Electrical Engineering; Computer Science; and Communications

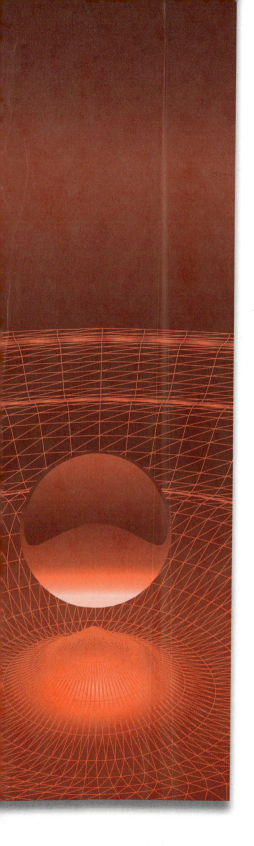

FIFTH EDITION

Probability and Statistics for Engineers

RICHARD L.
Scheaffer
University of Florida

•

MADHURI S.
Mulekar
University of South Alabama

•

JAMES T.
McClave
Infotech, Inc.

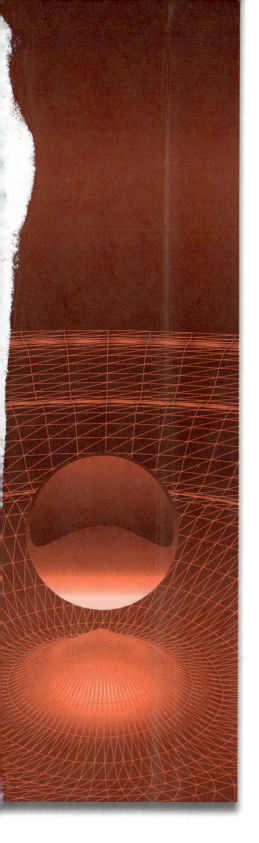

FIFTH EDITION

Probability and Statistics for Engineers

RICHARD L.
Scheaffer

University of Florida

•

MADHURI S.
Mulekar

University of South Alabama

•

JAMES T.
McClave

Infotech, Inc.

BROOKS/COLE
CENGAGE Learning™

Australia • Brazil • Japan • Korea • Mexico • Singapore • Spain • United Kingdom • United States

BROOKS/COLE
CENGAGE Learning™

Probability and Statistics for Engineers, Fifth Edition
Richard L. Scheaffer, Madhuri S. Mulekar, James T. McClave

Editor in Chief: Michelle Julet

Publisher: Richard Stratton

Senior Sponsoring Editor: Molly Taylor

Associate Editor: Daniel Seibert

Editorial Assistant: Shaylin Walsh

Associate Media Editor: Andrew Coppola

Marketing Manager: Ashley Pickering

Marketing Coordinator: Erica O'Connell

Marketing Communications Manager: Mary Anne Payumo

Associate Content Project Manager: Jill Clark

Art Director: Linda Helcher

Senior Manufacturing Buyer: Diane Gibbons

Text Rights Account Manager: Timothy Sisler

Production Service: MPS Limited, A Macmillan Company

Text Designer: Rokusek Design

Cover Designer: Rokusek Design

Compositor: MPS Limited, A Macmillan Company

For product information and technology assistance, contact us at
Cengage Learning Customer & Sales Support, 1-800-354-9706

For permission to use material from this text or product, submit all requests online at **www.cengage.com/permissions**
Further permissions questions can be emailed to
permissionrequest@cengage.com

Library of Congress Control Number: 2009932826

ISBN-13: 978-0-534-40302-7

ISBN-10: 0-534-40302-6

Brooks/Cole
20 Channel Center Street
Boston, MA 02210
USA

Cengage Learning is a leading provider of customized learning solutions with office locations around the globe, including Singapore, the United Kingdom, Australia, Mexico, Brazil, and Japan. Locate your local office at: **international.cengage.com/region**

Cengage Learning products are represented in Canada by Nelson Education, Ltd.

For your course and learning solutions, visit **www.cengage.com**.

Purchase any of our products at your local college store or at our preferred online store **www.cengagebrain.com**.

Printed in the USA
1 2 3 4 5 6 7 13 12 11 10

Contents

Preface

This book is a calculus-based introduction to probability and statistics for students in engineering, with emphasis on problem solving through the collection and analysis of data. The problems presented, which deal with topics such as energy, pollution, weather, chemical processes, global warming, production issues, and bridge operation patterns, show the breadth of societal issues that engineers and scientists are called upon to solve. Examples and exercises contain real scenarios, often with actual data.

The Fifth Edition of this book has been modified to offer a more modern approach to teaching of probability and statistics for engineers, incorporating many more examples relevant to the real world situations that engineers encounter. More emphasis is now given on making sense of the data, understanding concepts, and interpreting the results. Emphasis is taken away from the computations that can be easily accomplished using calculators or computers.

In its modern approach to data analysis, this book highlights graphical techniques to summarize data, look for patterns, and assess distributional assumptions. To make these plots efficiently, and to allow for easy handling of data in the context of making inferences and building models, technology is incorporated throughout the book, in the form of outputs from popular software like Minitab, JMP, and SAS as well as graphing calculators like TI-89. However, other statistical packages or graphing calculators with statistical capabilities available to students and teachers can also be used.

The motivating theme of almost any successful business or industry today is improving the quality of process. Whether the product is cars that transport us, medicine that heals us or movies that entertain us, customers expect high quality at a reasonable price. Engineers, who are key personnel in many of these industries, must become proficient in the use of statistical techniques of process improvement. The theme of quality improvement is central to the development of statistical ideas throughout this book. Discipline-appropriate activities for students are incorporated to engage students in hands-on discovery-based learning of different concepts.

Organization

The book is designed to be compact yet effectively cover the essentials of an introduction to statistics for those who must use statistical techniques in their work. Following a discussion in Chapter 1 and 2 of the use of data analysis techniques in quality improvement for univariate and bivariate data, Chapter 3 discusses commonly used techniques of data collection and different types of biases encountered due to inaccurate use of these methods. Chapters 4 through 7 develop the basic ideas of probability and probability distributions, beginning with applications to frequency data in one- and two-way tables and continuing to more defined univariate and multivariate probability distributions. Chapter 8, as a transition between probability and statistics, introduces sampling distributions for certain statistics in common use. Chapters 9 and 10 present the key elements of statistical inference through a careful look at estimation and hypothesis testing. Chapters 11 deals with the process of fitting models to data, making inference about the model parameters, and then using these models in making decisions. Chapter 12 discusses the importance of designing experiments to answer specific questions and makes use of the models from earlier chapters in analyzing data from designed experiments.

Suggested Coverage

This book organizes the most important ideas in introductory statistics so all can be covered in a one-semester course. One possibility is to cover all of Chapters 1, 2, and 3, followed by a more cursory coverage of Chapters 4, 5, and 6. Selections from Chapter 5 should include the binomial and Poisson distributions, while those from Chapter 6 should include the exponential and normal distributions. For a more applications-oriented course, Chapter 7 can be skipped. The ideas of expectation of linear functions of random variables and correlation are presented again in later chapters, within the context of their application to data analysis and inference. Chapters 8, 9, and 10 present the heart of statistical inference and its more formal connections to quality assurance, and should be covered rather thoroughly. The main ideas of Chapter 2 and later continued in Chapter 11 should receive much attention in an introductory course, even at the expense of skipping some of the probability material in Chapters 5 through 7, because building statistical models by regression techniques is an essential topic for scientific investigations. Design and analysis of experiments, the main ideas of Chapter 12, are key to statistical process improvement and should receive considerable attention. The use of computer software for statistics will enhance coverage of Chapters 8 through 12.

New to This Edition

Material in the Fifth Edition has been reorganized so that the book can be divided into five major categories. It begins with an emphasis on univariate and bivariate data analysis, moves to the data collection methods, continues with probability theory, develops estimation and inference methods, and ends with the model development. Different commonly used data collection methods, randomization techniques, and types of biases that can result due to incorrect use of techniques have been added. Because of this added material the book now has twelve chapters.

The Fifth Edition much more strongly encourages the use of technology. Examples are scattered throughout the book showing computer and calculator outputs that are used for obtaining solutions and for decision making. To take advantage of the power offered by statistical software and calculators, larger real-life datasets are incorporated in addition to smaller datasets to develop concepts and incorporate them into exercises.

For an improved flow of topics, some of the topics have been regrouped, such as coverage of bivariate data analysis right after univariate data analysis. New student activities that are continued through several chapters covering different concepts are also included to provide students with an understanding of how different topics are connected to each other.

Supplements

A **Student Solutions Manual** (*ISBN-10: 0-538-73595-3; ISBN-13: 978-0-538-73595-7*), available for separate purchase, contains fully worked-out solutions to the same selected exercises solved in the short-answer key in the back of the book.

Complete instructor-only solutions to all exercises from the book are available through Cengage's **Solution Builder** service. Adopting instructors can sign up for access at *www.cengage.com/solutionbuilder* and create custom printouts of the solutions directly matched to their homework assignments.

The book's **Companion Website**, at *www.cengage.com/statistics/scheaffer*, offers downloadable versions of all of the data sets used in the book in the native file formats of several of the most popular statistical software packages. Extra supplementary topics are also provided, along with other helpful resources.

Acknowledgments

The authors hope that students in engineering and the sciences find the book both enjoyable and informative as they seek to learn how statistics is applied to their field of interest. The original authors offer thanks to Madhuri Mulekar in appreciation of her fine work in expanding and modernizing the interaction of statistics and engineering in this edition of the text. We had much assistance in making the book a reality and wish to thank all of those who helped along the way. In particular, we want to thank the reviewers of this edition for their many insightful comments and suggestions for improvement: William Biles, University of Louisville; Leszek Gawarecki, Kettering University; James Guffey, Truman State University; Megan Mocko, University of Florida; Raja Nasser, Louisiana Tech University; Steven Rigdon, Southern Illinois University; and David Zeitler, Grand Valley State University. We also extend our continuing gratitude to the reviewers of all four previous editions of the book. A special thanks goes to Jill Pellarin for her detailed copyediting work, Laurel Chiappetta for thoroughly checking the material for accuracy, and Laura Lawrie for her project management and organization during the book's production. It was a real pleasure to work with our editors Molly Taylor and Daniel Seibert, and with Jill Clark during production of the book. They and the fine staff at Cengage Learning made the work on this textbook really pleasurable; we thank them for their dedication to quality with a personal touch.

Richard L. Scheaffer
Madhuri S. Mulekar
James T. McClave

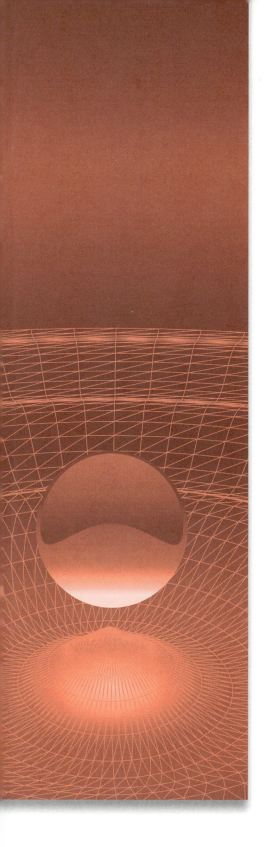

FIFTH EDITION

Probability and Statistics for Engineers

RICHARD L.
Scheaffer
University of Florida

•

MADHURI S.
Mulekar
University of South Alabama

•

JAMES T.
McClave
Infotech, Inc.

Data Collection and Exploring Univariate Distributions

How does air quality compare among different metropolitan areas? Is the air quality changing over the years? If yes, how? Is it improving or deteriorating? Data such as that pictured in Figure 1.1 can help answer these questions. The data values in the plots are the number of days a year that the air quality index (AQI) exceeds a limit standard for healthy air (air quality index > 100). The plot on the left looks at three selected years, whereas the plot on the right summarizes the data for the period 1993–2002. To answer questions such as those posed above, we have to carefully study this and other air quality data from different perspectives.

In today's information society, decisions are made on the basis of data. A family checks school districts before purchasing a house; a company checks the labor and transportation conditions before opening a new plant; a physician checks the outcomes of recent clinical trials before prescribing a medicine; and an engineer tests the tensile strength of wire before winding it into a cable. The decisions made from the data—correctly or incorrectly—affect us every day of our lives. This chapter presents basic ideas on how to systematically study data for the purpose of making decisions. These basic ideas will form the building blocks for the remainder of the book, as it builds a solid picture of modern data analysis along with a solid foundation in the design of statistical studies.

Figure 1.1
The number of AQI exceedences for selected metropolitan areas

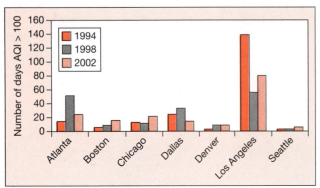

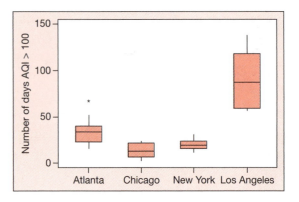

Introduction

Producing a product or a service of high quality while lowering cost is a complex matter. As a result, managers in business and industry think about such issues as how the quality of incoming materials affects the quality of outgoing product, how changing the flow of in-house activities might improve efficiency, or what reaction customers might have to the services provided. How do these managers make decisions? Among many possibilities, they might measure quality of products, compare the quality and prices of input materials from different providers, and provide customers with customer satisfaction surveys, among other things.

- The years 2008–09 saw a terrible downturn in the job market, but even during the period of rapid economic growth, not all people of working age had jobs. The U.S. Census Bureau collected data (Reasons people do not work, *U.S. Census Bureau*, Current Population Reports, July 2001, **http://www.census.gov**) that provides insight into the identification and potential availability of nonworkers during the period of tightening of labor markets.
- The Mobile County, Alabama, school system implemented several programs to help high school students prepare for a graduation exam. In an attempt to assess the success or failure of these programs, *The Mobile Register* (Feb. 15, 2004) collected data from the Mobile County High Schools. They reported that about 7% of seniors failed. That's an improvement over the 11% who failed the previous year.
- With all the advances in distillation technology (the process of separating chemical compounds), one would expect the failure rate in distillation towers to be on the decline. However, malfunction data collected by Kister (*Trans IchemE*, 1900, 1997, 2003) showed the reverse; the tower failure rate is on the rise and accelerating.

Virtually all the success stories in scenarios similar to those listed above have one thing in common. Decisions were and are made on the basis of objective, quantitative information—data!

Even in our daily lives, all of us are concerned about quality. We want to purchase high-quality goods and services, from automobiles to television sets, from medical treatment to the sound at local movie theater. Not only do we want maximum value for our dollar when we purchase goods and services, but we also want to live high-quality lives. Consequently, we think seriously about the food we eat, the amount of exercise we get, and the stresses that build up from our daily activities. How do we make decisions that confront us in our daily lives?

- We compare prices and value.
- We read food labels to determine calories and cholesterol.
- We ask our physician about possible side effects of a prescribed medication.
- We make mental notes about our weight gains and losses from day to day.

In short, we too are making daily decisions on the basis of objective, quantitative information—data!

Formally or informally, data provide the basis for many of the decisions made in our world. Use of data allows decisions to be made on the basis of factual information rather than subjective judgment, and the use of facts should produce better results. Whether the data lead to good results depends upon how the data were produced and how they were analyzed. For many of the simpler problems confronting us, data are readily available, but it may take some skill to analyze them properly. Thus, techniques for analyzing data will be presented first, with ideas on how to produce good data coming later in the book.

1.1 A Model for Problem Solving and Its Application

Whether we are discussing products, services, or lives, the goal is to improve quality. Along the way to improve quality, numerous decisions must be made on the basis of objective data. It is, however, possible that various decisions could affect one another, so the wise course of action is to review the set of decisions together in a wider problem-solving context. Because this procedure may involve the study of many variables and how they interrelate, a systematic approach to problem solving is essential to making good decisions efficiently.

For example, a civil engineer is to solve the problem of hampered traffic flow through a small town with only one main street. After collecting data on the volume of traffic, the immediate decision seems easy—widen the main street from two to four lanes. Careful consideration might suggest, though, that more traffic lights will be needed for the wider street so that traffic can cross it at other points. In addition, traffic now using the side streets will begin using the improved main street. A "simple" problem has become more complex once all the possible factors have been brought into the picture. The whole process of traffic flow can be improved only by taking a more detailed and careful approach to the problem.

The model outlined below contains the essential steps for solving problems in industry and business:

1. **State the problem or question:** Going from a loose idea or two about a problem to a careful statement of the real problem requires careful assessment of the situation and a clear idea of the goals of the study.

   I am pressed to get my homework assignments done on time and I do not seem to have adequate time to complete all my assignments. I still want to work out each day and spend some time with friends. Upon review of my study habits, it seems that I study rather haphazardly and tend to procrastinate. The real problem, then, is not to find more hours for study but to develop an efficient study plan.

2. **Collect and analyze data:** List all factors and collect data on factors thought to be important. A goal is to solve the specific problem addressed in the first step. Use data analysis techniques appropriate for the data collection methods used.

   The data show that I study 5 hours a day and work late into the evening. Then I go to the gym and find out the gym is crowded in the evening, slowing down my workout.

3. **Interpret the data and make decisions:** After analyzing the data, study the analysis carefully, and pose potential solutions to the original problems or questions.

   I realized that I often study late into evening because I watch television and visit my friends before studying. And I am tired when I begin to study. So, I have to study before watching television or visiting friends. Also, I will work out in the mornings to save time.

4. **Implement and verify the decisions:** After the solution is posed, put it into practice for a trial to see whether it is feasible. Collect new data on the revised process to see whether improvement is actually realized.

   I tried the new schedule with an earlier study time for 2 weeks and it worked fine. I seemed to have more time to complete my assignments without devoting more hours to study. The gym is just as busy in the morning, so no workout time was saved with the new schedule.

5. **Plan next actions:** The trial in step 4 may show that the earlier decision solved the problem. More likely, though, the decision may lead to partially satisfactory results. Then there is another problem: to tackle and plan a response while the process is still firm in mind and the data are still fresh.

   I would like to find time for pleasure reading. How can I fit that into my revised schedule?

Figure 1.2
A model for problem solving

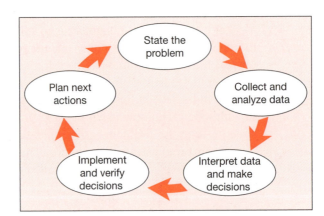

Figure 1.2 shows a typical problem-solving model. An industrial application of the problem-solving model is described below:

Office personnel responsible for the processing of customer payments to a large utility company noticed that they were receiving complaints both from customers ("Why hasn't my check cleared?") and from the firm's accounting division ("Why did it take so long to get these receipts deposited?"). The office staff decided it was time to put their quality-improvement training into action.

1. **State the problem or question:** After brainstorming the general problem, the team collected background data on elapsed time for processing payments and found that about 1.98% of the payments (representing 51,000 customers) took more than 72 hours to process. They suspected that many of these were "multis" (payment containing more than one check or bill) or "verifys" (payments in which the bill and the payment do not agree).

    Figure 1.3 is a *Pareto chart* that demonstrates the correctness of the team's intuition; "multis" account for 63% of the batches requiring more than 72 hours to process, but comprise only 3.6% of the total payments processed. "Verifys" are not nearly as serious as first suspected, but they are still the second leading contributor to the problem. The problem can now be made specific: Concentrating first on the "multis," reduce payments requiring more than 72 hours processing time to 1% of the total.

2. **Collect and analyze data:** What factors affect the processing of "multis"? Pareto charts and *scatterplots* can be used to identify such factors. The effect on elapsed time of processing payments can be seen in the scatterplot (Figure 1.4). It shows that cash carryover from the previous day affects elapsed time and thereby the entire payment processing system. An efficient handling of each day's bills must reduce the carryover to the next day.

3. **Interpret the data and make decisions:** With data in hand that clearly showed the computerized check processing system to be a major factor in the processing delays, a plan was developed to reprogram the machine so that it would accept multiple checks and coupons as well as both "multis" and "verifys" that were slightly out of balance.

Figure 1.3

A Pareto chart

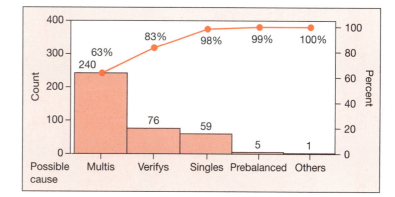

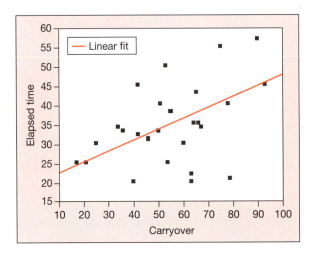

4. **Implement and verify the decisions:** The reprogramming was completed along with a few additional changes. It produced improvements that were close to the goal set forth in the problem statement. The record of processing times during a trial period is shown in the *time series plot* (Figure 1.5).

5. **Plan next actions:** The carryover problem remains. Along the way it was discovered that the mail-opening machine had an excessive amount of downtime.

   The distribution of downtimes is shown in the *histogram* (Figure 1.6). Resolution of these two problems led to an improvement that exceeded the goal set in the original problem statement.

Figure 1.5
A time series plot

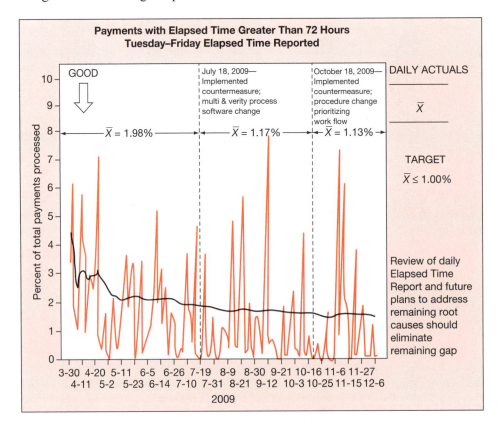

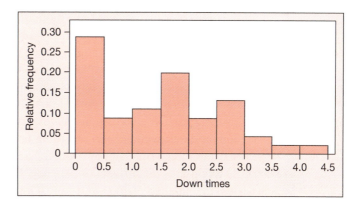

1.2   Types of Data and Frequency Distribution Tables

As we have seen in the earlier example, decisions are made using the information collected for a project or the measurements taken from the experiment. Statistics is a science of data. The collected data are used to estimate the unknown characteristics, make decisions, develop ideas, and implement policies. Here are some examples of data collected by engineers:

- Measuring lifetimes of transistors.
- Counting the number of drawbridge operations (openings and closings) per month.
- Measuring the ozone level of the air.
- Counting the number of days per year the Air Quality Index (AQI) exceeded 100.
- Classifying air as healthy or unhealthy based on the AQI index.
- Classifying the types of complaints seen in a consumer response program.
- Measuring the percent of chemical output of a process when the temperature of the process is varied.

These examples show that there are basically two different types of data: quantitative (numerical) and qualitative (categorical).

Data in the form of numerical measurements or counts are referred to as **quantitative (numerical) data**.

Data in the form of classifications into different groups or categories are referred to as **qualitative (categorical) data**.

The number of bridge operations per year, the ozone level, and percent of chemical yield are examples of quantitative (numerical) data. The type of complaint received and the classification of air (healthy or unhealthy) based on the AQI are examples of qualitative (categorical) data.

Sometimes data that appear to be numeric in nature can actually be categorical. For example, when program errors are recorded as "file not found" or "incorrect address," then the categorical nature of the data is obvious. However, if the errors are recorded using numerical codes such as 401 (for file not found) and 408 (for incorrect address), then its categorical nature is not obvious. But it is still categorical data, and we use only categorical data analysis tools to analyze such data. In sample survey data, yes/no responses are very commonly coded as 1/0.

Often for the convenience of recording or ease of analysis, quantitative data are converted into categorical data by grouping a range of values together. For example, the U.S. Environmental Protection Agency, Office of Air Quality and Standards, classifies air quality using AQI as "Good" (0–50), "Moderate" (51–100), "Unhealthy for sensitive group" (101–150), "Unhealthy" (151–200), "Very unhealthy" (201–300), and "Hazardous" (over 300). Although actual AQI levels are numerical, this classification of air quality results in ordered categorical data.

Sometimes we take only one measurement per unit and sometimes more, where a unit can be a person, an item, a location, a time slot, etc.

- If only one measurement is taken per unit, then the data are referred to as *univariate data*.
- If two measurements are taken on each unit, the result is *bivariate data*.
- If more than 2 measurements are taken per unit, the result is *multivariate data*.

For example,

- Measuring ozone level each day results in a univariate dataset of ozone levels.
- Measuring ozone and carbon monoxide level each day will result in a bivariate dataset consisting of ozone and carbon monoxide levels per day.
- Measuring ozone, carbon monoxide, particulate matter, and sulfur dioxide levels each day will result in a multivariate dataset consisting of four measurements per day.

Raw data collected in, say, a spreadsheet is not usually in a form useful for decision-making. To be useful in leading to reasoned conclusions, the data must be organized through the use of *descriptive tools* and *measures* that are useful for data presentation, data reduction, data exploration, and data summarization. Selection of an appropriate descriptive tool depends on the type of data collected, the number of measurements taken per unit, the total number of measurements, and the questions of interest.

There are two basic types of descriptive tools and measures—graphical descriptive tools and numerical descriptive measures—and each will be discussed in detail in the two subsequent sections. Before this, however, it is essential to understand the useful intermediate step of getting data from a list to a meaningful table.

Example 1.1

Table 1.1 summarizes the types of complaints received by an automobile paint shop over a period of time. This is a categorical variable, so the order of the rows is arbitrary. It is useful, however, to list the complaints in the order of their frequency, from high to low. Quite commonly, relative frequencies (proportions that can be easily turned into percentages) are more useful summaries than the frequencies themselves. These can be accumulated to form cumulative relative frequencies, another useful summary.

Table 1.1

Complaints Gathered at an Auto Paint Shop

Complaint	Frequency	Relative Frequency	Cumulative Relative Frequency
Dent	32	0.485	0.485
Chip	17	0.258	0.743
Peel	5	0.076	0.819
Mar	4	0.061	0.880
Bubble	3	0.045	0.925
Scratch	3	0.045	0.970
Burn	1	0.015	0.985
Streak	1	0.015	1.000

Source: Minitab

This tabular formation makes it easy to see that dents account for nearly half of the complaints, and dents, chips, and peels account for over 80% of the complaints.

The **frequency** of a measurement in a dataset is the number of times that measurement occurs in the dataset. Frequency is denoted by the letter f.

A **frequency distribution table** is a table giving all the measurements and their frequencies. The measurement can be reported in a grouped or ungrouped manner.

The **relative frequency** of an observation in a set with n measurements is the ratio of frequency to the total number of measurements, denoted by rf, where

$$rf = f/n$$

The **cumulative relative frequency** gives the proportion of measurements less than or equal to a specified value, and denoted by crf.

As mentioned above, quantitative data are sometimes grouped into meaningful categories so that they become categorical in nature. The following example illustrates this point.

Example 1.2

Table 1.2 compares the air quality indices for Los Angeles, California, and Orlando, Florida, for 2007 by looking at the EPA standard categories for air quality. With this summary table, it is easy to see, for example, that Orlando has moderate to good air quality more than 99% of the time, whereas the figure is only about 88% for Los Angeles.

Table 1.2

Air Quality Indices for Los Angeles, California, and Orlando, Florida, 2007

	Range of Indices	LA			Orlando		
		Number of Days	Relative Frequency	Cumulative Relative Frequency	Number of Days	Relative Frequency	Cumulative Relative Frequency
Good	0–50	155	0.464	0.464	233	0.850	0.850
Moderate	51–100	138	0.413	0.877	39	0.142	0.992
Unhealthy for sensitive groups	101–150	36	0.108	0.985	1	0.004	0.996
Unhealthy	151–200	5	0.015	1.000	1	.004	1.000
Total days monitored		**334**			**274**		

Source: **http://www.epa.gov/oar/data/**

Exercises

1.1 Classify the following data as quantitative or qualitative data.

    a The day of the week on which power outage occurred.

    b Number of automobile accidents at a major intersection in Pittsburgh on the fourth of July weekend.

    c Ignition times of material used in mattress covers (in seconds).

    d Breaking strength of threads (in ounces).

    e The gender of workers absent on a given day

    f Number of computer breakdowns per month.

    g Number of gamma rays emitted per second by a certain radioactive substance.

    h The ethnicity of students in a class.

    i Daily consumption of electric power by a certain city (in millions of kilowatt hours).

    j The level of ability to handle a spinner achieved by machinists at the end of training.

1.2 In each of the following studies, identify the measurements taken and classify them as quantitative or qualitative.

    a Since the discovery of the ozone hole over the South Pole in the 1980s, scientists have been monitoring it using satellites and ground stations. The National Oceanic & Atmospheric Administration has been tracking changes in ozone levels (in percent deviation from normal) and the ozone hole size (in million square miles).

    b A manufacturer of a particular heart surgery drug used to prevent excessive bleeding funded clinical trials. Physicians interested in the safety of drug studied incidence of kidney failure (yes or no), amount of blood loss (in cubic centimeters), length of recovery period (in days), serious complications (yes or no), and side effects (yes or no) for patients undergoing heart surgery.

    c After hurricane Katrina, the Federal Emergency Management Agency collected from the affected homeowners information on the amount of damage (in dollars), type of damage (wind or water), insurance status (insured or not insured), among other variables.

1.3 Table 1.3 displays the number of adults in the United States that fall into various income categories, according to the number of years of formal education.

    a Produce a companion table that shows relative frequencies for the income categories within each education column.

    b Comment on the relationship between income and years of education.

1.4 Table below gives a summary of all U.K. private finance initiative/public-private partnership

Capital Value (£ million)	Number of Projects Completed
Less than 50	145
50–100	58
101–150	20
151–200	10
201–250	5
251–300	3
301–350	2
351–400	1
401–450	3
451–500	2
501–700	1

Table 1.3

**Educational Attainment versus Income Category
(Counts in thousands for U.S population age 25 and over)**

| Income Category | Years of Formal Education | | | | | |
	12	14	16	18	20 +	Totals
0–29,999	24,270	5,505	8,115	1,936	259	**40,085**
30,000–59,999	12,273	4,750	10,000	3,671	522	**31,216**
60,000–89,999	2,100	1,082	4,389	1,996	472	**10,039**
90,000 and above	536	294	2,657	1,193	498	**5,178**
Totals	**39,179**	**11,631**	**25,161**	**8,796**	**1,751**	**86,518**

Source: U.S. Census Bureau

(PFI/PPP) projects completed by October 2002 (Bayley, 2003, *Proc. ICE*). Add a cumulative relative frequency column to the data table to help answer the following questions.

a What percent of completed projects had a capital value of at most 250 million pounds?

b What percent of completed projects had capital values exceeding 450 million pounds?

c How many completed projects had a capital value between 301 and 500 million pounds?

1.3 Tools for Describing Data: Graphical Methods

Compared to tables, graphs and charts are generally more appealing and effective for communicating information. Although modern technology makes generating beautiful graphs an easy task, we still have to determine which type of graph is most suitable for the data on hand and the questions to be answered.

- Qualitative data are popularly summarized using bar charts and Pareto charts.
- Quantitative data are commonly summarized using dotplots, histograms, boxplots, and cumulative frequency charts.
- Data collected over time are generally summarized using the time series plots (discussed in Chapter 2).
- Bivariate quantitative data are typically summarized using scatterplots (discussed in Chapter 2).

1.3.1 Graphing Categorical Data

Bar Chart: Bar charts are useful for summarizing and displaying the patterns in categorical data. A bar chart is created by plotting all the categories in the data on one axis and the frequency (or relative frequency or percentages) of occurrence of each category in the data on the other axis. Either horizontal or vertical bars of height (or length) equal to the frequency (or relative frequency or percentages) are drawn. Multiple bar charts are useful for comparing more than one dataset as long

Figure 1.7

Frequency of paint shop complaints

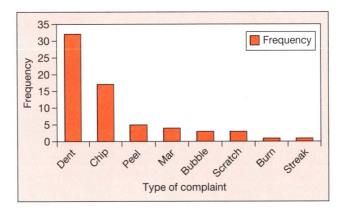

as all the datasets are sorted into the same categories. The bars for different datasets are identified using different colors or shading patterns.

Figure 1.7 shows a simple bar chart for the paint shop complaints of Example 1.1, giving the clear impression that dents and chips are by far the most common. Figure 1.8 shows the multiple bar chart for the air quality comparison of Example 1.2, showing clearly that Orlando has much better air quality on most days.

Pareto chart: In the late 19th century, the Italian economist V. Pareto showed that the distribution of income is very uneven; most of the world's wealth is controlled by a small percentage of people. Similarly, in most industrial processes, most defects arise from a relatively small number of causes. The process of sorting out the few vital causes from the trivial many is referred to as Pareto analysis. The *Pareto chart* is a simple device used for process improvement based on the idea: Sort out the few causes of most of your problems, and solve them first. A *Pareto chart* is a simple bar graph with the height of the bars proportional to the contributions from each source and the bars ordered from tallest to shortest. Figure 1.9 shows the bar chart of the paint shop complaints turned into a Pareto chart.

Figure 1.8

Relative frequencies of air quality categories

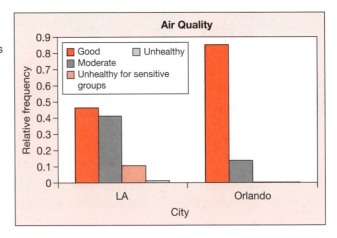

Figure 1.9

Pareto chart of the paint shop complaints

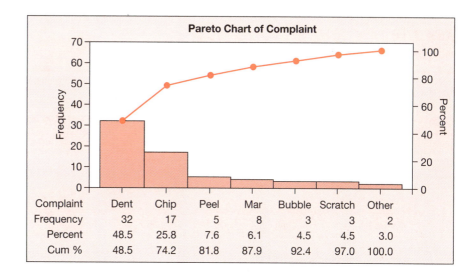

Complaint	Dent	Chip	Peel	Mar	Bubble	Scratch	Other
Frequency	32	17	5	8	3	3	2
Percent	48.5	25.8	7.6	6.1	4.5	4.5	3.0
Cum %	48.5	74.2	81.8	87.9	92.4	97.0	100.0

Example 1.3

What policies should be recommended in order to improve air quality in the United States? To assess the air quality and the effects of recent improvements on air quality, look at data on the Air Quality Index (AQI). The pollutants that contribute most to the AQI are carbon monoxide (CO), ozone (O_3), particulate matter (PM), and sulfur dioxide (SO_2). Table below gives emission estimates for years 1990 and 2000 for both carbon monoxide and ozone (which comes mainly from volatile organic compounds, or VOC). The source categories and changes over the years can be seen from the table below, but the important features of the data can be seen more clearly through a graphic display such as Figure 1.10. Notice that the measurements here are quantitative, not categorical, but the Pareto charting technique can still be used to make meaningful plots for comparing these data.

Source Category	1990	2000	Source Category	1990	2000
Carbon monoxide: (million short tons)			Volatile organic compounds: (million short tons)		
Fuel combustion	5.510	4.590	Fuel combustion	1.005	1.206
Industrial Process	5.582	7.521	Industrial process	10.000	8.033
Transportation	76.635	76.383	Transportation	8.988	8.396
Miscellaneous	11.122	20.806	Miscellaneous	1.059	2.710

The relative importance of the source categories for the various pollutants must be known before appropriate policy decisions can be made. Pareto charts show these relative contributions quite effectively.

Figure 1.10 provides the Pareto charts for carbon monoxide and ozone emissions comparing years 1990 and 2000. Clearly, transportation was the largest contributor to carbon monoxide, although its share was reduced from 1990 to 2000. Although industrial processes were the largest contributor to ozone in 1990, transportation was also a major contributor and fell right behind industrial processes. In 2000, these two contributors switched places; however, both together are still major sources of over 80% of ozone problems. Contributions by miscellaneous sources have increased. So a careful screening of miscellaneous sources is needed to identify any new factors contributing to the carbon monoxide level and institute measures to control them.

Figure 1.10

Carbon monoxide and ozone emissions

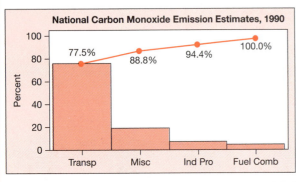

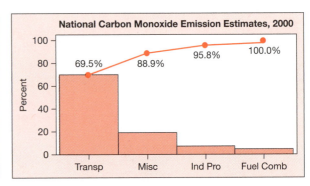

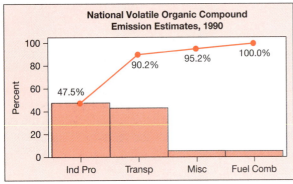

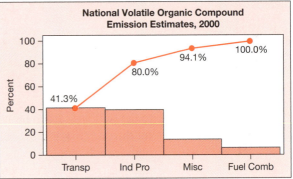

1.3.2 Graphing numerical data

AQI exceedences, the number of days the air quality index exceeds 100 at a particular site, provides one quantitative measure of air quality around the United States. Table 1.4 shows the number of days for which AQI was greater than 100 (i.e., unhealthy air) for each of 17 years in each of 15 metropolitan areas. It is not easy to see patterns in data arrayed like this, but some simple graphical devices will help us see patterns more clearly.

Dotplot: Perhaps the simplest of all plots of numerical data is the dotplot, which simply plots the data as points on a real number line. A dotplot is constructed by placing a dot over the number line in the position corresponding to each measurement. If more than one measurement has the same value, then dots are placed above each other forming a column. Dotplots work well for comparing multiple datasets but are not efficient for very large datasets.

Example 1.4

Refer to AQI exceedences data in Table 1.4. Figure 1.11 shows a dotplot for the AQI exceedences by year. Among other things, it shows that during 1990, all cities had less than 60 days of unhealthy air, except Los Angeles, which recorded unhealthy air for 161 days. In 2006, all the cities studied had less than 50 days of unhealthy air, including Los Angeles.

A comparison of three cities over 17 years (Figure 1.12) shows that Houston had more unhealthy days than New York, and Boston had fewer unhealthy days than New York.

Table 1.4

Air Quality Index, AQI Exceedences at All Trend Sites

Metropolitan Area	1990	1991	1992	1993	1994	1995	1996	1997	1998	1999	2000	2001	2002	2003	2004	2005	2006
Atlanta	42	23	20	36	15	35	25	31	50	73	39	24	20	12	12	11	18
Boston	0	0	0	0	0	0	0	0	0	4	0	3	9	8	1	4	1
Chicago	5	23	6	3	7	22	6	9	10	19	13	33	20	10	9	23	5
Dallas	0	0	5	10	24	20	8	20	24	16	20	14	7	5	9	10	13
Denver	9	6	11	6	2	3	0	0	7	3	2	8	7	17	0	1	6
Detroit	11	27	7	5	11	14	13	11	17	23	15	31	26	19	5	24	6
Houston	51	36	32	27	38	65	26	46	38	51	42	28	21	31	22	28	18
Kansas City	2	9	1	4	10	21	6	16	14	3	10	4	7	11	0	9	11
Los Angeles	161	156	166	128	127	100	75	40	49	54	63	81	81	88	65	43	34
New York	15	30	4	11	13	17	11	22	14	22	19	19	27	11	6	15	11
Philadelphia	39	49	24	51	26	30	22	32	37	33	21	34	35	19	9	21	18
Pittsburgh	18	21	9	13	19	25	11	20	39	40	32	50	50	37	39	48	36
San Francisco	0	0	0	0	0	2	0	0	0	10	4	12	17	1	4	5	2
Seattle	2	0	2	0	1	0	0	0	3	6	8	6	7	2	1	3	5
Washington, DC	14	40	12	42	21	32	14	25	44	41	20	27	33	12	10	18	18

Figure 1.11

Number of AQI exceedences for 1990, 1998, and 2006

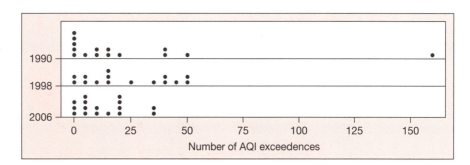

Figure 1.12

Number of AQI exceedences by city

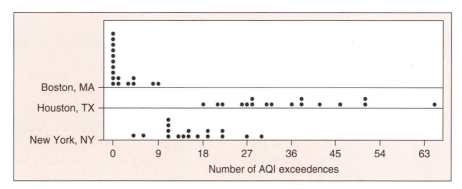

Histogram: Dotplots can become cumbersome to construct and study if the datasets are large. Histograms preserve some of the essential features of the dotplot with regard to displaying the shape of data distributions and their position on the real number line, but they are not limited by the size of the datasets. A histogram is formed by marking off intervals on the real number line on which the measurements fall and then constructing a bar over each interval with height equal to the frequency or relative frequency of the measurements falling into that interval.

Example 1.5

The Environmental Protection Agency collects data on the levels of contributors to the AQI from the entire nation. Figure 1.13 shows relative frequency histograms for two of the main contributors of AQI, VOC and CO, recorded during 1999 for different counties in Alabama. The distributions for both pollutants are right skewed (i.e., skewed toward higher values) with outliers on the higher end. Most of the volatile organic compound readings are below 20,000, with few counties recording higher values. Most counties in Alabama recorded carbon monoxide levels below 100,000. In both cases, there were a few counties recording high values of pollutants; these high values can be identified as outliers.

Figure 1.13

Pollutant readings in Alabama, 1999

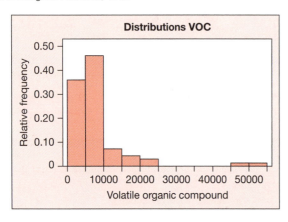 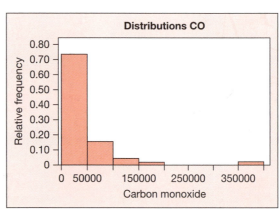

One note of caution is in order: When you are constructing histograms, the intervals that form the width of the bars should be of equal length. That allows the area of the bar, relative to the total area of all bars, to reflect the proportion of dataset that is actually inside the interval covered by the bar. A correct visual image of the data distribution is then portrayed.

Cumulative relative frequency chart: A cumulative relative frequency chart for quantitative data divided into categories is constructed by plotting cumulative relative frequencies against the upper class limit and then connecting the plotted points. Such charts are typically S-shaped, as seen in the following example:

Example 1.6

A company receives a supply of a particular component from two different manufacturers, A and B. To compare their lifetimes the company tested 200 components by each manufacturer. The experiment was stopped after 400 hours. Table 1.5 gives the frequency distribution of lifetimes measured for tested components from each manufacturer. The cumulative frequency table is given in Table 1.6 and the cumulative relative frequency chart is shown in Figure 1.14. The plot shows that components from manufacturer A have a higher failure rate with shorter lifetimes. About 50% of the A's fail by 100 hours, whereas less than 5% of the B's fail by that time. At the other end of the scale, over 90% of the B's last until 300 hours, whereas the percentage for the A's is well under 90%.

Table 1.5

Frequency Distribution of Lifetime

Lifetime (in hours)	Number of Components for Manufacturer	
	A	B
Lifetime ≤ 50	54	0
50 < Lifetime ≤ 100	47	6
100 < Lifetime ≤ 150	27	34
150 < Lifetime ≤ 200	19	66
200 < Lifetime ≤ 250	11	55
250 < Lifetime ≤ 300	15	24
300 < Lifetime ≤ 350	6	8
350 < Lifetime ≤ 400	6	5
400 < Lifetime	15	2

Table 1.6

Cumulative Frequency Distribution of Lifetime Data

Lifetime (in hours)	Number of Components for Manufacturer	
	A	B
Lifetime ≤ 50	54	0
Lifetime ≤ 100	101	6
Lifetime ≤ 150	128	40
Lifetime ≤ 200	147	106
Lifetime ≤ 250	158	161
Lifetime ≤ 300	173	185
Lifetime ≤ 350	179	193
Lifetime ≤ 400	185	198
Total	200	200

Figure 1.14

Cumulative relative frequency chart for lifetime data

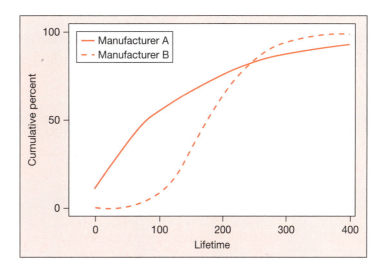

Boxplots: Dotplots and histograms are excellent graphic displays for focusing attention on key aspects of the shape of a distribution of data (symmetry, skewness, clustering, gaps, as explained below), but they are not good tools for making comparisons among multiple datasets. A simple graphical display that is ideal for making comparisons is the *boxplot*, which will be discussed in detail in Section 1.5.3.

1.3.3   Visualizing distributions

The overall pattern of a distribution of measurements is generally described by three components: center, spread, and shape.

- **Center** of a distribution describes a point near the middle of a distribution that might serve as a typical value or a balance point for the distribution.
- **Spread** of a distribution describes how the data points spread out around a center.
- **Shape** of a distribution describes the basic pattern of the plotted data along with any notable departures from that pattern. The most basic shapes can be described using the idea of symmetry. Although it is possible to identify shapes of distributions from the frequency distribution tables, it is easier to identify them from the graphs and charts.

Symmetric distribution: The upper half of the distribution is approximately a mirror image of the lower half of the distribution. For example, the distribution of SAT scores of students is approximately symmetric (see Figure 1.15a).

Left-skewed distribution: The lower half of the distribution extends farther out than the upper half of the distribution; in other words, the distribution has a longer left tail than the right tail. There are more measurements with higher values than there are with lower values. For example, the distribution of grades on an easy exam is typically left skewed because more students score higher grades and fewer students score lower grades (see Figure 1.15b).

Right-skewed distribution: The upper half of the distribution extends farther out than the lower half of the distribution; in other words, the distribution has a longer right tail than the left tail. There are more observations with lower values than higher values. For example, the distribution of incomes for employees in a large firm; few people in a given company earn high salaries (see Figure 1.15c).

In addition to the center, spread, and shape, striking departures from patterns are also used to describe the data.

- **Clusters and gaps:** Are any observations grouped together unusually? Are there any significant gaps in the data? For example, if you summarize the diameters of holes made by a stamping machine and the data clusters at two separate points, then a possible explanation for such clustering might be two machine operators are setting the machine differently during their shifts.

Figure 1.15

Shapes of distributions

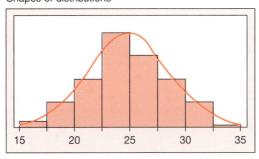

(a) Symmetric distribution

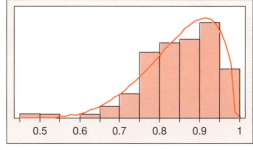

(b) Left-skewed distribution

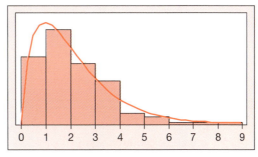

(c) A right-skewed distribution

- **Outliers:** Outlier is a measurement that appears to be different from the rest of the data. For example, when studying the lifetimes of batteries, most of the batteries had to be replaced within 100 hours of use, but one lasted over 120 hours. In a salary distribution of employees of any organization, typically the salary of a high official is an outlier (or a few higher ones are outliers). Later in this chapter, we'll study graphical and numerical methods of identifying outliers. An outlier does not necessarily mean an incorrectly recorded measurement and should not be discarded without an investigation.

Graphical tools are very attractive and good at conveying information. Then why don't we use just graphs and charts for decision making? For two major reasons:

- Graphs are very subjective. Graphs can look very different when created by different people or programs and are capable of giving wrong impressions. For example, a histogram could be constructed to show features like clusters and gaps or to hide these features, depending on the whim of the data analyst.
- Graphs are not amenable to precise mathematical descriptions and analysis. It is, for example, very difficult to describe in precise terms how much one histogram differs from another. By comparison, it is easy for you to compare your grade point average with that of another student.

Example 1.7

In the late 1800s, Lord Rayleigh studied the density of nitrogen compounds produced by chemical reactions in his laboratory as compared to those coming from the air outside his laboratory (cleansed from known impurities). Figure 1.16a shows a possible histogram for Rayleigh's data, from which no obvious conclusions could be drawn. The dotplot in Figure 1.16b shows quite clearly that the air samples have larger densities. The observation that the nitrogen densities produced from air gave larger values led to the conjecture that air samples might contain chemicals in addition to nitrogen, and this led to the discovery of inert gases like radon in the atmosphere. The lesson to be learned: Data plots are very helpful, but they must be used with care.

Figure 1.16
Distribution of nitrogen densities

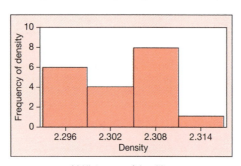

(a) Histogram of densities

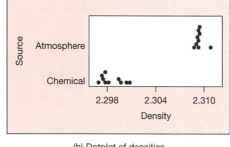

(b) Dotplot of densities

Exercises

1.5 Kennerley (*Proceedings of ICE*, 2003) analyzed complaints received by the Complaint Commissioner in the area where Channel Tunnel Rail Link construction was in progress. A variety of complaints received were categorized as listed below.

- **(1)** Property damage
- **(2)** Protection of livestock and machinery
- **(3)** Drainage systems
- **(4)** Water courses affected by construction
- **(5)** Cleaning of public highways
- **(6)** Plant noise and vibration
- **(7)** Blue routes & restricted times of use
- **(8)** Temporary and permanent diversions
- **(9)** TMP
- **(10)** Screening/fencing
- **(11)** Access
- **(12)** Working hours
- **(13)** Site operations

a Identify the major categories of complaints using the Pareto chart of complaints in figure below.

b Which categories comprise about 80% of complaints?

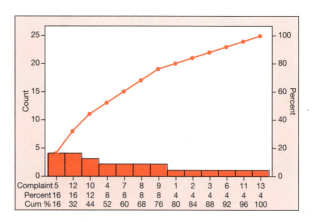

1.6 Sulfur dioxide is one of the major components of the AQI that measures air quality. Table below gives the sulfur dioxide estimates (in short tons) for years 1980 and 2000 (National Air Quality and Emissions Trends Report, **http://www.epa.gov/airtrends**).

SO₂ Source	1980	2000
Fuel Combustion		
Electrical Utilities	16.073	11.389
Industrial	2.951	2.894
Other	0.971	0.593
Industrial processes	3.807	1.498
Transportation	0.697	1.805
Total	**25.905**	**18.201**

a Produce and compare the Pareto charts of source categories for years 1980 and 2000.
b Identify important contributors of sulfur dioxide emission.

1.7 Lead is a major pollutant in both air and water, but is not part of the AQI. Table below gives the lead emission estimates (in short tons), and source categories for years 1980, 1990, and 2000 (National Air Quality and Emissions Trends Report, **http://www.epa.gov/airtrends**).

Lead Pollution Source	1980	1990	2000
Fuel combustion			
Electrical utilities	0.129	0.064	0.072
Industrial	0.060	0.018	0.017
Other	4.111	0.418	0.412
Industrial processes			
Chemical and allied product MFG	0.104	0.136	0.218
Metal processing	3.026	2.170	2.078
Other industrial processes	0.808	0.169	0.053
Waste disposal and recycling	1.210	0.804	0.813
Transportation			
On-road vehicles	60.501	0.421	0.020
Non-road engines and vehicles	4.205	0.776	0.545
Total	**74.153**	**4.975**	**4.228**

a Sketch a Pareto chart of source categories for years 1980, 1990, and 2000.
b Compare the Pareto charts in part (a).
c Has the quality of our environment improved in recent years? Justify your answer.

1.8 Top 10 suppliers of U.S. crude oil imports (in million barrels) for the year 2003 are listed below (U.S. Census Bureau, *Statistical Abstract of the United States,* 2005).

Saudi Arabia* 629	Venezuela* 439
Nigeria* 306	Iraq* 171
Kuwait* 75	Mexico 580
Canada 565	Angola 132
United Kingdom 127	Norway 60

a Construct a bar chart for these data. Comment on the U.S. crude oil import from different countries.

b Construct a dot plot of crude oil imports. Comment on the distribution of U.S. crude oil import from different countries. What information from a bar chart in part (a) is missing from a dot plot in part (b)?
c Suppose the countries marked by an asterisk (*) are OPEC countries. Compare crude oil amounts supplied by OPEC and non-OPEC countries, using the appropriate graph.

1.9 *The Mobile Register* (Nov. 16, 2003) reported information about Alabama's 203 aerospace companies. The companies were classified into different sectors. Table below shows the distribution of companies by sector classification and the numbers of employees in each sector.

Aerospace Sector	No. of Companies	No. of Employees
General manufacturing	30	1,070
Missile and space vehicle parts mfg.	18	2,367
Aircraft parts maintenance, repair, and overhaul	20	2,248
Engineering and RFD services	6	6,831
Information technology services	86	12,832
Missile space vehicle mfg.	5	5,095

a Construct a bar chart to summarize Alabama aerospace employment. Describe the results.
b Construct an appropriate plot to compare the number of employees per company in different sectors. Discuss the results.

1.10 Kister (*Trans IChemE*, 2003) studied malfunctioning of distillation towers. Table below lists the number of cases stratified by different causes. The data are also stratified by the year of failure ("Prior to 1991" or "1992 and Later").

Tower Malfunctioning Cause	Total cases	Prior to 1991	1992 and Later
Coking	26	4	22
Scale, corrosion products	22	11	11
Precipitation, salting out	17	4	13
Solids in feed	10	6	4
Polymer	9	3	5

a Construct an appropriate plot to describe this data.
b Compare "Prior to 1991" failures with "1992 and Later" failures.

1.11 The World Steel Organization reported that world crude steel production stands at 839.3 million short tons for the first 9 months of 2004, 8.7% higher compared with the same period of 2003.

The crude steel production (in millions of short tons) is reported in table below. Note that the Commonwealth of Independent States (CIS) includes Byelorussia, Kazakhstan, Moldova, Russia, Ukraine, and Uzbekistan.

	2004	2003
Asia/Oceania	393.3	352.0
European Union	159.0	150.6
North America	109.0	102.1
CIS	92.2	87.1
South America	37.7	35.5
Other Europe	23.8	21.0
Africa	13.5	13.4
Middle east	11.0	10.7

a Construct a bar chart to describe this data.
b Comment on the patterns observed, if any.

1.12 The Department of Energy (**http://www.energy.gov**) reported the production of top world oil producers and consumption of top world oil consumers (both in million barrels per day). The production is reported in Table 1.7 as average daily production for the first half of 2004, and the consumption is reported as 2004 estimates or forecasts.

a Make appropriate charts to describe world production and consumption of oil.
b Comment on your findings, with a particular reference to United States.

1.13 The Connecticut River Bridge is a railroad drawbridge. The Federal law requires Amtrak to leave it open except when trains are approaching.

Wilson (*Proceedings of ICE*, 2003) studied the number of bridge operations per month. This area has considerable commercial and navy traffic as well as summer activities. A histogram and dotplot for the number of operations of the Connecticut River Bridge are given in Figure below. Describe the strengths and weaknesses for each of the two types of plots.

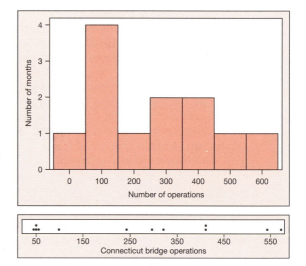

1.14 The National Aeronautics and Space Administration has sent many orbiter missions to space. The number of crew members on the orbiter for each flight between 1981 and 2003 are listed below (read by row): **http://www.ksc.nasa.gov/shuttle/missions/missions.html**):

Table 1.7

Oil Production by Producing Countries

Top World Oil Producers	Oil Production (million barrel/day)	Top world Oil Consumers	Oil Consumption (million barrels/day)
Saudi Arabia	10.6	United States	20.42
Russia	9.4	China	6.29
United States	8.74	Japan	5.45
Iran	4.03	Germany	2.68
Mexico	3.87	Russia	2.65
China	3.58	India	2.35
Norway	3.30	Canada	2.22
Canada	3.15	Brazil	2.20
Venezuela	2.84	South Korea	2.17
United Arab Emirates	2.80	France	2.07

```
2 2 2 2 4 4 5 5 6 5 5 6 7 5 5 7 7 7
7 5 5 8 7 7 7 5 5 5 5 5 5 5 5 5 5 5
5 7 5 7 7 5 5 6 7 7 7 7 7 7 6 5 5 5
7 6 5 7 7 6 5 6 7 6 6 6 6 7 7 5 5 7
5 6 7 6 6 6 7 6 5 6 7 7 7 7 5 7 6 7 7
6 7 6 7 5 7 6 7 7 7 5 5 7 7 5 7 8 7
7 7 7 6 7 7 7
```

a Construct a frequency histogram for the numbers of crew members.
b On how many flights were more than five crew members sent into space?
c What percent of flights were manned with at most three personnel?
d If the above data is in chronological order of flights (by row), what kind of long-term pattern do you notice in the number of crew sent into space with the orbiter?

1.15 A simple histogram is also a useful chart as a quality improvement tool. The frequency histogram of quality measurements on the diameters of 500 steel rods is given in Figure below. These rods had a lower specification limit (LSL) of 1.0 centimeter, and rods with diameters under the LSL were declared defective.

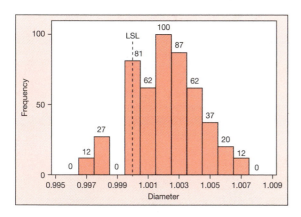

a What percent of rods were declared defective?
b Do you see anything unusual in this histogram? Can you offer a possible explanation of what is happening here?

1.16 The environmental lead impact studies on the quality of water often measure the level of chemical toxicants by the LC50, the concentration lethal to 50 percent of the organisms of a particular species under a given set of conditions and specified time period. Two common toxicants are methyl parathion and Baytex. LC50 measurements on water samples for each of these toxicants are as follows:

| Methyl parathion (MP) | 2.8, 3.1, 4.7, 5.2, 5.2, 5.3, 5.7, 5.7, 6.6, 7.1, 8.9, 9.0 |
| Baytex (B) | 0.9, 1.2, 1.3, 1.3, 1.4, 1.5, 1.6, 1.6, 1.7, 1.9, 2.4, 3.4 |

a Construct appropriate plots for these data. Comment on the shapes of these distributions.
b Compare the two distributions and describe the differences between two distributions.

1.17 The back-to-back histogram (often called a population pyramid) shown in Figure below displays the age distributions of the U.S. population in 1890 and 2005 (projected by U.S. Census Bureau, *Statistical Abstract of the United States*, 2001).

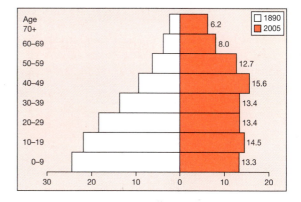

a Is the age distribution of the U.S. population changing? If so, what might that say about our quality of life in future years?
b Approximately what percentage of the population was under age 30 in 1890? In 2005?
c How would you describe the major changes in the age distribution over this century?

1.18 Ease of getting to work is one measure of quality of life in Metropolitan areas. Staggered starting times sometimes help alleviate morning traffic congestion and such a policy is often preferred by workers. Figure on next page shows the results of a study of work-start times in Pittsburgh. The plots are cumulative percentages of customers for official work-start times, arrival times at work, and desired work start times.

a Do official start times seem to bunch together more than actual arrival times?
b How would you compare the distribution of desired work start times to the distribution of arrival times? To that of official work start times?
c Give a concise explanation of what this plot shows.

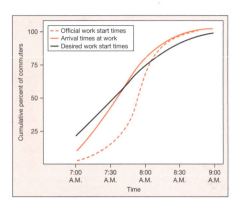

1.19 Table 1.8 gives the index of industrial production for selected countries (U.S. Census Bureau, *Statistical Abstract of the United States,* 2000).

a Construct appropriate plots to summarize the index of industrial production for the years 1990 and 1998. How would you describe these distributions?

b Calculate the difference between indices for 1990 and 1998. Summarize these differences using an appropriate plot. Comment on the result.

Table 1.8

Index of Industrial Production

Country	1990	1998	Country	1990	1998
Australia	92.9	106.7	Korea, South	66.4	106.1
Austria	92.3	115.9	Luxembourg	97.1	114.0
Belgium	99.3	108.8	Mexico	95.7	128.3
Canada	88.4	109.8	Netherlands	90.6	107.7
Czech Republic	144.9	109.8	New Zealand	87.0	105.0
Denmark	86.0	110.0	Norway	78.6	108.4
Finland	86.9	122.1	Poland	87.7	127.4
France	100.4	108.7	Portugal	97.0	114.1
Germany	103.2	108.5	Spain	96.6	111.4
Greece	101.9	109.8	Sweden	88.0	112.5
Hungary	113.6	129.1	Switzerland	96.0	108.0
Ireland	63.2	144.0	Turkey	81.3	120.5
Italy	93.2	102.9	United Kingdom	94.1	102.8
Japan	103.2	99.0	United States	86.4	114.8

1.4 Tools for Describing Data: Numerical Measures

As seen from examples in the previous section, graphs are useful, but they have certain deficiencies and must be interpreted carefully. At times, we have to summarize data using tools that will not be subjective and will give exactly the same answer regardless of who uses them. Numerical measures provide such summaries of quantitative data; such measures are useful in drawing meaningful conclusions from the data. Numerical data analysis provides a set of measures such that each measure produces a single numerical value to describe an important characteristic of the entire dataset. This helps in describing the key features of a data distribution and, perhaps more importantly, facilitates the comparison of different datasets. The numerical summaries of data are generally divided into *measures of center and measures of variation* (or *spread*).

1.4.1   Measures of center

As the name suggests, the measure of center determines the central point of the data, or the point around which all the measurements are scattered. Two basic measures of center are the arithmetic average, or *mean*, and the center value, or *median*. Although the mean is used more commonly, the median is quite useful, especially for skewed distributions. The grade point average of a student is a mean (weighted average to be exact) that attempts to measure where a student performance centers. Housing prices are often characterized by the median because of the extreme skewness in the distribution.

Mean: The mean of a dataset is the arithmetic average (commonly referred to as the average) of the numerical values. It is the most commonly used measure of center and measures the center of gravity, or the point at which the dataset balances.

Mean

If the dataset contains n measurements labeled $x_1, x_2, \ldots, x_n$, then the mean (read as "x bar") is defined as

$$\bar{x} = \frac{1}{n} \sum_{i=1}^{n} x_i$$

For example, Figure 1.17 shows a dotplot of mileage traveled by the SouthFlight helicopter in transporting accident victims to the University of South Alabama hospitals. Although miles traveled ranged from 2 to 69 miles, most flights were of shorter distances. As a result, the dotplot is balanced (as shown by the fulcrum), at a point located more toward the lower end (at 23.4 to be exact) than in the middle (35.5 miles) of the range of miles traveled.

Figure 1.17

A dotplot of SouthFlight mileage and the mean mileage

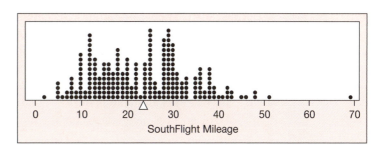

SouthFlight Mileage

Median: Another measure of the center is the median. Compared to the mean, the median is used less often for statistical purposes. The median is a measure of position. Fifty percent of the measurements have values at or below the median, whereas the remaining 50% measurements have values at or above the median. Median is denoted using the letter M. We can determine the median for a dataset containing n measurements as follows:

> ### Median
>
> For any dataset, the median is the middle of the ordered array of numerical values.

- Arrange n measurements in increasing order of value, in other words, from smallest to largest.
- Compute $l = (n + 1)/2$.
- Then the median M = the value of the lth measurement in the ordered array of measurements.

Note that if the dataset contains an odd number of measurements, then the median belongs to the dataset. But if the dataset contains an even number of measurements, then the median does not belong to the dataset. In fact, it is the midpoint between the middle two measurements. The typical price of a house reported for a city is usually the median price because the house prices are distributed right skewed, with possibly few extremely expensive houses.

Example 1.8

In the order of the cities in Table 1.4, the AQI data for year 2003 are

$$12, 8, 10, 5, 17, 19, 31, 11, 88, 11, 19, 37, 1, 2, 12$$

The mean of these data is

$$\bar{x} = \frac{1}{n}\sum_{i=1}^{n} x_i = \frac{1}{15}(12 + 8 + \cdots + 12) = 18.87$$

To compute the median, order the data by value as follows:

$$1, 2, 5, 8, 10, 11, 11, 12, 12, 17, 19, 19, 31, 37, 88$$

Because the data on 15 different cities are available, the median is the value of the $(15 + 1)/2 = 8$th measurement in the ordered array; therefore, $M = 12$. In other words, at least half of the cities included in the study recorded at most 12 AQI exceedences. Looking back at Table 1.6, note that Atlanta and Washington, DC, both have median AQI exceedences. Denver, Detroit, Houston, Los Angeles, Philadelphia, and Pittsburgh fall in the top half of the data; in other words, their number of AQI exceedences are greater than the median number of AQI exceedences.

Note that the median of these data is 12.00 and the mean is 18.87. Why are these measures of center so different? Study the dotplot of the data given below.

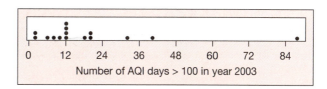

Note that Los Angeles (88) is an extremely large value, far away from the remainder of the data. Perhaps this "outlier" is responsible for the discrepancy between the mean and the median. If the 88 is removed from the data, the mean and the median of the reduced dataset of 14 measurements are 13.93 and 11.5, respectively. This shows that the mean is very sensitive to one or more extreme values, whereas the median is not. The important conclusions are as follows:

- The mean is affected by the extreme measurements. One large (or small) measurement, in comparison to the rest of the measurements in the dataset, can pull the mean to the high (or low) side of the data distribution.
- The median is little affected by the extreme measurements.

The upshot of this fact is that the median rather than the mean generally gives the more useful measure of center for datasets that are skewed or contain extremely large or small data values.

Example 1.9

Mean transportation time for accidents in rural Alabama from 3,133 cases is 13.67 min, whereas that in urban Alabama in 2,065 cases is 8.97 min. The transportation time is defined as the time to transport the vehicular accident victim from the site of accident to the emergency care medical facility by the EMS (emergency medical service) vehicle. What is the overall mean transportation time for the state of Alabama?

Solution   In computing the mean transportation time for the state of Alabama, we must take into account all the cases, rural as well as urban. The total transportation time in rural Alabama is $3,133(13.67) = 42,828$ min, and that for the urban Alabama is $2,065(8.97) = 18,523$ min. Therefore the total transportation time in Alabama is $42,828 + 18,523 = 61,351$ min in 5,198 cases. This gives the mean transportation time in Alabama as $61,523/5,198 = 11.84$ min.

Note that we cannot just average the rural and the urban mean transportation times to obtain the state mean. Because the two groups are of different sizes, we must make accommodations for the difference. This is also known as the *weighted average*, $\bar{x}_w$, and can be computed as

$$\bar{x}_w = \frac{n_1\bar{x}_1 + n_2\bar{x}_2}{n_1 + n_2}$$

where n_1 and n_2 are the respective sizes of two samples and $\bar{x}_1$ and $\bar{x}_2$ are the respective means of two samples.

1.4.2 Measures of position

Sometimes positions other than the center of a data distribution are of most interest. For example,

- A manufacturer of television sets might be interested in determining the length of warranty to be offered on the new sets. The manufacturer is willing to replace at most 5% of sets manufactured under the warranty. In other words, he needs to know the length of time by which at most 5% of the televisions are likely to fail. The measure of center would not be helpful because, for a fairly symmetric distribution of failure times, the mean would be close to the median, which would give us the time by which about 50% of the televisions are likely to fail. In this case, the manufacturer needs to know the 5th percentile of the distribution of the failure times.
- A student might be interested in knowing how his or her grade is compared to other students taking the same standardized test. If the student scores at 90th percentile, it means that at most 10% students scored better than him or her.

The measures used to describe position of a measurement with respect to the rest of measurements in the dataset are known as the *measures of position*. The *quartiles*, *percentiles*, and the *standardized scores* (*z-scores*) are the most commonly used measures of position.

Quartiles: Three quartiles, Q_L, Q_M, and Q_U, divide the dataset into four equal parts.

Q_L: First (or lower) quartile. Approximately 25% of the data values are at or below the lower quartile, i.e., it is the middle value of the lower half of the data.

Q_M: Second (or middle) quartile. It is same as the median. Approximately 50% of the data values are at or below the median.

Q_U: Third (or upper) quartile. Approximately 75% of the data values are at or below the upper quartile, that is, the middle value of the upper half of the data.

Percentiles: Ninety-nine percentiles divide the dataset into 100 equal parts. They are denoted by $P_1, P_2, \ldots, P_{99}$. Note that $P_{25} = Q_L$, $P_{50} = Q_M = M$, and $P_{75} = Q_U \cdot P_k =$ the kth percentile is the value such that approximately k percent of the data values are at or below it.

Example 1.10

Refer to the AQI data for year 2003 in Table 1.4. Determine the upper and lower quartiles of the AQI exceedances. Which cities are in the top 25% of the dataset? Which cities are in the bottom 25% of the dataset. How would you describe these groups?

Solution The AQI data for year 2003 ordered by value are as follows:

$$1, 2, 5, 8, 10, 11, 11, 12, 12, 17, 19, 19, 31, 37, 88$$

Earlier we saw that the median is 12. Therefore, the middle of the lower 7 measurements, the lower quartile, is 8. Similarly, the middle of the upper 7 measurements, the upper quartile, is 19.

- Dallas, San Francisco, and Seattle are the cities that are in the bottom 25% of the data. They are the cities with most clean air in this group of cities.
- Houston, Los Angeles, and Pittsburgh are the cities in the top 25% of the data. They are the cities with comparatively unhealthy air in this group of cities.

1.4.3   Measures of variation (or spread)

Either the mean or the median will do an adequate job of describing center of most distributions, but how can we measure variation (spread) in the data values? Three commonly used measures of variation are range, interquartile range, and standard deviation.

Range: The range (denoted by R) is the simplest of the measures of spread and is defined as the difference between the two extreme data values. It is easy to compute and understand. The range is not, however, a reliable measure because it depends only on two extreme measurements and does not take into account values of remaining measurements. For example, the range of AQI exceedences in year 2003 is $80 - 1 = 79$, but this can be misleadingly large as an indicator of spread because the 80 is far away from the bulk of the data.

> **Range**
>
> $R = $ Largest measurement $-$ smallest measurement.

Interquartile range: Interquartile range (IQR) is a measure of spread based on the measures of position and is defined as the difference between the upper quartile (Q_L) and the lower quartile (Q_U). If the median is used to measure center, then it is appropriate to use the IQR to measure variation in the data. Because it gives the range of the middle 50% of the data, the IQR is not affected by changes in the values of the extreme observations. Example 1.10 shows that the lower and the upper quartiles for the AQI data for the year 2003 are 8 and 19, respectively. That means the interquartile range is IQR $= 19 - 8 = 11$. Note that Los Angeles (80) is an "outlier," but the interquartile range would not change if this value were changed to, say, 55.

> **Interquartile Range**
>
> $$\text{IQR} = Q_U - Q_L$$

Example 1.11

Website **http://www.factmonster.com/ipka/** lists the home run records of American League (years 1901–2005) and National league (1876–2005) champions. The quartiles and the IQR for the home run data for both leagues are listed in table below.

City	Q_L	Median	Q_U	IQR
American League	32.00	41.00	47.5	15.50
National League	15.75	35.00	44.5	28.75

Comparison of IQR values shows that the number of home runs by the National League champions differed more from year to year than that by the American League Champions. In other words, the American League champions were more consistent from year to year than the National League champions in terms of the number of home run hits.

Variance and standard deviation: If the center of the dataset is defined by the mean, then the "typical" deviation from the mean is described by standard deviation. The variance is the typical squared deviation. It is denoted by s^2 and is measured in squared units. The contribution that a single data point makes to the variation of the dataset can be measured by the difference between the data value and the mean. Thus, the n data points will produce n deviations from the mean. How should we combine these n deviations into a single number that measures a typical deviation? Averaging the deviations will not work because the positive and negative deviations will cancel out. Averaging the absolute values of the deviations (known as *mean absolute deviation*) might work, but it is difficult to work with the absolute value functions. Alternatively, we can square the deviations and average the squared deviations into one number. This measure is known as the *variance*. For the reasons that will become clearer later in the text, the squared deviations are averaged using a divisor $(n - 1)$ instead of n.

Variance

If the dataset contains n measurements labeled $x_1, x_1, \ldots, x_n$, then the standard deviation is defined as

$$s^2 = \frac{\sum_{i=1}^{n}(x_i - \bar{x})^2}{n - 1}$$

The standard deviation is the square root of the variance.

Standard deviation:

$$s = \sqrt{s^2}$$

The positive square root of the variance, denoted by s, is known as the *standard deviation*. The standard deviation is measured in the same units as the data. Standard deviation and variance are better measures of variation than is the range because they take into account the contribution of each and every measurement. Besides measuring the spread of data points, the standard deviation is also used as a standard unit of distance. The distances between data points or between a data point and mean is often measured in terms of the number of standard deviations.

Example 1.12

Look at the AQI data for year 2003 in Example 1.10 (12, 8, 10, 5, 17, 19, 31, 11, 88, 11, 19, 37, 1, 2, 12). The mean for this data is 18.87. The variance can be computed as follows:

$$s^2 = \frac{(12 - 18.87)^2 + (8 - 18.87)^2 + \cdots + (12 - 18.87)^2}{15 - 1} = 462.25$$

The standard deviation is $s = \sqrt{462.24} = 21.5$ (see table below). This is a typical deviation of AQI exceedences from the mean for year 2003, as shown in Figure 1.18.

Mean	18.87
Std Dev	21.50
n	15

Figure 1.18

Typical deviation in AQI exceedences for 2003

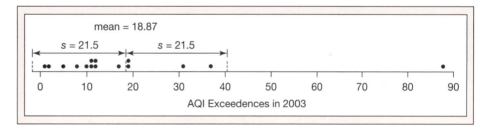

mean = 18.87

$s = 21.5$ $s = 21.5$

AQI Exceedences in 2003

Why is the standard deviation so large? Note that Los Angeles has an unusually large value in the dataset of Example 1.10. If Los Angeles is removed, the standard deviation for the AQI days of the remaining 14 cities is $s = 10.19$, which looks more reasonable than 21.50 as a typical deviation from the mean. This shows that the standard deviation is sensitive to changes in the extreme measurements in the dataset. Now look at the AQI data for New York and Washington, DC, listed in table below. The standard deviation for Washington, DC, is 11.70, and that for New York is 6.92. The dotplot in Figure 1.19 shows that the larger standard deviation indicates greater spread among the measurements.

City	Number of AQI Exceedences	Mean	Standard Deviation
New York	15, 30, 4, 11, 13, 17, 11, 22, 14, 22, 19, 19, 27, 11, 6, 15, 11	15.71	6.92
Washington, DC	14, 40, 12, 42, 21, 32, 14, 25, 44, 41, 20, 27, 33, 12, 10, 18, 18	24.88	11.70

Figure 1.19
Dotplot of AQI exceedences in New York and Washington, DC

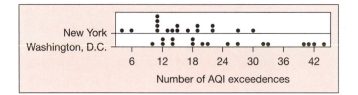

Example 1.13

For a sample of domestic and foreign used cars advertised for sale in *The Mobile Register* (May 31, 2004), the year of the car was recorded. The data are as follows:

Year of domestic cars advertised
1995 2001 2004 2000 2002 2002 2000 2001 1999 2002 2004 2001 2002
1996 1990 1995 1992 1995 1999 1996 1999 1999 1998 2001 2002 2004
2004 2004 2001 1997 2002 2001 2002 2001 2000 2002 1999 2001 2002
2000 2003 2001 2001 1999 2002 2001 2002 2001 2000 2002 2001 2002
2000 2000 2002 2001 2002 2002 2001 2001 2002 2002 2003 2003 2002
2001 2002 2001 2002 2003 2002

Year of foreign cars advertised
1997 2000 2002 2002 2001 2003 1995 1990 1992 1991 1997 2000 2000
1998 2000 1998 2001 2004 2001 2000 2001 2000 2002 2003 2003 2002
2002 2002 2003 1999 2000 2001 2003 2003 2000 2001

a Calculate the mean and median for each group. What do these values suggest about the data distribution?
b Calculate the standard deviation for each group. What do these values suggest about the data distribution?
c Describe the distributions of the year of domestic and foreign cars advertised.

Solution   Table below gives the summary statistics for the data.

	Domestic	Foreign
Mean	2000.592	1999.917
Std Dev	2.638	3.367
Median	2001.0	2001.0
Q_1	2000	1999.5
Q_3	2002	2002
n	71	36

a The fact that the mean is smaller than the median suggests that the distribution is skewed toward the smaller values (older cars) because the extremely small values pull the mean in that direction.
b The standard deviations are relatively large. In fact, adding two standard deviations to the mean takes us beyond 2004, in other words, off the scale for the data collected in 2004. This helps confirm the fact that the distributions must be skewed toward the smaller values (as explained later in this chapter).

c The relative frequency histograms in Figure 1.20a and Figure 1.20b show that both the distributions are left skewed, indicating more recently manufactured cars than older cars being advertised for sale. The means are about the same for the two distributions, but the foreign cars have a higher standard deviation. The IQRs, however, are about the same for the two distributions. Both of the distributions have a few "outliers" on the lower end, indicating a few comparatively older cars being advertised for sale.

Figure 1.20

Distribution of year of cars

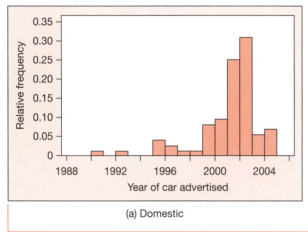

(a) Domestic

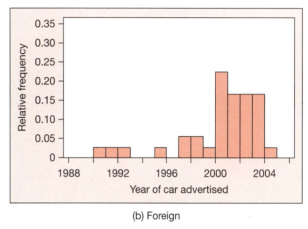

(b) Foreign

1.4.4   Reading computer printouts

Numerous computer software programs are available that can be used to compute numerical summary measures. Different programs report different summary measures. Printouts from JMP and Minitab for AQI data from year 2003 are given here:

(1) Mean
(2) Median
(3) Standard deviation
(4) Q_1

(5) Q_3
(6) n = number of observations
(7) Smallest data value
(8) Largest data value

```
JMP Printout

Distributions 2003

Quantiles                                     Moments
100.0%    maximum  (8)  88.000      Mean            (1) 18.866667
99.5%                   88.000      Std Dev         (3) 21.497065
97.5%                   88.000      Std Err Mean        5.5505184
90.0%                   57.400      upper 95% Mean      30.771345
75.0%    quartile (4)  19.000       lower 95% Mean       6.9619888
50.0%    median   (2)  12.000       N               (6) 15
25.0%    quartile (5)   8.000
10.0%                   1.600
2.5%                    1.000
0.5%                    1.000
0.0%     minimum  (7)   1.000
```

Minitab Printout

Descriptive Statistics: 2003

Variable	N	N*	Mean	SE Mean	StDev	Minimum	Q1	Median	Q3	Maximum
2003	15	0	18.87	5.55	21.50	1.00	8.00	12.00	19.00	88.00
	(6)		(1)		(3)	(7)	(4)	(2)	(5)	(8)

1.4.5 The effect of shifting and scaling of measurements on summary measures

For a simple scaffold structure made of steel, it is important to study the increase in length of tension members under load. For a load of 2,000 kg, the length increases (in cm) for 10 similar tension members were 2.5, 2.2, 3.0, 2.1, 2.7, 2.5, 2.8, 1.9, 2.2, and 2.6. The mean and standard deviation of lengths are given by

$$\bar{x} = \frac{1}{n}\sum_{i=1}^{n}x_i = \frac{1}{10}(2.5 + 2.2 + \cdots + 2.6) = 2.45 \text{ cm}$$

$$s = \sqrt{\frac{1}{n-1}\sum_{i=1}^{n}(x_i - \bar{x})^2} = \sqrt{\frac{1}{10-1}[(2.5 - 2.45)^2 + \cdots + (2.6 - 2.45)^2]}$$

$$= \sqrt{0.8183} = 0.34 \text{ cm}$$

However, the manager requires you to report the results in inches. How to get the mean and standard deviation of results in inches without transforming each value and repeating the calculations? Let x_i denote the measurement in centimeters and y_i the corresponding measurement in inches. Then $y_i = 0.4x_i$. Using this relation we get

$$\bar{y} = \frac{1}{n}\sum_{i=1}^{n}y_i = \frac{1}{n}\sum_{i=1}^{n}(0.4x_i) = 0.4\left[\frac{1}{n}\sum_{i=1}^{n}x_i\right] = 0.4\bar{x} = 0.4\,(2.45)$$

$$= 0.98 \text{ inches.}$$

$$s_y = \sqrt{\frac{1}{n-1}\sum_{i=1}^{n}(y_i - \bar{y})^2} = \sqrt{\frac{1}{10-1}[(0.4x_i - 0.4\bar{x})^2]}$$

$$= \sqrt{0.4^2\left[\frac{1}{n-1}\sum_{i=1}^{n}(x_i - \bar{x})^2\right]} = \sqrt{0.4^2 s_x^2}.$$

Thus, $s_y = \sqrt{0.4^2(0.34^2)} = 0.4\,(0.34) = 0.136$ inches.

The general rule for linearly transformed data is as follows. Suppose measurements $x_1, x_2, \ldots, x_n$ are transformed using constants a and b to new measurements $y_1, y_2, \ldots, y_n$ by

$$y_i = ax_i + b.$$

If $x_1, x_2, \ldots, x_n$ have mean $\bar{x}$ and variance s_x^2, then $y_1, y_2, \ldots, y_n$ have mean $\bar{y}$ and variance s_y^2 given by

$$\bar{y} = a\bar{x} + b \quad \text{and} \quad s_y^2 = a^2 s_x^2.$$

- If we shift the measurements by adding or subtracting a constant, then the measure of center gets shifted by the same amount, but the measure of variation is unaffected by any shift in measurements.
- If we scale measurements by multiplying or dividing by a constant, the measures of center and variation are both affected.

Example 1.14

A lawnmower manufacturer provides several different lines of lawnmowers to businesses. The mean and standard deviation of prices of their lawnmowers are $500.00 and $125.00, respectively. In response to increase in petroleum prices that led to increased transportation costs, the manufacturer decided to increase the prices of its products. Two different schemes were proposed as follows:

   A Increase the price of each lawnmower by $50.00.
   B Increase the price of each lawnmower by 10%.

     a For each of these two schemes, determine the mean and standard deviation of the new prices.
     b Explain the effect of each scheme on the prices of the lawnmowers.

Solution   Let x_i = current price of lawnmower, and y_i = the new price of lawnmower.

     a Using scheme A, $y_i = x_i + 50.0$. Therefore, $\bar{y} = \bar{x} + 50.0 = \$550.0$. The standard deviation will remain unchanged with this shift in prices.

        Using scheme B, $y_i = 1.10 \, x_i$. Therefore, $\bar{y} = 1.10\bar{x} = 1.10(500) = 550$. The standard deviation of the new prices will be

$$s_y = \sqrt{1.10^2 s_x^2} = \sqrt{1.10^2 (125.0)^2} = 1.10(125.0) = 137.5.$$

     b Although two schemes will result in the same mean increase in prices, they will have different effects on the spread of the prices. With scheme A, all the prices will shift up in the same amount, but the spread of prices from low-priced mower to high-priced mower will remain the same. On the other hand, with scheme B, the prices of mowers will shift up by different amounts. The low-priced mowers will have smaller amounts of increases in price, and the high-priced mowers will have comparatively larger amounts of increases, resulting in larger spread of prices.

Exercises

1.20 Table 1.9 gives the age distribution of the U.S. population in 1890 and 2010 (projected):

   a What age class contained the median age in 1890?

   b What age class contained the median age in 2010?

   c Compare the median ages and comment on the change in age distributions.

1.21 The data in table below show the percent change in the crude oil imports into the United States from January 2009 to June 2009 for each of the OPEC countries from which oil is imported: **(http://www.eia.doe.gov)**. The crude oil imports are recorded in thousand barrels per month.

Country	% Change	Country	% Change
Algeria	−38.39	Libya	56.29
Angola	−20.34	Nigeria	57.90
Ecuador	−46.42	Saudi Arabia	−31.89
Iraqu	−36.36	United Arab Emirates	92.35
Kuwait	−28.59	Venezuela	−11.54

   a Construct a dotplot for the percent change in the crude oil import.

   b Calculate the mean and the standard deviation of the percent change in the crude oil import.

   c Calculate the median and the interquartile range of the percent change in the crude oil import.

   d Do the mean and standard deviation offer good descriptions of center and variation? Why or why not? The United Arab Emirates contribution changed by 92% over 6 months. United Arab Emirates' contribution to the U.S. crude oil imports is very small (less than 3% in Jan. 2009), even with the increased supply (6% in June 2009). Eliminate this unusual measurement and repeat parts (b) and (c) with the reduced dataset. Comment on the effect this unusually large measurement has on the mean and standard deviation.

1.22 *The Los Angeles Times* of September 20, 1981, contains the following statement: "According to the latest enrollment analysis by age-categories, half of the [Los Angeles Community College] district's 128,000 students are over the age of 24. The average student is 29." Can this statement be correct? Explain.

1.23 "In the region we are traveling west of Whitney, precipitation drops off and the average snow depth on April 1 for the southern Sierra is a modest 5 to 6 feet. And two winters out of three the snow pack is below average" (Bowen, *The High Sierra*, 1972, p. 142). Explain how this statement can be true.

1.24 The air sulfur dioxide levels recorded in 67 counties of Alabama in 1999 **(http://www.epa.gov/air/airtrends/)** are listed here:

2,447 1,586 410 177 423 90 486 2,230 502 213 454 4,073 1,494 135 119 634 70,727 255 140 599 156 926 591 3,131 681 470 24,963 11,747 284 543 253 57,138 143 179 1,888 47,204 82,789 163 1,331 2,274 1,364 931 176 206 4,778 2,028 423 1,504 91,311 2,620 6,126 10,708 203 207 6,760 205 2,414 1,031 106,774 807 11,874 628 5,070 98,705 25,940 2,549 290

   a Construct a histogram for these data.

   b Calculate the mean and standard deviation for these data.

   c Calculate the median and IQR for these data.

   d Describe the shape center and spread of these data using the results of part (a) and either part (b) or part (c), whichever you think most appropriate.

1.25 Communication skills, both written and oral, remain key to professional success within the engineering profession. Bonk, Imhoff, and Cheng (*J. Professional Issues in Engineering Education and Practice,* 2002) studied effects of pooled efforts between engineering, business, and English departments.

   a The mean evaluation scores for the fourth-year capstone course in structural engineering during the fall 2000 and spring 2001 semesters were 4.0 and 4.2. During the fall and spring

---

Table 1.9

U.S. Population in 1890 and 2010 (projected)

Age	0–9	10–19	20–29	30–39	40–49	50–59	60–69	70+
1890	24%	22%	18%	13.5%	9.5%	6.5%	4%	3%
2010	13.2%	13.9%	13.7%	12.7%	14.2%	13.7%	9.5%	9.2%

Source: U.S. Census Bureau, National Population and Projections Summary Tables, January 2000 (**http://www.census.gov/**)

semesters, respectively, 30 and 33 students completed the evaluations. Determine the composite mean evaluation score for the entire academic year.

b During the spring 2001 semester, the mean scores (number of responses) in writing, environmental, transportation, structural, and civil engineering, respectively, were reported to be 4.2(30), 2.7(29), 3.0(29), 4.2(30), and 3.0(30). The numbers in the parenthesis are the sample sizes. Determine the mean score for the spring semester of 2001.

1.26 Table 1.10 gives the total revenue (in millions), taxes (per capita), and expenditure (per capita) for 50 states (U.S. Census Bureau, *Statistical Abstract of the United States*, 2000).

  a What numerical summaries best describe the total tax revenue?

  b What numerical summaries best describe the per capita tax revenue?

  c Write the brief description of the key features of each dataset.

d Compute the difference between the per capita tax revenue and per capita expenditure for each state. Plot the differences and describe the distribution. Identify outliers, if any.

e What numerical summaries best describe the difference in per capita tax revenue and per capita expenditure?

1.27 The environmental lead impact studies on the quality of water often measure the level of chemical toxicants by the LC50, the concentration lethal to 50% of the organisms of a particular species under a given set of conditions and specified time period. Two common toxicants are Methyl Parathion and Baytex. LC50 measurements on water samples for each of these toxicants are as follows:

| Methyl parathion (MP) | 2.8, 3.1, 4.7, 5.2, 5.2, 5.3, 5.7, 5.7, 6.6, 7.1, 8.9, 9.0 |
| Baytex (B) | 0.9, 1.2, 1.3, 1.3, 1.4, 1.5, 1.6, 1.6, 1.7, 1.9, 2.4, 3.4 |

Table 1.10

Revenue, Taxes, and Expenditure by State (in dollars)

State	Revenue	Taxes	Expenditures	State	Revenue	Taxes	Expenditures
AL	17,860	1,248	3,741	MT	4,224	3,027	4,473
AK	6,186	7,109	14,270	NE	5,944	1,046	3,553
AZ	15,489	699	3,230	NV	6,644	1,615	3,216
AR	10,330	1,054	3,932	NH	4,575	4,454	3,504
CA	1,76,081	1,802	4,927	NJ	42,788	3,493	4,425
CO	19,774	1,110	3,540	NM	9,099	2,351	5,010
CT	17,750	5,539	5,295	NY	1,12,439	4,212	5,586
DE	5,114	4,879	5,410	NC	32,203	1,218	3,854
FL	46,371	1,137	3,070	ND	3,373	2,432	4,549
GA	25,250	895	3,314	OH	52,809	1,646	4,204
HI	6,591	4,320	5,536	OK	12,746	1,725	3,290
ID	5,826	1,773	3,748	OR	18,219	1,848	4,699
IL	41,348	2,416	3,608	PA	45,887	1,565	4,185
IN	20,767	1,390	3,523	RI	5,483	5,503	5,048
IA	10,255	867	4,185	SC	16,865	2,354	4,451
KS	8,713	808	3,774	SD	3,171	2,923	3,549
KY	18,550	2,052	4,259	TN	17,344	589	3,198
LA	17,811	1,785	3,671	TX	65,525	787	3,027
ME	5,207	3,280	4,469	UT	9,132	1,765	4,060
MD	20,939	2,165	3,989	VT	3,143	3,794	5,499
MA	29,304	6,585	5,067	VA	22,760	1,801	3,722
MI	43,347	2,010	4,663	WA	23,646	2,104	4,643
MN	26,135	1,128	4,937	WV	8,297	2,272	4,054
MS	11,693	1,335	4,101	WI	18,826	2,252	4,598
MO	20,134	2,018	3,351	WY	2,880	2,725	5,355

a Calculate the mean and standard deviation for the two toxicants.

b Do they provide good measures of center and variation? Explain.

1.28 Is the system wearing out? Sometimes evidence of decreasing quality can be found by the careful study of measurements taken over time. Time intervals between successive failures of the air conditioning system on a Boeing 720 jet plane were recorded in order (*Source: Proschan, Technometrics*, 5(3), 1963, p. 376.). The 30 measures are divided into two groups of 15 to see whether the later time intervals seem shorter than the earlier ones (possibly indicating wearing out of the system). The data (in hours) are as follows:

Group I   23, 261, 87, 7, 120, 14, 62, 47, 225, 71, 246, 21, 20, 5, 42

Group II   12, 120, 11, 2, 14, 71, 14, 11, 16, i90, 1, 16, 52, 95

How would you analyze these data to build a case for (or against) the possibility of the air conditioning systems wearing out? What is your conclusion?

1.29 According to American society of Civil Engineers (**http://www.asce.org**), a structurally deficient bridge may either be closed or restricted to light vehicles because of its deteriorated structural components. Although not necessarily unsafe, these bridges must have limits for speed and weight. On the other hand, a functionally obsolete bridge has older design features and, although it is not unsafe for all vehicles, it cannot safely accommodate current traffic volumes, and vehicle sizes and weights. Table below shows the available data from the U.S. Department of Transportation about the condition of U.S. highway bridges by state as of August 13, 2007 (**http://www.bts.gov/**):

State	% Structurally Deficient Bridges	% Functionally Obsolete Bridges
AL	12	14
AK	8	15
FL	3	15
GA	7	13
KY	10	21
LA	13	16
MS	18	8
NC	13	16
SC	14	9
TN	7	14
VA	9	17
WV	15	22

a Compute the mean and standard deviation of the percentages of both types of bridges.

b Compare the mean and standard deviations for both types of bridges and comment on your findings.

1.5   Summary Measures and Decisions

For a data distribution that is fairly symmetric and mound-shaped, the center and spread are well described by the mean and standard deviation. Many datasets in practice are somewhat symmetric and mound-shaped (or can be transformed to be so) and so the mean and the standard deviation are, in fact, widely used measures. For a data distribution that is not shaped like a symmetric mound, the center and variation are better described by medians and the interquartile range, especially if the distribution is skewed. For either of these situations, certain basic properties and definitions allow data analysts to make objective decisions that can be unambiguously communicated to others.

1.5.1   The empirical rule

The mean and the standard deviation are a good way to summarize center and spread for a mound-shaped, nearly symmetric distribution of data. Armed with knowledge of only mean and standard deviation, we can determine something about the relative frequency behavior of the dataset.

> ### The Empirical Rule
> For any mound-shaped, nearly symmetric distribution of data,
>
> - The interval $\bar{x} \pm s$ contains approximately 68% of the data points.
> - The interval $\bar{x} \pm 2s$ contains approximately 95% of the data points.
> - The interval $\bar{x} \pm 3s$ contains almost all the data points.

Example 1.15

One of the measures of quality in baseball is a batting average. Look at the American (AL) and National (NL) League batting champions and their batting averages (**http://www.factmonster.com/ipka/**). Histograms in Figure 1.21a and Figure 1.21b show the distributions of batting averages of league champions from 1901 to 2005.

Figure 1.21
Batting averages of Champions (1901–2005)

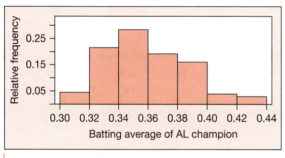

(a) American League

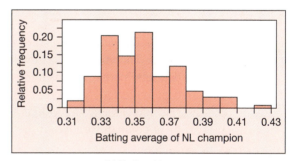

(b) National League

Both the distributions are mound-shaped and fairly symmetric. Although distribution of batting averages in the National league is slightly skewed to right, the skewness is not alarmingly prominent. Therefore, the empirical rule should apply to both the distributions. The mean and standard deviation of batting averages of champions from each league are listed in table below

83.4

44.6

47
15

	AL	NL
Mean	0.357	0.353
Std Dev	0.027	0.021
n	105	105

Table below gives the percentage of data points in each distribution that lie within one and two standard deviations of the mean. The results show that the empirical rule holds quite well for these data.

	American League		National League	
k	Interval $(\bar{x} - ks, \bar{x} + ks)$	Observed % Data in the Interval	Interval $(\bar{x} - ks, \bar{x} + ks)$	Observed % Data in the Interval
1	(0.330, 0.384)	68/105 = 65%	(0.332, 0.374)	75/105 = 71%
2	(0.303, 0.411)	101/105 = 100%	(0.311, 0.395)	99/105 = 100%

The empirical rule is a very useful result. It allows the standard deviation to be given a distance interpretation. If we know the mean and the standard deviation, then we know the interval within which most of the data will fall. Perhaps more importantly, we know that few observations will lie more than two standard deviations from the mean in mound-shaped, symmetric distributions.

1.5.2 Standardized values and z-scores

A z-score gives the distance between a measurement and the mean of the distribution from which it came in terms of the standard deviation of that distribution. A negative z-score indicates that the measurement is smaller than the mean, and a positive z-score indicates that the measurement is larger than the mean.

z-score

$$z\text{-score} = \frac{\text{measurement} - \text{mean}}{\text{standard deviation}}$$

Standardized scores are free of the units in which the data values are measured. Therefore, they are useful for comparison of observations from different datasets, perhaps measured on different scales. Imagine a set of environmental data collected for over 3,000 counties in the United States. The z-scores on selected variables for your county will help you decide how your county's environment compares to that of the rest of the counties in the nation.

Example 1.16

Summary statistics for the carbon monoxide and sulfur dioxide emission estimates measured at different locations in the United States in 1999 are given in table below. That year Mobile County, Alabama, recorded carbon monoxide emission estimates at 189,966.99 and sulfur dioxide emission estimates at 91,310.67. The z-scores, which allow comparisons of Mobile County to the United States as a whole, for each of the pollutants, are as follows:

Carbon Monoxide:

$$z\text{-score} = \frac{189,966.99 - 37,298.013}{84,369.21} = 1.81$$

Sulfur Dioxide:

$$z\text{-score} = \frac{91,310.67 - 5,616.0483}{18,869.243} = 4.54$$

	CO	SO$_2$
Mean	37,298.013	5,616.0483
Std Dev	84,369.21	18,869.243
N	3,143	3,143

A comparison of emission levels of two pollutants tells us that sulfur dioxide emission was about half of the carbon monoxide level. But such a comparison is not appropriate because of different centers and spreads of the two distributions. The *z*-scores indicate that Mobile County had considerably high emission levels of both the pollutants compared to the nation in 1999. However, the emission level of sulfur dioxide (4.54 standard deviations higher than the national average) was alarmingly higher than the emission level of carbon monoxide (1.81 standard deviations higher than the national average).

1.5.3   Boxplots

Boxplot: Dotplots and histograms are excellent graphic displays for focusing attention on key aspects of the shape of a distribution of data (symmetry, skewness, clustering, gaps), but they are not good tools for making comparisons among multiple datasets. A simple graphical display that is ideal for making comparisons is the *boxplot*. Also known as a *box-and-whiskers* plot, a boxplot is a graphical summary based on the measures of position. It is useful for identifying outliers. Here is how to construct one:

- Draw a measurement scale.
- Construct a rectangular box drawn from Q_L to Q_U.
- Draw a line at median dividing the box into two compartments.
- Draw whiskers from each end of the box to the farthest measurement, so long as that measurement is within 1.5 IQR from the end of the box.
- Identify any measurements beyond the length of the whiskers as *outliers*, using dots, stars, or any such symbols.

As shown in Figure 1.22, a box in the boxplot encloses the middle 50% of the measurements. The lower whisker covers the bottom 25% of the measurements, whereas the upper whisker covers the top 25% of the measurements. The length of the box is equal to the IQR and measures variation in the measurements. In comparing two boxplots on the same scale, a longer box is an indication of greater spread of the measurements. A skewed distribution is likely to result in a boxplot with one longer whisker and median toward one side of the box (rather than in the middle of the box). If both the whiskers are of about the same length, there are no

Figure 1.22
A boxplot

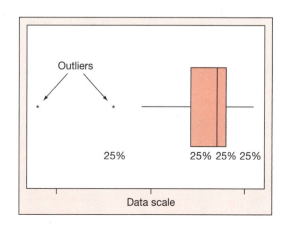

outliers, and the median lies approximately in the middle of the box, then the distribution is likely to be somewhat symmetric. However, boxplots fail to identify gaps in the data, so it is possible, for example, to get a symmetric boxplot for a distribution with two peaks and a gap in between (bimodal distribution). Therefore, a boxplot by itself should not be used to judge the symmetry of the distribution.

Example 1.17

Figure 1.23 shows the parallel boxplots of AQI data (Table 1.4) for Boston, Houston, and New York. What features of AQI exceedences in these cities stand out?

Solution   First, notice some of the same features that were apparent in the other plots.

- The New York data are fairly symmetric.
- Boston observed a lower number of AQI exceedences than New York and Houston. It also has an outlier on the higher end, indicating a year with comparatively high AQI exceedences.
- Houston has the highest AQI exceedences among the three cities.

However, the clusters and gaps that could be seen using the dotplots of Figure 1.12 cannot be seen in the boxplot.

Figure 1.23

Comparison of AQI days among cities

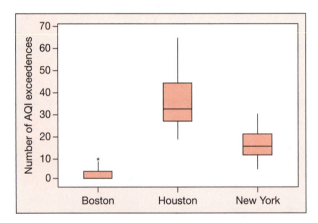

Example 1.18

BestTrack data **(http://www.bbsr.edu/rpi/research/demaria/demaria4.html)** contains different characteristics measured every 6 hours for all the tropical storms developed in the Atlantic Ocean from 1988 to 2002. Not all storms have eyes, but for storms that developed an eye, the eye radius (measured in km) was also recorded. Table below gives the summary of recorded eye radii.

	Eye Radius
100.0% (Maximum)	50.900
75.0% (Upper quartile	27.800
50.0% (Median)	18.500
25.0% (Lower quartile)	13.900
0.0% (Minimum)	4.600
Mean	21.742
Std Dev	9.263
n	463

- The lower quartile = 13.9 km; in other words, at least 25% of the eye radii recorded were at most 13.9 km.
- The median eye radius = 18.5 km; in other words, at least 50% of the eye radii recorded were at most 18.5 km.
- The upper quartile = 27.8 km; in other words, at least 75% of the eye radii recorded were at most 27.8 km.

The IQR = 27.8 − 13.9 = 13.9 km. Because 1.5 IQR = 1.5 (13.9) = 20.85, the maximum whisker length is 20.85. To make a boxplot, first draw a rectangle from Q_L = 13.9 to Q_U = 27.8. Draw a line inside the box at 18.5 to identify the center by median. Because $Q_L − 1.5\ IQR$ is negative (and eye radius cannot be negative), draw the lower whisker from the lower end of the box to the smallest eye radius recorded, 4.6. Because $Q_U + 1.5\ IQR$ = 48.65, draw the upper whisker to the largest radius recorded within 48.65 km. Plot all the measurements beyond whiskers individually.

Figure 1.24 shows only one measurement beyond the upper whisker. The largest radius recorded during this 15-year period was 50.9 km. Further investigation identified this as a radius recorded for hurricane Georges in 1998.

Figure 1.24

A boxplot of eye radius of hurricanes

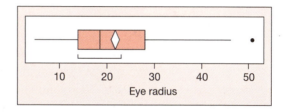

1.5.4   Detecting outliers

Data analysis at this exploratory level can be thought of as detective work with data, and one important bit of detection is to find outliers. An outlier might indicate something went wrong with the data collection process (as you would likely think if you saw a human height recorded as 9.5 feet), or something important caused an unusual value (as you might think if you saw a 20% increase in number of defectives recorded on one day). Outliers are not necessarily mistakes, and should not be removed from dataset without further investigation. A method of identifying an outlier is essential for data analysis. Two different methods in common use are described here, one based on the empirical rule and the other on boxplots.

Detecting an Outlier

Two different rules commonly used for detecting outliers are as follows:

1. An outlier will be any data point that lies below $\bar{x} − 3s$ or above $\bar{x} + 3s$.
2. An outlier will be any data point that lies below $Q_L − 1.5(IQR)$ or above $Q_U + 1.5(IQR)$.

Example 1.19

The summary statistics generated using Minitab for the 2001 AQI data is given below. Using these summary statistics, the IQR $= 33 - 8 = 25$, and $1.5(25) = 37.5$. Because the lower quartile is 8, then $Q_L - 1.5(IQR) = 8 - 37.5 < 0$. As no count of days could be negative, there could not be any outliers on the lower end. In other words, in 2001 no city in the study reported exceptionally clean air compared to the other cities in the study. On the other hand, $Q_U + 1.5(IQR) = 33 + 37.5 = 70.5$, and Los Angeles (81 AQI days) exceeds it. Thus the 2001 AQI data has one outlier, Los Angeles (Figure 1.35a). In other words, Los Angeles recorded extremely unhealthy air (high number of AQI exceedences) compared to the other cities in the study.

Descriptive Statistics: 2001

Variable	N	N*	Mean	SE Mean	StDev	Minimum	Q1	Median	Q3	Maximum
2001	15	0	24.93	5.28	20.45	3.00	8.00	24.00	33.00	81.00

Los Angeles is an outlier in years 1996 and 2001 (Figure 1.25a), two out of three years in the display. Extreme outliers like Los Angeles tend to dominate the display and make it difficult to see other patterns. Figure 1.25b shows what happens when Los Angeles is removed from the dataset. There are no more outliers in the data. Now we can see that variation in the number of AQI exceedences has decreased from 2001 to 2006, an indication that differences in air quality among cities are diminishing. In 1996, the distribution of AQI exceedences is highly skewed toward higher values, whereas in 2006 the distribution is only slightly skewed toward the higher values. The median AQI exceedences increased from 1996 to 2001, an indication of decreased air quality over 5 years. From 2001 to 2006, the decrease in median indicates an improvement in air quality.

Figure 1.25

Number of AQI exceedences at trend sites

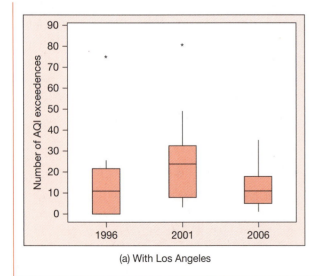

(a) With Los Angeles

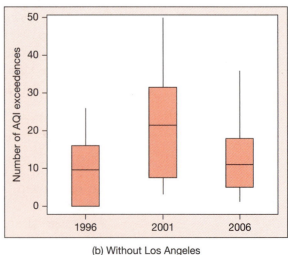

(b) Without Los Angeles

Exercises

1.30   Refer to the differences in per capita tax revenue and per capita expenditure in Table 1.10.
  a   Identify outliers using a boxplot. Comment on the differences between outliers on the higher end versus those on the lower end.
  b   Calculate z-score for the outlier on the lower end. Interpret this z-score.

1.31   Percent on-time arrivals and departures of flights during the fourth quarter of 2003 at major U.S. airports (**http://airconsumer.ost.dot.gov**) are listed in Table 1.11.
  a   Compare the distributions of on-time arrivals and on-time departures by constructing both histograms and boxplots for each dataset.
  b   What proportion of airports listed in Table 1.11 meet or exceed the goal of 90% on-time departures.
  c   What proportion of airports listed in Table 1.11 meet or exceed the goal of 90% on-time arrivals.
  d   Does the empirical rule hold better for arrival or departure data? Can you give a reason for your answer?
  e   What percent of airports had percent arrivals within 5% of the mean?
  f   What percent of airports had percent departures within 5% of the mean?
  g   Identify those airports that can be described as outliers for on-time arrivals. Do the same for on-time departures.

  h   Is Atlanta better on arrivals or departures, based on comparison to the other airports? What about Chicago O'Hare?

1.32   The National Bridge Inventory Program of the U.S. Federal Highway Administration, Office of Bridge Technology (**http://www.fhwa.dot.gov/bridge/britab.cfm**) recorded the percent of obsolete bridges in the year 2003 by state (See Table 1.12). Bridges are classified as functionally obsolete if they have deck geometry, load carrying capacity, clearance, or approach roadway alignment that no longer meet the criteria for the system of which the bridge is a part.
  a   Construct a histogram and box plot for the bridge data.
  b   Describe the data distributions, including any unusual characteristics observed. Identify the states considered to be outliers, if any.
  c   Describe any changes that occur if the District of Columbia and Puerto Rico are removed from the dataset.

1.33   Are highways getting safer? How can we begin to measure possible improvements in the travel by motor vehicle? One possibility is to study traffic safety data for any discernable patterns or trends. The motor vehicle death rates (deaths per 100 million miles driven) for the 50 states plus the District of Columbia are listed in the Table 1.13 (U.S. National Highway Safety Traffic Administration,

Table 1.11

Percent On-Time Arrivals (ARR) and Departures (DEP) at the Major Airports

Airport	ARR	DEP	Airport	ARR	DEP
Atlanta Hartsfield Int	80.5	83.5	Newark Int	71.7	81.7
Baltimore/Washington Int	83.9	83.4	New York Kennedy Int	82.7	83.4
Boston, Logan Int	79.8	82.6	New York LaGuardia	72.8	82.8
Charlotte Douglas	87.1	89.2	Orlando Intl	84.9	89.3
Chicago Midway	82.9	81.9	Philadelphia Int	78.2	80.4
Chicago O'Hare	67.0	73.4	Phoenix Sky Harbor Int	84.2	83.5
Cincinnati Int	83.4	86.0	Pittsburgh Greater Int	84.5	88.1
Dallas/Ft. Worth Reg	86.6	86.0	Portland Int	80.7	86.8
Denver Int	84.6	87.4	Ronald Reagan Nat	83.1	89.2
Detroit, Metro Wayne	85.4	86.9	St. Louis Lambert	86.4	89.3
Fort Lauderdale	82.9	87.0	Salt Lake City Int	81.9	84.7
Houston George Bush	84.2	90.4	San Diego Int	80.2	84.5
Las Vegas McCarran Int	79.5	82.9	San Francisco Int	74.1	84.5
Los Angeles Int	81.3	86.0	Seattle-Tacoma Int	77.4	83.9
Miami Int	82.3	86.1	Tampa Int	84.7	89.9
Minneapolis/St. Paul Int	81.4	87.2	Washington Dulles	82.2	85.5

Table 1.12

Percent of Obsolete Bridges in the United States

State	% Obsolete Bridges	State	% Obsolete Bridges	State	% Obsolete Bridges	State	% Obsolete Bridges
Al	14.6	IA	7.3	NJ	22.6	VT	17.5
AK	17.5	KS	10.6	NM	8.9	VA	17.2
AZ	7.8	KY	21.2	NY	25.4	WA	20.6
AR	15.7	LA	16.0	NC	16.5	WV	21.8
CA	16.1	ME	20.3	ND	5.7	WI	6.5
CO	11.5	MD	20.3	OH	14.5	WY	7.3
CT	24.0	MA	38.6	OK	6.5	PUERTO RICO	36.3
DE	9.8	MI	12.3	OR	16.7		
DC	57.1	MN	4.0	PA	17.5		
FI	15.7	MS	7.9	RI	27.8		
GA	12.1	MO	13.1	SC	8.9		
HI	32.5	MT	9.9	SD	8.1		
ID	10.5	NE	9.5	TN	15.1		
IL	7.8	NV	8.9	TX	15.6		
IN	11.1	NH	18.2	UT	8.8		

Table 1.13

Motor Vehicle Deaths by State, 1980, 2002

State	1980	2002	State	1980	2002	State	1980	2002
AL	3.2	1.8	KY	3.2	2.0	ND	2.9	1.3
AK	3.3	1.8	LA	5.0	2.0	OH	2.8	1.3
AZ	5.3	2.2	ME	3.5	1.5	OK	3.5	1.6
AR	3.6	2.1	MD	2.6	1.2	OR	3.4	1.3
CA	3.5	1.3	MA	2.5	0.9	PA	2.9	1.5
CO	3.2	1.7	MI	2.8	1.3	RI	2.4	1.0
CT	3.0	1.0	MN	3.0	1.2	SC	3.8	2.2
DE	3.6	1.4	MS	4.2	2.4	SD	3.7	2.1
DC	1.2	1.3	MO	3.4	1.8	TN	3.4	1.7
FL	3.6	1.8	MT	4.9	2.6	TX	3.8	1.7
GA	3.5	1.4	NE	3.5	1.6	UT	3.1	1.3
HI	3.3	1.3	NV	5.7	2.1	VT	3.7	0.8
ID	4.8	1.9	NH	3.0	1.0	VA	2.7	1.2
IL	3.0	1.3	NJ	2.2	1.1	WA	3.4	1.2
IN	3.0	1.1	NM	5.4	2.0	WV	4.9	2.2
IA	3.3	1.3	NY	3.4	1.1	WI	3.1	1.4
KS	3.4	1.8	NC	3.6	1.7	WY	4.9	2.0

Traffic Safety Facts, annual, **http://www.nhtsa.dot.gov/people/Crash/Index.html**).

a Construct histograms and boxplots for each year. Are the distributions better characterized by means and standard deviations or medians and interquartile ranges?

b Which states tend to be the outliers for the 1980 data? Why might this be the case?

c Does it appear that highway safety improved between 1980 and 2002? Construct an argument based on these data.

1.6   Summary

The modern industrial and business approach to quality improvement provides a philosophy for problem solving, as well as a model for problem solving and a set of statistical tools to accomplish the task. The philosophy centers on identification of the *real* problem, the use of simple tools to understand the problem and clearly convey the message to others, and the importance of teamwork in identifying the problem and proposing a solution. The simple tools include *Pareto charts* to identify the most important factors that should be improved first; *stemplots and histograms* to show the shapes of data distributions, observe symmetry and skewness, and spot clusters and gaps in the data; *boxplots* to graphically display medians, quartiles, and interquartile ranges and to conveniently compare essential features of two or more data sets; *scatterplots* to show association (correlation) between two measured variables; and *time series plots* to display trends overtime.

Refinements to the above tools include the *mean* as a measure of center to complement the *median* introduced with boxplots, and the *standard deviation* as a measure of variability to complement the *interquartile range*. The mean and standard deviation (most appropriate for mound-shaped, approximately symmetric data distributions) can be used to standardize values so that data measured on different scales can be compared. These two measures also lead to the empirical rule for connecting the mean and standard deviation to a relative frequency interpretation of the underlying data for mound-shaped distributions.

Supplementary Exercises

1.34 The September 19, 2004, issue of *Parade* reported results from the PARADE/Research! America poll (**http://www.researchamerica.org**). A cross-section of 1,000 Americans was asked, "Have you ever used the Internet to look for information on these topics: specific diseases, nutrition, health care providers (doctors/hospitals), exercise, health insurance, weight control)?" The percent responding "yes" was reported as shown in table at the right. Make a bar chart and a Pareto chart for these data. Comment on essential features of the data.

Category	Percentage
Specific diseases	58%
Nutrition	33%
Prescription drugs	32%
Health care providers	28%
Exercise	27%
Health insurance	22%
Weight control	21%

1.35 Table 1.14 lists (top to bottom) the annual temperatures, in degrees Fahrenheit, recorded at

Table 1.14

Annual Temperatures at Central Park and Newnan (1901–2000) °F

Central Park	Newnan	Central Park	Newnan	Central Park	Newnan	Central Park	Newnan
52.91	63.41	52.89	62.07	56.88	62.90	54.94	61.41
54.37	61.41	51.92	60.89	53.59	61.65	53.36	59.57
53.29	60.71	53.43	64.03	55.09	62.04	54.34	60.70
53.39	61.72	51.23	62.79	55.77	62.71	52.97	60.49
51.73	59.64	53.43	65.14	57.03	62.53	55.68	60.15
52.27	61.98	53.55	62.39	54.88	62.92	54.95	61.08
52.37	60.78	54.09	63.03	54.67	62.32	55.22	60.38
50.19	61.61	54.49	62.76	53.49	62.47	54.87	60.38
52.75	61.57	55.79	65.06	55.43	62.64	55.98	58.65
54.25	63.02	55.13	64.20	52.54	60.39	55.45	60.22
52.19	63.20	54.24	65.10	55.40	61.92	55.53	60.71
54.33	63.49	52.95	62.51	54.02	60.72	55.25	61.92
52.98	62.81	53.05	62.89	55.06	61.12	55.12	60.60
52.84	62.22	53.42	62.74	53.42	62.23	54.78	60.05
52.79	65.00	54.46	61.40	53.57	60.58	53.96	60.17
52.04	61.57	55.20	62.43	54.56	61.10	57.22	62.93
54.26	63.33	54.65	62.38	54.19	61.09	56.98	61.90
51.30	62.78	51.88	59.32	55.05	59.78	53.93	60.13
52.79	63.43	54.90	61.92	52.97	60.53	55.49	60.41
51.53	64.10	54.06	60.52	54.07	60.00	55.23	60.88
50.04	62.39	53.74	61.63	54.79	60.25	55.28	61.16
52.48	63.55	54.64	63.09	54.26	61.69	53.72	60.59
53.07	63.87	54.08	62.39	55.21	61.29	54.28	59.98
52.28	61.91	55.31	63.32	53.86	61.47	57.13	61.48
54.89	65.24	53.76	61.19	56.06	61.48	56.48	61.34
53.54	64.27	54.04	62.85	54.69	61.13	53.80	59.06

Central Park, New York, and Newnan, Georgia, for the years 1901–2000.

a Make a histogram of temperatures for each location.

b Make parallel boxplots for temperatures. Comment on the distributions of temperatures.

c Compare the distributions of temperatures at Central Park and Newnan.

1.36 A survey of 8,453 consumers was conducted by BIGresearch for the National Retail Federation. The consumers were asked, "When will you begin your back-to-school shopping?" Table below shows the results reported by *USA Today* (August 19, 2004).

When will you begin your back-to-school shopping?	Percentage
Two months before school starts	15
Three weeks to a month before school starts	41
One to two weeks before school starts	35
The week school starts	6
After school starts	3

a Construct a bar chart to display these data. Comment on your findings.

b What percent of consumers shop before school starts?

c How many consumers in the survey responded that they shop during the week school starts?

1.37 In the 2000 presidential elections, 60% of the eligible voters showed up to vote. A Census Bureau survey of November 2000 (**http://www.census.gov**) reported the percentage who voted, classified by household income is given below:

Household Income	Percentage
Less than $5,000	34
$5,000 to $9,999	41
$10,000 to $14,999	44
$15,000 to $24,999	51
$25,000 to $34,999	58
$35,000 to $49,999	62
$50,000 to $74,999	69
$75,000 and over	74

Make an appropriate plot to display these data. Comment on your findings.

1.38   Table below gives the age and gender of patients participating in a medical study at the University of South Alabama.

47	M	26	M	70	F	63	F	76	M
75	F	67	M	63	F	49	M	72	F
73	M	22	M	59	F	59	F	60	F
41	F	32	F	69	F	76	F	64	F
54	M	24	F	55	F	52	M	58	F
23	F	23	F	68	F	57	M	37	M
61	F	24	F	78	F	80	F	88	M
26	M	73	M	37	F	44	F	33	M
61	M	77	M	63	F	63	F	52	M
73	F	56	F	78	M	57	F	50	M
20	M	59	F	60	F	76	F	71	F
33	M	62	F	56	F	52	F	44	M
23	M	64	F	63	F	65	F	68	F
36	M	76	F	75	F	61	M	62	F
58	M	68	F	44	M	77	F	62	F
53	M	59	F	78	F	74	M	70	F
21	F	62	F	64	F	69	M	67	F
68	M	37	M	48	M	46	F	74	F
52	M	67	F	79	M	67	M	63	F
74	M	72	F	60	F	72	M	76	F
54	M	73	F	72	F	55	F	56	M
65	F	49	F	43	M	60	F	54	F

a   Construct a histogram for the age of patients, disregarding their gender. Describe the distribution.
b   Does the empirical rule work well for these data?
c   Suppose all outliers are removed from the data. Does the empirical rule now work better? Why or why not?
d   Construct histograms and boxplots for the age of patients by gender. Compare the age distributions of male and female patients based on information in these plots.

1.39   Wolmuth and Surters (*IEE Proc. Commun.*, 2003) studied bridge failures as they relate to the size of the crowd on the bridge at the time of failure. They reported the distribution of crowd size on the failed bridges as shown in table below. Make an appropriate plot of these data and describe the distribution, in the context of the problem, using center, spread, and shape.

Number Reported on the Bridge	Number of Bridge Collapses
<26	2
26–150	14
150–250	4
251–500	3
501–750	2
>750	2

1.40   Taylor, Wamuziri, and Smith (*Civil Engineering*, 2003) studied operation of Big-Canopy Systems used in building construction. A big canopy erected at the construction site provides a comprehensive overhead delivery system and improved working environment with more advanced construction manipulators and control systems. Such a system includes columns, beams, slabs, interior wall elements, drainpipes, air-conditioning ducts, and so on. Big construction corporations expect reduction in construction costs of large structures can be reduced and efficiency improved by use of a Big-Canopy system compared to the traditional methods. Table below shows the labor (man-days per m^2 floor area of superstructure erection work) required with different construction techniques for a superstructure erection. Make an appropriate graph and comment on your findings.

Technique	Labor Man-Days/m^2 Floor Space
Conventional method	100.0
System framework method	72.9
PC concrete and tower crane	38.6
Big-Canopy project Y	25.6
Big-Canopy project K	26.0
Big-Canopy project H	20.0

1.41   Fanny, Weise, and Henderson (2003) studied the impact of a photovoltaic system installed on the rooftop of the National Institute of Standards and Technology Building on the conservation of electricity usage. The energy savings were recorded monthly from November 2001 until October 2002.

Month	Energy Savings	Max Peak Demand Savings	Thermal Savings	Total Savings
Nov 01	71.49	1.77	5.09	78.35
Dec 01	61.43	37.39	9.57	108.38
Jan 02	54.47	50.10	8.52	113.09
Feb 02	94.84	56.71	8.47	160.01
Mar 02	104.19	75.28	6.56	186.03
Apr 02	132.77	63.33	3.17	199.27
May 02	166.18	79.92	1.66	247.75
Jun 02	164.24	38.40	0.60	279.19
Jul 02	154.17	81.12	0.87	433.12
Aug 02	148.62	56.71	0.81	343.81
Sep 02	140.58	56.97	0.16	336.02
Oct 02	67.35	31.09	4.35	192.92
Total	**1,360.32**	**628.76**	**49.83**	**2,677.94**

a Make a bar chart that shows savings in the three categories over time.

b Describe the patterns observed in the chart.

c Compute the mean and standard deviation of the total savings over the one-year period.

d Construct a boxplot for the total savings. Identify the outliers, if any.

1.42 McMullin and Owen (*J. Professional Issues in Eng. Edu. and Practice*, 2002) reported assessment of one engineering course taught using distance learning technology at San Jose State University and San Francisco State University. The mean grade achieved by 27 students at the home site was 2.9. The mean grade achieved by 16 students at the distant site was 2.6. Determine the mean grade of students enrolled in this program.

1.43 Sulfur oxide is one of the major components of the AQI measure of air quality. Table below shows sulfur dioxide emission estimates for the

Source Category	1990	1995	2000
Fuel combustion	20.3	16.2	14.9
Industrial processes	1.9	1.6	1.5
Transportation	1.5	1.3	1.8
Miscellaneous	0.0	0.0	0.0
Total	**23.7**	**19.2**	**18.2**

Note: Some columns may not sum to total due to rounding

years 1990, 1995, and 2000, along with major source categories (U.S. Environmental Protection Agency, Office of Air Quality Planning and Standards). Obviously, the most important source is fuel combustion from refineries and mills. Produce a Pareto chart of source categories for these three years. Comment on any similarities and differences you observe.

1.44 Lead is a major pollutant in both air and water, but it is not part of the AQI. Table below shows lead emission estimates and source categories for the years 1990, 1995, and 2000 (U.S. Environmental Protection Agency, Office of Air Quality Planning and Standards).

Source Category	1990	1995	2000
Fuel combustion	0.5	0.5	0.5
Industrial processes	3.3	2.9	3.2
Transportation	1.2	0.6	0.6
Total	**5.0**	**3.9**	**4.2**

a Sketch a Pareto chart of source categories for the years 1990, 1995, and 2000. How do these compare?

b Has the quality of our environment improved in recent years with regard to lead?

Exploring Bivariate Distributions and Estimating Relations

Do hurricanes with higher maximum low-level winds have lower minimum surface pressure? Is the change in maximum low-level winds dependent on minimum surface pressure?

   To answer such questions, we must study the relationships between different variables. In the previous chapter, we studied tools for analyzing univariate data (i.e., a set of measurements on a single variable), categorical as well numerical. Frequently, however, one variable is dependent upon one or more related variables. For example, Figure 2.1 shows that the minimum surface pressure of a hurricane tends to decrease as maximum wind speed increases. Similarly, the amount of energy required to heat a house depends on the size of the house, and the chemical yield of a process depends on the temperature of the chemical process. In this chapter, we will begin the study of tools for analyzing *bivariate data*, detecting patterns, and using the detected patterns for decision making. We introduce models in which the mean of one variable can be written as a linear function of another variable. This procedure, known as *regression analysis*, is one way of characterizing association between two variables. The strength of such a linear association is measured by the correlation coefficient.

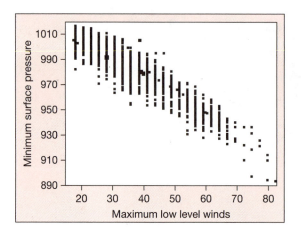

Figure 2.1
Scatterplot of pressure
versus wind speed

Introduction

Many engineering applications involve *modeling* the relationships among sets of measurements. For example, a chemical engineer might want to model the yield of a chemical process as a function of the temperature and pressure at which the reactions take place. An electrical engineer might be interested in modeling the daily peak load of a power plant as a function of the time of year and the number of customers. An environmental engineer might want to model the dissolved oxygen of samples from a large lake as a function of the algal and nitrogen content of the sample. Such modeling generally makes use of a statistical methodology called *regression analysis*, which will be introduced in this chapter.

In regression analysis, the data used to model relationships must be collected from independently selected items. For example, in studying the relationship between size of houses and heating bills, the houses used in the study should be selected independently from one another. If the data are collected over time, then there may be some dependence among successive observations. For example, the maximum daily temperatures recorded at the Mobile-Pascagoula Regional Airport will be dependent upon one another. The temperature in one place on one day depends on the temperature at the same place on the previous day. A technique known as the *time series analysis* is used to analyze such time-dependent data. This book will not go deeply into time series analysis, but we will distinguish between exploratory analyses of time series data and regression analyses of non–time series data.

This chapter first describes use of two-way tables and percentages to study relations between two categorical variables. Then it presents exploratory analyses of measurements depending on the time when they were taken, followed by the regression models for the association between two sets of measurements that are not time dependent. Chapter 11 extends the exploratory regression ideas to more than two sets of measurements and discusses the statistical inference of regression models.

2.1 Two-Way Table for Categorical Data

The clarity with which patterns in data can be seen often depends upon the manner in which the quantitative information is displayed. In the two examples that follow, observe how the *two-way table* adds information to the one-way tables and plots introduced earlier.

Example 2.1

Table 2.1 and the bar chart in Figure 2.2 display projections of net new workers by the Bureau of Labor Statistics for the years 2000–2010 in the United States. It shows that about 24% of net new workers will be white males, with lower percentages for other categories.

However simple the table and however attractive the bar chart, these are one-way summaries that are not very useful for interpreting data and showing associations among the variables. A closer look at the bar chart and the one-way table reveals that two categorical factors are actually defined here, gender and race. Thus, the data can be conveniently displayed in a two-way table (Table 2.2) that shows how the two variables are related.

Table 2.1

Projections on Net New Workers

Category	Percent
White men	24
White women	23
Black men	6
Black women	9
Asian men	6
Asian women	7
Hispanic men	12
Hispanic women	13

Figure 2.2

Projections on net new workers

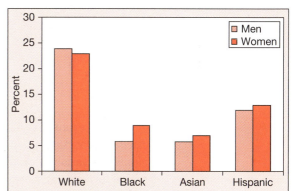

Table 2.2

Projections on Net New Workers by Gender and Race

	Women	Men	Total
White	23%	24%	**37%**
Black	9%	6%	**15%**
Asian	7%	6%	**13%**
Hispanic	13%	12%	**25%**
Total	**52%**	**48%**	**100%**

From this 4×2 table (i.e., a table with 4 rows and 2 columns of data), we see that many more women than men will be entering the workforce, and a quarter of the net new workers will be Hispanic. These features are more easily spotted on a two-way table than a one-way table.

Example 2.2

The Maine Department of Public Safety collected data on the seat belt use and injury (**http://www.maine.gov/portal/index.php**) for 71,552 passengers involved in car or light truck accidents. For each passenger involved in an accident, two types of data were recorded:

- Whether the passenger was wearing a seat belt
- Whether the passenger was injured

Note that both of these are categorical measurements, recorded as "yes" or "no." Just a list of yes and no responses for 71,552 passengers is of not much use for decision making. But the data summarized in a 2×2 table below reveals some informative patterns. The *marginal percentages* show that the majority of passengers were wearing seatbelts (85.4%), and the majority of passengers were not injured (91.2%). The *conditional percentages* show that the percent of passengers injured was almost four times higher among those not wearing seat belts $\left(\frac{2323}{10,461} (100) = 22.2\% \right)$ than that among those wearing seat belts $\left(\frac{3984}{61,091} (100) = 6.5\% \right)$. Thus, there appears to be an association between seat belt use and injuries.

Counts (percentages)	Injured?		
	Yes	**No**	**Total**
Seat belt? Yes	3,984 (5.6)	57,107 (79.8)	**61,091 (85.4)**
No	2,323 (3.2)	8,138 (11.4)	**10,461 (14.6)**
Total	**6,307 (8.8)**	**65,245 (91.2)**	**71,552 (100)**

These examples showed how to use two-way tables of data collected for categorical variables. However, in many experiments the data are collected for two numerical variables or for one numerical variable periodically over a certain time period. In such cases, we use regression analysis or time series techniques to analyze the collected data.

Exercises

2.1 Table below shows the data collected by the University of South Alabama Trauma Center from the Emergency Medical Services records for the years 2001 and 2002. For each person involved in an accident, two types of data were recorded: mortality/nonmortality and the seating position in the motor vehicle (front, middle, or rear seat).

Count	Mortality	Nonmortality	Total
Front	565	39,974	**40,539**
Middle	55	3,098	**3,153**
Rear	2	157	**159**
Total	**622**	**43,229**	**43,851**

a Make appropriate graph(s) to display these data.

b What can you say about the possible association between mortality and seating position?

2.2 Table below gives the percent of housing units in the United States in 2003 (**http://www.census.gov/hhes/www/housing/ahs/nationaldata.html**) by the type of housing and owner/renter classification.

Category	Percent
Single-family home, owner	59.82%
Single-family home, renter	10.10%
Multiple-unit home, owner	3.21%
Multiple-unit home, renter	20.40%
Mobile home/trailer, owner	5.21%
Mobile home/trailer, renter	1.26%

a Show a better way of displaying these data for comparison of owners and renters within different types of housing units.

b Make an appropriate chart to display these data.

c Compare and comment on the percent of owned or rented housing units by different types. Is there any apparent association between the type of housing unit and ownership?

2.3 Table below gives the number of housing units (in thousand) in United States in 2003 (**http://www.census.gov/hhes/www/housing/ahs/nationaldata.html**) by the type of sewage disposal and geographic location of the housing unit. For the geographic location, the United States was divided into four parts: Northeast (NE), Midwest (MW), South (S), and West (W).

Does there appear to be any association between the mode of sewage disposal used and the geographic location of the housing unit? Justify your answer using data provided in table below.

	NE	MW	S	W
Public Sewer	15,984	19,737	28,023	20,369
Septic tank	4,181	4,734	10,075	2,684
Other	6	16	43	16

2.4 Table below gives the electric energy retail sales (in billions of kilowatt-hours) by class of service and state for year 2001 (**http://www.ela.doe.gov**, January 2003).

State	Residential	Commercial	Industrial	Total
Alabama	27.8	18.9	31.8	79.2
Florida	101.4	74.0	18.8	199.7
Georgia	44.4	37.8	33.9	117.8
Louisiana	25.8	17.7	28.6	74.7
Mississippi	16.9	11.4	15.3	44.3

Is there an apparent association between the class of service and the state?

2.2 Time Series Analysis

Often data are collected over certain time period. For example:

- Number of industrial accidents recorded per week.
- Number of machine breakdowns per month.
- Quarterly production figures.
- Annual profits.
- Daily temperatures.
- Monthly unemployment rates.
- Number of service calls received per hour.

In Chapter 1, we studied air quality data by using various univariate summary tools. But two-way features of the background data on air quality remain to be studied. What are the trends in the emission of pollutants over time? Such trends are easily spotted on a time series plot.

Example 2.3

The total emission estimates for carbon monoxide and volatile organic compounds across the decade of the study are listed in table below. Figure 2.3 shows the time series plot of this data. Notice that carbon monoxide levels decreased slightly during the first part of the decade, but they seem to have taken an upward turn during the latter half of the decade. The volatile organic compounds show a decrease over the decade, but at a very slow rate.

Year	Carbon Monoxide	Volatile Organic Compound
1989	106.439	22.513
1990	99.119	21.053
1991	101.797	21.249
1992	99.007	20.862
1993	99.791	21.099
1994	103.713	21.683
1995	94.058	20.918
1996	104.600	19.924
1997	105.466	20.325
1998	101.246	19.278
1999	102.356	19.439
2000	109.300	20.384

Figure 2.3

CO and VOC emission (million short ton)

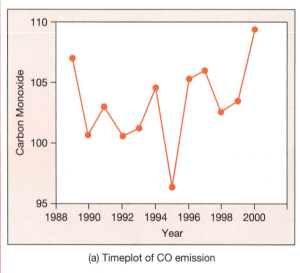

(a) Timeplot of CO emission

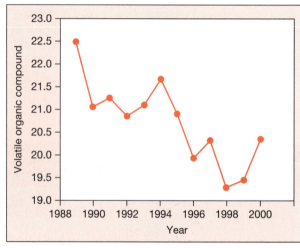

(b) Timeplot of VOC emission

Example 2.4

The average AQI days over 100 per city from Table 1.6 (average over 15 selected cities) over years 1990–2006 are

24.60, 28.00, 19.93, 22.40, 20.93, 25.73, 14.46, 18.13, 23.06, 26.53, 20.53, 24.93, 24.46, 18.86, 12.80, 17.53, 13.46.

The time series plot of the average AQI days over 100 is shown in Figure 2.4. Overall, there is a downward trend, but there also seems to be a cyclic pattern in the number of exceedences over time. A sharp decrease followed by a gradual increase occurs every four years. Can you think of any reason for this?

It would be necessary to look at data for more than one decade and conduct further investigations in order to establish the reasons for such a pattern, but notice that during this time period the drop coincides with the election year. It is possible that election year policies led to different measures being implemented resulting in a decrease in AQI days.

Figure 2.4

Average number of days
with AQI > 100

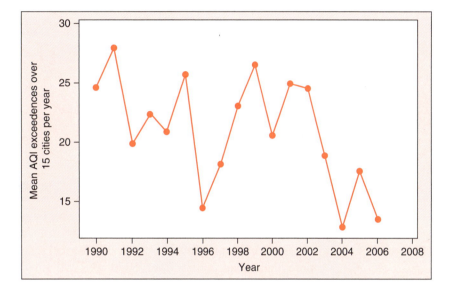

Example 2.5

Table 2.3 gives data on the annual fuel consumption (billion gallons) and the miles per gallon (mpg) by two types of vehicles: cars and vans. The classification "vans" includes vans, pickups, and SUVs. The U.S. Federal Highway Commission collected the data for years 1970–1999. The mpg for a vehicle determines the fuel efficiency of that vehicle. Studying how the fuel consumption and the fuel efficiency have changed over about a 30-year period is enhanced by use of time series plots (Figure 2.5).

Table 2.3

Fuel Consumption and Average MPG for Cars and Vans, 1970–1999

Year	Fuel Consumption (billion gallons)		Average Miles per Gallon (mpg)		Year	Fuel Consumption (billion gallons)		Average Miles per Gallon (mpg)	
	Car	Van	Car	Van		Car	Van	Car	Van
1970	67.8	12.3	13.5	10.0	1989	74.1	33.3	18.0	16.1
1975	74.3	19.1	14.0	10.5	1990	69.8	35.6	20.3	16.1
1980	70.2	23.8	16.0	12.2	1991	64.5	38.2	21.2	17.0
1981	69.3	23.7	16.5	12.5	1992	65.6	40.9	21.0	17.3
1982	69.3	22.7	16.9	13.5	1993	67.2	42.9	20.6	17.4
1983	70.5	23.9	17.1	13.7	1994	68.1	44.1	20.8	17.3
1984	70.8	25.6	17.4	14.0	1995	68.1	45.6	21.1	17.3
1985	71.7	27.4	17.5	14.3	1996	69.2	47.4	21.2	17.2
1986	73.4	29.1	17.4	14.6	1997	69.9	49.4	21.5	17.2
1987	73.5	30.6	18.0	14.9	1998	71.7	50.5	21.6	17.2
1988	73.5	32.7	18.8	15.4	1999	73.2	52.8	21.4	17.1

U.S. Federal Highway Commission, *Highway Statistics*, annual

Figure 2.5

Fuel consumption and efficiency of cars and vans

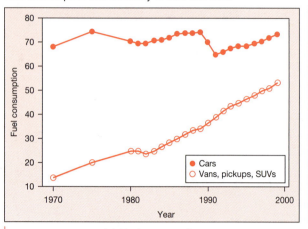

(a) Fuel consumption

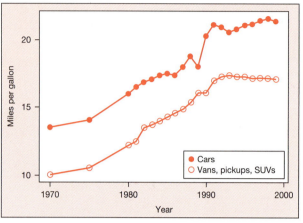

(b) Fuel efficiency

The fuel consumption by vans is increasing at a fairly steady rate, more than quadrupling from 1970 to 1999. In comparison to vans, the fuel consumption by cars has not changed much overall (Figure 2.5a). During the '80s, the fuel consumption by cars increased slightly; in 1990–91, it dropped considerably, but during the '90s it increased steadily at a very low rate. Increased fuel consumption by vans, at a higher rate than that of cars, may be the result of more people opting to drive vans, pickups, and trucks rather than cars. On the other hand, the fuel efficiency (as determined by mpg) has increased steadily for both cars and vans (Figure 2.5b). During the '80s, the fuel efficiency of vans increased at a higher rate than that during the '90s. In 1989, fuel efficiency of cars decreased slightly, but then it increased again over the next decade. This drop in fuel consumption in 1990 might be due to the increase in fuel efficiency.

Example 2.6

The mean annual temperatures (in °F) recorded at Newnan, Georgia, for years 1901 to 2000 (read from left to right) are as follows:

59.64 61.98 60.78 61.61 61.57 63.02 63.20 63.49 62.81 62.22
65.00 61.57 63.33 62.78 63.43 64.10 62.39 63.55 63.87 61.91
65.24 64.27 62.07 60.89 64.03 62.79 65.14 62.39 63.03 62.76
65.06 64.20 65.10 62.51 62.89 62.74 61.40 62.43 62.38 59.32
61.92 60.52 61.63 63.09 62.39 63.32 61.19 62.85 62.90 61.65
62.04 62.71 62.53 62.92 62.32 62.47 62.64 60.39 61.92 60.72
61.12 62.23 60.58 61.10 61.09 59.78 60.53 60.00 60.25 61.69
61.29 61.47 61.48 61.13 61.41 59.57 60.70 60.49 60.15 61.08
60.38 60.38 58.65 60.22 60.71 61.92 60.60 60.05 60.17 62.93
61.90 60.13 60.41 60.88 61.16 60.59 59.98 61.48 61.34 59.06

The time series plot of the mean annual temperatures at Newnan during the last century is given in Figure 2.6. Notice two types of temperature variations, long term and short term. Long-term variation is an overall trend in annual temperatures. It shows an increasing trend during the early part of the last century with annual temperatures reaching their peak around the end of first quarter and then a decreasing trend in annual temperatures. Short-term variations are the year-to-year fluctuations in temperatures, also known as "jitters." These jitters tend to mask the long-term variations.

One method of smoothing out these jitters is to compute and plot the moving averages (MA) of the temperatures. To get moving averages of length 3 (i.e., MA (3)), we can average temperatures for 3 consecutive years and plot the averages against the middle year of each trio. For example, average temperatures for years 1901, 1902, 1903, and plot the average against year 1902. Next, average temperatures for years 1902, 1903, 1904, and plot the average against year 1903. Continue this process until the last one, in other words, the average of temperatures for years 1998, 1999, and 2000 plotted against year 1999. Figure 2.7 shows that the moving average process has smoothed out some of the jitters in the temperature trend, making it easier to observe the trend in annual Newnan temperatures over the 20th century.

When the data are collected over time, the successive measurements, or the measurements separated by a certain time period, are often associated with each other. For example, each day's highest temperature is generally dependent on the previous day's highest temperature at the same location, or the soft-drink sale on Saturday is associated with that on the previous Saturday. Such dependence between measurements taken over time (measured by *autocorrelation*) sets time series analysis apart from the regression analysis introduced later in this chapter and taken up again in Chapter 11.

Figure 2.6

Annual temperatures (in °F) at Newnan 1902–2000

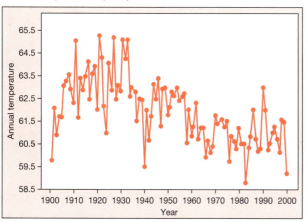

Figure 2.7

MA(3) for annual temperatures (in °F) at Newnan

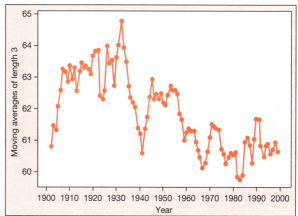

Exercises

2.5 Sulfur dioxide is one of the major components of the AQI measure of air quality. Table 2.4 gives the sulfur dioxide estimates (in short tons) for years 1980 and 2000. (National Air Quality and Emissions Trends Report, **http://www.epa.gov/airtrends/**).

    a Construct a time series plot for the total emission across the years. Comment on any trend you observe.

    b Construct a time series plot for each of the sources of sulfur dioxide emission across the years. Comment on any trend you observe.

2.6 Table 2.5 gives the vehicle accident data for different types of trucks (U.S. Census Bureau, *Statistical Abstract of the United States*, 2003).

    a Construct a time series plot for each type of vehicle.

    b Compare and comment on any trends observed.

2.7 Table 2.6 gives the average fuel prices (in dollars per Btu) for four different sectors from 1970 to 2004 (U.S. Census Bureau, *Statistical Abstract of the United States*, 2008).

    a Construct a time series plot for energy cost for each sector.

    b Compare and comment on any trends observed.

2.8 Table 2.7 gives data on research and development (R&D) expenditures (in millions of dollars) by industry and universities (U.S. Census Bureau, *Statistical Abstract of the United States*, 2008). Because there are other sources of funds, such as Federal government and nonprofit organizations not included in the table, the industry and university numbers do not add to the total.

    a Construct time series plots of R&D expenditures for industries and universities.

    b Compare and comment on any trends observed.

    c Compute R&D expenditures by other sources. Plot the data and compare the trends to the trends observed in the industry and university expenditures.

Table 2.4

Sulfur Dioxide Estimates for 1980–2000

Source	1989	1990	1991	1992	1993	1994	1995	1996	1997	1998	1999	2000
Fuel combustion	19.924	20.290	19.796	19.493	19.245	18.887	16.230	16.232	16.649	16.746	16.027	14.876
Electrical utilities	16.215	15.909	15.784	15.416	15.189	14.889	12.080	12.730	13.195	13.416	12.653	11.389
Industrial	3.086	3.550	3.256	3.292	3.284	3.218	3.357	2.863	2.805	2.742	2.788	2.894
Other	0.624	0.831	0.755	0.784	0.772	0.780	0.793	0.639	0.649	0.588	0.586	0.593
Industrial processes	2.010	1.900	1.720	1.758	1.723	1.675	1.638	1.408	1.458	1.463	1.457	1.498
Transportation	1.349	1.476	1.517	1.553	1.497	1.297	1.311	1.791	1.816	1.837	1.853	1.805
Total	**23.293**	**23.679**	**23.044**	**22.813**	**22.474**	**21.875**	**19.189**	**19.447**	**19.939**	**20.059**	**19.349**	**18.201**

Table 2.5

Vehicle Accident Frequencies for Trucks

Vehicle Type	1990	1994	1995	1996	1997	1998	1999	2000
Light trucks	8601	8904	9568	9932	10249	10705	11265	11418
Pickup	5979	5574	5939	5904	5887	5921	6127	5953
Utility	1214	1757	1935	2147	2380	2713	3026	3324
Van	1154	1508	1639	1832	1914	2042	2088	2104
Other	254	65	56	46	68	29	24	37

Table 2.6

Average Fuel Prices for 1970–2004 (dollars per Btu)

Year	Residential	Commercial	Industrial	Transportation	Year	Residential	Commercial	Industrial	Transportation
1970	2.10	1.98	0.84	2.31	1988	10.66	10.82	5.00	6.57
1971	2.24	2.16	0.92	2.37	1989	11.02	11.27	4.92	7.18
1972	2.37	2.33	0.99	2.38	1990	11.88	11.89	5.23	8.28
1973	2.72	2.56	1.10	2.57	1991	12.08	12.07	5.18	7.99
1974	3.38	3.41	1.78	3.70	1992	11.98	12.17	5.13	7.93
1975	3.81	4.08	2.20	4.02	1993	12.28	12.58	5.16	7.88
1976	4.13	4.39	2.43	4.21	1994	12.63	12.74	5.15	7.92
1977	4.77	5.13	2.78	4.48	1995	12.63	12.64	4.97	8.09
1978	5.13	5.51	3.03	4.59	1996	12.73	12.78	5.40	8.77
1979	6.00	6.28	3.63	6.19	1997	13.29	13.05	5.34	8.70
1980	7.46	7.85	4.71	8.61	1998	13.48	13.07	4.91	7.48
1981	8.82	9.49	5.52	9.84	1999	13.19	12.87	5.12	8.23
1982	9.78	10.37	6.05	9.43	2000	14.27	13.93	6.49	10.78
1983	10.66	10.94	6.21	8.45	2001	15.72	15.62	6.80	10.21
1984	10.68	11.10	6.12	8.26	2002	14.74	14.70	6.28	9.63
1985	10.91	11.65	6.03	8.27	2003	15.82	15.56	7.47	11.21
1986	10.75	11.22	5.36	6.22	2004	17.16	16.55	8.54	13.37
1987	10.71	10.98	5.17	6.59					

Table 2.7

R&D Expenditure 1960–2004 (in millions of dollars)

Year	Expenditure	Industry	University	Year	Expenditure	Industry	University
1960	13,711	4,516	67	1983	89,950	45,264	1,357
1961	14,564	4,757	75	1984	102,244	52,187	1,514
1962	15,636	5,124	84	1985	114,671	57,962	1,743
1963	17,519	5,456	96	1986	120,249	60,991	2,019
1964	19,103	5,888	114	1987	126,360	62,576	2,262
1965	20,252	6,549	136	1988	133,880	67,977	2,527
1966	22,072	7,331	165	1989	141,889	74,966	2,852
1967	23,346	8,146	200	1990	151,990	83,208	3,187
1968	24,666	9,008	221	1991	160,872	92,300	3,457
1969	25,996	10,011	233	1992	165,347	96,229	3,568
1970	26,271	10,449	259	1993	165,726	96,549	3,709
1971	26,952	10,824	290	1994	169,201	99,203	3,938
1972	28,740	11,715	312	1995	183,618	110,870	4,110
1973	30,952	13,299	343	1996	197,338	123,416	4,435
1974	33,359	14,885	393	1997	212,142	136,227	4,837
1975	35,671	15,824	432	1998	226,455	147,845	5,162
1976	39,435	17,702	480	1999	245,036	164,660	5,618
1977	43,338	19,642	569	2000	267,557	186,135	6,230
1978	48,719	22,457	679	2001	277,736	188,439	6,824
1979	55,379	26,097	785	2002	276,591	180,711	7,341
1980	63,224	30,929	920	2003	289,025	186,174	7,648
1981	72,292	35,948	1,058	2004	300,060	191,376	7,932
1982	80,748	40,692	1,207				

2.9   Table below gives the total U.S. consumption and conventional hydroelectric power consumption (in quadrillion Btu). The data were reported by the U.S. Census Bureau's *Statistical Abstract of the United States* (2001).

Year	Total U.S. consumption	Hydroelectric
1990	6.26	3.14
1994	6.39	2.97
1995	6.96	3.47
1996	7.45	3.92
1997	7.37	3.94
1998	6.99	3.55
1999	7.37	3.42

a   Construct a time series plot for total U.S. energy consumption.

b   Compare and comment on any trends observed.

c   Compute the hydroelectric power consumption as the percentage of total consumption. Construct a time series plot and comment on any trends observed.

2.10   Table below gives (in thousands) the number of state motor vehicle registrations for the United States. Construct a time series plot and describe the trend observed.

Year	Vehicles	Year	Vehicles
1925	20,069	1945	31,035
1930	26,750	1950	49,162
1935	26,546	1955	62,689
1940	32,453	1960	73,858
1965	90,358	1985	171,689
1970	108,418	1990	188,798
1975	132,949	1995	201,530
1980	155,796	2000	221,475

2.11   Table below gives data on the urban and rural federal-aid highways (in thousand miles) as reported by the U.S. Census Bureau's *Statistical Abstract of the United States* (2001). Federal-aid highways are the highways constructed using funding from the federal government.

Year	Urban	Rural
1980	624	3331
1985	691	3171
1990	757	3123
1993	803	3102
1994	814	3093
1995	814	3093
1996	834	3100
1997	843	3116
1998	842	3065

a   Construct time series plots for urban and rural highways.

b   Compare and comment on any trends observed.

2.12   Table below gives the crude oil prices in U.S. dollars per barrel for years 1975–2006 (U.S. Census Bureau, *Statistical Abstract of the United States*, 2008).

Year	Price	Year	Price	Year	Price
1975	10.38	1986	14.55	1997	19.04
1976	10.89	1987	17.90	1998	12.52
1977	11.96	1988	14.67	1999	17.51
1978	12.46	1989	17.97	2000	28.26
1979	17.72	1990	22.22	2001	22.95
1980	28.07	1991	19.06	2002	24.10
1981	35.24	1992	18.43	2003	28.53
1982	31.87	1993	16.41	2004	36.98
1983	28.99	1994	15.59	2005	50.24
1984	28.63	1995	17.23	2006	60.23
1985	26.75	1996	20.71		

a   Display this data using a time series plot.

b   Identify any trends in the crude oil prices.

c   Identify any peaks or troughs or sudden changes in the crude oil prices. Associate them with then current events that lead to such sudden price changes.

2.13   Table 2.8 gives the age-adjusted deaths per 100,000 population from 1960 to 2001 (U.S. Census Bureau, *Statistical Abstract of the United States*, 2003) for the leading disease causes. Construct a time series plot for each cause and comment on how the patterns compare.

Table 2.8

Death Rates per 100,000 Population, 1960–2001

Year	Heart Disease	Cancer	Diabetes	Influenza	Liver Disease
1960	559.0	193.9	22.5	53.7	13.3
1961	545.3	193.4	22.1	43.4	13.3
1962	556.9	193.3	22.6	47.1	13.8
1963	563.4	194.7	23.1	55.6	14.0
1964	543.3	193.6	22.5	45.4	14.2
1965	542.5	195.6	22.9	46.8	14.9
1966	541.2	196.5	23.6	47.9	15.9
1967	524.7	197.3	23.4	42.2	16.3
1968	531.0	198.8	25.3	52.8	16.9
1969	516.8	198.5	25.1	47.9	17.1
1970	492.7	198.6	24.3	41.7	17.8
1971	492.9	199.3	23.9	38.4	17.8
1972	490.2	200.3	23.7	41.3	18.0
1973	482.0	200.0	23.0	41.2	18.1
1974	458.8	201.5	22.1	35.5	17.9
1975	431.2	200.1	20.3	34.9	16.7
1976	426.9	202.5	19.5	36.8	16.4
1977	413.7	203.5	18.2	31.0	15.8
1978	409.9	204.9	18.3	34.5	15.2
1979	401.6	204.0	17.5	26.1	14.8
1980	412.1	207.9	18.1	31.4	15.1
1981	397.0	206.4	17.6	30.0	14.2
1982	389.0	208.3	17.2	26.5	13.2
1983	388.9	209.1	17.6	29.8	12.8
1984	378.8	210.8	17.2	30.6	12.7
1985	375.0	211.3	17.4	34.5	12.3
1986	365.1	211.5	17.2	34.8	11.8
1987	355.9	211.7	17.4	33.8	11.7
1988	352.5	212.5	18.0	37.3	11.6
1989	332.0	214.2	20.5	35.9	11.6
1990	321.8	216.0	20.7	36.8	11.1
1991	313.8	215.8	20.7	34.9	10.7
1992	306.1	214.3	20.8	33.1	10.5
1993	309.9	214.6	22.0	35.2	10.3
1994	299.7	213.1	22.7	33.9	10.2
1995	296.3	211.7	23.4	33.8	10.0
1996	288.3	208.7	24.0	33.2	9.80
1997	280.4	205.7	24.0	33.6	9.60
1998	272.4	202.4	24.2	34.6	9.50
1999	267.8	202.7	25.2	23.6	9.70
2000	257.6	199.6	25.0	23.7	9.50
2001	247.8	196.0	25.3	22.0	9.50

2.3

Scatterplots: Graphical Analysis of Association between Measurements

How should we explore the possible relationship between two measurement variables, neither of which is time? The simplest graphical tool used for detecting association between two variables is the *scatterplot*, which simply plots the ordered pairs of data points on a rectangular coordinate system, similar to the time series plot. Three examples follow: Figure 2.8a is the plot used in the introduction to this chapter showing minimum surface pressure of storms plotted against maximum low level winds; Figure 2.8b shows time to destination plotted against distance traveled for helicopters delivering victims of automobile accidents to a hospital; and Figure 2.8c is a plot of time on the scene versus time to reach the scene for an emergency medical service.

- If the plot shows a roughly elliptical cloud (instead of a curved, fan-shaped, or clustered cloud) with data points spread throughout the ellipse, then a conclusion of *linear association* is reasonable. For example, the narrow ellipses enclosing most of the data points in Figure 2.8a and Figure 2.8b indicate that there is evidence of linear association between the variables in each pair.
- If the ellipse tilts upward to the right, as in Figure 2.8b, the association is *positive*. If the ellipse tilts downward to the right, as in Figure 2.8a, the association is *negative*.
- If the ellipse is thin and elongated, the association is *strong*. If the ellipse is closer to a circle or is horizontal, the association is *weak*. The horizontal and somewhat "fat" and irregular ellipse of Figure 2.8c indicates very weak association between the time to the scene and the time at the scene.

The study of AQI data in Chapter 1 concentrated on plotting one variable at a time. But policymakers might be able to take advantage of association between the variables. For example, the same policies might be applicable to two cities with similar records, or a change in manufacturing operations might improve levels of

Figure 2.8

Scatterplots

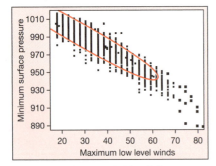

(a) Pressure versus wind speed of storms

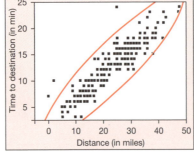

(b) Time to destination versus distance traveled

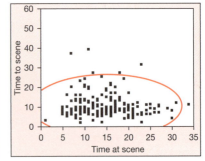

(c) Time to scene versus time at scene

two pollutants positively associated with each other. The decision-making process should begin with an exploration of which variables are associated with each other and of the nature of the relationships.

Consider 2006 AQI data for Chicago and Detroit (Table 1.6). Chicago had 5 and Detroit had 11 AQI days, which produces a pair (5, 11). Each year produces a similar ordered pair for Detroit and Pittsburgh. When we plot these ordered pairs, we get a scatterplot (see Figure 2.9a).

Notice that Detroit and Chicago tend to vary together; when Chicago has a high number of AQI days, so does Detroit. This fairly strong positive association between Detroit and Chicago might lead to the conclusion, after further investigation, that similar pollution control policies would be appropriate for both the cities. Looking at Figure 2.9b, we see that the number of AQI days for Pittsburgh and Chicago may be positively associated, but the association is much weaker than that between Detroit and Chicago. The solution to Chicago's problems may not have much in common with the solutions to Pittsburgh's problems.

Across the cities, how are the 2006 AQI days associated with 1996 AQI days? Can we display changes across the decade? Pairing on cities, Figure 2.10a shows the scatterplot of 2006 against 1996.

It looks as if there is a positive association between the 2006 and 1996 AQI days, as we would expect, but it is difficult to see what is happening for most of the cities because of the strong *influence* of Los Angeles. Putting Los Angeles on the graph with other cities causes a scaling problem; all the other cities cluster together in one corner. Removing Los Angeles produces a new scatterplot, as shown in Figure 2.10b. Now Pittsburgh seems to have more influence, but it is possible to see the pattern for the other cities. The association is still positive but appears to be a bit weaker.

Suppose we draw a $y = x$ line on the scatterplot, as shown in Figure 2.11. A point below that line indicates that 1996 AQI days exceed the 2006 AQI days, which is a good sign. There are 5 cities out of 14 (remember that we removed Los Angeles from the dataset) below this line: Atlanta, Chicago, Detroit, Houston, and

Figure 2.9

Association between AQI days in different cities in 2006

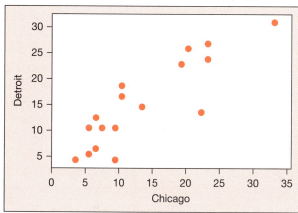

(a) AQI days at Detroit and Chicago

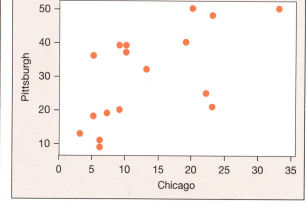

(b) AQI days at Pittsburgh and Chicago

Figure 2.10

Association between AQI days in 1996 and 2006, with and without Los Angeles

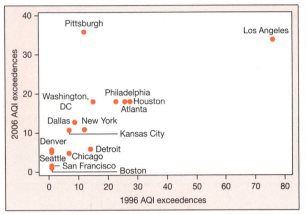

(a) With Los Angeles

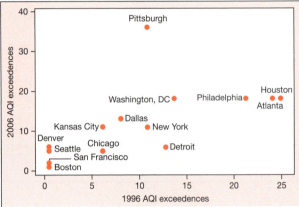

(b) Without Los Angeles

Figure 2.11

Association between AQI days in 1996 and 2006, without Los Angeles

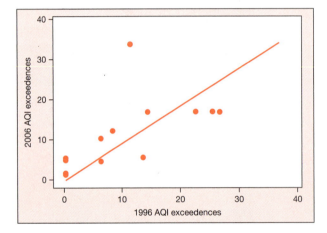

Philadelphia. A point above the line indicates worsening air quality and the need for new policies. Unfortunately, 9 out of 14 cities in the study fall in this category.

What can we say about a city such as Chicago that is close to the $y = x$ line? Looking at a scatterplot this way, can we identify those cities on which we should concentrate our efforts in improving air quality? Is Pittsburgh one of them?

Perhaps there are associations among the pollutants making up the AQI that would allow some economy in their treatment. The U.S. Environmental Protection Agency has set national air quality standards for six principal air pollutants: nitrogen dioxide (NO_2), ozone (O_3), sulfur dioxide (SO_2), particulate matter (PM), carbon monoxide (CO), and lead (Pb). Table 2.9 lists the maximum air quality concentrations of four pollutants measured in 2000 at the 15 cities studied earlier. The carbon monoxide concentration is measured in 1,000,000 ppm (parts per million); ozone and SO_2 are measured in ppm, and PM_{10} is measured in $\mu g/m^3$.

Figure 2.12 shows *a scatterplot matrix* of pollutant concentrations paired on the 15 sites around the country. Such scatterplot matrices are useful for reviewing pairwise relations among several variables. The scatterplot matrix is symmetric

Table 2.9

Maximum Concentrations of Air Pollutants in Year 2000

City	CO (ppm)	O_3 (ppm)	PM_{10} ($\mu g/m^3$)	SO_2 (ppm)
Atlanta	4.1	0.023	0.11	0.019
Boston	3.4	0.029	0.08	0.030
Chicago	8.3	0.032	0.08	0.075
Dallas	3.5	0.014	0.10	0.047
Denver	2.1	0.016	0.08	0.009
Detroit	4.4	0.024	0.08	0.043
Houston	4.2	0.021	0.12	0.031
Kansas City	1.8	0.017	0.09	0.039
Los Angeles	9.5	0.044	0.11	0.010
New York	9.3	0.038	0.09	0.046
Philadelphia	5.1	0.028	0.10	0.027
Pittsburgh	2.4	0.025	0.09	0.086
San Francisco	1.7	0.020	0.05	0.007
Seattle	2.4	0.021	0.07	0.011
Washington, D.C.	4.9	0.023	0.09	0.030

Source: U.S. Environmental Protection Agency, National Air Quality and Emissions Trends Report, 2003.

Figure 2.12

A scatterplot matrix

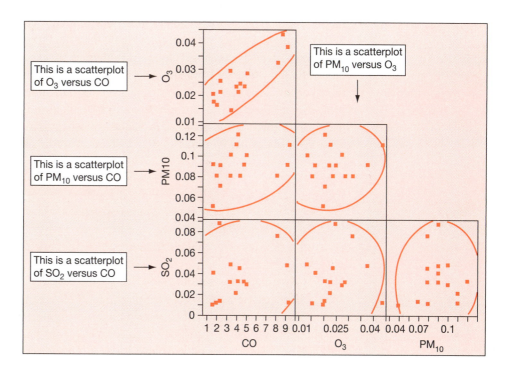

because the association between CO and O_3 is the same as the association between O_3 and CO. Thus scatterplots in the upper triangle (not shown here) are essentially the same as the scatterplots in the lower triangle, except that the axes are reversed. With data collected for four variables (CO, O_3, PM_{10}, and SO_2), we get six pairs, resulting in six different scatterplots.

Upon inspection, notice that

- The plot of ozone and carbon monoxide shows fairly strong positive association.
- The plot of particulate matter against carbon monoxide also shows some positive association, although not as strong as that between CO and O_3.
- The scatterplots suggest no strong association between SO_2 and other three pollutants.

Ozone is not directly emitted but is formed when nitrogen oxides and volatile organic compounds react in the presence of sunlight. On further checking, we discover that particulate matter comes mostly from factories, construction sites, and motor vehicles, whereas carbon monoxide comes mostly from motor vehicles. Because both involve motor vehicles and because construction sites are usually polluted areas, perhaps both of these pollutants can, to some extent, be controlled simultaneously.

Example 2.7

In heat exchangers, frequently old tubes have to be removed and replaced by new ones. This process often results in enlarged tubesheet holes (i.e., the hole in which the tube is inserted). When new tubes are installed, they are expanded into enlarged tubesheet holes using different procedures; however, they also result in thinning walls of tubes. Repeated enlargement of holes could result in thinning of the tubewalls beyond the material strain limit. Shuaib et al. (*J Press. Vessel Tech.*, 2003) conducted an experiment to evaluate the effect of one such procedure (roller expansion of heat exchanger tubes in enlarged tubesheet holes) on thinning of tubewalls (i.e., tubewall reduction). They investigated different levels of tube-to-tubesheet hole clearance (i.e., overtolerance, OT) to simulate the hole enlargement process, and recorded resulting percent wall reduction (WR). The WR of tube is useful in evaluating the adequacy of tube–tubesheet joints. The data for 25 roller-expanded tubes are given in Table 2.10 (read from the chart).

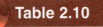

Table 2.10

OT and WR Data

OT	WR	OT	WR
−0.0010	1.30	0.0050	5.50
−0.0010	3.00	0.0062	4.20
−0.0005	1.60	0.0087	4.10
−0.0005	3.50	0.0105	6.00
−0.0005	4.25	0.0123	4.90
−0.0005	4.30	0.0127	6.80
0.0005	3.80	0.0127	6.90
0.0010	3.80	0.0132	5.20
0.0015	2.70	0.0137	7.20
0.0018	4.40	0.0140	6.70
0.0025	4.90	0.0148	5.80
0.0038	2.80	0.0148	5.90
0.0042	3.50		

Figure 2.13

Tubewall reduction vs. overtolerance

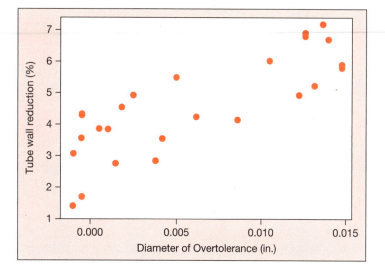

Figure 2.13 gives the scatterplot of expanded tube wall reduction versus overtolerance of the joints. This scatterplot shows a strong relation between the overtolerance and the tubewall reduction. Note that increasing overtolerance has an increasing effect on the roller expanded tubewall reduction. In general, wall reduction seems to increase linearly with overtolerance.

The maximum value of OT allowed by the Tubular Exchanger Manufacturer Association (TEMA) standard is 0.002 in. We can note from the scatterplot that within allowable TEMA standard, the wall reduction falls in the range of 1.3% to 4.4% only. The industry-accepted maximum wall reduction level is 12%. This experiment showed that even when the OT level was increased beyond seven times the TEMA standard, the percent wall reductions of all tested tubes were well below the industry-accepted level. Note that the maximum wall reduction recorded was only 7.2%.

Exercises

2.14   Are highways getting safer? How can we begin to measure possible improvements in the travel by motor vehicle? One possibility is to study traffic safety data for any discernable patterns or trends. The motor vehicle death rates per 100 million miles driven in the 50 states plus the District of Columbia are listed in Table 2.11 (U.S. National Highway Safety Traffic Administration, *Traffic Safety Facts*, annual (**http://www.nhtsa.dot.gov/ people/Crash/Index.html**).

    a Construct a scatterplot of the 2002 data versus the 1980 data. Describe the pattern. Which point deviates most extremely from the pattern?

    b Are there any interesting clustering patterns among the states?

    c What feature of the plot will help you say something about the improvement in highway safety over these years?

2.15   The longer you stay in school, the more you eventually earn! True? Explore the data of Table 2.12 to see whether it appears to support this claim. Educational attainment is turned into an approximate number to simplify the statistical analysis.

    a Construct a scatterplot of median weekly earnings versus years of schooling. Describe the pattern. Does a linear trend completely describe the pattern here?

    b Construct a scatterplot of unemployment versus years of schooling. Describe any patterns you see here.

Table 2.11

Motor Vehicle Deaths by State, 1980 and 2002

State	1980	2002	State	1980	2002	State	1980	2002
AL	3.2	1.8	KY	3.2	2.0	ND	2.9	1.3
AK	3.3	1.8	LA	5.0	2.0	OH	2.8	1.3
AZ	5.3	2.2	ME	3.5	1.5	OK	3.5	1.6
AR	3.6	2.1	MD	2.6	1.2	OR	3.4	1.3
CA	3.5	1.3	MA	2.5	0.9	PA	2.9	1.5
CO	3.2	1.7	MI	2.8	1.3	RI	2.4	1.0
CT	3.0	1.0	MN	3.0	1.2	SC	3.8	2.2
DE	3.6	1.4	MS	4.2	2.4	SD	3.7	2.1
DC	1.2	1.3	MO	3.4	1.8	TN	3.4	1.7
FL	3.6	1.8	MT	4.9	2.6	TX	3.8	1.7
GA	3.5	1.4	NE	3.5	1.6	UT	3.1	1.3
HI	3.3	1.3	NV	5.7	2.1	VT	3.7	0.8
ID	4.8	1.9	NH	3.0	1.0	VA	2.7	1.2
IL	3.0	1.3	NJ	2.2	1.1	WA	3.4	1.2
IN	3.0	1.1	NM	5.4	2.0	WV	4.9	2.2
IA	3.3	1.3	NY	3.4	1.1	WI	3.1	1.4
KS	3.4	1.8	NC	3.6	1.7	WY	4.9	2.0

Table 2.12

Educational Attainment, Income, and Unemployment*

Education Attained	Years of Schooling (approximate)	Median Weekly Earnings in 2006 (dollars)	Unemployment Rate in 2006 (percent)
Professional degree	20	1,474	1.1
Doctoral degree	20	1,441	1.4
Master's degree	18	1,140	1.7
Bachelor's degree	16	962	2.3
Associate degree	14	721	3.0
Some college, no degree	13	674	3.9
High school graduate	12	595	4.3
Less than a high school diploma	10	419	6.8

*Data are 2006 annual averages for persons ages 25 and over. Earnings are for full-time wage and salary workers.

Source: Bureau of Labor Statistics.

2.16 Refer to the index of industrial production data in Table 1.15.
  a Construct a scatterplot of 1998 indices versus 1990 indices.
  b Describe the nature of relationship. Comment on any patterns you see.

2.17 Refer to the tax revenue data from Table 1.10.
  a Construct a scatterplot of per capita expenditure versus per capita tax revenue.
  b Describe the nature of the relationship. Comment on any patterns you see.

  c Calculate the difference between the per capita expenditure and per capita tax revenue. Plot these differences against the per capita expenditure. Describe the observed pattern.
  d Note that Alaska is an outlier in the scatterplot of part (a). Remove Alaska from the dataset and redo the scatterplot of per capita expenditure versus per capita tax revenue. Comment on any patterns observed. Describe any differences from your answer in part (a).

2.18 Seamless steel and aluminum cylinders that are used to transport high-pressure gases are required to meet safety regulations promulgated by the U.S. Department of Transportation and other national authorities. Smith, Rana, and Hall (*J. Pressure Vessel Technology*, 2003) used "fitness for service" assessment procedures described in API Recommended Practice 579 to predict, by analysis, the cylinder burst pressure. They also conducted a number of hydrostatic burst tests on selected cylinders to verify the analytical results. The measured (experimental) and calculated (analytical) burst pressures (RSF, or remaining strength factor) are reported in table below

Flaw Description			
Longitudinal Notch		Rectangular local thin area (LTA)	
Measured RSF	Calculated RSF	Measured RSF	Calculated RSF
0.91	0.95	0.94	0.98
0.87	0.90	1.00	0.96
0.83	0.84	0.94	0.94
0.78	0.77	0.93	0.91
0.78	0.68	0.97	0.86
1.00	0.98	0.88	0.80
1.01	0.95	0.76	0.72
0.98	0.93		

a Construct a scatterplot of calculated RSF values versus the measured RSF values.

b Describe the nature of relationship. Do calculated RSF values provide good estimates for the measured RSF values?

c Construct separate scatterplots for the two different types of flaws. Do you observe any differences in association?

2.19 The importance of management to the long-term careers of practicing professional engineers has long been recognized. Palmer (*J. Professional Issues in Engineering Education and Practice*, 2003) reported results of a survey in which academicians in U.S. MBA programs and executives responsible for technology management in industry in the United States rated the importance of technology management knowledge and skills. The ranking was reported on a scale of 0–5 and averaged over all the respondents. The results are reproduced in Table 2.13.

a Construct a scatterplot of ranking by executives versus ranking by professors.

b Comment on any relationship observed.

Table 2.13

Ranking by Professors and Executives

Management Knowledge/Skill	Ranking by	
	Professors	Executives
Integration of technology and business strategy	4.56	4.25
Working across functional boundaries	4.47	4.34
Written communication skills	4.40	4.43
Strategic role of technology	4.36	4.25
Achieving implementation	4.33	4.59
Implementation of new technology	4.31	4.14
Identification of new technological opportunity	4.29	3.95
Oral communication skills	4.24	4.50
Management of complexity and ambiguity	4.24	4.32
Ability to apply analytical techniques	4.09	4.18
Solving problems on a timely basis	4.07	4.57
Gaining users' support	4.05	4.27
Facility in human relations	4.04	3.89
Transfer of technology within organizations	3.98	4.11
Ability to apply theoretical knowledge	3.95	3.66
New product development	3.93	4.20
Business strategy and competition	3.89	4.16
General business functions	3.82	4.11
Ability to manage technical professionals	3.82	4.36
Selection of technological projects	3.80	3.93
Transfer of technology between organizations	3.80	3.52
Management of risk	3.78	4.30
Timing of technological choice	3.73	3.98
Technology acquisition	3.73	3.66
Process of technological innovation	3.73	3.60
Producing clearly actionable results	3.72	4.36
Use of manufacturing technology	3.60	3.80
Evaluation of technical projects	3.47	3.60
Perform technological assessment/evaluation	3.44	3.77
Handling data gaps and conflicts	3.34	4.00
Management of research	3.20	3.48
Use of information technology	3.04	3.19
Environmental issues	3.02	3.09
Influence of government policy	2.98	2.74
Social aspects	2.91	3.09
Ethical aspects	2.89	3.14
Financing technical projects	2.84	3.52
Legal aspects	2.78	3.35
General engineering functions	2.55	3.80

2.20 Table below gives the average amount of refined oil products imported and exported per day over the years 1990–2000 (U.S. Census Bureau, *Statistical Abstract of the United States*, 2001). The averages are reported in 1,000 barrels (bbl) per day, where a barrel contains 42 gallons.

a Make time series plots for the import and export of oil products. Comment on any trends observed.

b Make a scatterplot of export versus import amounts. Comment on any trends observed.

Year	Import	Export
1990	2123	748
1991	1844	885
1992	1805	861
1993	1833	904
1994	1933	843
1995	1605	855
1996	1971	871
1997	1936	896
1998	2002	835
1999	2122	822
2000	2146	974

2.4 Correlation: Estimating the Strength of a Linear Relation

Once a linear association between two variables appears to be reasonable, the next step is to measure the strength of the association. This is accomplished through the *correlation coefficient*. Correlation between pairs of observations is measured by Pearson's correlation coefficient, calculated as follows for observed measurement pairs (x, y):

$$r = \frac{1}{n-1} \sum_{i=1}^{n} \left(\frac{x - \bar{x}}{s_x} \right) \left(\frac{y - \bar{y}}{s_y} \right) = \frac{1}{n-1} \sum_{i=1}^{n} z_x z_y$$

where z_x and z_y are the standardized x and y scores, respectively. Because terms in the denominator are always positive, the product of terms in the numerator will determine the sign (positive or negative) of r. Geometrically, the correlation coefficient describes how close the data points are to the line describing the relation between x and y.

- Note that $-1 \leq r \leq 1$ always.
- A value of r near or equal to zero implies little or no linear relationship between y and x.
- In contrast, the closer r is to 1 or -1, the stronger the linear relationship between y and x. If $r = \pm 1$, all the points fall exactly on the line.
- A positive value of r implies that y increases as x increases.
- A negative value of r implies that y decreases as x increases.

Each of these situations is portrayed in Figure 2.14.

Figure 2.14

Values of *r* and their implications

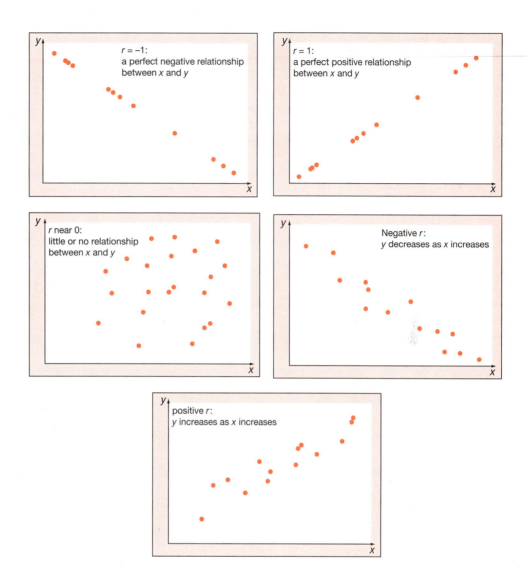

Example 2.8

Suppose the mean daily peak load (*y*) for a power plant and the maximum outdoor temperature (*x*) for a sample of 10 days is as given in Table 2.14.

The scatterplot of the collected data is given in Figure 2.15. It shows a positive association between the maximum daily temperature and the daily peak load. How strong is this relation? We can determine the strength from the correlation coefficient.

Using the mean and standard deviation given in table below, the data are standardized by

$$z = \frac{\text{measurement} - \text{mean}}{\text{standard deviation}}$$

	Mean	**Standard Deviation**
Temperature (*x*)	91.5	6.5021
Load (*y*)	194.8	46.2645

Table 2.14

Power Load and Temperature

Day	Maximum Temperature (x)	Peak Power Load (y)
1	95	214
2	82	152
3	90	156
4	81	129
5	99	254
6	100	266
7	93	210
8	95	204
9	93	213
10	87	150

Figure 2.15

Scatterplot of power load and temperature

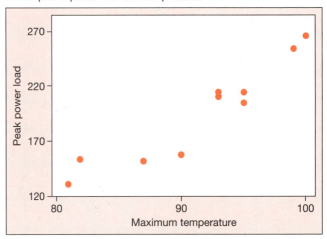

The standardized data appear in table below. The computed correlation coefficient is

$$r = \frac{1}{n-1}\sum_{i=1}^{n} z_x z_y = \frac{8.497}{10-1} = 0.944.$$

x	y	z_x	z_y	$z_x z_y$
95	214	0.538	0.415	0.223
82	152	−1.461	−0.925	1.351
90	156	−0.231	−0.839	0.193
81	129	−1.615	−1.422	2.296
99	254	1.153	1.279	1.476
100	266	1.307	1.539	2.011
93	210	0.230	0.328	0.075
95	204	0.538	0.198	0.107
93	213	0.230	0.393	0.090
87	150	−0.692	−0.968	0.670

$$\sum z_x z_y = 8.497$$

The fact that the value of r is positive and near 1 indicates that the peak power load is very strongly associated with the daily maximum temperature, and the peak power load tends to increase as the daily maximum temperature increases *for this sample of 10 days.*

How do individual points affect the correlation coefficient? Remember from Chapter 1 that the further the z-score is away from 0, the further away the data point is from the mean of the dataset. Look at the point (81, 129), which has large (and negative) z-scores compared to the other data points. The large $z_x z_y$ (2.296) indicates that this point contributes a considerable amount to the correlation coefficient. If this point is omitted from the dataset, the correlation coefficient will decrease. Computation shows that the correlation coefficient calculated from the remaining 9 points is 0.775, compared to 0.944 with all 10 data points.

In Chapter 1, you learned that outliers can have a tremendous effect on the mean and standard deviation. Outliers in scatterplots are points widely separated from the rest of the data in any direction, and different kinds of outliers have different effects on the correlation coefficient. It is important to remember that points separated from the rest of the data are *potentially influential*. The extent of influence of any point can be judged, in part, by computing the correlation coefficient with and without that point.

Example 2.9

Refer to O_3 and SO_2 data collected from 15 cities in the year 2000 (Table 2.19). The standardized values of this pollution data are shown in table below.

City	O_3	SO_2	z_{O_3}	z_{SO_2}	$z_{O_3} z_{SO_2}$
Atlanta	0.023	0.019	−0.245	−0.645	0.157
Boston	0.029	0.030	0.489	−0.172	−0.084
Chicago	0.032	0.075	0.856	1.762	1.508
Dallas	0.014	0.047	−1.345	0.558	−0.752
Denver	0.016	0.009	−1.101	−1.075	1.182
Detroit	0.024	0.043	−0.122	0.386	−0.047
Houston	0.021	0.031	−0.489	−0.129	0.063
Kansas City	0.017	0.039	−0.978	0.214	−0.210
Los Angeles	0.044	0.010	2.323	−1.032	−2.397
New York	0.038	0.046	1.589	0.515	0.820
Philadelphia	0.028	0.027	0.366	−0.301	−0.110
Pittsburgh	0.025	0.086	−0.000	2.235	−0.000
San Francisco	0.020	0.007	−0.611	−1.161	0.709
Seattle	0.021	0.011	−0.489	−0.989	0.483
Washington, DC	0.023	0.030	−0.245	−0.172	0.042

Figure 2.16

Scatterplot of O_3 versus SO_2, with and without LA

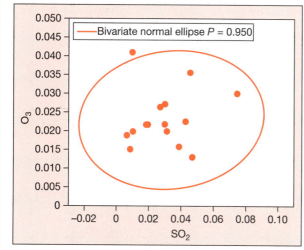

(a) Scatterplot of O_3 versus SO_2

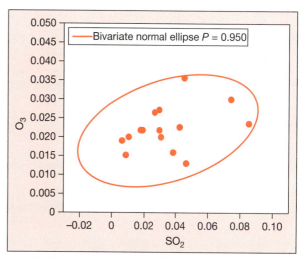

(b) Scatterplot of O_3 versus SO_2 without Los Angeles

The scatterplot of the data is given in Figure 2.16a. The cloud is almost circular and does not show any association between the O_3 and SO_2 levels measured. This conclusion is also supported by the fact that the correlation coefficient is equal to 0.097. Now notice that Los Angeles (with $z_{O_3} z_{SO_2} = -2.397$) is outside the cloud. Remember from Chapter 1 that Los Angeles was an outlier in terms of AQI days. With such a high negative z-score product, this data point exerts considerable influence on the correlation coefficient, as its large negative z-score product will reduce the correlation coefficient considerably.

Figure 2.16b shows the scatterplot of O_3 versus SO_2 without Los Angeles. Now the cloud is more oval shaped and the incline indicates positive association, although the cloud is not very compact and the association not very strong. Computations show that the correlation coefficient between O_3 and SO_2 without Los Angeles is 0.383.

Example 2.10

Refer to the pollutant data of 15 cities measured in year 2000 (Table 2.9). The matrix of correlation coefficients between different pollutants is given in table below.

	CO	O_3	PM_{10}	SO_2
CO	1.0000	0.8688	0.361	0.1722
O_3	0.8688	1.0000	0.2045	0.0976
PM_{10}	0.3610	0.2045	1.0000	0.0912
SO_2	0.1722	0.0976	0.0912	1.0000

The correlation coefficients show that all four pollutants are positively associated with each other; in other words, an increase in the level of one is associated with an increase in the level of the others.

However, the levels of carbon monoxide are most strongly associated with the ozone levels, followed by particulate matter levels. Very weak to almost nonexistent associations are observed among the levels of other pollutants. Refer to the scatterplot matrix of Figure 2.12. The differences in strengths of correlations are evident from the narrowness or wideness of ellipses enclosing the data points, and the positive or negative nature of correlation is evident from the inclination of the ellipse.

Example 2.11

Refer to the expanded tubewall reduction data from Example 2.7. The correlation coefficient between the tubewall reduction and overtolerance is equal to 0.811. This shows that there is a fairly strong positive association between the tubewall reduction and the overtolerance, as shown by the scatterplot in Figure 2.13.

A word of caution: Do not infer a causal relationship on the basis of high correlation. Although it is probably true that a high maximum daily temperature *causes* an increase in peak power demand, the same would not necessarily be true of the relationship between availability of oil and peak power load, even though they are probably positively correlated. That is, we would not be willing to state that low availability of oil *causes* a lower peak power load. Rather, the scarcity of oil tends to cause an increase in conservation awareness, which in turn tends to cause a decrease in peak power load. Thus, a large value for a sample correlation coefficient indicates only that the two variables tend to change together (are associated); causality must be determined after considering other factors in addition to the size of the correlation.

Exercises

2.21 Refer to the traffic safety data of Exercise 2.14.
  a Calculate the correlation between the number of motor vehicle deaths in 1980 and 2002.
  b Interpret this correlation in terms of the problem.
  c Refer to your scatterplot in Exercise 2.14. What conclusions can you make from the scatterplot that you cannot from the correlation coefficient?
  d Refer to your scatterplot in Exercise 2.14. What conclusions can you make from the correlation coefficient that you cannot from the scatterplot?

2.22 Refer to the index of industrial production data in Table 1.9.
  a Calculate correlation between the index of industrial production in 1990 and 1998.
  b Interpret this correlation in terms of the problem.
  c Refer to your scatterplot in Exercise 2.16. Does it support your conclusion from the correlation coefficient?

2.23 Refer to the tax revenue data of Table 1.10 and the plot of Exercise 2.17.
  a Compute a correlation matrix for revenue, taxes, and expenditure.
  b Interpret correlation coefficients in terms of the problem.
  c Which variables have the strongest association? Justify your answer.
  d Are there any outliers affecting the correlations? If yes, what is their effect?

2.24 Refer to the burst pressure (RSF) data for industrial cylinders reported in Exercise 2.18.
  a Calculate the correlation between the measured and calculated RSF.
  b Is there any association between the measured and calculated RSF? Justify your answer.

  c Does the nature or strength of association differ by the type of flaw? Justify your answer.

2.25 Refer to the managerial skill/knowledge data of Exercise 2.13. Is there any association between the professors' and executives' opinions about the importance of technology management knowledge and skills? Justify your answer.

2.26 Refer to the refined oil import and export data in Exercise 2.20.
  a Is there any association between the amount of import and export of refined oil products by the United States?
  b Make a time series plot showing trend in import and export. Discuss your findings from the time series plot.
  c Compare and discuss your findings from the correlation coefficient and the time series plot.

2.27 Zsaki et al. (*International J. for Numerical Methods in Fluids*, 2003) studied the solution to the vector Laplacian system (inner system) using the finite element tearing and interconnecting (FETI) method. They used different mesh sizes from 1/48 to 1/8 to search across the entire computation domain. Table 2.15 shows the number of inner FETI iterations with and without the reconjugation needed to obtain numerical solution to the system when using different mesh sizes.
  a Determine whether there is any association between the mesh size used and the inner FTEI iterations without reconjugation.
  b Determine whether there is any association between the mesh size and the inner FTEI iterations with reconjugation.
  c Determine whether there is any association between the mesh size and difference in the inner FTEI, without and with reconjugation.

Table 2.15

Mesh Size and Number of Iterations Needed

Mesh size	1/8	1/12	1/16	1/20	1/28	1/32	1/38	1/40	1/48
No. of inner FETI iterations (without reconjugation)	217	252	269	271	291	291	291	293	311
No. of inner FETI iterations (with reconjugation)	95	107	109	113	115	118	118	119	119

2.5 Regression: Modeling Linear Relationships

Once it is established that two variables are associated with each other, the next step is to quantify that relationship through a mathematical model. Such a quantification (model) is useful for measuring average rate of change and predicting values of one variable for known values of the other variable, among other things. For example, a problem facing every power plant is the estimation of the daily peak load. Suppose we wanted to model the peak power load as a function of the maximum temperature for the day.

The scatterplot (Figure 2.17) shows that a straight-line function might describe this relation well. However, notice that not all data points fall on this line. In fact, they will not fall on any one line mainly because of other sources of variation such as the geographical location of the plant and number of customers affecting the peak power load.

Figure 2.17

Straight-line fit to power load and temperature data

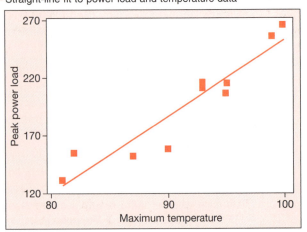

Figure 2.18

A simple linear regression model

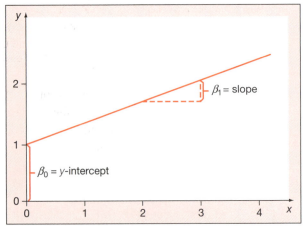

The statistical process of finding mathematical equations (or models) that represent the relationship between two variables is called **regression analysis**. The simplest of these models is a straight line, which will be referred to as a **simple linear regression model** or **simple linear regression equation**. Here, we consider only simple linear models; more complicated models will be discussed in Chapter 11.

In general, a linear relation between two variables is given by following equation:

$$y = \beta_0 + \beta_1 x$$

where

- y is the **response variable**.
- x is the **explanatory variable**.
- β_0 is the **y-intercept**. It is the value of y, for $x = 0$ (see Figure 2.18).
- β_1 is the **slope** of the line. It gives the amount of change in y for a unit change in the value of x (see Figure 2.18).

Fitting such a line (a regression line) to a set of data involves estimating the slope and intercept to produce a line that is denoted by $\hat{y} = \hat{\beta}_0 + \hat{\beta}_1 x$.

2.5.1 Fitting the model: The least-squares approach

In the example of daily peak power load described earlier, we saw that the peak power load has a strong positive association with the daily maximum temperature. A mathematical model for peak power load as a function of maximum temperature would allow us to estimate the peak power load for a given maximum temperature. From the scatterplot, the straight-line model seems appropriate to describe this relationship. However, which line should we use? If we model the relation by eyeballing a line, then it will result in a very subjective assessment of the relation. We want to calculate the line modeling the relationship in such a way that, regardless of who does the modeling, the same equation would result from a given set of measurements.

The scatterplot in Figure 2.19 shows that for the day with maximum temperature 90 °F, the measured peak power load is 156 megawatts. Suppose we are using the line plotted on the scatterplot to model the relation between the peak power load and the maximum temperature. Then, as marked, on the day with maximum temperature 90 °F, the peak load is predicted to be about 185 megawatts. The difference $(156 - 185) = -29$ megawatts is the *residual* between the observed power load value and the one on plotted line. That is, $(y_i - \hat{y}_i) = -29$ for the observation at $x = 90$.

Note that a few points are above the line and a few are below. For the points below the line, this line overpredicts the peak power, whereas for those above the line, the line underpredicts the peak power load. So there is a residual associated with each measured data point unless the point falls on the line, in which case the residual is zero. Some data points have a positive residual associated with them (i.e., underprediction), and some data points have a negative residual associated with them (i.e., overprediction). When all residuals add to zero, we can say that,

Figure 2.19

Residual in estimating
response for a given *x*

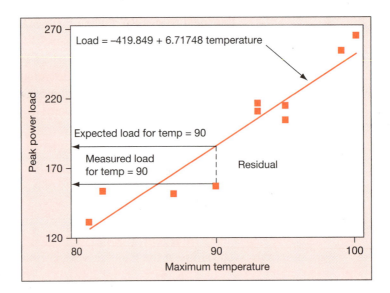

on the average, there is no residual in prediction. But what about the variation in residuals? To determine that, we compute the sum of the squared residuals, commonly called the *sum of squared errors* (SSE) as

$$SSE = \sum (y_i - \hat{y}_i)^2$$

Note the similarity between SSE and the formula for the numerator of a variance. Using this similarity, the SSE value can be interpreted as the total distance between the data points and the fitted line. But different lines will result in different values of SSE. How, then, do we decide which line is "best"? One approach is to select a line that minimizes the SSE as the *line of best fit*. Such a line is also known as the *least-squares regression line*. The least-squares estimates of slope and *y*-intercept are referred to as the *least-squares regression coefficients*.

The **least-squares regression line** is $\hat{y} = \hat{\beta}_0 + \hat{\beta}_1 x$ with

$$\text{slope } \hat{\beta}_1 = \frac{\sum (x - \bar{x})(y - \bar{y})}{\sum (x - \bar{x})^2} \text{ and } y\text{-intercept } \hat{\beta}_0 = \bar{y} - \hat{\beta}_1 \bar{x}$$

A line estimated using least-squares estimates of the *y*-intercept and the slope has the smallest possible SSE. This implies that, on the average, this line is likely to give predictions closer to the actual values than any other line.

Properties of a least-squares regression line:

- The sum (and the mean) of residuals is zero.
- The variation in residuals is as small as possible for the given dataset.
- The line of best fit will always pass through the point $(\bar{x}, \bar{y})$.

Example 2.12

Refer to Table 2.14 and Figure 2.15 showing data collected on the maximum temperature and the peak power load.

a Compute the least-squares estimates of y-intercept and slope. Interpret them.
b How do you expect the peak power load to change if the maximum temperature increases by 5 °F.
c Write the equation of the least-squares regression line.

Solution From the data we get the calculations shown in table below:

Temp x	Load y	$(x - \bar{x})$	$(y - \bar{y})$	$(x - \bar{x})^2$	$(x - \bar{x})(y - \bar{y})$
95	214	3.5	19.2	12.25	67.2
82	152	−9.5	−42.8	90.25	406.6
90	156	−1.5	−38.8	2.25	58.2
81	129	−10.5	−65.8	110.25	690.9
99	254	7.5	59.2	56.25	444.0
100	266	8.5	71.2	72.25	605.2
93	210	1.5	15.2	2.25	22.8
95	204	3.5	9.2	12.25	32.2
93	213	1.5	18.2	2.25	27.3
87	150	−4.5	−44.8	20.25	201.6
Total	**915**	**1,948**	**0**	**0**	**380.50**	**2,556**

a Using the totals from the above table, we can compute the least-squares estimates as

$$\hat{\beta}_1 = \frac{\sum (x - \bar{x})(y - \bar{y})}{\sum (x - \bar{x})^2} = \frac{2,556}{380.5} = 6.7175$$

$$\hat{\beta}_0 = \bar{y} - \hat{\beta}_1 \bar{x} = \left(\frac{1,948}{10}\right) - 6.7175\left(\frac{915}{10}\right) = -419.85$$

This line models the relation between the peak power load and maximum temperature only for the range of data collected (Figure 2.20). Note that all the data collected were for the temperatures that ranged from 81 °F to 100 °F. No measurements on the peak power load were taken for the maximum temperature of 0 °F. Therefore, we cannot meaningfully interpret the y-intercept value. From the data, it seems we are modeling the relationship for summer temperatures. The same relationship might not hold for winter temperatures.

However, for this situation, $\hat{\beta}_1 = 6.7175$ tells us that for every one degree (in °F) increase in the maximum temperature, the peak power load will increase on the average by 6.7175 megawatts.

b If the maximum temperature increases by 5 °F, then the peak power load is expected to increase by about $5(6.7175) = 33.5875$ megawatts.

c The least-squares regression line is $\hat{y} = -419.85 + 6.7175x$, where $\hat{y}$ = the predicted peak power load and x = the maximum daily temperature.

Figure 2.20

Straight-line fit to power load and temperature data

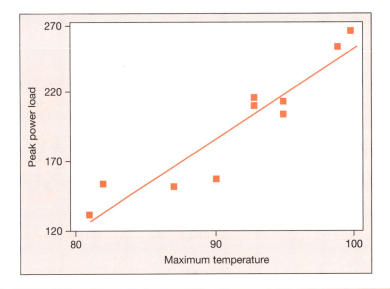

Remember that the residuals (or errors) are the vertical distances between the observed points and the regression line, in other words ($y_i - \hat{y}_i$).

For the least-squares regression line $\hat{y} = -419.85 + 6.7175x$, the predicted values $\hat{y}_i$, the residuals ($y_i - \hat{y}_i$), and the squared residuals ($y_i - \hat{y}_i)^2$ are shown in table below. The SSE for this regression line is 2,093.43. No other straight line will yield as small an SSE as this one.

x	y	$\hat{y} = \hat{\beta}_0 + \hat{\beta}_1 x$	$(x - \hat{x})^2$	$(y - \hat{y})^2$
95	214	218.31	−4.31	18.58
82	152	130.98	21.02	441.84
90	156	184.72	−28.72	824.84
81	129	124.27	4.73	22.37
99	254	245.18	8.82	77.79
100	266	251.90	14.10	198.81
93	210	204.88	5.12	26.21
95	204	218.31	−14.31	204.78
93	213	204.88	8.12	65.93
87	150	164.57	−14.57	212.28

SSE = 2,093.43

Example 2.13

Refer to Example 2.7 of overtolerance and expanded tubewall reduction. The Minitab output for a linear fit to these data is given below.

```
The regression equation is
WR = 3.23 + 215 OT

Predictor          Coef        SE Coef              T            P
Constant         3.2310         0.2725          11.86        0.000
OT               215.30         32.41            6.64        0.000

S = 0.9553       R-Sq = 65.7%        R-Sq(adj) = 64.2%
```

Determine the *y*-intercept and slope of the line of best fit. Interpret their values in the context of the problem.

Solution   The printout shows the relation between the overtolerance (OT) and the wall reduction (WR) as

$$WR = 3.23 + 215 \; OT$$

This equation shows that the *y*-intercept is $\hat{\beta}_0 = 3.23$ and the slope is $\hat{\beta}_1 = 215$. These values are also given in the table under the column titled Coef (short for coefficient). The value for Constant in this column gives the *y*-intercept, and the value for OT in this column gives the slope of the least-squares regression line.

- The *y*-intercept tells us that for zero overtolerance, the expected wall reduction is on the scale of 3.23.
- The slope tells us that for every one unit increase in overtolerance, the wall reduction is expected to increase by about 215 units.

2.5.2   Using the model for prediction

Once we have identified an appropriate model and estimated the slope and intercept from the data, we can use it for prediction. As seen earlier, the value of *y* on the least-squares line corresponding to the given *x* is referred to as the *predicted value*. It represents the most likely value of *y* at that particular *x*.

Example 2.14

Consider the example of peak power load. Predict the required peak power load, if tomorrow's maximum temperature is expected to be (a) 95 °F, (b) 98 °F, and (c) 102 °F.

Solution   **a**   We know that maximum temperature is not the only factor on which the peak power load depends. On different days with maximum temperature 95 °F, we are likely to observe different peak power load requirements. However, the relation that we developed earlier between the peak power load and the maximum temperature gives the typical value of peak power load for a given maximum temperature. Using that relation we get

$$\hat{y} = -419.85 + 6.7175x = 419.85 + 6.7175(95) = 218.3125 \text{ megawatts.}$$

The actual peak load is likely to differ from this predicted value, but the fitted relation tells us that the likely peak load will be around 218.3 megawatts.

Looking back at the data, we see that on the two days for which the data were collected, the maximum temperature was recorded to be 95 °F. Although the maximum temperature was the same, the peak power loads for those days (214 and 204, respectively) are not the same. Because the regression line gives the typical peak power load for the day with given maximum temperature, we can determine the residual associated with the prediction. On day 1, the residual $= 214 - 218.3 = -3.7$; for this day, the regression line overpredicted the peak power load by 3.7 megawatts. On day 8, the residual $= 204 - 218.3 = -13.7$; for this day, the regression line overpredicted the peak power load by 13.7 megawatts.

b Using $x = 98$ in the least-squares equation, we get,

$$\hat{y} = -419.85 + 6.7175x = -419.85 + 6.7175(98) = 238.465 \text{ megawatts.}$$

Looking back at the data, we see that no data were collected for the day with maximum temperature 98 °F. However, the data were collected for maximum temperatures ranging from 81 °F to 100 °F. Within this range of temperatures, we have seen that the straight-line model represents the data well. So assuming the peak power load is linearly related to the maximum temperature, we can use the least-squares line to interpolate between the values. It predicts the peak power load of 238.5 megawatts for the day with maximum temperature 98 °F.

c Now we are interested in predicting peak power load for the day with maximum temperature 102 °F. Using the regression line, we could say that the typical peak power load would be 265.335 megawatts. However, how reliable would this prediction be? Looking back at the data, we see that the maximum temperature of 102° is outside the range of the data collected. From the scatterplot, it is known that the linear relation holds well within the range of the data collected. However, the relation between the peak power load and the maximum temperature outside the range of the data is unknown. Therefore, use of the fitted model for prediction outside the range of the data is not recommended. Even when used, the predicted value must be used very carefully because a high amount of error might be associated with it.

Example 2.15

Consider Example 2.7 of overtolerance and tubewall reduction. Predict the wall reduction for the overtolerance 0.010.

Solution   Although no data were collected for OT $= 0.010$, this OT value is within the range of data collected. Therefore, as shown in Figure 2.21, using the least-squares line, we get

$$\text{WR} = 3.23 + 215 \text{ OT}$$
$$= 3.23 + 215 (0.010) = 5.38$$

A wall reduction of 5.38 is predicted for OT $= 0.010$.

Figure 2.21

Fitted least-squares regression line for WR-OT data

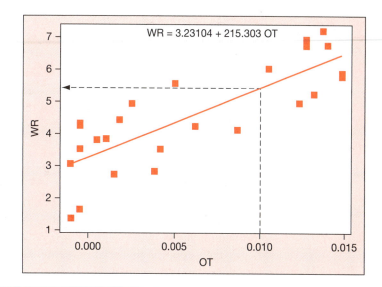

2.5.3   The relation between the correlation coefficient and the slope

The correlation coefficient and the slope of the regression line respectively were defined as

$$r = \frac{1}{n-1}\Sigma\left(\frac{x-\bar{x}}{s_x}\right)\left(\frac{y-\bar{y}}{s_y}\right) \quad \text{and} \quad \hat{\beta}_1 = \frac{\Sigma(x-\bar{x})(y-\bar{y})}{\Sigma(x-\bar{x})^2}$$

Note that $\Sigma(x-\bar{x})^2$ is the numerator of the variance of x-values. Thus, if we have n data points, we can write $\Sigma(x-\bar{x})^2 = (n-1)s_x^2$. Taking these relations into account, we can easily see that

$$r = \frac{1}{n-1}\Sigma\left(\frac{x-\bar{x}}{s_x}\right)\left(\frac{y-\bar{y}}{s_y}\right) = \frac{\Sigma(x-\bar{x})(y-\bar{y})}{[(n-1)s_x]s_y}$$

$$= \frac{\Sigma(x-\bar{x})(y-\bar{y})}{\Sigma(x-\bar{x})^2}\frac{s_x}{s_y} = \hat{\beta}_1\frac{s_x}{s_y}$$

Because both r and $\hat{\beta}_1$ provide information about the utility of the model, it is not surprising that there is a similarity in their computational formulas. Particularly note that $\Sigma(x-\bar{x})(y-\bar{y})$ appears in the numerator of both the expressions. Because the denominators of both terms are always positive, the sign of both r and $\hat{\beta}_1$ is determined by the sign of $\Sigma(x-\bar{x})(y-\bar{y})$. If the correlation coefficient is positive, then the slope of the best-fitting line is positive too.

Activity: Do taller persons have bigger feet?

Suppose you are interested in determining how a person's height is associated with foot size.

To do this, you will collect heights and foot sizes of several persons and determine whether there is any association between the two. If the association appears to be linear, you will estimate the nature of the relation.

- Mark a scale on the board. Measure the height of each student against this scale.
- Mark a scale on a large piece of paper. Have each student stand with back to the wall and place his or her foot on the piece of paper with the heel against the wall (see Figure 2.22). Measure the foot length.
- Use a scatterplot to determine whether there is a linear relation between the foot length and height.
- If yes, then estimate this relation using the least-squares technique.
- Use this relation to estimate the instructor's foot length using his or her height as a predictor.

Figure 2.22
Measuring technique

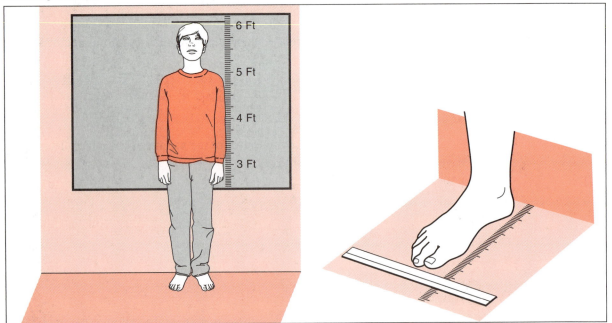

Exercises

2.28 Refer to traffic safety data given in Exercise 2.14.
   a Model the relationships between 2002 and 1980 traffic deaths using the least-squares technique, if the association between the two variables being studied appears to be linear.
   b Interpret the slope of the simple linear model in terms of the variables being measured.

2.29 Refer to the tax revenue data from Table 1.10 and Exercise 2.17.
   a Remove Alaska from the dataset. Model the relationship between per capita expenditure and per capita tax revenue using the least-squares technique and a simple linear model.

b Interpret the slope of the model in part (a) in terms of the variables in the model.

2.30 Refer to the RSF data from Exercise 2.18.

  a Find a simple linear model for the relationship(s) using the least-squares technique if the association between the measured and calculated values appears to be linear.

  b Should we use linear relation to estimate the relation between the measured and calculated RSF when considering two types of flaws separately? Explain why or why not.

2.31 Refer to the ranking data from Table 2.13.

  a Find a simple linear model for the relationship between the ranking by professors and that by executives.

  b Interpret the slope of the regression line in terms of the problem.

  c What do you expect the executive's rank to be if the professor gave a rank of 4.0?

2.32 Chapman and Demeritt (*Elements of Forest Mensuration,* 2nd ed., Albany, NY, J.B. Lyon Company [now Williams Press], 1936) reported diameters (in inches) and ages (in years) of oak trees. The data are shown in table below.

Age	Diameter	Age	Diameter	Age	Diameter
4	0.8	14	2.5	30	6.0
5	0.8	16	4.5	30	7.0
8	1.0	18	4.6	33	8.0
8	2.0	20	5.5	34	6.5
8	3.0	22	5.8	35	7.0
10	2.0	23	4.7	38	5.0
10	3.5	25	6.5	38	7.0
12	4.9	28	6.0	40	7.5
13	3.5	29	4.5	42	7.5

  a Make a scatterplot. Is there any association between the age of the oak tree and the diameter? If yes, discuss the nature of the relation.

  b Can the diameter of oak tree be useful for predicting the age of the tree? If yes, construct a model for the relationship. If no, discuss why not.

  c If the diameter of one oak tree is 5.5 inches, what do you predict the age of this tree to be?

2.33 Good measures of body fat are difficult to come by, but a good measure of body density can be made from underwater weighing. Table below lists the percent fat and body density of 15 women (*Source:* Pollock, unpublished research report, University of Florida, 1956).

  a Construct a scatterplot of fat versus density. Discuss the nature of relation between the

percent body fat and the body density of women.

  b Use the data in table below to find a good regression model to predict % body fat from body density for women.

  c A woman's body density is measured to be 1.050. Use the relation in part (b) to predict the % body fat for this woman?

Density	%Fat	Density	%Fat
1.053	19.94	1.001	44.33
1.053	20.04	1.052	20.74
1.055	19.32	1.014	38.32
1.056	18.56	1.076	9.91
1.048	22.08	1.063	15.81
1.040	25.81	1.023	34.02
1.030	30.78	1.046	23.15
1.064	15.33		

2.34 Table below lists the weight (in pounds) and length (in inches) for 11 largemouth bass (*Source: The Mathematics Teacher,* 1997, p. 666).

Length (inches)	Weight (pounds)
9.4	0.4
11.1	0.7
12.1	0.9
13.0	1.2
14.3	1.6
15.8	2.3
16.6	2.6
17.1	3.1
17.8	3.5
18.8	4.0
19.7	4.9

  a Find a simple linear regression model for predicting the weight of a largemouth bass from knowledge of its length. Do you see any problems in using this linear relationship as a model for these data?

  b If the length of a largemouth bass caught is 15.00 inches, what is its expected weight?

2.35 To measure the effect of certain toxicants found in water, concentrations that kill 50% of the fish in a tank over a fixed period of time (LC50s) are determined in laboratories. There are two methods for conducting these experiments. One method has water continuously flowing through the tanks, and the other has static water conditions. The Environmental Protection Agency (EPA) wants to adjust all results to the flow-through conditions.

a Given the data in table at the right on a sample of 10 toxicants, establish a model that will allow adjustment of the static values to corresponding flow-through values.

b Do you see any problems with using this prediction model for new static values anywhere between 0 and 40? Explain.

Flow-Through	Static
23.00	39.00
22.30	37.50
9.40	22.20
9.70	17.50
0.15	0.64
0.28	0.45
0.75	2.62
0.51	2.36
28.00	32.00
0.39	0.77

2.6   The Coefficient of Determination

Another way to measure the contribution of x in predicting Y is to consider how much the errors of prediction of Y were reduced by using the information provided by x. If you do not use x, the best prediction for any value of Y would by $\bar{y}$, and the sum of squares of the deviations of the observed y values about $\bar{y}$ is the familiar

$$SS_{yy} = \Sigma(y_i - \bar{y})^2$$

On the other hand, if you use x to predict Y, the sum of squares of the deviations of the y values about the least-squares line is

$$SSE = \Sigma(y_i - \hat{y}_i)^2$$

To illustrate, suppose a sample of data has the scatterplot shown in Figure 2.23a. If we assume that x contributes no information for the prediction of y, the best prediction for a value of y is the sample mean $\bar{y}$, which is shown as the horizontal line in Figure 2.23b. The vertical line segments in Figure 2.23b are the deviation of the points about the mean $\bar{y}$. Note that the sum of squares of deviations for the model $\hat{y} = \bar{y}$ is

$$SS_{yy} = \Sigma(y_i - \bar{y})^2$$

Now suppose you fit a least-square line to the same set of data and locate the deviations of the points about the line, as shown in Figure 2.23c. Compute the deviations about the prediction lines in parts (b) and (c) of Figure 2.23. You can see the following:

1. If x contributes little or no information for the prediction of y, the sum of squares of deviations for the two lines,

$$SS_{yy} = \Sigma(y_i - \bar{y})^2 \quad \text{and} \quad SSE = \Sigma(y_i - \hat{y})^2$$

will be nearly equal.

Figure 2.23
A comparison of the deviations for two models

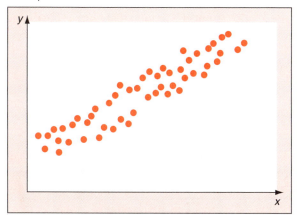

(a) Scatterplot of data

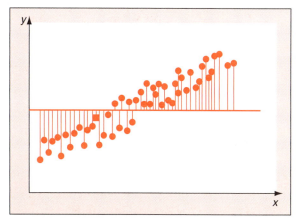

(b) Assumption: x contributes no information for predicting y, i.e., $\hat{y} = \bar{y}$

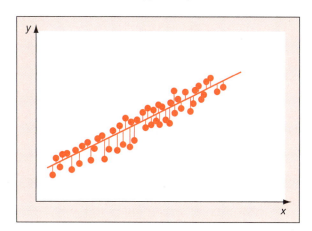

(c) Assumption: x contributes information for predicting y, i.e., $\hat{y} = \hat{\beta}_0 + \hat{\beta}_1 x$

2. If x does contribute information for the prediction of y, SSE will be smaller than SS_{yy}, as shown in Figure 2.23. In fact, if all the points fall on the least-squares line, then SSE = 0.

It can be shown that

$$\Sigma(y_i - \bar{y})^2 = \Sigma(y_i - \hat{y}_i)^2 + \Sigma(\hat{y}_i + \bar{y})^2$$

We label the third sum. $\Sigma(\hat{y}_i - \bar{y})^2$, the Sum of Squares for Regression and denote it by SSR. Thus,

$$SS_{yy} = SSE + SSR$$

and it follows that

$$SS_{yy} - SSE \geq 0$$

The SSR assesses differences between $\hat{y}_i$ and $\bar{y}$. If the regression line has slope near zero, then $\hat{y}_i$ and $\bar{y}$ will be close together for any i, and SSR will be small. If, on the other hand, the slope is far from zero. SSR will be large.

Then the reduction in the sum of squares of deviations that can be attributed to x, expressed as a proportion of SS_{yy}, is

$$\frac{SS_{yy} - SSE}{SS_{yy}} = \frac{SSR}{SS_{yy}}$$

It can be shown that this quantity is equal to the square of the simple linear coefficient of correlation.

The square of the coefficient of correlation is called the **coefficient of determination**. it represents the proportion of the sum of squares of deviations of the y values about their mean that can be attributed to a linear relation between y and x.

$$r^2 = \frac{SS_{yy} - SSE}{SS_{yy}} = 1 - \frac{SSE}{SS_{yy}} = \frac{SSR}{SS_{yy}}$$

Note that r^2 is always between 0 and 1. Thus $r^2 = 0.60$ means that 60% of the sum of squares of deviations of the observed y values about their mean is attributable to the linear relation between y and x.

Example 2.16

Calculate and interpret the coefficient of determination for the peak power load example.

Solution   We have previously calculated

$$SS_{yy} = 19{,}263.6 \quad \text{and} \quad SSE = 2{,}093.4$$

so we have

$$r^2 = \frac{SS_{yy} - SSE}{SS_{yy}} = \frac{19{,}263.6 - 2{,}093.4}{19{,}263.6} = 0.89$$

Note that we could have obtained $r^2 = (0.94)^2 = 0.89$ more simply by using the correlation coefficient calculated in the previous section. However, use of the above formula is worthwhile because it reminds us that r^2 represents the fraction reduction in variability from SS_{yy} to SSE. In the peak power load example, this fraction, is 0.89. which means that the sample variability of the peak load about their mean is reduced by 89% when the mean peak loads is modeled as a linear function of daily high temperature. Remember, though that this statement holds true only for the sample at hand. We have no way of knowing how much of the population variability can be explained by the linear model.

Exercises

2.36 Refer to the oak tree data in Exercise 2.32.
  a Compute r and r^2 values.
  b Interpret your answers in part (a) in terms of the variables being measured.

2.37 Refer to the body density data in Exercise 2.33.
  a Compute r and r^2 values.
  b Interpret your answers in part (a) in terms of the relationship between body fat and body density.

2.38 Refer to the largemouth bass data in Exercise 2.34.
  a Compute r and r^2 values.
  b Interpret your answers in part (a) in terms of the relationship between length and weight.

2.39 In the summer of 1981, the Minnesota Department of Transportation installed a state-of-the-art weigh-in-motion scale in the concrete surface of the eastbound lanes of Interstate 494 in Bloomington, Minnesota, The system is computerized and monitors traffic continuously. It is capable of distinguishing among 13 different types of vehicles (car, five-axle semi, five-axle twin trailer, etc.). The primary purpose of the system is to provide traffic counts and weights for use in the planning and design of future roadways. Following installation, a study was undertaken to determine whether the scale's readings corresponded with the static weights of the vehicles being monitored. Studies of this type are known as *calibration studies*. After some preliminary comparisons using a two-axle, six-tire truck carrying different loads (see table below), calibration adjustments were made in the software of the weigh-in-motion system, and the weigh-in-motion scales were reevaluated (Wright, Owen, and Pena, "Status of MN/DOT's" Weigh-in-Motion Program," Minnesota DOT, 1983).

  a Construct two scatterplots, one of y_1 versus x, and the other of y_2 versus x.
  b Use the scatterplots of part (a) to evaluate the performance of the weigh-in-motion scale both before and after the calibration adjustment.
  c Calculate the correlation coefficient for both sets of data and interpret their values. Explain how these correlation coefficients can be used to evaluate the weigh-in-motion scale.
  d Suppose the sample correlation coefficient for y_2 and x were 1. Could this happen if the static weights and the weigh-in-motion reading disagreed? Explain.

2.40 A problem of both economic and social concern in the United States is the importation and sale of illicit drugs. The data shown in Table 2.16 are a part of a larger body of data collected by the Florida Attorney General's Office in a attempt to relate the incidence of drug seizures and drug arrests to the characteristics of the Florida counties. They show the number y of drug arrests per county in 1982, the density x_1 of the county (population per square mile), and the number x_2 of law enforcement employees. In order to simplify the calculations, we show data for only ten counties.
  a Fit a least-squares line to relate the number y of drug arrests per county in 1982 to the country population density x_1.
  b We might expect the mean number of arrests to increase as the population density increases. Do the data support this theory? Test by using $\alpha = 0.05$.
  c Calculate the coefficient of determination for this regression analysis and interpret its value.

Trial No.	Static Weight of Truck x	Weight-in-Motion Reading	
		y_1 (before calibration adjustment)	y_2 (after calibration adjustment)
1	27.9	26.0	27.8
2	29.1	29.9	29.1
3	38.0	39.5	37.8
4	27.0	25.1	27.1
5	30.3	31.6	30.6
6	34.5	36.2	34.3
7	27.8	25.1	26.9
8	29.6	31.0	29.6
9	33.1	35.6	33.0
10	35.5	40.2	35.0

Note: All weights are in thousands of pounds.

Table 2.16

Number of Arrests and County Characteristics

County	1	2	3	4	5	6	7	8	9	10
Population Density x_1	169	68	278	842	18	42	112	529	276	613
Number of Law Enforcement Employees x_2	498	35	772	5,788	18	57	300	1,762	416	520
Number of Arrests in 1982 y	370	44	716	7,416	25	50	189	1,097	256	432

2.41   Repeat the instruction of Exercise 2.40, except let the independent variable be the number x_2 of county law enforcement employees.

2.42   Refer to Exercise 2.40

   a   Calculate the correlation coefficient r between the county population density x_1 and the number of law enforcement employees x_2.

   b   Does the correlation between x_1 and x_2 differ significantly from zero? Test by using $\alpha = 0.05$.

2.43   Note that the regression lines of Exercise 2.41 and 2.42 both have positive slopes. How might the value for the correlation coefficient, calculated in Exercise 2.42 part (a), be related to this phenomenon?

2.7   Residual Analysis: Assessing the Adequacy of the Model

In the earlier sections, we learned to fit a straight-line model to appropriately structured data and measure the contribution of x for predicting y. But if the data do not follow a linear pattern, then, obviously, fitting a linear model is not a good idea. One approach used to explore the adequacy of a model is referred to as *residual analysis*. Remember that the residual is the difference between the observed and predicted value of y for a given value of x, in other words, $(y_i - \hat{y}_i)$.

There are certain conditions that must hold among the residuals in order for the linear regression theory of later chapters to work well:

- The residuals appear to be nearly random quantities, unrelated to the value of x.
- The variation in residuals does not depend on x; in other words, the variation in the residuals appears to be about the same no matter which value of x is being considered.

These conditions can be checked using a plot of the residuals graphed against the values of x or $\hat{y}$. A plot of residuals versus x or $\hat{y}$ is known as a *residual plot*. In interpreting such plots,

- Lack of any trends or patterns can be interpreted as an indication of random nature of residuals.
- Nearly constant spread of residuals across all the values of x can be interpreted as an indication of variation in residuals not dependent on x.

If a model fits the data well, there will be no discernable pattern in the residual plot; the points will appear to be randomly scattered about the plane.

Example 2.17

Figure 2.24

Residual plot for power load data

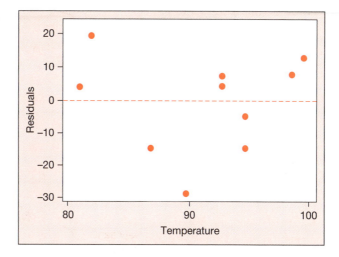

A residual plot for the peak power load of data is given in Figure 2.24.

- These residuals look fairly random, but there is a hint of a pattern in that residuals for the middle temperature values are mostly negative, whereas those at either end are positive. This is an indication that the linear model might not be a best fit for these data. As will be seen in Chapter 11, a quadratic model gives a better fit for these data than a linear model.
- The variation in residuals seems to be about the same for small x values as for large ones, but this is difficult to check here because of the small number of repeated x values.

Example 2.18

The least-squares fit and a residual plot for the overtolerance and wall reduction data are given in Figure 2.25.

Figure 2.25

Least-squares fit and residual plot for WR-OT data

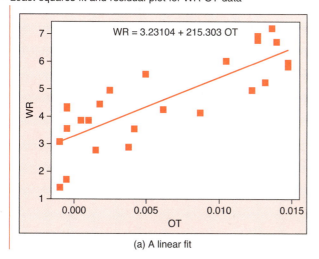

(a) A linear fit

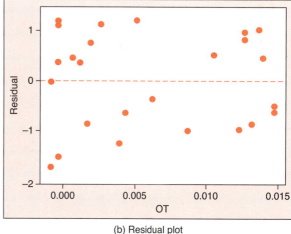

(b) Residual plot

A look at the least-squares fit (Figure 2.25a) indicates a fairly good fit of the linear model to the data. The residual plot (Figure 2.25b) identifies no pattern or trend. The residuals look fairly random. Also, the variation in residuals seems to be about the same across different values of x. It seems that the least-squares line fits data well and is a good model to describe the behavior of wall reduction as a function of overtolerance.

Exercises

2.44 For the relationship developed for the traffic safety data in Exercise 2.28,
  a Construct a residual plot.
  b Comment on the shape of the residual plot.
  c Comment on the appropriateness of the linear relation developed for these data.

2.45 Refer to the tax revenue data analyzed in Exercise 2.29.
  a Construct a residual plot.
  b Comment on the shape of the residual plot.
  c Comment on the appropriateness of the linear relation developed for these data.

2.46 Refer to the RSF data analyzed in Exercise 2.30.
  a Construct a residual plot.
  b Comment on the shape of the residual plot.
  c Comment on the appropriateness of the linear relation developed for these data.

2.47 Refer to the ranking data analyzed in Exercise 2.31.
  a Construct a residual plot.
  b Comment on the shape of the residual plot.
  c Comment on the appropriateness of the linear relation developed for these data.

2.48 Refer to data on age and diameter of oak trees analyzed in Exercise 2.32.
  a Construct a residual plot.
  b Comment on the shape of the residual plot.
  c Comment on the appropriateness of the linear relation developed for these data.

2.49 Refer to data on density and % body fat analyzed in Exercise 2.33.
  a Construct a residual plot.
  b Comment on the shape of the residual plot.
  c Comment on the appropriateness of the linear relation developed for these data.

2.50 Refer to data on the length and weight of largemouth bass analyzed in Exercise 2.34.
  a Construct a residual plot.
  b Comment on the shape of the residual plot.
  c Comment on the appropriateness of the linear relation developed for these data.

2.51 Table below lists the U.S. retail gasoline price (cents per gallon) and U.S. FOB cost of crude oil (dollars per barrel) as reported by the U.S. Energy Information Administration (**http://www.eia.doe.gov/**).

Year	Crude Oil	Gasoline	Year	Crude Oil	Gasoline
1984	27.60	91.6	1995	15.69	76.1
1985	25.84	91.9	1996	19.32	84.3
1986	12.52	63.7	1997	16.94	83.1
1987	16.69	67.7	1998	10.76	66.0
1989	16.89	76.8	1999	16.47	76.2
1990	20.37	89.9	2000	26.27	109.1
1991	16.89	81.8	2001	20.46	102.2
1992	16.77	78.7	2002	22.63	94.3
1993	14.71	75.3	2003	25.86	113.5
1994	14.18	72.9	2004	33.75	142.3

  a Construct a scatterplot of gasoline prices versus crude oil prices. Describe the nature of the relation.
  b Estimate the relation using the least-squares technique.
  c Construct a residual plot.
  d Comment on the appropriateness of the linear relation between gasoline and crude oil prices.

2.8   Transformations

What if the conditions for regression are not met or the simple regression model does not fit? One way to improve the fit of a model is to *transform* the data and fit a linear model to the transformed data, the subject of this section. Another way is to fit a more complex model, which will be done in Chapter 11.

Example 2.19

Table below shows the population density in the United States for each census year since 1790.

Year	Population Density	Year	Population Density
1790	4.5	1900	21.5
1800	6.1	1910	26.0
1810	4.3	1920	29.9
1820	5.5	1930	34.7
1830	7.4	1940	37.2
1840	9.8	1950	42.6
1850	7.9	1960	50.6
1860	10.6	1970	57.5
1870	10.9	1980	64.0
1880	14.2	1990	70.3
1890	17.8	2000	79.6

Because the population density experienced an increase over the years, we might start building a model of the relationship between population density and year by fitting a simple linear regression model. The Minitab output for a linear model is shown in Figure 2.26.

Figure 2.26

Minitab output for regression of population density versus year

```
Regression Analysis: PopDens versus Year
The regression equation is
PopDens = -623 + 0.343 Year

Predictor        Coef      SE Coef           T          P
Constant      -622.79        49.99      -12.46      0.000
Year          0.34335      0.02636       13.02      0.000

S = 7.846      R-Sq = 89.5%      R-Sq(adj) = 88.9%
```

To the casual observer, the linear model might seem to fit well, with a large r^2 value. However, a scatterplot of the data and a plot of residuals will show that the "good" fit is not nearly good enough. The actual data (Figure 2.27a) show a definite curvature; in other words, the rate of increase in the population density seems larger in

Figure 2.27

A scatterplot and a residual plot for population density by year

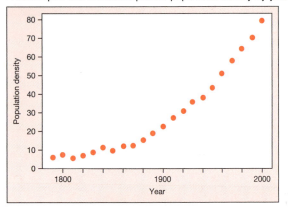

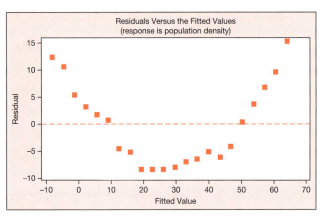

(a) Scatterplot   (b) Residual plot

later years. This curvature is magnified in the plot of residuals from the straight regression line (see Figure 2.27b). The residuals show a definite pattern; they cannot be regarded as random fluctuations (in part due to the fact that these are time series data).

Because the rate of increase in population seems to be growing larger with time, an exponential relationship between the average population and year might be appropriate. Exponential growth can be modeled by

$$y = ae^{bx}$$

Then $\ln(y) = \ln(a) + bx$, and there will be a linear relationship between $\ln(y)$ and x. The graph of $\ln(y)$ versus year for the population data is shown in Figure 2.28. This transformation does appear to "linearize" the data, although there seems to be more variation in the population size during the early half of the 19th century than in the rest of the time period.

Figure 2.28

A scatterplot and a residual plot for population density by year

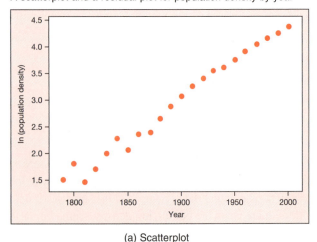

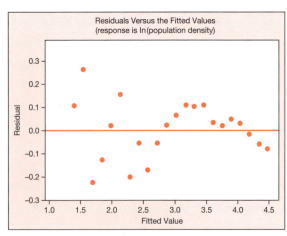

(a) Scatterplot   (b) Residual plot

Figure 2.29

Minitab output for regression model of ln (population density) versus year

```
Regression Analysis: ln(y) versus Year

The regression equation is
ln(y) = -24.7 + 0.0146 Year

Predictor            Coef      SE Coef        T        P
Constant         -24.7103       0.7940   -31.12    0.000
Year            0.0145871     0.0004188    34.83    0.000

S = 0.1246      R-Sq = 98.4%      R-Sq(adj) = 98.3%
```

A simple linear model fit to the data produced the Minitab results given in Figure 2.29.

This is a much better fitting straight line, as is shown by the scatterplot (Figure 2.28a) and the higher value of coefficient of determination (increased from 89.5% to 98.4%). Also, the residual plot (see Figure 2.28b) indicates that the pattern shown by the residuals in the linear fit is removed. The exponential model fits well, but the fit is not equally good for earlier years and later years. Can you suggest a historical reason for such a differential in the fit of the model to this data?

In the interpretation of exponential models, note that e^b is (1+ the growth rate). For these data, $e^b = 1.015$, so the population density has increased, on the average, at 1.5% per year over this time period.

Fortunately, a rather simple transformation solved the "linearization" problem in the population growth example. Sometimes, however, such a transformation on *y* or *x*, or both, is much more difficult to find, may not produce such drastic improvements, or may not even exist. Example 2.20 describes a case where a reasonable transformation makes a reasonable improvement.

Example 2.20

The Florida Game and Freshwater Fish Commission is interested in developing a model that will allow the accurate prediction of the weight of an alligator from more easily observed data on length. Length (in inches) and weight (in pounds) data for a sample of 25 alligators are shown in table below.

Weight	Length	Weight	Length
130	94	83	86
51	74	70	88
640	147	61	72
28	58	54	74
80	86	44	61
110	94	106	90
33	63	84	89
90	86	39	68
36	69	42	76
38	72	197	114
366	128	102	90
84	85	57	78
80	82		

A plot of weight versus length (Figure 2.30a) shows that a linear model will not fit these data well. If, however, we proceed to fit a simple linear model anyway, coefficient of determination is about 0.84 (from the Minitab printout in Figure 2.30c).

So as not to be completely misled, look at the residual plot (Figure 2.30b). This plot shows a definite curved pattern in the residuals, ample evidence of the fact that the straight line is not a good model.

Figure 2.30

Output for the linear growth model

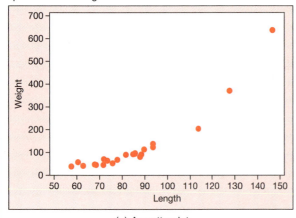

(a) A scatterplot

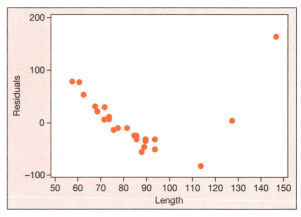

(b) A residual plot

```
Regression Analysis: Weight versus Length

The regression equation is
Weight = -393.264 + 5.90235 Length


Predictor         Coef       SE Coef          T          P
Constant       -393.26         47.53      -8.27      0.000
Length          5.9024        0.5448      10.83      0.000


S = 54.0115      R-Sq = 83.6%       R-Sq(adj)  = 82.9%
```

(c) A regression output

How should we proceed from here? The exponential growth model got us out of trouble before and could work again here. Plotting ln(weight) versus length (Figure 2.31a) does seem to produce a scatterplot that could be fit by a straight line. The fit of ln(weight) as a function of length is much better than the fit in the original data; Coefficient of determination increased from about 0.84 to 0.96 (see Minitab output in Figure 2.31c).

However, the residual plot (Figure 2.31b) shows that the variation in residuals is much greater at the shorter lengths than at the longer lengths, and this condition violates the condition of a nearly constant spread of residuals that a good model should produce.

Figure 2.31

Fit of ln(weight) versus length of alligators

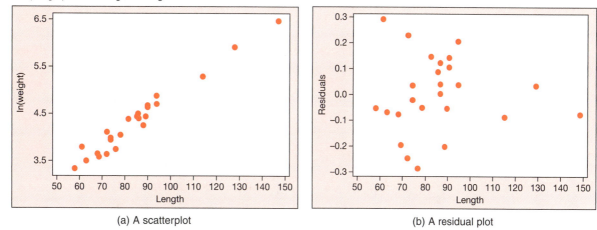

(a) A scatterplot

(b) A residual plot

```
Regression Analysis: ln-wt versus Length

The regression equation is
ln-wt = 1.33534 + 0.0354158 Length


Predictor        Coef     SE Coef        T         P
Constant       -1.3353      0.1314    10.16     0.000
Length        0.035416    0.001506    23.52     0.000


S = 0.149299      R-Sq = 96.0%      R-Sq(adj) = 95.8%
```

(c) A regression output

A little reflection on the nature of the physical problem might help at this point. Weight is related to volume, a three-dimensional feature. So it's not surprising that weight does not turn out to be a linear function of length. A model of the form

$$\text{weight} = a\,(\text{length})^{b}$$

allows weight to be modeled as a function of length raised to an appropriate power. Taking logarithms, we get a linear model as follows:

$$\ln(\text{weight}) = \ln(a) + b\ln(\text{length})$$

Fitting ln(weight) versus ln(length) will provide an estimate of b. The plot of ln(weight) versus ln(length) is shown in Figure 2.32a, and the scatterplot looks quite linear.

The fitted model is almost as good in terms of coefficient of determination (see Figure 2.32c) as the previous model, and the residuals look quite random (Figure 2.32b), although a little curvature might still remain.

Figure 2.32

Fit of ln(weight) versus ln(length) for alligators

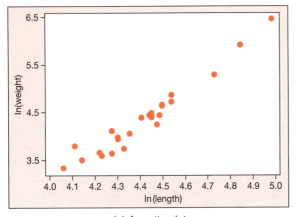

(a) A scatterplot

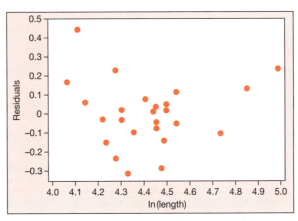

(b) A residual plot

```
Regression Analysis: in-wt versus ln-lgth

The regression equation is
ln-wt = -10.2 + 3.29 ln-lgth


Predictor          Coef      SE Coef         T          P
Constant       -10.1746       0.7316     13.91      0.000
ln-lgth          3.2860       0.1654     19.87      0.000


S = 0.1753       R-Sq = 94.5%       R-Sq(adj)  = 94.3%
```

(c) A regression output

To find the appropriate transformation, one must think of transformation that relates to the phenomenon under study. It should not be too surprising that populations grow exponentially (by a multiplicative factor), and the weight of the alligator is related to the power of its length (about 3). A few of the commonly used models and transformations to linearize the relations or stabilize the error variance are as follows:

- The *exponential transformation*: If an exponential model of the form $y = ae^{bx}$ seems appropriate, fit $\ln(y) = \hat{\beta}_0 + \hat{\beta}_1 x$. This may be a good model if the variation in errors (or y values) increases with x. In such situations, the distribution of errors will be positively skewed. This transformation reduces the dependence of variation of errors on the x values. It is also a good model if y appears to be increasing by a multiplicative factor (as in compound interest) rather than an additive factor.
- The *power transformation*: If a power model of the form $y = ax^b$ seems appropriate, fit $\ln(y) = \hat{\beta}_0 + \hat{\beta}_1 x \ln(x)$. This is often a good model for variables that have a geometric relationship, such as area to length or volume to length.
- The *square root transformation*: If the variation in y values is proportional to the mean of y, fit $\sqrt{y} = \hat{\beta}_0 + \hat{\beta}_1 x$. One such situation occurs when the response

variable (y) has the Poisson distribution (to be discussed in Chapter 5). This transformation is useful for count data, such as the number of items produced per unit of time or the number of defects per unit of area.

More discussion about transformations is available in Draper and Smith (1981); Kutner, Nachtsheim, and Neter (2004); and Box and Cox (1964).

Influential Observations

Interpreting the significance and the strength of a linear association through R^2 requires great care—and plots of data. Sometimes one or two data points may have great *influence* on the nature of a relation as measured by the correlation coefficient and the slope of the fitted line. If a measurement has a strong influence, then it can totally change the nature of relationship between two variables.

Consider the ozone and PM_{10} levels and the AQI days in year 2000 for 15 cities reported in Table 2.9 and Table 1.4, respectively. The correlation matrix for these variables is given in Table 2.17. The correlations show fairly strong positive association ($r = 0.7822$) between AQI_{2000} and PM_{10}, and mild positive ($r = 0.4713$) association between AQI_{2000} and O_3.

For the line fitted between AQI_{2000} and O_3, the JMP output in Figure 2.33a gives $r^2 = 0.22$, and the scatterplot (Figure 2.33b) identifies Los Angeles with a considerably high AQI_{2000} value. This point appears to be far away from the others, and from the line of best fit. For a straight line fit through this data, this point (measurements for Los Angeles) will produce a large residual and affect the slope of the line considerably.

How does the removal of Los Angeles from the data affect the relationship? If the relationship changes considerably by addition or removal of a measurement, then such a measurement is an *influential measurement*.

After removal of Los Angeles from the data, the correlation between O_3 and AQI_{2000} reduces to $r = -0.0568$ (see Table 2.18). In fact, from a fairly strong positive association the relation changes to practically no association (or an extremely weak positive association) between AQI_{2000} and O_3 (Figure 2.34b). The regression line shows no trend of any practical significance. Los Angeles was highly influential; its presence increased coefficient of determination and the slope of the line considerably.

Table 2.17

Correlation Matrix for 15 Cities' data with Los Angeles included

Correlations LA Included	O_3	PM_{10}	AQI_{2000}
O_3	1.0000	0.2045	0.4713
PM_{10}	0.2045	1.0000	0.7822
AQI_{2000}	0.4713	0.7822	1.0000

Table 2.18

Correlation Matrix with Los Angeles Excluded

Correlations LA Excluded	O_3	PM_{10}	AQI_{2000}
O_3	1.0000	−0.0074	0.0568
PM_{10}	−0.0074	1.0000	0.8104
AQI_{2000}	0.0568	0.8104	1.0000

Figure 2.33

JMP output for AQI vs. O_3

```
Linear Fit
AQI2000 = -4.226282 + 990.38462*O₃

Summary of Fit
RSquare                          0.222096
RSquare Adj                      0.162257
Root Mean Square Error            15.7276
Mean of Response                 20.53333
Observations (or Sum Wgts)             15

Parameter Estimates

Term        Estimate Std Error  t Ratio  Prob >|t|
Intercept  -4.226282  13.47812    -0.31     0.7588
O₃         990.38462  514.0727     1.93     0.0762
```

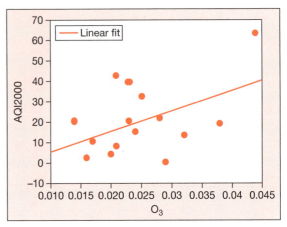

(a) A regression output   (b) Relation between AQI and O_3

Figure 2.34

JMP output for AQI vs. O_3 without Los Angeles

```
Linear Fit
AQI2000 = 14.809468 + 113.79893*O₃

Summary of Fit
RSquare                          0.003231
RSquare Adj                      -0.07983
Root Mean Square Error           13.52278
Mean of Response                     17.5
Observations (or Sum Wgts)             14

Parameter Estimates

Term        Estimate Std Error  t Ratio  Prob >|t|
Intercept  14.809468  14.11314    1.05     0.3147
O₃         113.79893  577.0257    0.20     0.8470
```

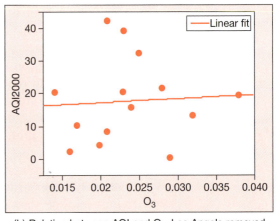

(a) A regression output   (b) Relation between AQI and O_3, Los Angels removed

In short, the linear model could have appeared to be highly significant, with high coefficient of determination, simply because of the influence of a single point. To guard against having one or two points unduly affect results, always plot the data and residuals, and study plots carefully.

Now consider the relation between AQI_{2000} and PM_{10} measurements (Figure 2.35b). Again Los Angeles turns out to be an influential measurement (top right-hand corner). Removal of measurements for Los Angeles improved the relationship between AQI_{2000} and PM_{10} (Figure 2.35b). The correlation coefficient increased from 0.7 to 0.82; that is, R^2 increased from 0.61 to 0.66, the slope dropped from 767.55 to 613.97 (Figure 2.35a and Figure 2.36), and the scatterplot has a more pronounced linear trend.

Figure 2.35

JMP output for AQI vs. PM₁₀

```
Linear Fit
AQI2000 = -48.03416 + 767.54658*PM10

Summary of Fit
RSquare                        0.611873
RSquare Adj                    0.582018
Root Mean Square Error         11.10929
Mean of Response               20.53333
Observations (or Sum Wgts)           15

Parameter Estimates

Term        Estimate  Std Error  t Ratio  Prob >|t|
Intercept   -48.03416   15.41538    -3.12     0.0082
PM10        767.54658  169.5465      4.53     0.0006
```

(a) A regression output

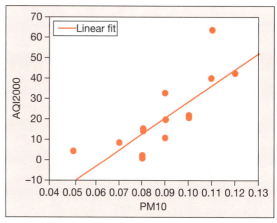

(b) Relation between AQI and PM10

Figure 2.36

JMP output for AQI vs. PM₁₀ without Los Angeles

```
Linear Fit
AQI2000 = -36.44134 + 613.96648*PM10

Summary of Fit
RSquare                        0.656775
RSquare Adj                    0.628173
Root Mean Square Error         7.935201
Mean of Response                  17.5
Observations (or Sum Wgts)           14

Parameter Estimates

Term        Estimate  Std Error  t Ratio  Prob >|t|
Intercept   -36.44134   11.45476    -3.18     0.0079
PM10        613.96648  128.1253      4.79     0.0004
```

(a) A regression output

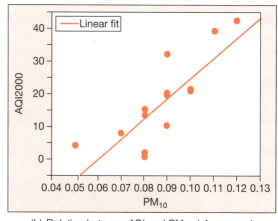

(b) Relation between AQI and PM₁₀, LA removed

Exercises

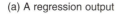

2.52 Refer to the largemouth bass data of Exercise 2.34. Find a model that fits better than the simple linear one and justify that it does, in fact, fit the data better.

2.53 If coolant in a reactor pressure vessel is lost or reduced by accident, then the emergency core cooling system starts operating and the coolant flows into the vessel, resulting in a steep temperature gradient in the walls of vessel. Under this condition, maintaining high pressure can change the stress distribution on the inner walls of vessel and develop cracks in the wall of vessel. The American Society of Mechanical Engineers has developed codes to evaluate integrity of such vessels measured by elastic hoop stress (9MPa). Kim, Koo, Choi, and Kim (*Transactions of ASME*, 2003) performed full 3-D elastic-plastic finite element analysis to investigate the present ASME code in evaluating the integrity of vessel. The elastic hoop stress (in MPa) measured after 700s for different parameter (x/a) values are reported in table below (read from a chart).

x/a	0	0.03	0.05	0.175	0.25	0.35	0.45	0.725	1.00
Stress	425	400	380	280	270	250	240	200	160

a Fit a simple linear regression model to the relation between the stress and the parameter x/a.

b Assess the strength of the estimated relation.

c Is a linear or nonlinear model more appropriate?

d Using the model you choose, predict the hoop stress level for $x/a = 0.30$.

2.54 Tidal force is the main hydrodynamic force in estuaries. Shi et al. (*Int. J. Numer. Meth. Fluids*, 2003) reported the recorded tidal elevations (in meters) at stations 16 and 18 at the North Passage of Changing Estuary (read from graphs). The pairs of tidal elevations recorded from 2:00 PM till 12:00 noon are given in table below.

St 16	St 18	St 16	St 18	St 16	St 18	St 16	St 18
0.4	0.6	0.8	0.5	0.1	0.7	0.5	0.1
1.2	1.0	0.4	0.1	0.5	1.0	0.0	−0.3
1.5	1.1	0.1	−0.3	1.0	1.1	−0.2	−0.7
1.5	1.1	−0.2	−0.7	1.2	0.8	−0.4	−0.7
1.4	1.1	−0.4	−0.5	1.0	0.6	−0.5	−0.8
1.2	1.1	−0.2	0.4	0.8	0.4		

a Make parallel time series plots. Comment on similarities and differences in the patterns in observed tidal elevations at stations 16 and 18.

b Make a scatterplot of tidal elevations at station 18 versus those at station 16. Comment on the nature of relationship.

c Assess the strength of the estimated relation.

2.55 Tidal force is the main hydrodynamic force in estuaries. Shi et al. (*Int. J. Numer. Meth. Fluids*, 2003) reported the recorded six different measurements of the relative depth of water (y) and tidal velocities (x) at stations 16 and 18 at 9:00 PM on August 12, 1978. The data reported (read from the graph) are listed in table below:

Station 16		Station 18	
Velocity	Relative Water Depth	Velocity	Relative Water Depth
0.6	0.02	0.55	0.02
0.88	0.2	0.85	0.2
1.0	0.6	0.95	0.8
1.1	0.4	1.05	0.4
1.15	0.8	1.05	1.0
1.2	1.0	1.2	0.6

a Construct scatterplots of depth versus velocity for stations 16 and 18 separately and then together.

b Comment on the nature of the relationship between depth and velocity. Does it appear to be linear?

c Fit an appropriate model to describe relationship between depth and velocity.

2.56 Table 2.19 gives the median annual earnings (in constant 2000 dollars) of all wage and salary workers of ages 25–34, by gender (*The Condition of Education*, 2000).

Table 2.19

Median Annual Earnings by Gender for 1971–2000

Year	Male	Female	Year	Male	Female	Year	Male	Female
1971	36,564	15,984	1981	31,617	18,449	1991	27,979	19,785
1972	38,118	16,686	1982	30,253	18,418	1992	27,274	20,268
1973	38,912	16,925	1983	30,317	18,781	1993	26,726	19,771
1974	36,833	16,873	1984	31,263	19,017	1994	26,764	19,370
1975	35,630	17,859	1985	31,345	19,032	1995	26,894	19,281
1976	35,788	18,006	1986	31,164	19,239	1996	27,497	19,940
1977	35,826	18,852	1987	31,204	20,025	1997	27,878	20,541
1978	36,572	18,196	1988	30,924	20,360	1998	31,028	21,642
1979	35,956	18,964	1989	30,261	20,149	1999	31,440	21,839
1980	33,278	18,747	1990	28,665	20,169	2000	31,175	22,447

a Plot male salaries against the year and female salaries against the year on the same plot. Describe the trends observed.

b Calculate the mean and standard deviation for each group.

c Calculate the appropriate z-score for each data point.

d Construct a time series plot of standardized salaries for each group on the same graph. Does the result show any interesting pattern?

e Construct a scatterplot of standardized salaries of male against female. Comment on any patterns you see.

f Suppose this year's average salary for males is $31,100. To what female average salary would this be comparable?

2.9 Reading a Computer Printout

The computations for a regression analysis are complicated and cumbersome. However, there are many software programs available that can easily do these computations and provide answers. In fact most of these computations including transformations and residual analysis can be performed using even handheld calculators such as TI-83 and TI-89. The standard computer printouts for a regression analysis of the peak power load data are provided here as a sample.

Estimated linear relation

(1) $\hat{y} = \hat{\beta}_0 + \hat{\beta}_1 x$

(2) Y-intercept $\hat{\beta}_0$

(3) Slope $\hat{\beta}_1$

(4) SS_{yy}

(4) SSE

(6) s^2

(7) S

(8) $r^2 = 1 - \dfrac{SSE}{SS_{yy}}$

JMP Printout

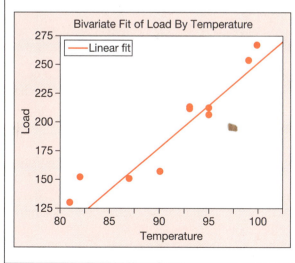

(1) Load = -419.8491 + 6.717477 Temperature

```
Summary of Fit
RSquare                           (8) 0.891312
RSquare Adj                           0.877726
Root Mean Square Error            (6) 16.17764
Mean of Response                       194.8
Observations (or Sum Wgts)                10
```

```
Analysis of Variance
Source     DF   Sum of Squares   Mean Square    F Ratio
Model       1         17169.871     17169.9    65.6049
Error       8   (5)   2093.729  (7)  261.7    Prob > F
C. Total    9   (4)  19263.600               <.0001
```

```
Parameter Estimates
Term              Estimate  Std Error  t Ratio  Prob >|t|
Intercept    (2) -419.8491   76.05778    -5.52     0.0006
Temperature  (3)  6.717477    0.82935     8.10    <.0001
```

Minitab Printout

```
The regression equation is
(1) Load = -420 + 6.72 Temperature

Predictor          Coef    SE Coef      T       P
Constant     (2) -419.85     76.06   -5.52   0.001
Temperat     (3)   6.7175    0.8294   8.10   0.000

(6) S = 16.18    (8) R-Sq = 89.1%    R-Sq(adj) = 87.8%


Analysis of Variance
Source            DF        SS       MS       F      P
Regression         1      17170    17170   65.60   0.000
Residual Error     8 (5)   2094  (7) 262
Total              9 (4)  19264
```

SAS Printout

```
                    The REG Procedure
                     Model: MODEL1
                 Dependent Variable: load

Analysis of Variance

                        Sum         Mean
Source       DF     Squares       Square   F Value   Pr > F
Model         1       17170        17170    65.60   <.0001
Error         8 (5) 2093.72878 (6) 261.71610
Corrected
Total         9     (4) 19264

      Root MSE      (7) 16.17764   R-Square  (8) 0.8913
      Dependent Mean    194.80000  Adj R-Sq      0.8777
      Coeff Var           8.30474


Parameter Estimates

                 Parameter   Standard
Variable   DF     Estimate      Error   t Value   Pr > |t|
Intercept   1 (2) -419.84915  76.05778    25.52    0.0006
temp        1 (3)    6.71748   0.82935     8.10    <.0001
```

2.10 Summary

Many engineering problems involve modeling the relationship between a dependent variable Y and one or more independent variables. In this chapter, we have discussed the case of a single independent variable x. The regression analysis presented here considers only the situation in which the value of Y is a linear function of x. The general procedure is to determine form of the relation using a Scatterplot, estimate the slope and Y-intercept of the relation, check the adequacy of assumption of linear relation, and when satisfied with the model's adequacy, use it for prediction. After a significant linear association has been found, the strength of the association is measured by the correlation r (or by coefficient of determination r^2). Since many subtleties affect such linear associations, data plots and residual plots should be studied carefully, and influential observations should be noted. Sometimes a transformation of one or both variables will help improve the fit of a linear model.

Supplementary Exercises

2.57 In the United States, the Environmental Protection Agency is charged with the job of monitoring the nation's environment, that is, air, water, and soil. Some of the major indicators of air quality are levels of carbon monoxide (CO), lead (Pb), nitrogen dioxide (NO_2), ozone, and sulfur dioxide (SO_2). Table below gives the national air quality data for years 1980–2000 (National Air Quality and Emissions Report, 2003).

Year	CO (ppm)	Pb (ppm)	NO$_2$ (ppm)	Ozone (ppm)	SO$_2$ (ppm)
1981	8.4	0.58	0.024	0.126	0.0102
1982	8.1	0.58	0.023	0.125	0.0095
1983	7.9	0.47	0.023	0.137	0.0093
1984	7.8	0.45	0.023	0.125	0.0095
1985	7.1	0.28	0.023	0.123	0.0090
1986	7.2	0.18	0.023	0.118	0.0088
1987	6.7	0.13	0.023	0.125	0.0086
1988	6.4	0.12	0.023	0.136	0.0087
1989	6.4	0.10	0.023	0.116	0.0085
1990	5.9	0.08	0.022	0.114	0.0079
1991	5.6	0.08	0.019	0.111	0.0081
1992	5.3	0.07	0.019	0.105	0.0076
1993	5.0	0.06	0.019	0.107	0.0074
1994	5.1	0.05	0.020	0.106	0.0072
1995	4.6	0.05	0.019	0.112	0.0057
1996	4.3	0.05	0.019	0.105	0.0057
1997	4.1	0.04	0.018	0.104	0.0056
1998	3.9	0.04	0.018	0.110	0.0055
1999	3.7	0.04	0.018	0.107	0.0053
2000	3.4	0.04	0.017	0.100	0.0051

a Are these different measures of air quality associated with each other?
b If yes, model relationships among them over these two decades.
c Plot each measure over time and discuss trends in different air quality measures over the two decades.

2.58 One of the problems in fluid dynamics is determining pressure in narrow and non-narrow regions through which fluid flows. It is a function of several parameters, including boundary conditions on mass and momentum flux. Stay and Barocas (*Int'l J. for Numerical Methods in Fluids,* 2003) demonstrated a localized reduction technique called "stitching" for solving such hydrodynamic problems. They compared peak pressure in a region calculated using the coupled Stokes-Raynolds model with the analytic solution. The difference between the analytic solution and the peak pressure calculated with the coupled Stokes-Raynolds model (i.e., error in pressure) and the CPU solution time for the coupled Stokes-Raynolds model are recorded in table below for different stitch lengths. Note that error measures the solution error or accuracy of solution by the coupled Stokes-Raynolds method and the CPU time indicates efficiency of the coupled Stokes-Raynolds method.

Stitch Length (cm)	Error in Pressure	CPU time (sec)
0.1	0.01	140
0.2	0.11	130
0.3	0.15	118
0.4	0.17	102
0.5	0.20	105
0.6	0.24	100
0.7	0.25	85
0.8	0.25	85
0.9	0.24	81
1.0	0.21	73
1.1	0.20	65
1.2	0.18	55
1.3	0.18	58
1.4	0.15	50

a Make a scatterplot of error in pressure versus stitch length. Comment on the trend observed.
b Make a scatterplot of CPU time versus stitch length. Comment on the trend observed.
c Put the information from both the scatterplots together to comment on the efficiency and accuracy of the coupled Stokes-Raynolds method.

2.59 Rigid structures suffer substantial damage due to effects of structural vibrations from natural disasters such as high winds or earth's tremors. The variable-orifice dampers are used in high-rise structures to provide a mechanically simple and low-power means to protect structures against natural disaster by reducing structural vibrations. Spencer and Nagarajaiah (*J. Structural Engg,* 2003) give data for buildings recently completed or currently under construction (see table).

Is there any association between the number of semi-active dampers used and the height of the building or the number of stories?

Stories	Height (m)	No. of Semi-Active Dampers
11	56.0	42
31	140.5	72
25	119.9	38
28	136.6	60
38	172.0	88
54	241.4	356
19	98.9	27
30	104.9	66
23	100.4	28

2.60   Moles et al. (*Chem Engg Research & Design,* 2003) studied how computation times of the stochastic methods are reduced using parallel implementation of different optimization methods or parallel processors. Their data about the number of parallel processors used and the corresponding speed-up are listed in table below. Note that speed-up measures the increase in computation speed. Model the relation between speed-up and the number of processors used.

Speed Up (y)	Number of Parallel Processors Used (x)
1	1
1.95	2
2.6	3
4.3	4
4.8	6

2.61   Timely assessment of cracks in steam generator tubes is important for extending the life of nuclear steam generators. Wang and Reinhardt (*J. Pressure Vessel Technology,* 2003) studied the failure assessment of a steam generator tube with defects, using a finite element method. For different percentage degraded areas (PDA) of the steam generator tube (10% to 90% degradation), they recorded the expected finite element limit load supported. Table below gives PDA and normalized finite element limit loads.

PDA (%)	Limit load
10	0.80
20	0.67
30	0.52
40	0.45
50	0.32
60	0.32
70	0.27
80	0.19
90	0.10

a   Describe the relation between PDA and limit load.

b   Fit an appropriate model to the relationship between PDA and the limit load, using regression technique.

2.62   Table 2.20 gives the median annual earnings (in constant 2000 dollars) of all wage and salary workers ages 25–34, by sex and educational attainment level (*The Condition of Education,* 2000).

a   For bachelor's degree or higher, plot earnings of male workers against the year and earnings of female workers against the year on the same plot. Describe and compare the trends observed.

b   For grades 9–11, plot earnings of male workers against the year and earnings of female workers against the year on the same plot. Describe and compare the trend observed.

c   Plot earnings of male workers against earnings of female workers for both the educational groups on the same plot. Describe and compare the trends observed.

d   Plot earnings of male workers with a bachelor's degree versus earnings of male workers with an education of grades 9–11. Describe the trend observed.

e   Plot earnings of female workers with a bachelor's degree versus earnings of female workers with an education of grades 9–11. Describe the trend observed.

2.63   Table below gives the number of households in the United States for the years 1970 to 2002 (*U.S. Census Bureau*).

Year	Number of Households (in thousands)
1790	558
1890	12,690
1900	15,964
1930	29,905
1940	34,949
1950	43,468
1955	47,788
1960	52,610
1965	57,251
1970	62,874
1975	71,120
1980	80,776
1985	86,789
1990	93,347
1995	98,990
2000	104,705
2002	109,297

Table 2.20

Median Annual Earnings by Gender for 1971–2000

Year	Bachelor's Degree or Higher Male	Female	Grades 9–11 Male	Female	Year	Bachelor's Degree or Higher Male	Female	Grades 9–11 Male	Female
1971	45,219	29,345	31,039	10,045	1986	41,608	29,437	19,204	10,690
1972	46,065	29,047	30,845	10,235	1987	41,473	30,164	20,305	11,404
1973	45,610	28,401	32,579	11,122	1988	40,720	30,131	19,469	9,305
1974	42,491	27,463	29,956	9,833	1989	40,656	30,889	19,559	10,037
1975	40,089	27,249	26,882	10,161	1990	38,770	30,503	18,628	9,139
1976	41,279	26,170	27,191	10,080	1991	39,019	29,516	16,471	9,910
1977	41,175	25,757	26,970	10,527	1992	39,070	30,684	16,596	11,724
1978	41,422	25,460	26,928	8,839	1993	38,014	30,245	16,201	8,905
1979	40,033	25,770	26,214	11,687	1994	37,437	29,822	16,588	9,248
1980	38,242	25,042	23,575	10,624	1995	37,553	29,328	17,847	9,436
1981	38,691	24,777	21,939	9,842	1996	38,593	28,940	16,928	9,789
1982	37,253	25,551	19,773	10,427	1997	38,410	31,024	18,191	10,279
1983	37,809	26,438	19,598	10,542	1998	41,695	31,789	18,569	10,989
1984	38,864	26,702	18,114	9,341	1999	42,341	32,145	18,582	10,147
1985	41,276	28,053	19,395	10,415	2000	42,292	32,238	19,225	11,583

a Make a time series plot. Comment on the trend observed over the years 1970 to 2002.

b Consider percent of households by size during years 1900 and 2000, listed in table below. Make appropriate plots to compare the distribution of household size in the years 1990 and 2000. Describe and compare the distributions. Comment on the changes observed in the distribution of household sizes over a century.

Household Size	1900	2000
1 person	5.1	25.5
2 persons	15.0	33.1
3 persons	17.6	16.4
4 persons	16.9	14.6
5 persons	14.2	6.7
6 persons	10.9	2.3
7 or more persons	20.4	1.4

c Now put the information obtained in parts (a) and (b) together and write a short note describing household and size distribution in the United States over the last century.

2.64 Pfeifer, Schubert, Liauw, and Emig (*Trans IchemE,* 2003) studied temperature measured when using different number of thermocouple holes. They set the reactor temperature to 310° C and measured the temperature on the top cover plate of the reactor, using two different electric controls (grouping and single control) of the heat cartridges. The number of thermocouple homes used and the temperatures recorded by the two controls are listed in table below.

a Make an appropriate plot to describe the temperatures recorded by the two controls when using different thermocouple holes.

b Describe any patterns observed from your plot.

c Compare the temperatures measured by the two different electric controls with each other as well as with respect to the temperature set.

Number of Thermocouple Holes	Temperature (°C) Grouping	Single Control
1	285	277
2	285	280
3	287	283
4	289	285
5	290	289
6	293	290
7	293	290
8	293	290
9	292	291
10	291	291
11	291	291

2.65 Industries use displacement washing with solvent to separate solutes from the voids in a porous medium. Tarleton, Wakeman, and Liang (*Trans IchemE,* 2003) developed a new technique to improve separation of solid–liquid mixture, using an electric field during washing. They recorded zeta potential of titania (TiO_2) in 0.01 or 0.001 M concentrations of $NaNO_3$ solutions at different pH levels. The data from their experiments are listed in table at the right. The zeta potential is the overall charge a particle acquires in a specific medium. It is used to understand and control colloidal suspensions, emulsions, and slurries. Its value is related to the stability of colloidal dispersions. Colloids with high zeta potential (negative or positive) are electrically stabilized. Colloids with low zeta potentials tend to coagulate or flocculate.

0.01M NaNO₃		0.001M NaNO₃	
Solution pH	Zeta Potential (mV)	Solution pH	Zeta Potential (mV)
3.5	45	3.5	43
4.5	35	4.0	33
5.4	15	4.75	28
5.8	−3	5.0	0
6.0	−5	5.5	−15
6.5	−22	6.1	−20
8.9	−62	6.8	−35
		7.8	−48
		8.8	−60

a Make a scatterplot of zeta potential against the solution pH for each $NaNO_3$ concentration.
b Describe the nature of the relationships.
c Fit the appropriate model to describe the predicted zeta potential.

Obtaining Data

What causes nitrate concentration to change during the catalytic oxidation of aniline, a compound used in the manufacture of drugs and plastics? (Oliviero et al., *Trans IChemE*, 2003)

Figure 3.1 illustrates the nitrate concentrations measured during the catalytic oxidation of aniline at different times from the start of reaction and when the reaction is run at different temperatures. The heating action illustrates one kind of question you can answer using statistical methods: How does the nitrate concentration change for reactions run at different temperatures?

So far, we have learned how to explore, summarize, display, and describe patterns in data. But how did we get that data in the first place? This chapter provides an introduction to four common ways of obtaining data; sample survey, census, experiment, and observational study. In this chapter, you'll learn about how to obtain data using these methods, about different types of biases that can get introduced due to inaccurately applying these methods, and about different types of conclusions that can be drawn from data obtained using these different methods.

Figure 3.1

Nitrate concentration by time and temperature

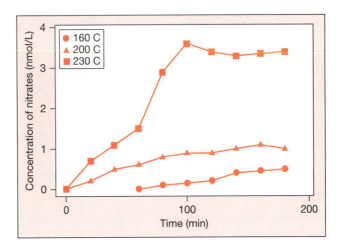

Introduction

The key to decision making is objective data; the key to good decision making is good objective data. It is not enough just to have data; the data must be valid and reliable in that they actually measure what they are supposed to measure and do so to a reasonable degree of accuracy.

Depending on the situation, engineers decide which data collection method would be most appropriate. Of course, the types of conclusions that can be made from the collected data depend on the data collection method used. For example, *sample surveys* allow generalization from the *sample* to the population, whereas *designed experiments* allow decisions about the cause of an observed pattern. We'll now study four different methods of data collection, namely, sample survey, census, experiment, and observational study.

3.1 Overview of Methods of Data Collection

3.1.1 Sample surveys and inference about populations

Some studies are designed for the purpose of estimating population characteristics, such as means or proportions. Before planning such a data collection activity, we need to identify the *population* involved. A population is the entire group of individuals or items in a study. A list (or comparable form of identification) of all members of a population is known as a *frame*. For example,

- a list of all students in the college of engineering
- a list all the equipment owned by a company
- a list of possible errors that can occur when a program is run
- a list of all addresses served by a power supplier
- a list of all bidders for a construction project
- a list all the trees in a particular plot.

Could serve as frames for various studies. For many populations, like residents of a state, a frame is not readily available.

A *sample* is a part of the population that is actually studied. For example,

- All the fish in Mobile Bay constitute a population, but the fish caught to measure mercury levels make up a sample.
- All the items produced in one run of an assembly line make up a population, but the items inspected for defects make up a sample.

> A **population** is the entire group of individuals or items in a study.
>
> A **sample** is the part of a population that is actually studied.

Sample Survey: A sample is collected and studied to gain information about a population. For example,

- During the process of negotiating an annual contract with a parking facility, the manager of a large company wants to know how many employees will need parking spaces next year. How can we get reliable information? One way is to question all employees, but this procedure would be somewhat inaccurate and very time-consuming. We could take the number of spaces in use this year and assume that the need for next year will be about the same, but this method would have inaccuracies as well. A simple technique that works well in many cases is to select a sample of employees not planning to retire at the end of this year and ask each selected employee if he or she will be requesting a parking space. From the proportion of "yes" answers, an estimate of the number of parking spaces required by the entire population of employees can be obtained.
- When Alabama was planning to offer tax incentives to Mercedes for building a plant in Alabama, the Mobile Register conducted a telephone survey of about 400 adult Alabama residents and asked them, "Should Alabama offer tax incentives to industries to relocate in the state?" People responded by saying "agree," "disagree," or "don't know." From those who agreed, an estimate of the percent of adult Alabama residents in favor of offering tax incentives to industries for relocation to the state was estimated.

The scenarios outlined above have all the elements of a typical sample survey. There is a question of "How many?" or "How much?" to be determined for a specific *target population*, the population to which we intend to apply the results of the study. The population from which the data is collected is known as the *sampled population*. It is desirable to have the target population the same as the sampled population, but in some circumstances they might differ. For example, random-digit-dialing telephone polls systematically leave out those without telephones and may miss those with cell phones.

An approximate answer for a population is derived from a sample of data extracted from the population of interest. Of key importance is the fact that the approximate answer will be a good approximation only if the sample truly represents the population under study. *Randomization* plays a vital role in the selection of samples that represent the population and hence produce good approximations. Virtually any sampling scheme that depends upon subjective judgments rather than randomization as to who (or which item) should be included in the sample will suffer from *judgmental bias*. As you will see in later chapters, randomization also forms the probabilistic basis for statistical inference.

Example 3.1

The Tennessee State board of Architectural and Engineering Examiners asked the Tennessee Society of Professional Engineers (TSPE) and the Consulting Engineers of Tennessee (CET) to look into various issues related to professional registration. One of the issues was the professional registration of engineering faculty in Tennessee. Between 1999 and 2003, they sent a survey to engineering deans to determine the registration rate for administrators (deans and department chairs) and full-time faculty. Also of interest were their opinions about the need for maintaining professional registration and whether they provide incentives to their faculty to obtain their PE certification. The survey questionnaire and results were reported by Madhaban and Malasri (*Journal of Professional Issues in Engineering Education and Practice*, 2003). The deans that received the survey questionnaire were not selected randomly from all available deans of engineering colleges. In fact, no specific scheme was used to select deans. Repeated rounds of mailings were used. What effect (if any) do you think this nonrandom selection will have on the outcome?

Census The United States conducts a census every 10 years; in other words, the government attempts to count everyone living in the United States and to measure various other features of the population. The information collected is used for future planning in such areas as taxation, building schools, planning retirement centers, and forecasting energy needs. A *census* means complete enumeration. It is a process of collecting information from every unit in the target population. In other words, a census is a big sample survey. Making a list of all the music CDs you own is taking a census of your music CDs. If a firm takes inventory, it is taking a census of everything in stock. The computerized record of all the employees of a firm is in fact a census of employees. So, the target population might be your CDs, the stock of the firm, or employees of a company, but the key that identifies census is that information is available on each element of that population. No randomization is used in the census data collected from all residents of the United States, but random sampling is used to augment these data on selected issues.

Example 3.2

U.S. News and World Report (September 2003) reported 50 top-rated doctoral universities in the country. They collected information on several important factors such as ACT or SAT scores, percent of freshmen in the top 10% of their high-school class, student/faculty ratio, graduation rate, freshmen retention rate, alumni giving rate, and so on, for all the doctoral universities in the country. Then, using statistical techniques, they ranked the universities. In this study, *U.S. News and World Report* collected information from all doctoral institutions in the country; no randomization was used in collecting these data. In other words, they conducted a census.

It is feasible to conduct a census if the population is small and the process of getting information does not destroy or modify units of the population. For example, the owner of a manufacturing firm might be interested in getting information about the stores to which his business supplies items produced. It is possible to gather this information even if there are 2,500 area stores to which he supplies items. However, in many situations census is not a method of choice to gather information. For large populations, a census can be a costly and time-consuming process of data gathering. Sometimes the process of measurements is destructive, as in testing an appliance for lifelength.

- One political advisor to a candidate for governor's position is interested in determining how much support his candidate has in the state. Suppose the state has 4,000,000 eligible voters. Then it will be too time-consuming to contact each and every voter to determine the amount of support. By the time the census is finished, the support level might change and the information collected may be useless.
- Suppose a Department of Fisheries is interested in determining the mercury level in the fish in Mobile Bay. Using a census will mean capturing all the fish in Mobile Bay and testing them for mercury level, which is not an advisable (or even possible) method of gathering information.
- A manufacturer of suspension cables is interested in determining the strength of cables produced by his factory. The strength test involves applying force till the cable breaks. Obviously, a census would leave no cables to use. So, a census would not be a practical method of gathering information in this situation.

3.1.2   Experiment and inference about cause and effect

An *experiment* is a planned activity designed to compare "treatments." In an experiment, the experimenter creates differences in the experimental units involved by subjecting them to different treatments and then observing the effects of such treatments on the measure of outcomes. For example,

- In laboratory testing, engineers at one car manufacturing facility run crash tests that involve running cars at different speeds (predetermined and controlled) and crashing them at a specific site. Then they measure the damage to the bumpers. In this example, the team of engineers creates the differences in environment by running cars at different speeds. (The cars are the experimental units and the speeds are the treatments.)
- Engineers interested in studying heat transfer use pipes of different sizes and control the direction in which water is flowing. In one study, the engineers create different environments by controlling the size of pipes and direction of water flow to determine the percent of heat transfer in those different environments. (The pipes are the experimental units and size-direction combinations are the treatments.)

As in sample surveys, randomization plays a vital role in designed experiments. By randomizing the assignment of different environments (treatments) to experimental units, biases that might result due to learning effects or specific orders can be avoided. Designed experiments are conducted not only to establish differences in outcomes under different environments, but also to establish cause and effect relations among outcomes and environments. In sample surveys, a sample is selected randomly from a population of interest to estimate some population characteristic; in designed experiments, different *experimental units* are assigned randomly to different treatments to study the treatment effects.

Example 3.3

Guo and Uea (*Trans IchemE*, 2003) conducted experiments to study effects of impregnation conditions on the textural and chemical characteristics of the prepared absorbents. They used three different concentrations (20%, 30%, and 40%) of three different solutions—zinc chloride ($ZnCl_2$), phosphoric acid (H_3PO_4), and potassium hydroxide (KOH)—and recorded the amounts of nitrogen dioxide (NO_2) and ammonia (NH_3) absorbed onto the oil-palm-shell absorbents. In this experiment, different treatments were created by using different concentrations of the solutions, and the effects of these different treatments were measured on the amount of nitrogen dioxide and ammonia absorbed. Nine different treatments created in this experiment can be listed as follows:

(1) 20% of $ZnCl_2$ (2) 30% of $ZnCl_2$ (3) 40% of $ZnCl_2$
(4) 20% of H_3PO_4 (5) 30% of H_3PO_4 (6) 40% of H_3PO_4
(7) 20% of KOH   (8) 30% of KOH   (9) 40% of KOH

3.1.3   Observational study

An *observational study* is a data collection activity in which the experimenter merely plays the role of an observer. The experimenter observes the differences in the conditions of units and observes the effects of these conditions on measurements taken on these units. The experimenter does not interject any treatment and does not contribute to the creation of observed differences. For example,

- One researcher collected information about the speed at which the car was traveling when a crash occurred and the amount of damage to the bumper from the accident reports filed by the local Police Department. In this example, the researcher has no control over the speed of the car. He did not contribute to creation of the differences among speeds. The researcher merely observed the differences in speeds and the result of them as measured by the amount of damage to the bumper.
- The weather station at the Mobile/Pascagoula Regional Airport recorded the wind speed, wind direction, and eye radius of the storm when Hurricane Danny stayed over Mobile Bay for three days. The meteorologists studied the relations among different factors in order to investigate reasons behind fluctuations in the eye of

the storm. Changing values of wind speed and wind direction had created different environments in the storm and such environments could be evaluated using differences in wind speeds and directions. However, the meteorologists did not control those scenarios, they merely observed those conditions created by nature.

Example 3.4

Wolmuth and Surters (*Proceedings of the ICE*, 2003) studied crowd-related failure of bridges in the world. They collected information on bridge failures from the years 1825–2000. For each failed bridge, they collected information on the age, use (road, footbridge, other), form (aluminum, chain, cable supports, concrete, deck structure, iron, steel, timber), span, width, occasion (cavalry or soldiers, sports gathering, religious gathering, river spectacle, toll dispute, other), crowd size, crowd action (walking from one end of the bridge to other, procession, crowd concentrated at one parapet, crowd going from one parapet to the other, queue, cavalry, soldiers or other military), number of deaths, number injured, and so forth. This was an observational study because the authors collected data from existing scenarios (they did not create differences in them) and analyzed collected data to answer specific questions about the bridge failure. Even an observational study such as this one provides very valuable information to engineers about planning bridge construction and proper use of bridges, but such studies do not allow cause-and-effect conclusions.

Although we might like to, it is not possible to conduct an experiment in all investigations that involve a comparison of treatments. Sometimes we must use an observational study instead of an experiment. For example:

- To study effects of asbestos on the health of workers in a certain industry that makes use of that product, an experiment will require a group of workers to be exposed to products containing asbestos while another group is not. It is unethical to expose somebody intentionally to possibly harmful chemicals so that damage to health can be measured.
- Certain inherited traits affect a worker's ability to perform certain tasks. It is not possible to randomly assign genetic traits to different workers; they are born with those traits.

In observational studies, results cannot be generalized to a population because observational studies use volunteers or samples of convenience, such as workers in the first shift instead of a random sample selected from all workers. However, we can sometimes check to see whether the results can reasonably be explained by chance alone.

Exercises

3.1   Engineers are interested in comparing the mean hydrogen production rates per day for three different heliostat sizes. From the past week's records, the engineers obtained the amount of hydrogen produced per day for each of three heliostat sizes. Then they computed and compared the sample means, which showed that the mean production rate per day increased with the heliostat sizes.

a Identify the type of study described here.

b Discuss the types of inference that can and cannot be drawn from this study.

3.2 To investigate reasons why people do not work, the Census Bureau interviewed a group of randomly selected individuals, from April–July 1996, in four separate rotation groups. Respondents were asked to select 1 of 11 categories consisting of economic and noneconomic reasons for not working, in response to the question, "What is the main reason you did not work at a job or business [in the last four months]?"

a Identify the type of study described here.

b Identify the population of interest.

3.3 Ariatnam, Najafi, and Morones (*Journal of Professional Issues in Engineering Education And Practice*, 2002) describe an overview of academies on horizontal directional drilling conducted to train engineers and inspectors for the California Department of Transportation. A pretest was administered on the first day prior to any instruction. Instruction and field experience were provided over a 3-day period, followed by a final test administered at the end of the last day. Although the average final test score of 75.27% was higher than the average pretest score of 55.61%, the difference was not significant.

a Identify the type of study described here.

b What is the purpose of administering a pretest?

3.4 A materials engineer wants to study the effects of two different processes for sintering copper (a process by which copper powder coalesces into a solid but porous copper) on two different types of copper powders. From each type of copper powder, she randomly selects two samples and then randomly assigns one of the two sintering processes to each sample by the flip of a coin. The response of interest measured is the porosity of the resulting copper. Explain what type of study this is and why.

3.5 A textile engineer is interested in measuring heat resistance of four different types of treads used in making fire-resistant clothing for firefighters. A random sample of 20 threads from each type was taken and subjected to a heat test to determine resistance (the length of time the fibers survive before starting to burn). Explain what type of study this is and why.

3.6 A manufacturer of "Keep It Warm" bags is interested in comparing the heat retention of bags when used at five different temperatures (100°F, 125°F, 150°F, 175°F, and 200°F). Thirty bags are selected randomly from last week's production and randomly assigned, six each, to five different groups. Items from group 1 at beginning temperature 100°F were kept in bags for an hour, and the temperatures of those items were recorded after an hour. Similarly, groups 2 to 5 were assigned items at 125°F, 150°F, 175°F, and 200°F, respectively.

a Identify the type of study used here.

b What type of inference is possible from this study?

3.2 Planning and Conducting Surveys

What are young people's career intentions and attitudes toward engineering? The Consortium of Engineering Colleges in one state was interested in the answer to this question so they could strategically plan to reverse the declining enrollments in engineering. To find an answer to such a specific question, they must conduct a carefully planned sample survey. The consortium could take a random sample of 11th and 12th graders from the state and ask them to complete a questionnaire that asks questions to determine their career intentions and attitudes toward engineering. Because the state has several school districts as well as different private schools, preparing a frame and getting a random sample might be a very time-consuming and expensive procedure. A more complex sampling plan (discussed later) could be used in this study, but random sampling is still the key to good data.

Once we know who is in the sample, we still need to get the pertinent information from them. The *method of measurement*, that is, the questions or measuring devices we use to obtain the data, should be designed to produce the most accurate data possible and should be free of *measurement bias*. Measurement bias occurs when variables are measured incorrectly. Measurement bias is the reason we refrain from asking leading questions in a poll and always check the zero on the scale before weighing anything in a laboratory. Because it is so difficult to get good information in a survey, every survey should be *pretested* on a small group of subjects similar to those that might appear in the final sample. The pretest not only helps improve the questionnaire or measurement procedure but also helps determine a good plan for *data collection* and *data management*. The data analysis should lead to clearly stated *conclusions* that relate to the original purpose of the study.

Key elements of the survey include the following:

1. State the *objectives* clearly.
2. Define the *target population* carefully.
3. Design the *sample selection* plan using *randomization*, so as to reduce *sampling bias*.
4. Decide on a *method of measurement* that will minimize *measurement bias*.
5. Use a *pretest* to try out the plan.
6. Organize the *data collection* and *data management*.
7. Plan for careful and thorough *data analysis*.
8. Write *conclusions* in light of the original objectives.

The following example shows how these points are realized in a real survey.

Example 3.5

A serious quality-of-life issue in today's world is the threat of HIV infection and AIDS. In a study reported in *Science*, 1992, a random sample of 2,673 residents of the United States between the ages of 18 and 75 was selected by random digit dialing (i.e., randomly dialing telephone numbers). A larger sample of 8,263 residents was randomly selected from high-risk cities. Some demographic characteristics of the resulting samples are shown in Table 3.1.

 Table 3.1

Demographic Characteristics of the National AIDS Behavioral Survey

	High-Risk Cities (%)	National (%)
Women	57.9	58.4
Men	42.1	41.6
African-American	33.8	13.2
Hispanic	20.7	8.3
White	42.7	75.8
Other	2.8	2.3

 **Table 3.2**

U.S. Population 1990 Census

	Percent
Women	51
Men	49
African-American	12
Hispanic	9
White	80

One of the main purposes of using randomization is to ensure that the sample is representative of the target population. For the target population of the United States as a whole, some aspects of this representation can be checked by comparing the demographic percentages of the national random sample with those in the population. Table 3.2 gives the percentages from the 1990 census. The national sample has a slightly higher percentage of women and a slightly lower percentage of whites, but overall it is quite representative. This agreement may allow us to have more confidence in the main results of the survey (*Science*, 1992)—the percentages of various HIV-related risk groups (see Table 3.3).

Table 3.3

Prevalence of HIV-Related Risk Groups

Risk Group	High-Risk Cities (%)	National (%)
Multiple partners	9.5	7.0
Risky partners	3.7	3.2
Transfusion recipient	2.1	2.3
Multiple partner and risky partner	3.0	1.7
Multiple partner and transfusion recipient	0.3	0.0
Risky partner and transfusion recipient	0.3	0.2
All others	0.7	0.7
No risk	80.4	84.9

For the nation as a whole, about 15% of those between the ages of 18 and 75 are at risk for HIV infection. That percentage increases to about 20% for high-risk cities. These percentages can be used as estimates of what might happen if a large national firm tries to hire, for example, 1,000 workers across the country. A good guess of the number at risk for HIV infection would be 15%, or 150 workers. What if the firm were to hire all 1,000 workers in high-risk cities?

3.2.1 Sampling methods

Once we decide to get information from the sample and not to conduct a census, the question that arises is how to get a sample. How do we decide who receives the questionnaire or which item gets tested? Does it matter how we select a sample? As you will see, different situations warrant different methods of sampling to address specific circumstances.

Some procedures use probabilistic approaches to determine which item of the population is to be selected in the sample, and some don't. For example, although the selection of a jury pool may be random, the actual jury selection from the available pool of jurors is not a random process. Using their judgment, lawyers from both parties question prospective jurors and decide who is to be selected as the jury for a given case. This may result in a biased jury, that is, selection of those with specific opinions. Such a procedure is known as *judgmental sampling*. This approach uses no probability and is entirely based on the judgment of the person selecting the sample. Judgmental sampling is not necessarily always a bad technique; at times, it can be advantageous. For example, a team of environmental engineers is checking a site for possible chemical pollution in the soil. Instead of randomly determining the spots from which the soil samples are taken, the engineers use their expertise in how pollution spreads and their knowledge about the layout of that particular site to select the most likely spots for testing. This could even achieve the required results with a smaller sample size, thus reducing the cost of the study.

Instead of using his expertise to select spots for getting soil samples, suppose an engineer takes soil samples from the easily accessible areas. Then the sample may not give representative results for the entire location. In fact, the results might be misleading due to missing some highly polluted spots in not-so-easy-to-reach areas. This sampling method results in a *sample of convenience*, another method that results in biased outcomes.

Volunteer sampling also results in biased outcomes. This method is popular with radio and television shows for conducting quick opinion polls. In volunteer sampling, it is up to individuals to decide whether they want to be part of the survey. For example, a company was viewing Mobile, Alabama, as a possible site for building a liquefied natural gas (LNG) processing center. Some residents were opposed to the idea, arguing that it poses a high risk to lives. A local television station conducted a survey to address the issue of whether Mobile should give permission to build a LNG processing center in the Mobile area. A telephone number was provided and respondents were asked to call and register their opinions. A computer responded to the calls and prompted callers to press 1 if they supported building the LNG center in Mobile area and 2 if they were against it. The television station then counted the number of 1's and 2's to determine the support for and opposition to the LNG center. But they might have inadvertently introduced *bias* into the results, because only those who feel very strongly about the issue (either in support or in opposition) are likely to call the advertised number and register their opinion.

> A sampling method is **biased** if it tends to give samples in which some characteristic of the population is underrepresented or overrepresented.

Biased sampling methods result in values that are systematically different from the population values, or systematically favor certain outcomes, resulting in systematic overestimation or underestimation. Judgmental sampling, samples of convenience, and volunteer samples are some of the methods that generally result in biased outcomes. Sampling methods, such as simple random, stratified random, systematic, and cluster, that are based on probabilistic selection of samples generally result in unbiased outcomes.

> **Simple random sampling** is a process of obtaining a sample from a population in which each sample of size *n* has an equal chance of being selected.

In a simple random sampling there is no preference for specific population items over the others, as all items have the same chance of being selected. Simple random samples, generally referred to as just "random samples," are obtained in two different ways: *sampling with replacement* and *sampling without replacement*.

- **Sampling with replacement from a finite population**—In selecting a sample of mother boards from a batch of 50 boards, suppose the first board selected at random and tested is added back to the population before selecting the second board. With this scheme, the chance of selection remains the same for both the boards, namely, 1 out of 50. When two boards are drawn with replacement, the probability of selecting two specific boards in order is $(1/50)(1/50) = 0.00040$ and is the same regardless of the boards specified.
- **Sampling without replacement from an infinite population**—In selecting the sample of mother boards, suppose the first sampled board is not added back to the population of 50 boards before the second sampled board is selected at random. Then, the probability of selecting two specified boards in order is $(1/50)(1/49) = 0.00041$, a slight increase.

It turns out that sampling without replacement from an infinite population (or from a very large population compared to the sample size) yields approximately the same result as sampling with replacement. For example, suppose you are selecting 2 computers from 200,000 manufactured this week. Here, the population size of 200,000 is quite large compared to the sample size of 2. When sampling without replacement, the available population size is decreasing as we keep on sampling, but the population size is so large compared to the sample size that the change in the chance of a specific computer getting selected is negligible for practical purposes. The chance of selecting the first specified computer is 1 out of 200,000 (i.e., 0.000005). Because sampling is without replacement, the chance of selecting the second specified computer increases to 1 out of 199,999 (i.e., 0.000005000025), not much different. The chance of selecting 2 specified computers out of 200,000 without replacement is $(0.000005)(0.000005000025) = 2.5000125\text{E-}11$, whereas that with replacement is $(0.000005)(0.000005) = 2.5\text{E-}11$. We can see that for all practical purposes, the chance of selection is the same. Thus, sampling without replacement, the usual practice, is treated theoretically like sampling with replacement.

When a quality control inspector inspects quality of an incoming lot, sampling is usually done without replacement. However, if the incoming lot size is very large compared to the sample selected for inspection, the effect is almost the same as that of getting a sample with replacement.

Suppose a quality control engineer is interested in checking the quality of items coming off the automated production line producing 20 items per minute. Based on the inspected items, he will decide whether the automated procedure is working as per specifications. Random sampling can become confusing and time-consuming, so he selects a random number from numbers between 1 and 20, say 15. Then he instructs the machine supervisor to inspect the 15th item at the beginning of the day and then every 20th item coming off the production line, that is,

Figure 3.2
Systematic sampling

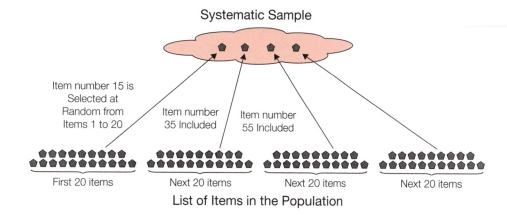

inspect item numbers 15, 35, 55, 75, 95, ... This method of sampling population is known as *systematic sampling*.

In systematic sampling, the frame is divided into consecutive segments, a random starting point is selected from the first segment, and then a sample is taken from the same point in each segment (see Figure 3.2). This method is popular among biologists, foresters, environmentalists, and marine scientists. It is also commonly used to sample people lined up to attend events.

Some populations can be divided into groups called *strata*. Ideally, strata are homogeneous groups of population units; that is, units in a given stratum are similar in some important characteristics, whereas those in different strata differ in those characteristics. For example, students in a university could be grouped by their year or major, or a city could be divided into geographic zones. If a population is divided into homogeneous strata and a simple random sample is selected from each stratum, the result is a *stratified random sample* (see Figure 3.3).

Figure 3.3
Stratified sampling

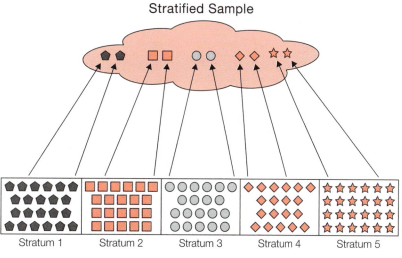

Suppose a company receives its supply of switches from three different suppliers (say, A, B, and C). Then each supplier forms a natural stratum. By taking separate random samples from different suppliers (strata), the industrial engineer can also measure differences among suppliers. Such a stratified sampling procedure could provide an accurate estimate within each stratum while leading to a more precise (reduced variation) combined estimate for the population as a whole.

If population items are packed 24 to a box, it would be time-consuming to take a random sample or a systematic sample that basically involves opening each box. We can consider each box as one *cluster*, select a few boxes at random from the entire shipment, and then inspect all items from the selected boxes. This sampling method is known as *cluster sampling* (see Figure 3.4). In cluster sampling, a population is divided into groups called clusters, and a simple random of clusters is selected. Then, either every unit is measured in each sampled cluster, or a second-stage random sample is selected from each sampled cluster. Cluster sampling methods may seem similar to stratified random sampling. In stratified sampling, a population is divided into homogeneous strata, whereas in cluster sampling, clusters are often nonhomogeneous. In stratified sampling, all strata are sampled, whereas in cluster sampling a census is conducted or random samples are taken from randomly selected clusters. Cluster sampling is often a cost-saving method because a sampling frame is not needed for each cluster, but only for the selected clusters. It is particularly cost saving for geographically dispersed clusters, such as plots of trees in a forest or rural towns scattered across a county.

All sampling designs are subject to *sampling error*, variation inherent in the sampling process. Even if the survey is repeated using the same sample size and the same questionnaires, the outcome will be different. This sampling error can be controlled to some extent by a good sampling design and can be measured by the inferential techniques that will be discussed in later chapters.

There are other types of errors, however, that are not so easily controlled or measured. One major problem in sample surveys is use of an incorrect sampling frame.

Figure 3.4

Cluster sampling

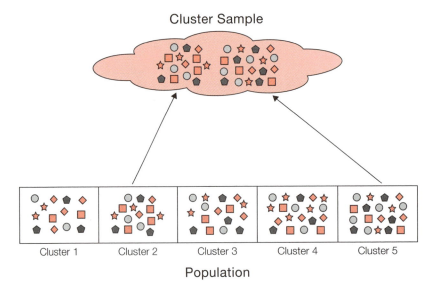

A telephone survey conducted using a telephone directory excludes those with unlisted phone numbers; however, computerized random digit dialing solves this problem. Another problem that arises in sample surveys is that interviewers sometimes substitute another person if the selected person is not available, an action that can introduce serious bias into the results. Under time and money constraints, reliable and accurate results are more likely to be forthcoming from a small sample survey with a good design than from a large survey that is hastily put together.

3.2.2 Selecting a simple random sample

Different kinds of random mechanisms are used for selecting samples in practice, from tossing a coin at a ball game to a very sophisticated random event generator at game shows on television. Here are some examples of using some kind of chance mechanism to select a simple random sample from the population.

- Prizes are often offered at large events. A portion of each ticket collected at the entrance is put in a large box. At the end, all the ticket stubs in the box are mixed thoroughly. Then a prespecified number of ticket stubs are picked from the box. The persons with the other half of the selected stubs receive prizes.
- Suppose there are 40 students in a class. In every class, the teacher asks one randomly selected student to go to the board and solve one homework problem. The teacher asks each student in the class to write his or her name on a separate (but identical) piece of paper and drop it in a box. At the beginning of each class, she mixes well all the pieces in the box and selects one piece at random, without looking at the name. The student whose name appears on the piece of selected paper is selected to solve the problem.
- Students are admitted to the magnet schools of the Mobile Area Public School System, using a lottery system. All those interested in attending magnet schools fill out applications. Their information is entered in the school system computer. Then the computer randomly selects students to be admitted to the magnet schools.

Example 3.6

Here is how the random selection process works. Table 1 in the Appendix gives random numbers. A portion of such a random number table is given in Table 3.4.

Table 3.4

Random Number Table

96410	96335	55249	16141	61826	57992	21382	33971	12082	91970
26284	92797	33575	94150	40006	54881	13224	03812	70400	45585
75797	18618	90593	54825	64520	78493	92474	32268	07392	73286
48600	65342	08640	78370	10781	58660	77819	79678	67621	74961
82468	15036	79934	76903	48376	09162	51320	84504	39332	26922

Suppose 951 students applied to the magnet schools, and 200 are to be selected for the next academic year. Number all the applicants, using three digit numbers such as 001, 002, 003, ..., 951. Start anywhere in the random number table and read sequentially, either horizontally or vertically. Note that the random number table will produce three-digit numbers from 000 to 999, but the school system is admitting only 200 children. So, ignore number 000 and numbers from 952 to 999. Suppose we start at the beginning of the random number table. Then we get numbers 964, 109, 633, 555, 249, 161, and so forth. Because there is no child numbered 964, ignore the number. Next, select the child numbered 109. Then select children numbered 633, 555, 249, 161, and so on, to be admitted to the magnet schools. Use each number only once and ignore any number that is repeated. Continue until 200 unique numbers are selected.

Activity

For this activity, use

- The random rectangles in Figure 3.5.
- The random number table (Table 1 in the Appendix) or a random number generator on the calculator or computer.

The sheet of random rectangles shows 100 shipments, numbered 1 to 100, of different sizes. The shipment size is determined by the number of rectangles in the shipment. Suppose we are interested in estimating the mean shipment size, using a sample.

Judgment Samples

1. Keep the sheet of rectangles covered until instructed to uncover it. At a signal from your instructor, take a look at the sheet and select five shipments that in your judgment represent the shipments on this page.
2. Determine the size of each selected shipment.
3. Compute the mean shipment size for your sample of size 5.
4. Repeat the process and get another sample of size 5. Compute the sample mean.
5. Compare the mean shipment sizes of two samples.
6. Combine two samples, and get a sample of size 10. Compute the sample mean.
7. Collect sample means for samples of size 5 from the entire class. Plot the data. Do the same with sample means for samples of size 10. Compare the two distributions.

Do you think the results of these judgment samples are biased? How?

Simple Random Samples

1. Note that all shipments are numbered using two-digit random numbers. We can consider shipment number 100 as 00. Using the random number table, select five shipments at random.
2. Determine the size of each selected shipment.
3. Compute the mean shipment size for your sample of size 5.
4. Repeat the process and get another sample of size 5. Compute the sample mean.

Figure 3.5
Random rectangles

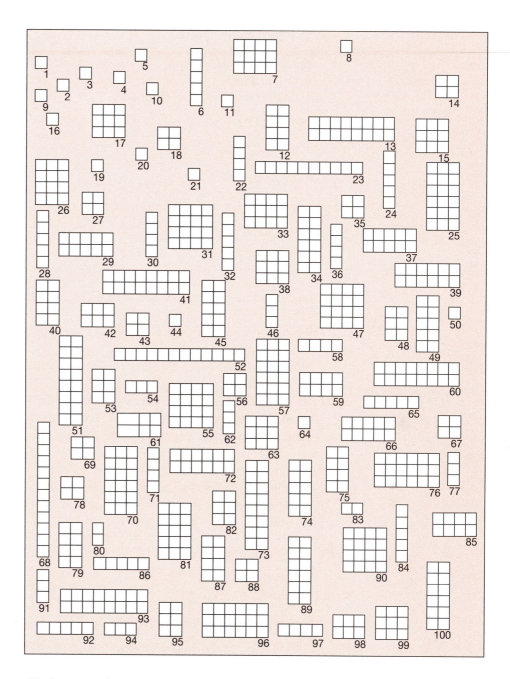

5. Compare the mean shipment sizes of the two samples.
6. Combine the two samples and get a sample of size 10.
7. Collect sample means for samples of size 5 from the entire class. Plot the data. Do the same with sample means for samples of size 10. Compare the two distributions.
8. Compare the distributions seen here with those seen for the judgment samples above; comment on the differences in patterns. Which method is better for estimating the population mean?

Will the small shipments get adequate representation compared to the large shipments?

Stratified Samples

1. Suppose the population is divided into two strata: shipments with sizes smaller than or equal to 6, and shipments with sizes larger than 6. Then the shipment numbers for two strata are as follows:

    STRATUM 1: Shipments with sizes ≤ 6:

    1 2 3 4 5 6 8 9 10 11 14 16 18 19 20 21 22 24
    27 28 30 32 35 36 42 43 44 46 48 50 53 54 56 58 62 64
    65 67 69 71 77 78 80 82 83 84 86 88 91 92 94 95 97 98

    STRATUM 2: Shipments with sizes > 6:

    7 12 13 15 17 23 25 26 29 31 33 34 37 38 39 40 41 45
    47 49 51 52 55 57 59 60 61 63 66 68 70 72 73 74 75 76
    79 81 85 87 89 90 93 96 99 100

2. Using the random number table, take a random sample of size 5 from each strata.
3. Determine the size of each selected shipment, and compute the mean shipment size for each sample.
4. Compare the mean shipment sizes of the two samples from the strata. Should these means differ?
5. Combine the two samples to get an estimate of the population mean based on a stratified random sample of size 10. To do this, observe that stratum 1 has 54% of the population data and stratum 2 only 46%. The combined (weighted) mean should then be (0.54) (mean of sample from stratum 1) + (0.46) (mean of sample from stratum 2).
6. Collect sample means for samples of size 5 from the entire class and make a plot. How do these estimates compare with those from simple random sampling above?

What if there was some pattern in the list of shipments?

Systematic Samples

1. List the shipments in two different ways.

    LIST1: Shipments listed by size, smallest to largest:

    1 2 3 4 5 8 9 10 11 16 19 20 21 44 50 64 80 83
    46 54 62 77 91 94 14 18 22 27 30 35 36 43 56 58 65 67
    69 71 78 88 97 6 24 28 32 84 86 92 42 48 53 82 95 98
    12 40 59 61 75 79 85 87 15 17 38 63 99 23 29 37 45 49
    66 74 7 26 33 34 39 52 68 72 89 100 81 13 31 41 47 51
    55 60 73 90 93 25 57 70 76 96

LIST2: Shipments listed by shipment number:

```
 1  2  3  4  5  6  7  8  9  10 11 12 13 14 15 16 17 18
19 20 21 22 23 24 25 26 27  28 29 30 31 32 33 34 35 36
37 38 39 40 41 42 43 44 45  46 47 48 49 50 51 52 53 54
55 56 57 58 59 60 61 62 63  64 65 66 67 68 69 70 71 72
73 74 75 76 77 78 79 80 81  82 83 84 85 86 87 88 89 90
91 92 93 94 95 96 97 98 99 100
```

2. Take a sample of 10 shipments from LIST1 by randomly selecting a starting point, using the random number table, and then taking every 10th shipment from the list.
3. Determine the size of each selected shipment, and compute the mean shipment size for your sample.
4. Take a sample of 10 shipments from LIST2 by randomly selecting a starting point, using the random number table, and then taking every 10th shipment from the list. Determine the size of each selected shipment, and compute the mean shipment size for your sample.
5. Collect the sample means from the class and, in separate plots, plot the sample means from LIST1 and LIST2.

Which seems to be the better situation for systematic sampling? How does systematic sampling compare with stratified random sampling and simple random sampling as a method of estimating the population mean?

Cluster Sampling

1. Listed in Table 3.5 are 20 clusters, containing five shipments each.
2. Randomly select two clusters, using the random number table.
3. Determine the size of each shipment in the selected clusters.
4. Combine the data from both the selected clusters and obtain a single estimate of the population mean.
5. Collect the estimates from the class and make a plot.

How does cluster sampling compare with the other three randomized methods used above as a way of estimating the population mean?

Table 3.5

Clusters

									Cluster Number										
1	2	3	4	5	6	7	8	9	10	11	12	13	14	15	16	17	18	19	20
1	8	19	64	30	58	88	91	6	84	12	79	38	37	7	52	81	51	93	96
2	9	20	14	35	67	97	77	28	86	40	87	63	45	26	68	13	55	25	98
3	10	21	18	36	69	80	62	32	92	59	85	99	49	33	72	31	60	57	42
4	11	44	22	43	71	83	54	24	95	61	15	23	66	34	89	41	73	76	82
5	16	50	27	45	78	94	46	65	53	75	17	29	74	39	100	47	90	70	48

Of the different sampling techniques used, discuss which one gave you the "best" estimates of the population mean and how you judged it to be the best.

- Make dotplots of the means from all the sampling methods, including judgmental. Discuss any similarities and differences you see in the patterns. Comment upon bias and variability.
- The true average (mean) value for the areas of rectangles is 7.42. Now discuss the notion of bias in judgmental samples as compared to random samples.

3.2.3 Sources of bias in surveys

When any controversial event occurs, newspapers, television stations, and radio stations, rush to conduct opinion polls and report results of these surveys. However as *The Wall Street Journal* reporter Cynthia Crossen (1992) notes, "These are most likely to be wrong because questions are hastily drawn and poorly pretested, and it is almost impossible to get a random sample in one night." It is possible that those people inaccessible on one night have systematically different opinions about the issue than those accessible people.

For a survey to produce reliable results,

1. The survey must be properly designed and conducted.
2. Samples must be selected using proper randomization techniques. Nonrandom selection of the sample will limit the generalizability of results.
3. The interviewers must be trained in proper interviewing techniques to avoid biases in responses. The attitude and behavior of the interviewer should not lead to any specific answers resulting in biased outcomes.
4. Wording of questions must be selected carefully so that leading questions and other forms of biased questions can be avoided, as wording of questions affects the response.

A survey is *biased* if it systematically favors certain outcomes. A bias can occur for many different reasons, such as improper wording of a question, the interviewer's behavior, nonrandom selection of sample, improper frame, and so on.

To estimate the percent of high school children that smoke, children are surveyed from one high school and asked whether they smoke cigarettes. If a parent is present at such an interview, the results will be biased because students may systematically tend to favor a negative answer over a positive one regardless of the truth. Such a bias that occurs due to the behavior of the interviewer or respondent is known as the *response bias*. It is possible to reduce response bias by carefully training interviewers and supervising the interview process. A *nonresponse bias* may occur if the person selected for the interview cannot be contacted or refuses to answer.

Most products come with a warranty card that a purchaser is expected to fill out and send to the manufacturer. One manufacturer decides to conduct a customer satisfaction survey, and he selects customers from the list of warranty cards received. By doing so, those who chose not to send the warranty card back or forgot to do so were totally excluded from the survey. If there is a systematic difference in the opinions of those who chose to send the warranty card back and those who chose not to send it back, then the results will be biased. This type of bias is known as an *undercoverage bias*, and it may occur if part of the population is left out of the selection process. For example, in a telephone survey, individuals without a telephone are left out of the selection process. In the United States, almost 98% of households have telephones, and a very small percent of the population is left out of the selection process. But in many African countries, for example, less than 4% of households have telephones, and a telephone survey would give seriously biased results. In every census, a certain percent of population is missed due to undercoverage. The undercoverage is higher in poorer sections of larger cities due to nonavailability of addresses.

A *wording effect bias* may occur if confusing or leading questions are asked. For example, an interviewer may first state, "The American Dental Association recommends brushing your teeth three times a day," and then ask, "How often do you brush your teeth every day?" To avoid looking bad, the respondent may feel compelled to answer "three or more times" even if he or she does not brush that often. The responses are likely to be a higher number than the truth represents. In this situation, the wording effect bias could be reduced or avoided by simply asking, "How often do you brush your teeth every day?"

Exercises

3.7 *Bias* is the tendency for a whole set of responses to read high or low because of some inherent difficulty with the measurement process. (A chipped die may have a bias toward 6's, in that it comes up 6 much more often than we would expect.) In the study of HIV-related risk groups discussed above, people in the survey were asked intimate questions about their personal lives as part of a telephone interview. It there a possibility of bias in the responses? If so, in which direction? Will randomization in selecting the respondents help reduce potential bias related to the sensitive questions? Will randomization in selecting the respondents help reduce any potential bias?

3.8 Readers of the magazine *Popular Science* were asked to phone in (on a 900 number) their responses to the following question: Should the United States build more fossil fuel–generating plants or the new so-called safe nuclear generators to meet the energy crisis of the '90s? Of the total call-ins, 86% chose the nuclear option. What do you think about the way the poll was conducted? What do you think about the way the question was worded? Do you think the result

are a good estimate of the prevailing mood of the country? (See *Popular Science*, August 1990, for details.)

3.9 "Food survey data all wrong?" This was the headline of a newspaper article on a report from the from the General Accounting Office of the U.S. Government related to the Nationwide Food Consumption Survey. The survey of 6,000 household of all incomes and 3,600 low-income households is intended to be the leading authority on who consumes what foods. Even though the original household were randomly selected, the GAO said the result were questionable because only 34% of the sampled household responded. Do you agree? What is the nature of the biases that could be caused by the low response rate? (The article was printed in the *Gainesville Sun,* September 11, 1991.)

3.10 "Why did they take my favorite show off the air?" The answer lies, no doubt, in low Nielsen ratings. What is the powerful rating system, anyway? Of the 92.1 million households in America, Nielsen Media Research randomly samples 4,000 on which to base their ratings. The sampling design is rather complex, but at the last two stages it involves randomly selecting city blocks (or equivalent units in rural areas) and the randomly selecting one household per block to be the Nielsen household. The *rating* for a program is the percentage of the sampled household that have a TV set on and turned to the program. The *share* for a program is the percentage of the *viewing households* that have a TV set turned to the program, where a viewing household is a household that has at least one TV set turned on.

a How many household are equivalent to one rating point?

b Is a share going to be larger or smaller than a rating point?

c For the week of April 19, 1992, "60 Minutes" was the top-rated show, with a rating of 21.7. Explain what this rating means.

d Discuss potential biases in the Nielsen ratings, even with the randomization on the selection of households carefully built in.

3.11 How does Nielsen determine who is watching what show? The determination comes from the data recorded in a journal by members of a Nielsen household. When a person begins to watch a show, he or she is supposed to log on. Computer vision researchers at the University of Florida are developing a peoplemeter that uses computer image recognition to passively, silently, and automatically record who is watching each show. Discuss the potential for this electronic device to reduce bias in the Nielsen ratings. Are they any new problems that might be caused by this device?

3.12 Decide whether these sampling methods produce a simple random sample of students from a class of 30 students. If not, explain why not.

a Select the first six students on the class roll sheet.

b Pick a digit at random and select those students whose phone number ends in that digit.

c If the classroom has six rows of chairs with five seats in each row, choose a row at random and select all students in that row.

d If the class consists of 15 males and 15 females, assign the males the numbers from 1 to 15 and the females the numbers from 16 to 30. Then use a random digit table to select six numbers from 1 to 30. Select the students assigned those numbers for your sample.

e If the class consists of 15 males and 15 females, assign the males the numbers from 1 to 15 and the females the numbers from 16 to 30. Then use a random digit table to select three numbers from 1 to 15 and three numbers from 16 to 30. Select the students assigned those numbers for your sample.

f Randomly choose a letter of the alphabet and select for the sample those students whose last name begins with that letter. If no last name begins with that letter, randomly choose another letter of the alphabet.

3.3   Planning and Conducting Experiments

Does listening to music on the assembly line help or hinder the assembly process? To answer such specific questions, we must conduct carefully planned *experiments*. The floor supervisor could let a few workers have their music on during assembly process and let a few workers work without any music. However, the time of the day when the assembly takes place could affect the outcomes as well. To address this concern, the supervisor could assign a few workers to work in the morning with music on and a few workers to work in the morning without music. Similarly, he could assign a few workers in the afternoon to work with music on and a few workers in the afternoon without music. This process will result in four *treatments*, namely, (1) morning with music, (2) morning without music, (3) afternoon with music, and (4) afternoon without music. The issue will be decided based on *measurements* taken, such as the number of assemblies completed per hour.

Males might produce different results from females, so perhaps we could *control* for gender by making sure both males and females are selected for each of the four treatment groups. That is, we could run the full experiment twice, once for males and once for females. On thinking about this *design* a bit deeper, we might be concerned that the native ability of the workers might also have some effect on the outcome. All the workers selected for this experiment have similar experiences and educational backgrounds, so differentiating on ability is difficult. Therefore, we'll *randomly* assign workers to the four treatment groups in the hope that any undetected differences in ability will be balanced out by the randomization process.

This study has most of the key elements of a *designed experiment*. The goal of an experiment is to measure the effect of one or more *treatments* on *experimental units* appropriate to the study to see whether some treatments are better than others. Here, four treatments are considered: the music (with and without) combined with the time of the day (morning or afternoon) that assembly takes place. The experimental unit in this case is the worker. Another *variable* of interest is the gender of the worker, but this is *controlled* in the design by *blocking*, that is, by making sure we have data from both sexes for all treatments. The variable "ability" cannot be controlled so easily, so we *randomize* the assignment of workers to treatments to reduce the possible biasing effect of ability on the response comparisons.

Key elements of any experiment are as follows:

1. Clearly define the *question* to be investigated.
2. Define the *treatments* to be used.
3. Identify other important variables that can be *controlled*, and possibly *block* experimental units by those variables.
4. Identify important *background (lurking) variables* that cannot be controlled but should be balanced by *randomization*.
5. Randomly assign treatments to the experimental units.
6. Decide on a method of measurement that will minimize *measurement bias*.
7. Organize the *data collection* and *data management*.
8. Plan for careful and thorough *data analysis*.
9. Write *conclusions* in light of the original question.
10. Plan a *follow-up study* to answer the question more completely or to answer the next logical question related to the issue on hand.

Experiments are made up of several different types of variables. The key variable is the response that is to be measured to answer the question around which the experiment is designed, such as productivity (in terms of assemblies completed per hour) in the music study. Other variables are chosen to be part of an experiment because of their potential to help produce or explain changes in the response, such as the presence or absence of music and the time of day in the above example.

Consider another example: An engineer is interested in studying the effect of pipe size and type on heat transfer. The engineer decides what types of pipes and what sizes of pipes (as measured by the diameter) to use in the experiment and then sets up an experiment to measure, as the response of interest, the amount of heat that is transferred through pipes of a specified length. In this experiment, diameter of the pipe is *one explanatory variable* and type of pipe is another, while the amount of heat transferred is the *response variable*.

> A **response variable** is the outcome of interest to be measured in an experiment. (This is sometimes called the dependent variable.)
>
> An **explanatory variable** is a variable that attempts to explain differences among responses. (This is sometimes called an independent variable.)

In the music experiment, each worker in the study is an *experimental unit*. In the pipe experiment, a time slot on the apparatus producing the heat is an experimental unit.

> An **experimental unit** is the smallest unit (person, animal, building, item, etc.) to which a treatment is applied.

When studying the effect of diameter of the pipe and type of pipe on heat transfer, the variable "diameter" is a *quantitative* factor, whereas "type of pipe" is a *qualitative* factor. An experiment can have one or more factors. The number of levels used in the experiment may differ from factor to factor. *Treatments* are the actual factor-level combinations used in the experiment.

> A **factor** is a variable whose effect on the response is of interest in the experiment.
>
> **Levels** are the values of a factor used in the experiment.
>
> **Treatments** are the factor-level combinations used in the experiment.

- In the example of music on the factory floor, two factors each at two levels are considered: music (with or without) and time of the day (morning or afternoon). Therefore, all four combinations (morning with music, morning without music, afternoon with music, and afternoon without music) considered in the experiment are referred to as treatments.
- In the experiment of heat transfer, suppose two types of pipes (copper and aluminum) in three different sizes (small, medium, and large) are used. Then there are $3 \times 2 = 6$ treatments of interest to the engineer: small copper pipe, medium copper pipe, large copper pipe, small aluminum pipe, medium aluminum pipe, and large aluminum pipe.

But how do we decide which worker will be assigned to which treatment? If we let workers decide under which condition they want to work, then those who like music will join one group and those who don't will join the other, confusing productivity with "love of music." Can we let the supervisor decide which worker should be assigned to which treatment? No!

That process might introduce some bias due to the opinion of the supervisor about a workers' productivity. In addition, the assignment should be made so that the workers with different skills are spread fairly evenly over four treatment groups, assuring that the skill effect on the productivity will be averaged out. Taking all of this into account, experience has shown that the best way to accomplish the necessary balance among treatments and uncontrolled variables is to assign treatments to experimental units at random. *Random assignment*, then, becomes one of the key marks of a good experiment. In the pipe experiment, the six treatments (combinations of size and type of pipe) would be assigned in random order to the apparatus that produces the heat transfer response.

Suppose the experimenter did decide to play music only in the morning and had no music in the afternoon. If the morning productivity turned out to exceed the afternoon productivity, then it would be impossible to tell whether the increased productivity were due to the music or the time of day. In this situation, music and time of day are said to be *confounded*. In comparative studies, *confounding* can lead to many false conclusions.

A **confounding variable** is a variable whose effect on the response cannot be separated from the effect of the treatments.

Randomizing helps protect against confounding effects. For example, in the heat transfer study, suppose the responses for all the aluminum pipes were measured before any of the copper pipes were tested. It might be that the apparatus gets hotter as the day wears on (or the operator gets tired) so that the effect of aluminum versus copper is confounded with heat of the machine (or operator fatigue). Randomization can help prevent this undesirable order from occurring.

Going back to the music on the factory floor, how can we control for variation in productivity due to different skill levels? Suppose there are some measures of skill level that allow classification of workers into groups such as "low skill level" and "high skill level." This is called *blocking*. Then some low-skill-level and some high-skill-level workers could be assigned to each treatment.

In summary, any experiment must have a well-defined response of interest and then must identify treatments that could affect the response in some important way. The goal of an experiment is to measure treatment effects so as to sort out which treatments are better than others. Often, extraneous factors get in the way and obscure the treatment effects. Some of these are easy to control, and some are not so easy to control. We use two different techniques to balance the effects of such factors, *randomization* and *blocking*.

- The technique of *randomization* is used to balance the effects of factors you cannot easily control. In other words, randomization is used to average out the effects of extraneous factors on responses. Which experimental unit receives which treatment is determined randomly. For example, in the experiment of the effect of music on productivity of workers, the floor supervisor will use a randomization mechanism to decide which participating worker should be assigned which one of the four treatments.

• The technique of *blocking* is used to control the effects of factors that you can control easily. For example, in the experiment on the effect of music on productivity of workers, the floor supervisor could separate workers by gender and conduct the experiment once for males and once for females. In this situation, the gender of a worker is used as a blocking factor to control for the effect of gender on productivity.

> A **block** is a group of homogeneous (similar in characteristics) experimental units.

Ideally, experimental units in a given block are similar in certain characteristics, whereas those in different blocks differ in those characteristics. It is the similarity in characteristics that we achieve by separating male workers from female workers and forming two blocks, one of male workers and another of female workers. In the example of music on the factory floor, two factors are considered. If production rate increases with music in the morning but decreases with music in the afternoon, then we would say that music and time of day interact to affect productivity. If the difference in the productivity between the music and no music groups is similar in the morning and in the afternoon, no *interaction effect* exists between factors music and time of day; otherwise, there is an interaction effect.

> An **interaction effect** is said to be present when one factor produces a different pattern of responses at one level of a second factor than it does at another level.

By plotting mean responses for all factor-level combinations, one can easily detect the presence of interaction graphically (see Figure 3.6). Think! While prescribing certain medicines, your doctor advises you not to take them with coffee. Why?

In general, when a chemical engineer runs experiments in the lab, she repeats the experiment more than once under exactly the same conditions. Why? Because even an experiment conducted under the same conditions can give slightly different results. Thus, *replication* is necessary to measure variation in responses at each treatment. This variation is also known as *within-treatment variation*. Replication reduces chance variation among responses while also allowing us to measure this chance variation. Treatments are usually compared by comparing their mean responses. As we'll see later (Chapter 10) the variation in these means decreases as the number of replication increases.

> **Replication** is the number of times a treatment appears in an experiment, or the number of experimental units to which each treatment is applied in an experiment.

Figure 3.6
Interaction effect

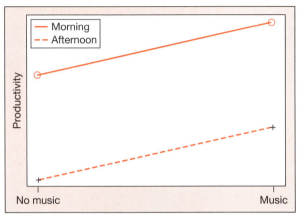

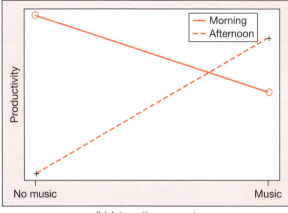

(a) No interaction present   (b) Interaction present

If all treatments are replicated exactly the same number of times in the experiment, then it is known as a *balanced design*. Suppose five workers are assigned to each of four treatments, resulting in the following data table (Table 3.6); then we say that each treatment was replicated five times. This replication will allow the floor supervisor to separate the *within-treatment variation* (i.e., the variation within the productivities of workers working under the same conditions) from the variation in productivity due to differences in working conditions (i.e., variation between the treatments made up of the factors "music" and "time of day").

Table 3.6

Replications in the Music Experiment

		Time of Day	
		Morning	**Afternoon**
Music	Yes	_____ _____ _____ _____ _____	_____ _____ _____ _____ _____
	No	_____ _____ _____ _____ _____	_____ _____ _____ _____ _____

3.3.1   Completely randomized design

Consider Table 3.6, showing replications in the music experiment. The differences among the average (mean) productivity of workers assigned to the four treatments may be a result of differences among these four treatments. The variation among these means measures *between-treatment variation*. Do all workers assigned to the same treatment have the same productivity? No, not likely. The differences among the productivities of workers assigned to the same treatment measure *within-treatment variation*, or worker variation. Without replication, it would not be possible to estimate variation within treatment. The goal of this experiment is to determine whether there is a significant variation between the mean productivity of workers in the four treatment groups. In other words, we want to determine whether the treatment variation is significantly larger than the worker variation. We can determine this by comparing between-treatment variation and within-treatment variation.

A protocol that describes exactly how the experiment is to be done is known as the *design of an experiment*. In order to develop an appropriate design for an experiment, first the engineer must identify major sources of variation that possibly affect the response. Different types of designs can be used depending on the number of factors involved and their relationships. Accordingly, we have to use different randomization schemes to control some factors and measure variation due to treatments.

> In a **completely randomized design**, treatments are assigned randomly to the experimental units, or experimental units are randomly assigned among the treatments.

This most basic design can compare any number of treatments. There are certain advantages in having an equal number of experimental units for each treatment, but it is not necessary. All the experimental units are expected to be as similar as possible because nonhomogeneous experimental units can confound effects due to experimental units and effects due to treatments.

Example 3.7 ——————————————————

A research and development department of a manufacturer of industrial drills has designed two new types of drill bits (A and B). The design engineers are interested in comparing the functioning of these two new types of bits with that of the currently marketed one (C). Metal sheets of specific thickness are available for the experiment. The drilling times to make holes of specific sizes will be recorded. Design an experiment to compare the three different types of bits.

Solution   The factor of interest is the bit type, with three levels (A, B, and C), which is an explanatory variable, because we are hoping that it would explain differences (if any) in drilling times. Therefore, bit types A, B, and C are three treatments in this experiment. The measured variable, drilling time, is the response variable. Using the available metal sheets, design the experiment as follows:

- Use some randomization scheme to divide the metal sheets into three groups. For example, throw a 6-sided die for each sheet. If a number 1 or 2 shows, then assign that sheet to group 1; if a number 3 or 4 shows, then assign that sheet to group 2; otherwise, assign the sheet to group 3.
- Drill a hole in each sheet in group 1, using a new bit A. Record the drilling times. Drill a hole in each sheet in group 2, using a new bit B. Record the drilling times. Drill a hole in each sheet in the group 3, using the currently marketed bit C. Record the drilling times. Randomize the order in which the drilling occurs.
- Compare the results.

This scheme can also be described using a schematic diagram, as shown in Figure 3.7.

In this experiment, we can identify two different sources of variation in the drilling times, one due to the use of different types of drill bits and the other among bits of the same type. The variation in drilling times using different drill bits is the between-treatment variation, whereas the variation in drilling times when using the same type of drill bit is within-treatment variation.

Figure 3.7

Schematic diagram of design of an experiment

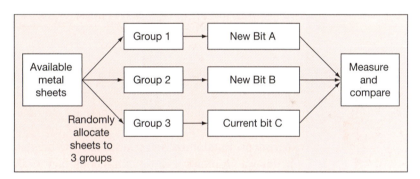

Example 3.8

Welded joints of pipes contain geometric patterns, which makes them likely to experience fatigue damage and develop cracks. Zhao (*Journal of Pressure Vessel Technology*, 2003) reported results of a designed experiment. They performed fatigue tests on a MTS 809 servo-hydraulic machine at a temperature of 240°C under total strain control. Six total strain amplitudes were used. Seven randomly assigned specimens were tested at each total strain amplitude level.

This is a balanced completely randomized design with six treatments (i.e., six total strain amplitudes) and seven replications (i.e., seven specimens). The experimenter is interested in comparing the mean fatigue life (measured in number of cycles) for six different strain amplitudes.

Note: Response variable: Fatigue life
Treatments: Six strain amplitudes
Number of replications = 7

Figure 3.8

Parallel dotplot of fatigue life by strain amplitudes

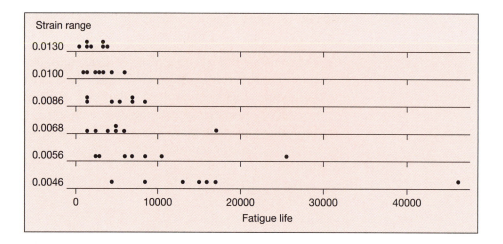

A total of 42 specimens was used in this experiment. Seven specimens were randomly assigned to each of six strain amplitudes. The resulting dotplot of the data, given in Figure 3.8, shows that the mean fatigue life differs for different strains. Those differences in mean fatigue life may be due to use of different strain amplitudes. However, there is also a considerable amount of variation in the fatigue lives of the specimens tested within the same strain amplitude.

3.3.2 Randomized block design

If treatments are the only systematic differences present in the experiment, then a completely randomized design is best for comparing responses. Often there are other factors affecting responses, such as the skill level of workers in the music experiment. Randomization can reduce the bias associated with such factors, but a better way to control those extraneous factors that can be measured before the experiment is conducted is by blocking. Groups of similar units, called *blocks*, are formed, and treatments are compared within each block. Blocking of experimental units allows removal of systematic differences in responses due to a known factor and leads to more precise conclusions from the experiment.

> In a **randomized block design**, all experimental units are grouped by certain characteristics to form homogeneous blocks, and a completely randomized design is applied within each block.

In the music experiment, to control for the differential effect of skill level on the productivity, the supervisor decided to separate workers by skill level (low and high) and assign workers within each skill-level group randomly to four treatment groups. So, the randomization was carried out within each block. It is like conducting a completely randomized design within each block. The resulting data table will look like Table 3.7.

Table 3.7

RBD Data Layout for Music Experiment

Low Skill Level				High Skill Level			
Music Morning	No Music Morning	Music Afternoon	No Music Afternoon	Music Morning	No Music Morning	Music Afternoon	No Music Afternoon
_____	_____	_____	_____	_____	_____	_____	_____
_____	_____	_____	_____	_____	_____	_____	_____
_____	_____	_____	_____	_____	_____	_____	_____
_____	_____	_____	_____	_____	_____	_____	_____
_____	_____	_____	_____	_____	_____	_____	_____

Now we can identify three different sources of variation.

1. Variation in productivity due to use of different treatments
2. Chance or within-treatment variation
3. Variation in productivity due to the skill level of workers

If there are only two treatments to be compared in each block, then the result is a *paired comparison design*. This can be designed in two different ways.

- Make each block of two experimental units, matched by some characteristic. Within each block, toss a coin to assign two treatments to two experimental units randomly. So, both treatments will be applied within each block, with each experimental unit receiving only one treatment. Because both experimental units are similar to each other except for the treatment received, the differences in responses may be attributed to the differences in treatments. For example, in a medical experiment comparing two treatments, subjects paired by matching background health conditions could have the treatments assigned at random within each pair.

- Alternatively, each experimental unit can be used as its own block. Assign both treatments to each experimental unit, but in random order. To control the effect of order of treatment, determine the order randomly. With each experimental unit, toss a coin to decide whether the order of treatments should be "treatment 1 → treatment 2," or "treatment 2 → treatment 1." Because both treatments are assigned to the same experimental unit, the individual effects of the experimental units are reduced so that differences in responses can more easily be attributed to differences in treatments. (A design of the type described here is sometimes called a *repeated measures design*.) For example, in a medical experiment comparing two treatments, it is sometimes possible to give each subject both treatments in random order, with a time delay between the administrations.

Example 3.9

Ermolova (*IEE Proceedings Communications*, 2003) designed an experiment to compare the gain using modified and conventional systems for the traveling wave tube amplifier (TWTA) model. For each of five different output back-off (OBO) values, the experimenter measured the gain for the TWTA model, using both systems. In this experiment, the OBO values serve as blocks, because the gain measured is matched or paired by the OBO values. The proposed modified system and the conventional system are the two treatments of interest. The measured gain is the response variable. The dotplot of the data is given in Figure 3.9.

From Figure 3.9, we can identify different sources of variation. As expected, the gain varies between the two systems, which is a treatment variation. The variation in gain at the different levels of OBO is the block variation. It is now easy to see that the conventional system has a higher gain than the modified system in each block. If the blocks had been ignored (envision sliding the two vertical rows of dots together), it would be much harder to say which of the two systems had the larger gains.

Figure 3.9

Dotplot of data for an RBD

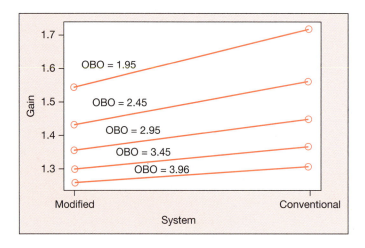

Example 3.10

A manufacturer is interested in comparing the time required to complete a task by two different sorters. Six different operators will be used to operate sorters in this experiment. The engineer in charge suspects that the experience level of each operator will affect the sorting time. Design an experiment to compare sorting times of the three sorters.

Solution   The response variable is sorting time to be measured. Two factors in this experiment are sorter (2) and operator (6), of which sorter is a treatment factor and operator is a blocking factor. To control for the operator effect, the engineer decides to block by operator, in other words, get each operator to operate both sorters. However, she suspects that the second sorter operated by each operator will get the advantage of a learning effect and will result in lower sorting times. Therefore, with each operator the engineer randomly decides the order of sorters operated (sorter 1 → sorter 2 or sorter 2 → sorter 1). By doing so, any advantage of a learning effect will be about the same for both the sorters, and it will get averaged out in the comparison of mean times.

Activity Look at the template of the paper helicopter given in Figure 3.10. Engineers are interested in studying the effect of wing length on the flight time of helicopters. The flight time is defined as the time for which the helicopter remains in flight. The *response* to be measured is flight time of the paper helicopter.

a The engineers know that flight times of helicopters vary with wing lengths. Consider helicopters with wing lengths 4.5, 5.5, and 6.5 inches. Design a completely randomized experiment to compare flight times of helicopters with the three different wing lengths.

b To increase the carrying capacity, the engineers are also interested in widening the body of the helicopter but suspect that it will have an effect on the flight times. Consider three different wing lengths (4.5, 5.5, and 6.5 inches), and two different body widths (1 inch and 2 inches). Design an experiment to compare the effect of wing length and body width on the flight times of helicopters.

c Are there any other factors affecting the flight times that you would like to control, such as the paper weight? Design a block experiment to compare flight times for different wing lengths. Discuss your choice of a blocking factor.

Figure 3.10

A template for a paper helicopter

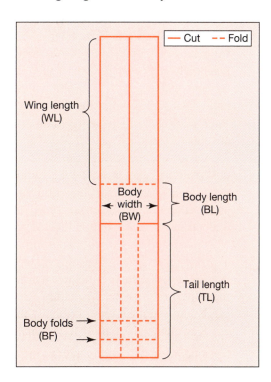

3.4   Planning and Conducting an Observational Study

"The economic conditions are deteriorating in the country; we need to plan for additional sessions of freshman engineering courses," the dean of an engineering college said to his faculty. The implication of this statement is that student enrollments are higher in economically bad years than in good years. Is this really the case? One way to answer this question is to collect data on the student enrollments for the two different groups, economically good years and bad years, and compare the results. Such an investigation looks somewhat like an experiment, but there is a key difference. In an experiment, treatments are randomly assigned to experimental units, and in the randomization process other factors of importance become balanced among the treatment groups. In the investigation suggested above, "good" and "bad" cannot be randomly assigned to years; therefore, the other factors that might affect student enrollments are not necessarily balanced in the study. All the dean can do is *observe* what is happening with the enrollment in his college. Thus, this will be an observational study.

Other important variables can and should be identified in the study. It might be possible to partially *control* some of them through *stratification*. Student enrollments vary by semester, so the data can be stratified according to the semester in which the student joined the program. In addition to the national economic condition, the regional economy has an effect on student enrollment. Enrollments might also differ by student status (students entering college directly from high school or returning from the workforce). No matter how much stratification is accomplished, some potentially important variables that cannot be balanced according to the national economy will always remain and will *confound* the results, in the sense that one never knows for sure whether the difference in enrollments is due to the economy or to some other factor. This is the main reason why *observational studies* are less desirable than experiments, but if done carefully, they are often better than no study at all.

Observational studies can be *retrospective* or *prospective*. In the retrospective observational studies, data collected in the past are used. In prospective studies, an experimenter plans data collection for future studies. However, the experimenter is an observer and exercises no control over the allocation of treatments to the experimental units. Using data collected from past hurricanes is a retrospective study, whereas a study to collect data from any future hurricanes (whenever it occurs) is a prospective study. One advantage of prospective studies over retrospective studies is that the experimenter has some freedom of deciding which variables to measure. In retrospective studies, whatever is available will be used; no new data can be collected, however critically they might be needed.

Key elements of an observational study are listed below:

1. Clearly define the *question* being investigated.
2. Identify the key variables to be observed.
3. Identify important variables that can be *controlled by stratification*.

4. Identify potential *confounding* variables that cannot be controlled.
5. Select *measurement units* in as balanced a way as possible, even though randomization is not used.
6. Decide on a *method of measurement* that will minimize measurement bias.
7. Organize the *data collection* and *data management*.
8. Plan a careful and thorough *data analysis*.
9. Write *conclusions* in light of the original question.
10. Plan a *follow-up study* to answer the question more completely or to answer the next logical question related to the issue at hand.

As in designed experiments, confounding can be a problem in observational studies too. In fact, it is more severe in observational studies because the experimenter has no control over the confounding variables. Observational studies often use samples of convenience, which generally are not representative of any population. Results of retrospective observational studies in which data are collected from people can be questionable because the memory of many people in recalling past events is not dependable. In the prospective studies, the researcher might think about measuring possible confounding variables; however, in the retrospective studies nothing can be done if confounding variables were not measured in the first place.

Example 3.11

Harbor Engineers are interested in developing algebraic models to describe patterns of sand shifting, using tidal elevations and tidal current velocities, because they affect the traffic of barges through the estuaries and deep-water navigations through the channels. To do so, they first need to identify possible factors affecting the response variable (patterns in sand shifting) and collect data. It is known that the tidal elevations also depend on the moon's phase and time of the day. Based on the locations and accessibility, the engineers identify four harbors for study, two used for deep-water navigation and two with shallower estuaries used for barge traffic. At each harbor, the engineers determine the location at which the measuring devices will be floated so that they will be in comparable locations for all four harbors. The engineers identify the following factors to be recorded: day, month, time of day, phase of moon, depth of channel, patterns in sand shifting, traffic intensity, and wind direction. Although the engineers made decisions on which sites to measure, this is still an observational study because there is neither randomization nor control of any of the explanatory variables.

Exercises

3.13 Suppose engineers at an environmental testing firm received a project to compare mean percolation rates (mm/yr) for four different cover types used in landfills. Twelve different landfills located in areas comparable in terms of atmospheric temperatures are available for their experiment. Four different cover types selected for this experiment are (1) conventional compacted clay,

(2) alternative monolithic, (3) conventional composite, and (4) alternative capillary barrier. Help these engineers design this experiment to compare the percolation rates.

3.14 An engineering department chair is interested in comparing the effect on the student grades of a well-established postgraduate module delivered using a traditional lecture mode and new distance-learning software. Fifty students enrolled in a postgraduate course are available for this study.

a Design an experiment to compare the two teaching methods.

b Suppose it is known that some of these students are full-time students and some are part-time students. Improve upon your design in part (a) to take into account the difference in part-time and full-time student achievements.

c Suppose the department chair decided to let students choose between classes with the two different modes of instruction (traditional or distance-learning). Discuss the consequences of his decision to let students choose the class.

3.15 A civil engineer is interested in comparing the wear rate of one type of propellant at 600°C, 800°C, and 100°C. Design an experiment to compare the mean wear rate at the three different temperatures.

3.16 Suppose it is possible to determine the wear rate analytically, that is, by using formulas that establish the relation between temperatures and wear rate. However, the engineer is not sure how closely the analytical method estimates the experimental wear rate. He is considering using a range of temperatures of 600°C to 1,200°C with an increment of 50°C for comparison (600(50)1200°C). Design an experiment to compare analytically estimated wear rates with the experimental wear rates.

3.17 A utility company is interested in comparing the effect of peak hour rates (dollars per kilowatt of usage) on the electricity usage by small manufacturing businesses.

a Describe a design for this experiment, using a sample of 30 small manufacturing businesses for a large industrial complex.

b Could your design be double blind? Explain.

3.18 In the 1990s, scientists became concerned about worldwide declines in the population sizes for amphibians. It was thought that one of the possible causes was the increasing amount of ultraviolet (UV) light to which the eggs were exposed, as many amphibians lay their eggs in shallow water fully exposed to sunlight. An experiment was designed for a shallow water creek in Oregon that is home to three amphibians, the toad, the tree frog, and the Cascade frog, which possess differing levels of an enzyme that protects eggs against the effects of UV light. Two types of UV filters were used, a blocking filter and a transmitting filter, along with a control group involving no filter (Ramsey and Schafer, *The Statistical Sleuth*, 1997).

a Describe the factors and levels for this experiment. How many treatments are there?

b The experimental units were 36 enclosures set up in the shallow water, each with room for 150 eggs. Describe how you would randomize the treatments to the experimental units.

Supplementary Exercises

3.19 The yield of a particular chemical reaction depends on the temperature at which the reaction is carried out. The investigators are interested in comparing the yield at temperatures 100°F, 150°F, and 200°F. They can conduct a reaction 12 times in an 8-hour shift.

a Identify the response variable.

b Identify treatments.

c Design a completely randomized experiment to compare yield at different temperatures.

d How many replications of each treatment are possible in an 8-hour shift?

3.20 The investigators from Exercise 3.19 suspected that in addition to temperature, the yield might also depend on the speed of the mixer. Along with temperatures 100°F, 150°F, and 200°F, they decided to compare two different speeds of the mixer, 1,200 and 1,800 rpm (revolutions per minute).

a Identify treatments.

b Design an experiment to compare yield at different temperatures and speeds.

3.21 According to an article in *Mobile Register* (Study: Breast-feeding may prevent diabetes in moms as well as babies), a new study published in the *Journal of the American Medical Association* on November 23, 2005, reported results of a long-running health study. In this study that began in 1976, 6.3% of women who breast-fed less than 1 year or not at all developed diabetes, compared with 5.5% of women who breast-fed for more than a year.

a Is this an observational study or an experiment? Why?

b Dr Schwartz of Dartmouth Medical School said the results may reflect the healthy lifestyles of women who breast-fed rather than breast-feeding itself. What do you think about this comment? If it is true, then what do you think is happening here?

3.22 The adult smoking rate declined, according to an article in *Mobile Register*, November 11, 2005. The reported results were based on a national household survey of 31,326 adults. People were defined as current smokers if they had smoked at least 100 cigarettes in their lifetime and said they still smoked on a daily or occasional basis. Among American adults, 22.5% in 2002, 21.6% in 2003, and 20.9% in 2004 described themselves as regular puffers.

a Are these results from a sample survey or an experiment? Why?

b Could 20.9% be used as a good estimate of the percent of American population that smokes?

3.23 Coffee may not affect blood pressure in women, reported an article in *Mobile Register*, November 9, 2005. This article was based on a study reported by the *Journal of the American Medical Association* (2005) that followed 155,594 mostly white female nurses, age 55 on the average, who took part in two long-running health studies. The participants were questioned periodically about their diets and health and followed for over 12 years.

a Is this an observational study or an experiment?

b Caffeine is a well-known beverage in both coffee and cola and has been shown to cause short-term increases in blood pressure. The article reported that drinking coffee doesn't seem to cause long-term high blood pressure. Can such a cause-and-effect conclusion be drawn from this study? Justify your response.

c The study also reported that women in the same study who drank four cans of cola or more each day seem to have a greater risk of high blood pressure. However, researchers cautioned that the study wasn't conclusive because sodium in cola might be a culprit. How would you describe these effects?

d Can these results be applied to all women in the United States? To all U.S. men?

3.24 With decreased natural fuel supplies and rising costs, in recent years more and more car manufacturers are switching to hybrid technology. The traditional vehicles draw power from either a gas or diesel engine. Hybrids, however, draw power from two energy sources, typically a gas or diesel engine combined with an electric motor. General Motors developed a hybrid technology that is used in mass-transit vehicles in Seattle. Although more fuel efficient, hybrid technology is also more expensive. Another city in the United States is considering switching to hybrid mass-transit vehicles and would like to compare fuel economy as well as emissions by traditional and hybrid mass-transit vehicles. Design a study to help them make a decision.

3.25 Too little sleep raises obesity risk, reported *Mobile Register*, November 17, 2005. A study involving about 18,000 adults participating in the federal government's National Health and Nutrition Examination Survey (NHANES) throughout the 1980s. Those who get less than 4 hours of sleep a night were 73% more likely to be obese than those who get the recommended 7–9 hours of rest, scientists discovered.

a Given the brief description presented above, is this a sample survey or an experiment? Why?

b Can we conclude that lack of sleep causes obesity? Justify your answer.

3.26 A pediatric surgeon has known from experience that children undergoing any planned surgeries tend to suffer from surgery jitters. Typically, children are given some sedatives or anti-anxiety drugs to calm them down. But this surgeon interested in some new ways to make surgery less frightening noticed how a friend's child became immersed with his Game Boy. The surgeon is interested in testing his theory that children often forget their fears of surgery as they play the games. Anxiety is measured using the Yale Pre-operative Anxiety Scale on a scale of 0–100.

a Design an experiment to test this surgeon's theory that anxiety level can be lowered by allowing children to play with Game Boys before surgery.

b Design an experiment to compare the effects of using anti-anxiety drugs and allowing use of Game Boy on anxiety level before surgery.

c Discuss how the conclusions from the two experiments described in parts (a) and (b) will differ.

Chapter

4

Probability

Are passengers wearing seat belts less likely to suffer injuries compared to those not wearing seatbelts?

In order to answer such a question we have to determine the likelihood of occurrence of such events. And to estimate the likelihood of such events, we have to get data as described by Figure 4.1. In this chapter, we'll study how randomness leads to different outcomes and how to use randomness to obtain data with a predictable pattern. We'll explore the concept and tools of probability, and the parallelism between probability and relative frequency of data summarized as frequency counts. A variety of examples of improvement of quality of life are used to illustrate the concepts.

Figure 4.1

Likelihood of injuries when wearing seat belts

Source: Data collected by the Maine Department of Public Safety. Do seat belts reduce injuries? Maine crash facts, 1996, **http://www.state.me.us/ dps/Bhs/seatbelts1996.htm** (Augusta, Maine)

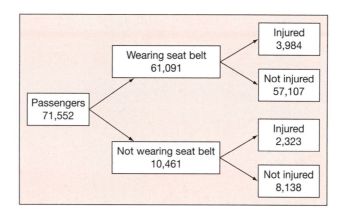

Introduction

Consider a coin used at the beginning of a ball game to assign roles to two teams. A question that arises at that point is: Is this a fair coin? In other words, what is the chance that the coin will land heads up? Most of us would say that this chance, or probability, is 0.5 or something very close to that. But what is the meaning of this number, 0.5? If the coin is tossed 10 times, will it come up heads exactly five times? Upon deeper reflection, most would agree that 10 tosses need not result in exactly five heads, but in repeated flipping the coin should land heads up approximately one-half of the time. From this point, the reasoning begins to get fuzzier. Will 50 tosses result in exactly 25 heads? Will 1,000 tosses result in exactly 500 heads? Not necessarily, but the fraction of heads should be close to one-half after "many" flips of the coin. Thus, 0.5 is regarded as a *long-run* or *limiting relative frequency*, as the number of flips gets large.

John Kerrich, a mathematician interned in Denmark during World War II, actually flipped a coin 10,000 times, keeping a careful tally of the number of heads. After 10 tosses, he had 4 heads, a relative frequency of 0.4; after 100 tosses, he had 44 heads (0.44); after 1,000 tosses, he had 502 heads (0.502); and after 10,000 tosses, he had 5,067 heads (0.5067). The relative frequency of heads remained very close to 0.5 after 1,000 tosses, although the actual figure at 10,000 tosses was slightly farther from 0.5 than was the figure at 1,000 tosses.

In the long run, Kerrich obtained a relative frequency of heads close to 0.5. For that reason, the number 0.5 can be called the *probability* of obtaining a head on the toss of a balanced coin. Another way of expressing this result is to say that Kerrich *expected* to see about 5,000 heads among the outcomes of his 10,000 tosses. He actually came close to his expectations, as have others who have repeated the coin-tossing study. The idea of a stabilizing relative frequency after many trials is at the heart of random behavior; another example that illustrates this idea comes from the study of properties of random digits.

A table of random digits (such as Table 1 in the Appendix), or random digits generated by computer, is produced according to the following model. Think of 10 equal-size chips numbered from 0 through 9, with one number per chip, thoroughly mixed in a box. Without looking, someone reaches into the box and pulls out a chip, recording the number on the chip. That process is a single draw of a random digit. Putting the chip back into the box, mixing the chips, and then drawing another chip produces a second random digit. A random number table is the result of hundreds of such draws, each from the same group of thoroughly mixed chips.

These examples describe the basic notion of probability as a long-run relative frequency. It works well in the situations when we have a good idea about what the relative frequency should be and therefore should work well when we don't.

4.1 Sample Space and Relationships among Events

Engineers are involved in many activities leading to measurable outcomes. For example:

- Measuring the force needed to bend a metal rod of specific thickness
- Measuring the widths of metal pieces cut by a rapid-cut industrial shear from metal sheets of different thickness
- Measuring the number of breakdowns of meat grinders per week

Suppose metal rods will be used in construction of a building, cut metal pieces will be used in making car bodies, and each breakdown of the meat grinder costs heavily in terms of repairs and business lost due to downtimes. Then an engineer will be interested in knowing the likelihood of these events:

- Metal rods sustaining more than the specified pressure
- The industrial shears cutting strips of width within 0.2 cm of required width
- No grinder breakdowns during the next week

In Chapter 2, you learned about distributions of data. Now, those distributions will be used to make probability statements about events that have not yet occurred, such as the probability of a rod being smaller than the specification limit or the probability of more than three breakdowns of a grinder in one week. Developing a probability model begins with a listing of the outcomes for the process being studied.

> A **sample space** is a set of all possible outcomes of a random process.

The sample space is generally denoted using the letter S. We can also think about a sample space as a set of all possible outcomes for a random selection from a specified population. The outcomes listed in a sample space do not overlap, and no outcome is omitted from the list.

For example, the sample space associated with a die toss can be written as $S = \{1, 2, 3, 4, 5, 6\}$. On the other hand, if we are only interested in acknowledging the occurrence of odd or even numbers in a die toss, we can write the sample space as $S = \{$even, odd$\}$. However, the former sample space provides all the necessary details, whereas some information is lost in the latter way of writing the sample space.

Suppose an inspector is measuring the length of a machined rod; a sample space could be listed as $S = \{1, 2, 3, \ldots, 50, 51, \ldots 75, 76, \ldots\}$, if the length is rounded to the closest integer. On the other hand, an appropriate sample space could be $S = \{x \mid x \in I, x > 0\}$, which is read as "the set of all integers numbers x such that $x > 0$."

As seen from these examples, sample spaces for a particular process are not unique; they must be selected so as to provide all pertinent information for a given situation. How to define the sample space depends on the nature of the measurement process and identified goal(s) of the experiment.

> An **event** is any subset of a sample space.

An event is a collection of elements from a sample space. Typically, events are denoted using capital letters such as A, B, C, $\ldots$ or E_1, E_2, E_3, $\ldots$

- In the process of checking the quality of manufactured items, a supervisor inspects one item and classifies it as defective or nondefective. Then the sample space is

$$S = \{\text{defective, nondefective}\}$$

  One event in this experiment can be defined as $A =$ Observing a defective item

- In the process of checking quality of a lot of incoming material, a supervisor inspects a sample of 25 items and classifies each as defective or nondefective. Then he counts the number of defective items in this sample. This experiment results in a sample space of

$$S = \{0, 1, 2, 3, 4, \ldots 25\}$$

  One event in this experiment can be defined as $A =$ Observing at most 2 defective items. Then $A = \{0, 1, 2\}$. Another event can be defined as $B =$ Observing at least 2 but no more than 7 defective items. Then $B = \{2, 3, 4, 5, 6, 7\}$.

- While checking the assembled computers, an inspector counts the number of defects, resulting in a sample space of

$$S = \{0, 1, 2, 3, \ldots\}$$

- A paper mill manufactures bathroom tissue, hand towels, tri-folds, paper rolls, packing paper, and wrapping paper. Every day, the company receives orders for a different product. Each product is ordered separately. This process results in a sample space of

$$S = \{\text{bathroom tissue, hand towels, tri-folds, paper rolls,}$$
$$\text{packing paper, and wrapping paper}\}$$

- A machine is set to cut rods of length 12 inches. An inspector is measuring the length of a machined rod and recording the deviation from the required length. The deviations are recorded to the nearest tenth of an inch. The machine arm extends only 2 inches on either side of the set length. Then the sample space is

$$S = \{-2.0, -1.9, \ldots, -0.3, -0.2, -0.1, 0, 0.1, 0.2, 0.3, \ldots, 1.9, 2.0\}$$

where the negative measurement indicates a rod shorter than the required length and a positive measurement indicates a rod longer than the required length. If $x =$ the deviation from the required length, then the sample space could be described as

$$S = \{x \mid x = 2.0(0.1)2.0\}$$

and it is read as "the set of all numbers between -2.0 and 2.0 inclusive with an increment of 0.1." Some possible events can be defined as follows:

$E_1 =$ Cutting rods of shorter than the required length
$E_2 =$ Cutting rods of longer than the required length
$E_3 =$ Cutting rods within 5/10 of an inch of the required length

We can write the sample spaces for these events as,

$E_1 = \{-2.0, -1.9, \ldots, -0.2, -0.1\}$
$E_2 = \{0.1, 0.2, \ldots, 1.9, 2.0\}$
$E_3 = \{-0.5, -0.4, -0.3, -0.2, -0.1, 0, 0.1, 0.2, 0.3, 0.4, 0.5\}$

Now the question that comes to mind is, "Are all possible outcomes listed in the sample space S equally likely to occur?"

- Consider the experiment of tossing a 6-sided fair die. The associated sample space is the list of numbers on six faces of this die, in other words, $S = \{1, 2, 3, 4, 5, 6\}$. We refer to this die as a fair die because the likelihood of each face when thrown is the same.
- Consider the experiment of measuring rod lengths described above. Are all deviations, -2.0 to 2.0 (in 0.10 inch increments) equally likely to occur? In other words, is the machine likely to cut about same number of rods with each possible deviation? Hopefully not. The machine is likely to cut more rods with smaller deviations and comparatively fewer rods with larger deviations from the

required length. In other words, too large and too small rods are less likely to be cut than the rods of lengths closer to 12 inches.

- Consider the process of classifying items as defective or nondefective. Any acceptable process is much more likely to produce nondefective items than defective items.

Two events are said to be *equally likely events* if one does not occur more often than the other. For example, the six possible outcomes for a fair die are equally likely, whereas possible deviations in rod lengths are not. In the process of classifying items as defective and nondefective, are the events A = defective item and B = nondefective item equally likely?

A tree diagram representation is useful in determining a sample space for an experiment.

Example 4.1

Suppose a firm is deciding to build two new plants, one in the east and one in the west. Four eastern cities (A, B, C, D) and two western cities (E, F) are being considered. Thus, there are 8 possibilities for locating the two plants as shown by the *tree diagram* in Figure 4.2:

Figure 4.2

A tree diagram

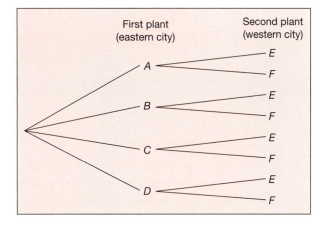

From the tree diagram in Figure 4.2, we can easily write the sample space as

$$S = \{AE, AF, BE, BF, CE, CF, DF, DF\}$$

where A, B, C, D are four eastern cities, and E and F are two western cities under consideration.

Venn diagrams are often convenient and effective for displaying sample spaces, events, and relations among different events. A Venn diagram is a pictorial representation of events and a sample space that makes use of circles.

Example 4.2

For the experiment of throwing a 6-sided die, let us define the following events:

A = An even number shows up
B = An odd number shows up
C = A number greater than 4 shows up
E_3 = Observing face with number 3

Then the sample space $S = \{1, 2, 3, 4, 5, 6\}$, and events $A = \{2, 4, 6\}$, $B = \{1, 3, 5\}$, $C = \{5, 6\}$, and $E_3 = \{3\}$ can be displayed using a Venn diagram, as shown in Figure 4.3.

Figure 4.3

Venn diagram for a die toss

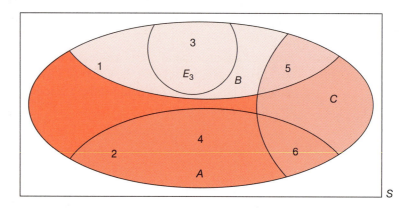

Example 4.3

Consider the following two situations:

- Prizes are to be given at a ballgame during half-time. The winners are to be selected by randomly choosing ticket stubs. Because each person has only one ticket stub, each person attending the game has an equal chance of winning. If we define events E_i as the ith attendee wins the prize, then the events E_i are equally likely.
- Raffle tickets are sold at the ballgame to benefit a charity organization. Attendees are offered tickets at a dollar apiece and allowed to buy as many as they want. As an incentive, a large-screen TV will be given to the owner of a ticket selected at random. If we define events E_i as the ith attendee wins the prize, then the events E_i are not equally likely. The more raffle tickets you buy, the higher your likelihood of winning.

Example 4.4

Suppose there are 100 majors in the department of Electrical Engineering at one university. Define events as follows:

A = Students enrolled in a calculus course
B = Students enrolled in a signal processing course

Of these 100 students, suppose 30 are enrolled in the calculus course, 25 are enrolled in the signal processing course, and 10 in both. Then, the sample space S is the list of all 100 students. Event A lists 30 students in the calculus course, and event B lists 25 students in the signal processing class. Figure 4.4 shows a Venn diagram representation of sample space and events.

Figure 4.4
Venn diagram of student enrollment

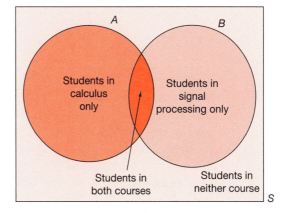

Events A and B are not equally likely. There are more students enrolled in the calculus course than the signal processing course. If a student is selected at random, then he/she is more likely to be enrolled in the calculus class than in the signal processing class.

Relationships among Events   In addition to the occurrence of events A and B, the Dean of the College of Engineering might also be interested in determining the likelihood of events that

- The student is not enrolled in a signal processing course.
- The student is enrolled in both calculus and signal processing.
- The student is enrolled in at least one of the two courses (calculus and signal processing).
- The student is enrolled in neither calculus nor signal processing.

Such events are determined by the complement of an event or by combining the events, using union and intersection.

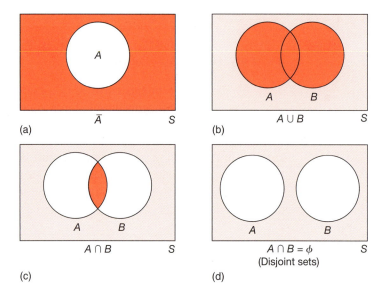

Figure 4.5

Venn diagram of set relations

- The *complement* of an event is the set of all possible outcomes in a sample space that do not belong to the event. In other words, the complement of A with respect to A is the set of all points in S that are not in A. The complement of an event A is denoted by $\bar{A}$. In Figure 4.5a, the shaded area shows the complement of event A.
- The *union* of events A and B is a set of all possible outcomes that belong to at least one of two events A and B. The union of events A and B is denoted by $(A \cup B)$ or (A or B). In Figure 4.5b, the shaded area shows the union of events A and B.
- The *intersection* of events A and B is a set of all possible outcomes that belong to both events A and B. The intersection of events A and B is denoted by $(A \cap B)$ or (A and B) or AB. In Figure 4.5c, the area overlapped by both the circles shows the intersection of events A and B.
- The *disjoint or mutually exclusive events* are events that have no outcome in common. Disjoint or mutually exclusive events cannot occur together. Figure 4.5d shows such events.

Figure 4.5 shows that

- $A \cup \bar{A} = S$ for any set A.
- $AB = \phi$ if events A and B are mutually exclusive. Here ϕ denotes the null set or the set consisting of no points.

Other important relationships among events are as follows:

- Distributive laws:

$$A(B \cup C) = AB \cup AC$$
$$A \cup (BC) = (A \cup B)(A \cup C)$$

- De Morgan's laws:

$$\overline{A \cup B} = \overline{A}\,\overline{B}$$
$$\overline{AB} = \overline{A} \cup \overline{B}$$

It is important to be able to relate descriptions of sets to their symbolic notation and to be able to list correctly or count the elements in sets of interest, as illustrated by the following examples.

Example 4.5

Suppose there are 100 majors in the department of electrical engineering at one university. A student is randomly selected from these 100 majors, then let us define events as follows:

A = Student is enrolled in a calculus course.
B = Student is enrolled in a signal processing course.

Of these 100 students, suppose 30 are enrolled in the calculus course, 25 are enrolled in the signal processing course, and 10 in both. Figure 4.6 shows the Venn diagram representation of the situation.

Figure 4.6

Venn diagram with counts of student enrollment

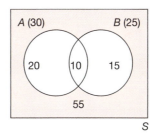

Then some of the events and their portions of the sample spaces are as follows:

- $\overline{A}$: Student is not enrolled in the Calculus course = {… list of 70 students not enrolled in the calculus course …}
- $A \cap B$: Student is enrolled in both the calculus and the signal processing course = {… list of 10 students enrolled in both the courses …}
- $A \cup B$: Student is enrolled in at least one of two courses (calculus and signal processing) = {… list of $20 + 10 + 15 = 45$ students enrolled in either calculus, or signal processing or both …}
- $\overline{A} \cap \overline{B}$ = Student is enrolled in neither course = {… list of 55 students not enrolled in either course …}

Example 4.6

The Bureau of Labor Statistics (net new workers, 2000–2010) is interested in the composition of the workforce in upcoming years. Their studies resulted in the information given in Table 4.1.

Table 4.1

Net New Workers, 2000–2010

	Women	Men	Total
White (non-Hispanic)	23%	24%	37%
Black	9%	6%	15%
Asian	7%	6%	13%
Hispanic	13%	12%	25%
Total	**52%**	**48%**	**100%**

The Venn diagram of key events in this study is given in Figure 4.7. The entire rectangle represents the number of net new workers. In Figure 4.7a, the circle labeled "Women" shows that the category of women is one of the subsets. Figure 4.7b shows schematically that the white (non-Hispanic) and nonwhite subsets do not overlap, whereas Figure 4.7c shows that the women and white subsets do overlap.

Figure 4.7
Venn diagrams for net new workers

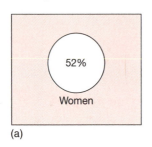

(a)

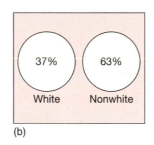

(b)

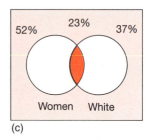

(c)

Example 4.7

Twenty electric motors are pulled from an assembly line and inspected for defects. Eleven of the motors are free of defects, eight have defects on the exterior finish, and three have defects in their assembly and will not run. Let A denote the set of motors having assembly defects and F the set having defects on their finish. Using A and F, write a symbolic notation for the following:

a the set of motors having both types of defects
b the set of motors having at least one type of defect
c the set of motors having no defects
d the set of motors having exactly one type of defect

Then give the number of motors in each set.

Solution **a** The motors with both types of defects must be in A and F; thus, this event can be written AF. Because only nine motors have defects, whereas A contains three and F contains eight motors, two motors must be in AF. (See Figure 4.8.)

b The motors having at least one type of defect must have either an assembly defect or a finish defect. Hence, this set can be written $A \cup F$. Because 11 motors have no defects, 9 must have at least one defect.

Figure 4.8

Number of motors for sets A and F

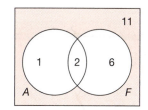

c The set of motors having no defects is the complement of the set having at least one defect and is written $\overline{A \cup F} = \overline{A}\,\overline{F}$ (by De Morgan's Law). Clearly, 11 motors fall into this set.

d The set of motors having exactly one type of defect must be either in A and not in F on in F and not in A. This set can be written $A\overline{F} \cup \overline{A}F$, and seven motors fall into the set.

Exercises

4.1 Of the 2.7 million engineers in the United States (as of 1988), 95.5% are male. In addition, 90.6% are white, 1.6% are black, 5.6% are Asian, 0.4% are Native American. Of the 0.7 million computer specialists, 69.1% are male, 88.3% are white, 3.7% are black, 6.6% are Asian, and 0.1% are Native American.

a Construct a meaningful table to compare number of engineers and number of computer scientists by sex.

b Construct a meaningful table to compare number of engineers and number of computer scientists by racial group.

4.2 Of 25 microcomputers available in a supply room, 10 have circuit boards for a printer, 5 have circuit boards for a modem, and 13 have neither type of board. Using P to denote those that have printer boards and M to denote those that have modem boards, symbolically denote the following sets, and give the number of microcomputers in each set.

a Those that have both boards.

b Those that have neither board.

c Those that have printer boards only.

d Those that have exactly one of the boards.

4.3 Construct a table to display the data in Exercise 4.2.

4.4 Five applicants (Jim, Don, Mary, Sue, and Nancy) are available for two identical jobs. A supervisor selects two applicants to fill these jobs.

a List all possible ways in which the jobs can be filled. (That is, list all possible selections of two applicants from the five.)

b Let A denote the set of selections containing *at least* one male. How many elements are in A?

c Let B denote the set of selections containing *exactly* one male. How many elements are in B?

d Write the set containing two females in terms of A and B.

e List the elements in $\overline{A}$, AB, $A \cup B$, and $\overline{AB}$.

4.5 Use Venn diagrams to verify the distributive laws.

4.6 Use Venn diagrams to verify De Morgan's Laws.

4.2 Definition of Probability

Consider the following situations:

- At the start of a football game, a balanced coin is flipped to decide who will receive the ball first. What is the chance that the coin will land heads up?
- You applied for a job at XYZ, Inc., and so did 23 other engineers. What is the likelihood of your getting that job?

At the start of a football game, a balanced coin is flipped to decide who will receive the ball first. Because the coin is balanced, there is an equal chance of the coin landing on heads or tails. So, we could say that the chance, or the probability, of this coin landing on heads is 0.5. What does that mean? If the coin is tossed 10 times, will it land exactly 5 times on heads? Not necessarily. Each run of 10 tosses will result in a different number of heads. However, in repeated flipping, the coin should land heads up approximately 50% of the time. Will 1,000 tosses result in exactly 500 heads? Not necessarily, but the fraction of heads should be close to 0.5 after many flips of a coin.

Words like "probably," "likely," and "chances" convey similar ideas. They convey some uncertainty about the happening of an event. In statistics, a numerical statement about the uncertainty is made using probability with reference to the conditions under which such a statement is true. For example, in the manufacturing business, the quality inspectors are uncertain about the quality of incoming lots, managers are uncertain about the demand for different items being manufactured, and the system manager is uncertain about exactly how long it will be before the next interruption in the communication system will occur.

Probability is a measure of the likelihood of an event. The probability of an event is generally denoted by P(event). If the events in a sample space are equally likely, then the intuitive idea of probability is related to the concept of relative frequency as

$$P(\text{event}) = \frac{\text{Number of outcomes favorable to the event}}{\text{Total number of possible outcomes}}$$

In the earlier example of engineering students (Example 4.5), a student is randomly selected from the 100 majors, and events A and B are defined as follows: A = Student is enrolled in a calculus course, and B = Student is enrolled in a signal processing course. Therefore,

$$P(A) = \frac{30}{100} \quad \text{and} \quad P(B) = \frac{25}{100}$$

Using similar logic, we can find

- The probability that a student is enrolled in both courses as 10/100
- The probability that a student is enrolled in neither course as 55/100

Suppose S is a sample space associated with a random process. A **probability** is a numerically valued function that assigns a number $P(A)$ to every event A such that

- $P(A) \geq 0$
- $P(S) = 1$
- If $A_1, A_2,\ldots$ is a sequence of mutually exclusive events, then

$$P\left(\bigcup_{i=1}^{\infty} A_i\right) = \sum_{i=1}^{\infty} P(A_i)$$

Probabilities are determined by constructing a theoretical model for a sample space or by collecting data from the random processes being studied.

Probabilities of events can be estimated from the observed data or simulated chance experiments.

Example 4.8

In the year 2002, was a randomly selected employed person more likely to be a man? Was a randomly selected employed person more likely to be a government employee?

To answer these questions, consider the estimates the Bureau of Labor Statistics provides on the employment rates during the year 2002, by gender and type of industry.

If a person employed in 2002 was selected at random, then

- P(selected person was a woman) = 0.46
- P(selected person was working for the government) = 0.53

Table 4.2 shows that, in the year 2002,

- The probability of being employed was higher for men (0.54) than women (0.46).
- About 53% of those employed were in the government compared to 38% in manufacturing and only 9% in information. In other words, the probability of being employed in the government was higher than the probability of being employed in the fields of manufacturing or information.

Table 4.2

Employment Rates

	Men	Women	Total
Manufacturing	26%	12%	**38%**
Government	23%	30%	**53%**
Information	5%	4%	**9%**
Total	**54%**	**46%**	**100%**

When no such data is available to assist us in estimating probabilities of events, we can sometimes simulate the events a large number of times and compute the relative frequencies of events. The following example shows the association between the probability of an event and the relative frequency of experimental outcomes.

Example 4.9

Take a standard paper cup, with the open end slightly larger than the closed end. Toss it in the air and allow it to land on the floor. What is the likelihood that it will land on its open end? On its side? We don't know. So let us experiment. The goal is to approximate the probability that the cup would land on the open end and the probability that it would land on its side.

- Toss the cup 100 times.
- Record the sequence of outcome (lands on open end, closed end, or side).
- Estimate the probability of landing on the open end as

$$P(\text{Landing on the open end}) = \frac{\text{Number of times cup landed on the open end}}{\text{Total number of tosses}}$$

- Plot this probability against the number of tosses.
- What do you observe about this probability from your graph?

When some students did this experiment in class, the sample paths of two of the trials came out as shown in Figure 4.8a (labeled as sample 1 and sample 2). Notice two important features of the graph:

- Both sample paths oscillate greatly for small numbers of observations and then settle down around a value close to 0.20.
- The variation between the two sample fractions is quite great for small sample sizes and quite small for larger sample sizes.

From the Figure 4.9a, we conclude that the probability of a tossed cup landing on its open end is approximately 0.20.

Figure 4.9

Proportions of cup landings

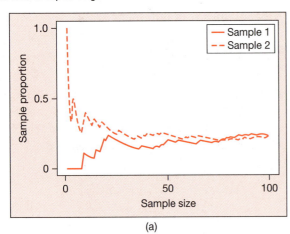

(a)

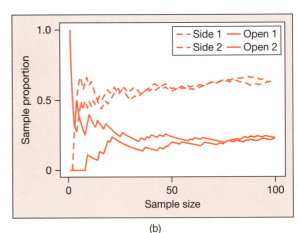

(b)

Now look at Figure 4.9b. It includes two sample paths for the outcome "landed on the side." Both paths seem to stabilize around 0.60. Using the notion of probability as a long-run relative frequency, we can say that the probability of a tossed cup landing on its side is about 0.6.

Because three events–landing on the open end, landing on the side, and landing on the closed end–make up the sample space, using these two results we can say that the probability of a tossed cup landing on its closed end is approximately 0.20.

Look at the following two results. They follow from the definition of probability.

Result 1: If events A and B are such that $A \subset B$, then $P(A) \leq P(B)$.

Because $A \subset B$, we can write event B as $B = A \cup \overline{A}B$. Events A and $\overline{A}B$ are mutually exclusive, hence by the definition, $P(B) = P(A) + P(\overline{A}B)$. Because $P(\overline{A}B) \geq 0$, by the definition it folows that $P(A) \leq P(B)$.

Result 2: The probability of an empty set is 0.

Because S and ϕ are mutually exclusive with $S \cup \phi = S$,
$1 = P(S) = P(S \cup \phi) = P(S) + P(\phi)$. Hence the result.

The definition of probability tells us only the rules such a function must obey; it does not tell us what numbers to assign to specific events. The actual assignment of numbers usually comes from empirical evidence or from careful thought about the selection process. If a die is balanced, we could toss it a few times to see whether the upper faces all seem equally likely to occur. Or we could simply assume this would happen and assign a probability of $1/6$ to each of the six elements in the sample space S as follows: $P(E_i) = 1/6$, $i = 1,2,3,4,5,6$. Once we have done this, the model is complete, becauseby definition we can now find the probability of any event. For example, if we define $A =$ an even number, $C = $ a number greater than 4, and $E_i = $ observing an integer $i(= 1,2,3,4,5,6)$, then

$$P(A) = P(E_2 \cup E_4 \cup E_6) = P(E_2) + P(E_4) + P(E_6) = \frac{1}{6} + \frac{1}{6} + \frac{1}{6} = \frac{1}{2}$$

and $$P(C) = P(E_5 \cup E_6) = P(E_5) + P(E_6) = \frac{1}{6} + \frac{1}{6} = \frac{1}{3}$$

Definition and the actual assignment of probabilities to events provide a probabilistic model for a random selection process. If $P(E_i) = 1/6$ is used for a die toss, the model is good or bad depending on how close the long-run relative frequencies for each outcome actually come to these numbers suggested by the theory. If the die is not balanced, then the model is bad, and other probabilities should be substituted for $P(E_i)$. No model is perfect, but many are adequate for describing real-world probabilistic phenomena.

The die-toss example assigns equal probabilities to the elements of a sample space, but such is not usually the case. If you have a quarter and a penny in your pocket and pull out the first one you touch, then the quarter may have a higher probability of being chosen because of its larger size. The data provide only approximations to the true probabilities, but these approximations are often quite good and are usually the only information we have on the events of interest, as shown in the example below.

Example 4.10

Distillation towers are used by chemical industries including petroleum refineries for distillation purposes. With all the advances in distillation technology, failure rate in distillation towers is expected to be on the decline. To identify the trend, Kister (*IChemE*, 2003) collected data on 900 cases and studied causes of malfunctions among such towers. The data identified plugging/coking as the undisputed leader of tower malfunction. Figure 4.10 describes the breakdown of causes of plugging stratified by the year of malfunction recorded as prior to 1992 (1992–) and 1992 or later (1992+).

Suppose a tower were selected at random from the records prior to 1992. What is the probability that the selected tower had plugging because of (a) coking, (b) precipitation, (c) polymer, (d) coking and polymer. What are these probabilities for a tower selected at random from records 1992 or later?

Solution   Assume that nothing else is known about the selected tower. We see from the pie chart for "1992–" that

a $P(\text{coking}) = 0.143$
b $P(\text{precipitation}) = 0.143$
c $P(\text{polymer}) = 0.107$
d $P(\text{coking and polymer}) = 0.143 + 0.107 = 0.250$

Figure 4.10
Causes of plugging resulting in tower malfunction

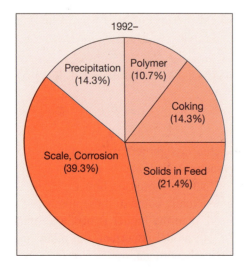
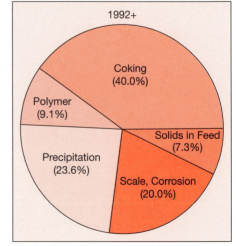

From the "1992+" pie chart, we get

a $P(\text{coking}) = 0.400$
b $P(\text{precipitation}) = 0.236$

   c $P(\text{polymer}) = 0.091$
   d $P(\text{coking and polymer}) = 0.400 + 0.091 = 0.491$

Notice that the likelihood of coking has tripled in the last decade compared to the prior 40 years, whereas the likelihood of precipitation has increased by about 50%. The likelihood of polymer as a cause of plugging has reduced very little. The likelihood of coking and polymer has almost doubled.

Example 4.11

Consider the national percentages for HIV-related risk groups (*Science*, 1992) given in Table 4.3. A firm hires a new worker after a national advertising campaign.

   a What is the probability that the worker falls in the "risky partner" category?
   b What is the probability that the worker is in at least one of the risk groups?
   c If the firm hires 1,000 workers, how many are expected to be at risk if the 1,000 came from the population at large?
   d If the firm hires 1,000 workers, how many are expected to be at risk if the 1,000 come from high-risk cities?

Table 4.3

Prevalence of HIV-Related Risk Groups, National and High-Risk Cities Samples

Risk Group	National Percent	National Number	High-Risk Cities Percent	High-Risk Cities Number
Multiple partners	7.0	170	9.5	651
Risky partner	3.2	76	3.7	258
Transfusion recipient	2.3	55	2.1	144
Multiple partner and risky partner	1.7	41	3.0	209
Multiple partner and transfusion recipient	0.0	1	0.3	20
Risky partner and transfusion recipient	0.2	4	0.3	19
All others	0.7	1	0.7	51
No risk	84.9	2,045	80.4	5,539

Solution   In order to answer these questions, we assume that prevalence of HIV-related risk groups has remained unchanged.

   a Solutions to practical problems of this type always involve assumptions. To answer parts (a) and (b) with the data given, we must assume that the new worker is randomly selected from the national population, which implies

$$P(\text{risky partner}) = 0.032$$

b Let's label the seven risk groups, in the order listed in the table, $E_1, E_2, \ldots, E_7$
Then the event of being in *at least one* of the groups can be written as the union of these seven, that is,

$$E_1 \cup E_2 \cup \cdots \cup E_7$$

The event "at least one" is the same as the event "E_1 or E_2 or E_3 or ... or E_7"
Because these seven groups are listed in mutually exclusive fashion,

$$
\begin{aligned}
P(E_1 \cup E_2 \cup \cdots \cup E_7) &= P(E_1) + P(E_2) + \cdots + P(E_7) \\
&= 0.070 + 0.032 + 0.023 + 0.017 \\
&\quad + 0.000 + 0.002 + 0.007 \\
&= 0.151
\end{aligned}
$$

Note that the mutually exclusive property is essential here; otherwise, we could not simply add the probabilities.

c We assume that all 1,000 new hires are randomly selected from the national population (or a subpopulation of the same makeup). Then 15.1% of the 1,000, or 151 workers, are expected to be at risk.

d We assume that the 1,000 workers are randomly selected from high-risk cities. Then 19.6%, or 196 workers, are expected to be at risk.

Do you think the assumptions necessary in order to answer the probability questions are reasonable here?

Exercises

4.7 For each of the following situations, discuss whether the term probability seems to be used correctly and, if so, how such a *probability* might have been determined.

a A physician prescribes a medication for an infection but says that there is a probability of 0.2 that it will cause an upset stomach.

b An engineer who has worked on the development of a robot says that it should work correctly for at least 1,000 hours with probability 0.99.

c A friend says that the probability of extraterrestrial life is 0.1.

4.8 How good is your intuition about probability? Look at the following situations and answer the questions without making any calculations. Rely on your intuition, and see what happens.

a A six-sided balanced die has four sides colored green and two colored red. Three sequences of tosses resulted in the following:

> RGRRR
> RGRRRG
> GRRRRR

Order the sequences from most likely to least likely.

b If a balanced coin is tossed six times, which of die the following sequences is more likely, HTHTTH or HHHTTT? Which of the following is more likely, HTHTTH or HHHHTH?

Activities like these were used by D. Kahneman, P. Slovic, and A. Tversky to study how people reason probabilistically, as reported in their book *Judgment Under Uncertainty: Heuristics and Biases.* Cambridge University Press, 1982.

4.9 A vehicle arriving at an intersection can turn left or continue straight ahead. Suppose an experiment consists of observing the movement of one vehicle at this intersection, and do the following.

a List the elements of a sample space.

b Attach probabilities to these elements if all possible outcomes are equally likely.

c Find the probability that the vehicle turns, under the probabilistic model of part (b).

4.10 A manufacturing company has two retail outlets. It is known that 30% of the potential customers buy products from outlet I alone, 50% buy from outlet II alone, 10% buy from both I and II, and 10% buy from neither. Let *A* denote the event that a potential customer, randomly chosen, buys from I and *B* denote the event that the customer buys from II. Find the following probabilities.

a $P(A)$ **b** $P(A \cup B)$
c $P(\bar{B})$ **d** $P(AB)$
e $P(A \cup \bar{B})$ **f** $P(\overline{AB})$
g $P(\overline{A \cup B})$

4.11 For volunteers coming into a blood center, 1 in 3 have O$^+$ blood, 1 in 15 have O$^-$, 1 in 3 have A$^+$, and 1 in 16 have A$^-$. What is the probability that the first person who shows up tomorrow to donate blood has each of these blood types

a Type O$^+$ blood
b Type O blood
c Type A blood
d Either type A$^+$ or O$^+$ blood

4.12 Information on modes of transportation for coal leaving the Appalachian region is shown in Figure 4.11. If coal arriving at a certain power plant comes from this region, find the probability that it was transported out of the region using each of these modes of transportation

a By track to rail
b By water only
c At least partially by truck
d At least partially by rail
e By modes not involving water. (Assume "other" does not involve water)

Figure 4.11
Modes of transportation for coal

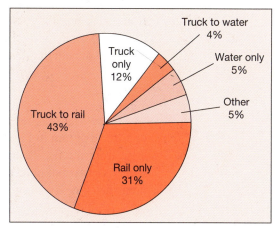

4.13 Hydraulic assemblies for landing gear coming from an aircraft rework facility are inspected for defects. History shows that 8% have defects in the shafts alone, 6% have defects in the bushings alone, and 2% have defects in both the shafts and the bushings. If a randomly chosen assembly is to be used on an aircraft, find the probability that it has each of these defects

a A bushing defect
b A shaft or bushing defect
c Only one of the two type of defects
d No defects in shafts or bushings

4.3 Counting Rules Useful in Probability

Let's look at the die toss from a slightly different perspective. Because there are six outcomes that should be equally likely for a balanced die, the probability of *A*, observe an even number, is

$$P(A) = \frac{3}{6} = \frac{\text{number of outcomes favorable to } A}{\text{total number of equally likely outcomes}}$$

This "definition" of probability will work for any random phenomenon resulting in a finite sample space with *equally likely* outcomes. Thus, it is important to be able to count the number of possible outcomes for a random selection. The number of outcomes can easily become quite large, and counting them is difficult without a few counting rules. Four such rules are presented in this section.

Example 4.12

Suppose a quality control inspector examines two manufactured items selected from a production line. Item 1 can be defective or nondefective, as can item 2. How many possible outcomes are possible for this experiment? In this case, listing the possible outcomes is easy. Using D_i to denote that the ith item is defective and N_i to denote that the ith item is nondefective, the possible outcomes are

$$D_1D_2 \qquad D_1N_2 \qquad N_1D_2 \qquad N_1N_2$$

These four outcomes could be placed in a two-way table as in Figure 4.12(a) or a tree diagram in Figure 4.12(b). This table helps us see that the four outcomes arise from the fact that the first item has two possible outcomes and the second item has two possible outcomes, and hence the experiment of looking at both items has $2 \times 2 = 4$ outcomes. This is an example of the *product rule*.

Figure 4.12

Possible Outcomes for inspecting two items

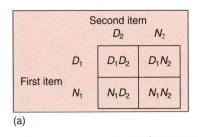

(a)

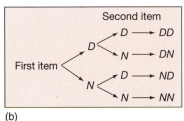

(b)

Product Rule

If the first task of an experiment can result in n_1 possible outcomes and for each such outcome, the second task can result in n_2 possible outcomes, then there are n_1n_2 possible outcomes for the two tasks together.

The product rule extends to more than two tasks in a sequence. If, for example, three items were inspected and each could be defective or nondefective, then there would be $2 \times 2 \times 2 = 8$ possible outcomes.

The product rule helps only in finding the number of elements in a sample space. We must still assign probabilities to these elements to complete our probabilistic model.

Example 4.13

Tree diagrams are also helpful in verifying the product rule and in listing possible outcomes. Suppose a firm is deciding where to build two new plants, one in the east and one in the west. Four eastern cities and two western cities are possibilities. Thus, there are $n_1n_2 = 4 \times 2 = 8$ possibilities for locating the two plants. Figure 4.13 shows the listing of these possibilities on a tree diagram.

Figure 4.13

Possible outcomes for locating two plants

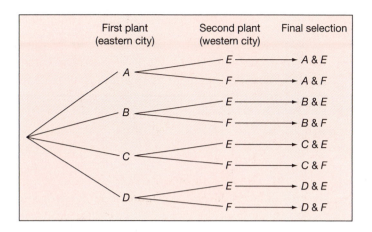

Example 4.14

In the case of the firm that plans to build two new plants, the eight possible outcomes are shown in Figure 4.13. If all eight choices are equally Likely (that is, one of the pairs of cities is selected at random), find the probability that city E gets selected.

Solution City E can be selected in four different ways, since there are four possible eastern cities to pair with it. Thus,

$$\{E \text{ gets selected}\} = \{AE\} \cup \{BE\} \cup \{CE\} \cup \{DE\}$$

Each of the eight outcomes has probability $1/8$, since the eight events are assumed to be equally likely. Because these eight events are mutually exclusive,

$$P(E \text{ gets selected}) = P(AE) + P(BE) + P(CE) + P(DE)$$

$$= \frac{1}{8} + \frac{1}{8} + \frac{1}{8} + \frac{1}{8} = \frac{1}{2}$$

Example 4.15

Five motors (numbered 1 through 5) are available for use, and motor 2 is defective. Motors 1 and 2 come from supplier I, and motors 3, 4, and 5 come from supplier II. Suppose two motors are randomly selected for use on a particular day. Let A denote the event that the defective motor is selected and B the event that at least one motor comes from supplier I. Find $P(A)$ and $P(B)$.

Solution The tree diagram in Figure 4.14 shows that there are 20 possible outcomes for this experiment, which agrees with our calculation using the product rule. That is, there are 20 events of the form $\{1, 2\}, \{1, 3\}$, and so forth. Because the motors are randomly selected, each of the 20 outcomes has probability $1/20$. Thus,

$$P(A) = P(\{1,2\} \cup \{2, 1\} \cup \{2, 3\} \cup \{2, 4\} \cup \{2, 5\} \cup \{3, 2\} \cup \{4, 2\} \cup \{5, 2\})$$

$$= \frac{8}{20} = 0.4$$

Similarly, we can show that B contains 14 of the 20 outcomes and that

$$P(B) = \frac{14}{20} = 0.7$$

Figure 4.14

Defective motor and two suppliers

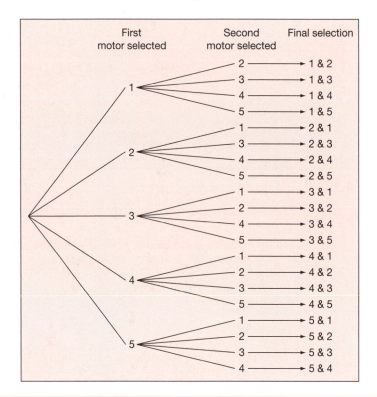

Suppose that from three pilots a crew of two is to be selected to form a pilot-copilot team. To count the number of ways this can be done, observe that the pilot's seat can be filled in three ways and the copilot's in two ways (after the pilot is selected), so there are $3 \times 2 = 6$ ways of forming the team. This is an example of a *permutation*.

> **Permutation** is the number of ordered arrangements, or permutations, of r objects selected from n distinct objects ($r \leq n$). It is given by
>
> $$P_r^n = n(n - 1) \cdots (n - r + 1) = \frac{n!}{(n - r)!}$$

The basic idea of a permutation can be thought of as filling r slots in a line with one object in each slot by drawing these objects one at a time from a pool of n distinct objects. The first slot can be filled in n ways, but the second can be filled in only $(n - 1)$ ways after the first is filled. Thus, by the product rule, the first two slots can be filled in $n(n - 1)$ ways. Extending this reasoning to r slots, we have that the number of ways of filling all r slots is

$$n(n - 1) \cdots (n - r + 1) = \frac{n!}{(n - r)!} = P_r^n$$

Example 4.16

From among 10 employees, three are to be selected for travel to three out-of-town plants, A, B, and C, with one employee traveling to each plant. Because the plants are in different cities, the order of assigning the employees to the plants is an important consideration. The first person selected might, for instance, go to plant A and the second to plant B. In how many ways can the assignments be made?

Solution Because order is important, the number of possible distinct assignments is

$$P_3^{10} = \frac{10!}{7!} = 10(9)(8) = 720$$

In other words, there are 10 choices for plant A but only nine for plant B and eight for plant C. This gives a total of $10(9)(8)$ ways of assigning employees to the plants.

Example 4.17

An assembly operation in a manufacturing plant involves four steps, which can be performed in any order. If the manufacturer wishes to experimentally compare the assembly times for each possible ordering of the steps, how many orderings will the experiment involve?

Solution The number of orderings is the permutation of $n = 4$ things taken $r = 4$ at a time. (All steps must be accomplished each time.) This turns out to be

$$P_4^4 = \frac{4!}{0!} = 4! = 4 \cdot 3 \cdot 2 \cdot 1 = 24$$

because $0! = 1$ by definition. (In fact, $P_r^r = r!$ for any integer r.)

Example 4.18

A manager is asked to rank four divisions of a firm with respect to their ability to adapt to new technology. If the divisions are labeled D_1, D_2, D_3, and D_4, what is the probability that D_2 gets ranked highest?

Solution The question has no answer as posited; there must be a random mechanism in operation in order to discuss probability. If the manager has no preferences and is merely ranking the firms randomly, then the question can be answered. There are now

$$P_4^4 = 4! = 24$$

equally likely rankings. If D_2 has the number one ranking, then there remain $3! = 6$ ways to rank the remaining divisions. Thus,

$$P(D_2 \text{ ranks first}) = \frac{3!}{4!} = \frac{6}{24} = \frac{1}{4}$$

Does this look reasonable? The question here could be extended to "What is the probability that D_2 ranks first and D_4 ranks second?" Once the first two ranks are taken, there remain only $2! = 2$ ways to fill out ranks three and four. Thus,

$$P(D_2 \text{ ranks first and } D_3 \text{ ranks second}) = \frac{2!}{4!} = \frac{2}{24} = \frac{1}{12}$$

Note that this answer is *not* $\left(\dfrac{1}{4}\right)\left(\dfrac{1}{4}\right)$.

Sometimes order of selection is not important, and we are interested only in the number of subsets of a certain size that can be selected from a given set.

> **Combination** is the number of distinct subsets, or combinations, of size r that can be selected from n distinct objects ($r \le n$). It is given by
>
> $$\binom{n}{r} = \frac{n!}{r!(n-r)!}$$

The number of ordered subsets of size r, selected from n distinct objects, is given by P_r^n. The number of unordered subsets of size r is denoted by $\binom{n}{r}$. Because any particular set of r objects can be ordered among themselves in $P_r^r = r!$ ways, it follows that

$$\binom{n}{r} r! = P_r^n$$

or

$$\binom{n}{r} = \frac{1}{r!} P_r^n = \frac{n!}{r!(n-r)!}$$

Example 4.19

Suppose that three employees are to be selected from ten to visit a new plant. (a) In how many ways can the selection be made? (b) If two of the ten employees are female and eight are male, what is probability that exactly one female gets selected among the three?

Solution   **a** Here, order is not important; we merely want to know how many subsets of size $r = 3$ can be selected from $n = 10$ people. The results is

$$\binom{10}{3} = \frac{10!}{3!(7!)} = \frac{10(9)8}{3(2)1} = 120$$

b We have seen that there are $\binom{10}{3} = 120$ ways to select three employees from 10. Similarly, there are $\binom{2}{1} = 2$ ways to select one female from the two available and $\binom{8}{3} = 28$ ways to select two males from the eight available. If selections are made at random (that is, if all subsets of three employees are equally likely to be chosen), then the probability of selecting exactly one female is

$$\frac{\binom{2}{1}\binom{8}{2}}{\binom{10}{3}} = \frac{2(28)}{120} = \frac{7}{15}$$

Example 4.20

Five applicants for a job are ranked according to ability, with applicant number 1 being the best, number 2 second best, and so on. These rankings are unknown to an employer, who simply hires two applicants at random. What is the probability that this employer hires exactly one of the two best applicants?

Solution   The number of possible outcomes for the process of selecting two applicants from five is

$$\binom{5}{2} = \frac{5!}{2!3!} = 10$$

If one of the two best is selected, the selection can be done in

$$\binom{2}{1} = \frac{2!}{1!(1!)} = 2$$

ways. The other selected applicant must come from among the three lowest-ranking applicants, which can be done in

$$\binom{3}{1} = \frac{3!}{1!(2!)} = 3$$

ways. Thus, the event of interest (hiring one of the two best applicants) can occur in $2 \times 3 = 6$ ways. The probability of this event is thus $6/10 = 0.6$.

The number of ways of **partitioning** n distinct objects into k groups containing $n_1, n_2, \ldots, n_k$ objects, respectively, is

$$\frac{n!}{n_1!n_2! \cdots n_k!}$$

where

$$\sum_{i=1}^{k} n_i = n$$

The partitioning of n objects into k groups can be done by first selecting a subset of size n_1 from the n objects, then selecting a subset of size n_2 from the $n - n_1$ objects that remain, and so on until all groups are filled. The number of ways of doing this is

$$\binom{n}{n_1}\binom{n - n_1}{n_2} \cdots \binom{n - n_1 - \cdots - n_{k-1}}{n_k}$$

$$= \frac{n!}{n_1!(n - n_1)!} \cdot \frac{(n - n_1)!}{n_2!(n - n_1 - n_2)!} \cdot \ldots \cdot \frac{(n - n_1 - \cdots n_{k-1})!}{n_k!0!}$$

$$= \frac{n!}{n_1!n_2! \cdots n_k!}$$

Example 4.21

Suppose that 10 employees are to be divided among three jobs, with three employees going to job I, four to job II, and three to job III. (a) In how many ways can the job assignment be made? (b) Suppose the only three employees of a certain ethnic group all get assigned to job I. What is the probability of this happening under a random assignment of employees to jobs?

Solution   **a** This problem involves a partitioning of the $n = 10$ employees into groups of size $n_1 = 3$, $n_2 = 4$, and $n_3 = 3$, and it can be accomplished in ways.

$$\frac{n!}{n_1!(n_2!)n_3!} = \frac{10!}{2!(4!)3!} = \frac{10 \cdot 9 \cdot 8 \cdot 7 \cdot 6 \cdot 5}{[3(2)1][3(2)1]} = 4,200$$

ways (Notice the large number of ways this task can be accomplished!)

b As seen in part (a) there are 4,200 ways of assigning the 10 workers to three jobs. The event of interest assigns three specified employees to job I. It remains to be determined how many ways the other seven employees can be assigned to jobs II and III, which is

$$\frac{7!}{4!3!} = \frac{7(6)(5)}{3(2)(1)} = 35$$

Thus, the chance of assigning three specific workers to job I is

$$\frac{35}{4,200} = \frac{1}{120}$$

which is very small, indeed!

Exercises

4.14   Two vehicles in succession are observed moving through the intersection of two streets.
  a   List the possible outcomes, assuming each vehicle can go straight, turn right, or turn left.
  b   Assuming the outcomes to be equally likely, find the probability that at least one vehicle turns left. (Would this assumption always be reasonable?)
  c   Assuming the outcomes to be equally likely, find the probability that at most one vehicle makes a turn.

4.15   A commercial building is designed with two entrances, door I and door II. Two customers arrive and enter the building.
  a   List the elements of a sample space for this observational experiment.
  b   If all elements in part (a) are equally likely, find the probability that both customers use door I; then find the probability that both customers use the same door.

4.16   A corporation has two construction contracts that are to be assigned to one or more of three firms (say I, II, and III) bidding for these contracts. (One firm could receive both contracts.)
  a   List the possible outcomes for the assignment of contracts to the firms.
  b   If all outcomes are equally likely, find the probability that both contracts go to the same firm.
  c   Under the assumptions of part (b), find the probability that one specific firm, say firm I, gets at least one contract.

4.17   Among five portable generators produced by an assembly line in one day, there are two defectives. If two generators are selected for sale, find the probability that both will be nondefective. (Assume the two selected for sale are chosen so that every possible sample of size two has the same probability of being selected.)

4.18   Seven applicants have applied for two jobs. How many ways can the jobs be filled if
  a   The first person chosen receives a higher salary than the second?
  b   There are no differences between the jobs?

4.19   A package of six light bulbs contains two defective bulbs. If three bulbs are selected for use, find the probability that none is defective.

4.20   How many four-digit serial numbers can be formed if no digit is to be repeated within any one number? (The first digit may be a zero.)

4.21   A fleet of eight taxis is to be randomly assigned to three airports A, B, and C, with two going to A, five to B, and one to C.

  a   In how many ways can this be done?
  b   What is the probability that the specific cab driven by Jones is assigned to airport C?

4.22   Show that
$$\binom{n}{r} = \binom{n-1}{r-1} + \binom{n-1}{r}, \ 1 \le r \le n.$$

4.23   Five employees of a firm are ranked from 1 to 5 based on their ability to program a computer. Three of these employees are selected to fill equivalent programming jobs. If all possible choices of three (out of the five) are equally likely, find the following probabilities
  a   The employee ranked number 1 is selected.
  b   The highest-ranked employee among those selected has rank 2 or lower.
  c   The employees ranked 4 and 5 are selected.

4.24   For a certain style of new automobile, the colors blue, white, black, and green are in equal demand. Three successive orders are placed for automobiles of this style. Find the following probabilities
  a   One blue, one white, and one green are ordered.
  b   Two blues are ordered.
  c   At least one black is ordered.
  d   Exactly two of the orders are for the same color.

4.25   A firm is placing three orders for supplies among five different distributors. Each order is randomly assigned to one of the distributors, and a distributor can receive multiple orders. Find the following probabilities
  a   All orders go to different distributors.
  b   All orders go to the same distributor.
  c   Exactly two of the three orders go to one particular distributor.

4.26   An assembly operation for a computer circuit board consists of four operations that can be performed in any order.
  a   In how many ways can the assembly operation be performed?
  b   One of the operations involves soldering wire to a microchip. If all possible assembly orderings are equally likely, what is the probability that the soldering comes first or second?

4.27   Nine impact wrenches are to be divided evenly among three assembly lines.
  a   In how many ways can this be done?
  b   Two of the wrenches are used, and seven are new. What is the probability that a particular line (line A) gets both used wrenches?

4.4 Conditional Probability and Independence

Conditional Probability

Sometimes the occurrence of one event alters the probability of occurrence of another event. Such events are called dependent events. Thus, it is necessary to discuss the occurrence of one event under the condition that another event has already occurred. The *conditional event A* given *B* is a set of outcomes for event *A* that occurs provided *B* has occurred. The conditional event is indicated by $(A \mid B)$ and read as "event *A* given *B*."

Example 4.22

Table 4.4 gives the seasonally unadjusted labor force in the U.S. (4th quarter of 2002) 25 years or older in age, as reported by the U.S. Bureau of Labor Statistics (figures in thousands). The total workforce is 123,052.

From the total workforce, the "employment rate," that is, the percent of employed workers, is 95%. But the overall employment rate does not tell us anything about the association between the education level and employment. To investigate this association, we calculate employment rates for each category separately. Narrowing the focus to a single row (like "education level") is often referred to as *conditioning on a row factor*. The conditional relative frequencies are given in Table 4.5.

Now it is apparent that the employment rate is associated with the education; those with less education have lower employment rates. In other words, the likelihood of unemployment is lowered with increasing education level.

Table 4.4

Civilian Labor Force in United States, 4th Quarter 2002

Education	Employed	Unemployed	Total
HS, 1–3 years	11,311	1,101	**12,412**
HS, 4 years	36,225	1,860	**38,085**
Associate degree	11,281	440	**11,721**
College, 1–3 years	21,300	1,094	**22,394**
College, 4 or more years	37,363	1,077	**38,440**
Total	**117,480**	**5,572**	**123,052**

Table 4.5

Employment Rates by Education, 2002

Education	Employed	Unemployed
HS, 1–3 years	91%	9%
HS, 4 years	95%	5%
Associate degree	96%	4%
College, 1–3 years	95%	5%
College, 4 or more years	97%	3%
Total	**95%**	**5%**

If 1,000 people are sampled from the national labor force, the expected number of unemployed workers is about 5%, that is, 50. If however, the 1,000 people all have a college education, the unemployment is expected to drop to 3%, that is, 30 people. On the other hand, if all 1,000 have only 1–3 years of highschool education, the unemployment is expected to be as high as 9%, that is, 90 persons.

Example 4.23

Refer to Example 4.10 about tower malfunction. The same data is reported in Table 4.6. How do the relative frequencies of causes compare for the two time frames?

Solution In this case, the focus is narrowed to a single column ("Prior to 1992" or "1992 and later"). This is often referred to as *conditioning on a column factor*. Note that prior to 1992, scale and corrosion was the major cause of plugging, but since 1992, coking has emerged as a major cause of plugging. The conditional relative frequencies also show that plugging due to precipitation is increasing and plugging due to solids in feed is on the decline.

Table 4.6

Causes of Plugging

Cause of Plugging	1992–	1992+
Coking	14.3	40.0
Scale, corrosion	39.3	20.0
Precipitation	14.3	23.6
Solids in feed	21.4	7.3
Polymer	10.7	9.1
Total	**100.0**	**100.0**

Example 4.24

Table 4.7 shows percentages of net new workers in the labor force. How do the relative frequencies for the three racial/immigrant categories compare between women and men?

Solution   Even though the data are given in terms of percentages rather than frequencies, the relative frequencies can still be computed. Conditioning on women, the total number represents 65% of the population, whereas the number of whites represents 42% of the population. Therefore, 42/65 represents the proportion of whites among the women. Proceeding similarly across the other categories produces *two* conditional distributions, one for women and one for men, as shown in Table 4.8.

Table 4.7

New Labor Force by Gender and Ethnicity

	Women	Men	Total
White	42%	15%	57%
Nonwhite	14%	7%	21%
Immigrant	9%	13%	22%
Total	**65%**	**65%**	**100%**

Table 4.8

New Labor Force by Ethnicity for gender

	Women	Men
White	65%	43%
Nonwhite	21%	20%
Immigrant	14%	37%
Total	**100%**	**100%**

Note that there is a fairly strong association between the two factors of sex and racial/immigration status. Among the new members of the labor force, women will be mostly white, and men will have a high proportion of immigrants.

Conditioning can also be represented in Venn diagrams. Suppose that out of 100 students completing an introductory statistics course, 20 were business majors. Ten students received A's in the course, and three of these students were business majors. These facts are easily displayed on a Venn diagram as shown in Figure 4.15, where A represents those student's receiving A's and B represents business majors.

Figure 4.15

Venn diagram of conditional event

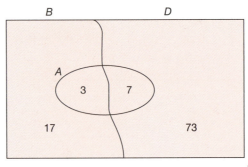

For a randomly selected student from this class, $P(A) = 0.1$ and $P(B) = 0.2$. But suppose we know that a randomly selected student is a business major. Then we might want to know the probability that the student received an A, *given* that she is a business major. Among the 20 business majors, 3 received A's. Thus, $P(A$ given $B)$, written as $P(A|B)$, is 3/20.

From the Venn diagram, we see that the conditional (or given) information reduces the effective sample space to just 20 business majors. Among them, 3 received A's. Note that

$$P(A|B) = \frac{3}{20} = \frac{P(AB)}{P(B)} = \frac{3/100}{20/100}$$

> If A and B are any two events, then the **conditional probability** of A given B, denoted by $P(A|B)$, is
>
> $$P(A|B) = \frac{P(AB)}{P(B)}$$
>
> provided $P(B) > 0$.

The events $(A \cap B)$ (intersection of events A and B) and $A|B$ (event A given that event B has occurred) are different from each other.

- With the event $(A \cap B)$, we are interested in the likelihood of occurrence of both events A and B in the entire sample space.
- With the event A/B, it is known that the event B has occurred, so there is no need to consider the entire sample space S. Instead, the sample space is reduced to B. We consider the likelihood of occurrence of event A in this reduced sample space.

Example 4.25

From five motors, of which one is defective, two motors are to be selected at random for use on a particular day. Find the probability that the second motor selected is nondefective, given that the first was nondefective.

Solution Let N_i denote the ith motor selected that is nondefective. We want to compute $P(N_2|N_1)$. From the above definition, we have, $P(N_2|N_1) = P(N_1N_2)/P(N_1)$. Looking at the 20 possible outcomes given in Figure 4.14, we can see that event N_1 contains 16 of these outcomes and N_1N_2 contains 12. Thus, because the 20 outcomes are equally likely,

$$P(N_2|N_1) = \frac{P(N_1N_2)}{P(N_1)} = \frac{12/20}{16/20} = \frac{12}{16} = 0.75$$

Does this result seem intuitively reasonable?

Conditional probabilities satisfy the three axioms of probability.

- Because $AB \subset B$, then $P(AB) \leq P(B)$. Also, $P(AB) \geq 0$ and $P(B) \geq 0$, so

$$0 \leq P(A|B) = \frac{P(AB)}{P(B)} \leq 1$$

- $P(S|B) = \dfrac{P(SB)}{P(B)} = \dfrac{P(B)}{P(B)} = 1$

- If $A_1, A_2, \ldots$ are mutually exclusive events, then so are $A_1B, A_2B, \ldots$ and

$$P\left(\bigcup_{i=1}^{\infty} A_i \,\middle|\, B\right) = \frac{P\left(\left(\bigcup_{i=1}^{\infty} A_i\right)B\right)}{P(B)} = \frac{P\left(\bigcup_{i=1}^{\infty}(A_iB)\right)}{P(B)} = \frac{\sum_{i=1}^{\infty} P(A_iB)}{P(B)}$$

$$= \sum_{i=1}^{\infty} \frac{P(A_iB)}{P(B)} = \sum_{i=1}^{\infty} P(A_i|B)$$

Conditional Probability and Diagnostic Tests Conditional probability plays a key role in many practical applications of probability. In these applications, important conditional probabilities are often drastically affected by seemingly small changes in the basic information from which probabilities are derived. Diagnostic tests, which indicate the presence or absence of fault/defect, are often used by technicians to detect defect. Virtually all diagnostic tests, however, have errors associated with their use. On thinking about the possible errors, it is clear that two different kinds of errors are possible:

- False positive: The test could show a machine/process/item to have the fault when it is in fact absent.
- False negative: The test could fail to show that a machine/process/item has the fault when in fact it is present.

Measures of these two types of errors depend on two conditional probabilities called *sensitivity* and *specificity*. Table 4.9 will help in defining and interpreting these measures. The true diagnosis may never be known, but often it can be

Table 4.9

True Diagnosis versus Test Results

| | | True Diagnosis | | |
		Fault Present	Fault Absent	Total
Test Results	Positive	a	b	a + b
	Negative	c	d	c + d
	Total	a + c	b + d	n

determined by more-intensive follow-up tests, which require investing more time and money.

where $n = a + b + c + d$. In this scenario, n items are tested and $a + b$ are shown by the test to have the fault. Of these, a items actually have the fault, and b do not have the fault (false positive). Of the $c + d$ which tested negative, c actually have the fault (false negative). Using these labels,

$$\text{Sensitivity} = \frac{a}{a + c} \quad \text{and} \quad \text{Specificity} = \frac{d}{b + d}$$

Obviously, a good test should have both sensitivity and specificity close to 1.

- If sensitivity is close to 1, then c (the number of false negatives) must be small.
- If specificity is close to 1, then b (the number of false positives) must be small.

Even when sensitivity and specificity are both close to 1, a screening test can produce misleading results if not carefully applied. To see this, we look at one other important measure, the predictive value of a test, given by

$$\text{Predictive value} = \frac{a}{a + b}$$

Sensitivity is the conditional probability of a positive diagnosis given that the item has the fault.

Specificity is the conditional probability of a negative diagnosis given that the item does not have the fault.

The **predictive value** is the conditional probability of the item actually having the fault given that diagnosis is positive.

Clearly a good test should have a high predictive value, but this is not always possible even for highly sensitive and specific tests. The reason that all three measures cannot always be close to 1 simultaneously lies in the fact that predictive value is affected by the *prevalence rate* of the fault. The prevalence rate of the fault is the proportion of population under study that actually has the fault, also known as the *defect rate* of production.

Example 4.26

Consider the three test situations, I, II, and III, described by Tables 4.10–4.12, respectively.

- Among the 200 items inspected in scenario I (Table 4.10), 100 have the fault, that is, a prevalence rate of 50%. The diagnostic test has sensitivity and specificity each equal to 0.90, and the predictive value is $90/100 = 0.90$. This is a good situation; the diagnostic test is good.

- In scenario II (Table 4.11), the prevalence rate changes to $100/1,100$, or 9%. Even though the sensitivity and specificity are still 0.90, the predictive value has dropped to $90/190 = 0.47$.
- In scenario III (Table 4.12), the prevalence rate is $100/10,100$, or about 1%, and the predictive value has dropped further to 0.08. Thus, only 8% of those tested positive actually have the fault, even though the test has high sensitivity and specificity.

What does this imply about the use of diagnostic (screening) tests on large populations in which the prevalence rate for the fault (disease) being studied is low? An assessment of the answer to this question involves a careful look at conditional probabilities.

Table 4.10

Test Situation I

		True Diagnosis		Total
Scenario I		Fault Present	Fault Absent	
Test Results	Positive	90	10	100
	Negative	10	90	100
	Total	100	100	200

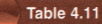

Table 4.11

Test Situation II

		True Diagnosis		Total
Scenario II		Fault Present	Fault Absent	
Test Results	Positive	90	100	190
	Negative	10	900	910
	Total	100	1000	1,100

Table 4.12

Test Situation III

		True Diagnosis		Total
Scenario III		Fault Present	Fault Absent	
Test Results	Positive	90	1,000	1,090
	Negative	10	9,000	9,010
	Total	100	10,000	10,100

Example 4.27

Receiving spam emails, also known as junk emails, along with legitimate emails is becoming a big problem for everybody. In response to increasing spam emails, a company developed an adaptive learning algorithm (ADA) that can be used for identifying spam emails from legitimate emails. The success rate of this algorithm depends on the spam percentage in the test set. Compute simulations indicated that when tested on a set of spam emails, this ADA correctly identified 98% of spam emails in the test sets. On the other hand, in cases involving legitimate emails, ADA declared 7% of the emails to be spam.

Suppose this firm receives about 10,000 emails a day, all to be screened by ADA. What can we say about the predictive value of such a screening process?

Solution   The result will depend on what is assumed about the rate of spam emails among the 10,000 emails. Suppose 15% emails were spam. With the ADA operating as in the simulation, the expected outcomes would be as shown in Table 4.13.

Table 4.13

Expected Count of Emails (15% spam)

		True Nature of Email		Total
		Spam	Legitimate	
ADA Classification	Spam	1,470	595	**2,065**
	Legitimate	30	7,905	**7,935**
	Total	**1,500**	**8,500**	**10,000**

Table 4.14

Expected Count of Emails (5% spam)

		True Nature of Email		Total
		Spam	Legitimate	
ADA Classification	Spam	490	665	**1,155**
	Legitimate	10	8,835	**8,845**
	Total	**500**	**9,500**	**10,000**

For this test, the predictive value is

$$\frac{1,470}{2,065} = 0.71$$

Of those emails classified as spam by ADA, 71% are expected to actually be spam emails.

A much more conservative view might be to select around 5% of the emails to be spam. The expected counts then become as shown in Table 4.14, and the predictive value for this test is only 0.42.

What is the danger of actually running a procedure with such a low predictive value? Less than half of the emails identified by ADA as spam are expected to be actually spam. Employees might delete a large percent of legitimate messages due to this misclassification.

Independence

In some situations, we expect events to behave independently, for example, a main battery and a backup battery of the system. We expect them to function independently, and the failure of the main battery should not lead to the failure of the backup battery. On the twin-engine planes, failure of one engine should not trigger the failure of the other. On the other hand, the functioning of a warning light should depend on the functioning of the system. At the first occurrence of malfunction, the warning light of the system should come on. Remember that we defined the occurrence of event A given that event B has occurred by $A|B$.

Probabilities are usually very sensitive to the conditioning information. Sometimes, however, a probability does *not* change when conditioning information is supplied. If the extra information derived from knowing that an event B has occurred does not change the probability of A—that is, if $P(A|B) = P(A)$—then events A and B are said to be *independent*. Because

$$P(A|B) = \frac{P(AB)}{P(B)}$$

the condition $P(A \mid B) = P(A)$ is equivalent to

$$\frac{P(AB)}{P(B)} = P(A) \qquad \text{or} \qquad P(AB) = P(A)\,P(B).$$

Sometimes the occurrence of one event alters the probability of occurrence of another event. Such events are called dependent events.

Independent Events

Two events A and B are said to be **independent** if and only if

$$P(A|B) = P(A) \text{ or } P(B|A) = P(B)$$

This is equivalent to stating that $P(AB) = P(A)P(B)$.

Example 4.28

Suppose a supervisor must select one worker for a special job from a pool of four available workers numbered 1, 2, 3, and 4. The supervisor selects the worker by mixing the four names and randomly selecting one. Let A denote the event that worker 1 or 2 is selected, B the event that worker 1 or 3 is selected, and C the event that worker 1 is selected. Are A and B independent? Are A and C independent?

Solution   Because the name is selected at random, a reasonable assumption for the probabilistic model is to assign a probability of $1/4$ to each individual worker. Then $P(A) = 1/2$, $P(B) = 1/2$, and $P(C) = 1/4$. Because the intersection AB contains only worker 1, $P(AB) = 1/4$. Now $P(AB) = 1/4 = P(A)P(B)$, so A and B are independent.

Because AC also contains only worker 1, $P(AC) = 1/4$. But $P(AC) = 1/4$ $\neq P(A)P(C)$, so A and C *are not* independent. A and C are said to be *dependent*, because the fact that C occurs changes the probability that A occurs.

Most situations in which independence issues arise are not like that portrayed in Example 4.28, in which events are well defined and one merely calculates probabilities to check the definition. Often, independence is *assumed* for two events in order to calculate their joint probability. For example,

- A = machine A does not break down today, and
- B = machine B will not break down today.

What is $P(AB)$, the probability that neither machine breaks down today?

$P(A)$ and $P(B)$ can be approximated from the repair records of the machines.

- If we assume independence, $P(AB) = P(A)P(B)$, a straightforward calculation.
- However, if we do not assume independence, we cannot calculate $P(AB)$ unless we form a model for their dependence structure or collect data on their joint performance.

Is independence a reasonable assumption? It may be, if the operation of one machine is not affected by the other, but it may not be if the machines share the same room, power supply, or job overseer. Independence is often used as a simplifying assumption and may not hold precisely in all cases for which it is assumed. Remember, probabilistic models are simply models, and they do not always precisely mirror reality. This should not be a major concern, because all branches of science make simplifying assumptions when developing their models, whether probabilistic or deterministic.

Exercises

4.28 Electric motors coming off two assembly lines are pooled for storage in a common stockroom, and the room contains an equal number of motors from each line. Motors are periodically sampled from that room and tested. It is known that 10% of the motors from line I are defective and 15% of the motors from line II are defective. If a motor is randomly selected from the stockroom and found to be defective, find the probability that it came from line I.

4.29 A diagnostic test for a certain disease is said to be 90% accurate in that, if a person has the disease, the test will detect it with probability 0.9. Also, if a person does not have the disease, the test will report that he or she doesn't have it with probability 0.9. Only 1% of the population has the disease in question. If a person is

chosen at random from the population and the diagnostic test reports him to have the disease, what is the conditional probability that he does, in fact, have the disease? Are you surprised by the size of the answer? Would you call this diagnostic test reliable?

4.30 A company produces two types of CDs, CD-R and CD-RW. The product CD-R constitutes 35% of total production. About 5% of the company's production is defective, of which 40% is CD-R. Suppose a CD is selected at random from today's production.

a Display these facts using a Venn diagram.

b What is the probability that the selected CD is defective?

c What is the probability that the selected CD is a CD-RW?

d What is the probability that a CD-R is selected given that a defective is the selected CD?

e Are the events "select CD-R" and "select a defective" independent? Justify your answer.

f What is the probability that a CD-RW is selected given that a defective is the selected CD?

4.31 The world is changing due to globalization, rapid changes, and technology integration. As a result, the university programs in civil engineering are facing challenges of preparing curriculums to meet the changing needs of industries. Once students obtain an understanding of civil engineering context, do they need a broader base of knowledge in management and financial issues to face challenges in the current professional environment? Chinowski (*J. Professional Issues in Engineering Education and Practice*, 2002) conducted a survey of top design and construction executives (vice presidents and above chosen from *Engineering News*' list of top firms) to ask about their educational background. The results are summarized in Table 4.15. Suppose one executive is selected at random.

a What is the probability that the selected executive has an engineering degree?

b If the selected executive is known to have an engineering degree, what is the probability that he/she has a master's degree?

c Are events "engineering degree" and "bachelor degree" independent of each other?

4.32 A small store installed a diesel-powered generator for emergency power outage because they function independently of electric power. The past history of the town indicates that on about 5% of days during summer the town loses electricity due to high winds. However, high winds do not affect functioning of generators. These generators have a failure rate of about 2%. On a given summer day, what is the probability that the store will be totally out of power?

4.33 Suppose that $P(A) = 0.6$, $P(B) = 0.3$, and $P(A \cap B) = 0.15$.

a Determine $P(A \mid B)$.

b Are events A and B independent?

4.34 A specific kind of small airplane used for dusting crops has two identical engines functioning inde-

Table 4.15

Educational Background of Executives

Degree Level	Engineering Degree	Business Degree
Bachelor	172	28
Masters	55	56
Ph.D.	7	1

pendently of each other, but can be flown with only one engine. The past records of this kind of engine indicate a 3% mid-air failure rate for each engine. One such plane crashed while dusting crops, and mid-air engine failure was suspected. What is the probability that such an airplane is likely to crash due to mid-air engine failure?

4.35 A manufacturer of computer keyboards has assembly plants at two different locations, one in Iowa and the other in South Carolina. The keyboard faults in general are classified into three different categories depending on the location: fault related to a letter key, a number key, or other function key. The summary of last year's inspection of production estimated the fault rates as follows (Table 4.16):

Table 4.16

Number of Defectives by Facility and Type of Fault

Production Facility	Location of Fault		
	Letter Key	Number Key	Other Function Key
Iowa	15	45	40
South Carolina	75	30	45

If a keyboard from this manufacturer is chosen at random, are the events "faulty letter key" and "produced in South Carolina facility" independent of each other?

4.5   Rules of Probability

Earlier we discussed relationships among different events. How can we use these relationships to determine the probabilities of required events? Rules that aid in the calculation of probabilities will now be developed. In each case, there is a similar rule for relative frequency data that will be illustrated using the *net* new worker data last visited in Example 4.24.

Categories like "women" and "men" are *complementary* in the sense that a worker must be in one or the other. Because the percentage of women among new workers is 65% (see Table 4.17), the percentage of men must be 35%; the two together must add to 100%.

Table 4.17

The Unconditional Percentages for Workforce

	Women	Men	Total
White	42%	15%	57%
Nonwhite	14%	7%	21%
Immigrant	9%	13%	22%
Total	**65%**	**35%**	**100%**

Table 4.18

The Conditional Percentages for Men and Women

	Women	Men
White	65%	43%
Nonwhite	21%	20%
Immigrant	14%	37%
Total	**100%**	**100%**

When we add percentages (or relative frequencies), we must take care to preserve the mutually exclusive character of the events. The percentage of whites, both men and women, is clearly 42% + 15% = 57% (see Table 4.17). However, the percentage of new workers that are either white *or* women is not 65% + 57%. To find the fraction of those who are white or women, the 42% that are *both* white and women must be subtracted from the above total. Otherwise, they will be counted twice, once with white workers and once with women workers. So, the total percentage of white or women can be obtained using information from Table 4.18 as

$$65\% + 57\% - 42\% = 80\%$$

which is same as accumulating the percentages among the appropriate mutually exclusive categories:

$$9\% + 14\% + 42\% + 15\% = 80\%$$

What about the 15% that are *both* men *and* white? Another way to view this type of question is to use the conditional percentages (see Table 4.18). If we know that 35% of the workers are men and 43% of men are white, then 43% of the 35%, that is, 15%, must represent the percentage of white men among the new workers.

Complementary Events

In probabilistic terms, recall that the complement $\overline{A}$ of an event A is the set of all outcomes in a sample space S that are not in A. Thus, $\overline{A}$ and A are mutually exclusive, and their union is S. That is, $A \cup \overline{A} = S$. It follows that

$$P(A \cup \overline{A}) = P(A) + P(\overline{A}) = P(S) = 1 \text{ or } P(\overline{A}) = 1 - P(A)$$

> If $\overline{A}$ is the **complement** of an event A in a sample space S, then
>
> $$P(\overline{A}) = 1 - P(A)\text{---}$$

Example 4.29

A quality-control inspector has ten assembly lines from which to choose products for testing. Each morning of a five-day week, she randomly selects one of the lines to work on for the day. Find the probability that a line is chosen more than once during the week.

Solution   It is easier here to think in terms of complements and first find the probability that no line is chosen more than once. If no line is repeated, five different lines must be chosen on successive days, which can be done in P_5^{10} ways. The total number of possible outcomes for the selection of five lines without restriction is 10^5, by an extension of the multiplication rule. Thus,

$$P(\text{no line is chosen more than once}) = \frac{P_5^{10}}{10^5} = \frac{10(9)(8)(7)(6)}{10^5} = 0.30$$

And

$$P(\text{a line is chosen more than once}) = 1.0 - 0.30 = 0.70.$$

Example 4.30

A manufacturing process produces about 2% of items that are defective. Suppose one item is selected at random from this manufacturing process. Let us define event A = selected item is defective. Then $P(A) = 0.02$. The complement of event A is $\overline{A}$ = selected item is not defective. Therefore, $P(\overline{A}) = 1 - P(A) = 1 - 0.02 = 0.98$, that is, 98% of the items manufactured are nondefective.

Additive Rule   The additive rule is useful in determining the probability of union of events.

Additive Rule

If A and B are any two events, then

$$P(A \cup B) = P(A) + P(B) - P(AB)$$

If A and B are mutually exclusive, then

$$P(A \cup B) = P(A) + P(B)$$

By extending this logic to more than two events, the formula for the probability of the union of k events $A_1, A_2, \ldots, A_k$ is given by

$$P(A_1 \cup A_2 \cup \cdots \cup A_k) = \sum_{i=1}^{k} P(A_i) - \sum\sum_{i<j} P(A_i A_j) + \sum\sum_{i<j<l} P(A_i A_j A_l)$$
$$- \cdots + \cdots - (-1)^k P(A_1 A_2 \ldots A_k)$$

Example 4.31

Suppose there are 50 students enrolled in the calculus course, 45 students enrolled in the signal processing course, and 10 in both. Select one student at random from this group of 100 students.

$A =$ Student is enrolled in a calculus course
$B =$ Student is enrolled in a signal processing course

The probability that the selected student is enrolled in a calculus class is $P(A) = 0.50$, the probability that the selected student is enrolled in a signal processing class is 0.45, and the probability that the selected student is enrolled in both the classes is 0.10. Then the probability that the student is enrolled in at least one of two classes is

$$P(A \cup B) = P(A) + P(B) - P(A \cap B),$$
$$= 0.50 + 0.45 - 0.10 = 0.85.$$

Note that two events A and B are not mutually exclusive because there are 10 students enrolled in both the courses. If we just add the probabilities of two events, then we'll be counting those 10 students twice, once in event A and once with event B. Therefore, we have to subtract it once.

The probability that the student is enrolled in none of these classes is

$$P(\overline{A \cup B}) = 1 - P(A \cup B) = 1 - 0.85 = 0.15.$$

Multiplicative Rule   The multiplication rule is useful to determine the probability of intersection of events. This rule is actually just a rearrangement of the definition of conditional probability for the case in which a conditional probability may be known and we want to find the probability of an intersection.

Multiplicative Rule

If A and B are any two events, then

$$P(AB) = P(A)P(B\,|\,A)$$
$$= P(B)P(A\,|\,B)$$

If A and B are independent, then

$$P(AB) = P(A)P(B)$$

Example 4.32

Records indicate that for the parts coming out of a hydraulic repair shop at an airplane rework facility, 20% will have a shaft defect, 10% will have a bushing defect, and 75% will be defect-free. For an item chosen at random from this output, find the probability of the following:

A: The item has at least one type of defect.
B: The item has only a shaft defect.

Solution   The percentages given imply that 5% of the items have both a shaft defect and a bushing defect. Let D_1 denote the event that an item has a shaft defect and D_2 the event that it has a bushing defect. Then $A = D_1 \cup D_2$ and

$$P(A) = P(D_1 \cup D_2) = P(D_1) + P(D_2) - P(D_1 D_2)$$
$$= 0.20 + 0.10 - 0.05 = 0.25$$

Another possible solution is to observe that the complement of A is the event that an item has no defects. Thus,

$$P(A) = 1 - P(\overline{A}) = 1 - 0.75 = 0.25$$

To find $P(B)$, note that the event D_1 that the item has a shaft defect is the union of the event that it has *only* a shaft defect (B) and the event that it has *both* defect ($D_1 D_2$). That is, $D_1 = B \cup D_1 D_2$ where B and $D_1 D_2$ are mutually exclusive. Therefore,

$$P(D_1) = P(B) + P(D_1 D_2)$$

or

$$P(B) = P(D_1) - P(D_1 D_2) = 0.20 - 0.05 = 0.15$$

You should sketch these events on a Venn diagram and verify the results derived above.

Example 4.33

A section of an electrical circuit has two relays in parallel, as shown in Figure 2.16. The relays operate independently, and when a switch is thrown, each will close properly with probability only 0.8. If the relays are both open, find the probability that current will flow from s to t when the switch is thrown.

Solution Let O denote an open relay and C a closed relay. The four outcomes for this experiment are shown by the following:

$$
\begin{array}{ccc}
 & \text{Relay} & \text{Relay} \\
 & 1 & 2 \\
E_1 = \{(O & , & O)\} \\
E_2 = \{(O & , & C)\} \\
E_3 = \{(C & , & O)\} \\
E_4 = \{(C & , & C)\}
\end{array}
$$

Figure 4.16

Two Relays in Parallel

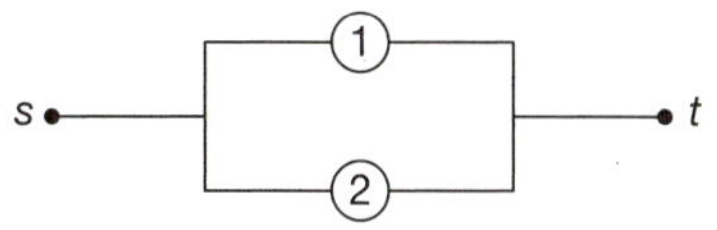

Because the relays operate independently, we can find the probability for these outcomes as follows:

$$
\begin{aligned}
P(E_1) &= P(O)P(O) = (0.2)(0.2) = 0.04 \\
P(E_2) &= P(O)P(C) = (0.2)(0.8) = 0.16 \\
P(E_3) &= P(C)P(O) = (0.8)(0.2) = 0.16 \\
P(E_4) &= P(C)P(C) = (0.8)(0.8) = 0.64
\end{aligned}
$$

If A denotes the event that current will flow from s to t, then

$$
A = E_2 \cup E_3 \cup E_4 \quad \text{or} \quad \overline{A} = E_1
$$

That is, at least one of the relays must close for current to flow. Thus,

$$
P(A) = 1 - P(\overline{A}) = 1 - P(E_1) = 1 - 0.04 = 0.96
$$

which is the same as $P(E_2) + P(E_3) + P(E_4)$.

Example 4.34

Three different orders are to be mailed to three supplier. However, an absent-minded secretary gets the orders mixed up and just sends them randomly to suppliers. If a match refers to the fact that a supplier gets the correct order, find the probability of (a) no matches (b) exactly one match.

Solution This problem can be solved by listing outcomes, since only three orders and suppliers are involved, but a more general method of solution will be illustrated. Define the following events:

A_1: match for supplier I
A_2: match for supplier II
A_3: match for supplier III

There are $3! = 6$ equally likely ways of randomly sending the orders to suppliers. There are only $2! = 2$ ways of sending the orders to suppliers if one particular supplier is required to have a match. Hence,

$$P(A_1) = P(A_2) = P(A_3) = 2/6 = 1/3$$

Similarly, it follows that

$$P(A_1A_2) = P(A_1A_3) = P(A_2A_3) = P(A_1A_2A_3) = 1/6$$

a $P(\text{no matches}) = 1 - P(\text{at least one match})$
$$= 1 - P(A_1 \cup A_2 \cup A_3)$$
$$= 1 - [P(A_1) + P(A_2) + P(A_3) - P(A_1A_2) - P(A_1A_3)$$
$$\quad - P(A_2A_3) + P(A_1A_2A_3)]$$
$$= 1 - [3(1/3) - 3(1/6) + (1/6)]$$
$$= 1/3$$

b $P(\text{exactly one match}) = P(A_1) + P(A_2) + P(A_3) - 2[P(A_1A_2) + P(A_1A_3)$
$$\quad + P(A_2A_3)] + 3P(A_1A_2A_3)$$
$$= 3(1/3) - 2(3)(1/6) + 3(1/6) = 1/2$$

Example 4.35 ——————————————————————————————

A company manufacturing ballpoint pens has two assembly lines. A random sample of produced ballpoint pens is taken periodically and inspected. The inspected pens are classified as defective–trash, defective–to be fixed, and nondefective. Table 4.19 shows the result of one such inspection.

Define events as follows:

$A1$ = Produced by assembly line 1
$A2$ = Produced by assembly line 2
DT = Defective–trash
DF = Defective–to be fixed
ND = Nondefective

Suppose a pen is selected at random.

a What is the overall defective–trash rate?
b What is the probability that it is a defective–trash pen manufactured by assembly line 1?
c Suppose a pen selected at random was manufactured by assembly line 1. Then what is the probability that it is defective–trash?

Table 4.19

Classification of Manufactured Pens

	Assembly Line 1	Assembly Line 2	Total
Defective–trash	8	2	**10**
Defective–to be fixed	13	27	**40**
Nondefective	59	91	**150**
Total	**80**	**120**	**200**

Solution

a Overall, 10 defective–trash items were recorded out of sampled 200. Therefore, the overall defective–trash rate is $10/200 = 0.05$; that is, overall 5% of the assembled pens are defective.

b Here we are interested in the co-occurrence of two events, namely, the pen is defective–trash and the pen was manufactured by line 1, that is, $DT \cap A1$. From Table 4.19 we get

$$P(DT \cap A1) = 8/200 = 0.04$$

c Here we know that the pen was assembled by line 1. So, instead of considering all 200 pens inspected, we can consider only the 80 pens manufactured by line 1. This knowledge about the manufacturing line has reduced the sample space under consideration from 200 to 80. Out of these 80 pens, 8 are defective–trash. Hence,

$$P(DT|A1) = 8/80 = 0.1$$

On the other hand, we can find this probability using the conditional probability formula given earlier as

$$P(DT|A1) = \frac{P(DT \cap A1)}{P(A1)} = \frac{0.04}{0.4} = 0.1$$

Similarly, we can determine $P(DT|A2) = 2/120 = 0.0167$. This shows that the defective rate changes depending on the assembly line. In other words, events "defective–trash" and "assembly line" are dependent.

Example 4.36

Let us refer to the question raised at the beginning of this chapter: "Are passengers wearing seat belts less likely to suffer injuries compared to those not wearing seatbelts?"

Data collected by the Maine Department of Public Safety [Do seat belts reduce injuries? Maine Crash Facts, 1996, **http://www.state.me.us/dps/Bhs/ seatbelts1996.htm** (Augusta, Maine)]. The data is given in Table 4.20.

Table 4.20

Seat Belt Usage versus Injury

	Count	Injured? Yes	No	Total
Seat belt? Yes	Yes	3,984	57,107	61,091
	No	2,323	8,138	10,461
	Total	6,307	65,245	71,552

Suppose a passenger is selected at random. Then

$$P(\text{Injury} \mid \text{seat belt}) = \frac{3{,}984}{61{,}091} = 0.065 \quad \text{and}$$

$$P(\text{Injury} \mid \text{No seat belt}) = \frac{2{,}323}{10{,}461} = 0.222$$

These conditional probabilities indicate that passengers not wearing seatbelts are three times likely to be injured in car accidents compared to those wearing seatbelts.

Bayes' Rule   The fourth rule we present in this section is based on the notion of a partition of a sample space. Events $B_1, B_2, \ldots, B_k$ are said to partition a sample space S if two conditions exist:

1. $B_i B_j = \phi$ for any pair i and j (ϕ denotes the null set)
2. $B_1 \cup B_2 \cup \ldots \cup B_k = S$

Figure 4.17
A Partition of S into B_1 and B_2

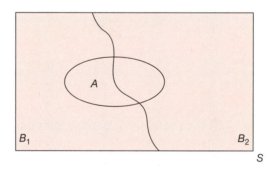

For example, the set of tires in an auto assembly warehouse might be partitioned according to suppliers, or employees of a firm might be partitioned according to level of education. We illustrate a partition for the case $k = 2$ in Figure 4.17. The key idea involving a partition is the observation that an event A can be written as the union of mutually exclusive events AB_1 and AB_2.

That is, $A = AB_1 \cup AB_2$ and thus $P(A) = P(AB_1) + P(AB_2)$

If conditional probabilities $P(A \mid B_1)$ and $P(A \mid B_2)$ are known, then $P(A)$ can be found by writing

$$P(A) = P(B_1)P(A \mid B_1) + P(B_2)P(A \mid B_2)$$

In problems dealing with partitions, it is frequently of interest to find probabilities of the form $P(B_1 \mid A)$, which can be written

$$P(B_1 \mid A) = \frac{P(B_1 A)}{P(A)} = \frac{P(B_1)P(A \mid B_1)}{P(B_1)P(A \mid B_1) + P(B_2)P(A \mid B_2)}$$

This result is a special case of *Bayes' Rule*.

Bayes' Rule

If $B_1, B_2, \ldots, B_k$ form a partition of S, and if A is any event in S then

$$P(B_j \mid A) = \frac{P(B_j)P(A \mid B_j)}{\sum\limits_{i=1}^{k} P(B_i)P(A \mid B_i)}$$

Example 4.37

A company buys tires from two suppliers, 1 and 2. Supplier 1 has a record of delivering tires containing 10% defectives, whereas supplier 2 has a defective rate of only 5%. Suppose 40% of the current supply came from supplier 1. If a tire is selected randomly from this supply and observed to be defective, find the probability that it came from supplier 1.

Solution Let B_i denote the event that a tire comes from supplier $i(i = 1,2)$, and note that B_1 and B_2 form a partition of the sample space for the experiment of selecting one tire. Let A denote the event that the selected tire is defective. Then

$$P(B_1 \mid A) = \frac{P(B_1)P(A \mid B_1)}{P(B_1)P(A \mid B_1) + P(B_2)P(A \mid B_2)}$$

$$= \frac{0.40(0.10)}{0.40(0.10) + 0.60(0.05)} = 0.5714.$$

Supplier 1 has a greater probability of being the party supplying the defective tire than does supplier 2.

Example 4.38

Table 4.21 shows a summary of employment in the United States for December 2003 provided by U.S. Bureau of Labor Statistics.

If an arbitrarily selected U.S. resident was asked, in Dec. 2003, to fill out a questionnaire on employment, find the probability that the resident would have been in the following categories:

a In the labor force
b Employed

Table 4.21

Employment Summary in the United States (December 2003)

Employment in the United States	(in millions)
Civilian noninstitutional population	223
Civilian labor force	147
Employed	139
Unemployed	8
Not in the labor force	76

   c Employed and in the labor force
   d Employed given that he or she was known to be in the labor force
   e Either not in the labor force or unemployed

Solution   Let L denote the event that the resident is in the labor force and E that he or she is employed.

   a Because 223 million people comprise the population under study and 147 million are in the labor force,

$$P(L) = \frac{147}{223}$$

   b Similarly, 139 million are employed; thus,

$$P(E) = \frac{139}{223}$$

   c The employed persons are a subset of those in the labor force; in other words, $EL = E$. Hence,

$$P(EL) = P(E) = \frac{139}{223}$$

   d Among the 147 million people known to be in the labor force, 139 million are employed. Therefore,

$$P(E|L) = \frac{139}{147}$$

Note that this value can also be found by using the definition of conditional probability as follows:

$$P(E\,|\,L) = \frac{P(EL)}{P(L)} = \frac{139/223}{147/223} = \frac{139}{147}$$

   e The event that the resident is not in the labor force, $\overline{L}$, is mutually exclusive from the event that he or she is unemployed. Therefore,

$$P(\overline{L} \cup \overline{E}) = P(\overline{L}) + P(\overline{E}) = \frac{76}{223} + \frac{84}{223} = \frac{160}{223}$$

Example 4.39

The ASF, NTSB accident database for year 2003 listed information on the flight accidents classified by the type of injury for different types of flights. This data is reported in Table 4.22. All figures are percentages reported according to the type of injury. For example, 36% of business flight accidents were fatal.

Table 4.22

Flight Accident Data by the Type of Injury

Type of flight	Type of Injury				All Types of Injuries
	Fatal	Minor	None	Serious	
Business	36	11	51	2	4
Instructional	13	12	66	9	18
Personal	22	14	54	10	78
All types of flights	21	14	56	9	100

If a flight accident is reported, find the probability that it is the following:

a An accident on a business flight
b An accident that resulted in a fatal injury
c An accident that resulted in a minor injury given that it was on a business flight
d An accident on a business flight that resulted in a minor injury
e An accident on a business flight given that it was fatal

Solution

a Of all accidents reported in Table 4.22, 4% are on business flights. Thus,

$$P(\text{Business flight}) = 0.04$$

b Of all accidents reported in Table 4.22, 21% are fatal. Thus,

$$P(\text{Fatal Accident}) = 0.21$$

c The first row of Table 4.22 deals only with accidents on business flights. Thus, 11% of all the accidents on business flights resulted in minor injuries. Note that this is a conditional probability and can be written as

$$P(\text{Minor} \mid \text{Business}) = 0.11$$

d In part (c), we know that the accident was on a business flight, and the probability in question was conditional on that information. Here (in part d), we must find the probability of an intersection between "minor injury" and "business flight."

$$P(\text{Minor} \cap \text{Business}) = P(\text{Business}) \, P(\text{Minor} \mid \text{Business})$$
$$= 0.04 \, (0.11) = 0.0044$$

e Remember the figures in Table 4.22 give probabilities of type of injury, given the type of flight. We are now asked to find the probability of a type of flight given the type of injury. This is exactly the type of situation to which Bayes' Rule applies. The type of flight partitions the set of all accidents into three different groups.

$$P(\text{Business}\,|\,\text{Fatal}) = \frac{P(\text{Business})P(\text{Fatal}\,|\,\text{Business})}{P(\text{Fatal})}$$

Now $P(\text{Fatal Accident}) = 0.21$ from Table 4.22. Then,

$$P(\text{Business}\,|\,\text{Fatal}) = \frac{(0.04)(0.36)}{0.21} = 0.069$$

In other words, 6.9% of fatal accidents occur in accidents involving business flights.

It will be interesting to see how $P(\text{Fatal Accident}) = 0.21$ is derived from the other information in the table.

$$P(\text{Fatal}) = P(\text{Fatal}\,|\,\text{Business})P(\text{Business}) + P(\text{Fatal}\,|\,\text{Instructional})P(\text{Instructional}) \\ + P(\text{Fatal}\,|\,\text{Personal})P(\text{Personal})$$

$$= (0.36)(0.04) + (0.13)(0.18) + (0.22)(0.78) = 0.2094$$

Example 4.40

Kister (*Trans IchemE,* 2003) studied causes of tower malfunctions and reported that plugging/coking is a cause of about 9.6% malfunctions. Often we use such data from past studies to anticipate results in the future. Suppose three new tower malfunctions are reported. Find the probability of each of the following:

a Exactly one tower malfunction is due to plugging.
b At least one tower malfunction is due to plugging.
c Exactly two malfunctions are due to plugging when it is known that at least one is due to plugging.

Assume that there is no association between the malfunctions reported at three towers.

Solution Assuming the three towers behave like a random sample from the population of old towers with regard to the malfunctions, we can assign a probability of 0.096 to the event that a tower malfunction will occur due to plugging.

Let T_i denote the event that tower i develops malfunction due to plugging ($i = 1,2,3$). A tree diagram (see Figure 4.18) can aid in showing the structure of outcomes for malfunctioning due to plugging. The probabilities along the branches can be multiplied because of the independence assumption.

a Exactly one tower malfunction due to plugging can occur in three possible ways; that is, only one of three events (T_1, T_2, T_3) occurs. Thus,

Figure 4.18
A tree diagram for independent malfunctions

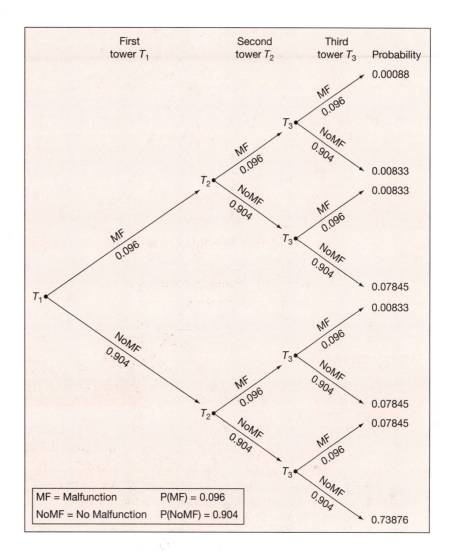

P(exactly one tower malfunction)

$$= P(T_1\overline{T}_2\overline{T}_3 \cup \overline{T}_1 T_2\overline{T}_3 \cup \overline{T}_1\overline{T}_2 T_3)$$
$$= P(T_1)P(\overline{T}_2)P(\overline{T}_3) + P(\overline{T}_1)P(T_2)P(\overline{T}_3) + P(\overline{T}_1)P(\overline{T}_2)P(T_3)$$
$$= (0.096)(0.904)(0.904) + (0.904)(0.096)(0.904)$$
$$\quad + (0.904)(0.904)(0.096)$$
$$= 3(0.096)(0.904)^2$$
$$= 0.2354$$

The probability of exactly one of three towers malfunctions due to plugging is 0.2354.

b At least one malfunction due to plugging can occur as "exactly two malfunctions due to plugging" and "exactly three malfunctions due to plugging." Thus, the probability of at least one malfunction due to plugging can be found by adding the probabilities of these two possibilities, which can be found by repeating the method described in part (a). An easier method is to use complement, so that

$$P(\text{at least one malfunction due to plugging})$$

$$= 1 - P(\text{no malfunction due to plugging})$$
$$= 1 - P(\bar{T}_1\bar{T}_2\bar{T}_3) = 1 - (0.904)^3 = 0.2612$$

c   This problem involves computation of conditional probability.

$$P(\text{exactly two malfunctions due to plugging} \mid \text{at least one due to plugging})$$

$$= \frac{P(\text{exactly two due to plugging})}{P(\text{at least one due to plugging})} = \frac{3(0.096)^2(0.904)}{0.2612} = 0.096$$

Note that the intersection of two events in this case is just the first event, "exactly two malfunctions due to plugging." The probability in the numerator is found in the manner analogous to that in part (a).

Exercises

4.36   Vehicles coming into an intersection can turn left or right or go straight ahead. Two vehicles enter an intersection in succession. Find the probability that at least one of the two vehicles turns left given that at least one of the two vehicles turns. What assumptions have you made?

4.37   A purchasing office is to assign a contract for computer paper and a contract for microcomputer disks to any one of three firms bidding for these contracts. (Any one firm could receive both contracts.) Find the probability that
   a   Firm I receives a contract given that both contracts do not go to the same firm.
   b   Firm I receives both contracts.
   c   Firm I receives the contract for paper given that it does not receive the contract for disks. What assumptions have you made?

4.38   The data in Table 4.23 give the number of accidental deaths overall and for three specific causes for the United States in 1984. You are told that a certain person recently died in an accident. Approximate the probability that
   a   It was a motor vehicle accident.
   b   It was a motor vehicle accident if you know the person to be male.
   c   It was a motor vehicle accident if you know the person to be between 15 and 24 years of age.
   d   It was a fall if you know the person to be over age 75.
   e   The person was male.

Table 4.23

Accidental death data in United States

	All Types	Motor Vehicle	Falls	Drowning
All ages	92,911	46,263	11,937	5,388
Under 5	3,652	1,132	114	638
5–14	4,198	2,263	68	532
15–24	19,801	14,738	399	1,353
25–44	25,498	15,036	963	1,549
45–64	15,273	6,954	1,624	763
65–74	8,424	3,020	1,702	281
75 and over	16,065	3,114	7,067	272
Male	64,053	32,949	6,210	4,420
Female	28,858	13,314	5,727	968

Source: The World Almanac, 1988.

4.39   The data in Table 4.24 show the distribution of arrival times at work by mode of travel for workers in the central business district of a large city. The figures are percentages. (The columns should add to 100, but some do not because of rounding.) A randomly selected worker is asked about his or her travel to work.

Table 4.24

Distribution of Arrival Times

Arrival Time	Transit	Drove Alone	Shared Ride with Family Member	Car Pool	All
Before 7:15	17	16	16	19	18
7:15–7:45	35	30	30	42	34
7:45–8:15	32	31	43	35	33
8:15–8:45	10	14	8	2	10
After 8:45	5	11	3	2	6

Source: C. Hendrickson and E. Plank, *Transportation Research*, 18A, no. 1, 1984.

a Find the probability that the worker arrives before 7:15 given that the worker drives alone.

b Find the probability that the worker arrives at or after 7:15 given that the worker drives alone.

c Find the probability that the worker arrives before 8:15 given that the worker rides in a car pool.

d Can you find the probability that the worker drives alone, using only these data?

4.40 An incoming lot of silicon wafers is to be inspected for defectives by an engineer in a microchip manufacturing plant. In a tray containing 20 wafers, assume four are defective. Two wafers are to be randomly selected for inspection. Find the probability that

a Both are nondefective.

b At least one of the two is nondefective.

c Both are nondefective given that at least one is nondefective.

4.41 In the setting of Exercise 4.40, find the same three probabilities if only two among the 20 wafers are assumed to be defective.

4.42 Among five applicants for chemical engineering positions in a firm, two are rated excellent and the other three are rated good. A manager randomly chooses two of these applicants to interview. Find the probability that the manager chooses the following:

a The two rated excellent

b At least one of those rated excellent

c The two rated excellent given that one of the chosen two is already known to be excellent

4.43 Resistors produced by a certain firm are marketed as 10-ohm resistors. However, the actual number of ohms of resistance produced by the resistors may vary. It is observed that 5% of the values are below 9.5 ohms and 10% are above 10.5 ohms. If two of these resistors, randomly selected, are used in a system, find the probability that

a Both have actual values between 9.5 and 10.5.

b At least one has an actual value in excess of 10.5.

4.44 Consider the following segment of an electric circuit with three relays. Current will flow from *a* to *b* if there is at least one closed path when the relays are switched to "closed." However, the relays may malfunction. Suppose they close properly only with probability 0.9 when the switch is thrown, and suppose they operate independently of one another. Let *A* denote the event that current will flow from *a* to *b* when the relays are switched to "closed."

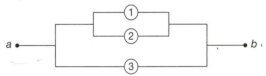

a Find $P(A)$.

b Find the probability that relay 1 is closed properly, given that current is known to be flowing from *a* to *b*.

4.45 With relays operating as in Exercise 4.44, compare the probability of current flowing from *a* to *b* in the series system with the probability of flow in the parallel system.

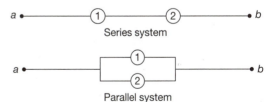

Series system

Parallel system

4.46 Electric motors coming off two assembly lines are pooled for storage in a common stockroom, and the room contains an equal number of motors from each line. Motors are periodically sampled from that room and tested. It is known that 10% of the motors from line I are defective and 15% of the motors from line II are defective. If a motor is randomly selected from the stockroom and found to be defective, find the probability that it came from line I.

4.47 Two methods, *A* and *B*, are available for teaching a certain industrial skill. The failure rate is 20% for *A* and 10% for *B*. However, *B* is more expensive and hence is used only 30% of the time. (*A* is used the other 70%.) A worker is taught the skill by one of the methods but fails to learn it correctly. What is the probability that he was taught by method *A*?

4.48 Table 4.25 gives estimated percentages of sports footwear purchased by those in various age groups across the United States.
a Explain the meaning of the 4.7 under aerobic shoes. Explain the meaning of the 67.0 under walking shoes.
b Suppose the percentages provided in the table are to be used as probabilities in discussing anticipated purchases for next year. Explain the 28.5 under jogging/running shoes as a probability.
c Your firm anticipates selling 100,000 pairs of gym shoes/sneakers across the country. How many of these would you expect to be purchased by children under 14 years of age? How many by 18–24-year-olds?
d You want to target advertising to 25–34-year-olds. If someone in this age group buys sports footwear next year, can you find the probability that he or she will purchase aerobic shoes?
e Write a brief description of the patterns you see in sales of sports footwear.

4.49 Study the data on employed scientists and engineers provided in Table 4.26. Think of the percentages as probabilities to aid in describing the anticipated employment pattern for scientists and engineers for this year.
a Explain the conditional nature of the columns of percentages.
b Suppose 3 million engineers are employed this year. How many would you expect to be in industry?
c Combine life scientists, computer specialists, physical scientists, environmental scientists, and mathematical scientists into one category. Find the percentage distribution across employment types for this new category. How does this distribution compare to that for engineers?
d Suppose industry adds 1 million new scientists this year. How many would you expect to be engineers?
e Describe any differences you see between the employment pattern of physical scientists and engineers and that of social scientists.

4.50 The segmented bar chart, as shown in Figure 4.19, is another way to display relative frequency data.
a Explain the percentages in the bar labeled Los Angeles.
b It appears that most whites live in Philadelphia. Is that a correct interpretation of these data?
c Describe in words the pattern of racial and ethnic diversity in these cities, based on these data.

Table 4.25

Consumer Purchases of Sports Footwear (in percent)

Characteristic	Total Households	Aerobic Shoes	Gym Shoes/ Sneakers	Jogging/ Running Shoes	Walking Shoes
Total	100.0	100.0	100.0	100.0	100.0
Age of user:					
Under 14 years old	20.3	4.7	34.6	8.3	2.8
14–17 years old	5.4	5.4	13.8	10.5	1.9
18–24 years old	10.7	11.5	9.7	10.0	3.3
25–34 years old	17.7	30.9	16.9	28.5	12.9
35–44 years old	14.7	22.5	10.3	24.0	19.7
45–64 years old	18.7	17.6	10.7	15.2	35.7
65 years old and over	12.5	7.4	4.0	3.5	23.7
Sex of user:					
Male	48.8	15.7	53.0	64.9	33.0
Female	51.2	84.3	47.0	35.1	67.0

Source: Statistical Abstract of the United States, 1991.

Table 4.26

Estimated Characteristics of Employed Scientists and Engineers, 1988

	Total	Engineers	Life Scientists	Computer Specialists	Physical Scientists	Social Scientists				Environmental Scientists	Mathematical Scientists
						Total	Economists	Other Social Scientists	Psychologists		
Total (in thousands)	5,286.4	2,718.6	458.6	708.3	312.0	531.0	219.8	311.2	275.9	113.4	168.6
Type of employer:											
Industry	68.0	79.5	37.9	78.7	56.1	52.5	64.0	44.4	40.7	58.4	42.0
Educ. institutions	13.5	4.3	35.2	6.6	25.4	24.5	19.0	28.5	31.0	16.0	44.1
Nonprofit organ.	3.5	1.5	7.3	2.4	3.3	6.0	2.3	8.7	17.6	1.2	2.0
Federal government	7.5	7.5	10.1	6.5	10.2	5.5	7.8	3.9	2.6	15.6	8.2
Military	0.8	1.0	0.4	0.7	0.4	0.4	0.2	0.5	0.2	1.4	0.8
Other government	5.1	4.6	7.5	3.6	9.6	9.6	5.2	12.7	6.1	5.6	1.9
Other and unknown	1.6	1.7	1.6	1.5	1.3	1.3	1.5	1.2	1.7	1.7	0.9

Figure 4.19

Segmented bar chart

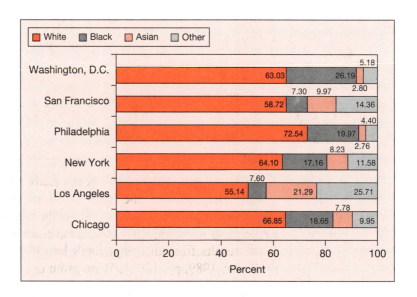

4.6 Odds and Odds Ratios

"What are the odds that our team will win today?"

This is a common way of talking about events whose unknown outcomes have probabilistic interpretations. The *odds* in *favor* of an event A are the ratio of the probability of A to the probability of $\overline{A}$. That is

$$\text{Odds in favor of } A = \frac{P(A)}{P(\overline{A})}$$

For example, the odds in favor of a balanced coin coming up heads when flipped are

$$P(H)/P(T) = 0.5/0.5 = 1$$

whereas, the odds in favor of an ace when a card is drawn from a well-shuffled pack are

$$\frac{P(\text{Ace})}{P(\text{Non–ace})} = \frac{4/52}{48/52} = \frac{1}{12} = 0.0833$$

Odds equal to 1 indicate that an event A, and its complement $\overline{A}$ have the same probability of occurrence. Odds larger than 1 indicate that an event A is more likely to occur than its complement, and odds smaller than 1 indicate than an event $\overline{A}$ is more likely to occur than A.

Often odds are written as $P(A) : P(\overline{A})$. In the case of a balanced coin, we would say the odds in favor of "heads" are 1:1 (read as "one to one"), and in the case of cards, we would say the odds in favor of an ace are 1:12 (read as "one to twelve").

The odds in favor of a randomly selected applicant getting admitted to college are $0.20/0.80 = 0.25$, or 1:4. The odds *against* that applicant getting admitted are $0.80/0.20 = 4$, or 4:1. The odds against an event A are same as the odds in favor of event $\overline{A}$.

Odds are not just a matter of betting and sports. They are a serious part of the analysis of frequency data, especially when categorical variables in a two-way frequency tables are being compared.

Example 4.41

During the 1980s, a long-term health study was conducted to determine whether aspirin helps lower the rate of heart attacks (myocardial infarctions). Approximately 22,000 physicians over the age of 40 participated in this study. The participants were randomly assigned to either an aspirin group or a placebo group. Some results from this physician's health study (*The New England Journal of Medicine,* 1989, pp. 129–135) are given in Table 4.27:

Table 4.27

Heart Attack Incidence by Treatment Group

	Myocardial Infarctions (M.I.)	No Myocardial Infarctions (no M.I.)	Total
Aspirin	139	10,898	11,037
Placebo	239	10,795	11,034
Total	**378**	**21,693**	**22,071**

For the aspirin group, the odds in favor of M.I. are

$$\frac{P(\text{M.I.})}{P(\overline{\text{M.I.}})} = \frac{139/11{,}037}{10{,}898/11{,}037} = \frac{139}{10{,}898} = 0.013$$

For the placebo group, the odds in favor of M.I. are

$$\frac{P(\text{M.I.})}{P(\overline{\text{M.I.}})} = \frac{239/11{,}034}{10{,}795/11{,}034} = \frac{239}{10{,}795} = 0.022$$

Thus, the above results show that the odds of heart attack with a placebo are considerably (almost 50%) higher than the odds with aspirin. More specifically, the ratio of the two odds is given by

$$\text{Odds ratio} = \frac{\text{odds of M.I. with aspirin}}{\text{odds of M.I. without aspirin}} = \frac{0.013}{0.022} = 0.59$$

The odds of an M.I. for the aspirin group are 59% of the odds for the placebo group. Odds ratios form a very useful single-number summary of the frequencies in a 2×2 table.

	I	II
A	a	b
B	c	d

The odds ratio has a simple form for any 2×2 table. The odds in favor of I for the A group are a/b, and the odds in favor of I for the B group are c/d. Therefore, the odds ratio is simply

$$\frac{a/b}{c/d} = \frac{ad}{bc}$$

which is the ratio of the product of the diagonal elements.

Example 4.42

For the 4th quarter of 2003, the U.S. Bureau of Labor Statistics reported following employment rates (in thousands) for 25 years old or older. Table 4.28 and Table 4.29 respectively show employment status by gender and education (High school graduates, no college; and college graduates). Interpret these tables using odds ratios.

Table 4.28

Employment Status by Gender

	Employed	Unemployed
Male	64,046	3,141
Female	55,433	2,556

Table 4.29

Employment Status by Education

	Employed	Unemployed
High school	36,249	1,962
College	39,250	1,165

Solution   From Table 4.28, the odds ratio is

$$\frac{(64{,}046)\,(2{,}556)}{(55{,}433)\,(3{,}141)} = 0.94$$

The odds in favor of being employed are about the same for males as for females (odds ratio close to 1). This result might be restated by saying that the risk of being unemployed is almost the same for males and females. The effect of gender on the employment status is negligible. For Table 4.29, the odds ratio is

$$\frac{(36{,}249)\,(1{,}165)}{(39{,}250)\,(1{,}962)} = 0.548 \approx 0.55.$$

Conditioning on these two education groups, the odds in favor of employment for the high school graduates is half of that for the college graduates. Education seems to have a significant impact on employability. In terms of risk, it is correct to say that the risk of unemployment for those with a high school education (no college) is $1/0.55 = 1.82$, or 82% higher than the risk of unemployment for those with college education.

In short, these easily computed odds and odds ratio serve a useful summary of tabled data for decision making.

Exercises

4.51 Cholesterol level seems to be an important factor in myocardial infarctions (M.I.). In Table 4.30 the data below give the number of M.I.'s over the number in the cholesterol group for each arm of the study. In this study, patients were randomly assigned to the aspirin or placebo group.

Table 4.30

Cholesterol Levels by Study Groups

Cholesterol Level (mg per 100 ml)	Aspirin Groups	Placebo Groups
≤159	2/382	9/406
160–209	12/1,587	37/1,511
210–259	26/1,435	43/1,444
≥260	14/582	23/570

a Did the randomization in the study seem to do a good job of balancing the cholesterol levels between the two groups? Explain.

b Construct a 2 × 2 table of aspirin or placebo versus M.I. response for each of the four cholesterol levels. Reduce the data in each table to the odds ratio.

c Compare the four odds ratios found in part (b). Comment on the relationship between the effect of aspirin on heart attacks and the cholesterol levels. Do you see why odds ratios are handy tools for summarizing data in a 2 × 2 table?

4.52 Is the race of a defendant associated with the defendant's chance of receiving the death penalty? This is a controversial issue that has been studied by many. One important data set was collected on 326 cases in which the defendant was convicted of homicide. The death penalty was given in 36 of these cases. Table 4.31 shows the defendant's race, the victim's race, and whether or not the death penalty was imposed.

a Construct a single 2 × 2 table showing penalty versus defendant's race, across all victims. Calculate the odds ratio and interpret it.

Table 4.31

Classification of Defendants and Victims by Race

	White Defendant		Black Defendant	
	Yes	**No**	**Yes**	**No**
White Victim	19	132	11	52
Black Victim	0	9	6	97

Source: M. Radelet, "Racial Characteristics and Imposition of the Death Penalty," *American Sociological Review,* 46, 1981, pp. 918–927.

b Decompose the table in (a) into two 2 × 2 tables of penalty versus defendant's race, one for while victims and one for black victims. Calculate the odds ratio for each table and interpret each one.

c Do you see any inconsistency between the results of part (a) and the results of part (b)? Can you explain the apparent paradox?

4.53 A proficiency examination for a certain skill was given to 100 employees of a firm. Forty of the employees were male. Sixty of the employees passed the examination, in that they scored above a preset level for satisfactory performance. The breakdown among males and females was as follows:

	Male M	Female F
Pass P	24	36
Fail $\overline{P}$	16	24

Suppose an employee is randomly selected from the 100 who took the examination.

a Find the probability that the employee passed given that he was a male.

b Find the probability that the employee was male given that the employee passed.

c Are events P and M independent?

d Are events P and F independent?

e Constructed and interpret a meaningful odds ratio for these data.

4.7 Summary

Data are the key to making sound, objective decisions, and *randomness* is the key to obtaining good data. The importance of randomness derives from the fact that relative frequencies of random events tend to stabilize in the long run; this long-run relative frequency is called *probability*. A more formal definition of probability allows for its use as a mathematical modeling tool to help explain and anticipate outcomes of events not yet seen. *Conditional probability, complements*, and rules for *addition* and *multiplication* of probabilities are essential parts of the modeling process. The relative frequency notion of probability and the resulting rules of probability calculations have direct parallels in the analysis of frequency data recorded in tables. Even with these rules, it is often more efficient to find an approximate answer through *simulation* than to labor over a theoretical probability calculation.

Supplementary Exercises

4.54 A coin is tossed four times, and the outcome is recorded for each toss.
 a List the outcomes for the experiment.
 b Let *A* be the event that the experiment yields three heads. List the outcomes in *A*.
 c Make a reasonable assignment of probabilities to the outcomes and find $P(A)$.

4.55 A hydraulic rework shop in a factory turns out seven rebuilt pumps today. Suppose three pumps are still defective. Two of the seven are selected for thorough testing and then classified as defective or nondefective.
 a List the outcomes for this experiment.
 b Let *A* be the event the selection includes no defectives. List the outcomes in *A*.
 c Assign probabilities to the outcomes and find $P(A)$.

4.56 The National Maximum Speed Limit (NMSL) of 55 miles per hour has been in force in the United States since early 1974. The data in Table 4.32 show the percentage of vehicles found to travel at various speeds for three types of highways in 1973 (before the NMSL), 1974 (the year the NMSL was put in force), and 1975.
 a Find the probability that a randomly observed car on a rural interstate was traveling less than 55 mph in 1973; less than 55 mph in 1974; less than 55 mph in 1975.
 b Answer the questions in part (a) for a randomly observed truck.
 c Answer the questions in part (a) for a randomly observed car on a rural secondary road.

4.57 An experiment consists of tossing a pair of dice.
 a Use the combinatorial theorems to determine the number of outcomes in the sample space *S*.
 b Find the probability that the sum of the numbers appearing on the dice is equal to 7.

4.58 **a** Show that $\binom{3}{0} + \binom{3}{1} + \binom{3}{2} + \binom{3}{3} = 2^3$.

 b Show that, in general, $\sum_{i=0}^{n} \binom{n}{i} = 2^n$.

4.59 Of the persons arriving at a small airport, 60% fly on major airlines, 30% fly on privately owned airplanes, and 10% fly on commercially owned airplanes not belonging to an airline. Of the persons arriving on major airlines, 50% are traveling for business reasons, while this figure is 60% for those arriving on private planes and 90% for those arriving on other commercially owned planes. For a person randomly selected from a group of arrivals, find the probability that
 a The person is traveling on business.
 b The person is traveling on business and on a private airplane.
 c The person is traveling on business given that he arrived on a commercial airliner.
 d The person arrived on a private plane given that he is traveling on business.

4.60 In how many ways can a committee of three be selected from 10 people?

4.61 How many different seven-digit telephone numbers can be formed if the first digit cannot be zero?

4.62 A personnel director for a corporation has hired ten new engineers. If three (distinctly different) positions are open at a particular plant, in how many ways can he fill the positions?

4.63 An experimenter wants to investigate the effect of three variables—pressure, temperature, and type of catalyst—on the yield in a refining process. If the experimenter intends to use three settings each for temperature and pressure and two types of catalysts, how many experimental runs will have to be conducted if she wants to run all possible combinations of pressure, temperature, and type of catalysts?

4.64 An inspector must perform eight tests on a randomly selected keyboard coming off an assembly line. The sequence in which the tests are conducted is important because the time lost between tests will vary. If an efficiency expert were to study all possible sequences to find the one that required the minimum length of time, how many sequences would be included in his study?

4.65 **a** Two cards are drawn from a 52-card deck. What is the probability that the draw will yield an ace and a face card in either order?

b Five cards are drawn from a 52-card deck. What is the probability that yield will be spades? that the draw will yield be of the same suit?

4.66 The quarterback on a certain football team completes 60% of his passes. If he tries three passes, assumed to be independent, in a given quarter, what is the probability that he will complete

a All three?

b At least one?

c At least two?

4.67 Two men each tossing a balanced coin obtain a "match" if both coins are heads or if both are tails. The process is repeated three times.

a What is the probability of three matches?

b What is the probability that all six tosses (three for each man) result in "tails"?

c Coin tossing provides a model for many practical experiments. Suppose the "coin tosses" represented the answers given by two students to three specific true-false questions on an examination. If the two students gave three matches for answers, would the low probability determined in part (a) suggest collusion?

4.68 Refer to Exercise 4.67. What is the probability that the pair of coins is tossed four times before a match occurs (that is, they match for the first time on the fourth toss)?

4.69 Suppose the probability of exposure to the flu during an epidemic is 0.6. Experience has shown that a serum is 80% successful in preventing an

inoculated person from acquiring the flu if exposed. A person not inoculated faces a probability of 0.90 of acquiring the flu exposed. Two persons, one inoculated and one not, are capable of performing a highly specialized task in a business. Assume that they are not at the same location, are not in contact with the same people, and cannot expose each other. What is the probability that at least one will get the flu?

4.70 Two gamblers bet $1 each on the successive tosses of a coin. Each has a bank of $6.

a What is the probability that they break even after six tosses of the coin?

b What is the probability that one player, say Jones, wins all the money on the tenth toss of the coin?

4.71 Suppose the streets of a city are laid out in a grid, with streets running north–south and east-west. Consider the following scheme for patrolling an area of 16 blocks by 16 blocks: A patrolman commences walking at the intersection in the center of the area. At the corner of each block he randomly elects to go north, south, east, or west.

a What is the probability that he will reach the boundary of his patrol area by the time he walks the first eight blocks?

b What is the probability that he will return to the starting point after walking exactly four blocks?

4.72 Consider two mutually exclusive events, A and B, such that $P(A) > 0$ and $P(B) > 0$. Are A and B independent? Give a proof with your answer.

4.73 An accident victim will die unless she receives in the next 10 minutes an amount of type A Rh$^+$ blood, which can be supplied by a single donor. It requires 2 minutes to "type" a prospective donor's blood and 2 minutes to complete the transfer of blood. A large number of untyped donors are available, and 40% of them have type A Rh$^+$ blood. What is the probability that the accident victim will be saved if only one blood-typing kit is available?

4.74 An assembler of electric fans uses motors from two sources. Company A supplies 90% of the motors, and company B supplies the other 10%. Suppose it is known that 5% of the motors supplied by company A are defective and 3% of the motors supplied by company B are defective. An assembled fan is found to have a defective motor. What is the probability that this motor was supplied by company B?

4.75 Show that for three events A, B, and C

$$P[(A \cup B)|C] = P(A|C) + P(B|C) - P[(A \cap B|C)]$$

4.76 If A and B are independent events, show that A and $\overline{B}$ are also independent.

Table 4.32

| Vehicle Speed (mph) | Rural Interstate | | | | | | Rural Primary | | | | | | Rural Secondary | | | | | |
| | Car | | | Truck | | | Car | | | Truck | | | Car | | | Truck | | |
	73	74	75	73	74	75	73	74	75	73	74	75	73	74	75	73	74	75
30–35	0	0	0	0	0	0	0	1	0	1	1	1	4	4	2	6	7	5
35–40	0	0	0	1	1	0	3	2	2	5	5	3	6	9	6	9	11	7
40–45	0	1	1	2	2	2	5	7	5	9	10	7	11	14	10	15	16	13
45–50	2	7	5	5	11	8	13	18	14	20	21	19	19	25	21	22	25	23
50–55	5	24	23	15	29	29	16	29	29	21	30	30	19	23	26	19	21	26
55–60	13	37	41	27	36	40	22	27	32	22	23	27	19	16	22	17	14	19
60–65	21	21	22	25	15	16	19	11	12	14	7	10	11	6	9	7	4	5
65–70	29	7	6	18	5	4	13	3	5	6	2	2	7	2	3	4	2	1
70–75	19	2	2	5	1	1	6	2	1	1	1	1	3	1	1	0	0	1
75–80	7	1	0	2	0	0	1	0	0	0	0	0	1	0	0	1	0	0
80–85	4	0	0	0	0	0	0	0	0	0	0	0	0	0	0	0	0	0

Source: D. B. Kamerud, *Transportation Research*, vol. 17A, no. 1, p.61.

4.77 Three events *A*, *B*, and *C* are said to be independent if

$$P(AB) = P(A)P(B)$$
$$P(AC) = P(A)P(C)$$
$$P(BC) = P(B)P(C)$$
and $$P(ABC) = P(A)P(B)P(C)$$

Suppose a balanced coin is independently tossed two times. Define the following events:

A: Heads appears on the first toss.
B: Heads appears on the second toss.
C: Both tosses yield the same outcome.

Are *A*, *B*, and *C* independent?

4.78 A line from *a* to *b* has midpoint *c*. A point is chosen at random on the line and marked *x* (the point *x* being chosen at random implies that *x* is equally likely to fall in any subinterval of fixed length *l*). Find the probability that the line segments *ax*, *bx*, and *ac* can be joined to form a triangle.

4.79 Eight tires of different brands are ranked from 1 to 8 (best to worst) according to mileage performance. If four of these tires are chosen at random by a customer, find the probabililty that the best tire among those selected by the customer is actually ranked third among the original eight.

4.80 Suppose that *n* indistinguishable balls are to be arranged in *N* distinguishable boxes so that each distinguishable arrangement is equally likely. If $n \geq N$, show that the probability that no box will be empty is given by

$$\frac{\binom{n-1}{N-1}}{\binom{N+N-1}{N-1}}$$

4.81 Relays in a section of an electrical circuit operate independently, and each one closes properly with probability 0.9 when a switch is thrown. The following two designs, each involving four relays, are presented for a section of a new circuit. Which design has the higher probability of current flowing from *a* to *b* when the switch is thrown?

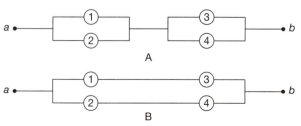

4.82 A blood test for hepatitis has the following accuracy:

	Test Result	
	+	−
Patient with Hepatitis	0.90	0.10
Patient without Hepatitis	0.01	0.99

The disease rate in the general population is 1 in 10,000.

a What is the probability that a person actually has hepatitis if he or she gets a positive blood test result?

b A patient is sent for a blood test because he has lost his appetite and has jaundice. The physician knows that this type of patient has probability 0.5 of having hepatitis. If this patient gets a positive result on his blood test, what is the probability that he has hepatitis?

4.83 Show that the following are true for any events *A* and *B*.

a $P(AB) \geq P(A) + P(B) - 1$

b The probability that exactly one of the events occurs is $P(A) + P(B) - 2P(AB)$.

4.84 Using Venn diagrams or similar arguments, show that for events *A*, *B*, and *C*

$$P(A \cup B \cup C) = P(A) + P(B) + P(C) - P(AB) - P(AC) - P(BC) + P(ABC)$$

4.85 Using the definition of conditional probability, show that

$$P(ABC) = P(A)P(B\,|\,A)P(C\,|\,AB)$$

4.86 There are 23 students in a classroom. What is the probability that at least two of them have the same birthday (day and month)? Assume that the year has 365 days. State your assumptions.

4.87 Does knowledge lead to the decision to be an engineer? To investigate this question, Hamill and Hodgkinson (*Proceedings of ICE*, 2003) surveyed 566, 11–16-year-olds, and classified them by their level of knowledge about engineering and the likelihood of becoming an engineer. The results are listed in Table 4.33.

a Find the probability that a randomly selected respondent said it was very unlikely that he or she will become an engineer.

b Find the probability that a randomly selected respondent knew nothing about engineering

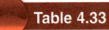

Table 4.33

Engineers and Their Level of Knowledge

Likelihood of Becoming an Engineer	Level of Knowledge about Engineering				
	Nothing at All	Not Much	A Little	Quite a Lot	A Lot
Very unlikely	179	132	53	8	4
Fairly unlikely	5	44	32	15	5
Fairly likely	3	4	16	16	5
Very likely	3	3	9	20	10

given that the respondent said it is very unlikely that he or she will become an engineer.

c Hamill and Hodgkinson concluded, "Lack of knowledge corresponds to engineering being an unlikely career, whereas knowledge makes it more likely." Can we make this conclusion from the data collected? Justify your answer.

Discrete Probability Distributions

One engineering firm enjoys 40% success rate in getting state government construction contracts. This month they have submitted bids on eight construction projects to be funded by the state government. How likely is this firm to get none of those contracts? Five out of eight contracts? All eight contracts?

To answer such questions, we obtain the probability distribution that describes the likelihood of all possible outcomes in a given situation (see Figure 5.1). In this chapter, we'll study the chance processes that generate numerical outcomes and the probabilities of occurrence of these outcomes. We'll also explore the uses of expected values of these outcomes. The numerical outcomes that take the form of counts such as the number of automobiles per family and the number of nonconforming parts manufactured per day are examples of *discrete random variables.* We'll study the probability distributions of such variables in this chapter. Variables that take the form of a continuous measurement, such as the diameter of a machined rod, are considered in the next chapter.

Figure 5.1

Likelihood of successfully bidding on projects

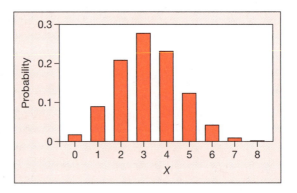

5.1 Random Variables and Their Probability Distributions

Most of the processes we encounter generate outcomes that can be interpreted in terms of real numbers, such as heights of children, number of voters favoring a certain candidate, tensile strengths of wires, and number of accidents at specified intersections. These numerical outcomes, with values that can change from trial to trial, are called *random variables*.

> A **random variable** is a real-valued function whose domain is a sample space.

Random variables are typically denoted by uppercase letters, such as $X, Y,$ and Z. The actual numerical values that a random variable can assume are denoted by lowercase letters, such as $x, y,$ and z. The probability that the random variable X takes a value x is denoted by $P(X = x)$ or $P(x)$. We encounter two types of random variables, *discrete* and *continuous*. In this chapter, we'll study discrete random variables and their probability distributions. In Chapter 6, we'll study continuous random variables and their probability distributions.

Example 5.1

A section of an electrical circuit has two relays, numbered 1 and 2, operating in parallel. The current will flow when a switch is thrown if either one or both of the relays close. The probability of a relay closing properly is 0.8 for each relay. We assume that the relays operate independently. Let E_i denote the event that relay i closes properly when the switch is thrown. Then $P(E_i) = 0.8$. A numerical event of some interest to the operator of this system is

X_i = the number of relays that close properly when the switch is thrown

Now X can take on only three possible values, 0, 1, and 2. We can find the probabilities associated with these values of X by relating them to the underlying events E_i. Thus, using the rules in Chapter 4, we get

$$P(X = 0) = P(\overline{E}_1 \cap \overline{E}_2) = P(\overline{E}_1) P(\overline{E}_2) = 0.2(0.2) = 0.04$$

because $X = 0$ means neither relay closes when the relays operate independently. Similarly,

$$P(X = 1) = P(E_1\overline{E}_2 \cup \overline{E}_1 E_2) = P(E_1) P(\overline{E}_2) + P(\overline{E}_1) P(E_2)$$
$$= 0.8(0.2) + 0.2(0.8) = 0.32$$

and

$$P(X = 2) = P(E_1 \cap E_2) = P(E_1) P(E_2) = 0.8(0.8) = 0.64$$

The values of the random variable X along with their probabilities are useful for keeping track of the operation of this system. The *number* of properly closing relays is the key to whether or not the system will work. The current will flow if X is at least 1, and this event has probability

$$P(X \geq 1) = P(X = 1) + P(X = 2) = 0.32 + 0.64 = 0.96$$

Note that we have mapped the outcomes of observing a process into a set of three meaningful real numbers and have attached a probability to each.

> A random variable X is said to be **discrete** if it can take on only a finite number, or a countably infinite number, of possible values x.

Following are some of the examples of a discrete random variable:

- Number of defective diodes in a sample of 64
- Number of switches in "On" position, out of 20 switches on a circuit board
- Number of nonconfirming assemblies out of 10
- Number of rolls of paper produced per day by a paper mill
- Number of defects per yard recorded for cable inspected
- Number of days recorded between consecutive failures of a network

In the relay example described earlier, the random variable X has only 3 possible values, and it is a relatively simple matter to assign probabilities to these values.

> The function $p(x)$ is called the **probability function** of X, provided
>
> 1. $P(X = x) = p(x) \geq 0$ for all values of x
> 2. $\sum_x P(X = x) = 1$

Table 5.1

Probability Distribution of Number of Closed Relays

x	P(x)
0	0.04
1	0.32
2	0.64
Total	**1.00**

Figure 5.2

Probability distribution of number of closed relays

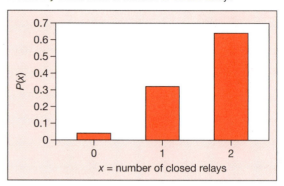

The probability function is sometimes called the *probability mass function* of X to denote the idea that a mass of probability is piled up at discrete points. It is often convenient to list the probabilities for a discrete random variable in a table or plot in a chart. With X defined as the number of closed relays, as in Example 5.1, the table and the graph of the probability mass function are given in Table 5.1 and Figure 5.2, respectively. The listing and the graph are two ways of representing the *probability distribution* of X.

Note that the probability function $p(x)$ satisfies two requirements:

1. $0 \le p(x) \le 1$ for $x = 0, 1, 2$
2. $p(0) + p(1) + p(2) = 1$

Example 5.2

The output of circuit boards from two assembly lines set up to produce identical boards is mixed into one storage tray. As inspectors examine the boards randomly, it is difficult to determine whether a board comes from line A or line B. A probabilistic assessment of this situation is often helpful.

Suppose a storage tray contains 10 circuit boards, of which 6 come from line A and 4 from line B. An inspector selects 2 of these identical-appearing boards for inspection. He is interested in X, the number of inspected boards from line A. Find the probability distribution for X.

Solution The experimenter consists of two selections, each of which can result in two outcomes. Let A_i denote the event that the ith board comes from line A, and B_i denote the event that it comes from line B. Then the probability of selecting two boards from line A (that is $X = 2$) is

$$P(X = 2) = P(A_1 A_2) = P(A \text{ on first}) \, P(A \text{ on second} \mid A \text{ on first})$$

The multiplicative rule of probability is used, and the probability of the second selection depends on what happened on the first selection. There are other possibilities for outcomes that will result in different values of X. These outcomes are conveniently listed using a tree diagram in Figure 5.3. The probabilities for the various selections are given on the branches of the tree.

It is easily seen that X has three possible outcomes, with the probabilities listed in Table 5.2.

Figure 5.3

Outcomes of circuit board selection

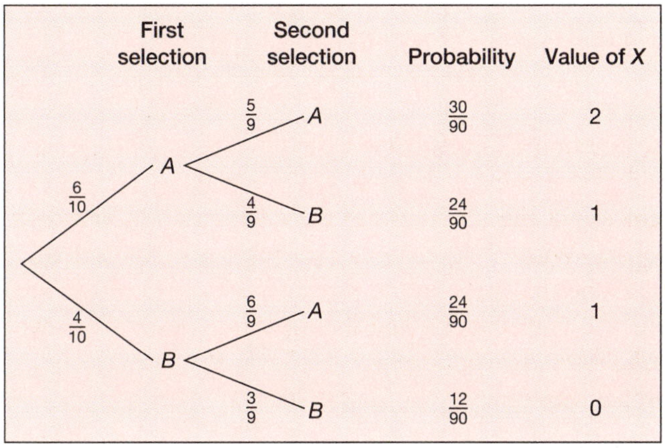

First selection	Second selection	Probability	Value of X
	$\frac{5}{9}$ A	$\frac{30}{90}$	2
A $\frac{6}{10}$	$\frac{4}{9}$ B	$\frac{24}{90}$	1
B $\frac{4}{10}$	$\frac{6}{9}$ A	$\frac{24}{90}$	1
	$\frac{3}{9}$ B	$\frac{12}{90}$	0

Table 5.2

Probability Distribution of Number of Boards Inspected from Line A

x	P(x)
0	12/90
1	48/90
2	30/90
Total	**1.00**

We sometimes study the behavior of random variables by looking at their *cumulative probabilities*. That is, for any random variable X, we may look at $P(X \leq b)$ for any real number b, which means the cumulative probability for X evaluated at b. Thus, we can define a function $F(b)$ as

$$F(b) = P(X \leq b).$$

The **distribution function** $F(b)$ for a random variable X is defined as

$$F(b) = P(X \leq b).$$

If X is discrete, $F(b) = \sum_{x=-\infty}^{b} p(x)$, where $p(x)$ is the probability function. The distribution function is also known as the **cumulative distribution function (cdf)**.

Example 5.3

The random variable X, denoting the number of relays closing properly (defined in Example 5.1) has the probability distribution given below:

$P(X = 0)$	$P(X = 1)$	$P(X = 2)$
0.04	0.32	0.64

Note that $P(X \leq 1.5) = P(X = 0) + P(X = 1) = 0.04 + 0.32 = 0.36$

The distribution function for this random variable (graphed in Figure 5.4) has the form

$$F(b) = \begin{cases} 0 & b < 0 \\ 0.04 & 0 \leq <1 \\ 0.36 & 1 \leq b < 2 \\ 1.00 & 2 \leq b \end{cases}$$

Figure 5.4

A distribution function for a discrete random variable

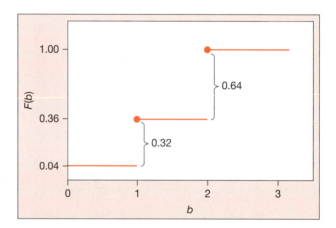

Exercises

5.1 Among 10 applicants for an open position, 6 are females and 4 are males. Suppose three applicants are randomly selected from the applicant pool for final interviews. Find the probability distribution for X, the number of female applicants among the final three.

5.2 The median annual income for household heads in a certain city is $28,000. Four such household heads are randomly selected for an opinion poll.
   a Find the probability distribution of X, the number (out of the four) that have annual incomes below $28,000.
   b Would you say that it is unusual to see all four below $28,000 in this type of poll? (What is the probability of this event?)

5.3 Wade Boggs of the Boston Red Sox hit 0.363 in 1987. (He got a hit on 36.3% of his official times at bat.) In a typical games, he was up to bat three official times. Find the probability distribution for X, the number of hits in a typical game. What assumptions are involved in the answer? Are the assumptions reasonable? Is it unusual for a good hitter like Wade Boggs to go zero for three in one game?

5.4 A commercial building has two entrances, numbered I and II. Three people enter the building at 9:00 A.M. Let X denote the number that select entrance I. Assuming the people choose entrances independently, find the probability distribution for X. Were any additional assumptions necessary in arriving at your answer?

5.5 Table 4.22 gives information on the type of injury sustained in accidents involving different types of flights. Suppose four independent flight accidents were reported in one day. Let X denote the number of accidents out of 4 that involved fatal injuries.

a Find the probability distribution for X and report it in a tabular form.

b Find the probability that at least one of the four accidents resulted in fatal injuries.

5.6 It was observed that 40% of the vehicles crossing a certain toll bridge are commercial trucks. Four vehicles will cross the bridge in the next minute. Find the probability distribution for X, the number of commercial trucks among the four, if the vehicle types are independent of one another.

5.7 Of the people entering a blood bank to donate blood, 1 in 3 have type O^+ blood, and 1 in 15 have type O^- blood. For the next three people entering the blood bank, let X denote the number with O^+ blood and Y the number with O^- blood. Assuming independence among the people with respect to blood type, find the probability distributions for X and Y. Also find the probability distribution for $X + Y$, the number of people with type O blood.

5.8 Daily sales records for a computer manufacturing firm show that it will sell 0, 1, or 2 mainframe computer systems with probabilities as listed:

Number of sales	0	1	2
Probability	0.7	0.2	0.1

a Find the probability distribution for X, the number of sales in a two-day period, assuming that sales are independent from day to day.

b Find the probability that at least one sale is made in the two-day period.

5.9 Four microchips are to be placed in a computer. Two of the four chips are randomly selected for inspection before assembly of the computer. Let X denote the number of defective chips found among the two chips inspected. Find the probability distribution for X if

a Two of the microchips were defective.

b One of the four microchips was defective.

c None of the microchips was defective.

5.10 When turned on, each of the three switches in the diagram below works properly with probability 0.9. If a switch is working properly, current can flow through it when it is turned on. Find the probability distribution for Y, the number of closed paths from a to b, when all three switches are turned on.

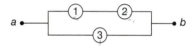

5.2   Expected Values of Random Variables

Because probability can be thought of as the long-run relative frequency of occurrence for an event, a probability distribution can be interpreted as showing the long-run relative frequency of occurrence for numerical outcomes associated with a random variable.

Game 1: Suppose that you and your friend are matching balanced coins. Each of you tosses a coin. If the upper faces match, you win \$1; if they do not match, you lose \$1 (i.e., your friend wins \$1).

The probability of a match is 0.5 and in the long run you will win about half the time. Thus, a relative frequency distribution of your winnings should look like Figure 5.5. Note that the negative sign indicates a loss to you.

On the average, how much will you win per game over the long run? If Figure 5.5 is a correct display of your winnings, you win -1 half the time and $+1$ half the time, for an average of

$$(-1)\left(\frac{1}{2}\right) + (+1)\left(\frac{1}{2}\right) = 0$$

This average is sometimes called your expected winnings per game or the *expected value* of your winnings. An expected value of 0 indicates that this is a fair game.

Figure 5.5

Relative frequency of winnings

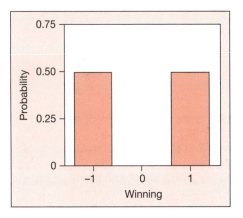

The expected value can be thought of as the mean value of your winnings over many games.

> The **expected value** of a discrete random variable X having probability function $p(x)$ is given by
>
> $$\mu = E(X) = \sum_x xp(x)$$
>
> The sum is over all values of x for which $p(x) > 0$.

Game 2: Now payday has arrived, and you and your friend up the stakes to $10 per game of matching coins. You now win -10 or $+10$ with equal probability. Your expected winnings per game are

$$(-10)\left(\frac{1}{2}\right) + (+10)\left(\frac{1}{2}\right) = 0$$

and the game is still fair.

The new stakes can be thought of as a function of the old in the sense that if X represents winnings per game when you were playing for $1, then $10X$ represents your winnings per game when you play for $10.

Game 3: You and your friend decide to play the coin–matching game by allowing you to win $1 if the match is tails and $2 if the match is heads. You lose $1 if the coins do not match.

Notice that this is not a fair game, because your expected winnings are

$$(-1)\left(\frac{1}{2}\right) + (+1)\left(\frac{1}{4}\right) + (+2)\left(\frac{1}{4}\right) = 0.25$$

So you are likely to win 25 cents per game in the long run. In other words, the game is favorable to you.

Game 4: In the coin-matching game, suppose you pay your friend $1.50 if the coins do not match. You'll still win $1 if the match is tails and $2 if the match is heads.

Now your expected winnings per game are

$$(-1.50)\left(\frac{1}{2}\right) + (+1)\left(\frac{1}{4}\right) + (+2)\left(\frac{1}{4}\right) = 0$$

and the game now is fair.

What is the difference between Game 4 and the original Game 1 in which the payoffs were $1?

The difference certainly cannot be explained by the expected value, because both games are fair. You can win more but also lose more with the new payoffs, so the difference between the two can be partially explained in terms of the variation of your winnings across many games. This increased variation can be seen in Figure 5.6, the relative frequency for your winnings in Game 4, which is more spread out than the one shown in Figure 5.3.

Figure 5.6

Relative frequency of winnings (Game 4)

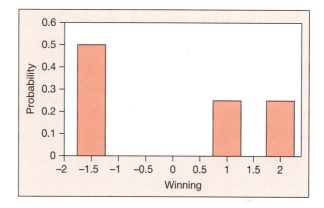

As described in Chapter 1, we often measure variation using the *variance* or the *standard deviation*.

> The **variance** of a random variable X with expected value μ is given by
> $$\sigma^2 = V(X) = E(X - \mu)^2$$
>
> The **standard deviation** of a random variable X is the square root of the variance given by
> $$\sigma = \sqrt{\sigma^2} = \sqrt{E(X - \mu)^2}$$

By expanding the square term in the definition of variance and solving we get

$$V(X) = E(X - \mu)^2 = E(X^2 - 2X\mu + \mu^2)$$
$$= E(X^2) - 2\mu E(X) + E(\mu^2) = E(X^2) - \mu^2$$

For the Game 1 depicted in Figure 5.5, the variance of your winnings (with $\mu = 0$) is

$$\sigma^2 = E(X - \mu^2) = (-1)^2\left(\frac{1}{2}\right) + (+1)^2\left(\frac{1}{2}\right) = 1$$

It follows that $\sigma = 1$ as well. For the Game 4 depicted in Figure 5.6, the variance of your winnings is

$$\sigma^2 = E(X - \mu)^2 = (-1.5)^2\left(\frac{1}{2}\right) + (+1)^2\left(\frac{1}{4}\right) + (+2)^2\left(\frac{1}{4}\right) = 2.375$$

and $\sigma = \sqrt{2.375} = 1.54$. Which game would you rather play? Game 1 or Game 4?

The standard deviation can be thought of as the size of a "typical" deviation between and observed outcome and the expected value.

- In Figure 5.5, each outcome (-1 or $+1$) does deviate precisely one standard deviation from the expected value.
- In Figure 5.6, the positive values deviate on the average 1.5 units from the expected value of 0, as does the negative value, which is approximately one standard deviation below the mean.

The mean and standard deviation provide a useful summary of the probability distribution for a random variable that can assume many values.

Example 5.4

Table below provides the age distribution of the population (U.S. Bureau of Census) for 1990 and 2050 (projected). The numbers are percentages.

Age Interval	Midpoint	1990	2050
Under 5	3	7.6	6.4
5–13	9	12.8	11.6
14–17	16	5.3	5.2
18–24	21	10.8	9.0
25–34	30	17.3	12.5
35–44	40	15.1	12.2
45–64	55	18.6	22.5
65–84	75	11.3	16.0
85 and over	92	1.2	4.6

Age is actually a continuous measurement, but when reported in categories, we can treat it as a discrete random variable. To move from continuous age intervals to discrete age classes, we assign each interval the value of its midpoint (rounded). The data in table above are interpreted as reporting that 7.6% of the 1990 population was around 3 years of age and 22.5% of the 2050 population is anticipated to be around 55 years of age. The open interval at the upper end was stopped at 100 years for convenience.

With the percentages interpreted as probabilities, the mean age for 1990 is approximated by

$$\mu = \sum xp(x) = 3(0.076) + 9(0.128) + \cdots + 92(0.012) = 35.5$$

For 2050, the mean age is approximated by

$$\mu = \sum xp(x) = 3(0.064) + 9(0.116) + \cdots + 92(0.046) = 41.2$$

The mean age is anticipated to increase rather markedly. The variation in the two age distributions can be approximated by the standard deviation. For 1990,

$$\sigma = \sqrt{\sum (x - \mu)^2 \, p(x)} = \sqrt{(3 - 35.5)^2(0.076) + \cdots + (92 - 35.5)^2(0.012)}$$
$$= 22.5$$

and employing similar calculation for 2050 data, we get $\sigma = 25.4$. These results are summarized in table below.

	1990	2050
Mean	35.5	41.2
Standard deviation	22.5	25.4

The population is not only getting older, on the average, but its variation is increasing too. What are some of the implications of these trends?

Example 5.5

The manager of a stockroom in a factory knows from her study of records that X, the daily demand (number of times used per day) for a certain tool, has the following probability distribution:

Demand	0	1	2
Probability	0.1	0.5	0.4

That is, 50% of the daily records show that the tool was used one time.
Find the expected daily demand for the tool and the variance.

Solution From the definition of expected value, we see that

$$E(X) = \sum xp(x) = 0(0.1) + 1(0.5) + 2(0.4) = 1.3$$

The tool is used an average of 1.3 times per day. From the definition of variance, we see that

$$V(X) = E(X - \mu)^2 = \sum_x (x - \mu^2) \, p(x)$$

$$= (0 - 1.3)^2(0.1) + (1 - 1.3)^2(0.5) + (2 - 1.3)^2(0.4) = 0.41$$

Often we use functions of random variables. The following results are useful in manipulating expected values.

> For any random variable X and constants a and b,
>
> 1. $E(aX + b) = aE(X) + b$
> 2. $V(aX + b) = a^2 V(X)$

Example 5.6

In Example 5.5, suppose it costs the factory $100 each time the tool is used. Find the mean and variance of the daily costs for use of this tool.

Solution Recall that $X =$ the daily demand. Then the daily cost of using this tool is $100X$. Using the above results we have,

$$E(100\,X) = 100\,E(X) = 100\,(1.3) = 130$$

and

$$V(100\,X) = 100^2\,V(X) = 10{,}000\,(0.41) = 4{,}100$$

The factory should budget $130 per day to cover the cost of using tool. The standard deviation of daily cost is $64.03.

Example 5.7

The operators of certain machinery are required to take a proficiency test. A national testing agency gives this test nationwide. The possible scores on the test are 0, 1, 2, 3, or 4. The test brochure published by the agency gave the following information about the scores and the corresponding probabilities:

Score	0	1	2	3	4
Probability	0.1	0.2	0.4	0.2	0.1

Compute the mean score and the standard deviation.

Solution The above probability distribution shows that about 10% of those who take the test score 4 on the proficiency test, or the probability of making 4 on the proficiency test is 0.1. The mean score or the expected score is

$$E(X) = \sum xp(x) = 0\,(0.1) + 1\,(0.2) + 2\,(0.4) + 3\,(0.2) + 4\,(0.1) = 2$$

Figure 5.7

Probability distribution of score

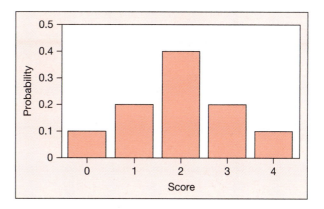

It is easy to see from the above distribution or the plot of the distribution in Figure 5.7 why the mean score is 2.

$$E(X^2) = \sum x^2 p(x) = 0^2(0.1) + 1^2(0.2) + 2^2(0.4) + 3^2(0.2) + 4^2(0.1) = 5.2$$

Because $\mu = E(X) = 2$, and $E(X^2) = 5.2$ we can compute the variance as

$$V(X) = E(X^2) - \mu^2 = 5.2 - (2^2) = 1.2$$

Therefore, the standard deviation of scores is $\sqrt{1.2} = 1.095$, which is approximately the average distance the scores are from the mean.

The mean and the variance give us some useful information about the center and spread of the probability mass. Recall that the standard deviation is often used as a measure of distance between data points and the mean of the distribution. Now suppose we know only the mean and the variance for a probability distribution. Then we can use the following two results to make some useful comments about the probabilities for certain intervals.

- The empirical rule: This rule was introduced in Chapter 2 as a data analysis tool. It still holds when the data distribution is interpreted as a probability distribution.

The Empirical Rule

For a mound-shaped, somewhat symmetric distribution of random variable X with mean μ and variance σ^2,

- $P(|X - \mu| < \sigma) \approx 0.68$
- $P(|X - \mu| < 2\sigma) \approx 0.95$
- $P(|X - \mu| < 3\sigma) \approx 1.00$

- Tchebysheff's theorem: This is another result (introduced in Chapter 1) relating μ, σ, and relative frequencies (probabilities). It gives a very rough but useful approximation that works in all cases.

> ### Tchebysheff's Theorem
>
> Let X be a random variable with mean μ and variance σ^2. Then for any positive k,
>
> $$P(|X - \mu| < k\sigma) \geq 1 - \frac{1}{k^2}$$

The inequality in the statement of Tchebysheff's theorem is equivalent to

$$P(\mu - k\sigma < X < \mu + k\sigma) \geq 1 - \frac{1}{k^2}$$

To interpret this result, let $k = 2$, for example. Then the interval from $\mu - 2\sigma$ to $\mu + 2\sigma$ must contain at least $1 - 1/k^2 = 1 - 1/4 = 3/4$ of the probability mass for the random variable.

Example 5.8

The daily production of electric motors at a certain factory averaged 120 with a standard deviation of 10.

a What fraction of days will have a production level between 100 and 140?
b Find the shortest interval certain to contain at least 90% of the daily production levels.

Solution   Because no information about the shape of the distribution of daily production levels is available, we'll use Tchebysheff's theorem to answer these questions.

a The interval 100 to 140 is $\mu - 2\sigma$ to $\mu + 2\sigma$, with $\mu = 120$ and $\sigma = 10$. Thus, $k = 2$, which gives $1 - 1/k^2 = 1 - 1/4 = 3/4$. At least 75% of the days will have total production in this interval.

This result could be closer to 95% if the daily production figures showed a mound-shaped, symmetric relative frequency distribution.

b To find k, we must set $1 - 1/k^2 = 0.90$ and solve for k.

$$1 - \frac{1}{k^2} = 0.90 \text{ gives } k = \sqrt{10} = 3.16$$

The interval is $\mu - (3.16)\sigma$ to $\mu + (3.16)\sigma$

i.e., $120 - 3.16(10)$ to $120 + 3.16(10)$

i.e., 88.4 to 151.6

Therefore, we can say that the interval $(88.4, 151.6)$ will contain at least 90% of the daily production levels.

Example 5.9

The cost for use of a certain tool varies depending on how frequently it is used. The frequency of use changes from day to day. The daily cost for use of this tool has a mean of $130 and standard deviation $64.03 (see Example 5.6). How often will this cost exceed $300?

Solution First we must find the distance between 300 and the mean (130), in terms of standard deviation of the distribution of costs as

$$k = \frac{300 - \mu}{\sigma} = \frac{300 - 130}{64.03} = 2.655$$

It shows that the cost of $300 a day is 2.655 standard deviations above the mean cost. Using Tchebysheff's theorem with $k = 2.655$, we get the interval

$$\mu - (2.655)\sigma \quad \text{to} \quad \mu + (2.655)\sigma$$

$$\text{i.e., } 130 - 2.655\,(64.03) \quad \text{to} \quad 130 + 2.655\,(64.03),$$

$$\text{i.e., } -40 \quad \text{to} \quad 300$$

with the associated probability at least $1 - 1/k^2 = 1 - 1/2.655^2 = 0.858$. In other words,

$$P(\$-40 < \text{daily cost} < \$300) \geq 0.858$$

Because the daily cost cannot be negative, at most 14.2% of the probability mass can exceed $300. Thus, the cost will exceed $300 no more than 14.2% of the time.

Example 5.10

Repeated here (from Example 5.4) are the mean and standard deviation of the age distribution of the U.S. population for years 1990 and 2050 (projected).

	1990	2050
Mean	35.5	41.2
Standard deviation	22.5	25.4

The 1- and 2-standard deviation intervals for these data and the percent of data within these intervals using the empirical rule and the actual percentages in these intervals are given in the following table.

		1990		2050	
Interval	Percentage by Empirical Rule	Interval	Actual Percentage	Interval	Actual Percentage
$\mu \pm \sigma$	68%	(13.0, 58.0)	67.1%	(15.8, 66.6)	61.4% to 77.4%
$\mu \pm 2\sigma$	95%	(0, 80.5)	<98.8*	(0, 92)	<100%

Note the following:

- The percentage of population inside the 1-standard deviation interval for 1990 cannot be determined precisely, but 13 is at the upper end of the 5–13 interval, and 58 is inside the 45–64 age interval. Adding the percentages above 13 and below 64 yields the percentage shown above.
- The upper limit of the 1-standard deviation interval for 2050, 66.6, is toward the low end of the 65–84 age interval, and depending on how we treat this category, the interval contains somewhere between 61.4% and 77.4% of the population.

As we can see from table on previous page, the percentages computed from the population are very close to the percentages expected using the empirical rule. In other words, the empirical rule does work for probability distributions as well as for datasets.

At this point, it may seem like every problem has its own unique probability distribution and we must start from the basics to construct such a distribution each time a new problem comes up. Fortunately, that is not the case. Certain basic probability distributions can be developed that will serve as models for a large number of practical problems. In the remainder of this chapter, we consider some fundamental discrete distributions.

Exercises

5.11 You are to pay $1 to play a game consisting of drawing one ticket at random from a box of numbered tickets. You win the amount (in dollars) of the number on the ticket you draw. Two boxes are available with numbered tickets as shown below:

I | 0, 1, 2   II | 0, 0, 0, 1, 4

5.12 The size distribution of U.S. families as estimated by the U.S. Bureau of Census is given in table below.

Number of Persons	Percentage
1	28.8
2	36.2
3	17.7
4	15.8
5	6.9
6	2.4
7 or more	1.4

a Calculate the mean and standard deviation of family size. Are these exact values or approximations?

b How does the mean family size compare to the median family size?

c Does the empirical rule hold for the family size distribution?

5.13 The graph in Figure 5.8 shows the age distribution for AIDS deaths in the United States through 1995. Approximate the mean and standard deviation of this age distribution. How does the mean age compare to the approximate median age?

Figure 5.8

Distribution of AIDS deaths by age, 1982 through 1995

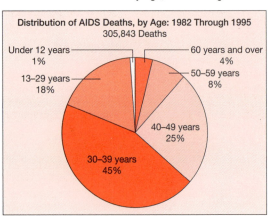

Distribution of AIDS Deaths, by Age: 1982 Through 1995
305,843 Deaths

Under 12 years 1%
13–29 years 18%
30–39 years 45%
40–49 years 25%
50–59 years 8%
60 years and over 4%

5.14 How old are our drivers? Table below gives the age distribution of licensed drivers in the United States.

Age	Number (in millions)
19 and under	9.4
20–24	16.2
25–29	17.0
30–34	19.1
35–39	20.4
40–44	21.1
45–49	19.5
50–54	17.4
55–59	13.4
60–64	10.2
65–69	8.4
70 and over	19.1
Total	191.0

a Describe this age distribution in terms of median, mean, and standard deviation.
b Does the empirical rule work well for this distribution?

5.15 Daily sales records for a computer manufacturing firm show that it will sell 0, 1, or 2 mainframe computer systems with probabilities as listed:

Number of sales	0	1	2
Probability	0.7	0.2	0.1

Find the expected value, variance, and standard deviation for daily sales.

5.16 Approximately 10% of the glass bottles coming off a production line have serious defects in the glass. If two bottles are randomly selected for inspection, find the expected value and variance of the number of inspected bottles with serious defects.

5.17 The number of breakdowns for a university computer system is closely monitored by the director of the computing center, because it is critical to the efficient operation of the center. The number averages 4 per week, with a standard deviation of 0.8 per week.
a Find an interval that must include at least 90% of the weekly figures on number of breakdowns.
b The center director promises that the number of breakdowns will rarely exceed 8 in a one-week period. Is the director safe in making this claim? Why or why not?

5.18 Keeping an adequate supply of spare parts on hand is an important function of the parts department of a large electronics firm. The monthly demand for microcomputer printer boards was studied for some months and found to average 28, with a standard deviation of 4. How many printer boards should be stocked at the beginning of each month to ensure that the demand will exceed the supply with probability less than 0.10?

5.19 An important feature of golf cart batteries is the number of minutes they will perform before needing to be recharged. A certain manufacturer advertises batteries that will run, under a 75-amp discharge test, for an average of 100 minutes, with a standard deviation of 5 minutes.
a Find an interval that must contain at least 90% of the performance periods for batteries of this type.
b Would you expect many batteries to die out in less than 80 minutes? Why or why not?

5.20 Costs of equipment maintenance are an important part of a firm's budget. Each visit by a field representative to check out a malfunction in a word processing system costs $40. The word processing system is expected to malfunction approximately five times per month, and the standard deviation of the number of malfunctions is 2.
a Find the expected value and standard deviation of the monthly cost of visits by the field representative.
b How much should the firm budget per month to ensure that the cost of these visits will be covered at least 75% of the time?

5.3 The Bernoulli Distribution

Consider the inspection of a single item taken from an assembly line. Suppose 0 is recorded if the item is nondefective and 1 is recorded if the item is defective. If X is the random variable denoting the condition of the inspected item, then

$X = 0$ with probability $(1 - p)$ and $X = 1$ with probability p, where p denotes the probability of observing a defective item. The probability distribution of X is given by

$$p(x) = p^x(1 - p)^{1-x} \qquad x = 0, 1$$

where $p(x)$ denotes the probability that $X = x$. Such a random variable is said to have a *Bernoulli distribution* or to represent outcome of a single Bernoulli trial. Any random variable denoting the presence or absence of a certain condition in an observed phenomenon will possess this type of distribution. Frequently, one of the outcomes is termed a "success" and the other a "failure." Note that $p(x) \geq 0$ and $\Sigma p(x) = 1$.

> ### The Bernoulli Distribution
>
> $$p(x) = p^x(1 - p)^{1-x} \qquad x = 0, 1 \qquad \text{for } 0 \leq p \leq 1$$
> $$E(X) = p \quad \text{and} \quad V(X) = p(1 - p)$$

Suppose we repeatedly observe items of this type and record a value of X for each item observed. Then the expected value of X is given by

$$E(X) = \sum_x xp(x) = 0\,p(0) + 1\,p(1) = 0\,(1 - p) + 1p = p$$

Thus, if 10% of the items are defective, we expect to observe an *average* of 0.1 defective per item inspected. In other words, we would expect to see one defective for every 10 items inspected.

Using the definition of variance, we get,

$$V(X) = E(X^2) - [E(X)]^2 = [0^2\,(1 - p) + 1^2 p] - p^2 = p(1 - p)$$

The Bernoulli distribution is a building block in forming other probability distributions, such as the binomial distribution.

5.4 The Binomial Distribution

Instead of inspecting a single item, as with the Bernoulli random variable, suppose we now independently inspect n items and record values for $X_1, X_2, \ldots, X_n$, where

$$X_i = \begin{cases} 1 & \text{if the } i\text{th inspected item is defective} \\ 0 & \text{otherwise} \end{cases}$$

We have in fact observed a sequence of n Bernoulli random variables. One especially interesting function of $X_1, X_2, \ldots, X_n$ is the sum $Y = \sum_{i=1}^{n} X_i =$ the number of defectives among the n sampled items $= 0, 1, \ldots, n$.

The probability distribution of Y under the assumption that $P(X_i = 1) = p$, where p remains constant over all trials, is known as the *binomial distribution*. We can easily find the probability of Y using basic probability rules. For simplicity, let us consider an example with three items, that is, $n = 3$ trials. Then the random variable Y can take on four possible values, 0, 1, 2, and 3. All possible outcomes of three trials can be described as follows

$Y = 0$	$Y = 1$	$Y = 2$	$Y = 3$
000	100	110	111
	010	101	
	001	011	

For $Y = 0$, all three X_is must be zero. Thus,

$$P(Y = 0) = P(X_1 = 0, X_2 = 0, X_3 = 0)$$
$$= P(X_1 = 0)\,P(X_2 = 0)\,P(X_3 = 0) = (1 - p)^3$$

For $Y = 1$, exactly one X_i is 1 and the other two are 0. But which X_i is 1? We don't know. So we need to take into account all possible ways of one $X_i = 1$ and the other two 0.

$$P(Y = 1) = P(X_1 = 1, X_2 = 0, X_3 = 0) \cup P(X_1 = 0, X_2 = 1, X_3 = 0)$$
$$\cup\, P(X_1 = 0, X_2 = 0, X_3 = 1)$$
$$= P(X_1 = 1)\,P(X_2 = 0)\,P(X_3 = 0) + P(X_1 = 0)P(X_2 = 1)\,P(X_3 = 0)$$
$$+\, P(X_1 = 0)\,P(X_2 = 0)\,P(X_3 = 1)$$
$$= 3p(1 - p)^2$$

For $Y = 2$, exactly two X_i are 1 and one must be 0. This can occur in three mutually exclusive ways. Hence,

$$P(Y = 2) = P(X_1 = 1, X_2 = 1, X_3 = 0) \cup P(X_1 = 0, X_2 = 1, X_3 = 1)$$
$$\cup\, P(X_1 = 1, X_2 = 0, X_3 = 1)$$
$$= P(X_1 = 1)\,P(X_2 = 1)\,P(X_3 = 0) + P(X_1 = 0)\,P(X_2 = 1)\,P(X_3 = 1)$$
$$+\, P(X_1 = 1)\,P(X_2 = 0)\,P(X_3 = 1)$$
$$= 3p^2(1 - p)$$

The event $Y = 3$ can occur only if all X_i are 1, so

$$P(Y = 3) = P(X_1 = 1, X_2 = 1, X_3 = 1)$$
$$= P(X_1 = 1)\,P(X_2 = 1)\,P(X_3 = 1) = p^3$$

Notice that the coefficient in each of the expressions is the number of ways of selecting y positions in the sequence in which to place 1s. Because there are three positions in the sequence, this number is $\binom{3}{y}$. Thus, we can write

$$P(Y = y) = \binom{3}{y} p^y (1 - p)^{3-y} \qquad y = 0, 1, 2, 3, \text{ where } n = 3$$

A random variable Y possesses a **binomial distribution** if

1. The experiment consists of a fixed number n of trials.
2. Each trial can result in one of only two possible outcomes, called "success" and "failure."
3. The probability of "success," p, is constant from trial to trial.
4. The trials are independent.
5. Y is defined as the number of successes among the n trials.

Many experimental and sample survey situations result in a random variable that can be adequately modeled by the binomial distribution. For example,

- Number of defectives in a sample of n items from a large population
- Counts of number of employees favoring a certain retirement policy out of n employees interviewed
- The number of pistons in an eight-cycle engine that are misfiring
- The number of electronic systems sold this week out of the n that are manufactured

The binomial distribution is defined by two parameters, n (number of trials) and p (probability of success on each trial). The binomial random variable $Y =$ the number of successes in n trials.

The Binomial Distribution

$$p(y) = \binom{n}{y} p^y (1 - p)^{n-y} \qquad y = 0, 1, 2, \ldots, n \qquad \text{for } 0 \leq p \leq 1$$

$$E(Y) = np \quad \text{and} \quad V(Y) = np(1 - p)$$

The figures on the next page show how the parameter values (n and p) affect the probability function for the binomial distribution.

Figure 5.9

The effect of n, and p on the binomial probability function

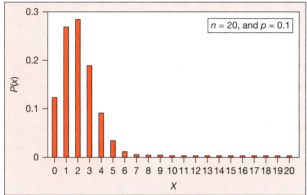

(a) $n = 20$, $p = 0.1$

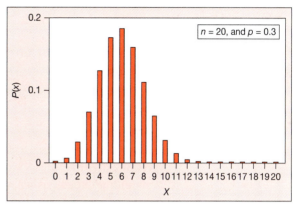

(b) $n = 20$, $p = 0.3$

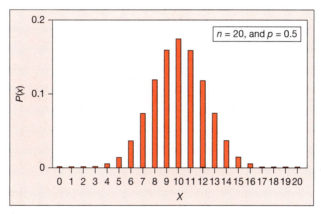

(c) $n = 20$, $p = 0.5$

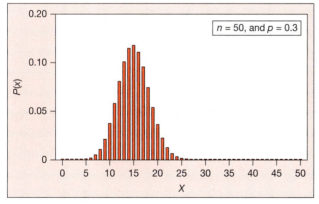

(d) $n = 50$, $p = 0.3$

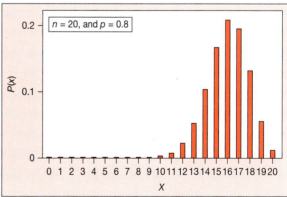

(e) $n = 20$, $p = 0.8$

From Figure 5.9 we see that

- For values of p near 0 or 1, the binomial distribution is skewed.
- As p gets closer to 0 (Figures 5.9a, and 5.9b), the binomial distribution becomes more right-skewed, while as p gets closer to 1 (Figure 5.9d), the binomial distribution becomes more left-skewed.
- For $p = 0.5$ (Figure 5.9c), the binomial distribution is symmetric and mound-shaped.
- The distribution becomes more symmetric as n increases (Figure 5.9b and 5.9e).

The mean and variance of the binomial distribution can be determined using the basic definition and solving the expression. This direct approach is tricky and requires regrouping of the terms in the summation. Later in Chapter 7, we'll see another approach using the sum of independent and identical random variables.

$$E(Y) = \sum_{y=0}^{n} y\, p(y) = \sum_{y=0}^{n} y \binom{n}{y} p^y (1-p)^{n-y} = np$$

and

$$V(y) = \sum_{y=0}^{n} (y-\mu)^2 p(y) = \sum_{y=0}^{n} (y-\mu)^2 \binom{n}{y} p^y (1-p)^{n-y} = np(1-p)$$

Example 5.11

Suppose a large lot contains 10% defective fuses. Four fuses are randomly sampled from the lot.

a Find the probability that exactly one fuse in the sample of four is defective.
b Find the probability that at least one fuse in the sample of four is defective.
c Suppose the four fuses sampled from the lot were shipped to a customer before being tested, on a guarantee basis. If any fuse is defective, the supplier will repair it without any change to the customer. Assume that the cost of making the shipment good is $C = 3Y^2$, where Y denotes the number of defectives in the shipment of four. Find the expected repair cost.

Solution We assume that the four fuses are sampled independently of each other and that the probability of being defective is the same (0.1) for each fuse. This will be approximately true if the lot is indeed *large*. If the lot is small, removal of one fuse would substantially change the probability of observing a defective on the second draw. For a large lot, the binomial distribution provides a reasonable model in this experiment with $n = 4$ and $p = 0.1$. Let

Y = the number of defective fuses out of 4 inspected

a The probability that exactly one fuse in the sample of four is defective is

$$P(Y = 1) = p(1) = \binom{4}{1}(0.1^1)(0.9^3) = 0.2916$$

b The probability that at least one fuse in the sample of four is defective is

$$P(Y \geq 1) = 1 - P(Y = 0) = 1 - \binom{4}{0}(0.1^0)(0.9^4) = 0.3439$$

c We know that $E(C) = E(3Y^2) = 3E(Y^2)$. For the binomial distribution, $\mu = E(Y) = np$ and $V(Y) = np(1 - p)$.

Because $V(Y) = E(Y - \mu)^2 = E(Y^2) - \mu^2$, we can get

$$E(Y^2) = V(Y) + \mu^2 = np(1 - p) + (np)^2$$

In this example, $n = 4$ and $p = 0.1$, and hence

$$\begin{aligned} E(C) = 3E(Y^2) &= 3[np(1 - p) + (np)^2] \\ &= 3[4(0.1)(0.9) + 4^2(0.1)^2] = 1.56 \end{aligned}$$

If the cost were originally in tens of dollars, we could expect to pay an average of $15.60 in repair costs for each shipment of four fuses.

Example 5.12

In a study of lifetimes for a certain type of battery, it was found that the probability of a lifetime X exceeding 4 hours is 0.135. If three such batteries are in use in independently operating systems, find the probability that

a Only one of the batteries lasts 4 hours or more.
b At least one battery lasts 4 hours or more.

Solution   Let $Y =$ the number of batteries (out of three) lasting 4 hours or more. We can reasonably assume that Y has a binomial distribution with $n = 3$ and $p = 0.135$. Hence,

a The probability that only one battery lasts 4 hours or more is

$$P(Y = 1) = p(1) = \binom{3}{1}(0.135^1)(0.865^2) = 0.303$$

b The probability that at least one battery lasts 4 hours or more is

$$P(Y \geq 1) = 1 - P(Y = 0) = 1 - \binom{3}{0}(0.135^0)(0.865^3) = 0.647$$

Example 5.13

An industrial firm supplies 10 manufacturing plants with a certain chemical. The probability that any one firm calls in an order on a given day is 0.2, and this probability is the same for all 10 plants. Find the probability that, on a given day, the number of plants calling in order is

 a Exactly 3   **b** At most 3   **c** At least 3

Solution Let Y = the number of plants (out of 10) calling in orders on the day in question. If the plants order independently, then Y can be modeled to have a binomial distribution with $n = 10$ and $p = 0.2$.

 a The probability of exactly 3 out of 10 plants calling in orders is

$$P(Y = 3) = \binom{10}{3}(0.2^3)(0.8^7) = 0.201$$

 b The probability of at most 3 out of 10 plants calling in orders is

$$P(Y \leq 3) = \sum_{y=0}^{3} p(y) = \sum_{y=0}^{3} \binom{10}{y}(0.2^y)(0.8^{10-y})$$

$$= 0.107 + 0.269 + 0.302 + 0.201 = 0.879$$

 c The probability of at least 3 out of 10 plants calling in orders is

$$P(Y \geq 3) = 1 - P(Y \leq 2) = 1 - \sum_{y=0}^{2} \binom{10}{y}(0.2^y)(0.8^{10-y})$$

$$= 1 - (0.107 + 0.269 + 0.302) = 0.322$$

Example 5.14

The guidance system for a rocket operates correctly with probability p when called upon. Independent but identical backup systems are installed in the rocket so that the probability that at least one system will operate correctly when called upon is no less than 0.99. Let n denote the number of guidance systems in the rocket. How large must n be to achieve the specified probability of at least one guidance system operating if

 a $p = 0.9$ **b** $p = 0.8$

Solution Let Y denote the number of correctly operating systems. If the systems are identical and independent, Y has a binomial distribution. Thus,

$$P(Y \geq 1) = 1 - P(Y = 0) = 1 - \binom{n}{0} p^0 (1 - p)^n = 1 - (1 - p)^n$$

The conditions specify that n must be such that $P(Y \geq 1) = 0.99$ or more.

a When $p = 0.9$, the condition

$$P(Y \geq 1) = 1 - (1 - 0.9)^n \geq 0.99$$

results in $(0.1)^n \leq 0.01$. So, $n = 2$. Thus, installing two systems will satisfy the specifications.

b When $p = 0.8$, the condition

$$P(Y \geq 1) = 1 - (1 - 0.8)^n \geq 0.99$$

results in $(0.2)^n \leq 0.01$. Now $0.2^2 = 0.04$ and $0.2^3 = 0.008$. Thus, we must have $n = 3$ systems so that $P(Y \geq 1) = 1 - (1 - 0.8)^n = 0.992 > 0.99$.

Example 5.15

One engineering firm enjoys 40% success rate in getting state government construction contracts. This month they have submitted bids on eight construction projects to be funded by the state government. The bids for different projects are assessed independently of each other.

a Find the probability that the firm will get none of those contracts?
b Find the probability that the firm will get five out of eight contracts?
c Find the probability that the firm will get all eight contracts?

Solution With each contract, there are two possible outcomes: The firm will get the contract or the firm will not get the contract. The probability that a firm will get the contract is $p = 0.4$. With $n = 8$ construction projects and $Y =$ the number contracts that the firm will get, we can use the binomial function to find the required probabilities.

a P(the firm will get none of those contracts) $= p(0) = \binom{8}{0} 0.4^0 \, 0.6^8 = 0.0168$.

There is 1.7% chance that the firm will get no contracts out of these eight projects. In other words, there is 98.3% chance that the firm will get at least one contract.

b P(the firm will get five out of eight contracts) $= p(5) = \binom{8}{0} 0.4^5 \, 0.6^3 = 0.1239$.

There is about a 12% chance that the firm will get five out of eight contracts.

c P(the firm will get all of those contracts) $= p(8) = \binom{8}{8} 0.4^8 \, 0.6^0 = 0.0007$.

There is very little chance (almost 0) that the firm will get all the contracts—not a very likely event.

Example 5.16

Virtually any process can be improved by use of statistics, including the law. A much publicized case that involved a debate about probability was the Collins case, which began in 1964. An incident of purse snatching in the Los Angeles area led to the arrest of Michael and Janet Collins. At their trial, an "expert" presented the probabilities listed in table below on characteristics possessed by a couple seen running from the crime scene:

Characteristic	Probability
Man with beard	1/10
Blond woman	1/4
Yellow car	1/10
Woman with ponytail	1/10
Man with moustache	1/3
Interracial couple	1/1,000

The chance that a couple had all these characteristics together is 1 in 12 million. Because the Collinses had these characteristics, the Collinses must be guilty. Do you see any problem with this line of reasoning?

Solution   First, no background data are given to support the probabilities used. Second, the six events are not independent of one another, and therefore the probabilities cannot be multiplied. Third and most interesting, the wrong question is being addressed. The question of interest is not "What is the probability of finding a couple with these characteristics?" Because one such couple has been found (the Collinses), the question is "What is the probability that another such couple exists, given that we have found one?"

Here we can use binomial distribution. In the binomial model, let

n = number of couples who could have committed the crime
p = probability that any one couple possesses the six listed characteristics
x = number of couples that posses the six characteristics

From the binomial distribution

$$P(X = 0) = (1 - p)^n$$

$$P(X = 1) = np(1 - p)^{n-1}$$

$$P(X \geq 1) = 1 - (1 - p)^{n-1}$$

Then the answer to the conditional question posed above is

$$P(X > 1 \mid X \geq 1) = \frac{1 - (1 - p)^n - np(1 - p)^{n-1}}{1 - (1 - p)^n}$$

Using $p = 1/12,000,000$ and $n = 12,000,000$, which are plausible but not well-justified guesses,

$$P(X > 1 \mid X \geq 1) = 0.42$$

So the probability of seeing another such couple, given that we've already seen one, is much larger than the probability of seeing such a couple in the first place. This holds true even if the numbers are dramatically changed. For instance, if n is reduced to 1 million, the conditional probability is 0.05, which is still much larger than $1/12,000,000$.

Exercises

5.21 Let X denote a random variable having a binomial distribution with $p = 0.2$ and $n = 4$. Find the following.
  a $P(X = 2)$ **b** $P(X \geq 2)$
  c $P(X \leq 2)$ **d** $E(X)$
  e $V(X)$

5.22 Let X denote a random variable having a binomial distribution with $p = 0.4$ and $n = 20$. Use Table 2 of the Appendix to evaluate the following.
  a $P(X \leq 6)$ **b** $P(X \geq 12)$
  c $P(X = 8)$

5.23 A machine that fills boxes of cereal underfills a certain proportion p. If 25 boxes are randomly selected from the output of this machine, find the probability that no more than two are underfilled when
  a $p = 0.1$ **b** $p = 0.2$

5.24 In testing the lethal concentration of a chemical found in polluted water, it is found that a certain concentration will kill 20% of the fish that are subjected to it for 24 hours. If 20 fish are placed in a tank containing this concentration of chemical, find the probability that after 24 hours.
  a Exactly 14 survive.   **b** At least 10 survive.
  c At most 16 survive.

5.25 Refer to Exercise 5.24
  a Find the number expected to survive out of 20
  b Find the variance of the number of survivors out of 20.

5.26 Among persons donating blood to a clinic, 80% are Rh$^+$ (that is, have the Rhesus factor present in their blood). Five people donate blood at the clinic on a particular day.
  a Find the probability that at least one of the five does not have the Rh factor.
  b Find the probability that at most four of the five have RH$^+$ blood.

5.27 Refer to Exercise 5.26. The clinic needs five Rh$^+$ donors on a certain day. How many people must donate blood to have the probability of at least five RH$^+$ donors over 0.90?

5.28 The *U.S. Statistical Abstract* reports that the median family income in the United States for 1989 was \$34,200. Among four randomly selected families, find the probability that

  a All four had incomes above \$34,200 in 1989.
  b One of the four had an income below \$34,200 in 1989.

5.29 According to an article by B.E. Sullivan in *Transportation Research* (18A, no. 2, 1984, p. 119), 55% of U.S. corporations say that one of the most important factors in locating a corporate headquarters is the "quality of life" for the employees. If five firms are contacted by the governor of Florida concerning possible relocation to that state, find the probability that at least three say the "quality of life" is an important factor in their decision. What assumptions have you made in arriving at this answer?

5.30 A. Goranson and J. Hall (*Aeronautical Journal,* November 1980, pp. 279–280) explain that the probability of detecting a crack in an airplane wing is the product of p_1, the probability of inspecting a plane with a wing crack; p_2, the probability of inspecting the detail in which the crack is located; and p_3, the probability of detecting the damage.
  a What assumptions justify the multiplication of these probabilities?
  b Suppose $p_1 = 0.9$, $p_2 = 0.8$, and $p_3 = 0.5$ for a certain fleet of planes. If three planes are inspected from this fleet, find the probability that a wing crack will be detected in at least one of them.

5.31 A missile protection system consists of n radar sets operating independently, each with probability 0.9 of detecting an aircraft entering a specified zone. (All radar sets cover the same zone.) If an airplane enters the zone, find the probability that it will be detected if
  a $n = 2$ **b** $n = 4$

5.32 Refer to Exercise 5.31. How large must n be if it is desired to have probability 0.99 of detecting an aircraft entering the zone?

5.33 A complex electronic system is built with a certain number of backup components in its subsystems. One subsystem has four identical components, each with probability 0.2 of failing in less than 1,000 hours. The subsystem will

operate if any two or more of the four components are operating. Assuming the components operate independently, find the probability that
a Exactly two of the four components last longer than 1,000 hours.
b The subsystem operates longer than 1,000 hours.

5.34 An oil exploration firm is to drill 10 wells, with each well having probability 0.1 of successfully producing oil. It costs the firm $10,000 to drill each well. A successful well will bring in oil worth $500,000.
a Find the firm's expected gain from the 10 wells.
b Find the standard deviation of the firm's gain.

5.35 A firm sells four items randomly selected from a large lot known to contain 10% defectives. Let Y denote the number of defectives among the four sold. The purchaser of the item will return the defective for repair, and the repair cost is given by

$$C = 3Y^2 + Y + 2$$

Find the expected repair cost.

5.36 From a large lot of new tires, n are to be sampled by a potential buyer, and the number of defectives X is to be observed. If at least one defective is observed in the sample of n tires, the entire lot is to be rejected by the potential buyer. Find n so that the probability of detecting at least one defective is approximately 0.90 if
a 10% of the lot is defective.
b 5% of the lot is defective.

5.37 Ten motors are packaged for sale in a certain warehouse. The motors sell for $100 each, but a "double-your-money-back" guarantee is in effect for any defectives the purchaser might receive. Find the expected net gain for the seller if the probability of any one motor being defective is 0.08. (Assume that the quality of any one motor is independent of the quality of the others.)

5.5 The Geometric and Negative Binomial Distributions

The geometric distribution is a special case of negative binomial distribution. Therefore, we'll study them together.

5.5.1 The geometric distribution

Suppose a series of test firings of a rocket engine can be represented by a sequence of independent Bernoulli random variables with

$$X_i = \begin{cases} 1 & \text{if the } i\text{th trial results in a successful firing} \\ 0 & \text{otherwise} \end{cases}$$

Assume that the probability of a successful firing (p) is constant for the trials. We are interested in $Y =$ the number of the trial on which the first successful firing occurs. The random variable Y follows a geometric distribution. The geometric distribution is defined by one parameter p. The probability distribution can be easily developed as follows:

Let Y denotes the number of the trial on which the first success occurs. The results of y successive trials are then as follows with $F =$ failure and $S =$ success:

$$\underbrace{F\,F\,F\,F\,\cdots\,F}_{(y-1)\,\text{trials}} \qquad \underbrace{S}_{\text{Last }(y\text{th})\,\text{trial}}$$

Therefore, the probability of observing failures on first $(y - 1)$ trials and success on the last $(y$th) trial (i.e., the probability of requiring y trials till first success is observed) is

$$\underbrace{(1 - p)(1 - p)(1 - p)(1 - p) \cdots (1 - p)}_{y - 1 \text{ trials}} \quad \underbrace{(p)}_{\text{Last } (y\text{th}) \text{ trial}} \quad = (1 - p)^{y-1} p$$

Success on Trial Number	Sequence	Probability
1	S	p
2	FS	$(1 - p)p$
3	FFS	$(1 - p)(1 - p)p$
4	$FFFS$	$(1 - p)(1 - p)(1 - p)p$
$\vdots$	$\vdots$	$\vdots$
y	$FF \ldots FS$	$(1 - p)(1 - p) \ldots (1 - p)p$

The Geometric Distribution

$$p(y) = p(1 - p)^{y-1} \qquad y = 1, 2, \ldots \qquad \text{for } 0 < p < 1$$

$$E(Y) = \frac{1}{p} \qquad \text{and} \qquad V(Y) = \frac{(1 - p)}{p^2}$$

Some examples of situations that result in a random variable whose probability can be modeled by a geometric distribution are as follows:

- The number of customers contacted before the first sale is made
- The number of years a dam is in service before it overflows
- The number of automobiles going through a radar check before the first speeder is detected

From the basic definition,

$$E(Y) = \sum_y yp(y) = \sum_{y=1}^{\infty} yp(1 - p)^{y-1} = p[1 + 2(1 - p) + 3(1 - p)^2 + \cdots]$$

This infinite series can be split into a triangular array of series as follows:

$$E(Y) = p[1 + (1 - p) + (1 - p)^2 + \cdots \\ + (1 - p) + (1 - p)^2 + \cdots \\ + (1 - p)^2 + \cdots \\ + \cdots]$$

Each line on the right side is an infinite, decreasing geometric progression with common ratio $(1 - p)$. Thus, the first line inside the bracket sums to $1/p$, the second to $(1 - p)/p$, the third to $(1 - p)^2/p$, and so on. On accumulating these totals and recalling that

$$a + ax + ax^2 + ax^3 + \cdots = \frac{a}{1 - x} \qquad \text{if } |x| < 1$$

we have

$$E(Y) = p\left[\frac{1}{p} + \frac{(1 - p)}{p} + \frac{(1 - p)^2}{p} + \cdots\right]$$

$$= 1 + (1 - p) + (1 - p)^2 + \cdots = \frac{1}{1 - (1 - p)} = \frac{1}{p}$$

This expected value should seem intuitively realistic. For example, if 10% of a certain lot of items is defective and if an inspector looks at randomly selected items one at a time, then he or she should expect to wait until the 10th trial to see the first defective. We can show that the variance of the geometric distribution is

$$V(Y) = \frac{(1 - p)}{p^2}$$

Example 5.17

A recruiting firm finds that 30% of the applicants for a certain industrial job have advanced training in computer programming. Applicants are selected at random from the pool and are interviewed sequentially.

a Find the probability that the first applicant having advanced training is found on the fifth interview.

b Suppose the first applicant with the advanced training is offered the position, and the applicant accepts. If each interview costs $300, find the expected value and variance of the total cost of interviewing incurred before the job is filled. Within what interval would this cost be expected to fall?

Solution Define Y as the number of the trial on which the first applicant having advanced training in programming is found. Then Y follows a geometric distribution with $p = 0.3$.

a The probability of finding a suitably trained applicant will remain constant from trial to trial if the pool of applicants is reasonably large. Thus,

$$P(Y = 5) = (0.7^4)(0.3) = 0.72$$

b The total cost of interviewing is $C = 300Y$. Now

$$E(C) = 300E(Y) = 300\left(\frac{1}{p}\right) = 300\left(\frac{1}{0.3}\right) = \$1{,}000$$

and

$$V(C) = (300^2)V(Y) = (300^2)\frac{(1 - p)}{p^2} = (300^2)\frac{(1 - 0.3)}{0.3^2} = 700{,}000$$

The standard deviation of the total cost is $\sqrt{700,000} = \$836.66$. Tchebysheff's theorem says that the total cost will lie within two standard deviations of its mean at least 75% of the time. Thus, it is quite likely that C will lie between

$$E(C) \pm 2\sqrt{V(C)}, \quad \text{i.e.,} \quad 1,000 \pm 2(836.66)$$

Because the lower bound is negative, that end of the interval is meaningless. However, we can still say that it is quite likely (at least 75% of the time) that the total cost of interviewing process before the position is filled up will be less than \$2,673.32.

5.5.2 The Negative binomial distribution

In Section 5.5.1 we saw that the geometric distribution models the probabilistic behavior of the number of the trial on which the *first success* occurs in a sequence of independent Bernoulli trials. But what if we were interested in the number of the trial for the second success, or the third success, or in general, the rth success? The distribution governing the probabilistic behavior in these cases is called the *negative binomial distribution*.

Let Y denote the number of the trial on which the rth success occurs in a sequence of independent Bernoulli trials with p denoting the common probability of "success." The negative binomial distribution is defined by two parameters, r and p.

> **The Negative Binomial Distribution**
>
> $$p(y) = \binom{y-1}{r-1} p^r (1-p)^{y-r} \quad y = r, r+1, \ldots \quad \text{for } 0 < p < 1$$
>
> $$E(Y) = \frac{r}{p} \quad \text{and} \quad V(Y) = \frac{r(1-p)}{p^2}$$

The expected value or the mean, and the variance for a negative binomially distributed random variable Y are found easily by analogy with the geometric distribution. Recall that Y denotes the number of the trial on which the first success occurs. Let

W_1 = the number of the trial on which the first success occurs

W_2 = the number of trials between the first success and the second success including the trial of the second success

W_3 = the number of trials between the second success and the third success including the trial of the third success and so forth. The results of the trials are then as follows with F = failure and S = success:

$$\underbrace{FF\cdots FS}_{w_1 \text{ trials}}\quad \underbrace{F\cdots FS}_{w_2 \text{ trials}}\quad \underbrace{F\cdots FS}_{w_3 \text{ trials}}\quad \cdots\quad \underbrace{F\cdots FS}_{w_r \text{ trials}}$$

It is easily observed that $Y = \sum_{i=1}^{r} W_i$, where W_i's are independent and each has a geometric distribution. By results to be discussed in Chapter 7,

$$E(Y) = \sum_{i=1}^{r} E(W_i) = \sum_{i=1}^{r} \frac{1}{p} = \frac{r}{p}$$

and

$$V(Y) = \sum_{i=1}^{r} V(W_i) = \sum_{i=1}^{r} \frac{(1-p)}{p^2} = \frac{r(1-p)}{p^2}$$

Example 5.18

Of a large stockpile of used pumps, 20% are unusable and need repair. A repairman is sent to the stockpile with three repair kits. He selects pumps at random and tests them one at a time. If a pump works, he goes on to the next one. If a pump doesn't work, he uses one of his repair kits to fix it. Suppose it takes 10 minutes to test a pump and 20 minutes to repair a pump that doesn't work. Find the expected value and variance of the total time it takes the repairman to use up his three kits.

Solution   Let Y be the trial on which the third defective pump is found. Then we see that Y has a negative binomial distribution with $p = 0.2$. The total time T taken to use up the three repair kits is

$$T = 10(Y - 3) + 3(30) = 10Y + 60$$

because each test takes 10 minutes, but the repair takes an extra 20 minutes, totaling 30 minutes for a pump that doesn't work. It follows that

$$E(T) = 10E(Y) + 60 = 10\left(\frac{3}{0.2}\right) + 60 = 210$$

and

$$V(T) = (10^2)V(Y) = (10^2)\left(\frac{3(0.8)}{0.2^2}\right) = 6,000$$

Thus, the total time to use up the kits has an expected value of 210 minutes and a standard deviation of $\sqrt{6,000} = 77.46$ minutes.

Example 5.19

As in Example 5.17, 30% of the applicants for a certain position have advanced training in computer programming. Suppose three jobs requiring advanced programming training are open. Find the probability that the third qualified applicant is found on the fifth interview, if the applicants are interviewed sequentially and at random.

Solution   Again, we assume independent trials, with the probability of finding a qualified candidate on any one trials being 0.3. Let Y denote the number of the trial on which the third qualified candidate is found. Then Y can reasonably be assumed to have a negative binomial distribution with $r = 3$ and $p = 0.3$, so

$$P(Y = 5) = p(5) = \binom{4}{2}(0.3^3)(0.7^2) = 6(0.3^3)(0.7^2) = 0.079$$

Exercises

5.38 Let Y denote a random variable having a geometric distribution, with probability of success on any trial denoted by p.
  a Find $P(Y \geq 2)$ if $p = 0.1$.
  b Find $P(Y > 4 \mid Y > 2)$ for general p. Compare the result with the unconditional probability $P(Y > 2)$.

5.39 Let Y denote a negative binomial random variable with $p = 0.4$. Find $P(Y \geq 4)$ if
  a $r = 2$ **b** $r = 4$

5.40 Suppose 10% of the engines manufactured on a certain assembly line are defective. If engines are randomly selected one at a time and tested, find the probability that the first nondefective engine is found on the second trial.

5.41 Refer to Exercise 5.40. Find the probability that the third nondefective engine is found
  a On the fifth trial.
  b On or before the fifth trial.

5.42 Refer to Exercise 5.40. Given that the first two engines are defective, find the probability that at least two more engines must be tested before the first nondefective is found.

5.43 Refer to Exercise 5.40. Find the mean and variance of the number of the trial on which
  a The first non-defective engine is found.
  b The third non-defective engine is found.

5.44 The employees of a firm that manufactures insulation are being tested for indications of asbestos in their lungs. The firm is requested to send three employees who have positive indications of asbestos to a medical center for further testing. If 40% of the employees have positive indications of asbestos in their lungs, find the probability that ten employees must be tested to find three positives.

5.45 Refer to Exercise 5.44. If each test costs $20, find the expected value and variance of the total cost of conducting the tests to locate three positives. Do you think it is highly likely that the cost of completing these tests would exceed $350?

5.46 If one-third of the persons donating blood at a clinic have O^+ blood, find the probability that
  a The first O^+ donor is the fourth donor of the day.
  b The second O^+ donor is the fourth donor of the day.

5.47 A geological study indicates that an exploratory oil well drilled in a certain region should strike oil with probability 0.2. Find the probability that
  a The first strike of oil comes on the third well drilled.
  b The third strike of oil comes on the fifth well drilled.
  What assumptions are necessary for your answers to be correct?

5.48 In the setting of Exercise 5.47, suppose a company wants to set up three producing wells. Find the expected value and variance of the number of wells that must be drilled to find three successful ones.

5.49   A large lot of tires contains 10% defectives. Four are to be chosen for placement on a car.
  a Find the probability that six tires must be selected from the lot to get four good ones.
  b Find the expected value and variance of the number of selections that must be made to get four good tires.

5.50   The telephone lines coming into an airline reservation office are all occupied about 60% of the time.
  a If you are calling this office, what is the probability that you complete your call on the first try? The second try? The third try?
  b If you and a friend both must complete separate calls to this reservation office, what is the probability that it takes a total of four tries for the two of you to complete your calls?

5.51   An appliance comes in two colors, white and brown, which are in equal demand. A certain dealer in these appliances has three of each color in stock, although this is not known to the customers. Customers arrive and independently order these appliances. Find the probability that
  a The third white is ordered by the fifth customer.
  b The third brown is ordered by the fifth customer.
  c All of the whites are ordered before any of the browns.
  d All of the whites are ordered before all of the browns.

5.6   The Poisson Distribution

The *Poisson distribution* occurs when we count the number of occurrences of an event over a given time period or length or area or volume. For example:

• The number of flaws in a square yard of fabric
• The number of bacterial colonies in a cubic centimeter of water
• The number of times a machine fails in the course of a workday

The Poisson Distribution

$$p(y) = \frac{\lambda^y e^{-\lambda}}{y!} \quad \text{for} \quad y = 0, 1, 2, \ldots$$

$$E(Y) = \lambda \quad \text{and} \quad V(Y) = \lambda$$

A number of probability distributions come about through the application of limiting arguments to other distributions. The *Poisson distribution* is one such distribution. It is used as the limit of the binomial distribution.

Consider the development of a probabilistic model for the number of accidents occurring at a particular highway intersection over a period of one week. We can think of the time interval as being split up into n subintervals such that

$P(\text{one accident in a subinterval}) = p$
$P(\text{no accidents in a subinterval}) = 1 - p.$

Note that we are assuming that the same value of p holds for all subintervals and that the probability of more than one accident in any one subinterval is zero. If the occurrence of accidents can be regarded as independent from one subinterval to another, then the total number of accidents in the time period (which equals the total number of subintervals containing one accident) will have a binomial distribution. There is no unique way to choose the subintervals and we therefore know neither n nor p. Thus, we want to look at the limit of the binomial probability distribution.

The Poisson distribution is defined by one parameter, namely, λ. The expected value of the binomial distribution is np, where the expected value of the Poisson distribution is λ. In order to keep the mean in the binomial case constant at the Poisson mean (i.e., $np = \lambda$), it is reasonable to assume that as n increases, p should decrease. With $np = \lambda$, that is, $p = \lambda/n$, we have

$$\lim_{n \to \infty} \binom{n}{y} p^y (1 - p)^{n-y} = \lim_{n \to \infty} \binom{n}{y} \left(\frac{\lambda}{n}\right)^y \left(1 - \frac{\lambda}{n}\right)^{n-y}$$

$$= \frac{\lambda^y}{y!} \lim_{n \to \infty} \left(1 - \frac{\lambda}{n}\right)^n \left(1 - \frac{\lambda}{n}\right)^{-y} \prod_{i=1}^{y-1} \left(1 - \frac{i}{n}\right)$$

$$= \frac{\lambda^y}{y!} e^{-\lambda}$$

We get this result using $\lim_{n \to \infty} \left(1 - \dfrac{\lambda}{n}\right)^n = e^{-\lambda}$

$$\lim_{n \to \infty} \left(1 - \frac{\lambda}{n}\right)^{-y} = 1 \quad \text{and}$$

$$\lim_{n \to \infty} \left(1 - \frac{i}{n}\right) = 1 \quad \text{for } i = 1, 2, \ldots, y - 1.$$

Figure 5.10 shows how the choices of λ affect the shape of a Poisson probability function. Note that the probability function is very asymmetric for $\lambda = 1$ (Figure 5.10a) but becomes fairly symmetric and mound-shaped as λ increases (Figures 5.10b and 5.10c).

We can intuitively determine what the mean and variance of a Poisson distribution should be by recalling the mean and variance of a binomial distribution and the relationship between the two distributions. A binomial distribution has mean np and variance $np(1 - p) = np - (np)p$. Now if n gets large and p remains at $np = \lambda$, the variance $np - (np)p = \lambda - \lambda p$ should tend toward λ. In fact, the Poisson distribution does have both mean and variance equal to λ.

The mean of the Poisson distribution is easily derived using the *Taylor series* expansion of e^x, namely,

$$e^x = 1 + x + \frac{x^2}{2!} + \frac{x^3}{3!} + \cdots$$

Then

$$E(Y) = \sum_y y p(y) = \sum_{y=0}^{\infty} y \frac{\lambda^y}{y!} e^{-\lambda} = \sum_{y=1}^{\infty} y \frac{\lambda^y}{y!} e^{-\lambda}$$

$$= \lambda e^{-\lambda} \sum_{y=1}^{\infty} \frac{\lambda^{(y-1)}}{(y - 1)!}$$

$$= \lambda e^{-\lambda} \left(1 + \lambda + \frac{\lambda^2}{2!} + \frac{\lambda^3}{3!} + \cdots\right)$$

$$= \lambda e^{-\lambda} (e^{\lambda}) = \lambda$$

Figure 5.10

The Poisson distributions

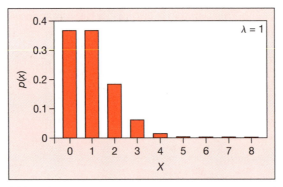

(a) $\lambda = 1$

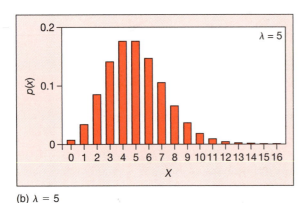

(b) $\lambda = 5$

(c) $\lambda = 8$

To find the variance, first find $E(Y^2) = E[Y(Y - 1)] + E(Y)$. Then derive the variance using similar arguments as in the expected value, and the relation $V(Y) = E(Y^2) - [E(Y)]^2$.

Example 5.20

For a certain Manufacturing industry, the number of industrial accidents averages three per week.

   a Find the probability that no accident will occur in a given week.
   b Find the probability that two accidents will occur in a given week.
   c Find the probability that at most four accidents will occur in a given week.
   d Find the probability that two accidents will occur in a given day.

Solution   **a** Using λ = mean number of accidents per week = 3, we get

$$P(\text{No accident in a given week}) = p(0) = \frac{3^0}{0!}e^{-3} = e^{-3} = 0.05$$

   b Using λ = mean number of accidents per week = 3, we get

$$P(\text{Two accidents in a given week}) = p(2) = \frac{3^2}{2!}e^{-3} = 0.224$$

c Using λ = mean number of accidents per week = 3, we get P(at most 4 accidents in a given week)

$$= p(0) + p(1) + p(2) + p(3) + p(4)$$

$$= \frac{3^0}{0!}e^{-3} + \frac{3^1}{1!}e^{-3} + \frac{3^2}{2!}e^{-3} + \frac{3^3}{3!}e^{-3} + \frac{3^4}{4!}e^{-3} = 0.815$$

d Now we are interested in the number of accidents on a given day. Thus, using λ = mean number of accidents per day = 3/7 = 0.2857, we get

$$P(\text{Two accidents on a given day}) = p(2) = \frac{0.2857^2}{2!}e^{-0.2857} = 0.031$$

Example 5.21

The manager of an industrial plant is planning to buy a new machine of either type A or B. For each day's operation, the number of repairs X that machine A requires is a Poisson random variable with mean $0.10t$, where t denotes the time (in hours) of daily operation. The number of daily repairs Y for machine B is a Poisson random variable with mean $0.12t$. The daily cost of operating A is $C_A(t) = 10t + 30X^2$; for B the cost is $C_B(t) = 8t + 30Y^2$. Assume that the repairs take negligible time and that the machines are to be cleaned each night so they operate like new machines at the start of each day. Which machine minimizes the expected daily cost if a day consists of (a) 10 hours? (b) 20 hours?

Solution   The expected cost for A and B are

$$E[C_A(t)] = 10t + 30E(X^2)$$
$$= 10t + 30[V(X) + (E(X))^2]$$
$$= 10t + 30[0.10t + 0.01t^2]$$
$$= 13t + 0.3t^2$$

$$E[C_B(t)] = 8t + 30E(Y^2)$$
$$= 8t + 30[0.12t + 0.0144t^2]$$
$$= 11.6t + 0.432t^2$$

a
$$E[C_A(10)] = 13(10) + 0.3(10)^2 = 160$$

and

$$E[C_B(10)] = 11.6(10) + 0.432(10)^2 = 159.2$$

which results in the choice of machine B.

b
$$E[C_A(20)] = 380$$

and

$$E[C_B(20)] = 404.8$$

which results in the choice of machine A.

In conclusion, B is more economical for short time periods because of its smaller hourly operating cost. However, for long time periods A is more economical because it needs to be repaired less frequently.

Exercises

5.52 Let Y denote a random variable having a Poisson distribution with mean $\lambda = 2$. Find the following.

 a $P(Y = 4)$ **b** $P(Y \geq 4)$
 c $P(Y < 4)$ **d** $P(Y \geq 4 | Y \geq 2)$

5.53 The number of telephone calls coming into the central switchboard of an office building averages four per minute.

 a Find the probability that no calls will arrive in a given 1-minute period.
 b Find the probability that at least two calls will arrive in a given 1-minute period.
 c Find the probability that at least two calls will arrive in a given 2-minute period.

5.54 The quality of computer disks is measured by sending the disks through a certifier that counts the number of missing pulses. A certain brand of computer disk has averaged 0.1 missing pulse per disk.

 a Find the probability that the next inspected disk will have no missing pulse.
 b Find the probability that the next inspected disk will have more than one missing pulse.
 c Find the probability that neither of the next two inspected disks will contain any missing pulse.

5.55 The National Maximum Speed Limit (NMSL) of 55 miles per hour has been in force in the United States since early 1974. The benefits of this law have been studied by D.B. Kamerud (*Transportation Research*, 17A, no. 1, 1983, pp. 51–64), who reports that the fatality rate for interstate highways with the NMSL in 1975 is approximately 16 per 10^9 vehicle miles.

 a Find the probability of at most 15 fatalities occurring in 10^9 vehicle miles.
 b Find the probability of at least 20 fatalities occurring in 10^9 vehicle miles.

(Assume that the number of fatalities per vehicle mile follows a Poisson distribution.)

5.56 In the article cited in Exercise 5.55, the projected fatality rate for 1975 if the NMSL had not been in effect was 25 per 10^9 vehicle miles. Under these conditions,

 a Find the probability of at most 15 fatalities occurring in 10^9 vehicle miles.
 b Find the probability of at least 20 fatalities occurring in 10^9 vehicle miles.
 c Compare the answers in parts (a) and (b) to those in Exercise 5.55.

5.57 In a time-sharing computer system, the number of teleport inquiries averages 0.2 per millisecond and follows a Poisson distribution.

 a Find the probability that no inquiries are made during the next millisecond.
 b Find the probability that no inquiries are made during the next 3 milliseconds.

5.58 Rebuilt ignition systems leave an aircraft rework facility at the rate of three per hour on the average. The assembly line needs four ignition systems in the next hour. What is the probability that they will be available?

5.59 Customer arrivals at a checkout counter in a department store have a Poisson distribution with an average of eight per hour. For a given hour, find the probability that

 a Exactly eight customers arrive.
 b No more than three customers arrive.
 c At least two customers arrive.

5.60 Refer to Exercise 5.59. If it takes approximately 10 minutes to service each customer, find the mean and variance of the total service time connected to the customer arrivals for 1 hour. (Assume that an unlimited number of servers are available, so no customer has to wait for service.) Is it highly likely that total service time would exceed 200 minutes?

5.61 Refer to Exercise 5.59. Find the probability that exactly two customers arrive in the 2-hour period of time

 a Between 2:00 PM and 4:00 PM (one continuous 2-hour period).
 b Between 1:00 PM and 2:00 PM and between 3:00 PM and 4:00 PM (two separate 1-hour periods for a total of 2 hours).

5.62 The number of imperfections in the weave of a certain textile has a Poisson distribution with a mean of four per square yard.

a Find the probability that a 1-square-yard sample will contain at least one imperfection.

b Find the probability that a 3-square-yard sample will contain at least one imperfection.

5.63 Refer to Exercise 5.62. The cost of repairing the imperfections in the weave is $10 per imperfection. Find the mean and standard deviation of the repair costs for an 8-square-yard bolt of the textile in question.

5.64 The number of bacteria colonies of a certain type in samples of polluted water has a Poisson distribution with a mean of two per cubic centimeter.

a If four 1-cubic-centimeter samples are independently selected from this water, find the probability that at least one sample will contain one or more bacteria colonies.

b How many 1-cubic-centimeter samples should be selected to have a probability of approximately 0.95 of seeing at least one bacteria colony?

5.65 Let Y have a Poisson distribution with mean λ. Find $E[Y(Y-1)]$ and use the result to show that $V(Y) = \lambda$.

5.66 A food manufacturer uses an extruder (a machine that produces bite-size foods such as cookies and many snack foods) that produces revenue for the firm at the rate of $200 per hour when in operation. However, the extruder breaks down an average of two times 10 hours of operation. If Y denotes the number of breakdowns during the time of operation, the revenue generated by the machine is given by

$$R = 200t - 50Y^2$$

where t denotes hours of operation. The extruder is shut down for routine maintenance on a regular schedule and operates like a new machine after this maintenance. Find the optimal maintenance interval t_0 to maximize the expected revenue between shutdowns.

5.67 The number of cars entering a parking lot is a random variable having a Poisson distribution with a mean of 4 per hour. The lot holds only 12 cars.

a Find the probability that the lot fills up in the first hour. (Assume all cars stay in the lot longer than 1 hour.)

b Find the probability that fewer than 12 cars arrive during an 8-hour day.

5.7 The Hypergeometric Distribution

The distributions already discussed in this chapter have as their basic building block a series of *independent* Bernoulli trials. The examples, such as sampling from large lots, depict situations in which the trials of the experiment generate, for all practical purpose, independent outcomes.

Suppose we have a relatively small lot consisting of N items, of which k are defective. If two items are sampled sequentially, then the outcome for the second draw is very much influenced by what happened on the first draw, provided that the first item drawn remains out of the lot. This is a situation of *dependent* trials.

In general, suppose a lot consists of N items, of which k are of one type (called successes) and $N - k$ are of another type (called failures). Suppose n items are sampled randomly and sequentially from the lot, with none of the sampled items being replaced, that is *sampling without replacement*. Let Y denote the total number of successes among the n sampled items (see Figure 5.11). Then the probability distribution of Y is described as the *hypergeometric distribution*. The hypergeometric distribution is defined by three parameters, n, k, and N. It can be easily developed using combinatorials from Chapter 4; for example, n items can be selected out of N without replacement in $\binom{N}{n}$ possible ways.

Figure 5.11

Hypergeometric sampling

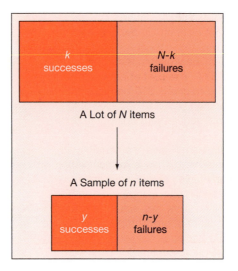

The Hypergeometric Distribution

$$p(y) = \frac{\binom{k}{y}\binom{N-k}{n-y}}{\binom{N}{n}} \quad y = 0, 1, \ldots, k, \text{ with } \binom{a}{b} = 0 \quad \text{if } a < b$$

$$E(Y) = n\frac{k}{N} \quad \text{and} \quad V(Y) = n\frac{k}{N}\left(1 - \frac{k}{N}\right)\left(\frac{N-n}{N-1}\right)$$

Experiments that result in a random variable possessing a hypergeometric distribution usually involve counting the number of "successes" in a random sample taken from a small lot. For example,

- Counting the number of males that show up on a committee of five randomly selected from among 20 employees
- Counting the number of Brand A alarm systems sold in three sales from a warehouse containing two Brand A systems and four Brand B systems.

The hypergeometric probability distribution arises from a situation quite similar to that of the binomial distribution, except that the trials are *dependent* in the hypergeometric distribution and *independent* in the binomial distribution. Observe that, because the probability of selecting a success on one draw is k/N, the mean of the hypergeometric distribution has the same form as the mean of the binomial distribution. Also, the variance of the hypergeometric distribution is like the variance of the binomial distribution, multiplied by $(N - n)/(N - 1)$, a correction factor for dependent samples arising from a finite population.

Example 5.22

A personnel director selects two employees for a certain job from a group of six employees, of which one is female and five are male. Find the probability that the female is selected for one of the jobs.

Solution If the selections are made at random and if Y denotes the number of females selected, the hypergeometric distribution would provide a good model for the behavior of Y. Here $N = 6$, $k = 1$, $n = 2$, and $y = 1$. Hence,

$$P(Y = 1) = p(1) = \frac{\binom{1}{1}\binom{5}{1}}{\binom{6}{2}} = \frac{1 \times 5}{15} = \frac{1}{3}$$

Using basic principles of probability and letting $X_1 = 1$ if the ith draw results in the female and $X_1 = 0$ otherwise, we can compute the same probability as follows:

$$P(Y = 1) = P(X_1 = 1, X_2 = 0) + P(X_1 = 0, X_2 = 1)$$

$$= P(X_1 = 1)\,P(X_2 = 0\,|\,X_1 = 1) + P(X_1 = 0)\,P(X_2 = 1\,|\,X_1 = 0)$$

$$= \left(\frac{1}{6}\right)\left(\frac{5}{5}\right) + \left(\frac{5}{6}\right)\left(\frac{1}{5}\right) = \frac{1}{3}$$

Example 5.23

In an assembly-line production of industrial robots, gearbox assemblies can be installed in 1 minute each if holes have been properly drilled in the boxes and in 10 minutes each if the holes must be redrilled. Twenty gearboxes are in stock, and it is assumed that two will have improperly drilled holes. Five gearboxes are to be randomly selected from the 20 available for installation in the next five robots in line.

   a Find the probability that all five boxes will fit properly.
   b Find the expected value, variance, and standard deviation of the time it takes to install these five gearboxes.

Solution In this problem, $N = 20$, and the number of nonconforming boxes is assumed to be $k = 2$, according to the manufacturer's standards. Let Y denote the number of nonconforming boxes (the number with improperly drilled holes) in the sample of five.

   a The probability that all five gearboxes will fit properly in the robots is

$$P(Y = 0) = \frac{\binom{2}{0}\binom{18}{5}}{\binom{20}{5}} = \frac{(1)\,(8{,}568)}{15{,}504} = 0.55$$

b The total time T taken to install the boxes (in minutes) is

$$T = 10Y + (5 - Y) = 9Y + 5$$

because each of Y nonconforming boxes takes 10 minutes to install, and the others take only one minute. To find $E(T)$ and $V(T)$, we first need $E(Y)$ and $V(Y)$.

$$E(Y) = n\left(\frac{k}{N}\right) = 5\left(\frac{2}{20}\right) = 0.5$$

and

$$V(Y) = n\left(\frac{k}{N}\right)\left(1 - \frac{k}{N}\right)\left(\frac{N - n}{N - 1}\right) = 5\left(\frac{2}{20}\right)\left(1 - \frac{2}{20}\right)\left(\frac{20 - 2}{20 - 1}\right)$$
$$= 0.355$$

It follows that

$$E(T) = 9E(Y) + 5 = 9(0.5) + 5 = 9.5$$

and

$$V(T) = 9^2 V(Y) = 81(0.355) = 28.755$$

Thus, installation time should average 9.5 minutes, with a standard deviation of $\sqrt{28.755} = 5.4$ minutes.

Exercises

5.68 From a box containing four white and three red balls, two balls are selected at random without replacement. Find the probability that
  a Exactly one white ball is selected.
  b At least one white ball is selected.
  c Two white balls are selected given that at least one white ball is selected.
  d The second ball drawn is white.

5.69 A warehouse contains 10 printing machines, 4 of which are defective. A company randomly selects five of the machines for purchase. What is the probability that all five of the machines are nondefective?

5.70 Refer to Exercise 5.69. The company purchasing the machines returns the defective ones for repair. If it costs $50 to repair each machine, find the mean and variance of the total repair cost. In what interval would you expect the repair costs

on these five machines to lie? [*Hint*: Use Tchebysheff's Theorem.]

5.71 A corporation has a pool of six firms, four of which are local, from which it can purchase certain supplies. If three firms are randomly selected without replacement, find the probability that
  a At least one selected firm is not local.
  b All three selected firms are local.

5.72 A foreman has 10 employees from whom he must select 4 to perform a certain undesirable task. Among the 10 employees, 3 belong to a minority ethnic group. The foreman selected all three minority employees (plus one other) to perform the undesirable task. The members of the minority group then protested to the union steward that they had been discriminated against by the foreman.

    The foreman claimed that the selection was completely at random. What do you think?

5.73 Specifications call for a type of thermistor to test out at between 9,000 and 10,000 ohms at 25°C. From ten thermistors available, three are to be selected for use. Let Y denote the number among the three that do not conform to specifications. Find the probability distribution for Y (tabular form) if

a The 10 contain two thermistors not conforming to specifications.

b The 10 contain four thermistors not conforming to specifications.

5.74 Used photocopying machines are returned to the supplier, cleaned, and then sent back out on lease agreements. Major repairs are not made, and, as a result, some customers receive malfunctioning machines. Among eight used photocopiers in supply today, three are malfunctioning. A customer wants to lease four of these machines immediately. Hence four machines are quickly selected and sent out, with no further checking. Find the probability that the customer receives

a No malfunctioning machines.

b At least one malfunctioning machine.

c Three malfunctioning machines.

5.75 An eight-cylinder automobile engine has two misfiring spark plugs. If all four plugs are removed from one side of the engine, what is the probability that the two misfiring ones are among them?

5.76 The "worst-case" requirements are defined in the design objectives for a brand of computer terminal. A quick preliminary test indicates that 4 out of a lot of 10 such terminals failed the "worst-case" requirements. Five of the 10 are randomly selected for further testing. Let Y denote the number, among the 5, that failed the preliminary test. Find the following.

a $P(Y \geq 1)$ **b** $P(Y \geq 3)$
c $P(Y \geq 4)$ **d** $P(Y \geq 5)$

5.77 An auditor checking the accounting practices of a firm samples three accounts from an accounts receivable list of eight accounts. Find the probability that the auditor sees at least one past-due account if there are

a Two such accounts among the eight.

b Four such accounts among the eight.

c Seven such accounts among the eight.

5.78 A group of six software packages available to solve a linear programming problem has been ranked from 1 to 6 (best to worst). An engineering firm selects two of these packages for purchase without looking at the ratings. Let Y denote the number of packages purchased by the firm that are ranked 3, 4, 5, or 6. Show the probability distribution for Y in tabular form.

5.79 Lot acceptance sampling procedures for an electronics manufacturing firm call for sampling n items from a lot of N items and accepting the lot if $Y \leq c$, where Y is the number of nonconforming items in the sample. From an incoming lot of 20 printer covers, 5 are to be sampled. Find the probability of accepting the lot if $c = 1$ and the actual number of nonconforming covers in the lot is

a 0   **b** 1   **c** 2   **d** 3   **e** 4

5.80 Given the setting and terminology of Exercise 5.79, answer parts (a) through (e) if $c = 2$.

5.81 Two assembly lines (I and II) have the same rate of defectives in their production of voltage regulators. Five regulators are sampled from each line and tested. Among the total of 10 tested regulators, there were 4 defectives. Find the probability that exactly two of the defectives came from line I.

5.8   The Moment-Generating Function

The moment-generating function (mgf) is a special function with many theoretical uses in probability theory. We denote it by $M(t)$.

Let Y be a random variable with probability distribution function $p(y)$. The **moment-generating function** of Y is the expected value of e^{ty}.

$$M(t) = E(e^{ty})$$

Moment-generating functions are useful in finding the expected values and determining the probability distributions of random variables. The expected values of powers of a random variable are often called *moments*. Thus, $E(Y)$ is the first moment and $E(Y^2)$ the second moment of Y. One use for the moment-generating function is that it does, in fact, generate moments of Y. When $M(t)$ exists, it is differentiable in a neighborhood of the origin $t = 0$, and derivatives may be taken inside the expectation. Thus,

$$M^{(1)}(t) = \frac{dM(t)}{dt} = \frac{d}{dt}E[e^{tY}] = E\left[\frac{d}{dt}e^{tY}\right] = E[Ye^{tY}]$$

Now if we set $t = 0$, we have

$$M^{(1)}(0) = E(Y)$$

Going on to the second derivative,

$$M^{(2)}(t) = E(Y^2 e^{tY})$$

which at $t = 0$ gives

$$M^{(2)}(0) = E(Y^2)$$

In general,

$$M^{(k)}(0) = E(Y^k)$$

It is often easier to evaluate $M(t)$ and its derivatives than to find the moments of the random variable directly.

Example 5.24

Evaluate the moment-generating function for the geometric distribution and use it to find the mean and variance of this distribution.

Solution We have, for the geometric random variable Y,

$$M(t) = E(e^{ty}) = \sum_{y=1}^{\infty} e^{ty} p(1-p)^{y-1}$$

$$= pe^t \sum_{y=1}^{\infty} (1-p)^{y-1} (e^t)^{y-1}$$

$$= pe^t \sum_{y=1}^{\infty} [(1-p)e^t]^{y-1}$$

$$= pe^t(1 + [(1-p)e^t] + [(1-p)e^t]^2 + \cdots$$

$$= pe^t\left[\frac{1}{1 - (1-p)e^t}\right]$$

because the series is geometric with common ratio e^t. To evaluate the mean, we have

$$M^{(1)}(t) = \frac{[1 - (1 - p)e^t]\, pe^t - pe^t\,[-(1 - p)e^t]}{[1 - (1 - p)\, e^t]^2} = \frac{pe^t}{[1 - (1 - p)e^t]^2}$$

and

$$M^{(1)}(0) = \frac{p}{[1 - (1 - p)]^2} = \frac{1}{p}$$

To evaluate the variance, we first need $E(Y^2) = M^{(2)}(0)$. Now

$$M^{(2)}(t) = \frac{[1 - (1 - p)e^t]^2\, pe^t - pe^t\{2[1 - (1 - p)e^t]\,(-1)(1 - p)e^t\}}{[1 - (1 - p)e^t]^4}$$

and

$$M^{(2)}(0) = \frac{p^3 + 2p^2(1 - p)}{p^4} = \frac{p + 2(1 - p)}{p^2}$$

Hence,

$$V(Y) = E(Y^2) - [E(Y)]^2 = \frac{p + 2(1 - p)}{p^2} - \frac{1}{p^2} = \frac{1 - p}{p^2}$$

Exercises

5.82 Find the moment-generating function for the Bernoulli random variable.

5.83 Show that the moment-generating function for the binomial random variable is given by

$$M(t) = [pe^t + (1 - p)]^n$$

Use this result to derive the mean and variance for the binomial distribution.

5.84 Show that the moment-generating function for the Poisson random variable with mean λ is given by

$$M(t) = e^{\lambda(e^t - 1)}$$

Use this result to derive the mean and variance for the Poisson distribution.

5.85 If X is a random variable with moment-generating function $M(t)$, and Y is a function of X given by $Y = aX + b$, show that the moment-generating function for Y is $e^{tb} M(at)$.

5.86 Use the result of Exercise 5.85 to show that

$$E(Y) = aE(X) + b \quad \text{and} \quad V(Y) = a^2 V(X)$$

5.9 Simulating Probability Distributions

Computers lend themselves nicely to use in the area of probability. They can be used not only to calculate probabilities but also to simulate random variables from specified probability distributions. A *simulation* can be thought of as an experiment performed on the computer. Simulation is used to analyze both theoretical and applied problems. A simulated model attempts to copy the behavior of the situation under consideration. Some practical applications of simulation might be to model inventory control problems, queuing systems, production lines, medical systems, and flight patterns of major jets. Simulation can also be used to determine the behavior of a complicated random variable whose precise probability distribution is difficult to evaluate mathematically.

Most mathematical software provide built-in random number generators for several probability distributions. Some calculators also provide random number generators for binomial and normal distributions.

Example 5.25

Suppose n_1, n_2, and n_3 items respectively are to be inspected from three production lines and the number of defectives are to be recorded. Let p_i be the probability of a defective item from line i ($i = 1, 2, 3$). What is the likelihood that the total number of defectives from three samples will be at least 5, for the following sample sizes and proportion defectives?

	Line 1	Line 2	Line 3
n	10	10	5
p	0.05	0.1	0.1

Solution Let X_i = the number of defective items out of n_i inspected. Then X_i follows a binomial distribution with parameters n_i and p_i, respectively. Suppose $Y = X_1 + X_2 + X_3$, the total number of defectives in three samples. Because the proportion defectives are different for different production lines, the distribution of Y is not binomial. When we don't know the distribution of random variable, we can simulate the situation and estimate the required probability. Here we describe this simulation as follows:

- Generate 100 samples of required size from each production line and record the number of defectives observed. This will give 100 observations for each production line.
- Compute the total number of defectives for each triplet of samples. This will give 100 values for the total Y.
- Generate a distribution of the variable Y using a histogram or a dotplot.

One such simulation using Minitab resulted in the accompanying histogram (Figure 5.12) and summary statistics (Figure 5.13).

The mean number of defective items in three samples together is 1.86 with a standard deviation of 1.378. The median is 2.0, indicating at least 50% of the time, two or more defectives are observed. The histogram shows that 4 out of 100 samples resulted in a total of 5 or more defective items. Thus, we can estimate the probability of observing total 5 or more defectives as $4/100 = 0.04$.

Figure 5.12

Simulated distribution of sum of binomial variables

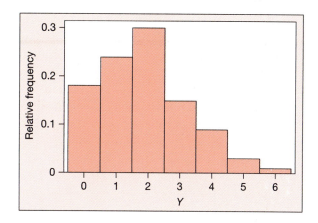

Figure 5.13

Summary statistics for simulation using Minitab

Variable	N	Mean	Median	TrMean	StDev	SE Mean
sum	100	1.860	2.000	1.789	1.378	0.138

Variable	Minimum	Maximum	Q1	Q3
sum	0.000	6.000	1.000	3.000

Example 5.26

Suppose six different types of coupons are randomly inserted in the cereal boxes by the manufacturer, of which only one type is specially marked. If a customer finds two specially marked coupons, then he/she can redeem them for a prize.

a What is the expected number of boxes the customer will have to buy in order to get 2 specially marked coupons?

b What is the probability of getting 2 specially marked coupons by opening less than 5 boxes?

Solution Let us assume that the manufacturer uses the same number of coupons of each type. Because the coupons are randomly inserted in the boxes, we can assume that the probability of finding a specially marked coupon in a given box is $p = 1/6$. We can simulate this experiment easily using a six-sided die.

- Designate any one side (say the side with number 1 on face) as the specially marked coupon.
- Toss the die successively and record whether the side with number 1 shows up or not. Stop when the side with number 1 shows up the second time.
- Record X = number to tosses to get number 1 twice (or number of tosses to get two specially marked coupons).
- Repeat this experiment 100 times. It results in 100 observations, each showing the number of tosses required to get two number 1s.
- Summarize the data using a histogram or a dotplot.

Results of one such simulation using Minitab are shown in Figure 5.14 and Figure 5.15.

Figure 5.14

Summary statistics for simulation using Minitab

Variable	N	Mean	Median	TrMean	StDev	SE Mean
x	100	10.460	10.000	10.144	5.858	0.586

Variable	Minimum	Maximum	Q1	Q3
x	2.000	31.000	5.250	14.000

Figure 5.15

Simulated negative binomial distribution

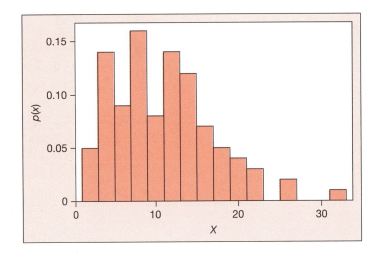

The distribution is mound-shaped and right-skewed.

a The expected number of boxes the customer will have to buy in order to get two specially marked coupons, in other words, the mean of the simulated distribution, is 10.46.

b The probability of getting two specially marked coupons by opening less than five boxes is

$$P(X = 2, 3, \text{ or } 4) = 19/100 = 0.19 \text{ approximately}$$

There is a 19% chance of getting two specially marked coupons by opening less than five boxes.

The theoretical distribution of X in this case is a negative binomial with $r = 2$ and $p = 1/6$.

Activity for Students Many situations occur in practice where we have questions that need to be answered with the help of probability theory. We can solve them using theory or get help from simulation technique. We can simulate the situation under consideration and study the possible outcomes to answer questions on hand. The following activity for students describes one such situation with a list of questions that must be answered.

Waiting for Blood   A local blood bank knows that about 40% of its donors have A^+ blood. Today, it needs $k = 3$ donors of A^+ blood type. How many donors do we expect to test in order to find three donors with A^+ blood type?

- Generate a distribution of values of X, the number of donors sequentially tested in order to find three A^+ donors.
- What are the approximate expected value and standard deviation of X from your generated data?
- Do these values agree with the theory?
- What is the estimated probability that the blood bank will need to test ten or more people to find the three A^+ donors?
- How will the answers to these questions change if k is increased to 4?

5.10   Summary

The outcomes of interest in most investigations involving random events are numerical in nature. The simplest type of numbered outcomes to model are counts, such as the number of nonconforming parts in a shipment, the number of sunny days in a month, or the number of water samples that contain a pollutant. One of the amazing results of probability theory is the fact that a small number of theoretical distributions can cover a wide array of applications. Six of the most useful discrete probability distributions were introduced in this chapter.

- The *Bernoulli* is simply an indicator random variable, using a numerical code to indicate the presence or absence of a characteristic.
- The *binomial* random variable counts the number of "successes" among a fixed number, n, of independent trials, each with the same probability of success.
- The *geometric* random variable counts the number of trials one needs to conduct, in sequence, until the first "success" is seen.
- The *negative binomial* random variable counts the number of trials one needs to conduct until the rth "success" is seen.
- The *Poisson* random variable arises from counts in a restricted domain of time, area, or volume and is most useful for counting fairly rare outcomes.
- The *hypergeometric* random variable counts the number of "successes" in sampling from a finite population, which makes the sequential selections dependent upon one another.

These theoretical distributions serve as models for real data that might arise in our quest to improve a process. Each involves assumptions that should be checked carefully before application is made to a particular case.

Supplementary Exercises

5.87 Construct probability histograms for the binomial probability distribution for $n = 5$ and $p = 0.1, 0.5$, and 0.9. (Table 2 of the Appendix will reduce the amount of calculation.) Note the symmetry for $p = 0.5$ and the direction of skewness for $p = 0.1$ and 0.9.

5.88 Use Table 2 of the Appendix to construct a probability histogram for the binomial probability distribution for $n = 20$ and $p = 0.5$. Note that almost all the probability falls in the interval $5 \leq y \leq 15$.

5.89 The probability that a single radar set will detect an airplane is 0.9. If we have five radar sets, what is the probability that exactly four sets will detect the plane? At least one set? (Assume that the sets operate independently of each other.)

5.90 Suppose that the four engines of a commercial aircraft were arranged to operate independently and that the probability of in-flight failure of a single engine is 0.01. What is the probability that, on a given flight,
  a No failures are observed?
  b No more than one failure is observed?

5.91 Sampling for defectives from large lots of a manufactured product yields a number of defectives Y that follows a binomial probability distribution. A sampling plan involves specifying the number of items to be included in a sample n and an acceptance number a. The lot is accepted if $Y \leq a$ and rejected if $Y > a$. Let p denote the proportion of defectives in the lot. For $n = 5$ and $a = 0$, calculate the probability of lot acceptance if
  a $p = 0$ **b** $p = 0.1$ **c** $p = 0.3$
  d $p = 0.5$ **e** $p = 1.0$
  A graph showing the probability of lot acceptance as a function of lot fraction defective is called the *operating characteristic curve* for the sample plan. Construct this curve for the plan $n = 5$, $a = 0$. Note that a sampling plan is an example of statistical inference. Accepting or rejecting a lot based on information contained in the sample is equivalent to concluding that the lot is either "good" or "bad", respectively. "Good" implies that a low fraction is defective and that the lot is therefore suitable for shipment.

5.92 Refer to Exercise 5.91. Use Table 2 of the Appendix to construct the operating characteristic curve for a sampling plan with
  a $n = 10$, $a = 0$ **b** $n = 10$, $a = 1$
  c $n = 10$, $a = 2$
  For each plan, calculate P(lot acceptance) for $p = 0, 0.05, 0.1, 0.3, 0.5$, and 1.0. Our intuition

suggests that sampling plan (a) would be much less likely to lead to acceptance of bad lots than plans (b) and (c). A visual comparison of the operating characteristic curves will confirm this supposition.

5.93 A quality-control engineer wants to study the alternative sampling plans $n = 5$, $a = 1$ and $n = 25$, $a = 5$. On a sheet of graph paper, construct the operating characteristic curves for both plans; use acceptance probabilities at $p = 0.05$, $p = 0.10$, $p = 0.20$, $p = 0.30$, and $p = 0.40$ in each case.
  a If you were a seller producing lots with fraction defective ranging from $p = 0$ to $p = 0.10$, which of the two sampling plans would you prefer?
  b If you were a buyer wishing to be protected against accepting lots with fraction defective exceeding $p = 0.30$, which of the two sampling plans would you prefer?

5.94 For a certain section of a pine forest, the number of diseased trees per acre Y has a Poisson distribution with mean $\lambda = 10$. The diseased trees are sprayed with an insecticide at a cost of \$3 per tree, plus a fixed overhead cost for equipment rental of \$50. Letting C denote the total spraying cost for a randomly selected acre, find the expected value and standard deviation for C. Within what interval would you expect C to lie with probability at least 0.75?

5.95 In checking river water samples for bacteria, water is placed in a culture medium so that certain bacteria colonies can grow if those bacteria are present. The number of colonies per dish averages 12 for water samples from a certain river.
  a Find the probability that the next dish observed will have at least 10 colonies.
  b Find the mean and standard deviation of the number of colonies per dish.
  c Without calculating exact Poisson probabilities, find an interval in which at least 75% of the colony count measurements should lie.

5.96 The number of vehicles passing a specified point on a highway averages 10 per minute.
  a Find the probability that at least 15 vehicles pass this point in the next minute.
  b Find the probability that at least 15 vehicles pass this point in the next 2 minutes.

5.97 A production line produces a variable number N of times each day. Suppose each item produced has the same probability p of not conforming to manufacturing standards. If N has a Poisson distribution with mean λ, then the number of nonconforming items in one day's production Y has a Poisson distribution with mean λp. The average number of resistors produced by a facility in one day has a Poisson distribution with mean 100. Typically, 5% of the resistors produced do not meet specifications.

a Find the expected number of resistors not meeting specifications on a given day.

b Find the probability that all resistors will meet the specifications on a given day.

c Find the probability that more than five resistors fail to meet specifications on a given day.

5.98 A certain type of bacteria cell divides at a constant rate λ over time. (That is, the probability that a cell will divide in a small interval of time t is approximately λt.) Given that a population starts out at time zero with k cells of this type and that cell divisions are independent of one another, the size of the population at time t, $Y(t)$, has the probability distribution

$$P[Y(t) = n] = \binom{N-1}{k-1} e^{-\lambda kt} (1 - e^{-\lambda t})^{n-k}$$

a Find the expected value of $Y(t)$ in terms of λ and t.

b If, for a certain type of bacteria cell, $\lambda = 0.1$ per second and the population starts out with two cells at time zero, find the expected population size after 5 seconds.

5.99 The probability that any one vehicle will turn left at a particular intersection is 0.2. The left-turn lane at this intersection has room for three vehicles. If five vehicles arrive at this intersection while the light is red, find the probability that the left-turn lane will hold all the vehicles that want to turn left.

5.100 Refer to Exercise 5.99. Find the probability that six cars must arrive at the intersection while the light is red to fill up the left-turn lane.

5.101 For any probability function $p(y)$, $\sum y p(y) = 1$ if the sum is taken over all possible values y that the random variable in question can assume. Show that this is true for the following:

a The binomial distribution

b The geometric distribution

c The Poisson distribution

5.102 The supply office for a large construction firm has three welding units of Brand A in stock. If a welding unit is requested, the probability is 0.7 that the request will be for this particular brand. On a typical day, five requests for welding units come to the office. Find the probability that all three Brand A units will be in use on that day.

5.103 Refer to Exercise 5.102. If the supply office also stocks three welding units that are not Brand A, find the probability that exactly one of these will be left immediately after the third Brand A unit is requested.

5.104 The probability of a customer arrival at a grocery service counter in any 1-second interval is equal to 0.1. Assume that customers arrive in a random stream and hence that the arrival at any one second is independent of any other.

a Find the probability that the fist arrival will occur during the third 1-second interval.

b Find the probability that the first arrival will not occur until at least the third 1-second interval.

5.105 Sixty percent of a population of customers is reputed to prefer Brand A toothpaste. If a group of consumers is interviewed, what is the probability the exactly five people must be interviewed before a consumer is encountered who prefers Brand A? At least five people?

5.106 The mean number of automobiles entering a mountain tunnel per 2-minute period is one. An excessive number of cars entering the tunnel during a brief period of time produces a hazardous situation.

a Find the probability that the number of autos n entering the tunnel during a 2-minute period exceeds three.

b Assume that the tunnel is observed during ten 2-minute intervals, thus giving 10 independent observations, Y_1, Y_2, ..., Y_{10}, on a Poisson random variable. Find the probability that $Y > 3$ during at least one of the ten 2-minute intervals.

5.107 Suppose that 10% of a brand of microcomputers will fail before their guarantee has expired. If 1,000 computers are sold this month, find the expected value and variance of Y, the number that have not failed during the guarantee period. Within what limit would Y be expected to fall? [*Hint*: Use Tchebysheff's theorem.]

5.108 a Consider a binomial experiment for $n = 20$, $p = 0.05$. Use Table 2 of the Appendix to calculate the binomial probabilities for $Y = 0, 1, 2, 3, 4$.

b Calculate the same probabilities as in (a), using the Poisson approximation with $\lambda = np$. Compare your results.

5.109 The manufacturer of a low-calorie dairy drink wishes to compare the taste appeal of a new formula (B) with that of the standard formula (A). Each of four judges is given three glasses in random order, two containing formula A and the other containing formula B. Each judge is asked to state which glass he or she most enjoyed.

Suppose the two formulas are equally attractive. Let Y be the number of judges stating a preference for the new formula.

a Find the probability function for Y.

b What is the probability that at least three of the four judges state a preference for the new formula?

c Find the expected value of Y.

d Find the variance of Y.

5.110 Show that the hypergeometric probability function approaches the binomial in the limit as $N \to \infty$ and $p = r/N$ remains constant. That is, show that

$$\lim_{N \to \infty} \frac{\binom{r}{y}\binom{N-r}{n-y}}{\binom{N}{n}} = \binom{n}{y} p^y q^{n-y}$$

for $p = r/N$ constant and $q = 1 - p$.

5.111 A lot of $N = 100$ industrial products contains 40 defectives. Let Y be the number of defectives in a random sample of size 20. Find $p(10)$ by using

a The hypergeometric probability distribution.

b The binomial probability distribution.

c Is N large enough so that the binomial probability function is a good approximation to the hypergeometric probability function?

5.112 For simplicity, let us assume that there are two kinds of drivers. The safe drivers, which comprise 70% of the population, have a probability of 0.1 of causing an accident in a year. The rest of the population consists of accident makers, who have a probability of 0.5 of causing an accident in a year. The insurance premium is $400 times one's probability of causing an accident in the following year. A new subscriber has an accident during the first year. What should be this driver's insurance premium for the next year?

5.113 A merchant stocks a certain perishable item. She knows that on any given day she will have a demand for either two, three, or four of these items with probabilities 0.1, 0.4, and 0.5, respectively. She buys the items for $1.00 each and sells them for $1.20 each. Any that are left at the end of the day represent a total loss. How many items should the merchant stock to maximize her expected daily profit?

5.114 It is known that 5% of a population have disease A, which can be detected by a blood test. Suppose that N (a large number) people are to be tested. This can be done in two ways: Either (1) each person is tested separately or (2) the blood samples of k people are pooled together and analyzed. (Assume that $N = nk$, with n an integer.) In method 2, if the test is negative, all the people in the pool are healthy (that is, just this one test is needed). If the test is positive, each of the k persons must be tested separately (that is, a total of $k + 1$ tests is needed).

a For fixed k, what is the expected number of tests needed in method 2?

b Find the k that will minimize the expected number of tests in method 2.

c How many tests does part (b) save in comparison with part (a)?

5.115 Four possible winning numbers for a lottery—AB-4536, NH-7812, SQ-7855, and ZY-3221—are given to you. You will win a prize if 1 of your numbers matches one of the winning numbers. You are told that one first prize of $100,000, two second prizes of $50,000 each, and 10 third prizes of $1,000 each will be awarded. The only thing you need to do is mail the coupon back. No purchase is required. From the structure of the numbers you have received, it is obvious that the entire list consists of all the permutations of two letters followed by four digits. Is the coupon worth mailing back for 44¢ postage?

Continuous Probability Distributions

Patients are transported by the ambulances from the scene of accident to the hospital for treatment. In what percent of cases are the patients transported within 20 minutes? How often does it take more than 45 minutes to get patients to the hospital?

To answer such questions, we look at the distribution of transportation times by ambulances for patients (see Figure 6.1). Random variables that are not discrete, such as transportation time, lifetime, or weight, are classified as *continuous*. In the last chapter, we discussed discrete random variables and their distributions. In this chapter, we'll discuss continuous random variables and the probability distributions used to describe their behavior. Applications of continuous random variables to reliability and other aspects of quality improvement are also presented.

Figure 6.1

Distribution of transportation time

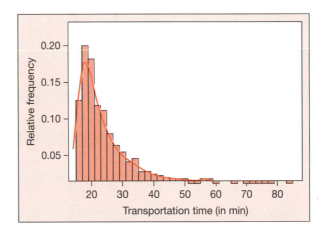

6.1 Continuous Random Variables and Their Probability Distributions

All the random variables discussed in Chapter 5 were discrete. Each could assume only specific values. But many random variables seen in practice have more than a countable collection of possible values. For example:

- Weights of patients coming into a clinic may be anywhere from, say, 80 to 300 pounds.
- Diameters of machined rods from a certain industrial process may be anywhere from 1.2 to 1.5 centimeters.
- Proportions of impurities in ore samples may run from 0.10 to 0.80.

These random variables can take on any value in an interval of real numbers. No one value can be ruled out as a possible observation in that interval. Because random variables of this type have a continuum of possible values, they are called *continuous random variables*.

> A random variable X is said to be **continuous** if it can take on the infinite number of possible values associated with intervals of real numbers.

Example 6.1

An experimenter is measuring the lifetime, X, of a battery. There are an infinite number of possible values that X can assume. We can assign positive probabilities to *intervals* of real numbers in a manner consistent with the axioms of probability. The experimenter measured the lifetimes of 50 batteries of a certain type, selected from a larger population of such batteries. The lifetimes (in hundreds of hours) are recorded in Table 6.1.

> ### Table 6.1
>
> **Lifetimes of Batteries (in hundreds of hours)**
>
> | | | | | |
> |---|---|---|---|---|
> | 0.406 | 0.685 | 4.778 | 1.725 | 8.223 |
> | 2.343 | 1.401 | 1.507 | 0.294 | 2.230 |
> | 0.538 | 0.234 | 4.025 | 3.323 | 2.920 |
> | 5.088 | 1.458 | 1.064 | 0.774 | 0.761 |
> | 5.587 | 0.517 | 3.246 | 2.330 | 1.064 |
> | 2.563 | 0.511 | 2.782 | 6.426 | 0.836 |
> | 0.023 | 0.225 | 1.514 | 3.214 | 3.810 |
> | 3.334 | 2.325 | 0.333 | 7.514 | 0.968 |
> | 3.491 | 2.921 | 1.624 | 0.334 | 4.490 |
> | 1.267 | 1.702 | 2.634 | 1.849 | 0.186 |

The relative frequency histogram for these data (Figure 6.2) shows clearly that, although the lifetimes ranged from 0 to 900 hours, most batteries lasted a low number of hours and very few lasted long hours. Here 32% of the 50 observations fall into the interval (0, 1) and another 22% fall into the interval (1, 2); in other words, about 54% batteries lasted at most 200 hours. This sample relative frequency histogram allows us to picture the sample behavior and gives us insight into a possible probabilistic model for the random variable X. The histogram of Figure 6.1 could be approximated closely by function.

$$f(x) = \frac{1}{2}e^{-x/2} \quad x > 0$$

We could take this function as a mathematical model for the behavior of the random variable X.

If we want to use a battery of this type in the future, we might want to know the probability that it will last longer than 400 hours. This probability can be approximated by the area under the curve to the right of the value 4, that is, by

$$\int_4^\infty \frac{1}{2}e^{-x/2}dx = 0.135.$$

The observed sample fraction of lifetimes that exceed 4 is $8/50 = 0.16$.

Figure 6.2

Relative frequency histogram of lifetimes of batteries

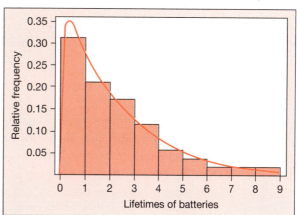

There are some questions for which the model would give more satisfactory answers than the sample fractions. For example, suppose we are interested in the probability that the lifetime exceeds 9. Then the model would suggest the answer

$$\int_{9}^{\infty} \frac{1}{2} e^{-x/2} dx = 0.011$$

whereas the sample shows no observations in excess of 9. Of course, the model suggests that we are likely to see about 1 in 100 lifetimes exceeding 9, and the experimenter tested only 50 batteries.

The function $f(x)$, which models the relative frequency behavior of X, is called the *probability density function*.

A function $f(x)$ of a random variable X is called the **probability density function** if

1. $f(x) \geq 0$ for all x

2. $\int_{-\infty}^{\infty} f(x)dx = 1$

3. $P(a \leq x \leq b) = \int_{a}^{b} f(x)dx$

Note that for a continuous random variable X,

$$P(X = a) = \int_{a}^{a} f(x)dx = 0$$

for any specific value of a. This does not rule out the possibility of occurrence of a specific lifetime; however, it means that the chance of observing a particular lifetime is extremely small.

Example 6.2

Refer to the random variable X of the lifetime example, which has associated with it a probability density function of the form

$$f(x) = \begin{cases} \dfrac{1}{2} e^{-x/2} & x > 0 \\ 0 & \text{elsewhere} \end{cases}$$

a Find the probability that the lifetime of a particular battery of this type is less than 200 hours.

b Find the probability that a battery of this type lasts more than 30 hours given that it has already been in use for more than 200 hours.

Solution **a** We are interested in determining the probability that $X < 2$.

$$P(X < 2) = \int_0^2 \frac{1}{2} e^{-x/2} dx = (1 - e^{-1}) = 1 - 0.368 = 0.767$$

b We are interested in $P(X > 3 \mid X > 2)$. Using the definition of conditional probability, and because $P([X > 3] \cap [X > 2]) = P(X > 3)$, we get

$$P(X > 3 \mid X > 2) = \frac{P(X > 3)}{P(X > 2)} = \frac{\int_0^3 \frac{1}{2} e^{-x/2} dx}{\int_0^2 \frac{1}{2} e^{-x/2} dx} = \frac{e^{-3/2}}{e^{-1}} = 0.606$$

Sometimes it is convenient to look at cumulative probabilities of the form $P(X \leq b)$, where b is some known constant. To do this, we use the distribution function.

If random variable X is continuous with the probability density function $f(x)$, then the **distribution function** X is

$$F(b) = P(X \leq b) = \int_{-\infty}^b f(x) \, dx$$

Note that $F'(x) = f(x)$ where $F'(x) = \dfrac{dF(x)}{dx}$

Example 6.3

In the lifetime example, X has a probability density function given by

$$f(x) = \begin{cases} \dfrac{1}{2} e^{-x/2} & x > 0 \\ 0 & \text{elsewhere} \end{cases}$$

Thus,

$$F(b) = P(X \leq b)$$

$$= \begin{cases} \int_0^b \frac{1}{2} e^{-x/2} dx = -e^{-x/2} \big|_0^b = 1 - e^{-b/2} & b > 0 \\ 0 & b \leq 0 \end{cases}$$

The function is shown graphically in Figure 6.3.

Figure 6.3

A distribution function for a continuous random variable

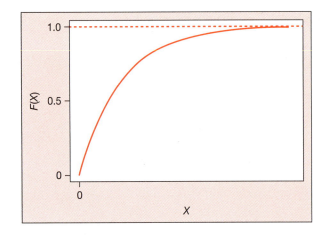

Example 6.4

A supplier of kerosene has a 200-gallon tank filled at the beginning of each week. His weekly demands show a relative frequency behavior that increases steadily up to 100 gallons and then levels off between 100 and 200 gallons. Let X denote weekly demand (in 100 gallons).

Then the relative frequencies for demand are modeled adequately by

$$f(x) = \begin{cases} 0 & x < 0 \\ x & 0 \le x \le 1 \\ 1/2 & 1 < x \le 2 \\ 0 & x > 2 \end{cases}$$

This function is graphed in Figure 6.4.

a Find $F(b)$ for this random variable.

b Use $F(b)$ to find the probability that demand will exceed 150 gallons on a given week.

Figure 6.4

$f(x)$ for Example 6.4

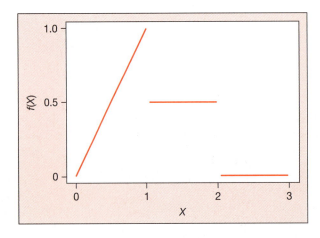

Solution   **a** From the definition,

$$F(b) = \int_{-\infty}^{b} f(x)dx,$$

In other words,

$$F(b) = \begin{cases} 0 & b < 0 \\ \int_{0}^{b} x\,dx = \dfrac{b^2}{2} & 0 \le b < 1 \\ \dfrac{1}{2} + \int_{1}^{b} \dfrac{1}{2}dx = \dfrac{1}{2} + \dfrac{b-1}{2} = \dfrac{b}{2} & 1 < b \le 2 \\ 1 & b > 2 \end{cases}$$

This function is graphed in Figure 6.5. Note that $F(b)$ is continuous over the whole real line even though $f(x)$ has two discontinuities.

b The probability that demand will exceed 150 gallons is given by

$$P(X > 1.5) = 1 - P(X \le 1.5) = 1 - F(1.5) = 1 - \frac{1.5}{2} = 0.25.$$

Figure 6.5

$F(b)$ for Example 6.5

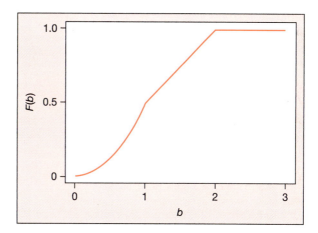

Exercises

6.1 For each of the following situations, define an appropriate random variable and state whether it is continuous or discrete.

   a An environmental engineer is looking at 10 field plots to determine whether they contain a certain type of insect.

   b A quality-control technician samples a continuously produced fabric in square-yard sections and counts the number of defects she observes for each sampled section.

   c A metallurgist counts the number of grains seen in a cross-sectional sample of aluminum.

   d The metallurgist of part (c) measures the area proportion covered by grains of a certain size rather than simply counting them.

6.2 Suppose a random variable X has a probability density function given by

$$f(x) = \begin{cases} kx(1-x) & 0 \le x \le 1 \\ 0 & \text{elsewhere} \end{cases}$$

a Find the value of k that makes this a probability density function.

b Find $P(0.4 \leq X \leq 1)$.

c Find $P(X \leq 0.4 \mid X \leq 0.8)$.

d Find $F(b) = P(X \leq b)$ and sketch the graph of this function.

6.3 The effectiveness of solar-energy heating units depends on the amount of radiation available from the sun. For a typical October, daily total solar radiation in Tampa, Florida, approximately follows the probability density function given below (units are hundreds of calories):

$$f(x) = \begin{cases} \dfrac{3}{32}(x-2)(6-x) & 2 \leq x \leq 6 \\ 0 & \text{elsewhere} \end{cases}$$

a Find the probability that solar radiation will exceed 300 calories on a typical October day.

b What amount of solar radiation is exceeded on exactly 50% of the October days, according to this model?

6.4 An accounting firm that does not have its own computing facilities rents time from a consulting company. The firm must plan its computing budget carefully and hence has studied the weekly use of CPU time quite thoroughly. The weekly use of CPU time approximately follows the probability density function given below (measurements in hours):

$$f(x) = \begin{cases} \dfrac{3}{64}x^2(4-x) & 0 \leq x \leq 4 \\ 0 & \text{elsewhere} \end{cases}$$

a Find the distribution function $F(x)$ for weekly CPU time X.

b Find the probability that CPU time used by the firm will exceed 2 hours for a selected week.

c The current budget of the firm covers only 3 hours of CPU time per week. How often will the budgeted figure be exceeded?

d How much CPU time should be budgeted per week if this figure is to be exceeded with probability only 0.10?

6.5 The pH, a measure of the acidity of water, is important in studies of acid rain. For a certain Florida lake, baseline measurements on acidity are made so any changes caused by acid rain can be noted. The pH of water samples from a lake is a random variable X with probability density function

$$f(x) = \begin{cases} \dfrac{3}{8}(7-x)^2 & 5 \leq x \leq 7 \\ 0 & \text{elsewhere} \end{cases}$$

a Sketch the curve of $f(x)$.

b Find the distribution function $F(x)$ for X.

c Find the probability that the pH will be less than 6 for a water sample from this lake.

d Find the probability that the pH of a water sample from this lake will be less than 5.5 given that it is known to be less than 6.

6.6 The "on" temperature of a thermostatically controlled switch for an air conditioning system is set at 60°F, but the actual temperature X at which the switch turns on is a random variable having probability density function

$$f(x) = \begin{cases} \dfrac{1}{2} & 59 \leq x \leq 61 \\ 0 & \text{elsewhere} \end{cases}$$

a Find the probability that it takes a temperature in excess of 60°F to turn the switch on.

b If two such switches are used independently, find the probability that both require a temperature in excess of 60°F to turn on.

6.7 The proportion of time, during a 40-hour workweek, that an industrial robot was in operation was measured for a large number of weeks, and the measurements can be modeled by the probability density function

$$f(x) = \begin{cases} 2x & 0 \leq x \leq 1 \\ 0 & \text{elsewhere} \end{cases}$$

If X denotes the proportion of time this robot will be in operation during a coming week, find the following:

a $P(X > 1/2)$

b $P(X > 1/2 \mid X > 1/4)$

c $P(X > 1/4 \mid X > 1/2)$

d Find $F(x)$ and graph this function. Is $F(x)$ continuous?

6.8 The proportion of impurities X in certain copper ore samples is a random variable having probability density function

$$f(x) = \begin{cases} 12x^2(1-x) & 0 \leq x \leq 1 \\ 0 & \text{elsewhere} \end{cases}$$

If four such samples are independently selected, find the probability that

a Exactly one has a proportion of impurities exceeding 0.5.

b At least one has a proportion of impurities exceeding 0.5.

6.2 Expected Values of Continuous Random Variables

As in the discrete case, we often want to summarize the information contained in a probability distribution by calculating expected values for the random variable and certain functions of random variable.

> For a continuous random variable X having probability density function $f(x)$,
>
> - The **expected value**, also known as the **mean**, of X is given by
>
> $$\mu = E(X) = \int_{-\infty}^{\infty} xf(x)dx$$
>
> - The **variance** of X is given by
>
> $$V(X) = E(X - \mu)^2 = \int_{-\infty}^{\infty} (x - \mu)^2 f(x)dx = E(X^2) - \mu^2$$

As seen in the previous chapter, if a random variable is shifted and scaled using a linear transformation, then the mean and the variance of the transformed variable are obtained as follows:

> For a continuous random variable and constants a and b,
>
> $$E(aX + b) = aE(X) + b \text{ and } V(aX + b) = a^2 V(X)$$

Example 6.5

For a lathe in a machine shop, let X denote the percentage of time out of a 40-hour workweek that the lathe is actually in use. Suppose X has a probability density function given by

$$f(x) = \begin{cases} 3x^2 & 0 \le x \le 1 \\ 0 & \text{elsewhere} \end{cases}$$

Find the mean and variance of X.

Solution   From the definition of expected value we have,

$$E(X) = \int_{-\infty}^{\infty} xf(x)dx = \int_0^1 x(3x^2)dx = \int_0^1 3x^3 dx = 3\left[\frac{x^4}{4}\right]_0^1 = \frac{3}{4} = 0.75$$

Thus, on the average, the lathe is in use 75% of the time.

To compute $V(X)$, we first find $E(X^2)$ as follows:

$$E(X^2) = \int_{-\infty}^{\infty} x^2 f(x)dx = \int_0^1 x^2(3x^2)dx = \int_0^1 3x^4 dx = 3\left[\frac{x^5}{5}\right]_0^1 = \frac{3}{5} = 0.60$$

Then

$$V(X) = E(X^2) - \mu^2 = 0.60 - (0.75^2) = 0.0375.$$

Example 6.6

The weekly demand X for kerosene at a certain supply station has a probability density function given by

$$f(x) = \begin{cases} x & 0 \le x \le 1 \\ 1/2 & 1 < x \le 2 \\ 0 & \text{elsewhere} \end{cases}$$

Find the expected weekly demand.

Solution   Using the definition to find $E(X)$, we must now carefully observe that $f(x)$ has different nonzero forms over two disjoint regions. Thus,

$$E(X) = \int_{-\infty}^{\infty} xf(x)dx = \int_0^1 x^2 dx + \int_1^2 (x/2)dx$$

$$= \left[\frac{x^3}{3}\right]_0^1 + \frac{1}{2}\left[\frac{x^2}{2}\right]_1^2 = \frac{1}{3} + \frac{1}{2}\left[2 - \frac{1}{2}\right] = \frac{13}{12} = 1.08$$

The expected weekly demand for kerosene is 108 gallons.

Tchebysheff's theorem, described in Chapter 5, holds for continuous random variables as well as for discrete random variables. Thus, if X is continuous with mean μ and standard deviation σ, then for any positive number k,

$$P(|X - \mu| < k\sigma) \ge 1 - \frac{1}{k^2}.$$

Example 6.7

The weekly amount Y spent for chemicals in a certain firm has a mean of \$445 and a variance of \$236. Within what interval would these weekly costs for chemicals be expected to lie at least 75% of the time?

Solution   To find an integral guaranteed to contain at least 75% of the probability mass for Y, we get $1 - (1/k^2) = 0.75$, which gives $k = 2$. Thus the interval $\mu - 2\sigma$ to $\mu + 2\sigma$ will contain at least 75% of the probability. This interval is given by

$$\left(445 - 2\sqrt{236},\ 445 + 2\sqrt{236}\right), \quad \text{i.e.,} \quad (414.28, 475.72)$$

Exercises

6.9 The temperature X at which a thermostatically controlled switch turns on has probability density function

$$f(x) = \begin{cases} \dfrac{1}{2} & 59 \le x \le 61 \\ 0 & \text{elsewhere} \end{cases}$$

Find $E(X)$ and $V(X)$.

6.10 The proportion of time X that an industrial robot is in operation during a 40-hour work week is a random variable with probability density function

$$f(x) = \begin{cases} 2x & 0 \le x \le 1 \\ 0 & \text{elsewhere} \end{cases}$$

a Find $E(X)$ and $V(X)$.
b For the robot under study, the profit Y for a week is given by

$$Y = 200X - 60$$

Find $E(Y)$ and $V(Y)$.
c Find an interval in which the profit should lie for at least 75% of the weeks that the robot is in use. [Hint: Use Tchebysheff's theorem.]

6.11 Daily total solar radiation for a certain location in Florida in October has probability density function

$$f(x) = \begin{cases} \dfrac{3}{32}(x-2)(6-x) & 2 \le x \le 6 \\ 0 & \text{elsewhere} \end{cases}$$

with measurements in hundreds of calories. Find the expected daily solar radiation for October.

6.12 Weekly CPU time used by an accounting firm has probability density function (measured in hours)

$$f(x) = \begin{cases} \dfrac{3}{64}x^2(4-x) & 0 \le x \le 4 \\ 0 & \text{elsewhere} \end{cases}$$

a Find the expected value and variance of weekly CPU time.
b The CPU time costs the firm \$200 per hour. Find the expected value and variance of the weekly cost for CPU time.
c Will the weekly cost exceed \$600 very often? Why or why not?

6.13 The pH of water samples from a specific lake is a random variable X with probability density function

$$f(x) = \begin{cases} \dfrac{3}{8}(7-x)^2 & 5 \le x \le 7 \\ 0 & \text{elsewhere} \end{cases}$$

a Find $E(X)$ and $V(X)$.
b Find an interval shorter than (5, 7) in which at least 3/4 of the pH measurements must lie.
c Would you expect to see a pH measurement blow 5.5 very often? Why?

6.14 A retail grocer has a daily demand X for a certain food sold by the pound, such that X (measured in hundreds of pounds) has probability density function

$$f(x) = \begin{cases} 3x^2 & 0 \le x \le 1 \\ 0 & \text{elsewhere} \end{cases}$$

(He cannot stock over 100 pounds.) The grocer wants to order $100k$ of food on a certain day. He buys the food at 6¢ per pound and sells it at 10¢ per pound. What value of k will maximize his expected daily profit?

6.3 The Uniform Distribution

We now move from a general discussion of continuous random variables to a discussion of specific models found in practice. Consider an experiment that consists of observing events occurring in a certain time frame, such as buses arriving at a bus stop or telephone calls coming into a switchboard. Suppose we know that one such event has occurred in the time interval (a, b), for example, a bus arrival between 8:00 and 8:10 AM. It may be of interest to place a probability distribution on the actual time of occurrence of the event under observation (say, X). A very simple model assumes that X is equally likely to lie in any small subinterval, say, of length d, no matter where that subinterval lies within (a, b). This assumption leads to the *uniform probability distribution*, the probability density function of which is given below and shown in Figure 6.6. One common application of this distribution is in random numbers. Note that random numbers have a uniform distribution on $(0, 1)$.

The Uniform Distribution

$$f(x) = \begin{cases} \dfrac{1}{b - a} & a \leq x \leq b \\ 0 & \text{elsewhere} \end{cases}$$

$$E(X) = \frac{a + b}{2} \quad \text{and} \quad V(X) = \frac{(b - a)^2}{12}$$

$$F(x) = \int_a^x \frac{1}{b - a}\, dx = \frac{x - a}{b - a} \quad a \leq x \leq b$$

Consider a subinterval $(c, c + d)$ contained entirely in (a, b). Then

$$P(c \leq x \leq c + d) = F(c + d) - F(c) = \frac{d}{b - a}$$

Note that this probability does not depend on the location of c, but only on the length d of the subinterval.

$$E(X) = \int_{-\infty}^{\infty} xf(x)dx = \int_a^b x\left(\frac{1}{b - a}\right)dx = \left(\frac{1}{b - a}\right)\left(\frac{b^2 - a^2}{2}\right) = \frac{a + b}{2}$$

It is intuitively reasonable that the mean value of a uniformly distributed random variable should lie at the midpoint of the interval. Now, to compute variance,

Figure 6.6

The uniform probability density function

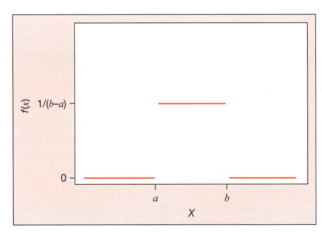

$$E(X^2) = \int_{-\infty}^{\infty} x^2(x)dx = \int_a^b x^2\left(\frac{1}{b-a}\right)dx = \left(\frac{1}{b-a}\right)\left(\frac{b^3-a^3}{3}\right)$$

$$= \frac{a^2+ab+b^2}{3}$$

gives

$$V(X) = E(X^2) - [E(X)]^2 = \frac{a^2+ab+b^2}{3} - \left(\frac{a+b}{2}\right)^2 = \frac{(b-a)^2}{12}$$

This result may not be intuitive, but we do see that the variance depends only on the length of the interval (a, b).

Example 6.8

The failure of a circuit board causes a computing system to shut down until a new board is delivered. The delivery time X is uniformly distributed over the interval of 1 to 5 days. The cost C of this failure and shutdown consists of a fixed cost, c_0, for the new part plus lost time and a cost (c_1) that increases proportional to X^2, so that

$$C = c_0 + c_1X^2$$

a Find the probability that the delivery time is two or more days.
b Find the expected cost of a single component failure.

Solution The delivery time X is uniformly distributed from 1 to 5 days, which gives

$$f(x) = \begin{cases} \dfrac{1}{4} & 1 \le x \le 5 \\ 0 & \text{elsewhere} \end{cases}$$

a The probability that the delivery time is two or more days (see Figure 6.7) is

$$P(X \ge 2) = \int_2^5 \left(\frac{1}{4}\right)dx = \left(\frac{1}{4}\right)(5-2) = \frac{3}{4}$$

Alternatively, we can use a calculator or some software to compute this probability. Use of the **cumulative probability** option from Minitab resulted in the following output (Figure 6.8).

Because we need $P(X \ge 2)$, subtract this number from 1.0, to get the answer.

$$P(X \ge 2) = 1 - P(X < 2) = 1 - 0.25 = 0.75$$

Figure 6.7

Probability of at least 2 days of delivery time

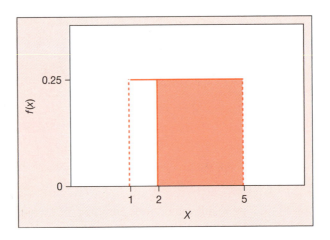

Figure 6.8

Minitab output for uniform CDF.

```
Cumulative Distribution Function

Continuous uniform on 1.00 to 5.00

        x           P(X <= x)
    2.0000            0.2500
```

b We know that $E(C) = c_0 + c_1E(X^2)$. Because $E(X^2) = V(X) + \mu^2$, we get

$$E(X^2) = \frac{(b-a)^2}{12} + \left(\frac{a+b}{2}\right)^2 = \frac{(5-1)^2}{12} + \left(\frac{1+5}{2}\right)^2 = \frac{31}{3}$$

Thus, $E(C) = c_0 + c_1E(X^2) = c_0 + c_1\left(\frac{31}{3}\right)$.

Exercises

6.15 Suppose X has a uniform distribution over the interval (a, b).
  a Find $F(x)$.
  b Find $P(X > c)$ for some point c between a and b.
  c If $a \le c \le d \le b$, find $P(X > d \mid X > c)$.

6.16 Upon studying low bids for shipping contracts, a microcomputer manufacturing firm finds that intrastate contracts have low bids that are uniformly distributed between 20 and 25, in units of thousands of dollars.
  a Find the probability that the low bid on the next intrastate shipping contract is below $22,000.

  b Find the probability that the low bid is in excess of $24,000.
  c Find the average cost of low bids on contracts of this type.

6.17 If a point is *randomly* located in an interval (a, b) and if X denotes its distance from a, then X will be assumed to have a uniform distribution over $(0, b - a)$.
  A plant efficiency expert randomly picks a spot along a 500-foot assembly line from which to observe work habits. Find the probability that she is
  a within 25 feet of the end of the line.
  b within 25 feet of the beginning of the line.
  c closer to the beginning than to the end of the line.

6.18 A bomb is to be dropped along a mile-long line that stretches across a practice target. The target center is at the midpoint of the line: The target will be destroyed if the bomb falls within a tenth of a mile on either side of the center. Find the probability that the target is destroyed if the bomb falls randomly along the line.

6.19 A telephone call arrived at a switchboard at a random time within a 1-minute interval. The switchboard was fully busy for 15 seconds into this 1-minute period. Find the probability that the call arrived when the switchboard was not fully occupied.

6.20 Beginning at 12:00 midnight, a computer center is up for 1 hour and down for 2 hours on a regular cycle. A person who doesn't know the schedule dials the center at a random time between 12:00 midnight and 5:00 AM. what is the probability that the center will be operating when he dials in?

6.21 The number of defective circuit boards among those coming out of a soldering machine follows a possion distribution. On a particular 8-hour workday, one defective board is found.
  a Find the probability that it was produced during the first hour of operation of that day.
  b Find the probability that it was produced during the last hour of operation for that day.
  c Given that no defective boards were seen during the first 4 hours of operation, find the probability that the defective board was produced during the fifth hour.

6.22 In determining the range of an acoustic source by triangulation, the time at which the spherical wave front arrives at a receiving sensor must be measured accurately. According to an article by J. Perruzzi and E. Hilliard (*Journal of the Acoustical Society of America*, 75(1), 1984, pp. 197–201), measurement errors in these times can be modeled as having uniform distributions. Suppose measurement errors are uniformly distributed from -0.05 to $+0.05$ microsecond.
  a Find the probability that a particular arrival-time measurement will be in error by less than 0.01 microsecond.
  b Find the mean and variance of these measurement errors.

6.23 In the setting of Exercise 6.22 suppose the measurement errors are uniformly distributed from -0.02 to $+0.05$ microsecond.

  a Find the probability that a particular arrival-time measurement will be in error by less than 0.01 microsecond.
  b Find the mean and variance of these measurement errors.

6.24 According to Y. Zimmels (*AIChE journal*, 29(4), 1983, pp. 669–676), the sizes of particles used in sedimentation experiments often have uniform distributions. It is important to study both the mean and variance of particle sizes, since in sedimentation with mixtures of various-size particles the larger particles hinder the movements of the smaller ones. Suppose spherical particles have diameters uniformly distributed between 0.01 and 0.05 centimeter. Find the mean and variance of the volumes of these particles. (Recall that the volume of a sphere is $\frac{4}{3}\pi r^3$.)

6.25 Arrivals of customers at a certain chekout counter follow a Poisson distribution. It is known that during a given 30-minute period one customer arrived at the counter. Find the probability that she arrived during the last 5 minutes of the 30-minute period.

6.26 A customer's arrival at a counter is uniformly distributed over a 30-minute period. Find the conditional probability that the customer arrived during the last 5 minutes of the 30-minute period given that there were no arrivals during the first 10 minutes of the period.

6.27 In tests of stopping distances for automobiles, those automobiles traveling at 30 miles per hour before the brakes are applied tend to travel distances that appear to be uniformly distributed between two points a and b. Find the probability that one of these automobiles
  a stops closer to a than to b.
  b stops so that the distance to a is more than three times the distance to b.

6.28 Suppose three automobiles are used in a test of the type discussed in Exercise 6.27. Find the probability that exactly one of the three travels past the midpoint between a and b.

6.29 The cycle times for trucks hauling concrete to a highway construction site are uniformly distributed over the interval 50 to 70 minutes.
  a Find the expected value and variance for these cycle times.
  b How many trucks would you expect to have to assign to this job so that a truckload of concrete can be dumped at the site every 15 minutes?

6.4 The Exponential Distribution

The lifetime data of Section 6.1 did not display a probabilistic behavior that was uniform but rather described one in which the probability over intervals of constant length decreased as the intervals moved farther and farther to the right. Such a behavior is described by the *exponential density*. The parameter θ of the exponential density is a constant that determines the rate at which the curve decreases. An exponential density function is given below. In Figure 6.9, an exponential density function with $\theta = 2$ is sketched.

The Exponential Distribution

$$f(x) = \begin{cases} \dfrac{1}{\theta}e^{-x/\theta} & x > 0 \\ 0 & \text{elsewhere} \end{cases}$$

$$E(X) = \theta \quad \text{and} \quad V(X) = \theta^2$$

$$F(x) = \begin{cases} 0 & x < 0 \\ 1 - e^{-x/\theta} & x \geq 0 \end{cases}$$

Figure 6.9

Exponential density function

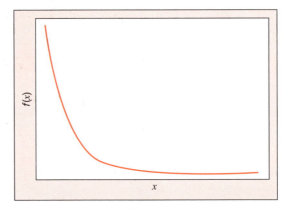

Many random variables occurring in engineering and the sciences can be approximately modeled as having exponential distributions. Figure 6.10 shows two examples of relative frequency distributions for times between arrivals (interarrival times) of vehicles at a fixed point on a one-directional roadway. Both of these relative frequency histograms can be modeled quite nicely by exponential functions. Note that the higher traffic density causes shorter interarrival times to be more frequent.

Finding expected value for the exponential function is simplified using the *gamma* (Γ) *function*. The function $\Gamma(\alpha)$ for $\alpha \geq 1$, is defined by

$$\Gamma(\alpha) = \int_0^\infty x^{\alpha-1}e^{-x}dx$$

Figure 6.10

Interarrival times of vehicles on a one-directional roadway

Source: D. Mahalel and A. S. Hakkert, *Transpotation Reasearch*, 17A, no. 4, 1983, p. 267.

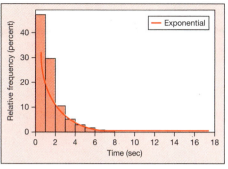

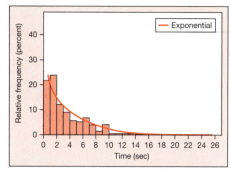

(a) Sample no. I (Vol 2,400 veh/h).

(b) Sample no. II (Vol 1,000 veh/h).

Integration by parts can be used to show that $\Gamma(\alpha + 1) = \alpha\Gamma(\alpha)$. It follows that $\Gamma(n) = (n - 1)!$ for any positive integer n. It is useful to note that $\Gamma(1/2) = 2\overline{\pi}$. The integral

$$\int_0^\infty x^{\alpha-1}e^{-x/\beta}dx \ \alpha, \beta > 0$$

can be evaluated by making the transformation $y = x/\beta$, or $x = \beta y$, $dx = \beta dy$. Then

$$\int_0^\infty (\beta y)^{\alpha-1}e^{-y}\beta dy = \beta^\alpha \int_0^\infty y^{\alpha-1}e^{-y}dy = \beta^\alpha\Gamma(\alpha)$$

Using this result, we see that for the exponential distribution,

$$E(X) = \int_0^\infty xf(x)dx = \frac{1}{\theta}\int_0^\infty xe^{-x/\theta} = \frac{1}{\theta}\Gamma(2)\theta^2 = \theta$$

Thus, the parameter θ is actually the mean of the distribution. To evaluate the variance,

$$E(X^2) = \int_{-\infty}^\infty x^2f(x)dx = \frac{1}{\theta}\int_0^\infty x^2e^{-x/\theta} = \frac{1}{\theta}\Gamma(3)\theta^3 = 2\theta^2$$

It follows that $V(X) = E(X^2) - \mu^2 = 2\theta^2 - \theta^2 = \theta^2$, and θ becomes the standard deviation as well as the mean.

Example 6.9

A sugar refinery has three processing plants, all receiving raw sugar in bulk. The amount of raw sugar (in tons) that one plant can process in one day can be modeled using an exponential distribution with mean of 4 tons for each of three plants. If each plant operates independently,

a Find the probability that any given plant processes more than 5 tons of raw sugar on a given day.

  b Find the probability that exactly two of the three plants process more than 5 tons of raw sugar on a given day.

  c How much raw sugar should be stocked for the plant each day so that the chance of running out of the raw sugar is only 0.05?

Solution   Let X = the amount of sugar processed in a day.

  a The probability that any given plant processes more than 5 tons is

$$P(X > 5) = \int_5^\infty f(x)dx = \int_5^\infty \frac{1}{4}e^{-x/4}dx = -e^{-x/4} \,|_5^\infty = e^{-5/4} = 0.2865$$

  Knowledge of the distribution function could allow us to evaluate this probability easily as

$$P(X > 5) = 1 - P(X \le 5) = 1 - F(5) = 1 - \left[1 - e^{-5/4}\right] = e^{-5/4}$$

  Alternatively we could use some software to compute this probability. The **cumulative probability** option from Minitab resulted in $P(X \le 5) = 0.7135$. Because we need $P(X \ge 5)$, subtract this number from 1.0 to get the answer. Thus,

$$P(X > 5) = 1 - P(X \le 5) = 1 - 0.7135 = 0.2865$$

  b Assuming that the three plants operate independently, the problem is to find the probability of two successes out of three tries, where 0.37 denotes the probability of success. This is a binomial problem, and the solution is

$$P(\text{exactly 2 use more than 5 tons}) = \binom{3}{2}0.2865^2(1 - 0.2865) = 0.1757$$

  c Let a = the amount of raw sugar to be stocked. The chance of running out of the raw sugar means there is a probability that more than the stocked amount is needed. We want to choose a so that

$$P(X > a) = \int_a^\infty \frac{1}{4}e^{-x/4}dx = e^{-a/4} = 0.05$$

  Solving this equation yields $a = 11.98$ tons.

   Alternatively, we can use some software to find the value of random variable. It is given that $P(X > a) = 0.05$. So, the cumulative probability at a, in other words, $P(X \le a)$, equals 0.95. Finding a such that $P(X > a) = 0.05$ is the same as finding a such that $P(X \le a) = 0.95$. Use the **inverse cumulative distribution function** option from Minitab to get this answer.

The Relationship between the Poisson and the Exponential Distributions Suppose events are occurring in time according to a Poisson distribution with a rate of λ events per hour. Thus, in t hours the number of events—say, Y—will have a Poisson distribution with mean value λt. Suppose we start at time zero and ask the question "How long do I have to wait to see the first event occur?" Let X denote the length of time until this first event. Then

$$P(X > t) = P[Y = 0 \text{ on the interval } (0, t)] = \frac{(\lambda t)^0 \, e^{-\lambda t}}{0!} = e^{-\lambda t}$$

and $P(X \leq t) = 1 - P(X > t) = 1 - e^{-\lambda t}$. We see that $P(X \leq t) = F(t)$, the distribution function of X, has the form of an exponential distribution function with $\lambda = (1/\theta)$. Upon differentiating, the probability density function of X is given by

$$f(t) = \frac{dF(t)}{dt} = \frac{d(1 - e^{-\lambda t})}{dt} = \lambda e^{-\lambda t} = \frac{1}{\theta} e^{-t/\theta} \quad t > 0$$

It shows that X has an exponential distribution.

   Actually, we need not start at time zero, for it can be shown that the waiting time from the occurrence of any event until the occurrence of the next event will have an exponential distribution for events occurring according to a Poisson distribution.

If

Y = number of occurrences of an event in time interval of length t
X = length of time between two successive occurrences of that event

then,

   • Y follows a Poisson distribution with parameter λ if and only if
   • X follows an exponential distribution with parameter θ.

where $\lambda = 1/\theta$.

Exercises

6.30 Suppose Y has an exponential density function with mean θ. Show that $P(Y > a + b \,|\, Y > a) = P(Y > b)$. This is referred to as the "memoryless" property of the exponential distribution.

6.31 The magnitudes of earthquakes recorded in a region of North America can be modeled by an exponential distribution with mean 2.4 as measured on the Richter scale. Find the probability that the next earthquake to strike this region will
  a Exceed 3.0 on the Richter scale
  b Fall between 2.0 and 3.0 on the Richter scale

6.32 Refer to Exercise 6.31. Of the next 10 earthquakes to strike this region, find the probability that at least one will exceed 5.0 on the Richter scale.

6.33 A pumping station operator observes that the demand for water at a certain hour of the day can be modeled as an exponential random variable with a mean of 100 cfs (cubic feet per second).
  a Find the probability that the demand will exceed 200 cfs on a randomly selected day

b What is the maximum water-producing capacity that the station should keep on line for this hour so that the demand will exceed this production capacity with a probability of only 0.01?

6.34 Suppose customers arrive at a certain checkout counter at the rate of two every minute.

a Find the mean and variance of the waiting time between successive customer arrivals.

b If a clerk takes 3 minutes to serve the first customer arriving at the counter, what is the probability that at least one more customer is waiting when the service of the first customer is completed?

6.35 The length of time X to complete a certain key task in house construction is exponentially distributed random variable with a mean of 10 hours. The cost C of completing this task is related to square of the time to completion by the formula

$$C = 100 + 40X + 3X^2$$

a Find the expected value and variance of C.

b Would you expect C to exceed 2,000 very often?

6.36 The inter-accident times (times between accidents) for all fatal accidents on scheduled American domestic passenger air flights, 1948–1961, were found to follow an exponential distribution with a mean of approximately 44 days (Pyke, *J. Royal Statistical Society* (B),27, 1968, p. 426).

a If one of those accidents occurred on July 1, find the probability that another one occurred in that same month.

b Find the variance of the inter accident times.

c What does the information given above suggest about the clumping of airline accidents?

6.37 The life lengths of automobile tires of a certain brand, under average driving conditions, are found to follow an exponential distribution with mean 30 (in thousands of miles). Find the probability that one of these tires bought today will last

a Over 30,000 miles

b Over 30,000 miles given that it already has gone 15,000 miles

l-up connections from remote terminals to a computing center at the rate of four te. The callers follow a Poisson distribu- call arrives at the beginning of a 1-minute nd the probability that a second call will in the next 20 seconds.

kdowns of an industrial robot follow n distribution with an average of

0.5 breakdowns per 8-hour workday. The robot is placed in service at the beginning of the day.

a Find the probability that it will not breakdown during the day.

b Find the probability that it will work for at least 4 hours without breaking down.

c Does what happened the day before have any effect on your answers above? Why?

6.40 One-hour carbon monoxide concentrations in air samples from a large city are found to have an exponential distribution with a mean of 3.6 ppm (Zamurs, *Air Pollution Control Association Journal, 34*(6), 1984, p. 637).

a Find the probability that a concentration will exceed 9 ppm.

b A traffic control strategy reduced the mean to 2.5 ppm. Now find the probability that a concentration will exceed 9 ppm.

6.41 The weekly rainfall totals for a section of the midwestern United States follow an exponential distribution with a mean of 1.6 inches.

a Find the probability that a weekly rainfall total in this section will exceed 2 inches.

b Find the probability that the weekly rainfall totals will not exceed 2 inches in either of the next two weeks.

6.42 The service times at teller windows in a bank were found to follow an exponential distribution with a mean of 3.2 minutes. A customer arrives at a window at 4:00 PM.

a Find the probability that he will still be there at 4:02 PM.

b Find the probability that he will still be there at 4:04 PM given that he was there at 4:02 PM.

6.43 In deciding how many customer service representatives to hire and in planning their schedules, it is important for a firm marketing electronic typewriters to study repair times for the machines. Such a study revealed that repair times have approximately an exponential distribution with a mean of 22 minutes.

a Find the probability that a repair time will last less than 10 minutes.

b The charge for typewriter repairs is $50 for each half hour or part thereof. What is the probability that a repair job will result in a charge of $100?

c In planning schedules, how much time should be allowed for each repair so that the chance of any one repair time exceeding this allowed time is only 0.10?

6.44 Explosive devices used in a mining operation cause nearly circular craters to form in a rocky-surface. The radii of these craters are exponentially distributed with a mean of 10 feet. Find the mean and variance of the area covered by such a crater.

6.5   The Gamma Distribution

Many sets of data, of course, will not have relative frequency curves with the smooth decreasing trend found in the exponential model. It is perhaps more common to see distributions that have low probabilities for intervals close to zero, with the probability increasing for a while as the interval moves to the right (in the positive direction), and then decreasing as the interval moves to the extreme positive side. In the case of electronic components, few will have short lifetimes, many will have something close to average, and very few will have extraordinarily long lifetimes.

For example, see Figure 6.11. It shows the distribution of transportation times for patients transported by emergency medical service vehicles from the scene of accident to the hospital in Alabama. It shows a mound-shaped distribution with a very long right tail and a short left tail. In other words, in cases of few patients, the transportation time was extraordinarily long. Often the transportation time is at most 30 min, but occasionally it goes beyond 30 min. Upon investigating, we found that many of these accidents were in rural Alabama where accident sites were some distance from the nearest hospital.

A class of functions that serve as a good model for this type of behavior is the *gamma* class. The gamma probability density function is given below.

The Gamma Distribution

$$f(x) = \begin{cases} \dfrac{1}{\Gamma(\alpha)\beta^{\alpha}}x^{\alpha-1}e^{-x/\beta} & x > 0 \\ 0 & \text{elsewhere} \end{cases}$$

$$E(X) = \alpha\beta \quad \text{and} \quad V(X) = \alpha\beta^2 \quad \alpha, \beta > 0$$

The parameters α and β determine the specific shape of the curve. Some typical gamma densities are shown in Figure 6.12. As described in the previous section, the function $\Gamma(\alpha)$ known as the gamma function is defined by

$$\Gamma(\alpha) = \int_0^\infty x^{\alpha-1}e^{-x}dx, \quad \alpha \geq 1$$

Some useful results:

- $\Gamma(\alpha + 1) = \alpha\Gamma(\alpha)$
- $\Gamma(n) = (n - 1)!$ for any positive integer n
- $\Gamma(1/2) = \sqrt{\pi}$

Figure 6.11

Transportation time (in min)

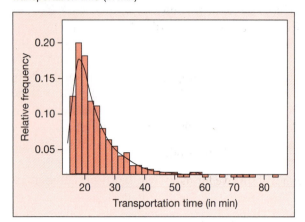

Figure 6.12

The gamma density function $\beta = 1$

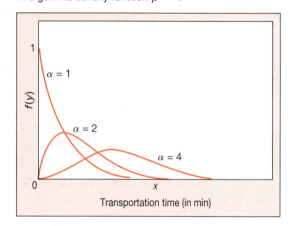

- For $\alpha, \beta > 0$, $\displaystyle\int_0^\infty x^{\alpha-1}e^{-x/\beta}dx = \beta^\alpha\Gamma(\alpha)$
- When $\alpha = 1$, the gamma density function reduces to the exponential.
- Suppose $X_1, X_2, \ldots, X_n$ represent independent, random gamma variables with parameters α and β. Then $Y = X_1 + X_2 + \cdots + X_n$ also has a gamma distribution with parameters $n\alpha$ and β. The derivation of this result is discussed in next chapter.

The derivation of the expected value and variance is very similar to the exponential case.

$$E(X) = \int_0^\infty x\frac{1}{\Gamma(\alpha)\beta^\alpha}x^{\alpha-1}e^{-x/\beta}dx = \frac{1}{\Gamma(\alpha)\beta^\alpha}\int_0^\infty x^\alpha e^{-x/\beta}dx$$

$$= \frac{1}{\Gamma(\alpha)\beta^\alpha}\Gamma(\alpha+1)\beta^{\alpha+1} = \alpha\beta$$

Similarly, we get $E(X^2) = \alpha(\alpha+1)\beta^2$, and hence,

$$V(X) = E(X^2) - [E(X)]^2 = \alpha(\alpha+1)\beta^2 - (\alpha\beta)^2 = \alpha\beta^2.$$

Example 6.10

A certain electronic system's primary component has lifetime X_1 that follows an exponential distribution with mean 400 hours. It is supported by an identical backup component with lifetime X_2. The backup component takes over immediately when the primary component fails, and the system fails when the back-up component fails. If the components operate independently, find

a The probability distribution for the lifetime of the system.
b Expected value for the lifetime of the system.
c The probability that the system will survive for more than 1000 hours.

Solution   Let $Y = X_1 + X_2$ be the lifetime of the system, where X_1 and X_2 are independent exponential random variables with $\beta = 400$.

a Remember that exponential is a gamma distribution with $\alpha = 1$. By the results stated above, Y will have a gamma distribution with $\alpha = 2$ and $\beta = 400$, i.e.

$$f_y(y) = \frac{1}{\Gamma(2)\,400^2}\, y e^{-y/400} \quad y > 0$$

b The expected lifetime of the system (i.e., the mean) is given by

$$E(Y) = \alpha\beta = 2(400) = 800.$$

c The probability that the system will survive for more than 1000 hours is equal to

$$P(Y > 1000) = \int_{1000}^{\infty} \frac{1}{\Gamma(2)\,400^2}\, y e^{-y/400}\, dy = 0.2873$$

which is intuitively reasonable.

Alternatively, we can use some software to compute this probability using gamma cumulative distribution function from Minitab $P(X \le 1{,}000) = 0.7127$. Because we need $P(X > 1{,}000)$, subtract this number from 1.0 to get the answer.

$$P(X > 1000) = 1 - P(X \le 1000) = 1 - 0.7127 = 0.2873$$

Example 6.11

Suppose that the length of time Y to conduct a periodic maintenance check (from previous experience) on office copiers follows a gamma-type distribution with $\alpha = 3$ and $\beta = 2$. Suppose a new repairman requires 20 minutes to check a machine. Does it appear that his time to perform a maintenance check disagrees with prior experiences?

Solution   The mean and variance for the length of maintenance times are

$$\mu = \alpha\beta = (3)(2) = 6, \quad \text{and} \quad \sigma^2 = \alpha\beta^2 = 3(2^2) = 12, \quad \text{i.e.,} \quad \sigma = \sqrt{12} = 3.46$$

The new repairman requires 20 min to check a machine. So the observed deviation from the mean is $(Y - \mu) = 20 - 6 = 14$ min and $y = 20$ min exceeds the mean $\mu = 6$ by $k = 14/3.46$ standard deviations. Then, from Tchebysheff's theorem,

$$P\big(|Y - 6| \ge 14\big) \le \frac{1}{k^2} = \left(\frac{3.46}{14}\right)^2 = 0.06$$

This probability is based on the assumption that the distribution of maintenance times has not changed from prior experience. Because $P(Y > 20\text{ min})$ is so small, we must conclude that our new maintenance man is somewhat slower than his predecessor.

Exercises

6.45 Four-week summer rainfall totals in a certain section of the midwestern United States have a relative frequency histogram that appears to fit closely to a gamma distribution with $\alpha = 1.6$ and $\beta = 2.0$.

   a Find the mean and variance of this distribution of 4-week rainfall totals.

   b Find an interval that will include the rainfall total for a selected 4-week period with probability at least 0.75.

6.46 Annual incomes for engineers in a certain industry have approximately a gamma distribution with $\alpha = 600$ and $\beta = 50$.

   a Find the mean and variance of these incomes.

   b Would you expect to find many engineers in this industry with an annual income exceeding $35,000?

6.47 The weekly downtime Y (in hours) for a certain industrial machine has approximately a gamma distribution with $\alpha = 3$ and $\beta = 2$. The loss, in dollars, to the industrial operation as a result of this downtime is given by

$$L = 30Y + 2Y^2$$

   a Find the expected value and variance of L.

   b Find an interval that will contain L on approximately 89% of the weeks that the machine is in use.

6.48 Customers arrive at a checkout counter according to a Poisson process with a rate of two per minute. Find the mean, variance, and probability density function of the waiting time between the opening of the counter and

   a The arrival of the second customer

   b The arrival of the third customer

6.49 Suppose two houses are to be built and each will involve the completion of a certain key task. The task has an exponentially distributed time to completion with a mean of 10 hours. Assuming the completion times are independent for the two houses, find the expected value and variance of

   a The total time to complete both tasks

   b The average time to complete the two tasks

6.50 The total sustained load on the concrete footing of a planned building is the sum of the deadload plus the occupancy load. Suppose the dead load X_1 has a gamma distribution with $\alpha_1 = 50$ and $\beta_1 = 2$, while the occupancy load X_2 has a gamma distribution with $\alpha_2 = 20$ and $\beta_2 = 2$. (Units are in kips, or thousands of pounds.)

   a Find the mean, variance, and probability density function of the total sustained load on the footing.

   b Find a value for the sustained load that should be exceeded only with probability less than $1/16$.

6.51 A 40-year history of maximum river flows for a certain small river in the United States shows a relative frequency histogram that can be modeled by a gamma density function with $\alpha = 1.6$ and $\beta = 150$. (Measurements are in cubic feet per second.)

   a Find the mean and standard deviation of the annual maximum river flows.

   b Within what interval will the maximum annual flow be contained with a probability of at least $8/9$?

6.52 The time intervals between dial-up connections to a computer center from remote terminals are exponentially distributed with a mean of 15 seconds. Find the mean, variance, and probability-distribution of the waiting time from the *opening* of the computer center until the fourth dial-up connection from a remote terminal.

6.53 If service times at a teller window of a bank are exponentially distributed with a mean of 3.2 minutes, find the probability distribution, mean, and variance of the time taken to serve three waiting customers.

6.54 The response times at an online terminal have approximately a gamma distribution with a mean of 4 seconds and a variance of 8. Write the probability density function for these response times.

6.6   The Normal Distribution

Perhaps the most widely used of all continuous probability distributions is the *normal distribution*. The normal probability density function is bell-shaped and centered at the mean value μ. Its spread is measured by the standard deviation σ. These two parameters, μ and σ, completely determine the location and spread of the normal density function (Figure 6.13). Its functional form is given below:

The Normal Distribution

$$f(x) = \frac{1}{\sigma\sqrt{2\pi}} e^{-(x-\mu)^2/2\sigma^2}$$

$$E(X) = \mu \quad \text{and} \quad V(X) = \sigma^2$$

Many naturally occurring measurements tend to have relative frequency distributions closely resembling the normal curve. For example, heights of adult American males tend to have a distribution that shows many measurements clumped around the mean height, with relatively few very short or very tall males in the population. Any time responses tend to be averages of independent quantities, the normal distribution tends to provide a reasonably good model for their relative frequency behavior. In contrast, lifetimes of biological organisms or electronic components tend to have relative frequency distributions that are neither normal nor close to normal.

Michelson (*Astronomical Papers*, 1881, p. 231) reported 100 measures of the speed of light. A histogram of measurements is given in Figure 6.14. The distribution is not perfectly symmetrical but still exhibits an approximately normal shape.

Some useful properties of the normal distribution:

- The mean μ determines the location of the distribution.
- The standard deviation σ determines the spread of the distribution. The normal distribution with larger σ is shorter and more spread (Figure 6.13).
- Any linear function of a normally distributed random variable is also normally distributed. That is, if X has a normal distribution with mean μ and variance σ^2 and $Y = aX + b$ for constants a and b, then Y is also normally distributed with mean $a\mu + b$ and variance $a^2\sigma^2$.

Figure 6.13

Normal density functions

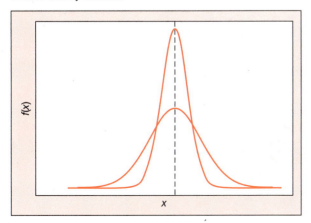

Figure 6.14

Michelson's measures of the speed of light in air

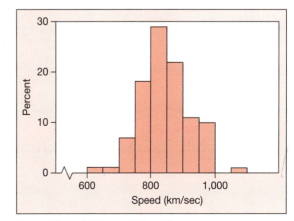

Source: Astronomical Papers, 1881, p. 231

Standard Normal Distribution

A normal distribution with $\mu = 0$ and $\sigma = 1$ is known as a standard normal distribution. The letter Z is used to indicate the standard normal variable.

$$f(z) = \frac{1}{\sqrt{2\pi}} e^{-z^2/2} \quad -\infty < z < \infty$$

$$E(Z) = 0 \quad \text{and} \quad V(Z) = 1$$

The last property leads to a very useful result. For any normally distributed random variable X, with parameters μ and σ^2, $Z = (X - \mu)/\sigma$ will have a standard normal distribution. Because any normally distributed random variable can be transformed to the standard normal, probabilities can be evaluated for any normal distribution simply by having a table of standard normal integrals available. Such a table, given in Table 4 of the Appendix, gives numerical values for the cumulative distribution function; that is,

$$P(Z \le z) = \int_0^z \frac{1}{\sqrt{2\pi}} e^{-x^2/2} \, dx$$

Values of this integral are listed in Table 4 of Appendix for z between -3.49 and 3.49.

Example 6.12

Find $P(Z \le 1.53)$ for a standard normal variable Z.

Solution   Remember that $P(Z \le 1.53)$ is equal to the shaded area in Figure 6.15.

Figure 6.15

A standard normal density function

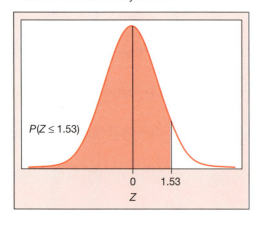

$P(Z \le 1.53)$

Figure 6.16

Using the z-table to find probability

z	.00	.01	.02	.03	.04
0.0	.5000	.5040	.5080	.5120	.5160
⋮	⋮	⋮	⋮	⋮	⋮
1.1	.8643	.8665	.8686	.8708	.8729
1.2	.8849	.8869	.8888	.8907	.8925
1.3	.9032	.9049	.9066	.9082	.9099
1.4	.9192	.9207	.9222	.9236	.9251
1.5	.9332	.9345	.9357	.9370	.9382
1.6	.9452	.9463	.9474	.9484	.9495
1.7	.9554	.9564	.9573	.9582	.9591
1.8	.9641	.9649	.9656	.9664	.9671
1.9	.9713	.9719	.9726	.9732	.9738

Now $P(Z \leq 1.53)$ is found in the cross-section of the row corresponding to 1.5 and the column corresponding to 0.03. Hence,

$$P(Z \leq 1.53) = 0.9370$$

Alternatively we can use a calculator or software to find this probability.

Example 6.13

If Z denotes a standard normal variable, find

a $P(Z \leq 1)$
b $P(Z < -1.5)$
c $P(Z > 1)$
d $P(-1.5 \leq Z \leq 0.5)$
e Find a value of Z, say, z_0, such that $P(Z \leq z_0) = 0.99$.

Solution   Use Table 4 in the Appendix or a calculator or a computer program to find these probabilities.

a $P(Z \leq 1) = 0.8413$, as shown by shaded area in Figure 6.17.
b $P(Z < -1.5) = 0.0668$, as shown by shaded area in Figure 6.18.
c $P(Z > 1) = 1.0 - P(Z \leq 1) = 1.0 - 0.8413 = 0.1587$, as shown by shaded area in Figure 6.19.
d $P(-1.5 \leq Z \leq 0.5) = P(Z \leq 0.5) - P(Z < -1.5) = 0.6915 - 0.0668 = 0.6247$, as shown by shaded area in Figure 6.20.
e To find the value of z_0 we must read the table inside out. We are given that the probability (or area under the curve) that Z is below some number z_0 is equal to 0.99 (see Figure 6.21), and we want to find the corresponding value of the

Figure 6.17
$P(Z \geq 1)$

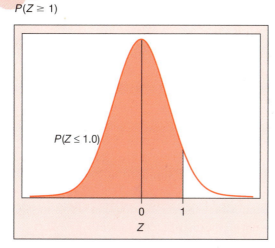

Figure 6.18
$P(Z < -1.5)$

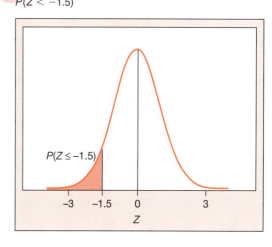

Figure 6.19
$P(Z > 1)$

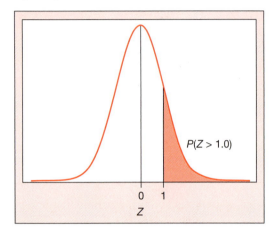

Figure 6.20
$P(-1.5 \leq z \leq 0.5)$

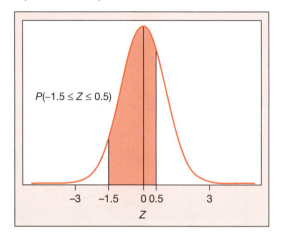

Figure 6.21
$P(Z \leq z_0) = 0.99$

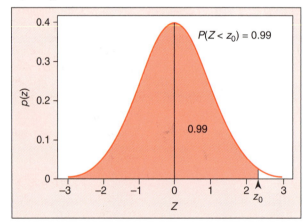

random variable Z. We must look for the given probability 0.99 on the area side of Appendix Table 4. The closest we can come is 0.9901, which corresponds to the z-value of 2.33. Hence, $z_0 = 2.33$.

Study of the table of normal curve areas for z-scores of 1, 2, and 3 shows how the percentages used in the empirical rule were determined. These percentages actually represent areas under the standard normal curve, as depicted in Figure 6.22.

The next example illustrates how the standardization works to allow Appendix Table 4 to be used for any normally distributed random variable.

Figure 6.22

Justification of the empirical rule

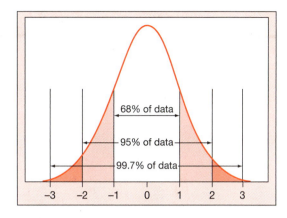

Example 6.14

A firm that manufactures and bottles apple juice has a machine that automatically fills 16-ounce bottles. There is, however, some variation in the amount of liquid dispensed (in ounces) into each bottle by the machine. Over a long period of time, the average amount dispensed into the bottles was 16 ounces, but there is a standard deviation of 1 ounce in these measurements. If the amount filled per bottle can be assumed to be normally distributed, find the probability that the machine will dispense more than 17 ounces of liquid into any one bottle.

Solution   Let X denote the ounces of liquid dispensed into one bottle by the filling machine. Then X is assumed to be normally distributed with mean 16 and standard deviation 1 (Figure 6.23). Hence,

$$P(X > 17) = P\left(\frac{X - \mu}{\sigma} > \frac{17 - \mu}{\sigma}\right)$$

$$= P\left(Z > \frac{17 - 16}{1}\right) = P(Z > 1) = 0.1587$$

The answer is found using Appendix Table 4, because $Z = (X - \mu)/\sigma$ has a *standard* normal distribution.

Figure 6.23

$P(X > 17)$

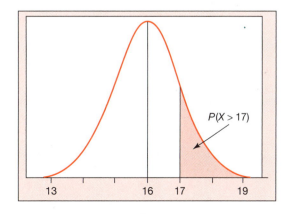

Example 6.15

Suppose that another machine, similar to the one of Example 6.14, operates so that ounces of fill have a mean equal to the dial setting for "amount of liquid" but have astandard deviation of 1.2 ounces. Find the proper setting for the dial so that 17-ounce bottles will overflow only 5% of the time. Assume that the amounts dispensed have a normal distribution.

Solution   Letting X denote the amount of liquid dispensed, we are now looking for a value of μ such that $P(X > 17) = 0.05$, as depicted in Figure 6.24. Now

$$P(X > 17) = P\left(\frac{X - \mu}{\sigma} > \frac{17 - \mu}{\sigma}\right) = P\left(Z > \frac{17 - \mu}{1.2}\right)$$

From Appendix Table 4, we know that if $P(Z > z_0) = 0.05$, then $z_0 = 1.645$. Thus, it must be that

$$\frac{17 - \mu}{1.2} = 1.645$$

and $\mu = 17 - 1.2(1.645) = 15.026$.

Figure 6.24

$P(X > 17) = 0.05$

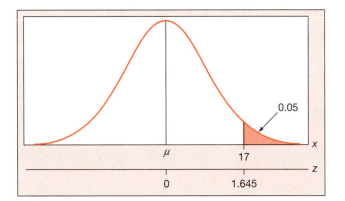

Example 6.16

The SAT and ACT college entrance exams are taken by thousands of students each year. The scores on the exam for any one year produce a histogram that looks very-much like a normal curve. Thus, we can say that the scores are approximately normally distributed. In recent years, the SAT mathematics scores have averaged around 480 with a standard deviation of 100. The ACT mathematics scores have averaged around 18 with a standard deviation of 6.

a An engineering school sets 550 as the minimum SAT math score for new students. What percent of students would score less than 550 in a typical year? (This percentage is called the *percentile* score equivalent for 550.)

b What would the engineering school set as a comparable standard on the ACT math test?

c What is the probability that a randomly selected student will score over 700 on the SAT math test?

Solution

a The percentile score corresponding to a raw score of 550 is the area under the normal curve to the left of 550, as shown in Figure 6.25. The z-score for 550 is

$$z = \frac{x - \mu}{\sigma} = \frac{550 - 480}{100} = 0.7$$

From Appendix Table 4 we get $P(X < 550) = 0.758$. It means the score of 550 is almost the 76th percentile, or about 75.8% of the students taking the SAT should have scores below this value.

b The question now is "What is the 75.8th percentile of the ACT math score distribution?" From the calculation above, percentile 75.8 for any normal distribution will be 0.7 standard deviations above the mean. Therefore, using the ACT distribution,

$$x = \mu + z\sigma = 18 + (0.7)6 = 22.2$$

A score of 22.2 on the ACT should be equivalent to a score of 550 on the SAT.

c A score of 700 corresponds to a z-score of

$$z = \frac{x - \mu}{\sigma} = \frac{700 - 480}{100} = 0.7$$

Thus,

$$P(SAT > 700) = P(z > 2.2) = 1.0 - 0.9861 = 0.0139$$

As can be seen from Figure 6.26, the chance of a randomly selected student scoring above 700 is quite small (around 0.014).

Figure 6.25

$P(X < 550)$

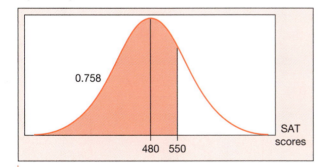

Figure 6.26

$P(X > 700)$

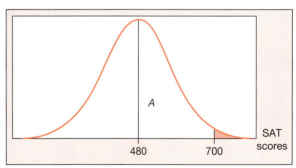

Example 6.17

The batting averages of the American League batting champions from 1901 through 2003 are graphed in the histogram in Figure 6.27. This graph looks somewhat normal in shape but has a little skewness toward the high values. The mean is 0.358 and the standard deviation is 0.027 for these data.

a Ted Williams batted 0.406 in 1941, and George Brett batted 0.390 in 1980. How would you compare these performances?

b Is there a good chance of anyone in the American League hitting over 0.400 in any one year?

Solution **a** Obviously, 0.406 is better than 0.390, but how much better? One way to compare these performances is to compare them to the remaining data using z-scores and percentile scores (see Figure 6.28).

Ted Williams:
$$z = \frac{0.406 - 0.358}{0.027} = 1.78$$

Percentile score $= P(Z \leq 1.78) = 0.9623$

George Brett:
$$z = \frac{0.390 - 0.358}{0.027} = 1.18$$

Percentile score $= P(Z \leq 1.19) = 0.8820$

The percentile score for Ted Williams is 0.96; in other words, Ted Williams hit as good as or better than 96% of the hitting leaders. The percentile score for George Brett is 0.88; that is, 88% of the League leaders hit as good as George Brett or lower. Clearly, both the performances are outstanding; however, Ted Williams did slightly better than George Brett.

b The chance of the League leader hitting over 0.400 in a given year can be approximated by looking at a z-score of

$$z = \frac{0.400 - 0.358}{0.027} = 1.56$$

This translates to a probability of hitting over 0.400 in a given year of 0.0599, or about 6 chances out of 100. (This is the probability of the League leader. What would be the chances for any other specified player?)

Figure 6.27

Batting averages of American League champions

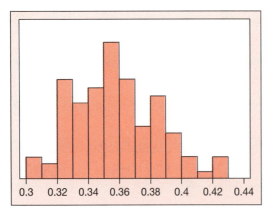

Figure 6.28

Comparison with Ted Williams' batting average

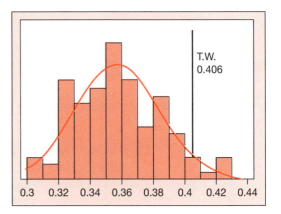

Example 6.18

The octane level of a multicomponent blend of gasoline depends on the volume fraction of components of the blend. Researchers at a gasoline processing plant were considering a different blend and were interested in knowing the likelihood of this blend producing an octane rating above 96. Because the distribution of octane ratings was unknown, they referred to the Foster and Klein data on the motor octane measured for multicomponent blends, reported by Snee (*Technometrics*, 1977) (reproduced in Table 6.2).

Because the distribution of octane ratings is unknown, plot the data using a histogram to look at the shape of the distribution. The histogram (Figure 6.29) shows fairly symmetric and mound-shaped distribution for the octane ratings. The summary statistics shows that the mean and median are about the same (see Figure 6.30).

From the histogram it seems reasonable to assume that the octane ratings are approximately normally distributed and we can estimate the probability of octane rating to be over 96 as,

$$P(\text{Octane rating} > 96) = P\left(z > \frac{96 - 90.671}{2.805}\right) = P(z > 1.90) = 0.0287$$

Table 6.2

Octane Ratings of Gasoline

88.5	95.6	88.3	94.2	89.2	93.3	89.8	91.8
90.4	92.2	87.7	93.3	87.6	92.7	88.3	91.8
89.6	91.6	89.3	92.2	83.4	94.7	84.3	93.2
85.3	92.3	87.4	90.4	89.7	91.2	86.7	91.1
86.7	91.0	87.9	90.4	88.9	91.1	90.3	91.0
87.5	91.0	88.2	90.3	88.6	90.1	91.2	92.6
91.6	92.2	91.5	94.2	90.8	93.4	90.9	93.0
89.3	89.8	90.5	90.0	88.6	87.8	88.3	88.5
89.0	88.7	94.4	90.6	90.7	100.3	93.7	89.9
98.8	90.1	96.1	89.9	92.7	91.1	92.7	

Figure 6.29

Distribution of octane ratings

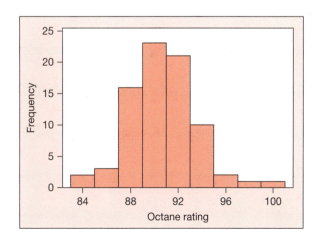

Figure 6.30
Summary statistics for octane ratings

Variable	N	Mean	Median	TrMean	StDev	SE Mean
octane	79	90.671	90.500	90.599	2.806	0.316
Variable	Minimum	Maximum	Q1	Q3		
octane	83.400	100.300	88.700	92.200		

What happens if the normal model is used to describe skewed distributions? To answer this question, let us look at two data sets on comparable variables and see how good the normal approximations are.

Example 6.19

Let us study cancer mortality rates (annual death rates per 100,000) for white males during 1996–2000. Figure 6.31 shows dotplots of cancer mortality rates for 67 counties of Florida and 78 reporting counties of Nebraska. The summary statistics for Florida and Nebraska are given in table below.

	Mean	Standard Deviation
Florida	261.65	49.74
Nebraska	226.22	37.29

On the average, Florida has a slightly higher mortality rate than Nebraska (see table above). Both the distributions look fairly symmetric and mound-shaped except for the possible outlier (a county with considerably higher mortality rate) in Florida (Figure 6.31). The variation in mortality rates is lower in Nebraska than in Florida. The key features of the differences in distributions can be demonstrated by looking at empirical versus theoretical relative frequencies as shown in table below.

Interval	Theoretical Proportion	Observed Proportion Florida	Observed Proportion Nebraska
$\bar{x} \pm s$	0.68	$57/67 = 0.806$	$53/78 = 0.679$
$\bar{x} \pm 2s$	0.95	$66/67 = 0.985$	$74/78 = 0.949$

Figure 6.31
Cancer mortality rates (per 100,000) for white males, 1996–2000

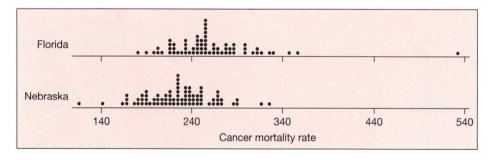

For Nebraska, the observed and the theoretical proportion of counties within one- or two-standard deviations of the mean are close together. However, for Florida, the observed proportions are slightly higher than the theoretical proportions. The intervals do seem to pick a few too many data points. In this case, the large outlier county in Florida inflates the standard deviation so that the one- and two-standard deviation intervals are too large to agree with the normal theory. The two-standard deviation interval still does not reach the outlier. This is typical of the performance of relative frequencies in skewed distributions with outliers. Be careful in interpreting standard deviation under skewed conditions because the standard deviation is affected by the outliers.

Quantile–Quantile (Q–Q) plots The normal model is popular for describing data distributions, and as seen above it often works well. How can we spot when it is not going to work?

Look carefully at histograms, dotplots, and stemplots to visually gauge symmetry and outliers. For example, look at the histograms for Florida and Nebraska (Figures 6.32a and 6.32b). The histogram for Florida shows an outlier on the higher end and skewness of distribution.

Another way to determine the normality of the distribution is to use Q–Q *plots*. If X has a normal (μ, σ) distribution, then $X = \mu + \sigma Z$ where Z has a standard normal distribution. The Q–Q plots make use of this property of a perfect linear relationship between X and Z.

- Order n measurements so that $x_1 \leq x_2 \leq x_3 \leq \cdots \leq x_n$.
- The value x_k has k/n values less than or equal to it, so it is the $(k/n)^{\text{th}}$ sample percentile.
- If the observations come from a normal distribution, they will be linearly related to the corresponding z-score, z_k.

Plots of the sample percentiles against the z-scores are known as *quantile–quantile* or *Q–Q plots*. The straight line indicates a theoretical normal distribution. If the

Figure 6.32

Mortality rates for counties in Florida and Nebraska

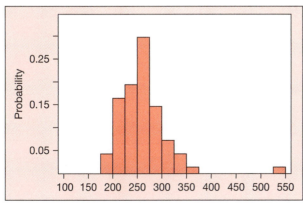

(a) Florida

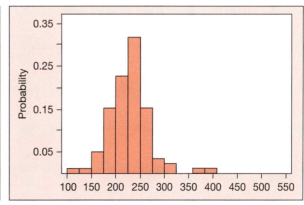

(b) Nebraska

sample percentile z-score points fall very nearly on the straight line, then it indicates a good fit of a normal distribution to the data. The slope of the line gives an estimate of the standard deviation from the data, and the intercept at $z = 0$ gives an estimate of the population mean.

Example 6.20

Let us look at the cancer mortality rates of white males for Florida and Nebraska. The sample percentiles and the corresponding z-scores for each state are given in table below:

Percentile	Florida	Nebraska	z–scores
5.0	180.448	165.275	−1.64485
10.0	198.383	178.736	−1.28155
20.0	220.102	195.036	−0.84162
30.0	235.762	206.79	−0.52440
40.0	249.144	216.832	−0.25335
50.0	261.651	226.219	0.00000
60.0	274.158	235.606	0.25335
70.0	287.539	245.649	0.52440
80.0	303.200	257.402	0.84162
90.0	324.918	273.702	1.28155
95.0	342.853	287.163	1.64485
99.9	414.209	340.716	

The 20th percentile for Florida is 220.102 (see table above). In other words, 20% of reporting Florida counties have cancer mortality rates at or below 220. Similarly, 95% of reporting Nebraska counties have cancer mortality rates at or below 287. For a normal distribution, 20% of the area will fall below z-score–0.84 and 95% of the area will fall below z-score 1.65.

The plots created by JMP using mortality data are shown in Figures 6.33a and 6.33b. The Q–Q plot for Florida (Figure 6.33a) shows that the distribution of cancer mortality rates in Florida is right-skewed. The point on the high end (below

Figure 6.33

Q–Q plots for cancer mortality rates

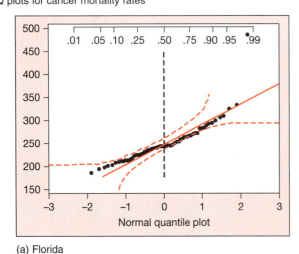

(a) Florida

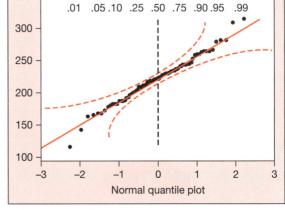

(b) Nebraska

scale 0.99) indicates a very long right tail. On the lower end too, the distribution deviates from normality and has a comparatively shorter left tail. In comparison, the distribution of cancer mortality rates in Nebraska is fairly symmetric. However, little bit of curvature on both ends and departure from the line indicates that the normal fit is not as good as in the middle of the distribution.

Example 6.21

Let us look at the batting averages of champions of National League and American League. Figure 6.34 shows histograms and boxplots for the batting averages.

Figure 6.34

Batting averages of champions

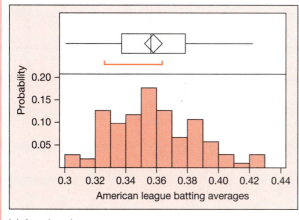

(a) American League

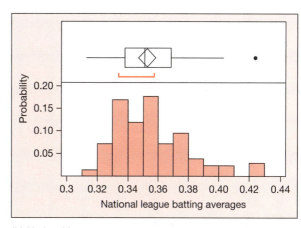

(b) National League

Figure 6.35

Quantile plots for batting averages

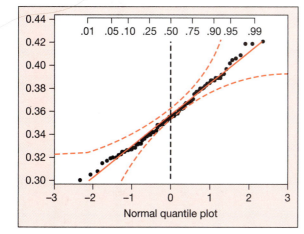

(a) American League

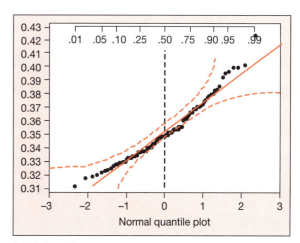

(b) National League

The stemplot of batting averages for the American League champions looks fairly symmetric without any outliers. However, the stemplot of batting averages for the National League champions looks more right-skewed with possibly an outlier on the higher end.

Figure 6.35 shows quantile plots of batting averages for American League and National League. The quantile plot for the National League shows considerable deviation from normality for the lower and higher quantiles with an outlier under scale 0.99.

When we are dealing with small samples, it is better to think of x_k as the $k/(n + 1)$ th sample percentile. The reason for this is that $x_1 \leq x_2 \leq x_3 \leq \cdots \leq x_n$ actually divides the population into $(n + 1)$ segments, with all segments expected to process roughly equal probability masses. Q–Q plots are cumbersome to construct by hand, especially if the dataset is large. Most computer programs for statistical analysis will generate Q–Q plots or normal probability plots quite easily.

Example 6.22

In the interest of predicting future peak particulate matter values and carbon monoxide levels, 15 observations were obtained from various sites across the United States (see table below). The ordered measurements are

CO: 1.7, 1.8, 2.1, 2.4, 2.4, 3.4, 3.5, 4.1, 4.2, 4.4, 4.9, 5.1, 8.3, 9.3, 9.5

Should the normal probability models be used to anticipate carbon monoxide levels in future?

i	x_i	$i/(n+1)$	z–scores
1	1.7	0.0625	−1.53412
2	1.8	0.1250	−1.15035
3	2.1	0.1875	−0.88715
4	2.4	0.2500	−0.67449
5	2.4	0.3125	−0.48878
6	3.4	0.3750	−0.31864
7	3.5	0.4375	−0.15731
8	4.1	0.5000	0.00000
9	4.2	0.5625	0.15731
10	4.4	0.6250	0.31864
11	4.9	0.6875	0.48878
12	5.1	0.7500	0.67449
13	8.3	0.8125	0.88715
14	9.3	0.8750	1.15035
15	9.5	0.9375	1.53412

Figure 6.36

Q–Q plot, peak CO values, 2000

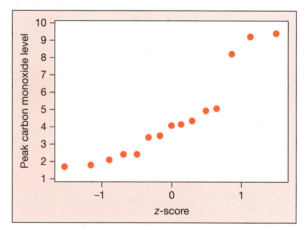

Solution   For $n = 15$, table on previous page provides the key component for the analysis. Recall that the z-scores are the standard normal values corresponding to the percentiles in the $i/(n + 1)$ column of Table 6.6. For example, about 75% of the normal curve's area is below 0.67449.

The Q–Q plot of x_i versus z_i in Figure 6.36 shows that the data depart from normality on the higher end. The lower values have too short a tail and the higher values have too long a tail. The data appear to come from a right-skewed distribution; predicting future values by using a normal distribution model would be a poor decision.

Exercises

6.55   Use Table 4 of the Appendix to find the following probabilities for a standard normal random variable Z.
a $P(0 \le Z \le 1.2)$
b $P(-0.9 \le Z \le 0)$
c $P(0.3 \le Z \le 1.56)$
d $P(-0.2 \le Z \le 0.2)$
e $P(-200 \le Z \le -1.56)$

6.56   For a standard normal random variable Z, use Table 4 of the Appendix to find a number z_0 such that
a $P(Z \le z_0) = 0.5$
b $P(Z \le z_0) = 0.8749$
c $P(Z \ge z_0) = 0.117$
d $P(Z \ge z_0) = 0.617$
e $P(-z_0 \le Z \le z_0) = 0.90$
f $P(-z_0 \le Z \le z_0) = 0.95$

6.57   The weekly amount spent for maintenance and repairs in a certain company has approximately a normal distribution with a mean of $400 and a standard deviation of $20. If $450 is budgeted to cover repairs for next week, what is the probability that the actual costs will exceed the budgeted amount?

6.58   In the setting of Exercise 6.57, how much should be budgeted weekly for maintenance and repairs so that the budgeted amount will be exceeded with probability only 0.1?

6.59   A machining operation produces steel shafts having diameters that are normally distributed with a mean of 1.005 inches and a standard deviation of 0.01 inch. Specifications call for diameters to fall within the interval 1.00 ± 0.02 inches. What percentage of the output of this operation will fail to meet specifications?

6.60   Refer to Exercise 6.59. What should be the mean diameter of the shafts produced to minimize the fraction not meeting specifications?

6.61   Wires manufactured for use in a certain computer system are specified to have resistances between 0.12 and 0.14 ohm. The actual measured resistances of the wires produced by Company A have a normal probability distribution with a mean of 0.13 ohm and a standard deviation of 0.005 ohm.
a   What is the probability that a randomly selected wire from Company A's production will meet the specifications?
b   If four such wires are used in the system and all are selected from Company A, what is the probability that all four will meet the specifications?

6.62   At a temperature of 25°C, the resistances of a type of thermistor are normally distributed with a mean of 10,000 ohms and a standard deviation of 4,000 ohms. The thermistors are to be sorted, with those having resistances between 8,000 and 15,000 ohms being shipped to a vendor. What fraction of these thermistors will be shipped?

6.63   A vehicle driver gauges the relative speed of a preceding vehicle by the speed with which the image of the width of that vehicle varies. This speed is proportional to the speed X of variation of the angle at which the eye subtends this width. According to Ferrani and others (*Transportation Research,* 18A, 1984, pp. 50–51), a study of many drivers revealed X to be normally distributed with a mean of zero and a standard deviation of $10(10^{-4})$ radians per second.
a   What fraction of these measurements is more than five units away from zero?
b   What fraction is more than 10 units away from zero?

6.64   A type of capacitor has resistances that vary according to a normal distribution with a mean of 800 megohms and a standard deviation of 200 megohms (see Nelson, *Industrial Quality Control,* 1967, pp. 261–268, for a more thorough discussion). A certain application specifies capacitors with resistances between 900 and 1,000 megohms.

a What proportion of these capacitors will meet this specification?

b If two capacitors are randomly chosen from a lot of capacitors of this type, what is the probability that both will satisfy the specification?

6.65 Sick-leave time used by employees of a firm in one month has approximately a normal distribution with a mean of 200 hours and a variance of 400.

a Find the probability that total sick leave for next month will be less than 150 hours.

b In planning schedules for next month, how much time should be budgeted for sick leave if that amount is to be exceeded with a probability of only 0.10?

6.66 The times of first failure of a unit of a brand of ink jet printers are approximately normally distributed with a mean of 1,500 hours and a standard deviation of 200 hours.

a What fraction of these printers will fail before 1,000 hours?

b What should be the guarantee time for these printers if the manufacturer wants only 5% to fail within the guarantee period?

6.67 A machine for filling cereal boxes has a standard deviation of 1 ounce of fill per box. What setting of the mean ounces of fill per box will allow 16-ounce boxes to overflow only 1% of the time? Assume that the ounces of fill per box are normally distributed.

6.68 Refer to Exercise 6.67. Suppose the standard deviation σ is not known but can be fixed at certain levels by carefully adjusting the machine. What is the largest value of σ that will allow the actual value dispensed to be within 1 ounce of the mean with probability at least 0.95?

6.69 The histogram in Figure 6.37 depicts the total points scored per game for every NCAA-tournament basketball game between 1939 and 1992 ($n = 1,521$). The mean is 143 and the standard deviation is 26.

a Does it appear that total points can be modeled as a normally distributed random variable?

b Show that the empirical rule holds for these data.

c Would you expect a total score to top 200 very often? 250? Why or why not?

d There are 64 games to be played in this year's tournament. How many would you expect to have total scores below 100 points?

6.70 Another contributor to air pollution is sulfur dioxide. Shown below are 12 peak sulfur dioxide measurements (in ppm) for 1990 from randomly selected U.S. sites:

0.003, 0.001, 0.014, 0.024, 0.024, 0.032,
0.038, 0.042, 0.043, 0.044, 0.047, 0.061

a Should the normal distribution be used as a model for these measurements?

b From a Q–Q plot for these data, approximate the mean and standard deviation. Check the approximation by direct calculation.

Figure 6.37

Histogram of total score per game for NCAA basketball games

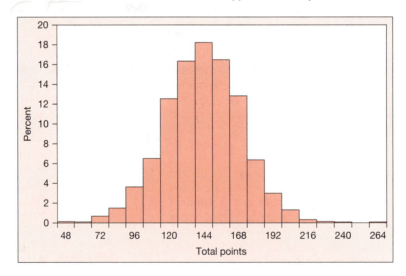

6.71 The cumulative percent distribution of population (20–29 years) by height and sex (1988–1994) is given in table below.

Height (inches)	Male	Female
56		0.6
57		0.7
58		1.2
59		3.1
60	0.1	6.3
61	0.1	11.5
62	0.5	21.8
63	1.3	34.3
64	3.4	48.9
65	6.9	62.7
66	11.7	74.0
67	20.8	84.7
68	32.0	92.4
69	46.3	96.2
70	58.7	98.6
71	70.1	99.5
72	81.2	100.0
73	87.4	100.0
74	94.7	100.0
75	97.9	100.0

a Produce a Q–Q plot for male and female heights.

b Discuss their goodness of fit for a normal distribution.

c Approximate the mean and standard deviation of these heights from the Q–Q plots.

6.72 The cumulative proportions of age groups of U.S. residents are shown in table below for years 1900 and 2000.

Age	Cumulative Proportion	
	1900	**2000**
5	0.121	0.068
15	0.344	0.214
25	0.540	0.353
35	0.700	0.495
45	0.822	0.576
55	0.906	0.709
65	0.959	0.795

a Construct a Q–Q plot for each year. Use these plots as a basis for discussing key differences between the two age distributions.

b Each of the Q–Q plots should show some departures from normality. Explain the nature of these departures.

6.7 The Lognormal Distribution

The idea of transformation was used in the regression analysis to imitate function with different shape. A similar idea is used in the most commonly used lognormal distribution. If Y is a normal random variable, then $X = e^Y$ follows a *lognormal distribution*. The distribution receives its name from the transformation used, $Y = \ln(X)$. The lognormal distribution is very commonly used to describe lifetime distributions, such as lifetime of products, components, and systems. It is a right-skewed distribution and takes values in the range $(0, \infty)$. The lognormal distribution has the following form, and the graph of the density function is shown in Figure 6.38.

Because the probabilities of a lognormal random variable are obtained by transforming a normal random variable, we can obtain the cumulative distribution function of lognormal using the cumulative distribution function of a standard normal variable, Z, as follows:

$$F(x) = P(X \le x) = P(e^y \le x)$$

$$= P(Y \le \ln x) = P\left(Z \le \frac{\ln x - \mu}{\sigma}\right)$$

Figure 6.38

The lognormal density function

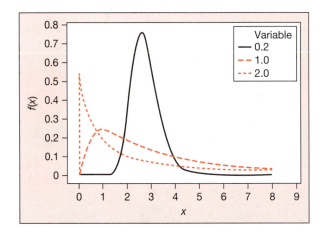

The Lognormal Distribution

If $Y = \ln(X)$ has a normal distribution with mean μ and variance σ^2, then X has a *lognormal* distribution with probability density function

$$f(x) = \begin{cases} \dfrac{1}{x\sigma\sqrt{2\pi}} \exp\left[\dfrac{-(\ln x - \mu)^2}{2\sigma^2}\right] & 0 < x < \infty \\\\ 0 & \text{elsewhere} \end{cases}$$

$$E(X) = \exp\left(\mu + \frac{\sigma^2}{2}\right) \quad \text{and} \quad V(X) = \exp(2\mu + \sigma^2)[\exp(\sigma^2) - 1]$$

Example 6.23 ─────────────────────────────────

The Department of Energy (DOE) has a Chronic Beryllium Disease Prevention Program. The focus of this program is to monitor equipment and buildings for beryllium contamination and exposure of workers to beryllium in the workplace. Data given in Table 6.3 were collected from a smelter facility used to recycle metal, where beryllium contaminants tend to deposit on all surfaces. Wipe samples were taken randomly from the surface of equipment in the workplace and tested for possible beryllium contamination (measured in μgm/100 cm^2).

The distribution of data (Figure 6.39) shows that lognormal distribution (given by the curve) is a reasonable choice to describe beryllium contamination at this smelter. From the samples, Frome and Wambach (2005, http://www.hss.energy.gov/HealthSafety/IIPP/sand/ORNLTM2005ab52.htm) estimated μ to be -2.291 and σ to be 1.276.

Table 6.3

Beryllium Contamination

0.015	0.070	0.270
0.015	0.075	0.290
0.015	0.095	0.345
0.025	0.100	0.395
0.025	0.125	0.395
0.040	0.125	0.420
0.040	0.145	0.495
0.040	0.145	0.840
0.045	0.150	1.140
0.050	0.150	
0.050	0.165	

Figure 6.39

Distribution of beryllium contamination

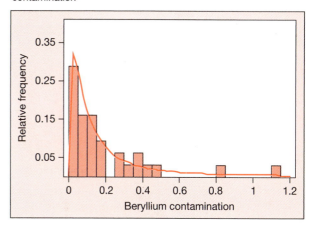

a If a wipe sample is taken randomly from this smelter, what is the probability that it will have beryllium contamination exceeding 0.50 μgm/100 cm^2?

Let X = the beryllium contamination; then $\ln X$ follows a normal distribution with mean -2.291 and standard deviation 1.276. Therefore, using the above stated result and standard normal cumulative distribution function, we get

$$P(X > 0.50) = 1 - P(X \leq 0.50)$$

$$= 1 - P[\ln(X) \leq \ln(0.50)]$$

$$= 1 - P\left[Z \leq \frac{\ln(0.50) - (-2.291)}{1.276}\right]$$

$$= 1 - P(Z \leq 1.25) = 1 - 0.89 = 0.11$$

b The DOE has established a safe limit for beryllium contamination to be $L_{0.95} = 0.20$ μgm/100 cm^2, where $L_{0.95}$ indicates 95th percentile. Use this data to determine the safety of workers' health at this smelter.

We need to determine $L_{0.95}$ value for this smelter such that $P(X \leq L_{0.95}) = 0.95$. Therefore, from the above result relating normal and lognormal distributions we get

$$P(X \leq L_{0.95}) = P\left[Z \leq \frac{\ln(L_{0.95}) - (-2.291)}{1.276}\right] = 0.95$$

Using the standard normal cumulative distribution function,

$$\frac{\ln(L_{0.95}) - (-2.291)}{1.276} = 1.645$$

Solving this equation, we get $L_{0.95} = \exp[-2.291 + 1.645(1.276)] = 0.8253 > 0.20$. Because this estimate of 95th percentile exceeds the DOE safety limit of 0.20, we can conclude that the beryllium contamination at this smelter is at an unhealthy level for workers.

c The expected amount of beryllium and the variance can be calculated as follows:

$$E(X) = \exp\left(\mu + \frac{\sigma^2}{2}\right) = \exp\left[(-2.291) + \frac{(1.276)^2}{2}\right] = 0.2283$$

and

$$V(X) = \exp\left(2\mu + \sigma^2\right)[\exp(\sigma^2) - 1]$$
$$= \exp\left(2(-2.291) + 1.276^2\right)\left[\exp(1.276^2) - 1\right] = 0.0977$$

Even the average amount of beryllium exceeds the safety limit.

Fitting a Lognormal Distribution It is easy to fit lognormal distribution using the fact that X has lognormal distribution if $Y = \ln(X)$ has normal distribution.

Consider beryllium contamination data from Table 6.3. The logarithms of these measurements (ln) are plotted in Figure 6.40. The histogram indicates that the distribution is fairly symmetric and mound-shaped. Also, the normal curve plotted on it resembles histogram fairly closely, which indicates that it is reasonable to assume ln(beryllium) values are approximately normally distributed. Therefore, using above result, we can say that beryllium measurements follow a lognormal distribution. The normal Q–Q plot for ln(beryllium) shown in Figure 6.41 also supports this conclusion.

Figure 6.40

Distribution of ln(beryllium)

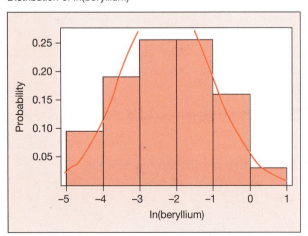

Figure 6.41

Normal Q–Q plot of ln(beryllium)

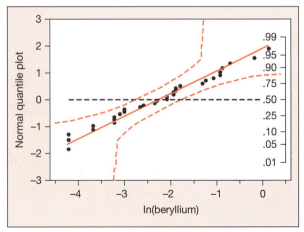

Exercises

6.73 The lifetime distributions of high-speed recordable optical disks is assumed to follow lognormal distribution (Irine and Okino, *IEEE Trans. on Magnetics*, 2007). Suppose $\mu = 2.5$ and $\sigma = 0.5$ weeks.

   a Compute the mean and variance of this distribution.

   b What percent of optical disks are expected to last over 20 weeks?

   c If 98% of disks fail before a_0 weeks, then determine the value of a_0.

6.74 In automobiles, rubber boot seals are used to protect velocity joints and steering mechanisms. Lifetimes of these boot seals depend on how much they come in contact with the metal shaft. The lognormal distribution has been used to describe the lifetime of boot seals (Guerin and Hambli, *J. Mechanical Designs*, 2007).

   a Using $\mu = 8$ and $\sigma = 4.0$ months, determine the percent of boots expected to fail within 5 months.

   b Determine the mean and variance of this distribution.

   c The manufacturer of these rubber boots is interested in replacing free-of-charge at most 2% of its product under warranty. How long of a warranty should the manufacturer offer?

6.75 One study of cancer patients showed that the survival times (in weeks) of cancer (Mesothelioma) patients since detection are distributed lognormal with $\mu = 3.5$ and $\sigma = 1.0$.

   a What percent of cancer patients are expected to survive for 6 months after detection?

   b What percent of patients are expected to survive beyond a year after detection?

   c At how many weeks will only 5% of cancer patients survive?

6.76 In a large-scale study to understand root causes of systems failures, Los Alamos National Laboratory collected data about 23,000 failures of different high-performance computing systems. Schroeder and Gibson (Technical Report CMU-PDL-05-112, 2005) showed that even for different systems, the repair times were well modeled by the lognormal distribution.

   a The mean repair time for failures caused by human error was estimated to be 163 min and standard deviation 418 min. Estimate μ and σ^2.

   b The mean repair time for failures due to environmental problems was estimated to be 572 min and standard deviation 808 min. Estimate μ and σ^2.

6.77 Because the use of CD-ROM and CD-R for storing data has increased considerably, successful retrieving of data from them becomes an important issue for their manufacturers. Eastman Kodak Company was interested in developing Kodak Writable CD and Kodak Photo CD with exceptional data life for CDs stored under different temperatures. To determine level of success in achieving this goal, the company tested a random sample of their product and found the lognormal distribution described the data lifetimes the best (Stinson, Ameli, and Zaino, http://www.cd-info.com/CDIC/Technology/CD-R/Media/Kodak.html). They concluded that 95% of the population of Kodak Writable CD Media will have a data lifetime of greater than 217 years under certain conditions.

   a Describe this statement using probability density.

   b If $\sigma = 5$, estimate the expected lifetime of Kodak Writable CD Media when used under similar conditions.

6.8 The Beta Distribution

Except for the uniform distribution of Section 6.3, the continuous distributions discussed thus far are defined as nonzero functions over an infinite interval. The beta distribution is very useful for modeling the probabilistic behavior of certain random variables, such as proportions, constrained to fall in the interval (0, 1). The beta distribution has the following functional form. The graphs of some of common beta density functions are shown in Figure 6.42.

The Beta Distribution

$$f(x) = \begin{cases} \dfrac{\Gamma(\alpha + \beta)}{\Gamma(\alpha)\Gamma(\beta)} x^{\alpha-1}(1-x)^{\beta-1} & 0 < x < 1 \\ 0 & \text{elsewhere} \end{cases}$$

For $\alpha, \beta > 0$ $E(X) = \dfrac{\alpha}{\alpha + \beta}$ and

$$V(X) = \dfrac{\alpha\beta}{(\alpha + \beta)^2(\alpha + \beta + 1)}$$

The following relation is used by the beta distribution. For $\alpha, \beta > 0$,

$$\int_0^1 x^{\alpha-1}(1-x)^{\beta-1}dx = \dfrac{\Gamma(\alpha)\Gamma(\beta)}{\Gamma(\alpha + \beta)}.$$

One measurement of interest in the process of sintering copper is the proportion of the volume that is solid, as opposed to the proportion made up of voids. The proportion due to voids is sometimes called the *porosity* of the solid. Figure 6.43 shows a relative frequency histogram of proportion of solid copper in samples from a sintering process. This distribution could be modeled with a beta distribution having a large α and small β.

Figure 6.42

The beta density functions

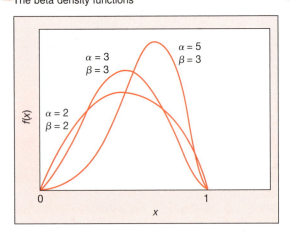

Figure 6.43

Solid mass of sintered linde copper

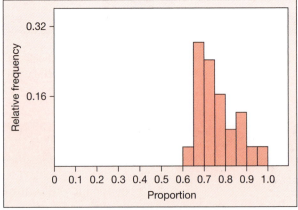

Source: Department of Material Science, University of Florida.

The expected value of the beta variable is computed as follows, recalling that $\Gamma(n + 1) = n\Gamma(n)$.

$$E(X) = \int_0^1 x \frac{\Gamma(\alpha + \beta)}{\Gamma(\alpha)\Gamma(\beta)} x^{\alpha-1}(1 - x)^{\beta-1} = \frac{\Gamma(\alpha + \beta)}{\Gamma(\alpha)\Gamma(\beta)} \int_0^1 x^{\alpha}(1 - x)^{\beta-1}$$

$$= \frac{\Gamma(\alpha + \beta)}{\Gamma(\alpha)\Gamma(\beta)} \frac{\Gamma(\alpha + 1)\Gamma(\beta)}{\Gamma(\alpha + \beta + 1)} = \frac{\alpha}{\alpha + \beta}$$

Similarly, we can show that

$$V(X) = \frac{\alpha\beta}{(\alpha + \beta)^2(\alpha + \beta + 1)}$$

Example 6.24

A gasoline wholesale distributor has bulk storage tanks holding a fixed supply. The tanks are filled every Monday. Of interest to the wholesaler is the proportion of this supply that is sold during the week. Over many weeks, this proportion has been observed to be modeled fairly well by a beta distribution with $\alpha = 4$ and $\beta = 2$.

a Find the expected value of this proportion.
b Is it highly likely that the wholesaler will sell at least 90% of the stock in a given week?

Solution Let X denote the proportion of the total supply sold in a given week.

a The expected proportion of supply sold in a given week is

$$E(X) = \frac{\alpha}{\alpha + \beta} = \frac{4}{4 + 2} = \frac{2}{3}$$

b We are interested in

$$P(X > 0.9) = \int_{0.9}^1 \frac{\Gamma(4 + 2)}{\Gamma(4)\Gamma(2)} x^3(1 - x)dx$$

$$= 20 \int_{0.9}^1 \frac{\Gamma(4 + 2)}{\Gamma(4)\Gamma(2)}(x^3 - x^4)dx = 20(0.004) = 0.08$$

It is not very likely that 90% of the stock will be sold in a given week.

Alternatively we can use some computer program to compute this probability. For example, the **cumulative probability** option for beta distribution from Minitab gives $P(X \leq 0.9) = 0.9185$. Therefore,

$$P(x > 0.9) = 1 - P(x \leq 0.9) = 1 - 0.9185 = 0.0815$$

Exercises

6.78 Suppose X has a probability density function given by

$$f(x) = \begin{cases} kx^3(1-x)^2 & 0 \le x \le 1 \\ 0 & \text{elsewhere} \end{cases}$$

  a Find the value of k that makes this a probability density function.
  b Find $E(X)$ and $V(X)$.

6.79 If X has a beta distribution with parameters α and β, show that

$$V(X) = \frac{\alpha\beta}{(\alpha + \beta)^2(\alpha + \beta + 1)}$$

6.80 During an 8-hour shift, the proportion of time X that a sheet-metal stamping machine is down for maintenance or repairs has a beta distribution with $\alpha = 1$ and $\beta = 2$. That is,

$$f(x) = \begin{cases} 2(1-x) & 0 \le x \le 1 \\ 0 & \text{elsewhere} \end{cases}$$

The cost (in hundreds of dollars) of this downtime, due to lost production and cost of maintenance and repair, is given by

$$C = 10 + 20X + 4X^2.$$

  a Find the mean and variance of C.
  b Find an interval in which C will lie with probability at least 0.75.

6.81 The percentage of impurities per batch in a certain type of industrial chemical is a random variable X having the probability density function

$$f(x) = \begin{cases} 12x^2(1-x) & 0 \le x \le 1 \\ 0 & \text{elsewhere} \end{cases}$$

  a Suppose a batch with more than 40% impurities cannot be sold. What is the probability that a randomly selected batch will not be sold?
  b Suppose the dollar value of each batch is given by

$$V = 5 - 0.5X,$$

Find the expected value and variance of V.

6.82 To study the disposal of pollutants emerging from a power plant, the prevailing wind direction was measured for a large number of days. The direction is measured on a scale of 0° to 360°, but by dividing each daily direction by 360, the measurements can be rescaled to the interval (0,1). These rescaled measurements X are found to follow a beta distribution with $\alpha = 4$ and $\beta = 2$. Find $E(X)$. To what angle does this mean correspond?

6.83 Errors in measuring the arrival time of a wave front from an acoustic source can sometimes be modeled by a beta distribution. (See Perruzzi and Hilliard, *J. Acoustical Society of America*, 1984) Suppose these errors have a beta distribution with $\alpha = 1$ and $\beta = 2$, with measurements in microseconds.

  a Find the probability that such a measurement error will be less than 0.5 microseconds.
  b Find the mean and standard deviation of these error measurements.

6.84 In blending fine and coarse powders prior to copper sintering, proper blending is necessary for uniformity in the finished product. One way to check the blending is to select many small random samples of the blended powders and measure the weight fractions of the fine particles. These measurements should be relatively constant if good blending has been achieved.

  a Suppose the weight fractions have a beta distribution with $\alpha = \beta = 3$. Find their mean and variance.
  b Repeat part (a) for $\alpha = \beta = 2$.
  c Repeat part (a) for $\alpha = \beta = 1$.
  d Which of the three cases, (a), (b), or (c), would exemplify the best blending?

6.85 The proportion of pure iron in certain ore samples has a beta distribution with $\alpha = 3$ and $\beta = 1$.

  a Find the probability that one of these randomly selected samples will have more than 50% pure iron.
  b Find the probability that two out of three randomly selected samples will have less than 30% pure iron.

6.9 The Weibull Distribution

The gamma distribution often serves as a probabilistic model for lifetimes of system components, but other distributions often provide better models for lifetime data. One such distribution is *Weibull*. A Weibull density function has the form shown here.

For $\gamma = 1$, this becomes an exponential density. For $\gamma > 0$, the function looks somewhat like the gamma function but has some different mathematical properties. The cumulative distribution function is given by

$$F(x) = \int_0^x \frac{\gamma}{\theta} t^{\gamma-1} e^{-t^\gamma/\theta} dt = -e^{-t^\gamma/\theta} \Big|_0^x = 1 - e^{-x^\gamma/\theta} \quad x > 0$$

The Weibull Distribution

$$f(x) = \begin{cases} \dfrac{\gamma}{\theta} x^{\gamma-1} e^{-x^\gamma/\theta} & x > 0 \\ 0 & \text{Otherwise} \end{cases}$$

$$\theta > 0 \quad \text{and} \quad \gamma > 0$$

$$E(X) = \theta^{1/\gamma} \Gamma\left(1 + \frac{1}{\gamma}\right)$$

$$V(X) = \theta^{2/\gamma} \left\{ \Gamma\left(1 + \frac{2}{\gamma}\right) - \left[\Gamma\left(1 + \frac{1}{\gamma}\right)\right]^2 \right\}$$

$$F(x) = 1 - e^{-x^\gamma/\theta} \quad x > 0$$

It is convenient to use transformation $Y = X^\gamma$ to study properties of Weibull. Using the above result, we see that the cumulative distribution function of Y is

$$F_Y(y) = P(Y \le y) = P(X^\gamma \le y) = P(X \le y^{1/\gamma}) = 1 - e^{-y/\theta} \quad y > 0$$

Hence, $f_Y(y) = \dfrac{dF_Y(y)}{dy} = \dfrac{1}{\theta} e^{-y/\theta}$ $y > 0$, which is the familiar exponential density. Therefore, if a random variable X has a Weibull distribution, then $Y = X^\gamma$ has the exponential distribution.

To find the expected value of the Weibull random variable, we use the following result with gamma type integrals:

$$E(X) = E\left(Y^{1/\gamma}\right) = \int_0^\infty y^{1/\gamma} \frac{1}{\theta} e^{-y/\theta} dy = \frac{1}{\theta} \Gamma\left(1 + \frac{1}{\gamma}\right) \theta^{(1+1/\gamma)} = \theta^{1/\gamma} \Gamma\left(1 + \frac{1}{\gamma}\right)$$

Letting $\gamma = 2$ in the Weibull density, we see that $Y = X^2$ has an exponential distribution. To reverse the idea, if we started out with an exponentially distributed random variable Y, then $\sqrt{Y}$ will have a Weibull distribution with $\gamma = 2$. We can illustrate this idea empirically.

Example 6.25

Let us look at the lifetimes of 50 randomly selected batteries listed in Table 6.4 Figure 6.44a shows that the lifetimes follow exponential distribution (given by the curve). Now let us take square roots of these lifetimes (listed in Table 6.4).

A relative frequency histogram of square roots is given in Figure 6.44b. Notice that the exponential form has now disappeared and that Weibull density with $\gamma = 2$ and $\theta = 2$ (given by the curve) is a much more plausible model for these measurements.

Table 6.4

Square Root of Lifetimes of Batteries

0.637	1.531	0.733	2.256	2.364	1.601	0.152	1.826	1.868	1.126
0.828	1.184	0.484	1.207	0.719	0.715	0.474	1.525	1.709	1.305
2.186	1.228	2.006	1.032	1.802	1.668	1.230	0.577	1.274	1.623
1.313	0.542	1.823	0.880	1.526	2.535	1.793	2.741	0.578	1.360
2.868	1.493	1.709	0.872	1.032	0.914	1.952	0.984	2.119	0.431

Figure 6.44
Relative frequency histograms and densities

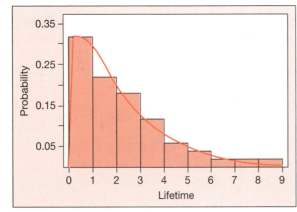

(a) Exponential density

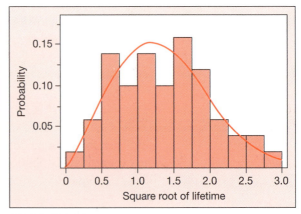

(b) Weibull density

Example 6.26

The length of service time during which a certain type of thermistor produces resistance within its specifications has been observed to follow a Weibull distribution with $\theta = 50$ and $\gamma = 2$. The measurements are in thousands of hours.

a Find the probability that one of these randomly selected thermistors, to be installed in a system today, will function properly for over 10,000 hours.

b Find the expected lifetimes for thermistors of this type.

Solution   Let X represent the lifetime of the thermistor in question.

a As seen earlier, the Weibull distribution has a closed-form expression for $F(x)$. Thus,

$$P(X > 10) = 1 - F(10) = 1 - [1 - e^{-10^2/50}] = e^{-2} = 0.14$$

b The expected lifetime for thermistors of this type is

$$E(X) = \theta^{1/\gamma}\Gamma\left(1 + \frac{1}{\gamma}\right) = \left(50^{1/2}\right)\Gamma\left(1 + \frac{1}{2}\right)$$

$$= \left(50^{1/2}\right)\Gamma\left(\frac{3}{2}\right) = \left(50^{1/2}\right)\left(\frac{1}{2}\right)\Gamma\left(\frac{1}{2}\right) = \left(50^{1/2}\right)\left(\frac{1}{2}\right)2\overline{\pi} = 6.27$$

Thus, the average service time for these thermistors is 6,270 hours.

Fitting Weibull Distribution   The Weibull distribution is versatile enough to be a good model for many distributions of data that are mound-shaped but skewed. Another advantage of the Weibull distribution over, say, the gamma distribution, is that a number of relatively straightforward techniques exist for actually fitting the model to data. Remember that for the Weibull distribution,

$$1 - F(x) = e^{-x^\gamma/\theta} \quad x > 0$$

Therefore, taking the logarithm (denoting natural logarithm by "ln") we get

$$LF(x) = \ln\left[\ln\left(\frac{1}{1 - F(x)}\right)\right] = \gamma \ln(x) - \ln(\theta)$$

Note that there is a linear relation between $LF(x)$ and $\ln(x)$ with slope γ and intercept $\ln(\theta)$.

Example 6.27

The New York City Department of Transportation hired the consulting firm of Steinman, Boynton, Gronquist, and Birdsall to estimate the strength of the Williamson Bridge's suspension cables. Wires from five different locations, numbered 1, 2, 3, 5, and 6, were used to establish properties of the cable wire. Perry

(*The American Statistician*, 1998) reported the cable strengths at location 6, as listed in Table 6.5:

Checking for outliers (Figure 6.45), the boxplot shows no outliers, and the histogram shows the distribution of strengths is skewed. Proceed as in the normal case to produce sample percentiles to approximate $F(x)$. As shown in Table 6.6, use $i/(n + 1)$ to approximate $F(x_i)$ in LF(x).

Figure 6.46 shows the plot of LF(x) versus ln(x) for the cable strength data. The points lie rather close to a straight line that has a slope about 16.56 and an intercept about -145. Thus, the cable strength seems to be adequately modeled by

Table 6.5

Strength of Suspension Bridge Cables

5800	6180	6470
5410	5700	6050
6190	6180	6250
5500	6780	5810
5900	5900	5710
5990	6380	6360
6300	6820	6240
6410	5750	6330
6590	6600	6370
6870	5870	6740
5300	6480	

Figure 6.45

Frequency histogram for cable strength data

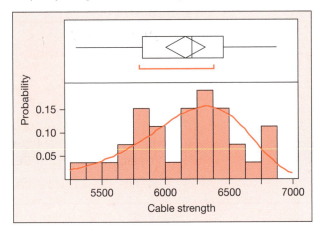

Table 6.6

Weibull Distribution to Cable Strength Data

i	x	ln(x)	$i/(n + 1)$	Approx LF(x)	i	x	ln(x)	$i/(n + 1)$	Approx LF(x)
1	5300	8.575	0.030	−3.481	17	6240	8.739	0.515	−0.323
2	5410	8.596	0.061	−2.772	18	6250	8.740	0.545	−0.238
3	5500	8.613	0.091	−2.351	19	6300	8.748	0.576	−0.154
4	5700	8.648	0.121	−2.046	20	6330	8.753	0.606	−0.071
5	5710	8.650	0.152	−1.806	21	6360	8.758	0.636	0.012
6	5750	8.657	0.182	−1.606	22	6370	8.759	0.667	0.094
7	5800	8.666	0.212	−1.434	23	6380	8.761	0.697	0.177
8	5810	8.667	0.242	−1.281	24	6410	8.766	0.727	0.262
9	5870	8.678	0.273	−1.144	25	6470	8.775	0.758	0.349
10	5900	8.683	0.303	−1.019	26	6480	8.776	0.788	0.439
11	5900	8.683	0.333	−0.903	27	6590	8.793	0.818	0.533
12	5990	8.698	0.364	−0.794	28	6600	8.795	0.848	0.635
13	6050	8.708	0.394	−0.692	29	6740	8.816	0.879	0.747
14	6180	8.729	0.424	−0.594	30	6780	8.822	0.909	0.875
15	6180	8.729	0.455	−0.501	31	6820	8.828	0.939	1.031
16	6190	8.731	0.485	−0.411	32	6870	8.835	0.970	1.252

Figure 6.46

Plot of LF(x) versus ln(x) for the cable strength data

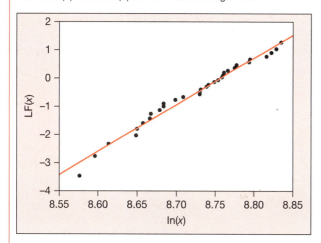

Figure 6.47

Weibull quantile plot for the cable strength data

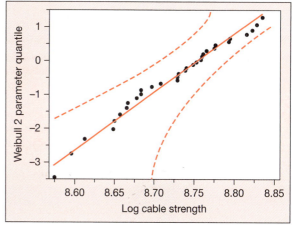

a Weibull distribution with γ approximately 15.56 and $\ln(\theta)$ approximately 145; that is, $\theta \approx 9.39 \times 10^{62}$.

Of course, in this era of technology, it is easier to fit the distribution using a computer program, as shown by the smooth curve in Figure 6.45, or as shown by the quantile plot (see Figure 6.47) generated using JMP.

Exercises

6.86 Fatigue life (in hundreds of hours) for a certain type of bearing has approximately a Weibull distribution with $\gamma = 2$ and $\theta = 4$.

   a Find the probability that a randomly selected bearing of this type fails in less than 200 hours.

   b Find the expected fatigue life for this type of bearing.

6.87 The maximum flood levels, in millions of cubic feet per second, for a certain U.S. river have a Weibull distribution with $\gamma = 1.5$ and $\theta = 0.6$. (See Cohen, Whitten, and Ding, *J. Quality Technology*, 1984, for more details.) Find the probability that the maximum flood level for next year

   a Will exceed 0.5

   b Will be less than 0.8

6.88 The time necessary to achieve proper blending of copper powders before sintering was found to have a Weibull distribution with $\gamma = 1.1$ and $\theta = 2$ (measurements in minutes). Find the probability that a proper blending in a randomly selected sample takes less than 2 minutes.

6.89 The ultimate tensile strength of steel wire used to wrap concrete pipe was found to have a Weibull distribution with $\gamma = 1.2$ and $\theta = 270$ (measure-

ments in thousands of pounds). Pressure in the pipe at a certain point may require an ultimate tensile strength of at least 300,000 pounds. What is the probability that a randomly selected wire will possess this required strength?

6.90 The yield strengths of certain steel beams have a Weibull distribution with $\gamma = 2$ and $\theta = 3,600$ (measurements in pounds per square inch). Two such randomly selected beams are used in a construction project that calls for yield strengths in excess of 70,000 psi. Find the probability that both beams meet the specifications for the project.

6.91 The pressure, in thousand psi, exerted on the tank of a steam boiler has a Weibull distribution with $\gamma = 1.8$ and $\theta = 1.5$. The tank is built to withstand pressures of 2,000 psi. Find the probability that this limit will be exceeded at a randomly selected moment.

6.92 Resistors being used in the construction of an aircraft guidance system have lifelengths that follow a Weibull distribution with $\gamma = 2$ and $\theta = 10$ (measurements in thousands of hours).

   a Find the probability that a randomly selected resistor of this type has a life length that exceeds 5,000 hours.

b If three resistors of this type are operating independently, find the probability that exactly one of the three resistors burns out prior to 5,000 hours of use.

c Find the mean and variance of the life length of such a resistor.

6.93 Failures to the bleed systems in jet engines were causing some concern at an air base. It was decided to model the failure-time distribution for these systems so that future failures could be anticipated a little better. Ten observed failure times (in operating hours since installation) were selected randomly from records and are given below (from *Weibull Analysis Handbook*, Pratt & Whitney Aircraft, 1983):

1,198, 884, 1,251, 1,249, 708, 1,082, 884, 1,105, 828, 1,013

Does a Weibull distribution appear to be a good model for these data? If so, what values should be used for γ and θ?

6.94 Maximum wind-gust velocities in summer thunderstorms were found to follow a Weibull distribution with $\gamma = 2$ and $\theta = 400$ (measurements in feet per second). Engineers designing structures in the areas in which these thunderstorms are found are interested in finding a gust velocity that will be exceeded only with probability 0.01. Find such a value.

6.95 The velocities of gas particles can be modeled by the Maxwell distribution, with probability density function given by

$$f(v) = 4\pi\left(\frac{m}{2\pi KT}\right)^{3/2} v^2 e^{-v^2(m/2KT)} \quad v > 0$$

where m is the mass of the particle, K is Boltzmann's constant, and T is the absolute temperature.

a Find the mean velocity of these particles.

b The kinetic energy of a particle is given by $(1/2)mV^2$. Find the mean kinetic energy for a particle.

6.10 Reliability

One important measure of the quality of products is their *reliability*, or probability of working for a specified period of time. We want products, whether they be cars, TV sets, or shoes, that do not breakdown or wear out for some time, and we want to know what this time period is expected to be. The study of reliability is a probabilistic exercise because data or models on component lifetimes are used to predict future behavior—how long a process will operate before it fails. In reliability studies, the underlying random variable of interest, X, is usually lifetime.

> If a component has lifetime X with distribution function F, then the **reliability** of the component is
>
> $$R(t) = P(X > t) = 1 - F(t)$$

The reliability functions for an exponential and a Weibull distributed lifetimes are as follows:

Exponential: $R(t) = e^{-t/\theta} \quad t \geq 0$

Weibull: $R(t) = e^{-t^\gamma/\theta} \quad t \geq 0$

Reliability functions of gamma and normal distributions do not exist in closed form. In fact, the normal model is rarely used to model lifetime data because length of time tends to exhibit positively skewed behavior.

Series and parallel systems A system is made up of a number of components, such as relays in an electrical system or check valves in a water system. The reliability of systems depends critically on how the components are networked into system. A *series* system (Figure 6.48a) fails as soon as any one of the components fails. A *parallel* system (Figure 6.48b) fails only when all components have failed.

Suppose each system component in Figure 6.48 has a reliability function $R(t)$, and suppose the components operate independently of one another. What are the system reliabilities, $R_s(t)$?

Figure 6.48

Series and parallel systems.

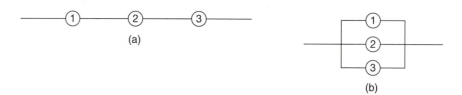

(a)

(b)

For a series system, lifetime will exceed t only if each component has lifetime exceeding t. Thus,

$$R_s(t) = R(t) \cdot R(t) \cdot R(t) = [R(t)]^3$$

For the parallel system, lifetime will be less than t only if all components fail before t. Thus,

$$1 - R_s(t) = [1 - R(t)] \cdot [1 - R(t)] \cdot [1 - R(t)] = [1 - R(t)]^3$$

$$\text{i.e., } \quad R_s(t) = 1 - [1 - R(t)]^3$$

Of course, most systems are combinations of components in series and components in parallel. The rules of probability must be used to evaluate the system reliability in each special case, but breaking down the system into series and parallel components often helps simplify this process.

• For a series of n independent components,

$$R_s(t) = [R(t)]^n$$

because $R(t) \leq 1$, $R_s(t)$ will decrease as n increases. Thus, adding more components in series will just make things worse!

• For n components operating in parallel,

$$R_s(t) = 1 - [1 - R(t)]^n$$

because $R(t) \leq 1$, $R_s(t)$ will increase as n increases. Thus, system reliability can be improved by adding backup components in parallel, a practice called *redundancy*.

How many components do we need to achieve a specified system reliability? If we have a fixed value for $R_s(t)$ in mind and have independently operating parallel components, then

$$1 - R_s(t) = [1 - R(t)]^n$$

Taking the natural logarithms of both sides and solving we get

$$n = \frac{\ln[1 - R_s(t)]}{\ln[1 - R(t)]}$$

Exercises

6.96 For a component with exponentially distributed lifetime, find the reliability to time $t_1 + t_2$ given that the component has already lived past t_1. Why is the constant failure rate referred to as the "memory-less" property?

6.97 For a series of n components operating independently, each with the same exponential life distribution, find an expression for $R_s(t)$. What is the mean lifetime of the system?

6.98 For independently operating components with identical life distributions, find the system reliabil-

ity for each of the following systems. Which has the higher reliability?

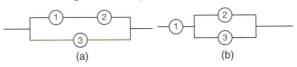

(a)   (b)

6.99 Suppose each relay in an electrical circuit has reliability 0.9 for a specified operating period of t hours. How could you configure a system of such relays to bring $R_s(t)$ up to 0.999?

6.11 The Moment-Generating Functions for Continuous Random Variables

As in the case of discrete distributions, the moment-generating functions of continuous random variables help in finding expected values and identifying certain properties of probability distributions.

> The **moment-generating function** of a continuous random variable X with probability density function $f(x)$ is given by
>
> $$M(t) = E(e^{tX}) = \int_{-\infty}^{\infty} e^{tX} f(x) dx$$
>
> when the integral exists.

We can use the moment-generating function to find the expected values of different powers of the random variable as

$$M^{(k)}(0) = \frac{d^{(k)}M(t)}{dt}\bigg|_{t=0} = E(X^k)$$

Two important properties of moment-generating functions are as follows:

1. If a random variable X has a moment-generating function $M_X(t)$, then $Y = aX + b$, for constants a and b, has a moment-generating function

$$M_Y(t) = e^{tb}M_X(at)$$

2. Moment-generating functions are unique to the distribution. That is, if two random variables have the same moment-generating functions, then they have the same probability distribution. If they have different moment-generating functions, then they have different distributions.

Example 6.28

For the exponential distribution, the moment-generating function is given by

$$M(t) = \int_0^{\infty} e^{tX}\left(\frac{1}{\theta}\right)e^{-x/\theta}dx = \frac{1}{\theta}\int_0^{\infty} e^{-x(1-\theta t)/\theta}dx$$

$$= \frac{1}{\theta}\Gamma(1)\left(\frac{\theta}{1 - \theta t}\right) = \frac{1}{(1 - \theta t)}$$

Because $M(t)$ needs to exist only in a neighborhood of zero, t can be small enough to make $(1 - \theta t)$ positive.

We can now use $M(t)$ to find $E(X)$, since $M'(0) = E(X)$. For exponential distribution, we have

$$E(X) = M'(0) = \left[\frac{-(-\theta)}{(1 - \theta t)^2}\right]_{t=0} = \theta$$

Similarly, we could find $E(X^2) = M''(0)$ and then $V(X) = M''(0) - [M'(0)]^2$.

Example 6.29

For the gamma distribution $M(t) = (1 - \beta t)^{-\alpha}$. Thus, if X has a gamma distribution, then

$$E(X) = M'(0) = \left[\frac{(-\alpha)(-\beta)}{(1 - \beta t)^{\alpha+1}}\right]_{t=0} = \alpha\beta$$

$$E(X^2) = M''(0) = \left[\frac{\alpha\beta(-\alpha - 1)}{(1 - \beta t)^{\alpha+2}} \right]_{t=0} = \alpha(\alpha + 1)\beta^2$$

$$E(X^3) = M'''(0) = \left[\frac{\alpha(\alpha + 1)\beta^2(-\alpha - 2)}{(1 - \beta t)^{\alpha+3}} \right]_{t=0} = \alpha(\alpha + 1)(\alpha + 2)\beta^3$$

Example 6.30

The force h exerted by a mass m moving at velocity v is

$$h = \frac{mv^2}{2}$$

Consider a device that fires a serrated nail into concrete at a mean velocity of 500 feet per second, where V, the random velocity, possesses a density function

$$f(v) = \frac{v^3 e^{-v/b}}{b^4 \Gamma(4)} \quad b = 500, \quad v > 0$$

If each nail possesses mass m, find the expected force exerted by a nail.

Solution Note that

$$E(H) = E\left(\frac{mV^2}{2} \right) = \frac{m}{2} E(V^2)$$

The density function for V is a gamma-type function with $\alpha = 4$ and $\beta = 500$. Thus,

$$E(V^2) = \alpha(\alpha + 1)\beta^2 = 4(4 + 1)(500^2) = 5,000,000$$

and therefore,

$$E(H) = \frac{m}{2} E(V^2) = \frac{m}{2}(5,000,000) = 2,500,000m$$

Example 6.31

Suppose X has a gamma distribution with parameters α and β. Then

$$M_X(t) = (1 - \beta t)^{-\alpha}.$$

Now let $Y = aX + b$ for constants a and b. From property 1 above,

$$M_Y(t) = e^{ib}[1 - \beta(at)]^{-\alpha} = e^{ib}[1 - (a\beta)t]^{-\alpha}$$

Because this moment-generating function is *not* of the gamma form (it has an extra multiplier of e^{ib}), Y will *not* have a gamma distribution (from property 2). If, however, $b = 0$ and $Y = aX$, then Y will have a gamma distribution with parameters α and $\alpha\beta$.

Example 6.32

Suppose X is normally distributed with mean μ and variance σ^2. To investigate its moment-generating function, let us first find the moment-generating function of $(X - \mu)$. Now

$$E\left(e^{t(X-\mu)}\right) = \int_{-\infty}^{\infty} e^{t(X-\mu)} \frac{1}{\sigma\sqrt{2\pi}} e^{-(X-\mu)^2/2\sigma^2} dx$$

Letting $y = x - \mu$, the integral becomes

$$E\left(e^{t(X-\mu)}\right) = E(e^{ty}) = \frac{1}{\sigma\sqrt{2\pi}} \int_{-\infty}^{\infty} e^{ty - y^2/2\sigma^2} dy$$

$$= \frac{1}{\sigma\sqrt{2\pi}} \int_{-\infty}^{\infty} \exp\left[-\frac{1}{2\sigma^2}(y^2 - 2\sigma^2 ty)\right] dy$$

Upon completing the square in the exponent, we have

$$-\frac{1}{2\sigma^2}(y^2 - 2\sigma^2 ty) = -\frac{1}{2\sigma^2}(y - \sigma^2 t)^2 + \frac{1}{2}t^2\sigma^2$$

So the integral becomes

$$E(e^{ty}) = \underbrace{e^{t^2\sigma^2/2} \frac{1}{\sigma\sqrt{2\pi}} \int_{-\infty}^{\infty} \exp\left[-\frac{1}{2\sigma^2}(y - \sigma^2 t)^2\right] dy}_{1} = e^{t^2\sigma^2/2}$$

because the remaining integrand forms a normal probability density that integrates to unity. Thus, the moment-generating function of Y is given by

$$M_Y(t) = e^{t^2\sigma^2/2}$$

Then $X = Y + \mu$ has a moment-generating function; using property 1 stated above,

$$M_X(t) = e^{t\mu} M_Y(t) = e^{t\mu} e^{t^2\sigma^2/2} = e^{t\mu + t^2\sigma^2/2}$$

For this same normal random variable X, let

$$Z = \frac{X - \mu}{\sigma} = \left(\frac{1}{\sigma}\right)X - \left(\frac{\mu}{\sigma}\right)$$

Then by property 1 stated above,

$$M_Z(t) = e^{-\mu t/\sigma} e^{\mu t/\sigma + t^2\sigma^2/2\sigma^2} = e^{t^2/2}$$

The moment-generating function of Z has the form of a moment-generating function for a normal random variable with mean 0 and variance 1. Thus, by property 2, variable Z must have a normal distribution with mean 0 and variance 1.

Exercises

6.100 Show that a gamma distribution with parameters α and β, has the moment-generating function

$$M(t) = (1 - \beta t)^{-\alpha}$$

6.101 Using the moment-generating function for the exponential distribution with mean θ, find $E(X^2)$. Use this result to show that $V(X) = \theta^2$.

6.102 Let Z denote a standard normal random variable. Find the moment-generating function of Z directly from the definition.

6.103 Let Z denote a standard normal random variable. Find the moment-generating function of Z^2. What does the uniqueness property of the moment-generating function tell you about the distribution of Z^2?

6.12 Simulating Probability Distributions

At times, theoretical distribution of the functions of random variables might not be known to the experimenters. In such situations, the probability distributions can be simulated for decision making. Most of the modern statistical software are capable of generating measurements from commonly used distributions.

Example 6.33

One particular assembly in bicycles requires inserting a solid bar into a hollow cylinder. The bicycle manufacturer receives bars and cylinders from two different contractors. The information provided by the contractors says that

- The bar diameters are approximately normally distributed with mean 2 cm and standard deviation 0.1 cm.
- The cylinder's inner diameters are approximately normally distributed with mean 2.1 cm and standard deviation 0.08 cm.

If the bar diameter is larger than the cylinder diameter, the assembly is not possible. Also bars with diameter more than 0.3 cm smaller than the cylinder are too loose to be used in the assembly. If one bar and one cylinder are selected at random, what is the probability that the assembly is not possible?

Solution Let $C =$ the diameter of the cylinder and $B =$ the diameter of the bar. Then

- B is approximately normally distributed with $\mu_B = 2$ cm and $\sigma_B = 0.1$ cm.
- C is approximately normally distributed with $\mu_C = 2.1$ cm and $\sigma_C = 0.08$ cm.

We are interested in the difference $(C - B)$. The assembly is not possible if $(C - B) < 0$ or $(C - B) > 0.3$. Let us simulate this situation. One such simulation using Minitab gives the following results:

Variable	N	Mean	SE Mean	StDev	Minimum	Q1	Median	Q3	Maximum
C-B	100	.0875	0.0135	0.1349	−0.2004	−0.0099	0.0853	0.1766	0.4189

Figure 6.49

Simulated distribution
of the difference in diameters

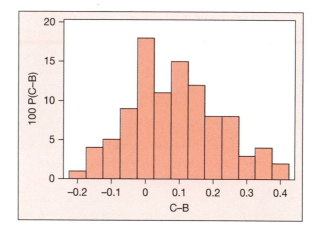

The simulated results show that in 24 pairs of bar and cylinder, the bar diameter is larger than the cylinder diameter resulting in infeasible assembly. Also in 5 pairs, the cylinder diameter is more than 0.3 cm larger than the bar diameter. Again this pairing results in useless assembly. Overall, in 29 out of 100 pairs the assembly is not possible. Therefore, we can conclude that the probability of useless pairing is about $29/100 = 0.29$.

This problem can be solved knowing that a linear combination of normal variates follows a normal distribution. The difference $(C - B)$ is normally distributed with mean and standard deviation,

$$\mu_{(C-B)} = 2.1 - 2.0 = 0.1 \quad \text{and}$$

$$\sigma_{(C-B)} = \sqrt{\sigma_C^2 + \sigma_B^2} = \sqrt{0.08^2 + 0.1^2} = 0.1281$$

Therefore, we can compute the probability of $(C - B) < 0$ or $(C - B) > 0.3$ using normal distribution as

$$P((C - B) < 0) + P((C - B) > 0.3)$$
$$= P\left(z < \frac{0 - 0.1}{0.1281}\right) + P\left(z > \frac{0.3 - 0.1}{0.1281}\right) = 0.2767$$

We can see that our simulated probability closely approximates the theoretical probability. Note that different runs of this simulation will result in different probabilities; however, they should be close to the theoretical value.

Example 6.34

Agricultural scientists at one research facility use standard handheld data collection instruments marketed by a manufacturer. The instrument works on a rechargeable battery and includes a backup battery. While the instrument is in use, the backup battery kicks in automatically when the rechargeable battery is depleted. The lifetime of rechargeable battery has a lognormal distribution with location parameter $\mu = 2.0$ and scale parameter $\sigma = 0.25$. The lifetime of backup battery has exponential distribution with mean 0.30. Suppose the researchers are planning a 5-hour data collection trip. What is the likelihood that the instrument will run out of charge and the scientist will not be able to complete data collection?

Solution Let R = the lifetime of a rechargeable battery and B = the lifetime of a backup battery. Then $T = R + B$ gives the length of time for which the instrument can be used on a field trip. It is known that R follows a lognormal distribution with parameters 2.0 and 0.25; and B follows an exponential distribution with mean 0.30. Using statistical software, we can easily generate (say, 100) random variables from each of these distributions and compute T for each pair. The distributions of measurements generated from one such simulation using JMP are given in Figure 6.50.

In this simulation, we have three measurements of T with values below 5. Therefore, we can say that there is about 0.03 probability that the instrument will not last long enough to complete a 5-hour data-collection trip. Note that different simulations will result in different answers; however, they should be fairly close to each other.

Figure 6.50

Simulated distributions of lifetimes

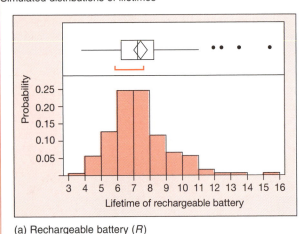

(a) Rechargeable battery (R)

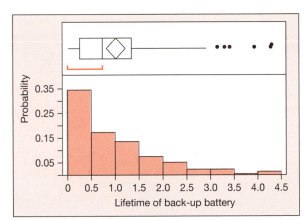

(b) Backup battery (B)

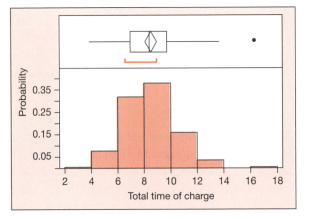

(c) Total (T)

Example 6.35

Consider a situation involving exponential random variables. In engineering, one is often concerned with a *parallel system*. A system is defined as a parallel system if it functions when at least one of the components of the system is working. Let us consider a parallel system that has two components. Define two independent random variables (X and Y) as follows:

X = Time until failure for component 1 (exponential with mean θ_x)
Y = Time until failure for component 2 (exponential with mean θ_y)

A variable of interest is the maximum of X and Y, i.e., say $W = \max\{X,Y\}$. Let us simulate the distribution of W to see its behavior. Figure 6.51 gives the results of three separate simulations for this situation. Note that the distribution of W does *not* look like an exponential distribution. All three distributions are mound-shaped

Figure 6.51

Relative frequency distributions of simulated *w* values

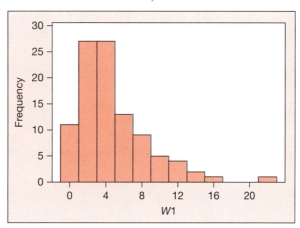

$$\theta_x = 4, \theta_y = 2$$
$$\overline{w} = 4.815, s = 3.835$$

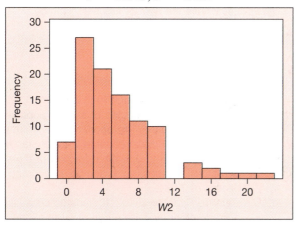

$$\theta_x = 1, \theta_y = 5$$
$$\overline{w} = 5.845, s = 4.427$$

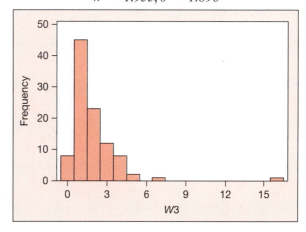

$$\theta_x = 0.5, \theta_y = 1.5$$
$$\overline{w} = 1.955, s = 1.896$$

and right-skewed. The distributions of X and Y (not shown here) are exponential, which means there is fairly high probability for time until failure to be very close to zero, whereas the probability of higher times to failure decreases rapidly with increasing time (refer to Section 6.4 for the shape of the exponential distribution). On the other hand, the maximum lifetime of two parallel components has relatively low probability of being close to zero and the maximum could also be quite large compared to most values of X and Y.

6.13 Summary

Improving the quality of processes often requires careful study of the behavior of observations on a continuum (time, weight, distance, etc.). Probability distributions for such measurements are characterized by *probability density functions* and cumulative *distribution functions*. The *uniform* distribution models the waiting time for an event that is equally likely to occur anywhere in a finite interval. The *exponential* distribution is a useful model for times between random events. The *gamma* model fits the behavior of sums of exponential variables. The most commonly used continuous probability model, *normal* distribution exhibits the symmetric and mound-shaped behavior often seen in real data. For the mound-shaped but skewed data, the *Weibull* provides a useful and versatile model. All models are approximations to reality, but the ones presented here are relatively easy to use while providing good approximations in a variety of settings.

Supplementary Exercises

6.104 Let Y possess a density function

$$f(y) = \begin{cases} cy & 0 \le y \le 2 \\ 0 & \text{elsewhere} \end{cases}$$

a Find c.
b Find $F(y)$.
c Graph $f(y)$ and $F(y)$.
d Use $F(y)$ in part (b) to find $P(1 \le Y \le 2)$.
e Use the geometric figure for $f(y)$ to calculate $P(1 \le Y \le 2)$.

6.105 Let Y have the density function given by

$$f(y) = \begin{cases} cy^2 + y & 0 \le y \le 2 \\ 0 & \text{elsewhere} \end{cases}$$

a Find c.
b Find $F(y)$.
c Graph $f(y)$ and $F(y)$.
d Use $F(y)$ in part (b) to find $F(-1)$, $F(0)$, and $F(1)$.

e Find $P(0 \le Y \le 0.5)$.
f Find the mean and variance of Y.

6.106 Let Y have the density function given by

$$f(y) = \begin{cases} 0.2 & -1 < y \le 0 \\ 0.2 + cy & 0 < y \le 1 \\ 0 & \text{elsewhere} \end{cases}$$

Answer parts (a) through (f) in Exercise 6.105.

6.107 The grade-point averages of a large population of college students are approximately normally distributed with mean equal to 2.4 and standard deviation equal to 0.5. What fraction of the students will possess a grade-point average in excess of 3.0?

6.108 Refer to Exercise 6.107. If students possessing a grade-point average equal to or less than 1.9 are dropped from college, what percentage of the students will be dropped?

6.109 Refer to Exercise 6.107. Suppose that three students are randomly selected from the student body. What is the probability that all three will possess a grade-point average in excess of 3.07?

6.110 A machine operation produces bearings with diameters that are normally distributed with mean and standard deviation equal to 3.0005 and 0.001, respectively. Customer specifications require the bearing diameters to lie in the interval 3.000 ± 0.0020. Those outside the interval are considered scrap and must be re-machined or used as stock for smaller bearings. With the existing machine setting, what fraction of total production will be scrap?

6.111 Refer to Exercise 6.110. Suppose five bearings are drawn from production. What is the probability that at least one will be defective?

6.112 Let Y have density function

$$f(y) = \begin{cases} cye^{-2y} & 0 \le y \le \infty \\ 0 & \text{Otherwise} \end{cases}$$

a Give the mean and variance for Y.
b Give the moment-generating function for Y.
c Find the value of c.

6.113 Find $E(X^k)$ for the beta random variable. Then find the mean and variance for the beta random variable.

6.114 The yield force of a steel-reinforcing bar of a certain type is found to be normally distributed with a mean of 8,500 pounds and a standard deviation of 80 pounds. If three such bars are to be used on a certain project, find the probability that all three will have yield forces in excess of 8,700 pounds.

6.115 The lifetime X of a certain electronic component is a random variable with density function

$$f(x) = \begin{cases} (1/100)e^{-x/100} & x > 0 \\ 0 & \text{elsewhere} \end{cases}$$

Three of these components operate independently in a piece of equipment. The equipment fails if at least two of the components fail. Find the probability that the equipment operates for at least 200 hours without failure.

6.116 An engineer has observed that the gap times between vehicles passing a certain point on a highway have an exponential distribution with a mean of 10 seconds.
a Find the probability that the next gap observed will be no longer than 1 minute.
b Find the probability density function for the sum of the next four gap times to be observed. What assumptions are necessary for this answer to be correct?

6.117 The proportion of time, per day, that all checkout counters in a supermarket are busy is a random variable X having probability density function

$$f(x) = \begin{cases} kx^2(1-x)^4 & 0 \le x < 0 \\ 0 & \text{elsewhere} \end{cases}$$

a Find the value of k that makes this a probability density function.
b Find the mean and variance of X.

6.118 If the lifelength X for a certain type of battery has a Weibull distribution with $\gamma = 2$ and $\theta = 3$ (with measurements in years), find the probability that the battery lasts less than four years given that it is now two years old;

6.119 The time (in hours) it takes a manager to interview an applicant has an exponential distribution with $\theta = 1/2$. Three applicants arrive at 8:00 AM, and interviews begin. A fourth applicant arrives at 8:45 AM. What is the probability that he has to wait before seeing the manager?

6.120 The weekly repair cost Y for a certain machine has a probability density function given by

$$f(y) = \begin{cases} 3(1-y)^2 & 0 \le y \le 1 \\ 0 & \text{elsewhere} \end{cases}$$

with measurements in hundreds of dollars. How much money should be budgeted each week for repair costs so that the actual cost will exceed the budgeted amount only 10% of the time?

6.121 A builder of houses has to order some supplies that have a waiting time for delivery Y uniformly distributed over the interval 1 to 4 days. Because she can get by without the supplies for 2 days, the cost of the delay is fixed at $100 for any waiting time up to 2 days. However, after 2 days the cost of the delay is $100 plus $20 per day for any time beyond 2 days. That is, if the waiting time is 3.5 days, the cost of the delay is $100 + $20(1.5) = $130. Find the expected value of the builder's cost due to waiting for supplies.

6.122 There is a relationship between incomplete gamma integrals and sums of Poisson probabilities, given by

$$\frac{1}{\Gamma(\alpha)} \int_k^\infty y^{\alpha-1} e^{-y} dy = \sum_{y=0}^{\alpha-1} \frac{\lambda^y e^{-\kappa}}{y!}$$

for integer values of α. If Y has a gamma distribution with $\alpha = 2$ and $\beta = 1$, find $P(Y > 1)$ by using the above equality and Table 3 of the Appendix.

6.123 The weekly downtime in hours for a certain production line has a gamma distribution with $\alpha = 3$

and $\beta = 2$. Find the probability that downtime for a given week will not exceed 10 hours.

6.124 Suppose that plants of a certain species are randomly dispersed over a region, with a mean density of λ plants per unit area. That is, the number of plants in a region of area A has a Poisson distribution with mean λA. For a randomly selected plant in this region, let R denote the distance to the nearest neighboring plant.

a Find the probability density function for R. [*Hint*: Note that $P(R > r)$ is the same as the probability of seeing no plants in a circle of radius r.]

b Find $E(R)$.

6.125 If X has a lognormal distribution with $\mu = 4$ and $\sigma^2 = 1$, find

a $P(X \leq 4)$

b $P(X > 8)$

6.126 The grains comprising polycrystalline metals tend to have weights that follow a lognormal distribution. For a certain type of aluminum, gram weights have a lognormal distribution with $\mu = 3$ and $\sigma = 4$ (in units of 10^{-2} gram).

a Find the mean and variance of the grain weights.

b Find an interval in which at least 75% of the grain weights should lie. [*Hint*: Use Tchebysheff's theorem.]

c Find the probability that a randomly chosen grain weighs less than the mean gram weight.

6.127 Let Y denote a random variable with probability density function given by

$$f(y) = (1/2)e^{-|y|} \quad -\infty < y < \infty$$

Find the moment-generating function of Y and use it to find $E(Y)$.

Chapter

7

Multivariate Probability Distributions

Do hurricanes with lower surface pressure and maximum low-level winds move forward more slowly? How does the forward speed of hurricane relate to the minimum surface pressure and the maximum low-level wind speed?

To answer such questions, we need to study the relationships between multiple variables. In the previous chapter, we studied tools for analyzing univariate data (i.e., a set of measurements on a single variable), and bivariate data (i.e., a set of two measurements on each item in the study). In many investigations, multiple types of observations are made on the same experimental units. Tensile and torsion properties are both measured on the same sample of wire, for example, or amount of cement and porosity ore both measured on the same sample of concrete. Physicians studying the effect of aspirin on heart attacks also want to understand its effect on strokes. This chapter provides techniques for modeling the joint behavior of two or more random variables being studied together. The ideas of *covariance* and *correlation* are introduced to measure the direction and strength of associations between variables.

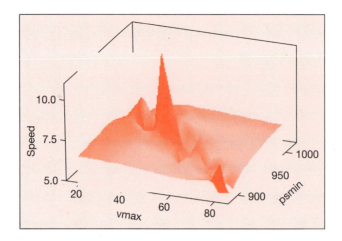

Figure 7.1

3-dimensional scatterplot of forward speed versus pressure and wind speed

7.1 Bivariate and Marginal Probability Distributions

Chapters 5 and 6 dealt with experiments that produced a single numerical response, or random variable, of interest. We discussed, for example, the lifetime X of a type of battery or the strength Y of a steel casing. Often, however, we want to study the joint behavior of two random variables, such as the joint behavior of lifetime *and* casing strength for these batteries. Perhaps, in such a study, we can identify a combination(s) of lifetime and casing strength that will balance cost of manufacturing with customer satisfaction. To proceed with such a study, we must know how to handle joint probability distributions. When only two random variables are involved, these joint distributions are called *bivariate distributions*. Some other examples of bivariate situations include these:

- A physician studies the joint behavior of pulse and exercise time.
- An educator studies the joint behavior of grades and time devoted to study, or the interrelationship of pretest and posttest scores.
- An economist studies the joint behavior of business volume and profits.

In fact, most real problems we come across will have more than one underlying random variable of interest.

Example 7.1

The National Highway Traffic Safety Administration is interested in the effect of seat belt use on saving lives. One study reported statistics on children under the age of 5 who were involved in motor vehicle accidents in which at least one fatality occurred. For 7,060 such accidents between 1985 and 1989, the results are shown in Table 7.1.

It is more convenient to analyze such data in terms of meaningful random variables than in terms of described categories. For each child, it is important to know

Table 7.1

Children Involved in Motor Vehicle Accidents

	Survivors	Fatalities	Total
No belt	1,129	509	**1,638**
Adult belt	432	73	**505**
Child seat	733	139	**872**
Total	**2,294**	**721**	**3,015**

whether or not he or she survived and what the seat belt situation was. For each child, define two random variables as follows:

$$X_1 = \begin{cases} 0 & \text{if child survived} \\ 1 & \text{if child did not survive} \end{cases} \quad \text{and} \quad X_2 = \begin{cases} 0 & \text{if no belt used} \\ 1 & \text{if adult belt used} \\ 2 & \text{if child seat used} \end{cases}$$

X_1 will keep track of the number of child fatalities and X_2 will keep track of the type of restraining device used for the child.

The frequencies from Table 7.1 are turned into the relative frequencies of Table 7.2 to produce the *joint probability distribution* of X_1 and X_2. In general, we write

$$P(X_1 = x_1, X_2 = x_2) = p(x_1, x_2)$$

and call $p(x_1, x_2)$ the *joint probability function* of (X_1, X_2). For example (see Table 7.2),

$$P(X_1 = 0, X_2 = 2) = p(0, 2) = \frac{733}{3,015} = 0.24$$

represents the approximate probability that a child will both survive *and* be in a child seat when involved in a fatal accident.

Table 7.2

Joint Probability Distribution

		X_1		
		0	**1**	**Total**
	0	0.38	0.17	**0.55**
X_2	1	0.14	0.02	**0.16**
	2	0.24	0.05	**0.29**
	Total	**0.76**	**0.24**	**1.00**

The probability that a child will be in a child seat is

$$P(X_2 = 2) = P(X_1 = 0, X_2 = 2) + P(X_1 = 1, X_2 = 2)$$
$$= 0.24 + 0.05$$
$$= 0.29$$

which is one of the *marginal probabilities* for X_2. The univariate distribution along the right margin is the *marginal distribution* for X_2 alone, and the one along the bottom row is the marginal distribution for X_1 alone.

The *conditional probability distribution* for X_1 given X_2 fixes a value of X_2 (a row of Table 7.2) and looks at the relative frequencies for values of X_1 in that row. For example, conditioning on $X_2 = 0$ produces

$$P(X_1 = 0 \mid X_2 = 0) = \frac{P(X_1 = 0, X_2 = 0)}{P(X_2 = 0)} = \frac{0.38}{0.55} = 0.69$$

and

$$P(X_1 = 1 \mid X_2 = 0) = 0.31$$

These two values show how survivorship relates to the situation in which no seat belt is used; 69% of children in such situations survived fatal accidents. That may seem high, but compare it to $P(X_1 = 0 \mid X_2 = 2)$. Does seat belt use seem to be beneficial?

There is another set of conditional distributions here, namely, those for X_2 conditioning on X_1. In this case,

$$P(X_2 = 0 \mid X_1 = 0) = \frac{P(X_1 = 0, X_2 = 0)}{P(X_1 = 0)} = \frac{0.38}{0.76} = 0.5$$

and

$$P(X_2 = 0 \mid X_1 = 1) = 0.17/0.24 = 0.71$$

What do these two probabilities tell you about the effectiveness of seat belts?

Let X_1 and X_2 be discrete random variables. The **joint probability distribution** of X_1 and X_2 is

$$p(x_1, x_2) = P(X_1 = x_1, X_2 = x_2)$$

defined for all real numbers x_1 and x_2.

The function $p(x_1, x_2)$ is called the **joint probability function** of X_1 and X_2.

The **marginal probability functions** of X_1 and X_2 respectively are given by

$$p_1(x_1) = \sum_{x_2} p(x_1, x_2) \quad \text{and} \quad p_2(x_2) = \sum_{x_1} p(x_1, x_2)$$

Example 7.2

There are three checkout counters in operation at a local supermarket. Two customers arrive at the counters at different times, when the counters are serving no other customers. Assume that the customers then choose a checkout station at random and independently of one another. Let X_1 denote the number of times counter A is selected and X_2 the number of times counter B is selected by the two customers.

a Find the joint probability distribution of X_1 and X_2.
b Find the probability that one of the customers visits counter B given that one of the customers is known to have visited counter A.

Solution **a** For convenience, let us introduce X_3, defined as the number of customers visiting counter C. Now the event ($X_1 = 0$ and $X_2 = 0$) is equivalent to the event ($X_1 = 0, X_2 = 0$, and $X_3 = 2$). It follows that

$$P(X_1 = 0, X_2 = 0) = P(X_1 = 0, X_2 = 0, X_3 = 2)$$
$$= P(\text{customer I selects counter } C \underline{\text{ and }} \text{customer II selects counter } C)$$
$$= P(\text{customer I selects counter}) \times P(\text{customer II selects counter } C)$$
$$= \frac{1}{3} \times \frac{1}{3} = \frac{1}{9}$$

because customers' choices are independent and each customer makes a random selection from among the three available counters. The joint probability distribution is shown in Table 7.3.

Table 7.3

Joint Distribution of X_1 and X_2

		X_1			Marginal Probabilities for X_2
		0	1	2	
X_2	0	1/9	2/9	1/9	4/9
	1	2/9	2/9	0	4/9
	2	1/9	0	0	1/9
Marginal Probabilities for X_1		4/9	4/9	1/9	1

It is slightly more complicated to calculate $P(X_1 = 1, X_2 = 0) = P(X_1 = 1, X_2 = 0, X_3 = 1)$. In this event, customer I could select counter A and customer II could select counter C, or customer I could select counter C and customer II could select counter A. Thus,

$$P(X_1 = 1, X_2 = 0) = P(\text{I selects } A)P(\text{II selects } C)$$
$$+ P(\text{I selects } C)P(\text{II selects } A)$$
$$= \frac{1}{3} \times \frac{1}{3} + \frac{1}{3} \times \frac{1}{3} = \frac{2}{9}$$

Similar arguments allow us to derive the results in Table 7.3.

b $P(X_2 = 1 \mid X_1 = 1) = \dfrac{P(X_2 = 1, X_1 = 1)}{P(X_1 = 1)} = \dfrac{2/9}{4/9} = \dfrac{1}{2}$

Does this answer agree with your intuition?

Before we move to the continuous case, let us first briefly review the situation in one dimension. If $f(x)$ denotes the probability density function of a random variable X, then $f(x)$ represents a relative frequency curve, and probabilities such as $P(a \leq X \leq b)$ are represented as areas under this curve. That is,

$$P(a \leq X \leq b) = \int_a^b f(x)\,dx$$

Now suppose we are interested in the joint behavior of two continuous random variables, say X_1 and X_2, where X_1 and X_2 might, for example, represent the amounts of two different hydrocarbons found in an air sample taken for a pollution study. The relative frequency behavior of these two random variables can be modeled by a bivariate function, $f(x_1, x_2)$, which forms a probability, or relative frequency, surface in three dimensions. Figure 7.2 shows such a surface. The probability that, X_1 lies in some interval and X_2 lies in another interval is then represented as a volume under this surface. Thus,

$$P(a_1 \leq X_1 \leq a_2, b_1 \leq X_2 \leq b_2) = \int_{b_1}^{b_2} \int_{a_1}^{a_2} f(x_1, x_2)\,dx_1\,dx_2$$

Note that the above integral simply gives the volume under the surface and over the shaded region in Figure 7.2.

Figure 7.2

A bivariate density function

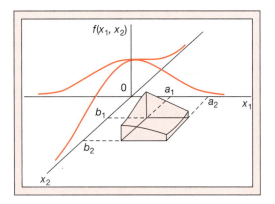

Example 7.3

A certain process for producing an industrial chemical yields a product containing two predominant types of impurities. For a certain volume of sample from this process, let X_1 denote the proportion of impurities in the sample and let X_2 denote the proportion of type I impurity among all impurities found. Suppose the joint distribution of X_1 and X_2, after investigation of many such samples, can be adequately modeled by the following function:

$$f(x_1, x_2) = \begin{cases} 2(1 - x_1) & 0 \le x_1 \le 1, 0 \le x_2 \le 1 \\ 0 & \text{elsewhere} \end{cases}$$

This function graphs as the surface given in Figure 7.3. Calculate the probability that X_1 is less than 0.5 and that X_2 is between 0.4 and 0.7.

Solution From the preceding discussion, we see that

$$P(0 \le x_1 \le 0.5, 0.4 \le x_2 \le 0.7) = \int_{0.4}^{0.7} \int_0^{0.5} 2(1 - x_1) dx_1 dx_2$$

$$= \int_{0.4}^{0.7} \left[-(1 - x_1)^2 \right]_0^{0.5} dx_2$$

$$= \int_{0.4}^{0.7} 0.75 \, dx_2$$

$$= 0.75(0.7 - 0.4) = 0.75(0.3) = 0.225$$

Thus, the fraction of such samples having less than 50% impurities and a relative proportion of type I impurities between 40% and 70% is 0.225.

Figure 7.3

Probability density function for Example 7.3

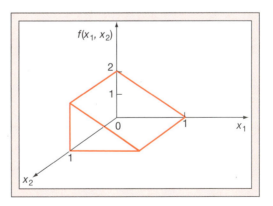

Just as the univariate, or marginal, probabilities were computed by summing over rows or columns in the discrete case, the univariate density function for X_1 in the continuous case can be found by integrating ("summing") over values of X_2.

The *marginal density function* of X_1, $f_1(x_1)$, is given by

$$f_1(x_1) = \int_{-\infty}^{\infty} f(x_1, x_2)\,dx_2$$

Similarly, the marginal density function of X_2, $f_2(x_2)$, is given by

$$f_2(x_2) = \int_{-\infty}^{\infty} f(x_1, x_2)\,dx_1$$

Example 7.4

For the situation in Example 7.3, find the marginal probability density functions for X_1 and X_2.

Solution   Let's first try to visualize what the answers should look like, before going through the integration. To find $f_1(x_1)$ we are to accumulate all probabilities in the x_2 direction. Look at Figure 7.3 and think of collapsing the wedge-shaped figure back onto the $(x_1, f(x_1, x_2))$ plane. Much more probability mass will build up toward the zero point of the X_1-axis than toward the unity point. In other words, the function $f_1(x_1)$ should be high at zero and low at 1. Formally,

$$f_1(x_1) = \int_{-\infty}^{\infty} f(x_1, x_2)\,dx_2 = \int_0^1 2(1 - x_1)\,dx_2 = 2(1 - x_1) \quad 0 \le x_1 \le 1$$

The function graphs as in Figure 7.4. Note that our conjecture is correct. Envisioning how $f_2(x_2)$ should look geometrically, suppose the wedge of Figure 7.3 is forced

Figure 7.4

Probability density function
$f_1(x_1)$ for Example 7.4

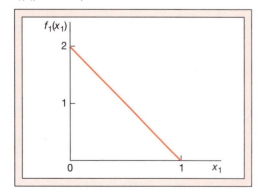

Figure 7.5

Probability density function
$f_2(x_2)$ for Example 7.4

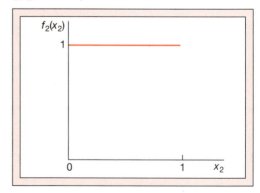

back onto the $(x_2, f_1(x_1, x_2))$ plane. Then the probability mass should accumulate equally all along the $(0,1)$ interval on the X_2-axis. Mathematically,

$$f_2(x_2) = \int_{-\infty}^{\infty} f(x_1, x_2)dx_1 = \int_0^1 2(1 - x_1)dx_1 = [-(1 - x_1)]_0^1 = 1 \quad 0 \le x_2 \le 1$$

and again our conjecture is verified, as seen in Figure 7.5.

Let X_1 and X_2 be continuous random variables. The joint **probability distribution function** of X_1 and X_2, if it exists, is given by a nonnegative function $f(x_1, x_2)$, which is such that

$$P(a_1 \le X_1 \le a_2, b_1 \le X_2 \le b_2) = \int_{b_1}^{b_2} \int_{a_1}^{a_2} f(x_1, x_2)dx_1 dx_2$$

The **marginal** probability density functions of X_1 and X_2, respectively, are given by

$$f_1(x_1) = \int_{-\infty}^{\infty} f(x_1, x_2)dx_2 \quad \text{and} \quad f_2(x_2) = \int_{-\infty}^{\infty} f(x_1, x_2)dx_1$$

Example 7.5

Gasoline is to be stocked in a bulk tank once each week and then sold to customers. To keep down the cost of inventory, the supply and demand are carefully monitored.

Let X_1 = the proportion of the tank that is stocked in a particular week
   X_2 = the proportion of the tank that is sold in that same week.

Due to limited supplies, X_1 is not fixed in advance but varies from week to week. Suppose a study over many weeks shows the joint relative frequency behavior of X_1 and X_2 to be such that the following density function provides an adequate model:

$$f(x_1, x_2) = \begin{cases} 3x_1 & 0 \le x_2 \le x_1 \le 1 \\ 0 & \text{elsewhere} \end{cases}$$

Note that X_2 must always be less than or equal to X_1. This density function is graphed in Figure 7.6. Find the probability that X_2 will be between 0.2 and 0.4 for a given week.

Figure 7.6
The joint density function for Example 7.5

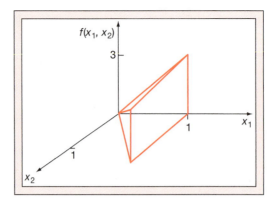

Solution   The question refers to the marginal behavior of X_2. Thus, it is necessary to find

$$f_2(x_2) = \int_{-\infty}^{\infty} f(x_1, x_2)dx_1 = \int_{x_2}^{1} 3x_1 dx_1 = \left[\frac{3}{2}x_1^2\right]_{x_2}^{1} = \frac{3}{2}\left(1 - x_2^2\right) \quad 0 \le x_2 \le 1$$

It follows directly that

$$P(0.2 \le X_2 \le 0.4) = \int_{0.2}^{0.4} \frac{3}{2}\left(1 - x_2^2\right)dx_2$$

$$= \frac{3}{2}\left(x_2 - \frac{x_2^3}{3}\right)\Bigg|_{0.2}^{0.4}$$

$$= \frac{3}{2}\left\{\left[0.4 - \frac{(0.4)^3}{3}\right] - \left[0.2 - \frac{(0.2)^3}{3}\right]\right\}$$

$$= 0.272$$

Note that the marginal density of X_2 graphs as a function that is high at $x_2 = 0$ and then tends to 0 as x_2 tends to 1. Does this agree with your intuition after looking at Figure 7.5?

7.2 Conditional Probability Distributions

Recall that in the bivariate *discrete* case the conditional probabilities for X_1 for a given X_2 were found by fixing attention on the particular row in which $X_2 = x_2$ and then looking at the relative probabilities within that row. That is, the individual cell probabilities were divided by the marginal total for that row in order to obtain conditional probabilities.

In the bivariate *continuous* case, the form of the probability density function representing the conditional behavior of X_1 for a given value of X_2 is found by slicing through the joint density in the X_1 direction at the particular value of X_2. This function then has to be weighted by the marginal density function for X_2 at that point. We will look at a specific example before giving the general definition of conditional density functions.

Example 7.6

Refer to the tank inventory problem of Example 7.5. Find the conditional probability that X_2 is less than 0.2 given that X_1 is known to be 0.5.

Solution Slicing through $f(x_1, x_2)$ in the x_2 direction at $X_1 = 0.5$ yields

$$f(0.5, x_2) = 3(0.5) = 1.5 \quad 0 \le x_2 \le 0.5$$

Thus, the conditional behavior of X_2 for a given X_1 of 0.5 is constant over the interval $(0, 0.5)$. The marginal value of $f_1(x_1)$ at $X_1 = 0.5$ is obtained as follows:

$$f_1(x_1) = \int_{-\infty}^{\infty} f(x_1, x_2)\,dx_2 = \int_0^{x_1} 3x_1\,dx_2 = 3x_1^2 \quad 0 \le x_1 \le 1$$

$$f_1(0.5) = 3(0.5)^2 = 0.75$$

Upon dividing, we see that the conditional behavior of X_2 for a given X_1 of 0.5 is represented by the function

$$f(x_2 \mid x_1 = 0.5) = \frac{f(0.5, x_2)}{f_1(0.5)} = \frac{1.5}{0.75} = 2 \quad 0 < x_2 < 0.5$$

This function of x_2 has all the properties of a probability density function and is the conditional density function for X_2 at $X_1 = 0.5$. Then

$$P(X_2 < 0.2 \mid X_1 = 0.5) = \int_0^{0.2} f(x_2 \mid x_1 = 0.5)\,dx_2 = \int_0^{0.2} 2\,dx_2 = 0.2(2) = 0.4$$

That is, among all weeks in which the tank was half full immediately after stocking, sales amounted to less than 20% of the tanks 40% of the time.

We see that the manipulations used above to obtain conditional density functions in the continuous case are analogous to those used to obtain conditional probabilities in the discrete case, except that integrals are used in place of sums.

Let X_1 and X_2 be jointly continuous random variables with joint probability density function $f(x_1, x_2)$ and marginal densities $f_1(x_1)$ and $f_2(x_2)$, respectively. Then the **conditional probability density function** of X_1 given $X_2 = x_2$ is given by

$$f(x_1 \mid x_2) = \begin{cases} \dfrac{f(x_1, x_2)}{f_2(x_2)} & f_2(x_2) > 0 \\ 0 & \text{elsewhere} \end{cases}$$

and the **conditional probability density function** of X_2 given $X_1 = x_1$ is given by

$$f(x_2 \mid x_1) = \begin{cases} \dfrac{f(x_1, x_2)}{f_1(x_1)} & f_1(x_1) > 0 \\ 0 & \text{elsewhere} \end{cases}$$

Example 7.7

A soft-drink machine has a random amount Y_2 in supply at the beginning of a given day and dispenses a random amount Y_1 during the day (with measurements in gallons). It is not resupplied during the day; hence, $Y_1 \leq Y_2$. It has been observed that Y_1 and Y_2 have joint density

$$f(y_1, y_2) = \begin{cases} 1/2 & 0 \leq y_1 \leq y_2, 0 \leq y_2 \leq 2 \\ 0 & \text{elsewhere} \end{cases}$$

That is, the points (Y_1, Y_2) are uniformly distributed over the triangle with the given boundaries. Find the conditional probability density of Y_1 given $Y_2 = y_2$. Evaluate the probability that less than $1/2$ gallon is sold, given that the machine contains 1 gallon at the start of the day.

Solution The marginal density of Y_2 is given by

$$f_2(y_2) = \int_{-\infty}^{\infty} f(y_1, y_2) \, dy_1 = \begin{cases} \int_0^{y_2} (1/2) \, dy_1 = (1/2)y_2 & 0 \leq y_2 \leq 2 \\ 0 & \text{elsewhere} \end{cases}$$

By Definition,

$$f_1(y_1 \mid y_2) = \frac{f(y_1, y_2)}{f_2(y_2)} = \begin{cases} \dfrac{1/2}{(1/2)y_2} = \dfrac{1}{y_2} & 0 \leq y_1 \leq y_2 \leq 2 \\ 0 & \text{elsewhere} \end{cases}$$

The probability of interest is

$$P(Y_1 < 1/2 \mid Y_2 = 1) = \int_{-\infty}^{1/2} f(y_1 \mid y_2 = 1)\,dy_1 = \int_0^{1/2} (1)\,dy_1 = 1/2$$

Note that if the machine had contained 2 gallons at the start of the day, then

$$P(Y_1 \leq 1/2 \mid Y_2 = 2) = \int_0^{1/2} \left(\frac{1}{2}\right) dy_1 = 1/4$$

Thus, the amount sold is highly dependent upon the amount in supply.

7.3 Independent Random Variables

Before defining independent random variables, we will recall once again that two events A and B are independent if $P(AB) = P(A)P(B)$. Somewhat analogously, two discrete random variables are independent if

$$P(X_1 = x_1, X_2 = x_2) = P(X_1 = x_1)P(X_2 = x_2)$$

for all real numbers x_1 and x_2. A similar idea carries over to the continuous case.

Discrete random variables X_1 and X_2 are said to be **independent** if

$$P(X_1 = x_1, X_2 = x_2) = P(X_1 = x_1)P(X_2 = x_2)$$

for all real numbers x_1 and x_2.

Continuous random variables X_1 and X_2 are said to be **independent** if

$$f(x_1, x_2) = f_1(x_1)f(x_2)$$

for all real numbers x_1 and x_2.

The concepts of joint probability density functions and independence extend directly to n random variables, where n is any finite positive integer. The n random variables $X_1, X_2, \ldots, X_n$ are said to be *independent* if their joint density function $f(x_1, \ldots, x_n)$ is given by

$$f(x_1, \ldots, x_n) = f_1(x_1)f_2(x_2)\ldots f_n(x_n)$$

for all real numbers $x_1, x_2, \ldots, x_n$.

Example 7.8

Show that the random variables having the joint distribution of Table 7.2 are not independent.

Solution   It is necessary to check only one entry in the table. We see that $P(X_1 = 0, X_2 = 0)$ = 0.38, whereas $P(X_1 = 0) = 0.76$ and $P(X_2 = 0) = 0.55$. Because

$$P(X_1 = 0, X_2 = 0) \neq P(X_1 = 0) P(X_2 = 0)$$

the random variables cannot be independent.

Example 7.9

Show that the random variables in Example 7.3, are independent.

Solution   Here,

$$f(x_1, x_2) = \begin{cases} 2(1 - x_1) & 0 \leq x_1 \leq 1, 0 \leq x_2 \leq 1 \\ 0 & \text{elsewhere} \end{cases}$$

We saw in Example 7.4 that

$$f_1(x_1) = 2(1 - x_1) \qquad 0 \leq x_1 \leq 1$$

and

$$f_2(x_2) = 1 \qquad\qquad 0 \leq x_2 \leq 1$$

Thus, $f(x_1, x_2) = f_1(x_1) f_2(x_2)$ for all real numbers x_1 and x_2; and X_1 and X_2 are independent random variables.

Exercises

7.1   Two construction contracts are to be randomly assigned to one or more of three firms (I, II, and III). A firm may receive more than one contract. Let
   X_1 = the number of contracts assigned to firm I
   X_2 = the number assigned to firm II.
   a Find the joint probability distribution for X_1 and X_2.
   b Find the marginal probability distribution for X_1.
   c Find $P(X_1 = 1 \mid X_2 = 1)$.

7.2   A radioactive particle is randomly located in a square area with sides one unit in length. Let X_1 and X_2 denote the coordinates of the particle. Since the particle is equally likely to fall in any subarea of a fixed size, a reasonable model for (X_1, X_2) is given by

$$f(x_1, x_2) = \begin{cases} 1 & 0 \leq x_1 \leq 1, 0 \leq x_2 \leq 1 \\ 0 & \text{elsewhere} \end{cases}$$

   a Sketch the probability density surface.
   b Find $P(X_1 \leq 0.2, X_2 \leq 0.4)$.
   c Find $P(0.1 \leq X_1 \leq 0.3, X_2 > 0.4)$.

7.3   The same study that produced the seat belt safety data of Table 7.1 also took into account the age of the child involved in a fatal accident. The results for those children wearing *no* seat belts are shown on the next page. Here, $X_1 = 1$ if the child did not survive and X_2 indicates the age in years. (An age of zero implies that the child was less

Age	Survivors	Fatalities
0	104	127
1	165	91
2	267	107
3	277	90
4	316	94

than 1 year old, an age of 1 implies the child was more than 1 year old but not yet 2, and so on.)

a Construct an approximate joint probability distribution for X_1 and X_2.

b Construct the conditional distribution of X_1 for fixed values of X_2. Discuss the implications of these results.

c Construct the conditional distribution of X_2 for fixed values of X_1. Are the implications the same as in part (b)?

7.4 An environmental engineer measures the amount (by weight) of particulate pollution in air samples (of a certain volume) collected over the smokestack of a coal-operated power plant. Let X_1 denote the amount of pollutant per sample when a certain cleaning device on the stack is not operating, and let X_2 denote the amount of pollutant per sample when the cleaning device is operating, under similar environmental conditions. It is observed that X_1 is always greater than $2X_2$, and the relative frequency behavior of (X_1, X_2) can be modeled by

$$f(x_1, x_2) = \begin{cases} k & 0 \le x_1 \le 2, 0 \le x_2 \le 1, 2x_2 \le x_1 \\ 0 & \text{elsewhere} \end{cases}$$

(That is, X_1 and X_2 are randomly distributed over the region inside the triangle bounded by $x_1 = 2$, $x_2 = 0$, and $2x_2 = x_1$.)

a Find the value of k that makes this a probability density function.

b Find $P(X_1 \ge 3X_2)$. (That is, find the probability that the cleaning device will reduce the amount of pollutant by one-third or more.)

7.5 Refer to Exercise 7.2.

a Find the marginal density function for X_1.

b Find $P(X_1 \le 0.5)$.

c Are X_1 and X_2 independent?

7.6 Refer to Exercise 7.4.

a Find the marginal density function of X_2.

b Find $P(X_2 \le 0.4)$.

c Are X_1 and X_2 independent?

d Find $P(X_2 \le 1/4 \mid X_1 = 1)$.

7.7 Let X_1 and X_2 denote the proportion of two different chemicals found in a sample mixture of chemicals used as an insecticide. Suppose X_1 and X_2 have joint probability density given by

$f(x_1, x_2)$

$$= \begin{cases} 2 & 0 \le x_1 \le 1, 0 \le x_2 \le 1, 0 \le x_1 + x_2 \le 1 \\ 0 & \text{elsewhere} \end{cases}$$

(Note that $X_1 + X_2$ must be at most unity since the random variables denote proportions within the same sample.)

a Find $P(X_1 \le 3/4, X_2 \le 3/4)$.

b Find $P(X_1 \le 1/2, X_2 \le 1/2)$.

c Find $P(X_1 \le 1/2 \mid X_2 \le 1/2)$.   *Same*

7.8 Refer to Exercise 7.5.

a Find the marginal density functions for X_1 and X_2.

b Are X_1 and X_2 independent?

c Find $P(X_1 > 1/2 \mid X_2 = 1/4)$.

7.9 Let X_1 and X_2 denote the proportions of time, out of one workweek, that employees I and II, respectively, actually spend performing their assigned tasks. The joint relative frequency behavior of X_1 and X_2 is modeled by the probability density function

$$f(x_1, x_2) = \begin{cases} x_1 + x_2 & 0 \le x_1 \le 1, 0 \le x_2 \le 1 \\ 0 & \text{elsewhere} \end{cases}$$

a Find $P(X_1 < 1/2, X_2 > 1/4)$.

b Find $P(X_1 + X_2 \le 1)$.

c Are X_1 and X_2 independent?

7.10 Refer to Exercise 7.9. Find the probability that employee I spends more than 75% of the week on her assigned task, given that employee II spends exactly 50% of the work week on his assigned task.

7.11 An electronic surveillance system has one of each of two different types of components in joint operation. Letting X_1 and X_2 denote the random life lengths of the components of type I and type II, respectively, we have the joint probability density function given by

$$f(x_1, x_2) = \begin{cases} (1/8)x_1 e^{-(x_1+x_2)/2} & x_1 > 0, x_2 > 0 \\ 0 & \text{elsewhere} \end{cases}$$

(Measurements are in hundreds of hours.)

a Are X_1 and X_2 independent?

b Find $P(X_1 > 1, X_2 > 1)$.

7.12 A bus arrives at a bus stop at a randomly selected time within a 1-hour period. A passenger arrives at the bus stop at a randomly selected time within the same hour. The passenger will wait for the bus up to one-quarter of an hour. What is the probability that the passenger will catch the bus? [*Hint*: Let X_1 denote the bus arrival time and X_2 the passenger arrival time. If these arrivals are independent, then

$$f(x_1, x_2) = \begin{cases} 1 & 0 \le x_1 \le 1, 0 \le x_2 \le 1 \\ 0 & \text{elsewhere} \end{cases}$$

Now find $P(X_2 \le X_1 \le X_2 + 1/4)$.]

7.13 Two friends are to meet at a library. Each arrives at independently and randomly selected times within a fixed 1-hour period. Each agrees to wait no more than 10 minutes for the other. Find the probability that they meet. Discuss how your solution generalizes to the study of the distance between two random points in a finite interval.

7.14 Each of two quality-control inspectors interrupts a production line at randomly, but independently, selected times within a given workday (of 8 hours). Find the probability that the two interruptions will be more than 4 hours apart.

7.15 Two telephone calls come into a switchboard at random times in a fixed 1-hour period. If the calls are made independently of each other,
  a find the probability that both calls are made in the first half hour.
  b find the probability that the two calls are made within 5 minutes of each other.

7.16 A bombing target is in the center of a circle with a radius of 1 mile. A bomb falls at a randomly selected point inside that circle. If the bomb destroys everything within 1/2 mile of its landing point, what is the probability that it destroys the target?

7.4 Expected Values of Functions of Random Variables

When we encounter problems that involve more than one random variable, we often combine the variables into a single function. We might be interested in the average lifelength of five different electronic components within the same system or the difference between two different strength-test measurements on the same section of cable. We now discuss how to find expected values of functions of more than one random variable.

The basic procedure for finding expected values in the bivariate case is given below. It can be generalized to higher dimensions.

Expected Values of Discrete and Continuous Random Variables

Suppose the discrete random variables (X_1, X_2) have a joint probability function given by $p(x_1, x_2)$. If $g(X_1, X_2)$ is any real-valued function of (X_1, X_2), then

$$E[g(X_1, X_2)] = \sum_{x_1} \sum_{x_2} g(x_1, x_2) p(x_1, x_2)$$

The sum is over all values of (X_1, X_2) for which $p(x_1, x_2) > 0$.

If (X_1, X_2) are continuous random variables with probability density function $f(X_1, X_2)$, then

$$E[g(X_1, X_2)] = \int_{-\infty}^{\infty} \int_{-\infty}^{\infty} g(x_1, x_2) f(x_1, x_2) dx_1 dx_2$$

If X_1 and X_2 are independent, then from the definition of expected value,

$$E[g(X_1)\,h(X_2)] = E[g(X_1)]\,E[h(X_2)]$$

A function of two variables that is commonly of interest in probabilistic and statistical problems is the *covariance*.

The **covariance** between two random variables X_1 and X_2 is given by

$$\mathrm{Cov}(X_1, X_2) = E[(X_1 - \mu_1)(X_2 - \mu_2)]$$

where $\mu_1 = E(X_1)$ and $\mu_2 = E(X_2)$.

The covariance helps us assess the relationship between two variables, in the following sense. If X_2 tends to be large when X_1 is large, and small when X_1 is small, then X_1 and X_2 will have a *positive covariance*. If, on the other hand, X_2 tends to be small when X_1 is large and large when X_1 is small, then the variables will have a *negative covariance*.

Whereas covariance measures the direction of the association between two random variables, *correlation* measures the strength of the association.

The **correlation coefficient** between two random variables X_1 and X_2 is given by

$$\rho = \frac{\mathrm{Cov}(X_1, X_2)}{\sqrt{V(X_1) \times V(X_2)}}$$

The correlation coefficient is a unitless quantity taking on values between -1 and $+1$. If $\rho = +1$ or $\rho = -1$, then X_2 must be a linear function of X_1. A simpler computational form for the covariance is given below without proof.

If X_1 has mean μ_1 and X_2 has mean μ_2, then

$$\mathrm{Cov}(X_1, X_2) = E(X_1 X_2) - \mu_1\mu_2$$

If X_1 and X_2 are independent random variables, then $E(X_1 X_2) = E(X_1)E(X_2)$. From the computational form of covariance, it is clear that independence between X_1 and X_2 implies that $\mathrm{Cov}(X_1, X_2) = 0$. The converse is not necessarily true. That is, zero covariance does not imply that the variables are independent. To see this, look at the joint distribution in Table 7.4

Table 7.4

Joint Probability Distribution of X_1 and X_2

		X_1			
		−1	**0**	**1**	**Total**
	−1	1/8	1/8	1/8	3/8
X_2	0	1/8	0	1/8	2/8
	1	1/8	1/8	1/8	3/8
Total		3/8	2/8	3/9	1

Clearly, $E(X_1) = E(X_2) = 0$, and $E(X_1 X_2) = 0$ as well. Therefore, $\text{Cov}(X_1, X_2) = 0$. On the other hand,

$$P(X_1 = -1, X_2 = -1) = 1/8 \neq P(X_1 = -1) \, P(X_2 = -1)$$

and so the random variables X_1 and X_2 are dependent.

Example 7.10

A firm that sells word processing systems keeps track of the number of customers who call on any one day and the number of orders placed on any one day. Let X_1 denote the number of calls, X_2 the number of orders placed, and $p(x_1, x_2)$ the joint probability function for (X_1, X_2); records indicate that

$$p(0, 0) = 0.04 \qquad p(2, 0) = 0.20$$
$$p(1, 0) = 0.16 \qquad p(2, 1) = 0.30$$
$$p(1, 1) = 0.10 \qquad p(2, 2) = 0.20$$

That is, for any given day; the probability of, say, two calls and one order is 0.30. Find $\text{Cov}(X_1, X_2)$.

Solution Let us first find $E(X_1, X_2)$, which is

$$E(X_1 X_2) = \sum_{x_1} \sum_{x_2} x_1 x_2 \, p(x_1, x_2)$$

$$= (0 \times 0) \, p(0, 0) + (1 \times 0) p(1, 0) + (1 \times 1) p(1, 1)$$
$$+ (2 \times 0) p(2, 0) + (2 \times 1) p(2, 1) + (2 \times 2) p(2, 2)$$

$$= 0(0.4) + 0(0.16) + 1(0.10) + 0(0.20) + 2(0.30) + 4(0.20)$$

$$= 1.50$$

let us find $\mu_1 = E(X_1)$ and $\mu_2 = E(X_2)$. The marginal distributions of X_1 and X_2 are as follows:

x_1	$p(x_1)$	x_2	$p(x_2)$
0	0.04	0	0.40
1	0.26	1	0.40
2	0.70	2	0.20

It follows that

$$\mu_1 = 0(0.04) + 1(0.26) + 2(0.70) = 1.66$$

and

$$\mu_2 = 0(0.40) + 1(0.40) + 2(0.20) = 0.80$$

Thus

$$\text{Cov}(X_1, X_2) = E(X_1 X_2) - \mu_1 \mu_2 = 1.50 - 1.66(0.80) = 0.172$$

From the marginal distribution of X_1 and X_2, it follows that

$$V(X_1) = E(X_1^2) - \mu_1^2 = 3.06 - (1.66)^2 = 0.30$$

and

$$V(X_2) = E(X_2^2) - \mu_2^2 = 1.2 - (0.8)^2 = 0.56$$

Hence,

$$\rho = \frac{\text{Cov}(X_1, X_2)}{\sqrt{V(X_1) \times V(X_2)}} = \frac{0.172}{\sqrt{(0.30)(0.56)}} = 0.42$$

There is a moderate positive association between the number of calls and the number of orders placed. Do the positive covariance and correlation agree with your intuition?

When a problem involves n random variables $X_1, X_2, \ldots, X_n$, we are often interested in studying linear combinations of those variables. For example, if the random variables measure the quarterly incomes for n plants in a corporation, we might want to look at their sum or average. If X_1 represents the monthly cost of servicing defective plants before a new quality-control system was installed and X_2 denotes that cost after the system was placed in operation, then we might want to study $(X_1 - X_2)$. A general result for the mean and variance of a linear combination of random variables is given below.

Mean and Variance of a Linear Combination of Random Variables

Let $Y_1, \ldots, Y_n$ and $X_1, \ldots, X_m$ be random variables with $E(Y_i) = \mu_i$ and $E(X_i) = \xi_i$. Define

$$U_1 = \sum_{i=1}^{n} a_i Y_i, \qquad U_2 = \sum_{j=1}^{m} b_j X_j$$

for constants $a_1, \ldots, a_n, b_1, \ldots, b_m$. Then

1. $E(U_1) = \sum_{i=1}^{n} a_i \mu_i$

2. $V(U_1) = \sum_{i=1}^{n} a_i^2 V(Y_i) + 2 \sum\sum_{i<j} a_i a_j \mathrm{Cov}(Y_i, Y_j)$

   where the double sum is over all pairs (i, j) with $i < j$

3. $\mathrm{Cov}(U_1, U_2) = \sum_{i=1}^{n} \sum_{j=1}^{m} a_i b_j \mathrm{Cov}(Y_i, X_j)$

When the random variables in use are independent, the variance of a linear function simplifies because the covariances become zero.

Example 7.11

Let $Y_1, Y_2, \ldots, Y_n$ be independent random variables with $E(Y_i) = \mu$ and $V(Y_1) = \sigma^2$. (These variables may denote the outcomes on n independent trials of an experiment.) Defining $\overline{Y} = (1/n) \sum_{i=1}^{n} Y_i$ show that $E(\overline{Y}) = \mu$ and $V(\overline{Y}) = \sigma^2/n$.

Solution   Note that $\overline{Y}$ is a linear function with all constants a_i equal to $1/n$. That is,

$$\overline{Y} = \left(\frac{1}{n}\right) Y_1 + \cdots + \left(\frac{1}{n}\right) Y_n$$

Using part (1) of above result

$$E(\overline{Y}) = \sum_{i=1}^{n} a_i \mu = \mu \sum_{i=1}^{n} a_i = \mu \sum_{i=1}^{n} \frac{1}{n} = \frac{n\mu}{n} = \mu$$

Using part (2) of above result

$$V(\overline{Y}) = \sum_{i=1}^{n} a_i^2 V(Y_i) + 2 \sum\sum_{i<j} a_{ij} \mathrm{Cov}(Y_i, Y_j)$$

but the covariance terms are all zero, because the random variables are independent. Thus,

$$V(\overline{Y}) = \sum_{i=1}^{n} \left(\frac{1}{n}\right)^2 \sigma^2 = \frac{n\sigma^2}{n^2} = \frac{\sigma^2}{n}.$$

Example 7.12

With X_1 denoting the amount of gasoline stocked in a bulk tank at the beginning of a week and X_2 the amount sold during the week, $Y = X_1 - X_2$ represents the amount left over at the end of the week. Find the mean and variance of Y if the joint density function of (X_1, X_2) is given by

$$f(x_1, x_2) = \begin{cases} 3x_1 & 0 \le x_2 \le x_1 \le 1 \\ 0 & \text{elsewhere} \end{cases}$$

Solution   We must first find the means and variances of X_1 and X_2. The marginal density function of X_1 is found to be

$$f_1(x_1) = \begin{cases} 3x_1^2 & 0 \le x_1 \le 1 \\ 0 & \text{elsewhere} \end{cases}$$

Thus,

$$E(X_1) = \int_0^1 x_1\left(3x_1^2\right)dx_1 = 3\left[\frac{x_1^4}{4}\right]_0^1 = \frac{3}{4}$$

Note that $E(X_1)$ can be found directly from the joint density function by

$$E(X_1) = \int_0^1\int_0^{x_1} x_1(3x_1)dx_2dx_1 = \int_0^1 3x_1^2[x_2]_0^{x_1}dx_1$$

$$= \int_0^1 3x_1^3dx_1 = \frac{3}{4}\left[x_1^4\right]_0^1 = \frac{3}{4}$$

The marginal density of X_2 is found to be

$$f_2(x_2) = \begin{cases} \dfrac{3}{2}\left(1 - x_2^2\right) & 0 \le x_2 \le 1 \\ 0 & \text{elsewhere} \end{cases}$$

Thus,

$$E(X_2) = \int_0^1 (x_2)\frac{3}{2}\left(1 - x_2^2\right)dx_2$$

$$= \frac{3}{2}\int_0^1 \left(x_2 - x_2^3\right)dx_2 = \frac{3}{2}\left\{\left[\frac{x_2^2}{2}\right]_0^1 - \left[\frac{x_2^4}{4}\right]_0^1\right\} = \frac{3}{2}\left\{\frac{1}{2} - \frac{1}{4}\right\} = \frac{3}{8}$$

Using similar arguments, it follows that

$$E\left(X_1^2\right) = 3/5 \qquad \text{and} \qquad V(X_1) = 3/5 - (3/4)^2 = 0.0375$$

Also $\qquad E\left(X_2^2\right) = 1/5 \qquad \text{and} \qquad V(X_2) = 1/5 - (3/8)^2 = 0.0594$

The next step is to find $\text{Cov}(X_1, X_2)$. Now

$$E(X_1 X_2) = \int_0^1 \int_0^{x_1} (x_1 x_2) 3x_1 \, dx_2 \, dx_1 = 3\int_0^1 \int_0^{x_1} x_1^2 x_2 \, dx_2 \, dx_1$$

$$= 3\int_0^1 x_1^2 \left[\frac{x_2^2}{2}\right]_0^{x_1} dx_1 = \frac{3}{2}\int_0^1 x_1^4 \, dx_1 = \frac{3}{2}\left[\frac{x_1^5}{5}\right]_0^1 = \frac{3}{10}$$

and

$$\text{Cov}(X_1, X_2) = E(X_1 X_2) - \mu_1 \mu_2$$

$$= \frac{3}{10} - \left(\frac{3}{4}\right)\left(\frac{3}{8}\right) = 0.0188$$

Thus,

$$E(Y) = E(X_1) - E(X_2)$$

$$= \frac{3}{4} - \frac{3}{8} = \frac{3}{8} = 0.375$$

and

$$V(Y) = V(X_1) + V(X_2) + 2(1)(-1)\text{Cov}(X_1, X_2)$$

$$= 0.0375 + 0.0594 - 2(0.0188)$$

$$= 0.0593$$

Example 7.13

A firm purchases two types of industrial chemicals. The amount of type I chemical, X_1, purchased per week has $E(X_1) = 40$ gallons with $V(X_1) = 4$. The amount of type II chemical, X_2, purchased has $E(X_2) = 65$ gallons with $V(X_2) = 8$. The type I chemical costs \$3 per gallon, whereas type II costs \$5 per gallon. Find the mean and variance of the total weekly amount spent for these types of chemicals, assuming that X_1 and X_2 are independent.

Solution   The dollar amount spent per week is given by $Y = 3X_1 + 5X_2$. This is a linear combination of two independent random variables, therefore

$$E(Y) = 3E(X_1) + 5E(X_2) = 3(40) + 5(65) = 445$$

and

$$V(Y) = (3)^2 V(X_1) + (5)^2 V(X_2) = 9(4) + 25(8) = 236$$

The firm can expect to spend \$445 per week on chemicals. The standard deviation of $\sqrt{236} = 15.4$ suggests that this figure will not drop below \$400 or exceed \$500 very often.

Example 7.14

Finite population sampling problems can be modeled by selecting balls from urns. Suppose an urn contains r white balls and $(N - r)$ black balls. A random sample of n balls is drawn without replacement, and Y, the number of white balls in the sample, is observed. From Chapter 5, we know that Y has a hypergeometric probability distribution. Find the mean and variance of Y.

Solution   We first observe some characteristics of sampling without replacement. Suppose that the sampling is done sequentially and we observe outcomes for $X_1, X_2, \ldots, X_n$, where

$$X_i = \begin{cases} 1 & \text{if the } i\text{th draw results in a white ball} \\ 0 & \text{otherwise} \end{cases}$$

Unquestionably, $P(X_1 = 1) = r/N$. But it is also true that $P(X_2 = 1) = r/N$, since

$$P(X_2 = 1) = P(X_1 = 1, X_2 = 1) + P(X_1 = 0, X_2 = 1)$$

$$= P(X_1 = 1)P(X_2 = 1 \mid X_1 = 1) + P(X_1 = 0)P(X_2 = 1 \mid X_1 = 0)$$

$$= \left(\frac{r}{N}\right)\left(\frac{r - 1}{N - 1}\right) + \left(\frac{N - r}{N}\right)\left(\frac{r}{N - 1}\right)$$

$$= \frac{r(N - 1)}{N(N - 1)} = \frac{r}{N}$$

The same is true for X_k; that is,

$$P(X_k = 1) = \frac{r}{N} \quad k = 1, \ldots, n$$

Thus, the probability of drawing a white ball on any draw, given no knowledge of the outcomes on previous draws, is r/N. In a similar way, it can be shown that

$$P(X_j = 1, X_k = 1) = \frac{r(r - 1)}{N(N - 1)} \quad j \neq k$$

Now observe that $Y = \sum_{i=1}^{n} X_i$ and hence

$$E(Y) = \sum_{i=1}^{n} E(X_i) = n\left(\frac{r}{N}\right)$$

To find $V(Y)$, we need $V(X_i)$ and Cov (X_i, X_j). Because X_i is 1 with probability r/N and 0 with probability $1 - (r/N)$, it follows that

$$V(X_i) = \frac{r}{N}\left(1 - \frac{r}{N}\right)$$

Also,

$$\text{Cov}(X_i, X_j) = E(X_i X_j) - E(X_i)E(X_j)$$

$$= \frac{r(r-1)}{N(N-1)} - \left(\frac{r}{N}\right)^2$$

$$= -\frac{r}{N}\left(1 - \frac{r}{N}\right)\left(\frac{1}{N-1}\right)$$

because $X_i X_j = 1$ if and only if $X_i = 1$ and $X_j = 1$. From Theorem 5.2, we have that

$$V(Y) = \sum_{i=1}^{n} V(X_i) + 2\sum\sum_{i<j} \text{Cov}(X_i, X_j)$$

$$= n\left(\frac{r}{N}\right)\left(1 - \frac{r}{N}\right) + 2\sum\sum_{i<j}\left[-\frac{r}{N}\left(1 - \frac{r}{N}\right)\left(\frac{1}{N-1}\right)\right]$$

$$= n\left(\frac{r}{N}\right)\left(1 - \frac{r}{N}\right) - n(n-1)\left(\frac{r}{N}\right)\left(1 - \frac{r}{N}\right)\left(\frac{1}{N-1}\right)$$

because there are $n(n-1)/2$ terms in the double summation. A little algebra yields

$$V(Y) = n\left(\frac{r}{N}\right)\left(1 - \frac{r}{N}\right)\left(\frac{N-n}{N-1}\right)$$

How strongly associated are X_i and X_j? It is informative to note that their correlation is

$$\rho = \frac{-\frac{r}{N}\left(1 - \frac{r}{N}\right)\left(\frac{1}{N-1}\right)}{\sqrt{\frac{r}{N}\left(1 - \frac{r}{N}\right) \times \frac{r}{N}\left(1 - \frac{r}{N}\right)}} = -\frac{1}{N-1}$$

which tends to zero as N gets large. If N is large relative to the sample size, the hypergeometric model becomes equivalent to the binomial model.

Exercises

7.17 Table 7.2 shows the joint distribution of fatalities and number of seat belts used by children under age 5. What is the "average" behaviour of X_1 and X_2, and how are they associated? Follow the steps below to answer this question.
  a Find $E(X_1)$, $V(X_1)$, $E(X_2)$, and $V(X_2)$.
  b Find $\text{Cov}(X_1, X_2)$. What does this result tell you?
  c Find the correlation coefficient between X_1 and X_2. How would you interpret this result?

7.18 In a study of particulate pollution in air samples over a smokestack, X_1 represents the amount of pollutant per sample when a cleaning device is not operating, and X_2 represents the amount when the cleaning device is operating. Assume that (X_1, X_2) has the joint probability density function

$$f(x_1, x_2) = \begin{cases} 1 & 0 \le x_1 \le 2, 0 \le x_2 \le 1, 2x_2 \le x_1 \\ 0 & \text{elsewhere} \end{cases}$$

The random variable $Y = X_1 - X_2$ represents the amount by which the weight of pollutant can be reduced by using the cleaning device.

a Find $E(Y)$ and $V(Y)$.

b Find an interval in which values of Y should lie at least 75% of the time.

7.19 The proportions X_1 and X_2 of two chemicals found in samples of an insecticide have the joint probability density function

$$f(x_1, x_2)$$
$$= \begin{cases} 2 & 0 \le x_1 \le 1, 0 \le x_2 \le 1, 0 \le x_1 + x_2 \le 1 \\ 0 & \text{elsewhere} \end{cases}$$

The random variable $Y = X_1 + X_2$ denotes the proportion of the insecticide due to both chemicals combined.

a Find $E(Y)$ and $V(Y)$.

b Find an interval in which values of Y should lie for at least 50% of the samples of insecticide.

c Find the correlation between X_1 and X_2 and interpret its meaning.

7.20 For a sheet-metal stamping machine in a certain factory, the time between failures, X_1, has a mean (MTBF) of 56 hours and a variance of 16. The repair time, X_2, has a mean (MTTR) of 5 hours and a variance of 4.

a If X_1 and X_2 are independent, find the expected value and variance of $Y = X_1 + X_2$, which represents one operation/repair cycle.

b Would you expect an operation/repair cycle to last more than 75 hours? Why or why not?

7.21 A particular fast-food outlet is interested in the joint behavior of the random variables Y_1, the total time between a customer's arrival at the store and his leaving the service window, and Y_2, the time that the customer waits in line before reaching the service window. Since Y_1 contains the time a customer waits in line, we must have $Y_1 \ge Y_2$. The relative frequency distribution of observed values of Y_1 and Y_2 can be modeled by the probability density function

$$f(y_1, y_2) = \begin{cases} e^{-y_1} & 0 \le y_2 \le y_1 < \infty \\ 0 & \text{elsewhere} \end{cases}$$

a Find $P(Y_1 < 2, Y_2 > 1)$.

b Find $P(Y_1 \ge 2Y_2)$.

c Find $P(Y_1 - Y_2 \ge 1)$. [*Note:* $Y_1 - Y_2$ denotes the time spent at the service windows.]

d Find the marginal density functions for Y_1 and Y_2.

7.22 Refer to Exercise 7.21. If a customer's total waiting time plus service time is known to be more than 2 minutes, find the probability that the customer waited less than 1 minute to be served.

7.23 Refer to Exercise 7.21. The random variable $Y_1 - Y_2$ represents the time spent at the service window.

a Find $E(Y_1 - Y_2)$.

b Find $V(Y_1 - Y_2)$.

c Is it highly likely that a customer would spend more than 2 minutes at the service window?

7.24 Refer to Exercise 7.21. Suppose a customer spends a length of time y_1 at the store. Find the probability that this customer spends less than half of that time at the service window.

7.5 The Multinomial Distribution

Suppose an experiment consist of n independent trials, much like the binomial case, but each trial cam result in any one of k possible outcomes. For example, a customer at a grocery store may choose any one of k checkout counters. Now suppose the probability that a particular trial results in outcome i is denoted by p_i, $(i = 1, \ldots, k)$ and p_i remains constant from trial to trial. Let Y_i, $i = 1, \ldots, k$, denote the number of the n trials resulting in outcome i. In developing a formula for $P(Y_1 = y_1, \ldots, Y_k = y_k)$, we first call attention to the fact that, because of independence of trials, the probability of having y_1 outcomes of type 1 through y_k outcomes of type k in a particular order will be

$$p_1^{y_1} p_2^{y_2} \cdots p_k^{y_k}$$

It remains only to count the number of such orderings, and this is the number of ways of partitioning the n trials into y_1 type 1 outcomes, y_2 type 2 outcomes, and so on through y_k type k outcomes, or

$$\frac{n!}{y_1! \, y_2! \cdots y_k!}$$

where

$$\sum_{i=1}^{k} y_i = n$$

Hence,

$$P(Y_1 = y_1, \ldots, Y_k = y_k) = \frac{n!}{y_1! \, y_2! \cdots y_k!} \, p_1^{y_1} p_2^{y_2} \cdots p_k^{y_k}$$

This is called the *multinomial probability distribution*. Note that if $k = 2$ we have the binomial case.

Example 7.15

Items under inspection are subject to two types of defects. About 70% of the items in a large lot are judged to be defect-free, whereas 20% have a type A defect alone and 10% have a type B defect alone. (None have both types of defects.) If six of these items are randomly selected from the lot, find the probability that three have no defects, one has a type A defect, and two have type B defects.

Solution   If we can assume that the outcomes are independent from trial to trial (item to item in our sample), which they nearly would be in a large lot, then the multinomial distribution provides a useful model. Letting Y_1, Y_2, and Y_3 denote the number of trials resulting in zero, type A, and type B defectives, respectively, we have $p_1 = 0.7, p_2 = 0.2$, and $p_3 = 0.1$. It follows that

$$P(Y_1 = 3, Y_2 = 1, Y_3 = 2) = \frac{6!}{(3!)(1!)(2!)} (0.7)^3 (0.2)(0.1)^2$$

$$= 0.041$$

Example 7.16

Find $E(Y_i)$ and $V(Y_i)$ for the multinomial probability distribution.

Solution   We are concerned with the marginal distribution of Y_i, the number of trials falling in cell i. Imagine all the cells, excluding cell i, combined into a single large cell. Hence, every trial will result in cell i or not with probabilities p_i and $1 - p_i$, respectively, and Y_i possesses a binomial marginal probability distribution. Consequently,

$$E(Y_i) = np_i \quad \text{and} \quad V(Y_i) = np_i q_i \qquad \text{where } q_i = 1 - p_i$$

Note: The same results can be obtained by setting up the expectations and evaluating. For example,

$$E(Y_i) = \sum_{y_1} \sum_{y_2} \cdots \sum_{y_k} y_i \frac{n!}{y_1! \, y_2! \cdots y_k!} \, p_1^{y_1} p_2^{y_2} \cdots p_k^{y_k}$$

Because we have already derived the expected value and variance of Y_i, we leave the tedious summation of this expectation to the interested reader.

Example 7.17

If $Y_1, \ldots, Y_k$ have the multinomial distribution given in Example 7.16, find $\text{Cov}(Y_s, Y_t)$, $s \neq t$.

Solution Thinking of the multinomial experiment as a sequence of n independent trials, we define

$$U_i = \begin{cases} 1 & \text{if trial } i \text{ results in class } s \\ 0 & \text{otherwise} \end{cases} \quad \text{and} \quad W_i = \begin{cases} 1 & \text{if trial } i \text{ results in class } t \\ 0 & \text{otherwise} \end{cases}$$

Then

$$Y_s = \sum_{i=1}^{n} U_i \quad \text{and} \quad Y_t = \sum_{j=1}^{n} W_j$$

To evaluate $\text{Cov}(Y_s, Y_t)$, we need the following results:

$$E(U_i) = p_s \quad \text{and} \quad E(W_j) = p_t$$

$$\text{Cov}(U_i, W_j) = 0 \quad \text{If } i \neq j \text{ (since the trials are independent)}$$

and

$$\text{Cov}(U_i, W_j) = E(U_i W_i) - E(U_i)E(W_i) = 0 - p_s p_t$$

since $U_i W_i$ always equals zero. From Theorem 5.2, we then have

$$\text{Cov}(Y_s, Y_t) = \sum_{i=1}^{n} \sum_{j=1}^{n} \text{Cov}(U_i, W_j)$$

$$= \sum_{i=1}^{n} \text{Cov}(U_i, W_i) + \sum_{i \neq j} \sum \text{Cov}(U_i, W_j)$$

$$= \sum_{i=1}^{n} (-p_s p_t) + 0$$

$$= -n p_s p_t$$

Note that the covariance is negative, which is to be expected since a large number of outcomes in cell s would force the number in cell t to be small.

Multinomial Experiment

1. The experiment consists of n identical trials.
2. The outcome of each trial falls into one of k classes or cells.
3. The probability that the outcome of a single trial will fall in a particular cell, say cell, i, is p_i $(i = 1, 2, \ldots, k)$ and remains the same from trial to trial. Note that $p_1 + p_2 + p_3 + \cdots + p_k = 1$.
4. The trials are independent.
5. The random variables of interest are $Y_1, Y_2, \ldots, Y_k$, where Y_i $(i = 1, 2, \ldots, k)$ is equal to the number of trials in which the outcome falls in cell i. Note that $Y_1 + Y_2 + Y_3 + \cdots + Y_k = n$

The Multinomial Distribution

$$P(Y_1 = y_1, \cdots, Y_k = y_k) = \frac{n!}{y_1! \, y_2! \cdots y_k!} p_1^{y_1} p_2^{y_2} \cdots p_k^{y_k}$$

where $\displaystyle\sum_{i=1}^{k} y_i = n$ and $\displaystyle\sum_{i=1}^{k} p_i = 1$

$$E(Y_i) = np_i \quad \text{and} \quad V(Y_i) = np_i(1 - p_i) \quad i = 1, \ldots, k$$

$$\text{Cov}(Y_i, Y_j) = -np_i p_j \quad i \neq j$$

Exercises

7.25 The National Fire Incident Reporting Service reports that, among residential fires, approximately 73% are in family homes, 20% are in apartments, and the other 7% are in other types of dwellings. If four fires are independently reported in one day, find the probability that two are in family homes, one is in an apartment, and one is in another type of dwelling.

7.26 The typical cost of damages for a fire in a family home is $20,000, the typical cost for an apartment fire is $10,000, and the typical cost for a fire in other dwellings is only $2,000. Using the information in Exercise 7.25, find the expected total damage cost for four independently reported fires.

7.27 Wing cracks, in the inspection of commercial aircraft, are reported as nonexistent, detectable, or critical. The history of a certain fleet shows that 70% of the planes inspected have no wing cracks, 25% have detectable wing cracks, and 5% have critical wing cracks. For the next five planes inspected, find the probability that

a One has a critical crack, two have detectable cracks, and two have no cracks.

b At least one critical crack is observed.

7.28 Given the recent emphasis on solar energy, solar radiation has been carefully monitored at various sites in Florida. For typical July days in Tampa, 30% have total radiation of at most 5 calories, 60% have total radiation of at most 6 calories, and 100% have total radiation of at most 8 calories. A solar collector for a hot water system is to be run for six days. Find the probability that three of the days produce no more than 5 calories, one of the days produces between 5 and 6 calories, and two of the days produce between 6 and 8 calories. What assumptions are you making in order for your answer to be correct?

7.29 The U.S. Bureau of Labor Statistics reports that approximately 21% of the adult population under age 65 is between 18 and 24 years of age, 28% is between 25 and 34, 19% is between 35 and 44, and 32% is between 45 and 64. An automobile manufacturer wants to obtain opinions on a new design from five randomly chosen adults from the group. Of the five so selected, find the approximate probability that two are between 18 and 24, two are between 25 and 44, and one is between 45 and 64.

7.30 Customers leaving a subway station can exit through any one of three gates. Assuming that each customer is equally likely to select any one of the three gates, find the probability that, among four customers,
  a Two select gate A, one selects gate B, and one selects gate C.
  b All four select the same gate.
  c All three gates are used.

7.31 Among a large number of applicants for a certain position, 60% have only a high school education, 30% have some college training, and 10% have completed a college degree. If five applicants are selected to be interviewed, find the probability that at least one will have completed a college degree. What assumptions are necessary for your answer to be valid?

7.32 In a large lot of manufactured items, 10% contain exactly one defect and 5% contain more than one defect. If 10 items are randomly selected from this lot for sale, the repair costs total $(Y_1 + 3Y_2)$, where Y_1 denotes the number

among the 10 having one defect and Y_2 denotes the number with two or more defects. Find the expected value and variance of the repair costs. Find the variance of the repair costs.

7.33 Refer to Exercise 7.32. If Y denotes the number of items containing at least one defect among the 10 sampled items, find the probability that
  a Y is exactly 2.
  b Y is at least 1.

7.34 Vehicles arriving at an intersection can turn right or left or can continue straight ahead. In a study of traffic patterns at this intersection over a long period of time, engineers have noted that 40% of the vehicles turn left, 25% turn right, and the remainder continue straight ahead.
  a For the next five cars entering the intersection, find the probability that one turns left, one turns right, and three continue straight ahead.
  b For the next five cars entering the intersection, find the probability that at least one turns right.
  c If 100 cars enter the intersection in a day, find the expected value and variance of the number turning left. What assumptions are necessary for your answer to be valid?

7.6   More on the Moment-Generating Function

The use of moment-generating functions to identify distributions of random variables works well when we are studying sums of independent random variables. For example, suppose X_1 and X_2 are independent exponential random variables, each with mean θ, and $Y = X_1 + X_2$. The moment-generating function of Y is given by

$$
\begin{aligned}
M_Y(t) = E(e^{tY}) &= E(e^{t(X_1+X_2)}) \\
&= E(e^{tX_1}e^{tX_2}) \\
&= E(e^{tX_1})E(e^{tX_2}) \\
&= M_{X_1}(t)M_{X_2}(t)
\end{aligned}
$$

because X_1 and X_2 are independent. From Chapter 6,

$$
M_{X_1}(t) = M_{X_2}(t) = (1 - \theta t)^{-1}
$$

and so

$$
M_Y(t) = (1 - \theta t)^{-2}
$$

Upon recognizing the form of this moment-generating function, we can immediately conclude that Y has a gamma distribution with $\alpha = 2$ and $\beta = \theta$. This result can be generalized to the sum of independent gamma random variables with common scale parameter β.

Example 7.18

Let X_1 denote the number of vehicles passing a particular point on the eastbound lane of a highway in 1 hour. Suppose the Poisson distribution with mean λ_1 is a reasonable model for X_1. Now let X_2 denote the number of vehicles passing a point on the westbound lane of the same highway in 1 hour. Suppose X_2 has a Poisson distribution with mean λ_2. Of interest is $Y = X_1 + X_2$, the total traffic count in both lanes per hour. Find the probability distribution for Y if X_1 and X_2 are assumed to be independent.

Solution   It is known from Chapter 6 that

$$M_{X_1}(t) = \exp[\lambda_1(e^t - 1)] \quad \text{and} \quad M_{X_2}(t) = \exp[\lambda_2(e^t - 1)]$$

By the property of moment-generating functions seen above,

$$M_Y(t) = M_{X_1}(t)M_{X_2}(t) = \exp[\lambda_1(e^t - 1)]\exp[\lambda_2(e^t - 1)]$$
$$= \exp[(\lambda_1 + \lambda_2)(e^t - 1)]$$

Now the moment-generating function for Y has the form of a Poisson moment generating function with mean $(\lambda_1 + \lambda_2)$. Thus, by the uniqueness property, Y must have a Poisson distribution with mean $(\lambda_1 + \lambda_2)$.

The fact that we can add independent Poisson random variables and still retain the Poisson properties is important in many applications.

Exercises

7.35   Find the moment-generating function for the negative binomial random variable. Use it to derive the mean and variance of that distribution.

7.36   There are two entrances to a parking lot. Cars arrive at entrance I according to a Poisson distribution with an average of three per hour and at entrance II according to a Poisson distribution with an average of four per hour. Find the probability that exactly three cars arrive at the parking lot in a given hour.

7.37   Let X_1 and X_2 denote independent normally distributed random variables, not necessarily having the same mean or variance. Show that, for any constants a and b, $Y = aX_1 + bX_2$ is normally distributed.

7.38   Resistors of a certain type have resistances that are normally distributed with a mean of 100 ohms and a standard deviation of 10 ohms. Two such resistors are connected in series, which causes the total resistance in the circuit to be the sum of the individual resistances. Find the probability that the total resistance

a Exceeds 220 ohms

b Is less than 190 ohms

7.39   A certain type of elevator has a maximum weight capacity X_1, which is normally distributed with a mean and standard deviation of 5,000 and 300 pounds, respectively. For a certain building equipped with this type of elevator, the elevator loading, X_2, is a normally distributed random variable with a mean and standard deviation of 4,000 and 400 pounds, respectively. For any given time that the elevator is in use, find the probability that it will be overloaded, assuming X_1 and X_2 are independent.

7.7 Conditional Expectations

Section 7.2 contains a discussion of conditional probability functions and conditional density functions; we now relate these functions to *conditional expectations*. Conditional expectations are defined in the same manner as univariate expectations, except that the conditional density function is used in place of the marginal density function.

> ## Conditional Expectation
>
> If X_1 and X_2 are any two random variables, the **conditional expectation** of X_1 given that $X_2 = x_2$ is defined to be
>
> $$E(X_1 | X_2 = x_2) = \int_{-\infty}^{\infty} x_1 f(x_1 | x_2) dx_1$$
>
> if X_1 and X_2 are jointly continuous and
>
> $$E(X_1 | X_2 = x_2) = \sum_{x_i} x_1 p(x_1 | x_2)$$
>
> if X_1 and X_2 are jointly discrete.

Example 7.19

Let's look again at supply and demand of soft drinks. Refer to the Y_1 and Y_2 of Example 7.7, which we shall relabel as X_1 and X_2.

$$f(x_1, x_2) = \begin{cases} 1/2 & 0 \le x_1 \le x_2, 0 \le x_2 \le 2 \\ 0 & \text{elsewhere} \end{cases}$$

Find the conditional expectation of amount of sales X_1 given that $X_2 = 1$.

Solution In Example 7.7, we found that

$$f(x_1 | x_2) = \begin{cases} 1/x_2 & 0 \le x_1 \le x_2 \le 2 \\ 0 & \text{elsewhere} \end{cases}$$

Thus, from definition of conditional expectation

$$E(X_1 | X_2 = 1) = \int_{-\infty}^{\infty} x_1 f(x_1 | x_2) \, dx_1 = \int_0^1 x_1 (1) \, dx_1 = \left. \frac{x_1^2}{2} \right|_0^1 = \frac{1}{2}$$

That is, if the soft-drink machine contains 1 gallon at the start of the day, then the expected sales for that day will be 1/2 gallon.

The conditional expectation of X_1 given $X_2 = x_2$ is a function of x_2. If we now let X_2 range over all its possible values, we can think of the conditional expectation as a function of the random variable X_2, and hence we can find the expected value of the conditional expectation. The result of this type of iterated expectation is given below.

> Let X_1 and X_2, denote random variables. Then
>
> $$E(X_1) = E[E(X_1 \mid X_2)]$$
>
> where, on the right-hand side, the inside expectation is with respect to the conditional distribution of X_1 given X_2, and the outside expectation is with respect to the distribution of X_2.

Example 7.20

A quality-control plan for an assembly line involves sampling $n = 10$ finished items per day and counting Y, the number of defectives. If p denotes the probability of observing a defective, then Y has a binomial distribution if the number of items produced by the line is large. However, p varies from day to day and is assumed to have a uniform distribution on the interval 0 to 1/4. Find the expected value of Y for any given day.

Solution   From the above result, we know that $E(Y) = E[E(Y \mid p)]$. For a given p, Y has a binomial distribution, and hence $E(Y \mid p) = np$. Thus,

$$E(Y) = E(np) = nE(p) = n \int_{0}^{1/4} 4p \, dp = n\left(\frac{1}{8}\right)$$

and for $n = 10$, $E(Y) = 10/8 = 5/4$.

This inspection policy should average 5/4 defectives per day, in the long run. The calculations could be checked by actually finding the unconditional distribution of Y and computing $E(Y)$ directly.

7.8 Summary

In Chapter 1, it was noted that process improvement requires careful study of all factors that might affect the process. In such a study, it is important to look not only at the *marginal distributions* of random variables by themselves but also at their *joint distributions* and the *conditional distributions* of one variable for fixed values of another. The *covariance* and *correlation* help assess the direction and strength of the association between two random variables.

Supplementary Exercises

7.40 Let X_1 and X_2 have the joint probability density function given by

$$f(x_1, x_2) = \begin{cases} Kx_1x_2 & 0 \le x_1 \le 1, 0 \le x_2 \le 1 \\ 0 & \text{elsewhere} \end{cases}$$

a Find the value of K that makes this a probability density function.
b Find the marginal densities of X_1 and X_2.
c Find the joint distribution of X_1 and X_2.
d Find $P(X_1 < 1/2, X_2 < 3/4)$.
e Find $P(X_1 \le 1/2 \mid X_2 > 3/4)$.

7.41 Let X_1 and X_2 have the joint density function given by

$$f(x_1, x_2) = \begin{cases} 3x_1 & 0 \le x_2 \le x_1 \le 1 \\ 0 & \text{elsewhere} \end{cases}$$

a Find the marginal density functions of X_1 and X_2.
b Find $P(X_1 \le 3/4, X_2 \le 1/2)$.
c Find $P(X_1 \le 1/2 \mid X_2 \ge 3/4)$.

7.42 From a legislative committee consisting of four Republicans, three Democrats, and two Independents, a subcommittee of three persons is to be randomly selected to discuss budget compromises. Let X_1 denote the number of Republicans and X_2 the number of Democrats on the subcommittee.
a Find the joint probability distribution of X_1 and X_2.
b Find the marginal distributions of X_1 and X_2.
c Find the probability $P(X_1 = 1 \mid X_2 \ge 1)$.

7.43 For Exercise 7.40, find the conditional density of X_1 given $X_2 = x_2$, Are X_1 and X_2 independent?

7.44 For Exercise 7.41.
a Find the conditional density of X_1 given $X_2 = x_2$.
b Find the conditional density of X_2 given $X_1 = x_1$.
c Show that X_1 and X_2 are dependent.
d Find $P(X_1 \le 3/4 \mid X_2 = 1/2)$.

7.45 Let X_1 denote the amount of a certain bulk item stocked by a supplier at the beginning of a week, and suppose that X_1 has a uniform distribution over the interval $0 \le X_1 \le 1$. Let X_2 denote the amount of this item sold by the supplier during the week, and suppose that X_2 has a uniform distribution over the interval $0 \le x_2 \le x_1$, where x_1 is a specific value of X_1.
a Find the joint density function for X_1 and X_2.
b If the supplier stocks an amount 1/2, what is the probability that he sells an amount greater than 1/4?

c If it is known that the supplier sold an amount equal to 1/4, what is the probability that he had stocked an amount greater than 1/2?

7.46 Let (X_1, X_2) denote the coordinates of a point selected at random inside a unit circle with center at the origin. That is, X_1 and X_2 have the joint density function given by

$$f(x_1, x_2) = \begin{cases} 1/\pi & x_1^2 + x_2^2 \le 1 \\ 0 & \text{elsewhere} \end{cases}$$

a Find the marginal density function of X_1.
b Find $P(X_1 \le X_2)$.

7.47 Let X_1 and X_2 have the joint density function given by

$$f(x_1, x_2) = \begin{cases} x_1 + x_2 & 0 \le x_1 \le 1, 0 \le x_2 \le 1 \\ 0 & \text{elsewhere} \end{cases}$$

a Find the marginal density functions of X_1 and X_2.
b Are X_1 and X_2 independent?
c Find the conditional density of X_1 given $X_2 = x_2$.

7.48 Let X_1 and X_2 have the joint density function given by

$$f(x_1, x_2)$$
$$= \begin{cases} K & 0 \le x_1 \le 2, 0 \le x_2 \le 1; 2x_2 \le x_1 \\ 0 & \text{elsewhere} \end{cases}$$

a Find the value of K that makes the function a probability density function.
b Find the marginal densities of X_1 and X_2.
c Find the conditional density of X_1 given $X_2 = x_2$.
d Find the conditional density of X_2 given $X_1 = x_1$.
e Find $P(X_1 \le 1.5, X_2 \le 0.5)$.
f Find $P(X_2 \le 0.5 \mid X_1 \le 1.5)$.

7.49 Let X_1 and X_2 have a joint distribution that is uniform over the region shaded in the diagram below.
a Find the marginal density for X_2.
b Find the marginal density for X_1.
c Find $P[(X_1 - X_2) \ge 0]$.

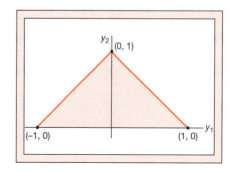

7.50 Refer to Exercise 7.40.
 a Find $E(X_1)$.
 b Find $V(X_1)$.
 c Find $\text{Cov}(X_1, X_2)$.

7.51 Refer to Exercise 7.41. Find Cov (X_1, X_2).

7.52 Refer to Exercise 7.42.
 a Find $\text{Cov}(X_1, X_2)$.
 b Find $E(X_1 + X_2)$ and $V(X_1 + X_2)$ by finding the probability distribution of $X_1 + X_2$.
 c Find $E(X_1 + X_2)$ and $V(X_1 + X_2)$ by using Theorem 5.2.

7.53 Refer to Exercise 7.47.
 a Find $\text{Cov}(X_1, X_2)$.
 b Find $E(3X_1 - 2X_2)$.
 c Find $V(3X_1 - 2X_2)$.

7.54 Refer to Exercise 7.48.
 a Find $E(X_1 + 2X_2)$.
 b Find $V(X_1 + 2X_2)$.

7.55 A quality control plan calls for randomly selecting three items from the daily production (assumed to be large) of a certain machine and observing the number of defectives. However, the proportion p of defectives produced by the machine varies from day to day and is assumed to have a uniform distribution on the interval (0,1). For a randomly chosen day, find the unconditional probability that exactly two defectives are observed in the sample.

7.56 The lifelength X of a fuse has a probability density

$$f(x) = \begin{cases} e^{-\lambda/\theta}/\theta & x > 0, \theta > 0 \\ 0 & \text{elsewhere} \end{cases}$$

Three such fuses operate independently. Find the joint density of their lifelengths, $X_1, X_2,$ and X_3.

7.57 A retail grocery merchant figures that her daily gain from sales X is a normally distributed random variable with $\mu = 50$ and $\sigma^2 = 10$ (measurements in dollars). X could be negative if she is forced to dispose of perishable goods. Also, she figures daily overhead costs Y to have a gamma distribution with $\alpha = 4$ and $\beta = 2$. If X and Y are independent, find the expected value and variance of the net daily gain. Would you expect the net gain for tomorrow to go above \$70?

7.58 Refer to Exercise 7.41 and 7.44.
 a Find $E(X_2 \mid X_1 = x_1)$.
 b Use the result about expected value expectation to find $E(X_2)$.
 c Find $E(X_2)$ directly from the marginal density of X_2.

7.59 Refer to Exercise 7.45. If the supplier stocks an amount equal to 3/4, what is the expected amount sold during the week?

7.60 Let X be a continuous random variable with distribution function $F(x)$ and density function $f(x)$. We can then write, for $x_1 \le x_2$,

$$P(X \le x_2 \mid X \ge x_1) = \frac{F(x_2) - F(x_1)}{1 - F(x_1)}$$

As a function of x_2 for fixed x_1, the right-hand side of this expression is called the conditional distribution function of X given that $X \ge x_1$. On taking the derivative with respect to x_2, we see that the corresponding conditional density function is given by

$$\frac{f(x_2)}{1 - F(x_1)} \qquad x_2 \ge x_1$$

Suppose a certain type of electronic component has lifelength X with the density function (lifelength measured in hours)

$$f(x) = \begin{cases} (1/200)e^{-x/200} & x \ge 0 \\ 0 & \text{elsewhere} \end{cases}$$

Find the expected lifelength for a component of this type that has already been in use for 100 hours.

7.61 The negative binomial variable X was defined as the number of the trial on which the rth success occurs in a sequence of independent trials with constant probability p of success on each trial. Let X_i denote a geometric random variable, defined as the number of the trial on which the first success occurs. Then we can write

$$X = \sum_{i=1}^{n} X_i$$

for independent random variables $X_1, \ldots, X_r$. Use results about mean and variance of a linear combination of random variables to show that $E(X) = r/p$ and $V(X) = r(1 - p)/p^2$.

7.62 A box contains four balls, numbered 1 through 4. One ball is selected at random from this box. Let

$X_1 = 1$ if ball 1 or ball 2 is drawn
$X_2 = 1$ if ball 1 or ball 3 is drawn
$X_3 = 1$ if ball 1 or ball 4 is drawn

and the X_i's are zero otherwise. Show that any two of the random variables $X_1, X_2,$ and X_3 are independent, but the three together are not.

Statistics, Sampling Distributions, and Control Charts

A manufacturing process is designed to produce bolts with a 0.5-inch diameter. Each hour, a random sample of five bolts is selected and their mean diameter computed. If the resulting sample mean is less than 0.48 or larger than 0.52, then the process is shut down. What is the probability that the manufacturing process will be shut down unnecessarily?

To answer such questions, we need to look at the distribution of the sample mean. In the process of making an inference from a sample to a population, we usually calculate one or more statistics, such as the mean, proportion, or variance. Because samples are randomly selected, the values of such statistics change from sample to sample. Thus, sample statistics themselves are random variables and their behavior can be modeled by probability distributions, known as *sampling distributions*. In this chapter, we'll study the properties of sampling distributions and the use of sampling distributions in control charts (see Figure 8.1), a decision-making tool employed for process improvement.

Figure 8.1

Control chart for a manufacturing process

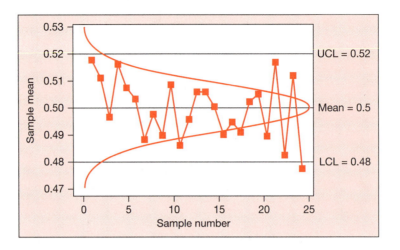

Introduction

Chapter 1 contained a discussion of a way to summarize data so that we can derive some useful and important information from them. We return to this idea of deriving information from data, but now we'll view each data set under study as a *sample* from a much larger set of data points that could be studied, called a *population*. For example,

- A sample of 50 household incomes could be selected from all the households in your community.
- A sample of five diameters could be taken from the population of machined rods.

We can visualize the nature of the population of values of which our sample data constitute only a small part. Our job is to describe the population using some summary measures as accurately as possible given only the data collected from the sample from that population. We are interested in estimating some characteristics of the population and want to decide how closely our sample mirrors the characteristics of the population from which it was selected.

- Numerical descriptive measures of a *population* are called *parameters*.
- Numerical descriptive measures of a *sample* are called *statistics*.

> A **statistic** is a function of sample observations that contains no unknown parameters.

Even though it is unknown, the mean annual income per household in your community is a parameter. A sample mean income calculated from a sample of household is a statistic. We hope that the sample mean will be close to the population mean.

The sample mean is only one of many possible statistics we study. To name a few, the median, the quartiles, the standard deviation, and the IQR are also important statistics. We concentrate on a careful study of the mean because of its central role in classical statistical inference procedures.

8.1 The Sampling Distributions

Consider a population with mean μ and variance σ^2. Suppose we take different samples of the same size from a given population and compute a mean for each selected sample, as shown in Figure 8.2.

Note that each sample will result in a different sample mean. If we are using a sample to make some inference about the population, then obviously our inference will depend on which sample gets selected. Naturally, we are not going to take all possible samples. This raises the following questions:

- How large is the difference between a sample mean and the population mean likely to be?
- Which values of sample mean are more likely to occur?
- Which values of sample mean are less likely to occur?

To answer these questions, we need to look at the probability distribution of sample mean, known as the *sampling distribution*.

> A **sampling distribution** is the probability distribution of a sample statistic.

Typically, the spread of the distribution is measured using *standard deviation*.

> The standard deviation of the sampling distribution of statistic is known as the **standard error** of statistic.

We'll use the notation *SE* to indicate the standard error. The following activities will help understand the concept of sampling distribution.

Figure 8.2

From the population to the sampling distribution

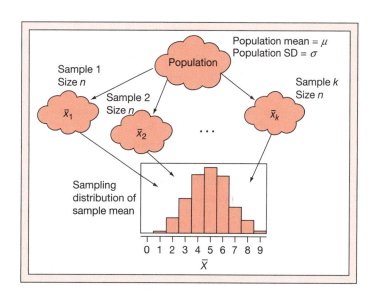

Activity   If we take different samples of the same size from the population of random rectangles, what will the distribution of sample means look like? How will the sample means differ from the population mean? Will it matter what size samples are taken?

For this activity, use

- Random rectangles sheet (Figure 3.5).
- Random number table (Table 1 of the Appendix) or a random number generator on TI-89.

Suppose 100 units pictured in random rectangles sheet (Figure 3.5) form a population of size 100. For each member of the population, the unit size is computed as the number of rectangles in that unit. The distribution of the unit size in the population is given in Figure 8.3. The mean and the standard deviation of this population are $\mu = 7.410$ and $\sigma = 5.234$, respectively.

Figure 8.3

Distribution of unit sizes

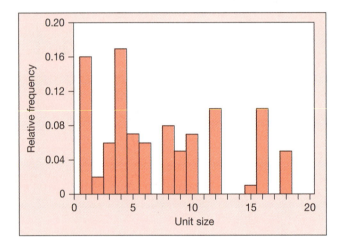

- Take a random sample of size $n = 5$ with replacement.
- Compute the sample mean.
- Repeat this process 100 times.
  This will result in a list of 100 sample means. (Alternatively, instead of each student taking 100 samples, each student in the class can take enough samples so that the whole class together will have approximately 100 samples.)
- Compute summary statistics for sample means.
- Make a histogram for sample means.
- Discuss the distribution of sample mean (i.e., the sampling distribution) through the histogram, using center, spread, and shape of the distribution.
- Repeat the process for $n = 20$.
- Compare the distributions of sample mean for $n = 5$ and $n = 20$ using center, spread, and shape of the distribution.

Such simulation can be done easily using computer software. Figure 8.4 gives the distributions of sample mean that resulted from one such simulation, and table on the next page gives the numerical summary.

What do we observe about the sampling distribution of $\overline{X}$?

	n = 5	n = 10	n = 20
Mean	7.440	7.456	7.486
Std Dev	2.127	1.703	1.169
Maximum	12.400	11.200	10.150
Q_3	8.800	8.875	8.250
Median	7.400	7.100	7.600
Q_1	5.850	6.200	6.900
Minimum	2.200	3.100	3.500

Figure 8.4

Sampling distribution of the mean of random rectangles

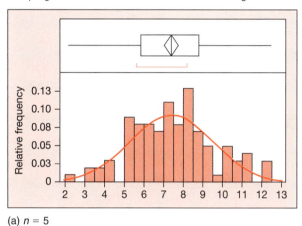

(a) n = 5

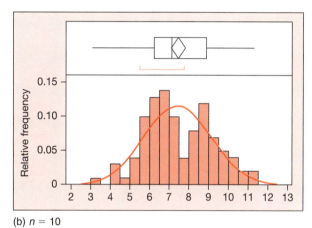

(b) n = 10

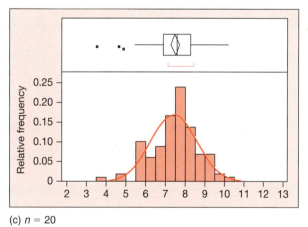

(c) n = 20

Notice that the sample means tend to be symmetric about the true population mean and tend to pile up more and more in the middle as the sample size increases. Also, the distribution of the sample mean becomes quite symmetric around the population mean with shrinking variability. In summary, we notice the following three characteristics:

- The center of the sampling distribution stayed in about the same place as the population mean and didn't change with the sample size.
- As the sample size increased, the spread of the sampling distribution decreased.
- As the sample size increased, the sampling distribution became more symmetric and mound-shaped.

The numerical summary for random rectangles shows that

- For $n = 5$, the standard error of $\overline{X}$ is approximately $SE(\overline{X}) = 2.13$.
- For $n = 10$, the standard error of $\overline{X}$ is approximately $SE(\overline{X}) = 1.70$.
- For $n = 20$, the standard error of $\overline{X}$ is approximately $SE(\overline{X}) = 1.17$.

This activity can be repeated using some other statistic, such as proportion, median, standard deviation, and variance to study their sampling distributions.

Let X_i denote the random outcome of the ith sample observation to be selected. Then $\overline{X} = \sum_{i=1}^{n} X_i/n$ and from the properties of expectations of linear functions,

$$E(\overline{X}) = \frac{1}{n}\sum_{i=1}^{n} E(X_i) = \frac{1}{n}\sum_{i=1}^{n}\mu = \mu$$

where $\mu = E(X_i)$, $i = 1, 2, \ldots, n$. Hence, $\overline{X}$ is an unbiased estimator of μ. If, in addition, the sampled observations are selected independently of one another,

$$V(\overline{X}) = \frac{1}{n^2}\sum_{i=1}^{n} V(X_i) = \frac{1}{n^2}\sum_{i=1}^{n} V(X_i) = \frac{1}{n^2}\sum_{i=1}^{n}\sigma^2 = \frac{\sigma^2}{n}$$

Thus, the standard deviation within any one sampling distribution of sample means should be $\sigma/\sqrt{n}$. To check this out for the data generated in Figure 8.4, calculate $\sigma/\sqrt{n}$ for each case and compare it to the sample standard deviation calculated from the 100 sample means. The results are given in Table 8.1.

We see that in each case the observed variability for the 100 sample means is quite close to the theoretical value of $\sigma/\sqrt{n}$.

In summary, the variability in the sample means decreases as the sample size increases, with the variance of a sample mean given by σ^2/n. The quantity $\sigma/\sqrt{n}$ is also referred to as the *standard error* of the mean. The variance σ^2 can be estimated by the sample variance s^2.

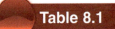

Table 8.1

Comparison of Theoretical and Computed ($n = 100$) Standard Deviation

Sample Size	Standard Deviation for Generated Data	$\sigma/\sqrt{n}$
5	2.1271	2.3401
10	1.7028	1.6551
20	1.1696	1.1704

The properties of the sample mean illustrated above hold only for *random samples*. That is, each observation in the sample must be independently selected from the sample probability distribution.

Suppose *n* resistors of the same type are placed into service and their lifetimes are measured. If the lifetimes are independent of one another, then the resulting measurements would constitute an observed random sample. The assumption of independence is difficult to justify in some cases, but careful thought should be given to the matter in every statistical problem.

- If the resistors are all in the same system and fail because of a power surge in the system, then their lifetimes are clearly not independent.
- If, however, they fail only from normal usage, then their lifetimes may be independent even if they are being used in the same system.

Most of the examples and problems in this book assume that sample measurements come from random samples.

Potential measurements of the lifetimes of resistors or random digits come from conceptual populations that are infinite. For finite populations, like the employees of a firm or the locks in a box, true independence of random samples is not generally achieved.

- If a lock is randomly selected from a box containing different kinds of locks and not replaced, then the probability distribution of the remaining locks in the box has changed. The second lock selected comes from a slightly different distribution.
- If, however, the finite population is large compared to the sample size, the sample observations from a series of random selections are nearly independent. The result, then, will still be called a random sample, and the properties of sample means presented above will still hold.

The term *random sample* suggests the manner in which items should be selected to ensure some degree of independence in the resulting measurements; that is, the items should be selected at *random*. One method of random selection involves the use of random number tables, which was described in more detail in Chapter 3.

Computer Simulation Activity

- Most commonly used statistical software give a list of several distributions. Select a distribution. Start with a symmetric distribution like a normal or a binomial with $p = 0.5$. Then repeat the process with more skewed distributions like a binomial with p close to 0 or 1, or a chi-square.
- Select 100 samples (i.e., 100 rows of data) each of size 5.
  This will result in data in 100 rows and 5 columns. Each row constitutes a sample.
- Compute the mean of each sample and save in another column.
  This will result in 100 sample means.
- Make a dotplot or a histogram to get the sampling distribution of $\overline{X}$.
- Compute summary statistics.
- Describe the sampling distribution using the center, spread, and shape of the distribution.

• Repeat the process using $n = 10, 20, 30$.
• Describe the effect of sample size on the sampling distributions using the center, spread, and shape of the distribution.

Results of one such simulation are described in the Example 8.1.

Example 8.1

Suppose we selected 100 samples each of size $n = 25$, from an exponential distribution with mean 10. The probabilistic model for this population is shown in Figure 8.5 and described below:

$$f(x) = \begin{cases} \dfrac{1}{10}e^{-x/10} & x > 0 \\ 0 & \text{elsewhere} \end{cases}$$

Figure 8.5

Exponential distribution with mean 10

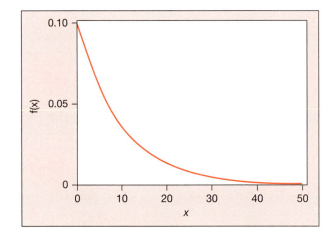

For this model,

$$E(X) = \mu = 10 \text{ and}$$

$$V(X) = \sigma^2 = (10)^2, \text{ or } \sigma = 10.$$

a Describe the distribution of means of samples of size 25.
b Describe the distributions of means of samples of sizes 5 and 10. Compare them to the sampling distribution in part (a).

Solution **a** The sample mean $\bar{x}$ was calculated for each of the 100 samples. The average of the 100 sample means turned out to be 10.24, reasonably close to the population mean 10. The standard deviation for the 100 values of $\bar{x}$ was calculated and found to be 2.17. Theoretically, the standard deviation of $\overline{X}$ is

$$\sqrt{V(\overline{X})} = \frac{\sigma}{\sqrt{n}} = \frac{10}{\sqrt{25}} = 2.0$$

which is not far from the observed value of 2.17.

A relative frequency histogram for the 100 values of $\bar{x}$ is shown in Figure 8.6. It looks mound-shaped and fairly symmetric, although we sampled from a very skewed distribution.

Figure 8.6

Frequency histogram for sample mean ($n = 25$)

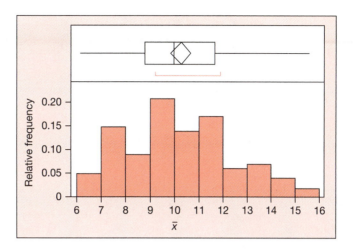

b To see what happens to the sampling distribution of $\bar{x}$ for smaller sample sizes, let us take 100 samples of sizes 5 and 10. The sampling distributions of $\bar{x}$ from one such simulation of 100 samples of sizes 5, 10, and 25 are shown in Figure 8.7. The summary statistic is reported in Table 8.2.

Figure 8.7

Sampling distributions of $\bar{x}$ for $n = 5$, 10, and 25

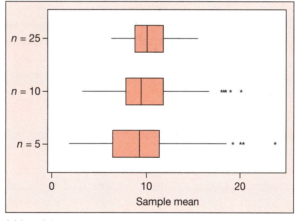

(a) boxplots

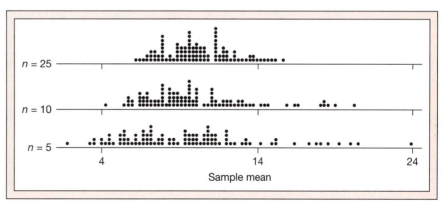

(b) dotplots

Table 8.2

Summary Statistic of Sample Means from Exponential Population

Variable	N	Mean	Median	TrMean	StDev	SE Mean
mean5	100	9.515	9.307	9.261	4.235	0.424
mean10	100	10.069	9.430	9.843	3.346	0.335
mean25	100	10.239	9.998	10.178	2.163	0.216

Variable	Minimum	Maximum	Q1	Q3
mean5	1.728	23.816	6.459	11.373
mean10	4.313	20.153	7.784	11.760
mean25	6.147	15.558	8.766	11.678

Comparison of three sampling distributions shows that the behavior of the sampling distribution differs for different sample sizes. However, notice a trend here. Although the population distribution is skewed,

- The mean of the sampling distribution remained close to 10, the population mean, regardless of the sample size.
- The spread of the sampling distribution was reduced as sample size increased.
- The shape of the sampling distribution became more symmetric and mound-shaped as the sample size increased.

Exercises

8.1   Concentrations of uranium 238 were measured in 12 soil samples from a certain region, with the following results in pCi/g (picoCuries per gram):

0.76, 1.90, 1.84, 2.42, 2.01, 1.77, 1.89, 1.56, 0.98, 2.10, 1.41, 1.32

a   Construct a boxplot.
b   Calculate $\bar{x}$ and s^2 for these data.
c   If another soil sample is randomly selected from this region, find an interval in which the uranium concentration measurement should lie with probability at least 0.75. [*Hint*: Assume the sample mean and variance are good approximations to the population mean and variance and use Tchebysheff's theorem.]

8.2   What's the average number of hours that students in your statistics class study in a week? Estimate this average by selecting a random sample of five students and asking each sampled student how many hours he or she studies in a typical week. Use a random-number table in selecting the sample. Collect answers from other students and construct a dotplot of the means

from samples of size 5. Describe the shape of the dotplot.

8.3   Obtain a standard six-sided die.

a   If the die is balanced, what will the outcomes of an ideal sample of 30 tosses look like?
b   Find the mean and variance of the ideal sample outcomes portrayed in part (a).
c   Toss your die 30 times and record the upper-face outcomes. How does the distribution of outcomes compare with the ideal sample in part (a)?
d   Calculate the sample mean and variance for the data in part (c). How do they compare with the results of part (b)?
e   Collect the sample means for 30 die tosses from other members of the class. Make a dotplot of the results and comment on its shape.

8.4   The dotplot of Figure 8.8 shows the median values of single-family housing (in thousands of dollars) in 152 U.S. metropolitan areas across the United States (http://www.realtor.org/research/research/metroprice). These data have a mean of 167.19, a median of 141.80, and a

Figure 8.8
Median prices of
single-family housing

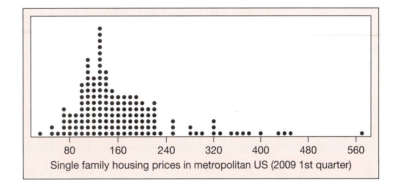

Single family housing prices in metropolitan US (2009 1st quarter)

standard deviation of 85.96. Four random samples, each of size $n = 5$, were selected from the 152 values; the sample data are displayed below.

Sample 1	Sample 2	Sample 3	Sample 4
97.3	429.9	219.1	108.6
155.3	145.2	116.3	109.1
85.9	132.8	169.3	290.7
111.4	323.2	209.8	129.9
121.1	402.0	109.1	216.5

Calculate the sample mean and standard deviation for each sample. Compare them to the population mean and standard deviation. Comment on the problems of estimating μ and σ for the type of population shown here.

8.5 Referring to the home-sale data of Exercise 8.4, calculate the median for each of the four random samples. Compare the sample medians to the population median. Do you think the sample median might be a better estimator of the "center" of the population in this case? Why or why not?

8.2 The Sampling Distribution of $\overline{X}$ (General Distribution)

We have seen that the approximated sampling distributions of a sample mean tend to have a particular shape (Figures 8.4, 8.6, and 8.7), being somewhat symmetric with a single mound in the middle. This result is not coincidental or unique to those particular examples. The fact that sampling distributions for sample means always tend to be approximately normal in shape is described by the *Central Limit Theorem*.

The Central Limit Theorem

If a random sample of size n is drawn from a population with mean μ and variance σ^2, then the sample mean $\overline{X}$ has approximately a normal distribution with mean μ and variance σ^2/n. That is, the distribution function of

$$\frac{(\overline{X} - \mu)}{\sigma/\sqrt{n}}$$

is approximately a standard normal. The approximation improves as the sample size increases.

Figure 8.9 shows three different populations from which random samples of sizes $n = 2, 5, 10, 25$ were taken. The sampling distributions of sample means computed from these samples are also shown. Notice that, when sampling from normally distributed population, the sampling distribution of sample means is

Figure 8.9

Sampling distributions of sample mean

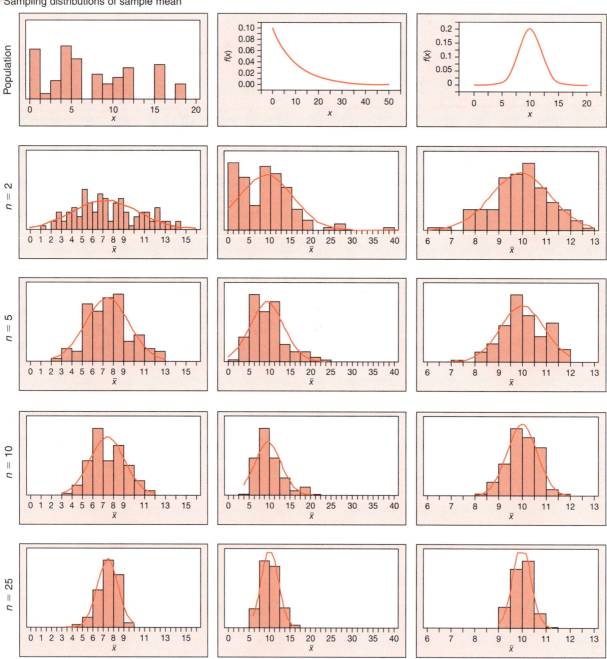

exactly normal regardless of sample sizes (even for small samples). For distributions that are neither mound-shaped nor symmetric, the sampling distributions are skewed for small samples, but for $n = 25$, they are fairly symmetric and mound-shaped (approximately normal).

The Central Limit Theorem provides a very useful result for statistical inference. Note that $\overline{X}$ is a random variable and has a probability distribution. The practical importance of this result is that, for large n, the sampling distribution of $\overline{X}$ can be closely approximated by a normal distribution. More precisely, for large n, we can find

$$P(\overline{X} \le b) = P\left(\frac{\overline{X} - \mu}{\sigma/\sqrt{n}} \le \frac{b - \mu}{\sigma/\sqrt{n}}\right) \approx P\left(Z \le \frac{b - \mu}{\sigma/\sqrt{n}}\right)$$

where Z is a standard normal random variable.

One thousand samples of size $n = 25$, 50, and 100 were drawn from a population having the probability density function

$$f(x) = \begin{cases} \dfrac{1}{10}e^{-x/10} & x > 0 \\ 0 & \text{elsewhere} \end{cases}$$

For this model,

$$E(X) = \mu = 10 \quad \text{and} \quad \sigma = 10$$

From the central limit theorem, we know not only that for large sample size, $\overline{X}$ has mean and variance σ^2/n if the population has mean μ and variance σ^2 but also that the probability distribution of $\overline{X}$ is approximately normal. We can confirm this tendency of sample means by looking at the distributions of means (Figure 8.10) for samples generated from the exponential distributions with mean 10. Although all the relative frequency histograms have a sort of bell shape, notice the tendency toward a symmetric normal curve is better for larger n.

Figure 8.10

Sampling distributions of sample mean for $n = 25$, 50, and 100

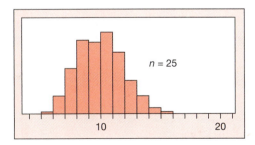

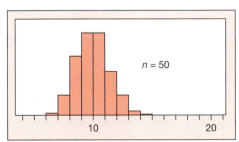

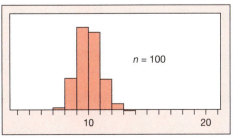

Example 8.2

Suppose a random sample of size $n = 25, 50,$ and 100 is drawn from a population having the exponential probability density function with mean 10. For this model, $\sigma = 10$ too. Find an interval (a, b) for each n such that $P(a \leq \overline{X} \leq b) = 0.95$.

Solution   From the Central Limit Theorem, we know that the distribution of $\overline{X}$ is approximately normal with mean $\mu_{\overline{X}} = \mu$ and standard deviation $\sigma_{\overline{X}} = \sigma/\sqrt{n}$ shown in this table.

n	$\mu_{\overline{x}}$	$\sigma_{\overline{x}}$
25	10	2.00
50	10	1.41
100	10	1.00

Now $P(a \leq \overline{X} \leq b) = 0.95$ is equivalent to

$$P\left(\frac{a - \mu}{\sigma/\sqrt{n}} \leq \frac{\overline{X} - \mu}{\sigma/\sqrt{n}} \leq \frac{b - \mu}{\sigma/\sqrt{n}} \right) = 0.95$$

Because $Z = \sqrt{n}(\overline{X} - \mu)/\sigma$ has approximately a standard normal distribution, this inequality can be approximated by

$$P\left(\frac{a - \mu}{\sigma/\sqrt{n}} \leq Z \leq \frac{b - \mu}{\sigma/\sqrt{n}} \right) = 0.95$$

However, we know that $P(-1.96 \leq Z \leq 1.96) = 0.95$. Comparing these two expressions, we get

$$\frac{a - \mu}{\sigma/\sqrt{n}} = -1.96 \quad \text{and} \quad \frac{b - \mu}{\sigma/\sqrt{n}} = 1.96$$

It gives $a = \mu - 1.96\sigma/\sqrt{n}$ and $b = \mu + 1.96\sigma/\sqrt{n}$. The accompanying table gives the intervals for different sample sizes. Note that the spread of sampling distributions decreased as the sample size increased. Consequently, interval (a, b) became shorter as n increased.

n	(a, b)
25	(6.08, 13.92)
50	(7.23, 12.77)
100	(8.04, 11.96)

Example 8.3

The fracture strengths of a certain type of glass average 14 (in thousands of pounds per square inch) and have a standard deviation of 2.

a What is the probability that the average fracture strength for 100 randomly selected pieces of this glass exceeds 14.5?

b Find an interval that includes the average fracture strength for 100 randomly selected pieces of this glass with probability 0.95.

Solution   According to the Central Limit Theorem, the average strength $\bar{X}$ has approximately a normal distribution with mean 14 and standard deviation $\sigma/\sqrt{n} = 2/\sqrt{100} = 0.2$.

a

$$P(\bar{X} > 14.5) = P\left(\frac{\bar{X} - \mu}{\sigma/\sqrt{n}} > \frac{14.5 - \mu}{\sigma/\sqrt{n}}\right)$$

$$= P\left(Z > \frac{14.5 - 14}{0.2}\right)$$

$$= P(Z > 2.5) = 0.0062$$

The probability of getting an average value (for sample of size 100) more than 0.5 units above the population mean is very small (see Figure 8.11).

Figure 8.11
$P(\bar{X} > 14.5)$

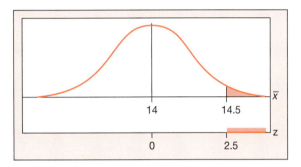

b   We are interested in determining interval (a, b) such that $P(a \le \bar{X} \le b) = 0.95$. Earlier we saw that such a and b are obtained as

$$a = 14 - 1.96(0.2) = 13.6 \quad \text{and} \quad b = 14 + 1.96(0.2) = 14.4$$

As illustrated by Figure 8.12, approximately 95% of the sample mean fracture strengths, for samples of size 100, should lie between 13.6 and 14.4.

Figure 8.12
$P(a \le \bar{X} \le b) = 0.95$

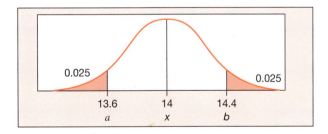

Example 8.4

A certain machine that is used to fill bottles with soda has been observed over a long period of time, and the variance in the amounts filled is found to be approximately $\sigma^2 = 1$ sq. ounce. However, the mean amount filled μ depends on an adjustment that may change from day to day, or from operator to operator.

a If 25 observations on the amount dispensed (in ounces) are taken on a given day (all with the same machine setting), find the probability that the sample mean will be within 0.3 ounces of the true population mean for that setting.

b How many observations should be taken in the sample so that the sample mean will be within 0.3 ounces of the population mean with probability 0.95?

Solution   **a** We assume that $n = 25$ is large enough for the sample mean $\overline{X}$ to have approximately a normal distribution with mean μ and standard deviation $\sigma/\sqrt{25} = 1/\sqrt{25} = 0.2$. Then, as shown in Figure 8.13

Figure 8.13
$P(|\overline{X} - \mu| \le 0.3)$

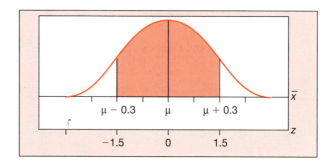

$$P(|\overline{X} - \mu| \le 0.3) = P[-0.3 \le (\overline{X} - \mu) \le 0.3]$$

$$= P\left[\frac{-0.3}{\sigma/\sqrt{n}} \le \frac{\overline{X} - \mu}{\sigma/\sqrt{n}} \le \frac{0.3}{\sigma/\sqrt{n}}\right]$$

$$= P\left[\frac{-0.3}{0.02} \le \frac{\overline{X} - \mu}{\sigma/\sqrt{n}} \le \frac{0.3}{0.2}\right]$$

$$= P[-1.5 \le Z \le 1.5] = 0.8664$$

b Now we want to find n such that

$$P(|\overline{X} - \mu| \le 0.3) = P[-0.3 \le (\overline{X} - \mu) \le 0.3] = 0.95$$

Because $\sigma = 1$, the standard error is $\sigma/\sqrt{n} = 1/\sqrt{n}$. Therefore, standardizing we get

$$P\left[-0.3\sqrt{n} \le \frac{(\overline{X} - \mu)}{\sigma/\sqrt{n}} \le 0.3\sqrt{n}\right] = 0.95$$

Thus, $P[-0.3\sqrt{n} \le Z \le 0.3\sqrt{n}] = 0.95$. From the normal probability table (Table 4) in the Appendix, we get $P[-1.96 \le Z \le 1.96] = 0.95$. Comparing these two expressions, we get $0.3\sqrt{n} = 1.96$ and it follows that

$$n = \left(\frac{1.96}{0.3}\right)^2 = 42.68$$

Thus, 43 observations will be needed for the sample mean to have 95% chance of being within 0.3 ounce of the population mean.

Exercises

8.6 Shear-strength measurements for spot welds of a certain type have been found to have a standard deviation of approximately 10 psi. If 100 test welds are to be measured, find the approximate probability that the sample means will be within 1 psi of the true population mean.

8.7 If shear-strength measurements have a standard deviation of 10 psi, how many test welds should be used in the sample if the sample mean is to be within 1 psi of the population mean with probability approximately 0.95?

8.8 The soil acidity is measured by a quantity called pH, which may range from 0 to 14 for soils ranging from low to high acidity. Many soils have an average pH in the 5–8 range. A scientist wants to estimate the average pH for a large field from n randomly selected core samples by measuring the pH in each sample. If the scientist selects $n = 40$ samples, find the approximate probability that the sample mean of the 40 pH measurements will be within 0.2 unit of the true average pH for the field.

8.9 Suppose the scientist of Exercise 8.8 would like the sample mean to be within 0.1 of the true mean with probability 0.90. How many core samples should she take?

8.10 Resistors of a certain type have resistances that average 200 ohms, with a standard deviation of 10 ohms. Twenty-five of these resistors are to be used in a circuit.

  a Find the probability that the average resistance of the 25 resistors is between 199 and 202 ohms.

  b Find the probability that the *total* resistance of the 25 resistors does not exceed 5,100 ohms. [*Hint:* Note that

$$P\left(\sum_{i=1}^{\infty} X_i > a\right) = P(n\overline{X} > a) = P(\overline{X} > a/n).]$$

  c What assumptions are necessary for the answers in parts (a) and (b) to be good approximations?

8.11 One-hour carbon monoxide concentrations in air samples from a large city average 12 ppm, with a standard deviation of 9 ppm. Find the probability that the average concentration in 100 samples selected randomly will exceed 14 ppm.

8.12 Unaltered bitumens, as commonly found in lead-zinc deposits, have atomic hydrogen/carbon (H/C) ratios that average 1.4, with a standard deviation of 0.05. Find the probability that 25 samples of bitumen have an average H/C ratio below 1.3.

8.13 The downtime per day for a certain computing facility averages 4.0 hours, with a standard deviation of 0.8 hour.

  a Find the probability that the average daily downtime for a period of 30 days is between 1 and 5 hours.

  b Find the probability that the total downtime for the 30 days is less than 115 hours.

  c What assumptions are necessary for the answers in parts (a) and (b) to be valid approximations?

8.14 The strength of a thread is a random variable with mean 0.5 pound and standard deviation 0.2 pound. Assume the strength of a rope is the sum of the strengths of the threads in the rope.

  a Find the probability that a rope consisting of 100 threads will hold 45 pounds.

  b How many threads are needed for a rope that will hold 50 pounds with 99% assurance?

8.15 Many bulk products, such as iron ore, coal, and raw sugar, are sampled for quality by a method that requires many small samples to be taken periodically as the material is moving along a conveyor belt. The small samples are then aggregated and mixed to form one composite sample. Let Y_i denote the volume of the ith small sample from a particular lot, and suppose $Y_1, \ldots, Y_n$ constitutes a random sample with each Y_i having mean μ and variance σ^2. The average volume of the samples μ can be set by adjusting the size of the sampling device. Suppose the variance of sampling volumes σ^2 is known to be approximately 4 for a particular situation (measurements are in cubic inches). It is required that the total volume of the composite sample exceed 200 cubic inches with probability approximately 0.95 when $n = 50$ small samples are selected. Find a setting for μ that will allow the sampling requirements to be satisfied.

8.16 The service times for customers coming through a checkout counter in a retail store are independent random variables with a mean of 1.5 minutes and a variance of 1.0. Approximate the probability that 100 customers can be serviced in less than 2 hours of total service time by this one checkout counter.

8.17 Refer to Exercise 8.16. Find the number of customers n such that the probability of servicing all n customers in less than 2 hours is approximately 0.1.

8.18 Suppose that $X_1, \ldots, X_{n_1}$ and $Y_1, \ldots, Y_{n_2}$ constitute independent random samples from populations with means μ_1 and μ_2 and variances σ_1^2 and σ_2^2, respectively. Then the Central Limit Theorem can be extended to show that $(\overline{X} - \overline{Y})$ is approximately normally distributed for large n_1 and n_2, with mean $(\mu_1 - \mu_2)$ and variance

$(\sigma_1^2/n_1) + (\sigma_2^2/n_2)$. Water flow through soils depends, among other things, on the porosity (volume proportion due to voids) of the soil. To compare two types of sandy soil, $n_1 = 50$ measurements are to be taken on the porosity of soil A, and $n_2 = 100$ measurements are to be taken on soil B. Assume that $\sigma_1^2 = 0.01$ and $\sigma_2^2 = 0.02$. Find the approximate probability that the difference between the sample means will be within 0.05 unit of the true difference between the population means, $(\mu_1 - \mu_2)$.

8.19 Refer to Exercise 8.18. Suppose samples are to be selected with $n_1 = n_2 = n$. Find the value of n that will allow the difference between the sample means to be within 0.04 unit of $(\mu_1 - \mu_2)$ with probability approximately 0.90.

8.20 An experiment is designed to test whether operator A or operator B gets the job of operating a new machine. Each operator is timed on 50 independent trials involving the performance of a certain task on the machine. If the sample means for the 50 trials differ by more than 1 second, the operator with the smaller mean gets the job. Otherwise, the experiment is considered to end in a tie. If the standard deviations of times for both operators are assumed to be 2 seconds, what is the probability that operator A gets the job even though both operators have equal ability?

8.3 The Sampling Distribution of $\overline{X}$ (Normal Distribution)

For large samples, the Central Limit Theorem tells us that the sampling distribution of $\overline{X}$ is approximately normal. What if we are dealing with samples not large enough to apply CLT? In that case, the population must satisfy some conditions such as approximate normality of the population sampled.

If population is normally distributed, then

- $Z = \dfrac{\overline{X} - \mu}{\sigma/\sqrt{n}}$ follows a standard normal distribution if σ is known.

- $T = \dfrac{\overline{X} - \mu}{S/\sqrt{n}}$ follows a t-distribution with $(n - 1)$ degrees of freedom if σ is unknown.

Because $E(\overline{X}) = \mu$, the sampling distribution of T centers at zero. Because σ (a constant) was replaced by S (a random variable), the sampling distribution of T shows more variability than the sampling distribution of Z. The sampling distribution of T is known as the *t-distribution*. It was first derived by William Gossett in the early 20th century. Because he published his results under the pseudo name "Student," this distribution also came to be known as the *Student's t-distribution*. A t-probability density function is pictured in Figure 8.14.

The *t*-probability density function is symmetric about zero and has wider tails (is more spread out) than the standard normal distribution. The shape of the distribution is similar to the Z-distribution, but unlike Z, the shape of the

Figure 8.14

A *t* distribution (probability density function)

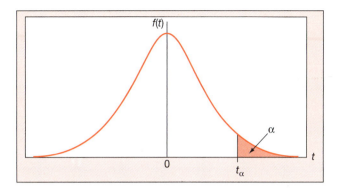

distribution of *t* depends on the degrees of freedom. As degrees of freedom increase, the *t*-distribution approaches the *Z*-distribution. For large degrees of freedom (at least 25), the difference between *t*-distribution and the *Z*-distribution becomes negligible, allowing us to use *Z*-distribution in place of *t*-distribution. Parallel to z_α, let us define

t_α = the *t*-value that cuts off an area α in the right tail of the *t*-distribution (see Figure 8.14)

Table 5 of the Appendix lists values of t_α for different values of α. Only upper tail values are given in the table because of the symmetry of the density function. Because the *t*-distribution approaches the standard normal distribution as *n* gets larger, note that t_α is approximately equal to z_α for large *n*.

Example 8.5

Concrete-filled tube columns are used as structural support due to their excellent seismic resistant properties. A structural engineer conducted an experiment and recorded failure strengths of 12 samples. The standard deviation of measurements was 989 kN (kilonewton, a measure of force). It is known that the failure strengths are normally distributed with mean 3,000 kN for this type of columns. What is the likelihood of the mean failure strength of 12 samples being lower than (a) 2,000 kN (b) 2,500 kN?

Solution   Because the failure strengths are normally distributed, but the variance of failure strengths in the population is not known, we know that the distribution of $\sqrt{n}(\overline{X} - \mu)/s$ is *t* with $(n - 1) = 11$ degrees of freedom. Using this result we can estimate the required probabilities as follows:

a $$P(\overline{X} < 2000) = P\left(t < \frac{2000 - \mu}{s/\sqrt{n}} \right)$$

$$= P\left(t < \frac{2000 - 3000}{989/\sqrt{12}} \right) = P(t < -3.50)$$

Using the *t*-table (Table 5 of the Appendix), we can see that for 11 degrees of freedom, $3.50 > 3.106 = t_{0.005}$. Therefore, we can conclude that $P(t < -3.50) < 0.005$.

Alternatively, using technology (calculator or computer software), we can get the exact required probability as $P(t < -3.50) = 0.0025$.

b
$$P(\overline{X} < 2500) = P\left(t < \frac{2500 - \mu}{s/\sqrt{n}} \right)$$
$$= P\left(t < \frac{2500 - 3000}{989/\sqrt{12}} \right) = P(t < -1.75)$$

Using the *t*-table (Table 5 of the Appendix), we can see that for 11 degrees of freedom, $t_{0.10} = 1.363 < 1.75 < 1.796 = t_{0.05}$. Therefore, we can conclude that $0.05 < P(t < -1.75) < 0.10$. In fact, 1.75 is very close to $1.796 = t_{0.05}$, which tells us that $P(t < -1.75)$ is just slightly more than 0.05.

Alternatively, using technology, we can get the required probability as $P(t < -1.75) = 0.054$.

Exercises

8.21 Shear-strength measurements for spot welds of a certain type have been found to have an approximate normal distribution with standard deviation 10 psi. If 10 test welds are to be measured, find the probability that the sample mean will be within 1 psi of the true population mean.

8.22 It is known from past samples that pH of water in Bolton Creek tends to be approximately normally distributed. The average water pH level of water in the creek is estimated regularly by taking 12 samples from different parts of the creek. Assuming they represent random samples from the creek, find the approximate probability that the sample mean of the 12 pH measurements will be within 0.2 unit of the true average pH for the field. The most recent sample measurements were as follows:

6.63, 6.59, 6.65, 6.67, 6.54, 6.13,
6.62, 7.13, 6.68, 6.82, 7.62, 6.56

8.23 Resistors of a certain type have resistances that are approximately normally distributed with mean 200 ohms. Fifteen of these resistors are to be used in a circuit. The standard deviation of 12 measurements is 10 ohms.

a Find the probability that the average resistance of the 15 resistors is between 199 and 202 ohms.

b Find the probability that the *total* resistance of the 15 resistors does not exceed 5,100 ohms.

8.24 One-hour carbon monoxide concentrations in air samples from a large city average 12 ppm. The standard deviation of carbon monoxide concentrations measured from the last five air samples was 9 ppm. Find the probability that the average concentration in next five samples selected randomly will exceed 14 ppm.

8.25 The downtime per day for a certain computing facility has averaged 4.0 hours for the last year. The standard deviation of downtimes of the last 20 failures is 0.8 hours.

a Find the probability that the average daily downtime for a period of 20 days is between 1 and 5 hours.

b Find the probability that the total downtime for the 30 days is less than 115 hours.

c What assumptions are necessary for the answers in parts (a) and (b) to be valid approximations?

8.26 The service times for customers coming through a checkout counter in a retail store are independent random variables with a mean of 1.5 minutes. The standard deviation of a random sample of 12 customers served is estimated to be 1.0 minute. Approximate the probability that 12 customers can be serviced in less than 15 minutes of total service time by this one checkout counter.

8.4

The Sampling Distribution of Sample Proportion Y/n (Large Sample)

We saw in Chapter 5 that a binomially distributed random variable Y can be written as a sum of independent Bernoulli random variables X_i. That is, if $X_i = 1$ with probability p and $X_i = 0$ with probability $(1 - p)$, $i = 1, 2, \ldots, n$, then $Y = X_1 + \cdots + X_n$ can represent the number of successes in a sample of n trials or measurements, such as the number of thermistors conforming to standards in a sample of n thermistors. Now the *fraction* of successes in n trials is $Y/n = \overline{X}$, the sample mean.

> For large n, using the Central Limit Theorem, Y/n has approximately normal distribution with a mean of $E(Y/n) = p$ and $V(Y/n) = p(1 - p)/n$.
>
> $np(1-p)$

Because $Y = n\overline{X}$, we get that Y has an approximately normal distribution with mean np and variance $np(1 - p)$. Because the approximate inference is easier than the exact for large n, this normal approximation to the binomial distribution is useful in practice.

Figure 8.15 shows the histogram of a binomial distribution for $n = 20$ and $p = 0.6$. The heights of bars represent the respective binomial probabilities. For this distribution, the mean is $np = 20(0.6) = 12$ and the variance is $np(1 - p) = 20(0.6)(0.4) = 4.8$. Superimposed upon the binomial distribution is a normal distribution with mean $\mu = 12$ and variance $\sigma^2 = 4.8$. Notice how the normal curve closely approximates the binomial histogram.

The normal approximation to the binomial distribution works well for even moderately large n, as long as p is not close to zero or 1. A useful rule of thumb is to make sure n is large enough so that $p \pm 2\sqrt{p(1 - p)/n}$ lies within the interval $(0, 1)$ before the normal approximation is used. Otherwise, the binomial distribution may be so asymmetric that the symmetric normal distribution cannot provide a good approximation. In general, this approximation is adequate if both $np \geq 10$ and $n(1 - p) \geq 10$.

Figure 8.15

A binomial distribution, $n = 20$ and $p = 0.6$, and a normal distribution, $\mu = 12$ and $\sigma^2 = 4.8$

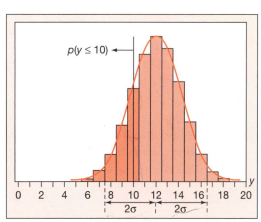

Example 8.6

Silicon wafers coming into a microchip plant are inspected for conformance to specifications. From a large lot of wafers, $n = 100$ are inspected at random. If the number of nonconformances Y is no more than 12, the lot is accepted. Find the approximate probability of acceptance if the proportion of nonconformances in the lot is $p = 0.2$.

Solution The number of nonconformances Y has a binomial distribution if the lot is indeed large. Before using the normal approximation, compute interval

$$p \pm 2\sqrt{\frac{p(1 - p)}{n}} \quad \text{i.e.,} \quad 0.2 \pm 2\sqrt{\frac{0.2(.8)}{100}} \quad \text{i.e.,} \quad 0.2 \pm 0.08$$

This interval is entirely within the interval $(0, 1)$. Thus, the normal approximation should work well. Alternatively, we could check $np = 100(0.2) = 20 > 10$ and $n(1 - p) = 100(1 - 0.2) = 80 > 10$. Now the probability of accepting the lot is $P(Y \leq 12) = P(Y/n \leq 0.12)$, where Y/n is an approximately normally distributed random variable for $n = 100$ with $\mu = p = 0.2$ and $\sigma = \sqrt{p(1 - p)/n} = 0.04$. Therefore, using the normal cumulative distribution function, it follows that

$$P(Y/n \leq 0.12) = P\left(\frac{Y/n - \mu}{\sigma} \leq \frac{0.12 - 0.2}{0.04}\right) = P(Z \leq -2.0) = 0.0228$$

There is only a small probability of 0.0228 of accepting any lot that has 20% nonconforming wafers.

Activity How many "Heads" in 50 tosses of a penny? What does the sampling distribution of sample proportion look like?

For this activity, use

- Pennies
- Stirling's sheet given in Figure 8.16.

Note that, Stirling's sheet marks "Heads" on one side and "Tails" on the other.

- Start at the top of the sheet and move down the sheet along the curves with each toss.
- Move one step toward side marked "Heads" if heads shows up on that toss, and opposite if tails shows up.
- Continue till you reach the bottom of the sheet. Total 50 tosses are required to reach the bottom.
- Read the percent of heads from the bottom scale. Divide by 100 to get a proportion of heads in 50 tosses.
- Collect data from all students in the class. Plot all the proportions to get a sampling distribution of sample proportion.
- Describe the shape of the sampling distribution using center, spread, and shape of the distribution.

Figure 8.16
Stirling's sheet

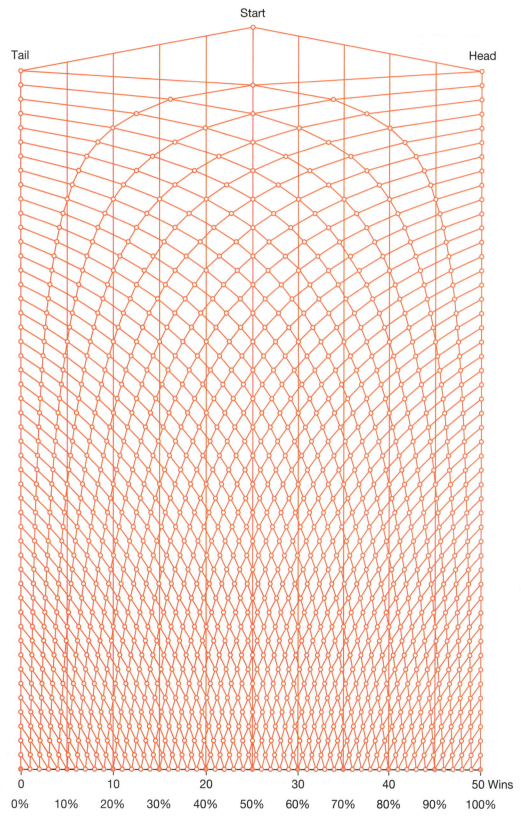

Exercises

8.27 The median age of residents of the United States is 33 years. If a survey of 100 randomly selected U.S. residents is taken, find the approximate probability that at least 60 of them will be under 33 years of age.

8.28 A lot acceptance sampling plan for large lots, similar to that of Example 8.6, calls for sampling 50 items and accepting the lot if the number of nonconformances is no more than 5. Find the approximate probability of acceptance if the true proportion of nonconformances in the lot is
   a 10%
   b 20%
   c 30%

8.29 Of the customers entering a showroom for stereo equipment, only 30% make purchases. If 40 customers enter the showroom tomorrow, find the approximate probability that at least 15 make purchases.

8.30 The quality of computer disks is measured by the number of missing pulses. For a certain brand of disk, 80% are generally found to contain no missing pulses. If 100 such disks are inspected, find the approximate probability that 15 or fewer contain missing pulses.

8.31 The capacitances of a certain type of capacitor are normally distributed with a mean of 53 mf (microfarads) and a standard deviation of 2 mf. If 64 such capacitors are to be used in an electronic system, approximate the probability that at least 12 of them will have capacitances below 50 mf.

8.32 The daily water demands for a city pumping station exceed 500,000 gallons with probability only 0.15. Over a 30-day period, find the approximate probability that demand for over 500,000 gallons per day occurs no more than twice.

8.33 At a specific intersection, vehicles entering from the east are equally likely to turn left, turn right, or proceed straight ahead. If 500 vehicles enter this intersection from the east tomorrow, what is the approximate probability that
   a 150 or fewer turn right?
   b At least 350 turn?

8.34 Waiting times at a service counter in a pharmacy are exponentially distributed with a mean of 10 minutes. If 100 customers come to the service counter in a day, approximate the probability that at least half of them must wait for more than 10 minutes.

8.35 A large construction firm has won 60% of the jobs for which it has bid. Suppose this firm bids on 25 jobs next month.
   a Approximate the probability that it will win at least 20 of these.
   b Find the exact binomial probability that it will win at least 20 of these. Compare this probability to your answer in part (a).
   c What assumptions are necessary for your answers in parts (a) and (b) to be valid?

8.36 An auditor samples 100 of a firm's travel vouchers to check on how many of these vouchers are improperly documented. Find the approximate probability that more than 30% of the sampled vouchers will be found to be improperly documented if, in fact, only 20% of all the firm's vouchers are improperly documented.

8.5 The Sampling Distribution of S^2 (Normal Distribution)

The beauty of the Central Limit Theorem lies in the fact that $\overline{X}$ will have an approximately normal sampling distribution, no matter what the shape of the probabilistic model for the population, so long as n is large and σ^2 is finite. For many other statistics, additional conditions are needed before useful sampling distributions can be derived. A common condition is that the probabilistic model for the population is itself normal. In real-world experimentation, we might expect that the distribution of population of measurements of interest is fairly symmetric and mound-shaped, that is, like a normal distribution.

Suppose a random sample of size n is taken from a normally distributed population with mean μ and variance σ^2. Then $U = (n - 1)S^2/\sigma^2$ will have a chi-square (χ^2) distribution with $(n - 1)$ degrees of freedom. The probability density function of χ^2 is

$$f(u) = \begin{cases} \dfrac{1}{\Gamma\left(\dfrac{n-1}{2}\right)2^{(n-1)/2}} u^{\frac{(n-2)}{2}-1} e^{-u/2} & u > 0 \\ \\ 0 & \text{elsewhere} \end{cases}$$

$$E(u) = n - 1 \quad \text{and} \quad V(u) = 2(n - 1)$$

The χ^2 density function is a special case of gamma density function. A typical χ^2 density function is shown in Figure 8.17. Specific values that cutoff certain right-hand tail areas under the χ^2 density function are given in Table 6 of the Appendix.

$\chi_\alpha^2(\nu) = $ the χ^2 value cutting off a tail area α under the χ^2 density function with ν degrees of freedom.

Figure 8.17

A χ^2 distribution (probability density function)

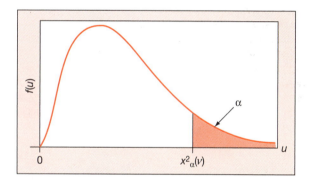

Simulation Activity What does the sampling distribution of variance look like?

- Using some software, take 100 random samples of sizes $n = 5, 10, 25$ from a normally distributed population with mean 10 and standard deviation 2.
- Compute sample variances.
- Make a histogram of sample variances.
- Discuss the distribution of sample variance (i.e., the sampling distribution of sample variance) using center, spread, and shape of the distribution.
- Compare the distributions of sample variance for $n = 5, 10, 25$ using center, spread, and shape of the distribution.

Results of one such simulation gave the summary statistics in table on the next page and distributions in Figure 8.18.

Figure 8.18

Sampling distributions of sample variance

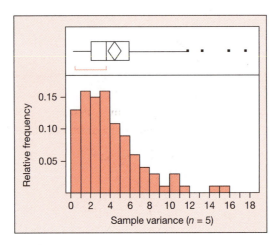

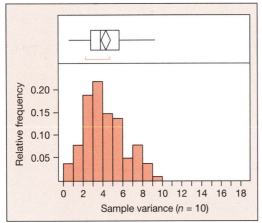

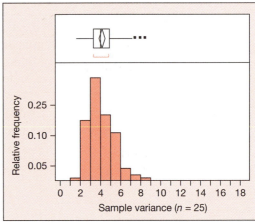

We can make the following observations from the distributions (Figure 8.18) and summary statistics in the table below.

	$n = 5$	$n = 10$	$n = 25$
Mean	3.98	4.23	4.08
Std Dev	3.01	2.06	1.27
Maximum	15.75	9.00	8.13
Q_3	5.30	5.58	4.82
Median	3.20	3.77	3.94
Q_1	1.83	2.81	3.24
Minimum	0.16	0.63	1.62

• The mean of the sampling distribution remained about the same regardless of the sample size; in other words, the sample variances were centered around the same point (marked by the population variance, $\sigma^2 = 4$), even when the sample size changed.

• The spread of the values of sample variances decreased as the sample size increased.

• Although we are sampling from a normally distributed population, the distribution of sample variances for small sample sizes is right-skewed. Tendency toward a symmetric bell-shaped curve is better for larger sample sizes.

Example 8.7

For a machine dispensing liquid into bottles, the variance σ^2 of the amount dispensed (in ounces) was known to be approximately 0.8. For a sample of 10 bottles, find two positive numbers a and b such that the sample variance S^2 of the amounts dispensed will be between a and b with probability 0.90, that is, $P(a \leq S^2 \leq b) = 0.90$. Assume that the population of amount dispensed per bottle is approximately normally distributed.

Solution Under the assumption of normality, $(n - 1)S^2/\sigma^2$ has a chi-square $\chi^2(n - 1)$ distribution. Because

$$P(a \leq S^2 \leq b) = P\left(\frac{(n - 1)a}{\sigma^2} \leq \frac{(n - 1)S^2}{\sigma^2} \leq \frac{(n - 1)b}{\sigma^2}\right)$$

the desired values can be found by setting $(n_2 - 1)b/\sigma^2$ equal to the value that cuts off an area of 0.05 in the upper tail of the χ^2 distribution, and $(n - 1)a/\sigma^2$ equal to the value that cuts off an area of 0.05 in the lower tail. Using Table 6 of the Appendix for $n - 1 = 9$ degrees of freedom yields

$$\frac{(n - 1)b}{\sigma^2} = x_{0.05}^2(9) = 16.919 \quad \text{and} \quad \frac{(n - 1)a}{\sigma^2} = x_{0.95}^2(9) = 3.32511$$

Thus, $b = 16.919(0.8)/9 = 1.504$ and $a = 3.32511(0.8)/9 = 0.296$. There is a 90% chance that the sample variance will fall between 0.296 and 1.504. Note that this is not the only interval that would satisfy the desired condition $P(a \leq S^2 \leq b) = 0.90$. But it is the convenient interval that has 5% area in each tail of the χ^2 distribution.

We saw earlier that a χ^2 distribution has a mean equal to its degrees of freedom and a variance equal to twice its degrees of freedom. In other words, if a random sample of size n is taken from a normally distributed population, then $E[(n - 1)S^2/\sigma^2] = (n - 1)$. Therefore, we get

$$\frac{(n - 1)}{\sigma^2}E(S^2) = (n - 1), \text{ which result in } E(S^2) = (n - 1)\frac{\sigma^2}{(n - 1)} = \sigma^2$$

Hence we can say that S^2 is an unbiased estimator of σ^2. Similarly, we can compute the variance of S^2 as follows:

$$V\left[\frac{(n - 1)S^2}{\sigma^2}\right] = 2(n - 1) \text{ gives } \left(\frac{n - 1}{\sigma^2}\right)^2 V[S^2] = 2(n - 1), \text{ which results in}$$

$$V(S^2) = 2(n - 1)\left(\frac{\sigma^2}{n - 1}\right)^2 = \frac{2\sigma^4}{n - 1}$$

Example 8.8

In designing mechanisms for hurling projectiles such as rockets at targets, it is very important to study the variance of distances by which the projectile misses the target center. Obviously, this variance should be as small as possible. For a certain launching mechanism, these distances are known to have a normal distribution with variance 100 m^2. An experiment involving 25 launches is to be conducted. Let S^2 denote the sample variances of the distances between the impact of the projectile and the target center.

  a Approximate $P(S^2 > 50)$.
  b Approximate $P(S^2 > 150)$.
  c Find $E(S^2)$ and $V(S^2)$.

Solution Let $U = (n - 1)S^2/\sigma^2$ have a $\chi^2(24)$ distribution for $n = 25$.

  a
$$P(S^2 > 50) = P\left(\frac{(n - 1)S^2}{\sigma^2} > \frac{(24 - 1)50}{100}\right) = P(U > 12)$$

Looking at the row for 24 degrees of freedom in Table 6 of the Appendix, we get

$$P(U > 12.4011) = 0.975 \quad \text{and} \quad P(U > 10.8564) = 0.990$$

Thus, $0.975 < P(U > 12) < 0.990$, i.e., $0.975 < P(S^2 > 50) < 0.990$

Therefore, $P(S^2 > 50)$ is a little larger than 0.975. We cannot find the exact probability using Table 6 of the Appendix, because it provides only selected tail areas. However, we can easily find the exact probability using a calculator or a computer program.

  b
$$P(S^2 > 150) = P\left(\frac{(n - 1)S^2}{\sigma^2} > \frac{(24 - 1)150}{100}\right) = P(U > 36)$$

Looking at the row for 24 degrees of freedom in Table 6 of the Appendix, we get

$$P(U > 36.4151) = 0.05 \quad \text{and} \quad P(U > 33.1963) = 0.10$$

Thus, $0.05 < P(U > 36) < 0.10$, i.e.,

$$0.975 < P(S^2 > 50) < 0.990$$

Thus, $P(S^2 > 150)$ is a little larger than 0.05

  c We know that

$$E(S^2) = \sigma^2 = 100 \quad \text{and} \quad V(S^2) = \frac{2\sigma^4}{n - 1} = \frac{2(100)^2}{24} = \frac{(100)^2}{12}$$

Following up on this result, the standard deviation of S^2 is $100/\sqrt{12} \approx 29$. Then by Tchebysheff's theorem, at least 75% of the values of S^2, in repeated sampling, should lie between 100 ± 29, that is, between 42 and 158. As a practical consideration, this much range in distance variances might cause the engineer to reassess the design of the launching mechanism.

Exercises

8.37 The efficiency ratings (in lumens per watt) of light bulbs of a certain type have a population mean of 9.5 and a standard deviation of 0.5, according to production specifications. The specifications for a room in which eight of these bulbs are to be installed call for the average efficiency of the eight bulbs to exceed 10. Find the probability that this specification for the room will be met, assuming efficiency measurements are normally distributed.

8.38 In the setting of Exercise 8.37, what should the mean efficiency per bulb equal if the specification for the room is to be met with probability approximately 0.90? (Assume the standard deviation of efficiency measurements remains at 0.5.)

8.39 The Environmental Protection Agency is concerned with the problem of setting criteria for the amount of certain toxic chemicals to be allowed in freshwater lakes and rivers. A common measure of toxicity for any pollutant is the concentration of the pollutant that will kill half of the test species in a given amount of time (usually 96 hours for fish species). This measure is called the LC50 (lethal concentration killing 50% of the test species). Studies of the effects of copper on a certain species of fish (say, species A) show the variance of LC50 measurements to be approximately 1.9, with concentration measured in milligrams per liter. If $n = 10$ studies on LC50 for copper are to be completed, find the probability that the sample mean LC50 will differ from the true population mean by no more than 0.5 unit. Assume that the LC50 measurements are approximately normal in their distribution.

8.40 If, in Exercise 8.39, it is desired that the sample mean differ from the population mean by no more than 0.5 with probability 0.95, how many tests should be run?

8.41 Suppose $n = 20$ observations are to be taken on normally distributed LC50 measurements, with $\sigma^2 = 1.9$. Find two numbers a and b such that $P(a \le S^2 \le b) = 0.90$. ($S^2$ is the sample variance of the 20 measurements.)

8.42 Ammeters produced by a certain company are marketed under the specification that the standard deviation of gauge readings be no larger than 0.2 amp. Ten independent readings on a test circuit of constant current, using one of these ammeters, gave a sample variance of 0.065. Does this suggest that the ammeter used does not meet the company's specification? [*Hint*: Find the approximate probability of a sample variance exceeding 0.065 if the true population variance is 0.04.]

8.43 A certain type of resistor is marketed with the specification that the variance of resistances produced is around 50 ohms. A sample of 15 of these resistors is to be tested for resistances produced.

a Find the approximate probability that the sample variance S^2 will exceed 80.

b Find the approximate probability that the sample variance S^2 will be less than 20.

c Find an interval in which at least 75% of such sample variances should lie.

d What assumptions are necessary for your answers above to be valid?

8.44 Answer the questions posed in Exercise 8.43 if the sample size is 25 rather than 15.

8.45 In constructing an aptitude test for a job, it is important to plan for a fairly large variance in test scores so the best applicants can be easily identified. For a certain test, scores are assumed to be normally distributed with a mean of 80 and a standard deviation of 10. A dozen applicants are to take the aptitude test. Find the approximate probability that the sample standard deviation of the scores for these applicants will exceed 15.

8.46 For an aptitude test for quality-control technicians in an electronics firm, history shows scores to be normally distributed with a variance of 225. If 20 applicants are to take the test, find an interval in which the sample variance of test scores should lie with probability 0.90.

8.6 Sampling Distributions: The Multiple-Sample Case

So far we have studied sampling distributions of a statistic when we are sampling from only one population. Often we sample from more than one population. Sometimes we are interested in estimating difference in two population means or proportions, and sometimes we are interested in estimating the ratio of variances, for example, in these comparisons:

- Mean lifetimes of two comparable brands of batteries
- Mean downtimes at two different factory locations
- Proportion defectives produced by 3 different assembly lines
- Variation in resistance measurements for rubber made using two different coupling agents

In each of the above examples, we are interested in comparing parameters for two or more populations. To do so, we must take a random sample from each of the populations and compare the statistics computed from the samples.

8.6.1 The sampling distribution of $(\overline{X}_1 - \overline{X}_2)$

Suppose your company is interested in introducing new modified batteries to the market. To show that new batteries on the average last longer than the current batteries, you might want to estimate the difference in mean lifetimes of current and new batteries produced by the company.

If we are interested in comparing two population means, we can do it easily by estimating the difference between the two. For comparing more than two population means, we have to use techniques that will be introduced in Chapter 12. Suppose two populations of interest have means μ_1 and μ_2, and variances σ_1^2 and σ_2^2, respectively. We are interested in estimating the difference $(\mu_1 - \mu_2)$, and intuitively we choose $(\overline{X}_1 - \overline{X}_2)$ to estimate $\mu_1 - \mu_2$. Suppose two random samples (of sizes n_1 and n_2 respectively), one from each population, are taken independently of each other. For each selected item, a specific variable of interest is measured (such as lifetime), and sample means are computed. The difference $(\overline{X}_1 - \overline{X}_2)$ is an unbiased estimator of $(\mu_1 - \mu_2)$. Because different random samples of the same size from the same population result in different sample means, $(\overline{X}_1 - \overline{X}_2)$ is a random variable. For independently taken samples, the sampling distribution of $(\overline{X}_1 - \overline{X}_2)$ is given on the next page.

General distribution, large samples: If both samples are large (i.e., $n_1 \geq 25$ and $n_2 \geq 25$), then the distribution function of

$$\frac{(\overline{X}_1 - \overline{X}_2) - (\mu_1 - \mu_2)}{\sqrt{\dfrac{\sigma_1^2}{n_1} + \dfrac{\sigma_2^2}{n_2}}}$$

is approximately a standard normal. The approximation improves as the sample size increases.

Small samples case, equal variances: If both the population distributions are normal with unknown variances that can be assumed to be equal (i.e., $\sigma_1^2 = \sigma_2^2$) (as in Figure 8.19), then the distribution function of

$$\frac{(\overline{X}_1 - \overline{X}_2) - (\mu_1 - \mu_2)}{S_p\sqrt{\dfrac{1}{n_1} + \dfrac{1}{n_2}}}$$

is a t-distribution with $n_1 + n_2 - 2$ degrees of freedom, where

$$S_p^2 = \frac{(n_1 - 1)S_1^2 + (n_2 - 1)S_2^2}{(n_1 - 1) + (n_2 - 1)}$$

is the pooled variance estimate of the common population variance.

Figure 8.19

Sampling from two populations with equal variances

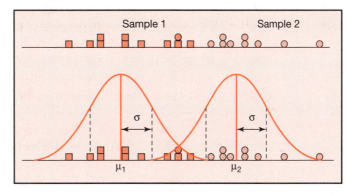

Small samples case, unequal variances: If both the populations are normal with unknown variances that are not equal (or the assumption of equal variances does not seem reasonable, i.e., $\sigma_1^2 \neq \sigma_2^2$), as shown in Figure 8.20, then the distribution function of T' is approximately a *t*-distribution with degrees of freedom equal to ν, where

$$T' = \frac{(\overline{X}_1 - \overline{X}_2) - (\mu_1 - \mu_2)}{\sqrt{\dfrac{S_1^2}{n_1} + \dfrac{S_2^2}{n_2}}} \quad \text{and} \quad \nu = \frac{\left(\dfrac{S_1^2}{n_1} + \dfrac{S_2^2}{n_2}\right)^2}{\dfrac{(S_1^2/n_1)^2}{n_1 - 1} + \dfrac{(S_2^2/n_2)^2}{n_2 - 1}}$$

In most cases, this formula gives a fractional value for the required degrees of freedom. To find the corresponding *t*-value from tables, use the next larger degrees of freedom. For example, if the above formula gives degrees of freedom equal to 17.35, then use 18 degrees of freedom in the table of *t*-distribution (Table 5 of appendix). Statistical software and calculators do this automatically.

Figure 8.20

Sampling from two populations with unequal variances

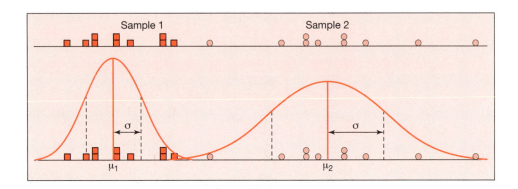

Example 8.9

A chemical manufacturer has two machines that are used to fill 1-liter bottles with certain chemicals. From long time observations, the variance of the amount filled per bottle was observed to be 200 ml for both machines. Suppose 25 observations are taken randomly on the amount dispensed by each machine on a given day. Find the probability that the difference in sample means for two machines will be at most 10 ml.

Solution   Assume that the mean amount filled per bottle by each machine is 1 L. Sample sizes are large enough for the difference in sample means $(\overline{X}_1 - \overline{X}_2)$ to have approximately normal distribution with mean $\mu_1 - \mu_2 = 1 - 1 = 0$ L, and standard deviation

$$\sqrt{\frac{\sigma_1^2}{n_1} + \frac{\sigma_2^2}{n_2}} = \sqrt{\frac{200}{25} + \frac{200}{25}} = \sqrt{16} = 4 \text{ L}$$

Thus,

$$P(|\overline{X}_1 - \overline{X}_2| \le 10) = P\left(\frac{-10 - 0}{4} \le Z \le \frac{10 - 0}{4}\right)$$

$$= P(-2.5 \le Z \le 2.5) = 0.9875$$

So, we are almost sure that the sample means are likely to have a difference of at most 10 ml.

Example 8.10

Young and Hancock (*J. Structural Engineering,* 2003) used tensile coupon tests to determine the material properties of the test specimens. The coupons were taken from the center of the web plate in the longitudinal direction of the finished specimen and tested according to Australian Standards. The samples were taken from two different groups. Six measurements of static 0.2% tensile-proof stress values were recorded from group 1 with mean population stress 450 MPa. Four measurements were recorded from group 2 with mean population stress 250 MPa. If sample standard deviations are 17.795 MPa and 9.129 MPa, respectively, estimate the probability that the sample mean tensile proof stress for group 1 is at least 150 MPa larger than that for group 2. It is reasonable to assume that the population variances of the tensile proof stress values for two groups are about the same.

Solution We are interested in estimating the probability $P(\overline{X}_1 - \overline{X}_2 \geq 150)$. Because the population variances are equal, estimate the pooled standard deviation as

$$S_p = \sqrt{\frac{(n_1 - 1)S_1^2 + (n_2 - 1)S_2^2}{(n_1 - 1) + (n_2 - 1)}} = \sqrt{\frac{5(17.795)^2 + 3(9.129)^2}{5 + 3}} = 15.138$$

If we assume that the tensile proof stress values are normally distributed then we know that the distribution function of

$$\frac{(\overline{X}_1 - \overline{X}_2) - (\mu_1 - \mu_2)}{\sqrt{S_p^2\left(\frac{1}{n_1} + \frac{1}{n_2}\right)}}$$

is a t-distribution with $n_1 + n_2 - 2 = 8$ degrees of freedom. Using this result we get

$$P(\overline{X}_1 - \overline{X}_2 \geq 150) = P\left(\frac{(\overline{X}_1 - \overline{X}_2) - (\mu_1 - \mu_2)}{S_p\sqrt{n_1^{-1} + n_2^{-1}}} \geq \frac{150 - (450 - 250)}{15.138\sqrt{6^{-1} + 4^{-1}}}\right)$$

$$= P(t(8) \geq -5.117)$$

Using the cumulative distribution function for t-distribution from TI-89, we get

$$P(\overline{X}_1 - \overline{X}_2 \geq 150) = 0.9995$$

Exercises

8.47 Shear-strength measurements for spot welds of two different types have been found to have approximate standard deviations of 10 and 12 psi, respectively. If 100 test welds are to be measured for each type, find the approximate probability that the two sample means will differ by at most 1 psi. Assume that both types of welds have the same mean shear strengths.

8.48 Soil acidity is measured by a quantity called pH. A scientist wants to estimate the difference in the average pH for two large fields using pH measurements from randomly selected core samples. If the scientist selects 20 core samples from field I and 15 core samples from field 2, independently of each other, find the approximate probability that the sample mean of the 40 pH measurements for field 1 will be larger than that for field 2 by at least 0.5. The sample variances for pH measurements for fields I and II are 1 and 0.8, respectively. In the past, both fields have shown approximately the same mean soil acidity levels.

8.49 One-hour carbon monoxide concentrations in air samples from a large city average 12 ppm, with a standard deviation of 9 ppm. The same from its twin city averages about 10 ppm, with a standard deviation of 9 ppm.

a Find the probability that the average concentration in 10 samples selected randomly and independently will exceed 1.4 ppm.

b What conditions are necessary to estimate the probability in part (a)?

8.50 The downtime per day for the university's computing server averages about 40 minutes and for the email server averages about 20 minutes. The downtimes were recorded for five randomly selected days, which showed standard deviations of 8 and 10 minutes, respectively. The director of computer services is interested in estimating the probability of the sample average for computing server is larger by at least 15 minutes than that for email server. The director uses t-distribution with mean 20 and pooled standard deviation to compute the probability of interest. Although the assumption of equal population variances is reasonable, which condition is violated here, making this probability calculation invalid?

8.51 The strengths of a type of thread manufactured by two machines are random variables with the same mean. The floor supervisor takes 10 random thread samples from each machine independently. The standard deviations of measured strengths are 0.2 pound and 0.18 pound, respectively. It is reasonable to assume that both machines are producing threads with same standard deviation of strength. Find the probability that there will be at least a 1-pound difference in the average strengths of two samples.

8.52 The service times for customers coming through a checkout counters in a retail store are independent random variables with a mean of 15 minutes and a variance of 4. At the end of the work day, the manager selects independently a random sample of 100 customers each served by checkout counters A and B. Approximate the probability that the sample mean service time for counter A is lower than that for counter B by 5 minutes.

8.53 An experiment is designed to test whether operator A or operator B gets the job of operating a new machine. Each operator is timed on 75 independent trials involving the performance of a certain task on the machine. If the sample means for the 75 trials differ by more than 5 seconds, the operator with the smaller mean gets the job. Otherwise, the experiment is considered to end in a tie. If the standard deviations of times for both operators are assumed to be 2 seconds, what is the probability that operator A gets the job even though both operators have equal ability?

8.6.2 The sampling distribution of $\overline{X}_D$

So far, we have considered the estimation of difference in population means when the samples are independent. But in many situations the two samples are dependent. For example:

- To determine whether one caliper is systematically reading higher than the other, the diameters of randomly selected components are measured using both the calipers. Then we have two measurements for each component.
- To compare the lifetimes of two brands of tires, both types of tires are mounted on rear wheels of each of 25 cars, one brand on one side and the other on the other side. Now we have two measurements for each car.

These examples show that by dependence, we mean there is some kind of matching involved between the first and the second sample, either by taking two measurements on the same experimental unit or by matching two experimental units through some common characteristics.

Consider an experiment described by Hu, Asle, Huang, Wu, and Wu (*J. Structural Engineering*, 2003). The experiment involved concrete-filled tube columns used for seismic resistance in structures. The experimenters were interested in comparing the failure strength measured through experimental data against that computed using a nonlinear finite element program. Obviously, these measurements are matched or paired. For each material specification, two measurements were taken, one experimental and one analytical. If we average measurements in the experimental and analytical groups separately, we would be averaging measurements taken under different conditions. However, the difference in each pair of measurements taken under the same conditions will lead to lower variation among the differences. Look at the data reported for columns in Table 8.3.

Table 8.3

Comparison of Failure Strengths

Experiment	Analysis	Difference
1666.0	1628.0	38.0
2016.9	2024.0	−7.1
893.0	860.0	33.0
3025.2	3029.0	−3.8
2810.0	2835.0	−25.0
2607.6	2618.0	−10.4

Let $(X_1, Y_1), (X_2, Y_2), \ldots, (X_n, Y_n)$ be n pairs of measurements. As seen from the examples, we are interested in the difference in two measurements taken on each item, that is, $d_1 = X_1 - Y_1, d_2 = X_2 - Y_2, \ldots, d_n = X_n - Y_n$. Now we consider the differences $d_1, d_2, \ldots, d_n$ as a random sample from a population of differences with mean μ_D. Then $\bar{d} = \sum d_i/n$ is an *unbiased estimator* (to be defined in Chapter 9) of μ_D, and the problem is reduced from a two-samples case to a one-sample case. Applying results of the sampling distributions for a single-samples case, we see that the distribution of

$$\frac{\bar{d} - \mu_D}{S_d/\sqrt{n}}$$

is a t-distribution with $(n - 1)$ degrees of freedom, where n is the number of pairs of measurements and S_d is the standard deviation of differences. The condition is that the population of differences must be normally distributed.

If the differences $d_1, d_2, \ldots, d_n$ are randomly selected from the population of normally distributed differences, the sampling distribution of

$$\frac{\bar{d} - \mu_D}{S_d/\sqrt{n}}$$

is a t-distribution with $(n - 1)$ degrees of freedom, where n is the number of pairs of measurements (X_i, Y_i) such that $d_i = X_i - Y_i$, and S_d is the standard deviation of differences.

Example 8.11

Two different standards/specifications are used by civil engineers for calculating the design strength of cold-formed steel structures subjected to concentrated bearing load. To compare design rules in the North American Specification (NAS) and the Australian/New Zealand Standard (AS/NZS) for cold-formed steel structures, Young and Hancock (*J. Structural Engineering*, 2003) conducted a series of tests on cold-formed steel channels. The design strengths were calculated under two loading conditions, end-one-flange (EOF) and end-two-flange (ETF), using the NAS specification (X_{NAS}) and the AS/NZS Standard ($X_{AS/NZS}$). The differences ($d = X_{NAS} - X_{AS/NZS}$) in design strengths were calculated for those two loading conditions. The following Minitab output gives the mean ($\bar{d}$) and the standard deviation (S_d) of differences for the two loading conditions.

Variable	N	Mean	StDev
d-EOF	12	10.18	12.19
d-ETF	12	-14.71	4.40

Assuming that both specifications are predicting the same strength, estimate the probability of observing a mean difference of

a At least 10.18 kN for EOF measurements
b At most –14.71 kN for ETF measurements

Solution In this experiment, the design strength was measured using two different specifications (NAS, and AS/NZS) for each of $n = 12$ tests, resulting in 12 paired measurements. Because we are assuming that both specifications are predicting the same strength, we get $\mu_D = 0$; that is, in the population, on the average there is no difference in the strengths measured by two specifications. Because the population variance of differences is unknown, we know that the sampling distribution of $\sqrt{n}(\bar{d} - \mu_D)/S_d$ is t with $(n - 1) = 11$ degrees of freedom. Using this result, we can compute the required probabilities.

a
$$P(\bar{d} \geq 12.19 \text{ kN}) = P\left(\frac{\bar{d} - \mu_D}{S_d/\sqrt{n}} \geq \frac{10.18 - 0}{12.19/\sqrt{12}}\right) = P(t(11) \geq 2.893)$$

$$= 0.0073 \text{ (using TI-89)}$$

There is less than 1% chance of observing a mean difference of at least 10.18 kN between the EOF strengths measured using NAS and AS/NZS, assuming that on the average there is no difference in the strengths measured by two specifications.

b
$$P(\bar{d} \leq -14.71 \text{ kN}) = P\left(\frac{\bar{d} - \mu_D}{S_d/\sqrt{n}} \geq \frac{-14.71 - (0)}{4.40/\sqrt{12}}\right) = P(t(11) \geq -11.58)$$

$$= 0.0000000838 \approx 0.0 \text{ (using TI-89)}$$

There is almost no chance of observing a mean difference of at most –14.71 kN between the ETF strengths measured using NAS and AS/NZS standards.

8.6.3 Sampling distribution of $(\hat{p}_1 - \hat{p}_2)$

Consider two populations with proportions of successes equal to p_1 and p_2, respectively. We are interested in the difference in population proportions $(p_1 - p_2)$. For example, we might want to study the difference in proportion defectives produced by factories at two different locations. Suppose we take independent random samples of sizes n_1 and n_2, respectively, from these two populations and compute the respective sample proportions, $\hat{p}_1 = x_1/n_1$ and $\hat{p}_2 = x_2/n_2$. Note that $(\hat{p}_1 - \hat{p}_2)$ is an unbiased estimator of $(p_1 - p_2)$. Different samples of sizes n_1 and n_2 will result in different $(\hat{p}_1 - \hat{p}_2)$ values, making $(\hat{p}_1 - \hat{p}_2)$ a random variable. Some values of $(\hat{p}_1 - \hat{p}_2)$ are more likely to occur than others. For sufficiently large sample sizes, the distribution of

$$\frac{(\hat{p}_1 - \hat{p}_2) - (p_1 - p_2)}{\sqrt{\dfrac{p_1(1 - p_1)}{n_1} + \dfrac{p_2(1 - p_2)}{n_2}}} \quad \text{is approximately standard normal.}$$

Example 8.12

Suppose a manufacturer has plants at two different locations. From the long-time observations, it is known that approximately 4% and 2.5% percent defectives are produced at these two locations. Random samples of 200 items are selected from one week's production at each location. What is the likelihood that the sample proportions will differ by more than 2 percentage points?

Solution It is known that $p_1 = 0.04$ and $p_2 = 0.025$. Also, we are taking samples of sizes $n_1 = n_2 = 200$. Then, using the earlier result, we get the standard deviation of the difference in sample proportions:

$$\sqrt{\frac{p_1(1 - p_1)}{n_1} + \frac{p_2(1 - p_2)}{n_2}} = \sqrt{\frac{0.04(0.96)}{200} + \frac{0.025(0.975)}{200}} = 0.0177$$

Therefore,

$$P(|\hat{p}_1 - \hat{p}_2| > 0.02) = P\left(|Z| > \frac{0.02 - (0.04 - 0.025)}{0.0177}\right)$$

$$= P(Z < -0.2825) + P(Z > 0.2825) = 0.3888$$

There is a fairly high chance (38.88%) of observing a difference of at most 2 percentage points between the sample proportion defectives.

8.6.4 The sampling distribution of S_1^2/S_2^2

When two samples from normally distributed populations are available in an experimental situation, it is sometimes of interest to compare the population variances, for example, comparison of variation in amount of sugar per can of peaches prepared using two different processes. We generally do this by estimating the ratio of the population variances, σ_1^2/σ_2^2.

> Suppose two independent random samples from normal distributions with respective sizes n_1 and n_2 yield sample variances of S_1^2 and S_2^2. If the populations in question have variances σ_1^2 and σ_2^2, respectively, then
>
> $$F = \frac{S_1^2/\sigma_1^2}{S_2^2/\sigma_2^2}$$
>
> has a known sampling distribution, called an *F-distribution*, with $\nu_1 = n_1 - 1$ and $\nu_2 = n_2 - 1$ degrees of freedom.

A probability density function of a random variable having an F-distribution is shown in Figure 8.21. The F-distribution is right-skewed and falls entirely in the first quadrant. The shape of the distribution depends on both sets of degrees of freedom. Let $F_\alpha(\nu_1, \nu_2)$ = the value of an F-variate that cut off the right-hand tail area of size α (see Figure 8.21).

The values of $F_\alpha(\nu_1, \nu_2)$ are given in Tables 7 and 8 of the Appendix for $\alpha = 0.05$ and 0.01, respectively. Left-hand tail values of F can be found by the equation

$$F_{1-\alpha}(\nu_1, \nu_2) = \frac{1}{F_\alpha(\nu_2, \nu_1)}$$

where the subscripts still denote the right-hand areas.

Figure 8.21

An *F*-distribution (probability density function)

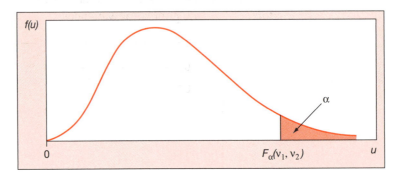

Example 8.13

Goldenberg (*IEEE Transactions on Electrical Insulation*, 1965) collected data on the lifetimes (in weeks) of wire specimen tested for thermal endurance according to AIEE Standard 57 in several laboratories. We are interested in the precision of results from two laboratories where tests were conducted at 200°C. Ten results were recorded at each laboratory. The results from both laboratories are supposed to have the same precision. Assume that the thermal endurances recorded using this test are normally distributed.

  a What is the likelihood that one sample variance is at least 2 times larger than the other, even when the population of test results at both laboratories have the same precision?
  b What is the likelihood that one sample variance is at least 4 times larger than the other, even when the population of test results at both laboratories have the same precision?

Solution Let σ_i^2 = the variance of population of test results at the ith lab, and s_i^2 = the variance of sample of test results from the ith lab ($i = 1, 2$). Because the results from both laboratories are supposed to have the same precision, let's assume $\sigma_1^2 = \sigma_2^2$. Using the above stated result and equality of population variances, we know that $\dfrac{S_1^2/\sigma_1^2}{S_2^2/\sigma_2^2} = \dfrac{S_1^2}{S_2^2}$ follows an F-distribution with $\nu_1 = n_1 - 1 = 9$ and $\nu_2 = n_2 - 1 = 9$ degrees of freedom.

  a The probability of observing one sample variance at least 2 times larger than the other is shown by the sum of the shaded areas in Figure 8.22. It is computed as follows:

$$P\left(\frac{s_1^2}{s_2^2} < \frac{1}{2}\right) + P\left(\frac{s_1^2}{s_2^2} > 2\right) = P(F(9, 9) < 0.5) + P(F(9, 9) > 2)$$

$$= 0.1582 + 0.1582 = 0.3164$$

We can get these probabilities using the technology. There is approximately 32% chance that one sample variance will be at least 2 times larger than the other, even if the population variances are equal.

Figure 8.22

$P(F < 0.5)$ and $P(F > 2)$ with (9, 9) degrees of freedom

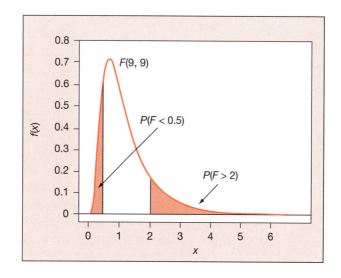

b Similarly, the probability of observing one sample variance at least 4 times larger than the other is

$$P\left(\frac{s_1^2}{s_2^2} < \frac{1}{4}\right) + P\left(\frac{s_1^2}{s_2^2} > 4\right) = P(F(9, 9) < 0.25) + P(F(9, 9) > 4)$$

$$= 0.0255 + 0.0255 = 0.051$$

We get these probabilities easily using technology (computer program or TI-89). There is an approximately 5% chance that one sample variance will be at least 4 times larger than the other, even if the population variances are equal.

8.7 Control Charts

The modern approach to improving the quality of a process, due mainly to the work of W. Edwards Deming, J. M. Juran, and their colleagues, emphasizes building quality into every step of the process. Screening out items that do not meet specifications after their completion has proven to be too costly and too ineffective in improving quality. It is far more economical to build quality into a process. Action on the process itself to improve quality is future oriented; it changes the process so that future items will have a better chance of exceeding specifications. Action on the output of a process, as in screening for quality at the factory door, is past oriented; it gives evidence of the quality of items already produced but does not help to assure better quality in the future.

All phases of a process are subject to variation, and, statistical techniques can help sort and quantify this variation. Consider two types of variations.

1. Variation arising from *common causes:* Common causes are the routine factors that affect a process and cause some variation, even though the process itself might be in "statistical control."
2. Variation arising from *special causes:* Special causes, or assignable causes, are not a regular part of the process and may not affect the entire process, but they do cause abnormally high variation when they show up.

Consider the time it takes you to get to school each day. Even though you may travel the same route at about the same time each day, there is some variation in travel times due to common causes like traffic, traffic lights, weather conditions, and how you feel. If, however, you arrive late for class on a particular day, it might be due to an extraordinarily long travel time as a result of a special cause such as a traffic accident, a flat tire, or an unusually intense storm.

The special causes are the main focus of attention in this chapter because statistical techniques are especially good at detecting them. Common-cause variation can be studied by the same techniques, but the process changes required to reduce this variation are not usually obvious or simple. Figure 8.23 diagrams process outputs ("size") for a situation involving only common causes and for another

Figure 8.23

Variation due to common and special causes

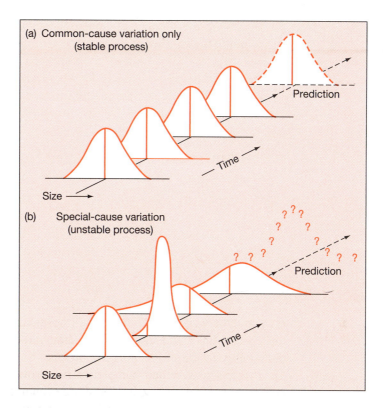

(a) Common-cause variation only (stable process)

Prediction

Time

Size

(b) Special-cause variation (unstable process)

? ? ?
? ?
?
? ? ? ? ?
? ? ?

Prediction

Time

Size

showing special causes. Stable output distributions, or systems "in control," are the goal; statistical techniques can help us achieve that goal by weeding out special causes of variation. The most common tool for assessing process variation is the *control chart*, first introduced by Dr. Walter Shewhart in the 1920s. Commonly used control charts are presented in the subsections that follow.

Imagine an industrial process that is running continuously. A standard quality improvement plan would require sampling one or more items from this process periodically and making the appropriate quality measurements. Usually, more than one item is measured at each time point to increase accuracy and measure variability. The objective in this section is to develop a technique that will help the engineer decide whether the center (or average location) of the measurements has shifted up or down and whether variation has changed substantially.

There are several different types of control charts. In this chapter, we'll discuss a few of them. The type of control chart used depends on the probability distribution of the quality measurements.

- Sampling quality measurements that possess a continuous probability distribution is referred to as "sampling by variables." For example, measuring amount of flour (per 5 lb bag) dispensed by the filling machine (measured in pounds), and diameter of rods cut by the machine (measured in mm). When sampling by variables we use the following charts:

    - $\overline{X}$-chart
    - $\overline{X}$- and R-charts
    - $\overline{X}$- and S-charts

- Observing the number of nonconforming items from a sample of items or counting the number of defects per item inspected is referred to as "sampling by attributes." For example, counting the number of overfilled cans per batch, and counting the number of defects in the finished surface of a table. When sampling by attributes we use following charts:

  - p-chart
  - c-chart
  - u-chart

8.7.1   The $\overline{X}$-chart: Known μ and σ

Suppose n observations are to be made at each time point in which the process is checked. Let X_i denote the ith observation ($i = 1, \ldots, n$) at the specified time point, and let $\overline{X}_j$ denote the average of the n observations at time point j. If $E(X_i) = \mu$ and $V(X_i) = \sigma^2$, for a process in control, then $\overline{X}_j$ should be approximately normally distributed with $E(\overline{X}_j) = \mu$ and $V(\overline{X}_j) = \sigma^2/n$. As long as the process remains in control, where will most of the $\overline{X}_j$ values lie? Because we know that $\overline{X}_j$ has approximately a normal distribution, the interval $\mu \pm 3(\sigma/\sqrt{n})$ has probability 0.9973 of including $\overline{X}_j$, as long as the process is in control.

If μ and σ were known or specified, we could use $\mu - 3(\sigma/\sqrt{n})$ as a lower control limit (LCL) and $\mu + 3(\sigma/\sqrt{n})$ as an upper control limit (UCL). If a value of $\overline{X}_j$ was observed to fall outside of these limits, we would suspect that the process might be out of control (the mean might have shifted to a different value) because this event has a very small probability of occurring (0.0027) when the process is in control. Figure 8.24 shows schematically how the decision process would work.

Note that if a process is declared to be out of control because $\overline{X}_j$ fell outside of the control limits, there is a positive probability of making an error. That is, we could declare that the mean of the quality measurements has shifted when, in fact, it has not. However, this probability of error is only 0.0027 for the three-standard deviation limits used above.

The $\overline{X}$-chart when μ and σ are known

$$UCL = \mu + 3\sigma/\sqrt{n}$$

$$CL = \mu$$

$$LCL = \mu - 3\sigma/\sqrt{n}$$

The $\overline{X}$-chart discussed here is used when the population mean and variance are known, which is very rarely the case. In most situations, the population mean and variance are unknown. In such cases, we need to estimate population mean and variance. The population mean is estimated using the sample mean. However,

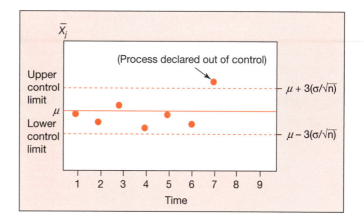

Figure 8.24

Schematic representation of $\overline{X}$-chart

we have a choice of estimators when estimating the population variance, which gives rise to two different charts.

- The $\overline{X}$- and S-charts: The sample variance S^2 is an unbiased estimator of the population variance σ^2. Of course, along with $\overline{X}$, the value of S^2 changes from sample to sample too. So when μ and σ are unknown and σ is estimated using S, then we construct the $\overline{X}$- and S-charts simultaneously.

- The $\overline{X}$- and R-charts: Although, the sample variance S^2 is an unbiased estimator of the population variance σ^2, it requires some computation. A simpler estimate of variation is given by the range of the data. So, we can use an approximate relation between range and the variance to estimate σ. This estimator, although not unbiased, gives simpler estimates for σ than S does. Although it is not an unbiased estimator, it gives fairly accurate estimates for σ. Using this result, we can construct the $\overline{X}$- and R-charts simultaneously.

8.7.2   The $\overline{X}$- and R-charts: Unknown μ and σ

If a control chart is to be started for a new process, μ and σ will not be known and hence must be estimated from the data. For establishing control limits, it is generally recommended that measurements be taken at least at $k = 20$ time points before the control limits are calculated. We will now discuss the details of estimating $\mu \pm 3\sigma/\sqrt{n}$ from k independent samples.

- For each of k random samples, calculate the sample mean, $\overline{X}_j$, and the range, R_j. Calculate the average of the means and ranges as

$$\overline{\overline{X}} = \frac{1}{k}\sum_{j=1}^{k}\overline{X}_j \quad \text{and} \quad \overline{R} = \frac{1}{k}\sum_{j=1}^{k}R_j$$

Now $\overline{\overline{X}}$ is a good measure of the center of the process output because $E(\overline{\overline{X}}) = \mu$. However, $\overline{R}$ by itself is not a good approximation for σ. It can be shown that

$$E(\overline{R}) = d_2\sigma$$

The values of constant d_2 are listed in Table 9 of the Appendix.

- Approximate the control limits $\mu \pm 3\sigma/\sqrt{n}$ by $\overline{\overline{X}} \pm A_2\overline{R}$, where

$$A_2 = \frac{3}{d_2\sqrt{n}}$$

The values of A_2 are also listed in Table 9 of the Appendix.

$\overline{X}$- and R-charts where μ and σ are known

$\overline{X}$-Chart	R-Chart
UCL $= \overline{\overline{X}} + A_2\overline{R}$	UCL $= D_4\overline{R}$
CL $= \overline{\overline{X}}$	CL $= \overline{R}$
LCL $= \overline{\overline{X}} - A_2\overline{R}$	LCL $= D_3\overline{R}$

- The sample ranges serve as a check on sample variation, in addition to their role in estimating σ for control limits on the process mean. It can be shown that $V(R_j) = d_3^2\sigma^2$. The values of d_3 are listed in Table 9 of the Appendix. The three-standard-deviation interval about the mean of R_j then becomes

$$d_2\sigma \pm 3d_3\sigma, \quad \text{i.e.,} \quad \sigma(d_2 \pm 3d_3).$$

For any sample, there is a high probability (approximately 0.9973) that its range will fall inside this interval if the process is in control.

- As in case of the $\overline{X}$-chart, if σ is not specified, it must be estimated from the data. The best estimator of σ based on $\overline{R}$ is $\overline{R}/d_2$, and the estimator of $\sigma(d_2 \pm 3d_3)$ then becomes

$$\frac{\overline{R}}{d_2}(d_2 \pm 3d_3), \quad \text{i.e.,} \quad \overline{R}\left(1 \pm 3\frac{d_3}{d_2}\right)$$

Letting $D_3 = 1 - 3d_3/d_2$ and $D_4 = 1 + 3d_3/d_2$, the control limits take on the form $(D_3\overline{R}, D_4\overline{R})$. The values of D_3 and D_4 are listed in Table 9 of the Appendix.

Example 8.14

A control chart is to be started for a new machine that fills boxes of cereal by weight. Five observations on amount of fill are taken every hour until 20 such samples are obtained. The data are given in Table 8.4. Construct a control chart for means and another for variation, based on these data. Interpret the results.

Solution The sample summaries of center and spread are given in Table 8.4. From these data,

$$\overline{\overline{x}} = \frac{1}{20}\sum_{j=1}^{20}\overline{x}_j = 16.32 \quad \text{and} \quad \overline{r} = \frac{1}{20}\sum_{j=1}^{20}r_j = 0.82$$

Table 8.4

Amount of Cereal Dispensed by Filling Machines

Sample No.	Readings 1	2	3	4	5	Summary $\bar{x}_j$	s_j	r_j (range)
1	16.1	16.2	15.9	16.0	16.1	16.06	0.114	0.3
2	16.2	16.4	15.8	16.1	16.2	16.14	0.219	0.6
3	16.0	16.1	15.7	16.3	16.1	16.04	0.219	0.6
4	16.1	16.2	15.9	16.4	16.6	16.24	0.270	0.7
5	16.5	16.1	16.4	16.4	16.2	16.32	0.164	0.4
6	16.8	15.9	16.1	16.3	16.4	16.30	0.339	0.9
7	16.1	16.9	16.2	16.5	16.5	16.44	0.313	0.8
8	15.9	16.2	16.8	16.1	16.4	16.28	0.342	0.9
9	15.7	16.7	16.1	16.4	16.8	16.34	0.451	1.1
10	16.2	16.9	16.1	17.0	16.4	16.52	0.409	0.9
11	16.4	16.9	17.1	16.2	16.1	16.54	0.439	1.0
12	16.5	16.9	17.2	16.1	16.4	16.62	0.432	1.1
13	16.7	16.2	16.4	15.8	16.6	16.34	0.358	0.9
14	17.1	16.2	17.0	16.9	16.1	16.66	0.472	1.0
15	17.0	16.8	16.4	16.5	16.2	16.58	0.319	0.8
16	16.2	16.7	16.6	16.2	17.0	16.54	0.344	0.8
17	17.1	16.9	16.2	16.0	16.1	16.46	0.503	1.1
18	15.8	16.2	17.1	16.9	16.2	16.44	0.541	1.3
19	16.4	16.2	16.7	16.8	16.1	16.44	0.305	0.7
20	15.4	15.1	15.0	15.2	14.9	15.12	0.192	0.5
Total						**326.4**	**6.74**	**16.4**
Mean						$\bar{\bar{x}} = 16.32$	$\bar{s} = 0.337$	$\bar{r} = 0.82$

From Table 9 of the Appendix (with $n = 5$), $A_2 = 0.577$, $D_3 = 0$, and $D_4 = 2.114$. Thus, the $\overline{X}$-chart has boundaries

$$\bar{\bar{x}} \pm A_2 \bar{r}, \quad \text{i.e.,} \quad 16.32 \pm (0.577)(0.82), \quad \text{i.e.,} \quad 16.32 \pm 0.47, \quad \text{or} \quad (15.85, 16.79)$$

The R-chart boundaries are

$$(D_3 \bar{r}, D_4 \bar{r}), \quad \text{i.e.,} \quad (0, 2.114[0.82]), \quad \text{or} \quad (0, 1.73)$$

These boundaries and the sample paths for both $\bar{x}_j$ and r_j are plotted in Figure 8.25. Note that no points are "out of control" among the ranges; the variability of the process appears to be stable. The mean for sample 20 falls below the lower control limit, indicating that the machine seems to be underfilling the boxes by a substantial amount. The operator should look for a special cause, such as a changed setting, a change in product flow into the machine, or a blocked passage.

If a sample mean or range is found to be outside the control limits and a special cause is found and corrected, new control limits are calculated on a reduced data set that does not contain the offending sample. Based on the first 19 samples from Example 8.14, the control limits for center become

$$16.38 \pm (0.577)(0.84) \quad \text{i.e.,} \quad 16.38 \pm 0.48$$

Now all sample means are inside the control limits.

Figure 8.25
$\bar{X}$-chart and R-chart for data in Example 8.14

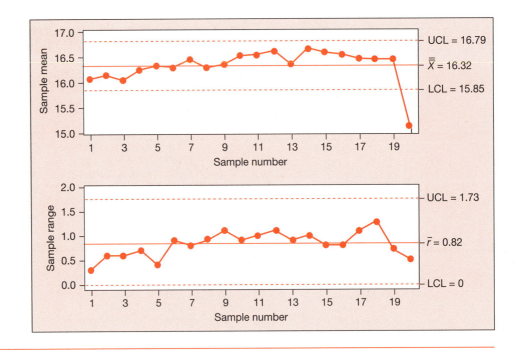

8.7.3   The $\bar{X}$- and S-charts: Unknown μ and σ

The sample standard deviation, S_j, is a more efficient estimator of σ than is a multiple of R, even though it is more cumbersome to calculate. Now

$$\bar{S} = \frac{1}{k}\sum_{j=1}^{k} S_j$$

can be adjusted by a constant, c_4, to approximate σ, because $E(\bar{S}) = E(S) = c_4\sigma$, where c_4 is listed in Table 9 of the Appendix. Thus, the control limits $\mu \pm 3\sigma/\sqrt{n}$ can be approximated by

$$\bar{\bar{X}} \pm \frac{3}{\sqrt{n}}\left(\frac{\bar{S}}{c_4}\right) \quad \text{or} \quad \bar{\bar{X}} \pm A_3\bar{S}$$

with A_3 values listed in Table 9.

What about using $\bar{S}$ as the basis of a chart to track process variability? In the spirit of three-standard-deviation control limits, $\bar{S} \pm 3\sqrt{V(S)}$ would provide control limits for sample standard deviations. Now

$$V(S) = E(S^2) - [E(S)]^2 = \sigma^2 - (c_4\sigma)^2 = \sigma^2(1 - c_4^2)$$

Because σ can be estimated by $\bar{S}/c_4$, the control limits become

$$\bar{S} \pm 3\frac{\bar{S}}{c_4}\sqrt{1 - c_4^2} \quad \text{i.e.,} \quad \bar{S}\left(1 \pm 3\frac{\sqrt{1 - c_4^2}}{c_4}\right)$$

These limits are usually written as $(B_3\bar{S}, B_4\bar{S})$, where B_3 and B_4 are listed in Table 9 of the Appendix.

$\overline{X}$- and *S*-chart where μ and σ are unknown

$\overline{X}$-Chart	*R*-Chart
UCL $= \overline{\overline{X}} + A_3\overline{S}$	UCL $= B_4\overline{S}$
CL $= \overline{\overline{X}}$	CL $= \overline{S}$
LCL $= \overline{\overline{X}} - A_3\overline{S}$	LCL $= B_3\overline{S}$

Example 8.15

Compute an $\overline{X}$-chart and an *S*-chart for the data of Table 8.4. Interpret the results.

Solution As before, $\overline{\overline{x}} = 16.32$, so $\overline{\overline{x}} \pm A_3\overline{s}$ becomes

$$16.32 \pm (1.427)(0.337) = 16.32 \pm 0.48, \text{ i.e., } (15.84, 16.80).$$

This is practically identical to the result found above based on ranges, and the 20th sample mean is still out of control.

For assessing variation, $(B_3\overline{s}, B_4\overline{s})$ becomes, $(0, 2.089[0.337])$, that is, $(0, 0.704)$. Note from Figure 8.26 that all 20 sample standard deviations are well within this range; the process variation appears to be stable.

Figure 8.26

$\overline{X}$- and *S*-charts for cereal data

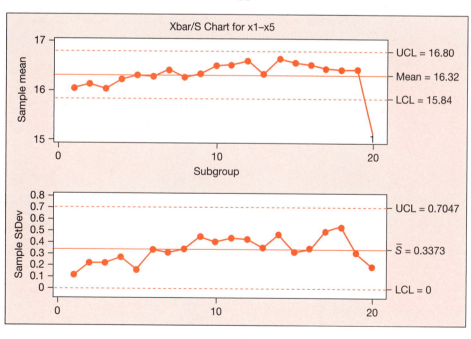

Points outside the control limits are the most obvious indicator of a potential problem with a process. More subtle features of the control chart, however, can serve as warning signs. For the $\overline{X}$-chart, roughly two-thirds of the observed sample means should lie within one standard deviation of the mean and one-third should lie beyond that. That is, on the chart itself, two-thirds of the plotted points should lie within the middle one-third of the region between the control limits. If too many points lie close to the center line, perhaps the limits are incorrectly calculated or the sample data are misrepresenting the process, If too few points lie close to the center line, perhaps the process is going "out of control," but a sample mean has not yet crossed the boundary.

In a similar spirit, long runs of points above or below the center line could indicate a potential problem. Ideally, the sample means should move above and below the center line rather frequently. If seven sample means in a row are above (or below) the center line, trouble may be brewing.

8.7.4   The *p*-chart

The control-charting procedures outlined in the preceding sections of this chapter depend on quality measurements that possess a continuous probability distribution. Sampling such measurements is commonly referred to as "sampling by variables." In many quality-control situations, however, we merely want to assess whether or not a certain item conforms to specifications. We now describe control charts for "sampling by attributes."

As in previous control-charting problems, suppose a series of k independent samples, each of size n, is selected from a production process. Let p denote the proportion of nonconforming (defective) items in the population (total production for a certain time period) for a process in control, and let X_i denote the number of defectives observed in the ith sample. Then X_i has a binomial distribution, assuming random sampling from large lots, with $E(X_i) = np$ and $V(X_i) = np(1 - p)$. We will usually work with sample fractions of defectives (X_i/n) rather than the observed number of defectives, Where will these sample fractions tend to lie for a process in control? As argued previously, most sample fractions should lie within three standard deviations of their mean, or in the interval

$$p \pm 3\sqrt{\frac{p(1 - p)}{n}}$$

because $E(X_i/n) = p$ and $V(X_i/n) = p(1 - p)/n$.

Because p is unknown, we estimate these control limits, using the data from all k samples, by calculating

$$\overline{P} = \frac{\sum_{i=1}^{k} X_i}{nk} = \frac{\text{total number of defectives observed}}{\text{total sample size}}$$

Because $E(\overline{P}) = p$, approximate control limits are given by

$$\overline{P} \pm 3\sqrt{\frac{\overline{P}(1 - \overline{P})}{n}}$$

p-chart

$$\text{UCL} = \overline{P} + 3\sqrt{\overline{P}(1 - \overline{P})/n}$$

$$\text{CL} = \overline{P}$$

$$\text{LCL} = \overline{P} - 3\sqrt{\overline{P}(1 - \overline{P})/n}$$

Example 8.16

A process that produces transistors is sampled every 4 hours. At each time point, 50 transistors are randomly sampled, and the number of defectives x_i is observed. The data for 24 samples are given in table below. Construct a control chart based on these samples.

Sample	x_i	x_i/n	Sample	x_i	x_i/n
1	3	0.06	13	1	0.02
2	1	0.02	14	2	0.04
3	4	0.08	15	0	0.00
4	2	0.04	16	3	0.06
5	0	0.00	17	2	0.04
6	2	0.04	18	2	0.04
7	3	0.06	19	4	0.08
8	3	0.06	20	1	0.02
9	5	0.10	21	3	0.06
10	4	0.08	22	0	0.00
11	1	0.02	23	2	0.04
12	1	0.02	24	3	0.06

Solution From the above data, the observed value of $\overline{P}$ is

$$\overline{p} = \frac{\sum\limits_{i=1}^{k} x_i}{nk} = \frac{52}{50(24)} = 0.04$$

Thus,

$$\overline{p} \pm 3\sqrt{\frac{\overline{p}(1 - \overline{p})}{n}}$$

becomes

$$0.04 \pm 3\sqrt{\frac{(0.04)(0.96)}{50}} \quad \text{i.e.,} \quad 0.04 \pm 0.08$$

This yields the interval (−0.04, 0.12), but we set the lower control limit at zero since a fraction of defectives cannot be negative, and we obtain (0, 0.12).

All of the observed sample fractions are within these limits, so we would feel comfortable in using them as control limits for future samples. The limits should be recalculated from time to time as new data become available.

The control limits and observed data points are plotted in Figure 8.27.

Figure 8.27

A p-chart for the data of table

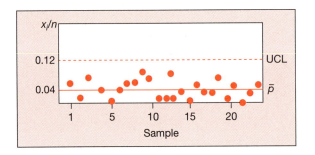

Figure 8.27

A p-chart for the data of table

8.7.5 The c-chart

In many quality-control problems, the particular items being subjected to inspection may have more than one defect, so we may wish to count defects instead of merely classifying an item as defective or nondefective. If C_i denotes the number of defects observed on the ith inspected item, a good working model for many applications is to assume that C_i has a Poisson distribution. We will let the Poisson distribution have a mean of λ for a process in control.

Most of the C_i's should fall within three standard deviations of their mean if the process remains in control. Since $E(C_i) = \lambda$ and $V(C_i) = \lambda$, the control limits become

$$\lambda \pm 3\sqrt{\lambda}$$

If k items are inspected, then λ is estimated by

$$\overline{C} = \frac{1}{k}\sum_{i=1}^{k} C_i$$

The estimated control limits then become

$$\overline{C} \pm 3\sqrt{\overline{C}}$$

c-chart

$$UCL = \overline{c} + 3\sqrt{\overline{c}}$$
$$CL = \overline{c}$$
$$LCL = \overline{c} - 3\sqrt{\overline{c}}$$

Example 8.17

Twenty rebuilt pumps were sampled from the hydraulics shop of an aircraft rework facility. The numbers of defects recorded are given in table on the next page. Use these data to construct control limits for defects per pump.

Pump	No. of Defects (C_i)	Pump	No. of Defects (C_i)
1	6	11	4
2	3	12	3
3	4	13	2
4	0	14	2
5	2	15	6
6	7	16	5
7	3	17	0
8	1	18	7
9	0	19	2
10	0	20	1

Solution   Assuming that C_i, the number of defects observed on pump i, has a Poisson distribution, we estimate the mean number of defects per pump by $\overline{C}$ and use

$$\overline{C} \pm 3\sqrt{\overline{C}}$$

as the control limits. Using the data in the table above, we have

$$\overline{c} = \frac{58}{20} = 2.9$$

and hence

$$\overline{c} \pm 3\sqrt{\overline{c}}$$

becomes

$$2.9 \pm 3\sqrt{2.9} \quad \text{i.e.,} \quad 2.9 \pm 5.1 \quad \text{i.e.,} \quad (0, 8.0)$$

Again the lower limit is raised to zero, since a number of defects cannot be negative.

All of the observed counts are within the control limits, so we could use these limits for future samples. Again, the limits should be recomputed from time to time as new data become available.

The control limits and data points are plotted in Figure 8.28.

Figure 8.28

A c-chart for the data of table

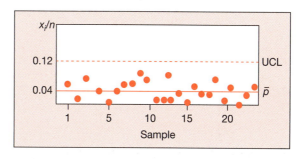

8.7.6 The *u*-chart

The *c*-chart introduced above works well when counts are being made of units of equal size, such as the number of paint defects per automobile or the number of bacteria per cubic centimeter in standard test tube samples of water. If the size of the sampling unit changes, adjustments must be made to the control limits, resulting in a *u*-chart. Such might be the case, for example, in counting the number of accidents per plant or the number of soldering errors per circuit board. If the plants have varying numbers of employees or the circuit boards have varying numbers of connections to be soldered, this information must be worked into the procedure for calculating control limits.

Suppose the *i*th sampled unit has size h_i (where h_i could be number of employees, number of electrical connections, volume of water sampled, area of glass examined, and so on) and produces a count of nonconformances C_i. We assume C_i has a Poisson distribution with mean λh_i. That is,

$$E(C_i) = \lambda h_i \quad \text{and} \quad V(C_i) = \lambda h_i$$

It follows that $U_i = C_i/h_i$ has the properties $E(U_i) = \lambda$ and

$$V(U_i) = \left(\frac{1}{h_i}\right)^2 (\lambda h_i) = \frac{\lambda}{h_i}$$

If a quality-improvement plan calls for *k* independently selected samples of this type, then

$$\overline{U} = \frac{\sum_{j=1}^{k} C_j}{\sum_{j=1}^{k} h_j}$$

is a good approximation to λ in the sense that $E(\overline{U}) = \lambda$. For the *i*th sample, the three-standard-deviation control limits around λ are given by

$$E(U_i) \pm 3\sqrt{V(U_i)} \quad \text{or} \quad \lambda \pm 3\sqrt{\frac{\lambda}{h_i}}$$

which can be approximated by

$$\overline{U} \pm 3\sqrt{\frac{\overline{U}}{h_i}}$$

Notice that the control limits might change with each sample if the h_i's change. For simplicity in constructing control limits, h_i may be replaced by

$$\overline{h} = \frac{1}{k} \sum_{j=1}^{k} h_j$$

as long as no individual h_i is more than 25% above or below this average.

u-chart

$$\text{UCL} = \bar{U} + 3\sqrt{\bar{U}/h_i}$$
$$\text{CL} = \bar{U}$$
$$\text{LCL} = \bar{U} - 3\sqrt{\bar{U}/h_i}$$

Example 8.18

The hydraulics shop of an aircraft rework facility was interested in developing control charts to improve its ongoing quality of repaired parts. The difficulty was that many different kinds of parts, from small pumps to large hydraulic lifts, came through the shop regularly, and each part may have been worked on by more than one person. The decision, ultimately, was to construct a *u*-chart based on number of defects per hours of labor to complete the repairs. The data for a sequence of 12 sampled parts are given in table below.

Part	Defects	Hours	Part	Defects	Hours
1	1	58.33	7	3	380.20
2	4	80.22	8	5	527.70
3	1	209.24	9	3	319.30
4	2	164.70	10	1	340.20
5	1	253.70	11	2	78.70
6	4	426.90	12	1	67.27

Source: Naval Air Rework Facility, Jacksonville, Florida.

Construct a *u*-chart for these data and interpret the results.

Solution Following the construction presented above,

$$\bar{u} = \frac{\sum c_i}{\sum h_i} = \frac{28}{2{,}906.46} = 0.0096$$

and the control limits are given by

$$\bar{u} \pm 3\sqrt{\bar{u}/h_i}$$

(The h_i values are too variable to use $\bar{h}$.) For $i = 1$ and 2, the upper control limits are

$$0.0096 + 3\sqrt{0.0096/58.33} = 0.048$$

and

$$0.0096 + 3\sqrt{0.0096/80.22} = 0.042$$

Now

$$u_1 = 1/58.33 = 0.017$$

and

$$u_2 = 4/80.22 = 0.049$$

Thus, u_1 is inside its control limit, but u_2 is outside its control limit. All other points are inside the control limits, as shown in Figure 8.29. Note that all lower control limits are rounded up to zero because they are calculated to be negative.

The second sampled part should be examined carefully for a special cause, such as the type of part, the training of the worker, the nature of the repairs to be made, the time and date of the repair, and so on.

Figure 8.29

A *u*-chart for the data of table

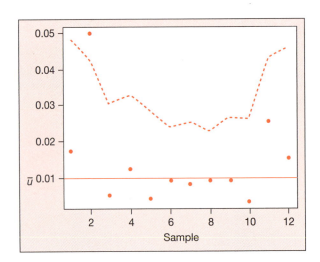

8.8   Process Capability

After a process is brought to a state of statistical control, questions may still be asked as to how well the process meets customer demands. Such questions deal with the *capability* of the process to meet specifications. There are many ways to assess process capability, but one of the more common ones is the use of the C_{pk} statistic.

Suppose a process has an upper specification limit (USL) and a lower specification limit (LSL). In other words, acceptable output values must lie between LSL and USL. In situations for which an $\overline{X}$-chart is appropriate, the center of the process is measured by $\overline{x}$. The process capability index C_{pk} measures how far the specification limits are from the process center in terms of process standard deviation. Specifically, we define

$$z_{\text{USL}} = \frac{\text{USL} - \overline{x}}{\hat{\sigma}} \quad \text{and} \quad z_{\text{LSL}} = \frac{\overline{x} - \text{LSL}}{\hat{\sigma}}$$

Where $\hat{\sigma}$ is an estimate of σ, the standard deviation of the outcome measure. Defining z_{min} as the smaller of z_{USL} and z_{LSL}, the capability index C_{pk} is given by

$$C_{pk} = \frac{z_{\text{min}}}{3}$$

All else being equal, a large value of $C_{pk}(C_{pk} > 1)$ tells us that the process is doing well because the standard deviation is small relative to the distance between specification limits. A small value of C_{pk} may be cause for concern.

In order to calculate C_{pk}, we must estimate σ from the data. This may be done directly from the input to the construction of control limits by either

$$\hat{\sigma}_1 = \bar{r}/d_2 \quad \text{or} \quad \hat{\sigma}_2 = \bar{s}/c_4$$

Example 8.19

The cereal weight data of Table 8.4 come from a machine that fills boxes labeled as weighing 16 ounces. Thus, LSL = 16 ounces for this study. To keep a reasonable margin of profit per box, the actual amount of cereal dispensed should not exceed 17 ounces. That is, USL = 17 ounces. Find C_{pk} for this process. Find the total proportion out of specification.

Solution The estimate of σ for this process should come from the data of Table 8.4 after removal of sample 20. (Recall that this sample mean was outside the control limits.) For the remaining 19 samples, $\bar{r} = 0.84$ and $\bar{s} = 0.345$. It follows that

$$\hat{\sigma}_1 = \bar{r}/d_2 = 0.84/2.326 = 0.361$$

and
$$\hat{\sigma}_2 = \bar{s}/c_4 = 0.345/0.940 = 0.367$$

Either estimate will do; we choose to use the estimate based on $\bar{s}$. Then

$$z_{\text{USL}} = \frac{\text{USL} - \bar{x}}{\hat{\sigma}} = \frac{17 - 16.38}{0.367} = 1.69$$

$$z_{\text{LSL}} = \frac{\bar{x} - \text{LSL}}{\hat{\sigma}} = \frac{16.38 - 16}{0.367} = 1.04$$

and

$$z_{\min} = \min(z_{\text{LSL}}, z_{\text{USL}}) = 1.04$$

It follows that

$$C_{pk} = \frac{z_{\min}}{3} = \frac{1.04}{3} = 0.35$$

This is quite a small C_{pk}, and it shows that the process is not meeting specifications very well. To make this statement more precise, the actual proportion of product not meeting specifications can be approximated from the normal curve. The area below $z_{\text{LSL}} = 1.04$ is 0.1492. The area above $z_{\text{USL}} = 1.69$ is 0.0455. Thus,

$$\text{proportion out of specification} = 0.1492 + 0.0455 = 0.1947$$

Nearly 20% of the boxes are filled outside the specification limits, a situation that is either making customers unhappy or hurting the profit margin. This process should be studied to see how the common-cause variation can be reduced.

Another commonly used index of process capability, C_p, is defined as

$$C_p = \frac{USL - LSL}{6\hat{\sigma}}$$

The idea behind this index is that most outcome measurements will lie within an interval of six standard deviations. If this six-standard-deviation interval fits inside the specification limits, then the process is meeting specifications most of the time. (C_p is greater than 1.) If the process output is centered at the midpoint between USL and LSL, then $C_p = C_{pk}$. Otherwise, C_p can be a poor index, because a poorly centered process, relative to USL and LSL, could still have a large C_p, value.

Exercises

8.54 Production of ammeters is controlled for quality by periodically selecting an ammeter and obtaining four measurements on a test circuit designed to produce 15 amps. The data in table below were observed for 15 tested ammeters.

	Reading			
Ammeter	**1**	**2**	**3**	**4**
1	15.1	14.9	14.8	15.2
2	15.0	15.0	15.2	14.9
3	15.1	15.1	15.2	15.1
4	150	14.7	15.3	15.1
5	14.8	14.9	15.1	15.2
6	14.9	14.9	15.1	15.1
7	14.7	15.0	15.1	15.0
8	14.9	15.0	15.3	14.8
9	14.4	14.5	14.3	14.4
10	15.2	15.3	15.2	15.5
11	15.1	15.0	15.3	15.3
12	15.2	15.6	15.8	15.8
13	14.8	14.8	15.0	15.0
14	14.9	15.1	14.7	14.8
15	15.1	15.2	14.9	15.0

a Construct control limits for the mean by using the sample standard deviations.

b Do all observed sample means lie within the limits found in part (a)? If not, recalculate the control limits, omitting those samples that are "out of control."

c Construct control limits for process variation and interpret the results.

8.55 Refer to Exercise 8.54. Repeat parts (a), (b), and (c) using sample ranges instead of sample standard deviations.

8.56 Refer to Exercise 8.54, part (a). How would the control limits change if the probability of the sample mean falling outside the limits when the process is in control is to be 0.01?

8.57 Refer to Exercises 8.54 and 8.55. The specifications for these ammeters require individual readings on the test circuit to be within 15.0 ± 0.4. Do the ammeters seem to be meeting this specification? Calculate C_{pk} and the proportion out of specification.

8.58 Bronze castings are controlled for copper content. Ten samples of five specimens each gave the measurements on percentage of copper that are listed in table on the next page.

Sample	Specimen				
	1	2	3	4	5
1	82	84	80	86	83
2	90	91	94	90	89
3	86	84	87	83	80
4	92	91	89	91	90
5	84	82	83	81	84
6	82	81	83	84	81
7	80	80	79	83	82
8	79	83	84	82	82
9	81	84	85	79	86
10	81	92	94	79	80

a Construct control limits for the mean percentage of copper by using the sample ranges.

b Would you use the limits in part (a) as control limits for future samples? Why or why not?

c Construct control limits for process variability. Does the process appear to be in control with respect to variability?

8.59 Filled bottles of soft drink coming off a production line are checked for underfilling. One hundred bottles are sampled from each half-day's production. The results for 20 such samples are listed in table below.

Sample	Defectives	Sample	Defectives
1	10	11	15
2	12	12	6
3	8	13	3
4	14	14	2
5	3	15	0
6	7	16	1
7	12	17	1
8	10	18	4
9	9	19	3
10	11	20	6

a Construct control limits for the proportion of defectives.

b Do all the observed sample fractions fall within the control limits found in part (a)? If not, adjust the limits for future use.

8.60 Refer to Exercise 8.59. Adjust the control limits so that a sample fraction will fall outside the limits with probability 0.05 when the process is in control.

8.61 Bolts being produced in a certain plant are checked for tolerances. Those not meeting tolerance specifications are considered to be defective. Fifty bolts are gauged for tolerance every hour. The results for 30 hours of sampling are listed in table below.

Sample	Defective	Sample	Defective	Sample	Defective
1	5	11	3	21	6
2	6	12	0	22	4
3	4	13	1	23	9
4	0	14	1	24	3
5	8	15	2	25	0
6	3	16	5	26	6
7	10	17	5	27	4
8	2	18	4	28	2
9	9	19	3	29	1
10	7	20	1	30	2

a Construct control limits for the proportion of defectives based on these data.

b Would you use the control limits found in part (a) for future samples? Why or why not?

8.62 Woven fabric from a certain loom is controlled for quality by periodically sampling 1-square-meter specimens and counting the number of defects. For 15 such samples, the observed number of defects is listed in table below.

Sample	Defects	Sample	Defects
1	3	9	6
2	6	10	14
3	10	11	4
4	4	12	3
5	7	13	1
6	11	14	4
7	3	15	0
8	9		

a Construct control limits for the average number of defects per square meter.

b Do all samples observed seem to be "in control"?

8.63 Refer to Exercise 8.62. Adjust the control limits so that a count for a process in control will fall outside the limits, with an approximate probability of only 0.01.

8.64 Quality control in glass production involves counting the number of defects observed in samples of a fixed size periodically selected from the production process. For 20 samples, the observed counts of numbers of defects are listed in table on the next page.

Sample	Defects	Sample	Defects
1	3	11	7
2	0	12	3
3	2	13	0
4	5	14	1
5	3	15	4
6	0	16	2
7	4	17	2
8	1	18	1
9	2	19	0
10	2	20	2

a Construct control limits for defects per sample based on these data.

b Would you adjust the limits in part (a) before using them on future samples?

8.65 Suppose variation among defect counts is important in a process like that in Exercise 8.63. Is it necessary to have a separate method to construct control limits for variation? Discuss.

8.66 An aircraft subassembly paint shop has the task of painting a variety of reworked aircraft parts. Some parts are large and require hours to paint, whereas some are small and can be painted in a few minutes. Rather than looking at paint defects per part as a quality-improvement tool, it was decided to group parts together and look at paint defects over a period of hours worked. Data in this form are given in table below for a sequence of 20 samples. Construct an appropriate control chart for these data. Does it look as if the painting process is in control?

Sample	Defects	Hours	Sample	Defects	Hours
1	4	51.5	11	1	38.8
2	1	51.5	12	3	57.8
3	0	47.1	13	0	39.1
4	0	47.1	14	2	49.1
5	2	53.1	15	0	41.6
6	3	40.4	16	1	51.5
7	0	57.7	17	1	42.1
8	0	51.5	18	2	54.3
9	0	34.8	19	1	36.6
10	1	51.5	20	0	36.6

8.67 Refer to the capability index discussion in Example 8.19. Suppose the process can be "centered" so that $\bar{\bar{x}}$ is midway between LSL and USL. How will this change C_{pk} and the proportion out of specification?

8.68 **a** Show that $C_p = C_{pk}$ for a centered process.

b Construct an example to show that C_p can be a poor index of capability for a process that is not centered.

8.69 For a random sample of size n from a normal distribution, show that $E(S) = c_4\sigma$ where

$$c_4 = \sqrt{\frac{2}{n-1}}\left(\frac{\Gamma(n/2)}{\Gamma((n-1)/2)}\right)$$

[*Hint*: Make use of proportions of the gamma distribution from Chapter 6.]

8.9   Summary

For random samples, sample *statistics* (like the mean and variance) assume different values from sample to sample, but these values collectively form predictable distributions, called *sampling distributions*. Most important, sample means and sample proportions have sampling distributions that are approximately normal in shape.

One of the principal applications of sampling distributions to process improvement comes in the construction of *control charts*. Control charts are basic tools that help separate special cause variation from common-cause variation in order to bring process into statistical control. Then *process capability* can be assessed.

It is important to remember that some of the sampling distributions we commonly deal with are exact and some are approximate. If the random sample under consideration comes from a population that can be modeled by a normal distribution, then $\overline{X}$ will have a normal distribution, and $(n - 1)S^2/\sigma^2$ will have a χ^2 distribution with *no* approximation involved. However, the Central Limit Theorem states that $\sqrt{n}(\overline{X} - \mu)/\sigma$ will have *approximately* a normal distribution for large n, under almost any probabilistic model for the population itself.

Sampling distributions for statistics will be used throughout the remainder of the text for the purpose of relating sample quantities (statistics) to parameters of the population.

Supplementary Exercises

8.70 The U.S. Census Bureau reports median ages of the population and median incomes for households. On the other hand, the College Board reports average scores on the Scholastic Aptitude Test. Do you think each group is using an appropriate measure? Why or why not?

8.71 Twenty-five lamps are connected so that when one lamp fails, another takes over immediately. (Only one lamp is on at any one time.) The lamps operate independently, and each has a mean life of 50 hours and a standard deviation of 4 hours. If the system is not checked for 1,300 hours after the first lamp is turned on, what is the probability that a lamp will be burning at the end of the 1,300-hour period?

8.72 Suppose that $X_1, \ldots, X_{40}$ denotes a random sample of measurements on the proportion of impurities in samples of iron ore. Suppose that each X_1 has the probability density function

$$f(x) = \begin{cases} 3x^2 & 0 \le x \le 1 \\ 0 & \text{elsewhere} \end{cases}$$

The ore is to be rejected by a potential buyer if $\overline{X} > 0.7$. Find the approximate probability that the ore will be rejected, based on the 40 measurements.

8.73 Suppose that $X_1, \ldots, X_{10}$ and $Y_1, \ldots, Y_2$ are independent random samples, with the X_i's having mean μ_1 and variance σ_1^2 and the Y_i's having mean μ_2 and variance σ_2^2. The difference between the sample means, $\overline{X} - \overline{Y}$, will again be approximately normally distributed, because

the Central Limit Theorem will apply to this difference.
a Find $E(\overline{X} - \overline{Y})$.
b Find $V(\overline{X} - \overline{Y})$.

8.74 A study of the effects of copper on a certain species (say, species A) of fish shows the variance of LC50 measurements (in milligrams per liter) to be 1.9. The effects of copper on a second species (say, species B) of fish produce a variance of LC50 measurements of 0.8. If the population means of LC50s for the two species are equal, find the probability that, with random samples of 10 measurements from each species, the sample mean for species A exceeds the sample means for species B by at least one unit.

8.75 If Y has an exponential distribution with mean θ, show that $U = 2Y/\theta$ has a χ^2 distribution with 2 degrees of freedom.

8.76 A plant supervisor is interested in budgeting for weekly repair costs (in dollars) for a certain type of machine. These repair costs over the past 10 years tend to have an exponential distribution with a mean of 20 for each machine studied. Let $Y_1, \ldots, Y_5$ denote the repair costs for five of these machines for the next week. Find a number c such that

$$P\left(\sum_{i=1}^{5} Y_i > c\right) = 0.05,$$

assuming the machines operate independently.

8.77 If Y has a $X^2(n)$ distribution, then Y can be represented as

$$Y = \sum_{i=1}^{n} X_i$$

where X_i has a $X^2(1)$ distribution and the X_i's are independent. It is not surprising, then, that Y will be approximately normally distributed for large n. Use this fact in the solution of the following problem.

A machine in a heavy-equipment factory products steel rods of length Y, where Y is a normally distributed random variable with a mean of 6 inches and a variance of 0.2. The cost C (in dollars) of repairing a rod that is not exactly 6 inches in length is given by

$$C = 4(Y - \mu)^2$$

where $\mu = E(Y)$. If 50 rods with independent lengths are produced in a given day, approximate the probability that the total cost for repairs exceeds \$48.

Estimation

The sulfate levels (micrograms per cubic meter) were measured in a state park near the factory location before and after installing a new pollution control device in the factory (see Figure 9.1). By how much has the pollution level decreased as a result of the use of a new pollution control device?

To answer this question, we need to use a statistical technique known as *estimation.* In many situations, the population characteristics (known as parameters) such as means or proportions are unknown, and sample statistics are employed to estimate these unknown parameters. The knowledge of the sampling distribution of the statistic employed allows us to construct intervals that include an unknown parameter value with high probability.

Figure 9.1

Sulfate levels before and after pollution control measures

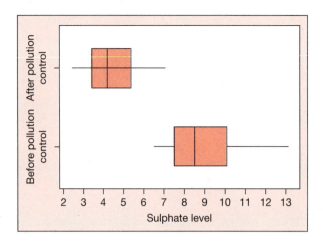

Introduction

Typically, all the information about a population of interest is not available. In many situations, it is also not practical to obtain information from each and every member of the population due to time, monetary, or experimental constraints. For example, to estimate the breaking strength of manufactured bricks, pressure must be applied until the brick breaks. In this case, if the manufacturer tries to collect information from the population, there will be no bricks left to use. Then how do we get information about the population? We take a random sample from the population and obtain information from the sampled units. We then use this information from the sample to estimate or to make inference about the unknown population characteristics. In this example, to estimate μ, the mean strength of the batch of bricks manufactured, the manufacturer would probably test a few bricks from this batch for breaking strength, compute the sample mean $\overline{X}$ from the measurements, and use the computed sample mean to estimate the mean breaking strength of the bricks from this whole batch.

We use the estimation technique frequently for decision making.

- Environmental engineers take a few soil samples and estimate chemical levels at the proposed construction site.
- Marine biologists sample a few fish from Mobile Bay and estimate the mercury level of the fish.
- Chemical engineers design experiments with a few trial runs and estimate the percentage of yield of specific chemicals from the process.
- A team of engineers from a construction company takes a few measurements and estimates the cost of a new highway construction project that involves two overpasses.
- Insurance adjusters take a few measurements and estimate the amount of damage to a building due to a hurricane.
- Pecan growers inspect a few pecan trees and estimate the damage due to late frost.
- Car mechanics run a few diagnostic tests and estimate the extent of damage and resulting repair costs.

The estimation processes use *estimators* to estimate unknown population characteristics.

> An **estimator** is a statistic that specifies how to use the sample data to estimate an unknown parameter of the population.

In an article describing a city's effort to revamp its transit system, *Mobile Register* (July 14, 2004, p. 4A) reported, "Metro transit has about 3,500 riders during each work week." The estimate reported in the newspaper gives a *point estimate*, that is, a single number (3,500 riders) describing an approximate number of users of the transit system per week.

Mobile Register (July 15, 2004, p. 4A) reported, "A few decades ago, before Goliath grouper became popular targets for the commercial fish market, biologists surveyed mating aggregation that averaged more than 100 fish per cluster. By the late 1980s, the average was less than 10 per aggregation.… Commercial fishing for Goliath grouper was halted in 1990, and in the years since, aggregations have climbed somewhat to 30 to 40 fish per aggregation." The recent estimate reported gives an *interval estimate*, that is, a range of values (30–40 fish per aggregation) describing an approximate aggregate size.

There are basically two types of estimators: point estimators and interval estimators

- A **point estimator** gives a single number as an estimate of the unknown parameter value.
- An **interval estimator** gives a range of possible values (i.e., an interval of possible values) as an estimate of the unknown parameter value.

What type of estimates are the ones reported for a few decades ago and for late 1980s Goliath grouper aggregate size?

A specific value of an estimator computed from the sample is known as an *estimate*. For example, the brick manufacturer could estimate the mean strength of bricks manufactured using a sample of bricks as:

- The mean strength of bricks is 150 lb per inch.
- The mean strength of bricks is between 135 and 165 lb per sq inch; or the mean strength of bricks is 150, give or take 15 lb, per sq inch.

A chemical engineer could estimate the yield of a chemical process from a few trial runs as:

- The yield of chemical A from this process is 28%.
- The yield of chemical A from this process is 25% to 31%; or the yield of chemical A from this process is 28%, give or take 3%.

In each of these two examples, the first option gives a point estimate of the unknown population characteristic (mean strength or chemical yield), whereas the second option gives a range of plausible values for the unknown population characteristic. In this chapter, we'll learn about computing and using different point and interval estimators for decision making.

9.1 Point Estimators and Their Properties

As described above, *Mobile Register* (July 14, 2004, p. 4A) reported, "Metro transit has about 3,500 riders during each work week." The estimate was obtained using data from a few weeks. If a sample of different weeks were used, would we get the same estimate? How close would it be to the actual number of passengers using the transit system?

Note that an estimator is a random variable. It seems obvious that the sample mean $\overline{X}$ could be used as an *estimator* of μ. But $\overline{X}$ will take different numerical values from sample to sample, so some questions remain. What is the magnitude of difference between $\overline{X}$ and μ likely to be? How does this difference behave as we take larger and larger samples (i.e., as n gets larger and larger)? In short, we need to study the behavior of $\overline{X}$, that is, the properties of $\overline{X}$ as an estimator of μ. Depending on the parameter to be estimated, we choose different statistics. Some of the most commonly used parameters and their point estimators are listed in Table 9.1. A carat symbol (^) is placed on top of the parameter to indicate its estimator. For example, if μ indicates a population mean, then $\hat{\mu}$ indicates its estimator, in other words, a sample mean.

The statistics used as estimators have sampling distributions. Sometimes the sampling distribution is known, at least approximately, as in the case of the mean of large samples. In other cases, the sampling distribution may not be completely specified. Even when the sampling distribution is unknown, we can often calculate the mean and variance of the estimator.

Table 9.1

Commonly Used Parameters and Their Estimators

Parameter	Estimator
μ = Population mean	$\hat{\mu} = \overline{X}$ = Sample mean
σ^2 = Population variance	$\hat{\sigma}^2 = S^2$ = Sample variance
p = Population proportion	$\hat{p} = X/n$ = Sample proportion
$\mu_1 - \mu_2$ = Difference in population means	$\widehat{\mu_1 - \mu_2} = \overline{X}_1 - \overline{X}_2$ = Differences in sample means
$p_1 - p_2$ = Difference in population proportions	$\widehat{p_1 - p_2} = \dfrac{X_1}{n_1} - \dfrac{X_2}{n_2}$ = Differences in sample proportions
$\dfrac{\sigma_1^2}{\sigma_2^2}$ = Ratio of two population variances	$\widehat{\sigma_1^2/\sigma_2^2} = S_1^2/S_2^2$ = Ratio of sample variances

We might then logically ask, "What properties would we like this estimator to possess?"

Let the symbol θ denote an arbitrary population parameter, such as μ or σ^2, and let $\hat{\theta}$ (read as "theta hat") denote an estimator of θ. We are then concerned with values of $E(\hat{\theta})$ and $V(\hat{\theta})$ as representing two different properties of point estimators: unbiasedness and precision.

- The *unbiasedness* of an estimator is determined by the expected value of the estimator.
- The *precision* of an estimator is determined by the variance of the estimator.

Unbiased estimator: Unbiasedness is a property associated with the sampling distribution of an estimator. Different samples result in different numerical values of $\hat{\theta}$, but we would hope that some of these values underestimate θ and others overestimate θ, so that the average value of $\hat{\theta}$ is close to θ.

> An estimator $\hat{\theta}$ is an **unbiased** estimator of θ if $E(\hat{\theta}) = \theta$.

In sampling distributions simulated in Chapter 8, we saw that the values of $\overline{X}$ tend to cluster around μ, the true population mean, when random samples are selected from the same population repeatedly. It demonstrates that $\overline{X}$ is an unbiased estimator of μ. We also showed it mathematically in Chapter 8.

More precise estimator: For an unbiased estimator $\hat{\theta}$, the sampling distribution of the estimator has mean value θ. How do we want the possible values of $\hat{\theta}$ to spread out to either side of θ for this unbiased estimator? Intuitively, it would be desirable to have all possible values of $\hat{\theta}$ be very close to θ; that is, we want the variance of $\hat{\theta}$ to be as small as possible.

> If $\hat{\theta}_1$ and $\hat{\theta}_2$ are both unbiased estimators of θ, then the one with the smaller variance is a **more precise** estimator of θ.

Suppose $\hat{\theta}_1$, $\hat{\theta}_2$, and $\hat{\theta}_3$ are the three estimators of parameter θ. Figure 9.1 shows the distributions of these three estimators.

- Figure 9.2a shows that the distribution of $\hat{\theta}_1$ is centered at θ, whereas the distribution of $\hat{\theta}_2$ is not. So, we know that $\hat{\theta}_1$ is an unbiased estimator of θ, but $\hat{\theta}_2$ is a biased estimator of θ. In fact, $\hat{\theta}_2$ is systematically giving higher estimates. Both the distributions have the same spread. So both the estimators are equally precise.

Figure 9.2

Properties of estimators: Unbiasedness and precision

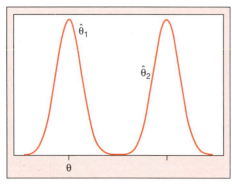

(a) Unbiased

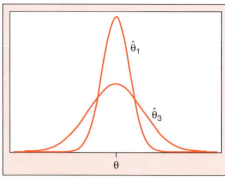

(b) More precise (or better)

- Figure 9.2b shows that both estimators, $\hat{\theta}_1$ and $\hat{\theta}_3$, are unbiased estimators because their distributions are centered at θ. However, estimator $\hat{\theta}_1$ is a more precise estimator than $\hat{\theta}_3$ because its distribution has a smaller spread than the spread of the distribution of $\hat{\theta}_3$.

Example 9.1

Suppose $X_1, \ldots, X_5$ denotes a random sample from some population with $E(X_i) = \mu$ and $V(X_i) = \sigma^2$, $i = 1, \ldots, 5$. The following are suggested as estimators for μ:

$$\hat{\mu}_1 = X_1 \quad \hat{\mu}_2 = \frac{1}{2}(X_1 + X_5) \quad \hat{\mu}_3 = \frac{1}{2}(X_1 + 2X_5) \quad \hat{\mu}_4 = \overline{X} = \frac{1}{5}(X_1 + \cdots + X_5)$$

Which estimator would you use and why?

Solution Looking at the expected values of these estimators, we have

$$E(\hat{\mu}_1) = \mu \quad E(\hat{\mu}_2) = \mu \quad E(\hat{\mu}_3) = \frac{3}{2}\mu \quad E(\hat{\mu}_4) = \mu$$

Thus, $\hat{\mu}_1$, $\hat{\mu}_2$, and $\hat{\mu}_4$ are all unbiased. We eliminate $\hat{\mu}_3$ because it is biased and tends to overestimate μ. Looking at the variances of the unbiased estimators, we have

$$V(\hat{\mu}_1) = V(X_1) = \sigma^2, \quad V(\hat{\mu}_2) = \frac{1}{4}[V(X_1) + V(X_5)] = \frac{\sigma^2}{2}$$

and

$$V(X_4) = V(\overline{X}) = \frac{\sigma^2}{5}$$

Thus, $\hat{\mu}_4$ would be chosen as the best estimator because it is unbiased and has the smallest variance among the three unbiased estimators.

Earlier in this section we discussed unbiased estimators. But not all estimators are unbiased. Estimators that are not unbiased are referred to as *biased estimators*. Suppose θ is our target to be estimated and $\hat{\theta}$ is its estimator. In the repeated attempts, if on the average we hit the target using $\hat{\theta}$, then $\hat{\theta}$ is an unbiased estimator of θ, that is, $E(\hat{\theta}) = \theta$. However, what if on the average we hit some point other than θ? In other words, with some estimator there might be a systematic tendency to overestimate or underestimate θ, that is, $E(\hat{\theta}) \neq \theta$. Then $\hat{\theta}$ is a *biased estimator* of θ. The amount of bias for the unbiased estimator is equal to 0. For the biased estimators, we can compute the amount of bias. Bias can be considered as an average deviation from the target.

> The **bias** is defined as the difference in the expected value of statistic and the parameter to be estimated.
>
> $$\text{Bias} = E(\hat{\theta}) - \theta$$

For example, before using a laboratory scale to weigh chemicals, you make sure that it is zeroed in (the needle points to 0 or the digital display shows 0). Why? Suppose the digital display of the empty scale shows 5 mg. What will happen if you go ahead and weigh chemicals anyway? Weight of each chemical recorded will be in excess of actual weight by 5 mg. So there is a systematic tendency to record extra weight of 5 mg for every item weighed on this scale. We would refer to this scale as a biased scale and the amount of bias would equal 5 mg.

Obviously, we would always like to use estimators that are unbiased and precise. Unfortunately not all estimators are unbiased and not all unbiased estimators are precise. So, we combine the measure of bias and the measure of spread into one measure called *Mean Squared Error* (*MSE*). It can be used to compare performance of different statistics for the same parameter.

> ### Mean Squared Error (MSE)
> If $\hat{\theta}$ is an estimator of θ, then the mean squared error is given by
>
> $$E(\hat{\theta} - \theta)^2 = MSE(\hat{\theta}) = (\text{Bias})^2 + V(\hat{\theta})$$

We know that, when the statistics are unbiased, we can select a better statistic by comparing variances. The MSE is useful in selection of a better statistic even when the statistics are biased.

> ### A Better Estimator
> If $\hat{\theta}_1$ and $\hat{\theta}_2$ are two estimators of θ, then the estimator $\hat{\theta}_1$ is considered a better estimator than $\hat{\theta}_2$ if
>
> $$MSE(\hat{\theta}_1) \leq MSE(\hat{\theta}_2)$$

Example 9.2

Suppose $X_1, X_2, \ldots, X_n$ is a random sample from a normal population with variance σ^2. Consider the following estimator for σ^2,

$$S^{*2} = \frac{\sum_{i=1}^{n}(X_i - \overline{X})^2}{n}$$

a Show that S^{*2} is a biased estimator of σ^2.
b Estimate the amount of bias when S^{*2} is used to estimate σ^2.

Solution **a** From Chapter 8, we know that statistic $(n-1)S^2/\sigma^2$ has a χ^2 distribution with $(n-1)$ degrees of freedom, where

$$S^2 = \frac{\sum_{i=1}^{n}(X_i - \overline{X})^2}{n-1}$$

From Chapter 6, we know that a χ^2 distribution has a mean equal to its degrees of freedom. Therefore, we get $E[(n-1)S^2/\sigma^2] = (n-1)$, and $S^{*2} = (n-1)S^2/n$. From this information, we can find the mean of S^{*2} as

$$E(S^{*2}) = E\left(\frac{(n-1)S^2}{n}\right) = \frac{\sigma^2}{n}E\left(\frac{(n-1)S^2}{\sigma^2}\right) = \frac{\sigma^2}{n}(n-1)$$

In other words, in repeated sampling, S^{*2} values tend to center around a value slightly lower than σ^2, namely, $\sigma^2(n-1)/n$. Therefore, we conclude that S^{*2} is a biased estimator of σ^2.

b The amount of bias when S^{*2} is used to estimate σ^2 is

$$\text{Bias} = E(S^{*2}) - \sigma^2 = \frac{(n-1)}{n}\sigma^2 - \sigma^2 = \frac{-1}{n}\sigma^2$$

Because $\sigma^2 > 0$, the negative bias indicates that S^{*2} systematically underestimates σ^2.

Note that S^{*2} defined with $(n-1)$ instead of n in the denominator is an unbiased estimator of σ^2 and therefore is more commonly used to estimate σ^2 than S^{*2}. It is also commonly referred to as the sample variance.

The above result can be shown easily using a simulation. Consider a normal population with mean 10 and variance 4.

- Take a random sample of size $n = 10$ and compute s^2 and s^{*2}.
- Repeat this process 1,000 times.
  This will result in two lists of 1,000 numbers.
- Compute summary statistics for s^2 and s^{*2}.
- Make histograms for s^2 and s^{*2}.

Figure 9.3

The sampling distributions of sample variances

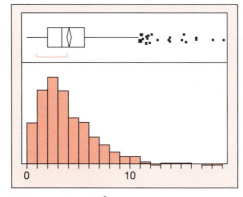

(a) Distribution of S^2

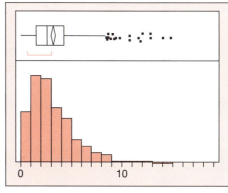

(b) Distribution of S^{*2}

One such simulation resulted in the histograms shown in Figure 9.3. The sampling distributions of both the estimators are right skewed as expected (chi-square). The summary statistics in table below shows that the $E(s^2)$ is about 4, whereas the $E(s^{*2})$ is lower than 4. As seen earlier, $E(s^{*2}) = 4(5 - 1)/5 = 3.2$, about what we got from the simulation. For large samples, this difference is negligible.

	S^2	S^{*2}
Mean	3.9261603	3.1409293
Std Dev	2.7578582	2.2062879
Std Err Mean	0.0872111	0.0697689
Upper 95% Mean	4.0972983	3.2778398
Lower 95% Mean	3.7550223	3.0040188
N	1000	1000

Example 9.3

Suppose $\hat{\theta}_1$ and $\hat{\theta}_2$ are two different estimators of θ. The estimator $\hat{\theta}_1$ is unbiased, while $E(\hat{\theta}_2) = 7\theta/2$. We also know that $V(\hat{\theta}_1) = 8$ and $V(\hat{\theta}_2) = 5$. Which estimator is a better estimator of θ? In what sense is it better?

Solution   Because $E(\hat{\theta}_2) = 7\theta/2$, we know that $\hat{\theta}_2$ is a biased estimator of θ. Therefore, we cannot compare the variances to determine which one, $\hat{\theta}_1$ or $\hat{\theta}_2$, is a better estimator of θ. Instead, we need to compare the mean squared errors.

Because $\hat{\theta}_1$ is an unbiased estimator, $MSE(\hat{\theta}_1) = V(\hat{\theta}_1) = 8$.
Now $Bias(\hat{\theta}_2) = E(\hat{\theta}_2) - \theta = (7\theta/2) - \theta = 5\theta/2$. Therefore,

$$MSE(\hat{\theta}_2) = V(\hat{\theta}_2) + [Bias(\hat{\theta}_2)]^2 = 5 + (5\theta/2)^2$$

We can say that $\hat{\theta}_1$ is a better estimator than $\hat{\theta}_2$ if $MSE(\hat{\theta}_1) < MSE(\hat{\theta}_2)$, that is, if $8 < 5 + (5\theta/2)^2$, in other words, if $\theta^2 > 12/25$. So in the sense of mean squared errors, we conclude that

- $\hat{\theta}_1$ is a better estimator of θ than $\hat{\theta}_2$ for $\theta < -\sqrt{12/25}$ or $\theta > \sqrt{12/25}$.
- $\hat{\theta}_2$ is a better estimator of θ than $\hat{\theta}_1$ for $-\sqrt{12/25} < \theta < \sqrt{12/25}$.

Exercises

9.1 Suppose X_1, X_2, X_3 denotes a random sample from the exponential distribution with density function

$$f(x) = \begin{cases} \dfrac{1}{\theta}\, e^{-x/\theta} & x > 0 \\ 0 & elsewhere \end{cases}$$

Consider the following four estimators of θ:

$$\hat{\theta}_1 = X_1 \qquad \hat{\theta}_2 = \frac{X_1 + X_2}{2}$$

$$\hat{\theta}_3 = \frac{X_1 + 2X_2}{3} \qquad \hat{\theta}_4 = \bar{X}$$

a Which of the above estimators are unbiased for θ?
b Among the unbiased estimators of θ, which has the smallest variance?

9.2 The reading on a voltage meter connected to a test circuit is uniformly distributed over the interval $(\theta, \theta + 1)$, where θ is the true but unknown voltage of the circuit. Suppose $X_1, \ldots, X_n$ denotes a random sample of readings from this voltage meter.
a Show that $\bar{X}$ is a biased estimator of θ.
b Find a function of $\bar{X}$ that is an unbiased estimator of θ.

9.3 The number of breakdowns per week for a certain minicomputer is a random variable X having a Poisson distribution with mean λ. A random sample $X_1, \ldots, X_n$ of observations on the number of breakdowns per week is available.

a Find an unbiased estimator of λ.
b The weekly cost of repairing these breakdowns is

$$C = 3Y + Y^2$$

Show that $E(C) = 4\lambda + \lambda^2$

c Find an unbiased estimator of $E(C)$ that makes use of the entire sample, $X_1, \ldots, X_n$.

9.4 The *bias* B of an estimator $\hat{\theta}$ is given by

$$B = |\theta - E(\hat{\theta})|$$

The *mean squared error*, or MSE, of an estimator $\hat{\theta}$ is given by

$$MSE = E(\hat{\theta} - \theta)^2$$

Show that $MSE(\hat{\theta}) = V(\hat{\theta}) + B^2$

[*Note:* $MSE(\hat{\theta}) = V(\hat{\theta})$ if $\hat{\theta}$ is an unbiased estimator of θ. Otherwise, $MSE(\hat{\theta}) > V(\hat{\theta})$.]

9.5 Refer to Exercise 9.2. Find $MSE(\bar{X})$ when $\bar{X}$ is used to estimate θ.

9.6 Suppose $X_1, \ldots, X_n$ is a random sample from a normal distribution with mean μ and variance σ^2.
a Show that $S = \sqrt{S^2}$ is a biased estimator of σ.
b Adjust S to form an unbiased estimator of σ.

9.7 For a certain new model of microwave oven, it is desired to set a guarantee period so that only 5% of the ovens sold will have a major failure in this length of time. Assuming the length of time until the first major failure for an oven of this type is normally distributed with mean μ and variance σ^2, the guarantee period should end at $\mu - 1.645\sigma$. Use the results of Exercise 9.6 to find an unbiased estimator of $(\mu - 1.645\sigma)$ based on a random sample $X_1, \ldots, X_n$ of measurements of time until the first major failure.

9.8 Suppose $\hat{\theta}_1$ and $\hat{\theta}_2$ are each unbiased estimators of θ, with $V(\hat{\theta}_1) = \sigma_1^2$ and $V(\hat{\theta}_2) = \sigma_2^2$. A new unbiased estimator for θ can be formed by

$$\hat{\theta}_3 = a\hat{\theta}_1 + (1 - a)\hat{\theta}_2$$

$(0 \leq a \leq 1)$. If $\hat{\theta}_1$ and $\hat{\theta}_2$ are independent, how should a be chosen so as to minimize $V(\hat{\theta}_3)$?

9.2   Confidence Intervals: The Single-Sample Case

In the last section, we learned about how to use point estimators $\hat{\theta}$ to estimate the unknown parameters θ and how to use the mean and variance of $\hat{\theta}$ to select a better estimator when more than one estimator is available. But different samples will result in different values of $\hat{\theta}$, giving rise to the sampling distribution of $\hat{\theta}$. However, once the selection of a better estimator is completed, the point estimation process does not take into account the variability in the values of $\hat{\theta}$. Also, it would be nice to know how small the distance between $\hat{\theta}$ and θ is likely to be. This is a probability question that requires knowledge of the sampling distribution of $\hat{\theta}$ beyond the behavior of $E(\hat{\theta})$ and $V(\hat{\theta})$.

An interval estimate uses the sampling distribution of the estimator of θ to determine an interval of plausible values for θ, called a confidence interval. Suppose $\hat{\theta}$ is an estimator of θ with a known sampling distribution, and we can find two quantities that depend on $\hat{\theta}$, say, $g_1(\hat{\theta})$ and $g_2(\hat{\theta})$, such that

$$P[g_1(\hat{\theta}) \leq \theta \leq g_2(\hat{\theta})] = 1 - \alpha$$

for some small positive number α. Then we can say that $(g_1(\hat{\theta}), g_2(\hat{\theta}))$ forms an interval that has probability $(1 - \alpha)$ of capturing the true θ. This interval is referred to as a confidence interval with confidence coefficient $(1 - \alpha)$. The quantity $g_1(\hat{\theta})$ is called the *lower confidence limit* (LCL), and $g_2(\hat{\theta})$ the *upper confidence limit* (UCL). We generally want $(1 - \alpha)$ to be near unity and $(g_1(\hat{\theta}), g_2(\hat{\theta}))$ to be as short an interval as possible for a given sample size. We illustrate the construction of confidence intervals for some common parameters here.

9.2.1   Confidence interval for a mean: General distribution

Suppose we are interested in estimating a mean μ for a population with variance σ^2 assumed to be known. We select a random sample $X_1, \ldots, X_n$ from this population and compute $\overline{X}$ as a point estimator of μ. If n is large, then by the Central

Limit Theorem, $\overline{X}$ has approximately a normal distribution with mean μ and variance σ^2/n. Thus, we can say that the interval $\overline{X} \pm 2\sigma/\sqrt{n}$ contains about 95% of the $\overline{X}$ values that could be generated in repeated random samplings from the population under study. Suppose we are to observe a single sample producing a single $\overline{X}$. A question of interest is "What possible values of μ would allow this $\overline{X}$ to lie in the middle 95% likely range of possible sample means?" This set of possible values for μ is the confidence interval with a confidence coefficient of approximately 0.95.

Under these conditions, $Z = \sqrt{n}(\overline{X} - \mu)/\sigma$ has approximately a standard normal distribution. Now, as shown in Figure 9.4, for any prescribed α we can find a value $z_{\alpha/2}$ such that

$$P(-z_{\alpha/2} \leq Z \leq + z_{\alpha/2}) = 1 - \alpha$$

Rewriting this probability statement, we have

$$1 - \alpha = P\left(-z_{\alpha/2} \leq \frac{\overline{X} - \mu}{\sigma/\sqrt{n}} \leq + z_{\alpha/2}\right)$$

$$= P\left(\overline{X} - z_{\alpha/2}\frac{\sigma}{\sqrt{n}} \leq \mu \leq \overline{X} + z_{\alpha/2}\frac{\sigma}{\sqrt{n}}\right)$$

Figure 9.4
Area enclosed by an interval

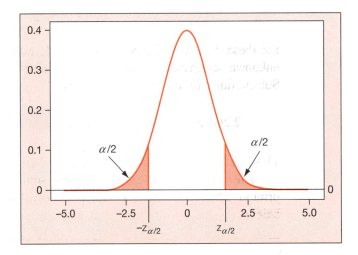

Large-Sample Confidence Interval for a Mean

A large random sample confidence interval for μ with confidence coefficient approximately $(1 - \alpha)$ is given by

$$\left(\overline{X} - z_{\alpha/2}\frac{\sigma}{\sqrt{n}}, \ \overline{X} + z_{\alpha/2}\frac{\sigma}{\sqrt{n}}\right)$$

If σ is unknown, replace it with s, the sample standard deviation, with no serious loss of accuracy.

The interval is centered at $\overline{X}$ and extends out the same distance on each side. This distance is known as the *margin of error* and is denoted by B.

$(1 - \alpha)\ 100\%$ **margin of error** to estimate μ from a large sample is

$$B = Z_{\alpha/2}\frac{\sigma}{\sqrt{n}}$$

The margin of error gives the maximum distance between μ and $\overline{X}$ with $(1 - \alpha)\ 100\%$ confidence. In other words, our estimated value $\overline{X}$ will be at most B units away from the unknown population mean μ with $(1 - \alpha)\ 100\%$ confidence.

Example 9.4

Consider once again the 50 lifetime observations in Table 6.1. Using these observations as the sample, find a confidence interval for the mean lifetime of batteries of this type, with confidence coefficient 0.95.

Solution   We are interested in estimating μ = the mean lifetime of batteries of this type. The sample size ($n = 50$) is large enough to use a large sample Z-interval to estimate μ as

$$\overline{x} \pm z_{\alpha/2}\frac{\sigma}{\sqrt{n}}$$

For these data, $\overline{x} = 2.268$ and $s = 1.935$, with units in 100 hours. Because σ is unknown, estimate it using s. For $(1 - \alpha) = 0.95$, we get $z_{\alpha/2} = z_{0.025} = 1.96$. Substituting s for σ, this interval yields

$$2.268 \pm 1.96\frac{1.932}{\sqrt{50}}, \quad \text{i.e.,} \quad 2.268 \pm 0.536, \quad \text{i.e.,} \quad (1.730, 2.802)$$

Thus, we are 95% confident that the true mean lies between 1.730 and 2.802 hours.

This can be easily obtained using some statistical software. The Minitab printout showing a 95% confidence interval for the population mean is shown below:

One-Sample Z: Lifetime

```
The assumed sigma = 1.932

Variable   N    Mean    StDev    SE Mean       95.0 % CI
lifetime  50    2.268   1.932      0.273    (1.732, 2.803)
```

Essentially, the computation of a confidence interval involves the following four steps:

1. Identifying the parameter of interest and selecting the appropriate confidence interval
2. Checking required conditions
3. Computing the interval
4. Interpreting the interval in the context of the problem

Checking Conditions   As with any other statistical procedure, the confidence intervals are derived under certain conditions. A large sample Z-interval can be used under the following conditions:

- A random sample is obtained from the population of interest.
- Sample size is large enough to approximate the distribution of sample mean by the normal distribution.

As discussed in Chapter 8, randomness of the selected sample is an important condition. For most statistical procedures discussed here, you must have a random sample. No inference procedure works without randomization. However, it is rarely satisfied fully in many practical situations. What effect do you think it will have on the inference drawn from such a sample?

How do we know if we have a large enough sample to use normal approximation? Refer to Figure 9.5:

- If the sample size is less than 15, be very careful. The measurements (or transformed measurements) must look like they came from a normal distribution. That is, there must not be any outliers and the distribution should be almost nearly symmetric.
- If the sample size is between 15 and 40, proceed with caution. Transform measurements, if the distribution is strongly skewed and/or there are outliers in the sample. Do not throw out the outliers. Compute the interval with and without the outliers. Do not rely on conclusions that depend on the inclusion or exclusion of outliers.
- If the sample size is over 40, it is safe to assume that the sample size is large enough to proceed safely with the use of normal approximation. The effect of skewness of the distribution on the significance level will be minimal. If the distribution is markedly skewed, it is worth employing some transformation to achieve symmetry of the distribution.

The transformation of measurements and the effect on symmetry is discussed in Chapters 2 and 11 in regression situations.

Figure 9.5

When is it safe to use normal approximation?

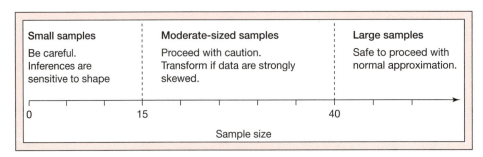

Example 9.5

The manager of a large manufacturing firm is interested in estimating the average distance traveled to work by its employees, using a 95% confidence interval. The manufacturing facility has a large employee parking lot, and the company issues parking permits to its employees.

The manager selected a random sample of 15 employees from the computerized list of parking permit holders and recorded the distance traveled from the odometer readings of cars of selected employees. Some statistics resulting from this random sample are listed below.

	Distance traveled
Mean	11.61
Std Dev	8.72
Maximum	84.00
75%	15.00
Median	10.00
25%	6.0
Minimum	1.00
n	15

a What conditions are necessary in order for this confidence interval to be appropriate?

b Discuss whether each of the conditions listed appears to be satisfied in this situation.

Solution **a** The population of interest is all employees of this manufacturing firm and the conditions are as follows:

- Fifteen selected employees constitute a random sample of all employees of this manufacturing firm.
- The distance traveled to work by employees is normal (or approximately normal), or the sample size is large enough to get approximate normal distribution for the sample mean.

b Let us discuss each assumption as follows:

- The condition of random sample is not fulfilled. Because the manager took a random sample of parking permit holders, the sample was not taken from the population of interest (i.e., the population of *all* employees). The employees that walk, carpool, or use local transportation (such as buses and trains) to get to the facility were excluded from the selection process.
- The sample size $n = 15$ may not be large enough to apply the central limit theorem. Therefore, the distribution of the sample mean might not be approximately normal. If the population distribution were normal, then the distribution of the sample mean would be normal even for a small sample of $n = 15$. But the statistics listed in table above show that the normality assumption is

not reasonable because the distribution is right-skewed. The asymmetric nature of the distribution is noticeable through the following:

- The range of the data is 83, which is 9.5 standard deviations, much larger than would be expected of a normal distribution (about 6 standard deviations).
- The minimum value is only 1.22 standard deviations below the mean, which is smaller than would be expected for a normal distribution.
- The maximum value is 8.30 standard deviations above the mean, which is larger than would be expected for a normal distribution.

What do we mean by "We are 95% confident …"?

Ninety-five percent confidence level means that, in repeated sampling, about 95% of the intervals formed by this method will contain μ.

Note that the interval computed from the sample is random, and the parameter is fixed.

- Before sampling, there is a probability of $(1 - \alpha)$ that an interval produced by the method used will include the true parameter value.
- After sampling, the resulting realization of the confidence interval either includes the parameter value or fails to include it, but we are quite confident that it will include the true parameter value if $(1 - \alpha)$ is large.

The following computer simulation illustrates the behavior of a confidence interval for a mean in repeated sampling.

Computer Simulation Start the simulation by selecting random samples of size $n = 100$ from a population having an exponential distribution with a mean value of $\mu = 10$. That is,

$$f(x) = \begin{cases} \dfrac{1}{10}e^{-x/10} & x > 0 \\ 0 & \text{elsewhere} \end{cases}$$

For each sample, construct a 95% confidence interval for μ by calculating $\bar{x} \pm 1.96s/\sqrt{n}$. Here is how you can do this activity with the help of Minitab (see Figure 9.6).

One such simulation showed that seven intervals do not include true mean value. Thus, we found that 93 out of 100 intervals include the true mean in one simulation (see Figure 9.7), whereas the theory says that 95 out of 100 should include μ. This is fairly good agreement between theory and application. Other such simulations will result in slightly different percentages; however, the sample percentages should remain about 95.

Figure 9.6

Minitab instructions for simulation

> **Minitab**
>
> - **Cal → Random data → Exponential**
> - Fill in the required fields.
>   - Generate **100** rows of data.
>   - Store in column(s): **c1–c25**.
>   - Mean: **10.0**
>   - Standard deviation: **2.0**
> - Click **OK** (twenty-five numbers in each row constitute a random sample of size 25 from an exponential population with mean 10).
> - Use **Cal → Row statistics** to compute mean and standard deviation from each sample, and save them in columns c26 and c27.
> - Use **Cal → Calculator** to compute LCL and UCL
>
> ```
> LCL = c26 - 1.96 * c27 / SQRT(25)
> UCL = c26 + 1.96 * c27 / SQRT(25)
> ```
>
> - Use **Cal → Calculator** to check which intervals contain true mean = 10. Fill in required fields as follows. Use keypad provided to enter expression.
>
>   > Store results in variable: **c28**
>   > Expression: 'LCL' <= 10 And 10 <= 'UCL'
>
> This step will create a new column of "1" and "0," where "1" will indicate the condition is satisfied (i.e., interval contains 10) and "0" will indicate failure to meet the condition (i.e., the interval does not contain 10).
> - Use **Stat → Tables → Tally** to count the number and percent of intervals containing population mean.

Figure 9.7

Confidence intervals from repeated sampling

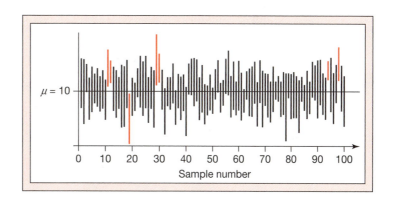

Determining Sample Size Often, in practice the first problem we face is that of deciding how many measurements to take. Because we know that a confidence interval for a mean μ will be of the form $\bar{X} \pm z_{\alpha/2}\sigma/\sqrt{n}$, we can choose the sample size to ensure certain accuracy before carrying out an experiment. Suppose the goal is to achieve an interval of length $2B$ so that the resulting confidence interval is of the form $\bar{X} \pm B$. We might, for example, want an estimate of the average assembly time for microcomputers to be within 3 minutes of the true average, with high probability. This tells us that B must be at most 3 or the total length of the interval must not exceed 6 in the final result. If $\overline{X} \pm z_{\alpha/2}\sigma/\sqrt{n}$ must result in $\overline{X} \pm B$, then we must have

$$B \leq z_{\alpha/2}\frac{\sigma}{\sqrt{n}}, \quad \text{which gives} \quad n \geq \left[\frac{z_{\alpha/2}\sigma}{B}\right]^2$$

Determining Sample Size to Estimate Mean

The sample size for establishing a confidence interval of the form $\overline{X} \pm B$ with confidence coefficient $(1 - \alpha)$ is given by

$$n \geq \left[\frac{z_{\alpha/2}\sigma}{B}\right]^2$$

Once the confidence coefficient $(1 - \alpha)$ is established, $z_{\alpha/2}$ is known. The only remaining problem is to find an approximation for σ. This can often be accomplished by prior experience, or by taking a preliminary sample, and using s to estimate σ. Even if σ is not known exactly, the above equation will give better results than arbitrarily guessing what the sample size should be. If a sample size determined this way is too large, making it too costly to sample, then either B must be increased or $(1 - \alpha)$ must be decreased.

Example 9.6

A large manufacturing firm is interested in estimating the average distance traveled to work by its employees. Past studies of this type indicate that the standard deviation of these distances should be in the neighborhood of 2 miles. How many employees should be sampled if the estimate is to be within 0.1 mile of the true average, with 95% confidence?

Solution   The resulting interval is to be of the form $\overline{X} \pm 0.1$ with $1 - \alpha = 0.95$. Thus, $B = 0.1$ and $z_{0.025} = 1.96$. It follows that

$$n \geq \left[\frac{z_{\alpha/2}\sigma}{B}\right]^2 = \left[\frac{1.96(2)}{0.1}\right]^2 = 1{,}536.64, \text{ i.e., } n \geq 1{,}537$$

Thus, at least 1,537 employees should be sampled to achieve the desired results.

The confidence intervals discussed so far have been two-sided intervals. Sometimes we are interested in only a lower or upper limit for a parameter, but not both. A one-sided confidence interval can be formed in such cases. For example, in the study of mean lifetimes of batteries (as in Example 9.4) we wanted only to establish a lower limit to this mean. That is, we want to find a statistic $g(\overline{X})$ so that

$$P[\mu > g(\overline{X})] = 1 - \alpha$$

Again, we know that $P\left(\dfrac{\overline{X} - \mu}{\sigma/\sqrt{n}} \le z_\alpha\right) = 1 - \alpha$, which gives

$$P\left(\overline{X} - z_\alpha \frac{\sigma}{\sqrt{n}}\right) = 1 - \alpha$$

Therefore,

$$\overline{X} - z_\alpha \frac{\sigma}{\sqrt{n}}$$

forms a lower confidence limit for μ, with $(1 - \alpha)$ 100% confidence.

Example 9.7

Consider once again the 50 lifetime measurements of Table 6.1. Using these measurements as a sample, find a lower limit for the mean lifetime of batteries of this type, with 95% confidence.

Solution   Once again using arguments similar to those used in Example 9.4, we get an approximate lower confidence limit for μ as

$$\overline{X} - z_{0.05} \frac{s}{\sqrt{n}}$$

Thus,

$$2.266 - (1.645) \frac{1.935}{\sqrt{50}} = 2.266 - 0.450 = 1.816$$

We are 95% confident that the mean battery lifetime exceeds 181.6 hours. Note that this lower confidence limit in the one-sided case is larger than the lower limit in the corresponding two-sided confidence interval.

9.2.2   Confidence interval for a mean: Normal distribution

If the samples we are dealing with are not large enough to ensure an approximately normal sampling distribution for $\overline{X}$, then the above results are not valid. What can we do in that case?

- One approach is to use techniques that do not depend upon distributional assumptions and use distribution-free techniques (see Chapter 12).
- Another approach is to make an additional assumption concerning the nature of the probabilistic model for the population.

In many situations, it is appropriate to assume that the random variable under study has a normal distribution. Under this assumption, we can develop an exact confidence interval for the mean μ using a method parallel to that used to develop the large-sample interval. We must find the function of $\overline{X}$ and μ for which we know the sampling distribution.

If $X_1, \ldots, X_n$ are independently normally distributed random variables, each with unknown mean μ and unknown variance σ^2, then we know that

$$T = \frac{\overline{X} - \mu}{S/\sqrt{n}}$$

follows a t-distribution with $(n-1)$ degrees of freedom. Using this result, we can construct a confidence interval for μ based on the t-distribution. From Table 5 of the Appendix, we can find a value $t_{\alpha/2}$, which cuts off an area of $\alpha/2$ in the right-hand tail of a t-distribution with known degrees of freedom. Thus,

$$P\left(-t_{\alpha/2} \le \frac{\overline{X} - \mu}{S/\sqrt{n}} \le +t_{\alpha/2}\right) = 1 - \alpha$$

for some prescribed α. Reworking as in the large-samples case, we get

$$P\left(\overline{X} - z_{\alpha/2}\frac{\sigma}{\sqrt{n}}, \ \overline{X} + z_{\alpha/2}\frac{\sigma}{\sqrt{n}}\right)$$

Normal Distribution: Confidence Interval for a Mean Based on t-Distribution

If $X_1, \ldots, X_n$ denotes a random sample from a normal distribution, with unknown variance σ^2, then a confidence interval for μ with confidence coefficient $(1 - \alpha)$ is

$$\left(\overline{X} - t_{\alpha/2}\frac{s}{\sqrt{n}}, \ \overline{X} + t_{\alpha/2}\frac{s}{\sqrt{n}}\right)$$

The degrees of freedom are given by $(n-1)$.

Conditions:

1. A random sample is obtained from the population of interest.
2. The measurements are normally distributed in the population.

The assumption of normality can be easily verified using the Q–Q plot described in Chapter 6, if the sample size is sufficiently large. Alternatively, we can plot the data using a dotplot or a histogram and visually verify whether the distribution significantly deviates from normality. None of these approaches are reliable for small samples.

Robustness of Statistical Procedure Each statistical procedure is derived under certain conditions. But not all conditions must be strictly satisfied for the application of the procedure. Some assumptions need only be loosely satisfied, that is, slight deviations from these assumptions do not affect the outcome of the procedure. Then the procedure is considered *robust* against such a condition. Some assumptions, unless satisfied, invalidate the outcome of the procedure. Then the procedure is *not robust* against such a condition.

Consider the two conditions listed for *t*-intervals:

- The *t*-interval is *not robust* against the condition of random sample. The confidence interval constructed from a non-random sample will not be valid.
- The *t*-interval is very *robust* against the condition of normality of the population sampled. In other words, some deviation from normality will not affect the confidence interval adversely. So we do not have to worry about strict adherence to the normality of the population. As long as the distribution is fairly symmetric, unimodal, and without any outliers, the *t*-confidence interval provides valid results.

Example 9.8

Prestressing wire for wrapping concrete pipe is manufactured in large rolls. A quality-control inspection required five specimens from a roll to be tested for ultimate tensile strength (UTS). The UTS measurements (in 1,000 psi) turned out to be 253, 261, 258, 255, and 256. Use these data to construct a 95% confidence interval estimate of the true mean UTS for the sampled roll.

Solution It is assumed that, if many wire specimens were tested, the relative frequency distribution of UTS measurements would be nearly normal. A confidence interval based on the *t*-distribution can then be employed. With $1 - \alpha = 0.95$ and $n - 1 = 4$ degrees of freedom, $t_{0.025} = 2.776$. From the observed data, $\bar{x} = 256.60$ and $s = 3.05$. Thus,

$$\bar{X} \pm t_{\alpha/2} \frac{s}{\sqrt{n}} \quad \text{gives} \quad 256.60 \pm (2.776) \frac{3.05}{\sqrt{5}}, \quad \text{i.e.,} \quad 256.60 \pm 3.79$$

We are 95% confident that the interval (252.81, 260.39) includes the true mean UTS for the roll.

Often we use available statistical packages to analyze data. See below a printout from Minitab.

One-Sample T: UTS

Variable	N	Mean	StDev	SE Mean	95.0 % CI
UTS	5	256.60	3.05	1.36	(252.81, 260.39)

Activity The goal is to estimate the mean flight time of a helicopter.

The **response** to be measured is flight time of the helicopter, defined as the time for which the helicopter remains in flight.

For this activity use

- A paper helicopter template from Chapter 3.
- A stop watch

1. Use the template given in Figure 3.10. Cut along the solid lines and fold along the dotted lines.
2. Release the helicopter from above your head while standing alone on a step stool or a ladder. Start timer at the drop of the helicopter. Stop timer when the helicopter hits floor.
3. Measure the time in flight (in sec) for 15 runs of this helicopter.
4. Record measurements in Table 9.2.
5. Repeat the process to get a second sample of 15 runs.
6. Construct a 95% confidence interval for the mean flight time for each sample.

Table 9.2

Data Recording Sheet

Sample 1				Sample 2			
Run No.	Time	Run No.	Time	Run No.	Time	Run No.	Time
1	___	9	___	1	___	9	___
2	___	10	___	2	___	10	___
3	___	11	___	3	___	11	___
4	___	12	___	4	___	12	___
5	___	13	___	5	___	13	___
6	___	14	___	6	___	14	___
7	___	15	___	7	___	15	___
8	___			8	___		

- Collect confidence intervals from the entire class.
- Plot them and observe variation in the estimates.
- Use your plot of confidence intervals to discuss the meaning of the term *confidence level*.

Activity The goal is to estimate the mean shipment size.
For this activity, use

- Random rectangles sheet from Chapter 3.
- Random number table (Table 1 in Appendix) or a random number generator on a calculator/computer

The sheet of random rectangles shows 100 shipments, numbered 1 to 100, of different sizes. The shipment size is determined by the number of grid squares in the shipment. Suppose we are interested in estimating the mean shipment size using a sample.

1. Note that all shipments are numbered using two-digit random numbers. We can consider shipment number 100 as 00. Using the random number table (Table 1 in Appendix), select 5 shipments at random.
2. Determine the size of each selected shipment.

3. Compute the mean shipment size for your sample of size 5. Construct a 95% confidence interval to estimate the mean shipment size.
4. Repeat the process and get another sample of size 5. Compute the sample mean. Construct a 95% confidence interval to estimate the mean shipment size.
5. Combine two samples, and get a sample of size 10. Construct a 95% confidence interval to estimate the mean shipment size.

- Collect confidence intervals from the entire class.
- Plot them and observe variation in the estimates.
- Use your plot of confidence intervals to discuss the meaning of the term *confidence level*.

[*Note:* The mean shipment size in the population is 7.41.]

9.2.3 Confidence interval for a proportion: Large sample case

Estimating the parameter p for the binomial distribution is analogous to the estimation of a population mean. As we saw in Chapter 5, the random variable Y having a binomial distribution with n trials can be written as $Y = X_1 + \cdots + X_n$, where X_i, $i = 1, \ldots, n$ are independent Bernoulli random variables with common mean p. Thus, $E(Y) = np$ or $E(Y/n) = p$ and Y/n is an unbiased estimator of p. Recall that $V(Y) = np(1 - p)$, and hence $V(Y/n) = p(1 - p)/n$.

Note that $Y/n = \overline{X}$, with the X_i's having $E(X_i) = p$ and $V(X_i) = p(1 - p)$. Then, applying the Central Limit Theorem to Y/n, the sample fraction of successes, we have that Y/n is approximately normally distributed with mean p and variance $p(1 - p)/n$. The large-sample confidence interval for p is then constructed by comparing with the corresponding result for μ. The observed fraction of successes, $\hat{p} = y/n$, will be used as the estimate of p.

Large-Sample Confidence Interval for a Proportion

A large sample confidence interval for p with confidence coefficient approximately $(1 - \alpha)$ is

$$\hat{p} \pm z_{\alpha/2}\sqrt{\frac{\hat{p}(1 - \hat{p})}{n}}$$

where $\hat{p} = y/n$

Conditions:

1. A random sample of size n is obtained from the population of interest.
2. The sample size is large enough to get an approximate normal distribution for the sampling distribution of $\hat{p}$.

Remember that almost all (99.7%) of the normal distribution falls within 3 standard deviations of the mean and the proportion is always between 0 and 1. Using these two facts, we can easily check condition (2) with the sample. If $\hat{p} \pm 3\sqrt{\hat{p}(1 - \hat{p})/n}$ interval falls completely in the interval $(0, 1)$, then it is safe to assume that the sample size is large enough to use the Z-interval. Alternatively, if $n\hat{p} \geq 10$, and $n(1 - \hat{p}) \geq 10$, then it is safe to assume that the second condition is satisfied.

Determining sample size Before carrying out an experiment, we want to determine the number of measurements to take in order to ensure certain accuracy. The goal is to achieve an interval with margin of error at least B (or of length at least $2B$). For example, we might be interested in estimating the proportion of voters in favor of the President within 5 percentage points of the true proportion, with high probability. We must have

$$B \leq z_{\alpha/2}\sqrt{\hat{p}(1 - \hat{p})/n}$$

which gives

$$n \geq \left(\frac{z_{\alpha/2}}{B}\right)^2 \hat{p}(1 - \hat{p})$$

- If $\hat{p}$ is known from some prior experiment, then use it to estimate the sample size. For example, an estimate of the President's popularity from last week's poll. Many engineering experiments are repeated regularly (maybe with some modification), which provides prior estimate of proportion.
- If no estimate of proportion is available, then obtain a conservative estimate of sample size using $\hat{p} = 1/2$. Note that $p(1 - p)$ is a quadratic function of p $(0 < p < 1)$ with maxima at $p = 1/2$. Using $\hat{p} = 1/2$ will give the largest sample size necessary to ensure the desired results.

Example 9.9

In certain water-quality studies, it is important to check for the presence or absence of various types of microorganisms. Suppose 20 out of 100 randomly selected samples of a fixed volume show the presence of a particular microorganism. Estimate the true probability p of finding this microorganism in a sample of the same volume, with a 90% confidence interval.

Solution   We assume that the number of samples showing the presence of the microorganism, out of n randomly selected samples, can be reasonably modeled by the binomial distribution. We are interested in estimating $p =$ probability of finding this microorganism in a sample. From a sample of size 100, we get $\hat{p} = 20/100 = 0.20$. Then $n\hat{p} = 20 > 10$ and $n(1 - \hat{p}) = 80 > 10$. Thus, we can use a large sample Z-interval to estimate $\hat{p}$ as

$$\hat{p} \pm z_{\alpha/2}\sqrt{\frac{\hat{p}(1 - \hat{p})}{n}}$$

For $(1 - \alpha) = 0.90$, $z_{\alpha/2} = z_{0.05} = 1.645$, the interval is

$$0.20 \pm 1.645\sqrt{\frac{0.2(0.8)}{100}}, \quad \text{i.e.,} \quad 0.20 \pm 0.066, \quad \text{i.e.,} \quad (0.134, 0.266)$$

We are about 90% confident that the true probability p of finding this microorganism in a sample is somewhere between 0.134 and 0.266.

Example 9.10

A firm wants to estimate the proportion of production-line workers who favor a revised quality-assurance program. The estimate is to be within 0.05 of the true proportion favoring the revised program, with a 90% confidence coefficient. How many workers should be sampled?

Solution Because no prior knowledge of p is available, use $p = 0.5$ in the calculation to determine the largest sample size necessary to ensure the desired results. In this problem, $B = 0.05$ and $(1 - \alpha) = 0.90$, so $z_{\alpha/2} = z_{0.05} = 1.645$. Thus,

$$n \geq \left(\frac{1.645}{0.05}\right)^2 (0.05)(1 - 0.05) = 270.6$$

A random sample of at least 271 workers is required in order to estimate the true proportion favoring the revised policy to within 0.05.

Activity For this activity, use

- Pennies
- Stirling's sheet from Chapter 8.

Goal To estimate the $P(\text{Heads})$ for your penny using a 95% confidence interval. Note that Stirling's sheet marks "Heads" on one side and "Tails" on the other.

1. Start at the top of the sheet and move down the sheet along the curves with each toss.
2. Move one step toward the side marked "Heads" if heads shows up on that toss, and in the opposite direction if tails shows up.
3. Continue till you reach the bottom of the sheet. A total of 50 tosses is required to reach the bottom.
4. Read the percent of heads from the bottom scale. Divide by 100 to get a proportion of heads in 50 tosses.

- Collect confidence intervals from all students in the class and plot them.
- Assuming the pennies are fair (unbiased) and hence $P(\text{Heads}) = 0.5$, discuss the concept of confidence level.

9.2.4 Confidence interval for a variance

Still under the assumption that the random sample $X_1, \ldots, X_n$ comes from a normal distribution, we know from Chapter 8 that $(n-1)S^2/\sigma^2$ has a χ^2 distribution, with $\nu = n - 1$ degrees of freedom. We can employ this fact to establish a confidence interval for σ^2. Using Table 6 of the Appendix, we can find two values, $\chi^2_{\alpha/2}(\nu)$ and $\chi^2_{1-\alpha/2}(\nu)$ such that

$$1 - \alpha = P\left[\chi^2_{1-\alpha/2}(\nu) \leq \frac{(n-1)S^2}{\sigma^2} \leq \chi^2_{\alpha/2}(\nu) \right]$$

$$= P\left[\frac{(n-1)S^2}{\chi^2_{\alpha/2}(\nu)} \leq \sigma^2 \leq \frac{(n-1)S^2}{\chi^2_{1-\alpha/2}(\nu)} \right]$$

Normal Distribution: Confidence Interval for a Variance

If $X_1, \ldots, X_n$ denotes a random sample from a normal distribution, then a confidence interval for σ^2 with confidence coefficient $(1 - \alpha)$ is

$$\left(\frac{(n-1)S^2}{\chi^2_{\alpha/2}(\nu)}, \frac{(n-1)S^2}{\chi^2_{1-\alpha/2}(\nu)} \right)$$

The degrees of freedom are $(n - 1)$.

Conditions:

1. A random sample is obtained from the population of interest.
2. The measurements are normally distributed in the population.

The assumption of normality can be easily verified using Q–Q plot described in earlier chapter if sample size is reasonably large. Alternatively, we can plot the data using a dotplot or a histogram and visually verify whether the distribution significantly deviates from normality.

Example 9.11

In laboratory work, it is desirable to run careful checks on the variability of readings produced on standard samples. In a study of the amount of calcium in drinking water undertaken as part of a water-quality assessment, the same standard was run through the laboratory six times at random intervals. The six readings (in parts per million [ppm]) were 9.54, 9.61, 9.32, 9.48, 9.70, and 9.26. Estimate σ^2, the population variance for readings on this standard, using a 90% confidence interval.

Solution From the data collected, $n = 6$ and $s^2 = 0.0285$. Using Table 6 of the Appendix, we have that $\chi^2_{0.95}(5) = 1.1455$ and $\chi^2_{0.05}(5) = 11.0705$. Thus the confidence interval for σ^2 is

$$\left(\frac{(n-1)S^2}{\chi^2_{0.05}(\nu)}, \frac{(n-1)S^2}{\chi^2_{0.95}(\nu)} \right), \quad \text{i.e.,} \quad \left(\frac{0.1427}{11.0705}, \frac{0.1427}{1.1455} \right), \quad \text{i.e.,} \quad (0.0129, 0.1246)$$

Thus, we are 90% confident that σ^2 is between 0.0129 and 0.1246 ppm^2. Note that this is a fairly wide interval, primarily because n is so small.

Exercises

9.9 *USA TODAY* (August 3, 2004) reported results of USA/TODAY/CNN/Gallop Poll from 1,129 likely voters. Before the Democratic Convention, 47% of likely voters opted for Kerry. After the Democratic Convention, 45% of likely voters opted for Kerry. The results are reported with a margin of error of 3 percentage points. Can we say that "Kerry's popularity dropped after the Democratic Convention"? Justify your answer.

9.10 For a random sample of 50 measurements on the breaking strength of cotton threads, the mean breaking strength was found to be 210 grams and the standard deviation 18 grams. Obtain a confidence interval for the true mean breaking strength of cotton threads of this type, with confidence coefficient 0.90.

9.11 A random sample of 40 engineers was selected from among the large number employed by a corporation engaged in seeking new sources of petroleum. The hours worked in a particular week were determined for each engineer selected. These data had a mean of 46 hours and a standard deviation of 3 hours. For that particular week, estimate the mean hours worked for all engineers in the corporation, with a 95% confidence coefficient.

9.12 An important property of plastic clays is the percent of shrinkage on drying. For a certain type of plastic clay, 45 test specimens showed an average shrinkage percentage of 18.4 and a standard deviation of 1.2. Estimate the true average percent of shrinkage for specimens of this type in a 98% confidence interval.

9.13 The breaking strength of threads has a standard deviation of 18 grams. How many measures on breaking strength should be used in the next experiment if the estimate of the mean breaking strength is to be within 4 grams of the true mean breaking strength, with confidence coefficient 0.90?

9.14 In the setting of Exercise 9.11, how many engineers should be sampled if it is desired to estimate the mean number of hours worked to within 0.5 hour with confidence coefficient 0.95?

9.15 Refer to Exercise 9.12. How many specimens should be tested if it is desired to estimate the percent of shrinkage to within 0.2 with confidence coefficient 0.98?

9.16 Upon testing 100 resistors manufactured by Company A, it is found that 12 fail to meet the tolerance specifications. Find a 95% confidence interval for the true fraction of resistors manufactured by Company A that fail to meet the tolerance specification. What assumptions are necessary for your answer to be valid?

9.17 Refer to Exercise 9.16. If it is desired to estimate the true proportion failing to meet tolerance specifications to within 0.05, with confidence coefficient 0.95, how many resistors should be tested?

9.18 Careful inspection of 70 precast concrete supports to be used in a construction project revealed 28 with hairline cracks. Estimate the true proportion of supports of this type with cracks in a 98% confidence interval.

9.19 Refer to Exercise 9.18. Suppose it is desired to estimate the true proportion of cracked supports to within 0.1, with confidence coefficient 0.98. How many supports should be sampled to achieve the desired accuracy?

9.20 In conducting an inventory and audit of parts in a certain stockroom, it was found that, for 60 items sampled, the audit value exceeded the book value on 45 items. Estimate, with confidence coefficient 0.90, the true fraction of items in the stockroom for which the audit value exceeds the book value.

9.21 The Environmental Protection Agency has collected data on the LC50 (concentration killing 50% of the test animals in a specified time interval) measurements for certain chemicals likely to be found in freshwater rivers and lakes. For a certain species of fish, the LC50 measurements (in parts per million) for DDT in 12 experiments yielded the following:

16, 5, 21, 19, 10, 5, 8, 2, 7, 2, 4, 9

Assuming such LC50 measurements to be approximately normally distributed, estimate the true mean LC50 for DDT with confidence coefficient 0.90.

normal

9.22 The warpwise breaking strength measured on five specimens of a certain cloth gave a sample mean of 180 psi and a standard deviation of 5 psi. Estimate the true mean warpwise breaking strength for cloth of this type in a 95% confidence interval. What assumption is necessary for your answer to be valid?

9.23 Answer Exercise 9.22 if the same sample data resulted from a sample of
a 10 specimens
b 100 specimens

9.24 Fifteen resistors were randomly selected from the output of a process supposedly producing 10-ohm resistors. The 15 resistors actually showed a sample mean of 9.8 ohms and a sample standard deviation of 0.5 ohm. Find a 95% confidence interval for the true mean resistance of the resistors produced by this process. Assume resistance measurements are approximately normally distributed.

9.25 The variance of LC50 measurements is important because it may reflect an ability (or inability) to reproduce similar results in identical experiments. Find a 95% confidence interval for σ^2, the true variance of the LC50 measurements for DDT, using the data in Exercise 9.21.

9.26 In producing resistors, variability of the resistances is an important quantity to study, as it reflects the stability of the manufacturing process. Estimate σ^2, the true variance of the resistance measurements, in a 90% confidence interval, if a sample of 15 resistors showed resistances with a standard deviation of 0.5 ohm.

9.27 The increased use of wood for residential heating has caused some concern over the pollutants and irritants released into the air. In one reported study in the northwestern United States, the total suspended particulates (TSP) in nine air samples had a mean of 72 $\mu g/m^3$ and a standard deviation of 23. Estimate the true mean TSP in a 95% confidence interval. What assumption is necessary for your answer to be valid? (See Core, Cooper, and Neulicht, *J. Air Pollution Control Association*, 31, no. 2,1984, pp. 138–143, for more details.)

9.28 The yield stress in steel bars (Grade FeB 400 HWL) reported by Booster (*J. Quality Technology*, 15, no. 4, 1983, p. 191) gave the following data for one test:

$$n = 150 \quad \bar{x} = 477 \text{ N/mm}^2 \quad s = 13$$

Estimate the true yield stress for bars of this grade and size in a 90% confidence interval. Are any assumptions necessary for your answer to be valid?

9.29 In the article quoted in Exercise 9.28, a sample of 12 measurements on the tensile strength of steel bars resulted in a mean of 585 N/mm^2 and a standard deviation of 38. Estimate the true mean tensile strength of bars of this size and type using a 95% confidence interval. List any assumptions you need to make.

9.30 Fatigue behavior of reinforced concrete beams in seawater was studied by Hodgkiess et al. (*Materials Performance*, July 1984, pp. 27–29). The number of cycles to failure in seawater for beams subjected to certain bending and loading stress was as follows (in thousands):

normal

n=9

774, 633, 477, 268, 407, 576, 659, 963, 193

Construct a 90% confidence interval estimate of the average number of cycles to failure for beams of this type.

9.31 Using the data given in Exercise 9.30, construct a 90% confidence interval for the variance of the number of cycles to failure for beams of this type. What assumptions are necessary for your answer to be valid?

9.32 The quantity

$$z_{\alpha/2}\frac{\sigma}{\sqrt{n}} \quad \left(\text{or } z_{\alpha/2}\sqrt{\frac{p(1-p)}{n}} \right)$$

used in constructing confidence intervals for μ (or p) is sometimes called the *sampling error*. A *Time* magazine (April 5, 1993) article on religion in America reported that 54% of those between the ages of 18 and 26 found religion to be a "very important" part of their lives. The article goes on to state that the result comes from a poll of 1,013 people and has a sampling error of 3%. How is the 3% calculated and what is its interpretation? Can we conclude that a majority of people in this age group find religion to be very important?

9.33 The results of a Louis Harris poll state that 36% of Americans list football as their favorite sport. A note then states: "In a sample of this size (1,091 adults) one can say with 95% certainty that the results are within plus or minus three percentage points of what they would be if the entire adult population had been polled." Do you agree? (*Source: Gainesville Sun*, May 7, 1968.)

9.34 A. C. Nielsen Co. has electronic monitors hooked up to about 1,200 of the 80 million American homes. The data from the monitors provide estimates of the proportion of homes tuned to a particular TV program. Nielsen offers the following defense of this sample size:

Mix together 70,000 white beans and 30,000 red beans and then scoop out a sample of 1,000. The mathematical odds are that the number of red beans will be between 270 and 330, or 27 to 33 percent of the sample, which translates to a "rating" of 30, plus or minus three, with a 20 to 1 assurance of statistical reliability. The basic statistical law wouldn't change even if the sampling came from 80 million beans rather than 100,000 (*Sky*, October 1982).

Interpret and justify this statement in terms of the results of this chapter.

9.3 Confidence Intervals: The Multiple Samples Case

All methods for confidence intervals discussed in the preceding section considered only the case in which a single random sample was selected to estimate a single parameter. In many problems, more than one population, and hence more than one sample, is involved. For example,

- We might want to estimate the difference between mean daily yields for two industrial processes for producing a certain liquid fertilizer.
- We might want to compare the rates of defectives produced on two or more assembly lines within a factory.

9.3.1 Confidence interval for linear functions of means: General distributions

It is commonly of interest to compare two population means, and this is conveniently accomplished by estimating the *difference* between the two means. Suppose the two populations of interest have means μ_1 and μ_2, respectively, and variances σ_1^2 and σ_2^2, respectively. The parameter to be estimated is $(\mu_1 - \mu_2)$. A large sample of size n_1 from the first population has a mean of $\overline{X}_1$ and a variance of S_1^2. An *independent* large sample from the second population has a mean of $\overline{X}_2$ and a variance of S_2^2. Remember from Chapter 8 that $(\overline{X}_1 - \overline{X}_2)$ is an unbiased estimator of $(\mu_1 - \mu_2)$ with variance $(\sigma_1^2/n_1) + (\sigma_2^2/n_2)$. Also, $(\overline{X}_1 - \overline{X}_2)$ will have approximately a normal distribution for large samples. Thus, a confidence interval for $(\mu_1 - \mu_2)$ can be formed in the same manner as the one for μ.

> **Confidence Interval for a Difference in Means: General Distribution**
>
> A large sample confidence interval for $(\mu_1 - \mu_2)$ with confidence coefficient $(1 - \alpha)$ is
>
> $$(\overline{X}_1 - \overline{X}_2) \pm z_{\alpha/2}\sqrt{\frac{\sigma_1^2}{n_1} + \frac{\sigma_2^2}{n_2}}$$

Conditions:

1. Both samples are taken at random from the respective populations of interest.
2. Samples are taken independently of each other.
3. Both the sample sizes are large enough to get an approximate normal distribution for the difference in sample means.

Note that the z-interval for general distributions is not robust for the conditions about random and independent samples, which means they must be satisfied for valid results. If $n_1 \geq 40$ and $n_2 \geq 40$, then it is safe to proceed with the use of normal approximation. If sample sizes are between 15 and 40, then proceed with caution. Refer to Section 9.2 (Figure 9.5) for a discussion on checking the sample size condition.

Example 9.12

A farm-equipment manufacturer wants to compare the average daily downtime for two sheet-metal stamping machines located in factories at two different locations. Investigation of company records for 100 randomly selected days on each of the two machines gave the following results:

$n_1 = 100$	$\bar{x}_1 = 12$ min	$s_1^2 = 6$ min^2
$n_2 = 100$	$\bar{x}_2 = 9$ min	$s_2^2 = 4$ min^2

Construct a 90% confidence interval estimate for the difference in mean daily downtimes for sheet-metal stamping machines located at two locations.

Solution   Let μ_i = the mean daily downtimes for sheet-metal stamping machines at location $i (i = 1, 2)$. We are interested in estimating $(\mu_1 - \mu_2)$. Estimating unknown σ_i^2 with s_i^2 $(i = 1, 2)$ we have

$$(\bar{x}_1 - \bar{x}_2) \pm z_{0.05}\sqrt{\frac{s_1^2}{n_1} + \frac{s_2^2}{n_2}}$$

which gives $(12 - 9) \pm (1.645)\sqrt{\dfrac{6}{100} + \dfrac{4}{100}}$,   i.e., 3 ± 0.52

We are about 90% confident that the difference in mean daily downtimes for sheet-metal stamping machines at two locations, that is $(\mu_1 - \mu_2)$, is between 2.48 and 3.52 min. This evidence suggests that μ_1 must be larger than μ_2.

Suppose we are interested in three populations, numbered 1, 2, and 3, with unknown means $\mu_1, \mu_2,$ and μ_3, respectively, and variances $\sigma_1^2, \sigma_2^2,$ and σ_3^2. If a random sample of size n_i is selected from the population with mean μ_i, $i = 1, 2, 3$, then any linear function of the form

$$\theta = a_1\mu_1 + a_2\mu_2 + a_3\mu_3 \text{ can be estimated by } \hat{\theta} = a_1\bar{X}_1 + a_2\bar{X}_2 + a_3\bar{X}_3$$

where $\overline{X}_i$ is the mean of the sample from population i. Also, if the samples are independent of one another,

$$V(\hat{\theta}) = a_1^2 V(\overline{X}_1) + a_2^2 V(\overline{X}_2) + a_3^2 V(\overline{X}_3)$$

$$= a_1^2 \left(\frac{\sigma_1^2}{n_1} \right) + a_2^2 \left(\frac{\sigma_2^2}{n_2} \right) + a_3^2 \left(\frac{\sigma_3^2}{n_3} \right)$$

and as long as all of the n_i's are reasonably large, $\hat{\theta}$ will have approximately a normal distribution.

Confidence Interval for a Linear Function of Means

A large-sample confidence interval for θ with confidence coefficient approximately $(1 - \alpha)$ is

$$\hat{\theta} \pm z_{\alpha/2} \sqrt{V(\hat{\theta})}$$

where θ is a linear function of population means.

Conditions:

1. All samples are taken at random from the respective populations of interest.
2. All samples are taken independently of each other.
3. All the sample sizes are large enough to get an approximate normal distribution for the linear function of sample means.

Example 9.13

A company has three machines for stamping sheet metal in factories located in three different cities. Let μ_i denote the average downtime (in minutes) per day for the ith machine, $i = 1, 2, 3$. The expected daily cost for downtimes on these three machines is

$$C = 3\mu_1 + 5\mu_2 + 2\mu_3$$

Investigation of company records for 100 randomly selected days on each of these machines showed the mean and standard deviations listed in table below. Estimate C using a 95% confidence interval.

$n_1 = 100$	$\overline{x}_1 = 12$ min	$s_1^2 = 6$ min^2
$n_2 = 100$	$\overline{x}_2 = 9$ min	$s_2^2 = 4$ min^2
$n_3 = 100$	$\overline{x}_2 = 4$ min	$s_2^2 = 5$ min^2

Solution We have seen above that an unbiased estimator of C is $\hat{C} = 3\overline{X}_1 + 5\overline{X}_2 + 2\overline{X}_3$ with variance

$$V(\hat{C}) = 9\left(\frac{\sigma_1^2}{100}\right) + 25\left(\frac{\sigma_2^2}{100}\right) + 4\left(\frac{\sigma_3^2}{100}\right)$$

The estimated daily cost is $\hat{C} = 3(12) + 5(9) + 2(14) = 109$. Because the σ_i^2's are unknown, we can estimate them with s_i^2's and estimate variance of $\hat{C}$ as

$$\hat{V}(\hat{C}) = 9\left(\frac{6}{100}\right) + 25\left(\frac{4}{100}\right) + 4\left(\frac{3}{100}\right) = 1.74$$

This approximation works well only for large samples. Thus, the 95% confidence interval for C is $109 \pm 1.96\sqrt{1.74}$, that is, 109 ± 2.58. We are 95% confident that the mean daily cost of downtimes on these machines is between \$106.42 and \$111.58.

9.3.2 Confidence interval for a linear function of means: Normal distributions

We must make some additional assumptions for the estimation of linear functions of means in small samples, where the Central Limit Theorem does not ensure approximate normality. If we have k populations and k independent random samples, we assume that all populations have *approximately normal distributions* with either

- A common unknown variance, or
- Different unknown variances, namely $\sigma_1^2, \sigma_2^2, \dots, \sigma_k^2$, respectively.

Equal variances case: As in the large-sample case of Section 9.2.1, the most commonly occurring function of means that we might want to estimate is a simple difference of the form $(\mu_1 - \mu_2)$. Suppose two populations of interest are normally distributed with means μ_1 and μ_2, and common variance σ^2. This condition of common variance should be considered carefully. If it does not seem reasonable, then the methods outlined below should not be used. Independent random samples from these populations yield sample statistics of $\overline{X}_1$ and S_1^2 from the first sample and $\overline{X}_2$ and S_2^2 from the second sample.

The obvious unbiased estimator of $(\mu_1 - \mu_2)$ is still $(\overline{X}_1 - \overline{X}_2)$. From Chapter 8, we know that

$$T = \frac{(\overline{X}_1 - \overline{X}_2) - (\mu_1 - \mu_2)}{S_p\sqrt{\dfrac{1}{n_1} + \dfrac{1}{n_2}}}, \text{ where } S_p^2 = \frac{(n_1 - 1)S_1^2 + (n_2 - 1)S_2^2}{n_1 + n_2 - 2}$$

has a t-distribution with $(n_1 + n_2 - 2)$ degrees of freedom. We can find $t_{\alpha/2}$ such that

$$P(-t_{\alpha/2} \leq T \leq t_{\alpha/2}) = 1 - \alpha$$

Reworking the inequality, we have

$$P\left[(\overline{X}_1 - \overline{X}_2) - t_{\alpha/2} S_p \sqrt{\frac{1}{n_1} + \frac{1}{n_2}} \leq \mu_1 - \mu_2 \leq (\overline{X}_1 - \overline{X}_2) + t_{\alpha/2} S_p \sqrt{\frac{1}{n_1} + \frac{1}{n_2}} \right] = 1 - \alpha$$

which yields the confidence interval for $(\mu_1 - \mu_2)$.

Confidence Interval for a Difference in Means: Normal Distributions with Common Variance

If both random samples come from independent normal distributions with common variance, then a confidence interval for $(\mu_1 - \mu_2)$, with confidence coefficient $(1 - \alpha)$ is

$$(\overline{X}_1 - \overline{X}_2) \pm t_{\alpha/2} S_p \sqrt{\frac{1}{n_1} + \frac{1}{n_2}}, \text{ where } S_p^2 = \frac{(n_1 - 1)S_1^2 + (n_2 - 1)S_2^2}{n_1 + n_2 - 2}$$

Degrees of freedom $= n_1 + n_2 - 2$.

Conditions:

1. Both samples are taken at random from the respective populations of interest.
2. Samples are taken independently of each other.
3. Both samples are taken for normally distributed populations.
4. Both the populations have same (unknown) variance, that is, $\sigma_1^2 = \sigma_2^2$.

Note that this inference procedure is not robust under the condition about random samples and independent samples.

Example 9.14

Copper produced by sintering (heating without melting) a powder under certain conditions is then measured for porosity (the volume fraction due to voids) in a certain laboratory. Two different batches of copper were produced and the batches are checked for consistency by measuring their porosity. The porosity measurements on four random samples from batch 1 show a mean of 0.22 and a variance of 0.0010. The porosity measurements on five random samples from batch 2 show a mean of 0.17 and a variance 0.0020. Estimate the difference between the (population) mean porosity measurements for these two batches.

Solution First, we assume that the porosity measurements in either batch could be modeled by a normal distribution. Because the sample variances are so close to each other, it is not unreasonable to assume that the population variances are approximately equal. Therefore, the pooled estimate of the common variance can be obtained as

$$S_p^2 = \frac{(n_1 - 1)S_1^2 + (n_2 - 1)S_2^2}{n_1 + n_2 - 2} = \frac{(4 - 1)(0.001) + (5 - 1)(0.002)}{4 + 5 - 2} = 0.0016$$

Thus, $s_p = \sqrt{0.0016} = 0.04$. Because the random samples are drawn independently of each other, we can construct a confidence interval based on the t-distribution using pooled estimate of variance. For a 95% confidence interval based on samples of sizes 4 and 5 we have $t_{0.025} = 2.365$ for 7 degrees of freedom. Using

$$(\overline{X}_1 - \overline{X}_2) \pm t_{\alpha/2} S_p \sqrt{\frac{1}{n_1} + \frac{1}{n_2}}$$

we get

$$(0.22 - 0.17) \pm (2.365)(0.04)\sqrt{\frac{1}{4} + \frac{1}{5}}, \quad \text{i.e.,} \quad 0.05 \pm 0.06$$

Thus, we are 95% confident that the difference in the mean porosity measurements for two batches is between -0.01 and 0.11. Because the interval contains zero, we would say that there is not much evidence of any difference between the two population means, that is, zero is a plausible value for $(\mu_1 - \mu_2)$, and if $\mu_1 - \mu_2 = 0$ then $\mu_1 = \mu_2$.

Example 9.15

One industrial engineer at a nylon manufacturing facility is interested in comparing the mean denier of nylon yarn spun by two spinning machines. (Denier is a unit of measurement of linear mass-density of textile fiber and is calculated as one gram per 9,000 meters.) He randomly selected samples of nylon yarn from two machines independently and recorded their denier as shown in Table 9.3.

Construct a 95% confidence interval to estimate the difference in mean denier of nylon yarn spun by two machines.

Solution We are interested in estimating $(\mu_1 - \mu_2)$, where $\mu_1 = $ the mean denier of nylon yarn spun by machine 1. $\mu_2 = $ the mean denier of nylon yarn spun by machine 2.

Parallel dotplot and boxplots in Figure 9.8 indicate the assumption of equal population variances would not be unreasonable. Therefore use pooled variances procedure to estimate the difference in population means.

Using sampled data, a confidence interval based on the t-distribution can be computed. The Minitab output for the t-interval is given in Figure 9.9. It shows that 95% confidence interval for the difference in mean denier computed from the data collected is $(-7.31, -2.23)$. We are 95% confident that the difference in mean denier is between -7.31 and -2.23. The interval does not contain zero, and falls entirely below zero indicating that the mean denier of yarn produced by machine 2 is significantly higher than that of machine 1.

Table 9.3

Linear–Mass Density of Textile Fibers from Two Spinning Machines

Machine 1	Machine 2
9.17	18.86
12.85	8.86
5.16	17.11
6.37	17.38
6.64	9.38
8.42	11.64
7.33	11.25
8.91	15.00
9.45	12.77
11.39	18.89
10.99	16.88
6.34	12.43
10.46	
14.30	
13.70	

Figure 9.8

Distribution of denier measured for threads

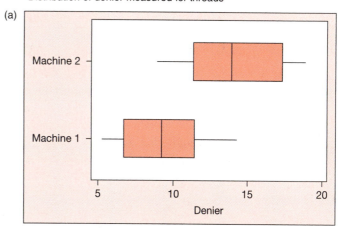

Figure 9.9

Minitab printout for Example 9.15

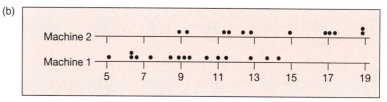

Two-Sample T-Test and CI: machine1, machine2

```
Two-sample T for machine1 vs machine2

            N     Mean    StDev    SE Mean
machine1   15     9.43    2.82      0.73

machine2   12    14.21    3.60      1.0


Difference = mu machine1 - mu machine2
Estimate for difference: -4.77
95% CI for difference: (-7.31, -2.23)
T-Test of difference = 0 (vs not =): T-Value = -3.87
P-Value = 0.001 DF = 25
Both use Pooled StDev = 3.19
```

Example 9.16

The sulfate levels (micrograms per cubic meter) were measured in a state park near a chemical factory before and after installing a new pollution control device in the factory. By how much has the pollution level decreased as a result of use of new pollution control device? Use a 95% confidence level. The sulfate data is given in table on the next page.

Before installing pollution control device	13.18, 9.42, 10.55, 10.11, 7.28, 8.53, 7.52, 8.04, 8.34, 6.91, 10.70, 9.21, 7.84, 9.46, 6.49
After installing pollution control device	5.31, 5.77, 3.36, 5.26, 2.43, 6.08, 3.77, 3.20, 3.49, 3.39, 2.99, 4.79, 6.99, 4.81, 3.99, 4.41, 7.12, 3.83, 3.57, 5.41

Solution   Let μ_B and μ_A respectively be the mean sulfate level at the nearby park before and after installing pollution control device. We are interested in estimating the difference $(\mu_B - \mu_A)$ using a 95% confidence interval. The parallel boxplot at the beginning of the chapter (Figure 9.1) indicates that the assumption of equal variances will be reasonable. The Minitab output for the data is shown in Figure 9.10.

The output indicates that a 95% confidence interval for $(\mu_B - \mu_A)$ is (3.347, 5.462). Therefore, we are 95% confident that the decrease in mean sulfate level is between 3.347 and 5.462 micrograms per cubic meter. Note that the entire interval falls above zero, indicating $\mu_B > \mu_A$.

Figure 9.10

Minitab printout for Example 9.16

```
Two-Sample T-Test and CI: Before, After

Two-sample T for Before vs After

            N      Mean     StDev     SE Mean
Before      15     8.90     1.75      0.45
After       20     4.50     1.32      0.30


Difference = mu Before - mu After
Estimate for difference:  4.405
95% CI for difference:  (3.347, 5.462)
Pooled StDev = 1.52
```

Suppose we have three populations of interest, numbered 1, 2, and 3, all assumed to be approximately normal in distribution. The populations have means μ_1, μ_2, and μ_3, respectively, and a common variance σ^2. We want to estimate a linear function of the form $\theta = a_1\mu_1 + a_2\mu_2 + a_3\mu_3$. A good estimator is $\hat{\theta} = a_1\overline{X}_1 + a_2\overline{X}_2 + a_3\overline{X}_3$, where $\overline{X}_i$ is the mean of the random sample from population i. Now

$$V(\hat{\theta}) = a_1^2 V(\overline{X}_1) + a_2^2 V(\overline{X}_2) + a_3^2 V(\overline{X}_3) = \sigma^2\left(\frac{a_1^2}{n_1} + \frac{a_2^2}{n_2} + \frac{a3}{n_3}\right)$$

The problem is now to choose the best estimator of the common variance σ^2 from three sample variances. It can be shown that an unbiased minimum variance estimator of σ^2 is

$$S_p^2 = \frac{\sum_{i=1}^{3}(n_i - 1)S_i^2}{\sum_{i=1}^{3}(n_i - 1)}$$

As previously used, the subscript p denotes "pooled" estimator. Because $\hat{\theta}$ will have a normal distribution, $\sum_{i=1}^{3}(n_i - 1)S_p^2/\sigma^2$ will have a χ^2 distribution with $\sum_{i=1}^{3}(n_i - 1)$ degrees of freedom. We can then use this statistic to derive a confidence interval for θ.

Confidence Interval for a Linear Function of Means: Normal Distributions with a Common Variance

If all random samples come from normal distributions with equal variances, a confidence interval for θ, with a confidence coefficient $(1 - \alpha)$ is

$$\hat{\theta} \pm t_{\alpha/2}S_p\sqrt{\frac{a_1^2}{n_1} + \frac{a_2^2}{n_2} + \frac{a_3^2}{n_3}} \quad \text{where} \quad S_p^2 = \frac{\sum_{i=1}^{3}(n_i - 1)S_i^2}{\sum_{i=1}^{3}(n_i - 1)}$$

Here $\theta = a_1\mu_1 + a_2\mu_2 + a_3\mu_3$, $\hat{\theta} = a_1\overline{X}_1 + a_2\overline{X}_2 + a_3\overline{X}_3$, and degrees of freedom $= \sum_{i=1}^{3}(n_i - 1)$.

Conditions:

1. All samples are taken at random from the respective populations.
2. Samples are taken independently of each other.
3. Samples are from normally distributed populations.
4. All populations have the same unknown variance.

Example 9.17

Refer to example 9.14. If a random sample of size ten is taken independently from a third batch of copper and porosity is measured for each selected sample. These yield a mean of 0.12 and a variance of 0.0018. Compare the mean porosity of this batch with the mean porosity of the batches from example 9.14 using a 95% confidence interval.

Solution   Again, we assume that the porosity measurements of the third batch follow a normal distribution. Because three sample variances (0.0010, 0.0020, and 0.0018) are very close to each other, it is not unreasonable to assume that the variances of these porosity measurements would be approximately equal for all three batches. The first two batches did not have significantly different means, therefore we assume that they have a common population mean μ. Because the two samples are of different sizes, we estimate this mean using a weighted average of the first two sample means as

$$\hat{\mu} = \frac{n_1\overline{x}_1 + n_2\overline{x}_2}{n_1 + n_2} = \frac{4(0.22) + 5(0.17)}{4 + 5} = 0.1922$$

Note that this weighted average is just the mean of two samples combined. This weighted average is unbiased and will have a similar variance than the simple average $(\bar{x}_1 + \bar{x}_2)/2$. To estimate $(\mu - \mu_3)$, we take

$$\hat{\mu} - \hat{\mu}_3 = 0.1922 - 0.12 = 0.0722$$

The pooled estimate of variance using all 3 samples is

$$s_p^2 = \frac{\sum_{i=1}^{3}(n_i - 1)s_i^2}{\sum_{i=1}^{3}(n_i - 1)}$$

$$= \frac{(4 - 1)(0.0010) + (5 - 1)(0.0020) + (10 - 1)(0.0018)}{4 + 5 + 9 - 3} = 0.0017$$

and $s_p = 0.0412$. Rewriting the function $(\mu - \mu_3)$, we can see that constants $a_1 = n_1/(n_1 + n_2)$, $a_2 = n_2/(n_1 + n_2)$, and $a_3 = -1$. Now the confidence interval estimate with $\alpha/2 = 0.025$ and 16 degrees of freedom is

$$(\hat{\mu} - \hat{\mu}_3) \pm t_{0.025}\, s_p \sqrt{\frac{a_1^2}{n_1} + \frac{a_2^2}{n_2} + \frac{a_3^2}{n_3}},$$

i.e., $0.0722 \pm (2.120)(0.0412)\sqrt{\dfrac{(4/9)^2}{4} + \dfrac{(5/9)^2}{5} + \dfrac{(-1)^2}{10}}$

yielding 0.0722 ± 0.0401. This means, we are 95% confident that the difference between the mean of the third batch and the average of means of the first two batches is between $(0.0321, 0.1123)$. This interval would suggest that the third batch of powder seems to give a lower mean porosity than the first because the interval does not contain zero and the entire interval is above zero.

Normal Distributions: Confidence Interval for a Difference in Means, Unequal Variances Case When two independent samples are taken from normally distributed populations, but the assumption of equal population variances does not seem reasonable, then from the chapter on sampling distributions we know that, statistic T follows an approximate t-distribution with ν degrees of freedom, where

$$T' = \frac{(\bar{X}_1 - \bar{X}_2) - (\mu_1 - \mu_2)}{\sqrt{\dfrac{S_1^2}{n_1} + \dfrac{S_2^2}{n_2}}} \quad \text{and} \quad \nu = \frac{\left(\dfrac{S_1^2}{n_1} + \dfrac{S_2^2}{n_2}\right)^2}{\dfrac{(S_1^2/n_1)^2}{(n_1 - 1)} + \dfrac{(S_2^2/n_2)^2}{(n_2 - 1)}}$$

In this situation, we would use what is known as *Smith-Satterthwaite's procedure* to estimate the difference in population means. The difference $(\bar{X}_1 - \bar{X}_2)$ still provides an unbiased estimator for $(\mu_1 - \mu_2)$. Now following the earlier procedure for constructing confidence intervals, we can find $t_{\alpha/2}$ such that

$$P(-t_{\alpha/2} \leq T' \leq t_{\alpha/2}) = 1 - \alpha$$

Reworking the inequality we have

$$P\left((\overline{X}_1 - \overline{X}_2) - t_{\alpha/2}\sqrt{\frac{S_1^2}{n_1} + \frac{S_2^2}{n_2}} \le (\mu_1 - \mu_2) \right.$$

$$\left. \le (\overline{X}_1 - \overline{X}_2) + t_{\alpha/2}\sqrt{\frac{S_1^2}{n_1} + \frac{S_2^2}{n_2}} \right) = 1 - \alpha$$

which gives the confidence interval for $(\mu_1 - \mu_2)$.

Confidence Interval for a Difference in Means: Normal Distributions with Unequal Variances

If random samples are taken independently from two normal populations with unequal variances, then a confidence interval for $(\mu_1 - \mu_2)$ with confidence coefficient $(1 - \alpha)$ is

$$(\overline{X}_1 - \overline{X}_2) \pm t_{\alpha/2}\sqrt{\frac{S_1^2}{n_1} + \frac{S_2^2}{n_2}} \quad \text{with} \quad \frac{\left(\dfrac{S_1^2}{n_1} + \dfrac{S_2^2}{n_2}\right)^2}{\dfrac{(S_1^2/n_1)^2}{(n_1 - 1)} + \dfrac{(S_2^2/n_2)^2}{(n_2 - 1)}} \quad \text{degrees of freedom.}$$

Conditions:

1. Both samples are taken at random from the respective populations of interest.
2. Samples are taken independently of each other.
3. Both the samples are taken from normally distributed populations.
4. The populations have different variances, that is, $\sigma_1^2 \ne \sigma_2^2$.

Example 9.18

Refer to spinning machine data (Example 9.15). Suppose the industrial engineer decided to compare machine 1 and machine 3. The denier recorded samples of yarn from machine 1 and 3 are shown in Table 9.4.

Construct a 95% confidence interval to estimate the difference in mean denier of nylon yarn spun by two machines.

Solution We are interested in estimating $(\mu_1 - \mu_3)$, where

μ_1 = the mean denier of nylon yarn spun by machine 1
μ_3 = the mean denier of nylon yarn spun by machine 3

The parallel dotplot and boxplots (Figure 9.11) indicate the assumption of equal population variances would not be reasonable. Therefore, use Smith-Satterthwaite's procedure to estimate the difference in population means.

Using sampled data, a confidence interval based on the t-distribution can be computed. The Minitab output for the t-interval is given in Figure 9.12. It shows that a 95% confidence interval for the difference in mean denier computed from

Table 9.4

Denier Measurements

Machine 1	Machine 3
9.17	12.17
12.85	11.22
5.16	11.42
6.37	11.73
6.64	12.33
8.42	12.21
7.33	12.21
8.91	10.93
9.45	12.16
11.39	11.61
10.99	10.41
6.34	11.91
10.46	
14.30	
13.70	

Figure 9.11

Distribution of denier

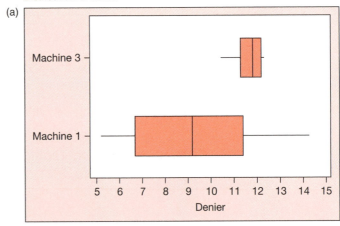

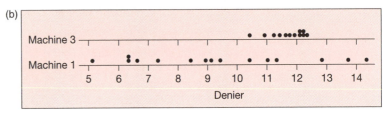

Figure 9.12

Minitab output for Example 9.19

Two-Sample T-Test and CI: machine1, machine3

```
Two-sample T for machine1 vs machine3

            N     Mean    StDev    SE Mean
machine1   15     9.43     2.82       0.73
machine3   12   11.693    0.599       0.17


Difference = mu machine1 - mu machine3
Estimate for difference: -2.259
95% CI for difference: (-3.857, -0.662)
T-Test of difference = 0 (vs not = ): T-Value = -3.01
P-Value = 0.009  DF = 15
```

the data collected is $(-3.857, -0.662)$. We are 95% confident that the difference in mean denier is between -3.857 and -0.662. The interval does not contain zero, indicating a significant difference between the mean denier by two machines. The entire interval falls below zero, indicating that the mean denier of yarn produced by machine 3 is significantly higher than that of machine 1.

Example 9.19

Lord Rayleigh was one of the earliest scientists to study the density of nitrogen. In his studies, he noticed something peculiar. The nitrogen densities produced from chemical compounds tended to be smaller than the densities of nitrogen produced from the air. He was working with fairly small samples, but was he correct in his conjecture? (*Proceedings, Royal Society* [London], 55, 1894, pp. 340–344).

Lord Rayleigh's data for chemical and atmospheric measurements are given below (Table 9.5), along with a dotplot (Figure 9.13).

Solution The units here are the mass of nitrogen filling a certain flask under specified pressure and temperature. The dotplot indicates that the assumption of equal variances is not unreasonable. The Minitab printout of a two-sample confidence interval is given in Figure 9.14. The printout gives a 95% confidence interval of $(-0.011507, -0.009515)$, which means, we are 95% confident that the difference in the mean chemical and atmospheric nitrogen density is between -0.011507 and -0.009515. This confidence interval is on the negative side of zero, indicating that the mean density of atmospheric nitrogen seems to be larger than the mean density of chemical nitrogen. Rather than suppressing this difference, Lord Rayleigh emphasized it, and that emphasis led to the discovery of the inert gases in the atmosphere.

Table 9.5

Nitrogen Density Measured by Two Methods

Chemical	Atmospheric
2.30143	2.31017
2.29890	2.30986
2.29816	2.31010
2.30182	2.31001
2.29869	2.31010
2.29940	2.31024
2.29849	2.31028
2.29889	2.31163
2.30074	2.30956
2.30054	

Figure 9.13

Dotplots of nitrogen densities

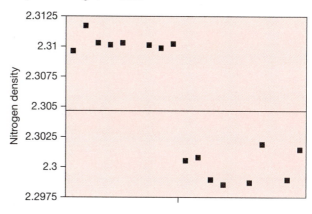

Figure 9.14

Minitab printout for
Example 9.20

Two-Sample T-Test and CI: Chemical, Atmospheric

```
Two-sample T for Chemical vs Atmospheric

                N        Mean        StDev      SE Mean
Chemical       10      2.29971      0.00131     0.00042
Atmosphe        9      2.310217     0.000574    0.00019


Difference = mu Chemical - mu Atmospheric
Estimate for difference: -0.010511
95% CI for difference: (-0.011507, -0.009515)
T-Test of difference = 0(vs not =): T-Value = -22.99
P-Value = 0.000 DF = 12
```

Activity The goal of this activity is to estimate the difference in mean flight times of helicopters with different wing lengths.

The **response** to be measured is flight time of the helicopter. It is defined as the time for which the helicopter remains in flight.

For this activity use,

- Paper helicopters (see Chapter 3 for a template) of two different wing lengths
- a stop watch

1. Form two teams.
2. Prepare two paper helicopters, one with 5.0 inches wing length and one with 6.0 inches wing length.
3. Let one team use helicopter with wing length 5.0 inches and the other team the helicopter with wing length 6.0 inches.

Table 9.6

Data Recording Sheet

	WL = 5 inches				WL = 6 inches			
Run No.	Time	Run No.	Time		Run No.	Time	Run No.	Time
1	___	9	___		1	___	9	___
2	___	10	___		2	___	10	___
3	___	11	___		3	___	11	___
4	___	12	___		4	___	12	___
5	___	13	___		5	___	13	___
6	___	14	___		6	___	14	___
7	___	15	___		7	___	15	___
8	___				8	___		

4. Release the helicopter from above your head while standing alone on a step stool or a ladder. Start the timer at the drop of the helicopter. Stop it when the helicopter hits the floor.
5. Record the flight times in the data sheet given in Table 9.6.
6. Measure the time in flight (in sec) for 15 runs of each helicopter.
7. Construct a 95% confidence interval for the difference in mean flight times.

9.3.3   Large-sample confidence interval for a linear function of proportions

Because we have seen that sample proportions behave like sample means and that for large samples the sample proportions will tend to have approximate normal distributions, we can adapt the above methodology to include estimation of linear functions of sample proportions. Again, the most common practical problem concerns the estimation of the difference between two proportions. If Y_i has a binomial distribution with sample size n_i and probability of success p_i, then $(p_1 - p_2)$ can be estimated by $(Y_1/n_1 - Y_2/n_2)$. After the samples are taken, we estimate p_1 by $\hat{p}_1 = y_1/n_1$ and p_2 by $\hat{p}_2 = y_2/n_2$.

Large Sample Confidence Interval for a Difference in Proportions

If large random samples are taken from two populations independently, then the confidence interval for $(p_1 - p_2)$ with confidence coefficient $(1 - \alpha)$ is

$$(\hat{p}_1 - \hat{p}_2) \pm z_{\alpha/2}\sqrt{\frac{\hat{p}_1(1 - \hat{p}_1)}{n_1} + \frac{\hat{p}_2(1 - \hat{p}_2)}{n_2}}$$

where $\hat{p}_1 = y_1/n_1$ and $\hat{p}_2 = y_2/n_2$.

Conditions:

1. Both samples are obtained randomly from the population of interest.
2. Both samples are taken independently of each other.
3. Both the sample sizes are large enough to get an approximately normal distribution for the sampling distribution of $(\hat{p}_1 - \hat{p}_2)$.

As in one sample case of estimating a proportion, we can check condition (3) easily from the samples. If

$$n_1\hat{p}_1 \geq 10, n_1(1 - \hat{p}_1) \geq 10, n_2\hat{p}_2 \geq 10, \quad \text{and} \quad n_2(1 - \hat{p}_2) \geq 10$$

then it is safe to assume that the last condition is satisfied.

Example 9.20

We want to compare the proportion of defective electric motors turned out by two shifts of workers. From the large number of motors produced in a given week, 250 motors were selected from the output of shift I and 200 motors were selected from the output of shift II. The sample from shift I revealed 25 to be defective and the sample from shift II showed 30 faulty motors. Estimate the true difference between the proportions of defective motors produced in two shifts using a 95% confidence interval.

Solution Let p_1 be the proportion of defective motors produced by workers in shift I and p_2 be the proportion of defective motors produce by workers in shift II. To estimate $(p_1 - p_2)$, using $n_1 = 250$, $n_2 = 200$, $y_1 = 25$, and $y_2 = 24$, we get $\hat{p}_1 = 25/250 = 0.10$ and $\hat{p}_2 = 24/200 = 0.15$. Because $n_1\hat{p}_1 = 25 > 10$, $n_1(1 - \hat{p}_1) = 225 > 10$, $n_2\hat{p}_2 = 24 > 10$, and $n_2(1 - \hat{p}_2) = 176 > 10$, the sample sizes are large enough to use normal approximation. At $\alpha = 0.05$, $z_{\alpha/2} = z_{0.025} = 1.96$, the confidence interval is computed as

$$(\hat{p}_1 - \hat{p}_2) \pm 1.96\sqrt{\frac{\hat{p}_1(1 - \hat{p}_1)}{n_1} + \frac{\hat{p}_2(1 - \hat{p}_2)}{n_2}}$$

Substituting sample proportions and sample sizes, we get

$$(0.10 - 0.15) \pm 1.96\sqrt{\frac{0.10(0.90)}{250} + \frac{0.15(0.85)}{200}}, \quad \text{i.e.,} \quad -0.05 \pm 0.062$$

We are 95% confident that the true difference in proportion of defective motors produced by two shifts is between -0.112 and 0.012. Because the interval contains zero, there does not appear to be any significant difference between the rates of defectives for the two shifts. That is, zero cannot be ruled out as a plausible value of the true difference between proportions of defective motors, which implies that these proportions may be equal.

More general linear functions of sample proportions are sometimes of interest, as shown in the following example.

Example 9.21

The personnel director of a large company wants to compare two different aptitude tests that are supposed to be equivalent in terms of which aptitude they measure. He suspects that the degree of difference in test scores is not the same for women and men. Thus, he sets up an experiment in which 200 men and 200 women of nearly equal aptitude are selected to take the tests. One hundred men take test I, and 100 independently selected men take test II. Likewise, one hundred women take test I, and 100 independently selected women take test II. The proportions, out of 100 receiving passing scores, are recorded for each of the four groups, and the data are as follows:

	Test I	Test II
Male	0.7	0.9
Female	0.8	0.9

Let p_{ij} denote the probability of a person of gender i passing test j. The director wants to estimate $(p_{12} - p_{11}) - (p_{22} - p_{21})$ to see whether the change in probability for males is different from the corresponding change for females. Estimate this quantity using a 95% confidence interval.

Solution   If $\hat{p}_{ij}$ denotes the observed proportion of a person of gender i passing test j (i.e., in row i and column j), then $(p_{12} - p_{11}) - (p_{22} - p_{21})$ is estimated by

$$(\hat{p}_{12} - \hat{p}_{11}) - (\hat{p}_{22} - \hat{p}_{21}) = (0.9 - 0.7) - (0.9 - 0.8) = 0.1$$

Because of the independence of the samples, the estimated variance of $(\hat{p}_{12} - \hat{p}_{11}) - (\hat{p}_{22} - \hat{p}_{21})$ is

$$\frac{\hat{p}_{12}(1 - \hat{p}_{12})}{100} + \frac{\hat{p}_{11}(1 - \hat{p}_{11})}{100} + \frac{\hat{p}_{22}(1 - \hat{p}_{22})}{100} + \frac{\hat{p}_{21}(1 - \hat{p}_{21})}{100}$$

$$= \frac{1}{100}[0.9(0.1) + 0.7(0.3) + 0.9(0.1) + 0.8(0.2)]$$

$$= 0.0055$$

Using normal approximation, we get $z_{0.025} = 1.96$. The interval estimate is then $0.1 \pm 1.96\sqrt{0.0055}$, that is, 0.1 ± 0.145. Therefore, we are 95% confident that the difference in the change in probability for males and females is between -0.045 and 0.245. Because this interval contains zero, there is really no reason to suspect that the change in probability for males is different from the change in probability for females.

9.3.4   Confidence interval for a ratio of variances: Normal distribution case

This procedure should be used very carefully in practice because it requires that both the samples come from normally distributed populations, and the procedure is *not* robust to this condition. In other words, results can be misleading if the procedure is applied when the populations are not normal.

Sometimes we are interested in comparing variances of two populations, for example comparing the precision of two procedures or comparing the variation in amounts filled by two filling machines.

From Chapter 8, we know that, if two independent random samples of sizes n_1 and n_2, respectively, are taken from normal distributions with variances σ_1^2 and σ_2^2, then $F = \dfrac{S_1^2/\sigma_1^2}{S_2^2/\sigma_2^2}$ follows an F-distribution with $\nu_1 = n_1 - 1$ and $\nu_2 = n_2 - 1$ degrees of freedom. Here S_1^2 and S_2^2 respectively are two sample variances.

Knowledge of the F-distribution allows us to construct a confidence interval for the ratio of two variances. We can find two values $F_{1-\alpha/2}(\nu_1, \nu_2)$ and $F_{\alpha/2}(\nu_1, \nu_2)$ as shown in Figure 9.15 so that

$$P(F_{1-\alpha/2}(\nu_1, \nu_2) \le F \le F_{\alpha/2}(\nu_1, \nu_2)) = 1 - \alpha$$

which yields

$$P\left(F_{1-\alpha/2}(\nu_1, \nu_2) \le \frac{S_1^2/\sigma_1^2}{S_2^2/\sigma_2^2} \le F_{\alpha/2}(\nu_1, \nu_2)\right) = 1 - \alpha$$

i.e., $$P\left(\frac{S_2^2}{S_1^2}F_{1-\alpha/2}(\nu_1, \nu_2) \le \frac{\sigma_2^2}{\sigma_1^2} \le \frac{S_2^2}{S_1^2}F_{\alpha/2}(\nu_1, \nu_2)\right) = 1 - \alpha$$

Using the identity for the lower-tailed F-values, $F_{1-\alpha}(\nu_1, \nu_2) = 1/F_\alpha(\nu_2, \nu_1)$, the confidence interval for σ_2^2/σ_1^2 is established.

Figure 9.15

The F-distribution

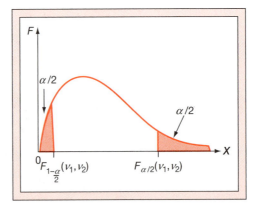

Confidence Interval for a Ratio of Population Variances: Normal Distributions

If both random samples are taken independently from normal distributions, then $(1 - \alpha)\,100\%$ confidence interval for σ_2^2/σ_1^2 is

$$\left(\frac{s_2^2}{s_1^2}\frac{1}{F_{\alpha/2}(\nu_2, \nu_1)}, \frac{s_2^2}{s_1^2}F_{\alpha/2}(\nu_1, \nu_2)\right)$$

Degrees of freedom $\nu_1 = n_1 - 1$ and $\nu_2 = n_2 - 1$.

> **Conditions:**
> 1. Both samples are obtained randomly from the population of interest.
> 2. Both samples are taken independently of each other.
> 3. The measurements are normally distributed in each population.

Again remember that, this procedure is *not* robust to the condition of normality and should be used only when both population distributions are known to be normal.

Example 9.22

A random sample of 10 measurements on the breaking strengths (measured in pounds per square inch) of a type of glass on one machine gave a variance of 2.31. An independent random sample of 16 measurements on a second machine, but with the same type of glass, gave a variance of 3.68. Estimate the true variance ratio, σ_2^2/σ_1^2, using a 90% confidence interval. Past experience has shown that the breaking strengths of glass on both machines are normally distributed.

Solution It is, of course, necessary to assume that the measurements in both populations can be modeled be normal distributions. With samples of sizes 10 and 16, we get 9 and 15 degrees of freedom, respectively. Then the confidence interval for confidence level 0.90 will yield

$$\left(\frac{s_2^2}{s_1^2} \frac{1}{F_{0.05}(15,\,9)}, \frac{s_2^2}{s_1^2} F_{0.05}(9,15) \right)$$

Now, using Table 7 of the Appendix, we get

$$\left[\frac{3.68}{2.31}\left(\frac{1}{3.01} \right), \frac{3.68}{2.31}(2.59) \right], \quad \text{i.e.,} \quad (0.53,\, 4.13)$$

Thus, we are 90% confident that the ratio σ_2^2/σ_1^2 is between 0.53 and 4.13. Equality of variances would imply $\sigma_2^2/\sigma_1^2 = 1.0$. Because our confidence interval includes 1.0, there is no real evidence that σ_2^2 differs from σ_1^2. Remember, variance measures precision. So we can conclude that the two machines do not differ significantly in precision (or have about the same precision of measuring the breaking strength of glass).

Exercises

9.35 The abrasive resistance of rubber is increased by adding a silica filler and a coupling agent to chemically bond the filler to the rubber polymer chains. Fifty specimens of rubber made with a type I coupling agent gave a mean resistance measure of 92, the variance of the measurements being 20. Forty specimens of rubber made with a type II coupling agent gave a mean of 98 and a variance of 30 on resistance measurements. Estimate the true difference between mean resistances to abrasion in a 95% confidence interval.

9.36 Refer to Exercise 9.35. Suppose a similar experiment is to be run again with an equal number of specimens from each type of coupling agent. How many specimens should be used if we want to estimate the true difference between mean resistances to within 1 unit, with a confidence coefficient of 0.95?

9.37 Two different types of coating for pipes are to be compared with respect to their ability to aid in resistance to corrosion. The amount of corrosion on a pipe specimen is quantified by measuring the maximum pit depth. For coating A, 35 specimens showed an average maximum pit depth of 0.18 cm. The standard deviation of these maximum pit depths was 0.02 cm. For coating B, the maximum pit depths in 30 specimens had a mean of 0.21 cm and a standard deviation of 0.03 cm. Estimate the true difference between mean depths in a 90% confidence interval. Do you think coating B does a better job of inhibiting corrosion?

9.38 Bacteria in water samples are sometimes difficult to count, but their presence can easily be detected by culturing. In 50 independently selected water samples from a certain lake, 43 contained certain harmful bacteria. After adding a chemical to the lake water, another 50 water samples showed only 22 with the harmful bacteria. Estimate the true difference between the proportions of samples containing the harmful bacteria, with a 95% confidence coefficient. Does the chemical appear to be effective in reducing the amount of bacteria?

9.39 In studying the proportion of water samples containing harmful bacteria, how many samples should be selected before and after a chemical is added if we want to estimate the true difference between proportions to within 0.1 with a 95% confidence coefficient? (Assume the sample sizes are to be equal.)

9.40 A large firm made up of several companies has instituted a new quality-control inspection policy. Among 30 artisans sampled in Company A, only 5 objected to the new policy. Among 35 artisans sampled in Company B, 10 objected to the policy. Estimate the true difference between the proportions voicing objection to the new policy for the two companies, with a confidence coefficient 0.98.

9.41 A measurement of physiological activity important to runners is the rate of oxygen consumption. *Research Quarterly*, May 1979, reports on the differences between oxygen consumption rates for college males trained by two different methods, one involving continuous training for a period of time each day and the other involving intermittent training of about the same overall duration. The sample sizes, means, and standard deviations are as follows (measurements in ml/kg • min):

Continuous Training	Intermittent Training
$n_1 = 9$	$n_2 = 7$
$\bar{y}_1 = 43.71$	$\bar{y}_2 = 39.63$
$s_1 = 5.88$	$s_2 = 7.68$

If the measurements are assumed to come from normally distributed populations, estimate the difference between the population means with a confidence coefficient 0.95. Also find a 90% confidence interval for the ratio of the true variances for the two training methods. Does it appear that intermittent training gives more variable results?

9.42 For a certain species of fish, the LC50 measurements (in parts per million) for DDT in 12 experiments were as follows, according to the EPA:

$$16, 5, 21, 19, 10, 5, 8, 2, 7, 2, 4, 9$$

Another common insecticide, Diazinon, gave LC50 measurements of 7.8, 1.6, and 1.3 in three independent experiments.

a Estimate the difference between the mean LC50 for DDT and the mean LC50 for Diazinon in a 90% confidence interval.

b What assumptions are necessary for your answer to be valid?

c Estimate the true variance ratio in a 90% confidence interval.

9.43 *Research Quarterly*, May 1979, reports on a study of impulses applied to a ball by tennis rackets of various construction. Three measurements on ball impulses were taken on each type of racket. For a Classic (wood) racket, the mean was 2.41 and the standard deviation was 0.02. For a Yamaha (graphite) racket, the mean was 2.22 and the standard deviation was 0.07. Estimate the difference between true mean impulses for the two rackets with a confidence coefficient 0.95. What assumptions are necessary for the method used to be valid?

9.44 Seasonal ranges (in hectares) for alligators were monitored on a lake outside Gainesville, Florida, by biologists from the Florida Game and Fish Commission. Six alligators monitored in the spring showed ranges of 8.0, 12.1, 8.1, 18.1, 18.2, and 31.7. Four different alligators monitored in the summer had ranges of 102.0, 81.7, 54.7, and 50.7. Estimate the difference between mean spring and summer ranges on a 95% confidence interval, assuming the data to come from normally distributed populations with common variance.

9.45 Unaltered, altered, and partly altered bitumens are found in carbonate-hosted lead-zinc deposits and may aid in the production of sulfide necessary to precipitate ore bodies in carbonate rocks. (See Powell, *Science*, April 6, 1984, p. 63.) The atomic hydrogen/carbon (H/C) ratios for 15 samples of altered bitumen had a mean of 1.02 and a standard deviation of 0.04. The ratios for 7 samples of partly altered bitumen had a mean of 1.16 and a standard deviation of 0.05. Estimate the difference between true mean H/C ratios for altered and partly altered bitumen, in a 98% confidence interval. Assume equal population standard deviations.

9.46 One-hour carbon monoxide concentrations in 45 air samples from a section of a city showed an average of 11.6 ppm and a variance of 82.4. After a traffic control strategy was put into place, 19 air samples showed an average carbon monoxide concentration of 6.2 ppm and a variance of 38.1. Estimate the true difference in average carbon monoxide concentrations in a 95% confidence interval. What assumptions are necessary for your answer to be valid? (See Zamurs, *Air Pollution Control Association Journal*, 34, no. 6, 1984, p. 637, for details of the measurements.)

9.47 The total amount of hydrogen chloride (HCl) in columns above an altitude of 12 kilometers was measured at selected sites over the El Chichon volcano both before and after an eruption. (See Mankin and Coffey, *Science*, October 12, 1984, p. 170.) Before-and-after data for one location were as follows (units are 1,015 molecules per square centimeter):

Preeruption	Posteruption
$n_1 = 40$	$n_2 = 20$
$\bar{x}_1 = 1.26$	$\bar{x}_2 = 1.40$
$s_1 = 0.14$	$s_2 = 0.04$

Estimate the difference between mean HCl amounts, with a confidence coefficient of 0.98.

9.48 Acid gases must be removed from other refinery gases in chemical production facilities in order to minimize corrosion of the plants. Two methods for removing acid gases produced the corrosion rates (in mm/yr) are listed below in experimental tests:

Method A: 0.3, 0.7, 0.5, 0.8, 0.9, 0.7, 0.8
Method B: 0.7, 0.8, 0.7, 0.6, 2.1, 0.6, 1.4, 2.3

Estimate the difference in mean corrosion rates for the two methods, using a confidence coefficient of 0.90. What assumptions must you make for your answer to be valid? (The data come from Kosseim et al., *Chemical Engineering Progress*, 80, 1984, p. 64.)

9.49 Using the data presented in Exercise 9.48, estimate the ratio of variances for the two methods of removing acid gases in a 90% confidence interval.

9.50 The number of cycles to failure for reinforced concrete beams was measured in seawater and in air. The data (in thousands) are as follows:

Seawater: 774, 633, 477, 268, 407, 576, 659, 963, 193
Air: 734, 571, 520, 792, 773, 276, 411, 500, 672

Estimate the difference between mean cycles to failure for seawater and air, using a confidence coefficient of 0.95. Does seawater seem to lessen the number of cycles to failure? (See Hodgkiess et al., *Materials Performance*, July 1984, pp. 27–29.)

9.51 The yield stresses were studied for two sizes of steel rods, with results as follows:

10-mm-diameter Rods	14-mm-diameter Rods
$n_1 = 51$	$n_2 = 12$
$\bar{x}_1 = 485$ N/mm^2	$\bar{x}_2 = 499$ N/mm^2
$s_1 = 17.2$	$s_2 = 26.7$

Estimate the average increase in yield stress of the 14-mm rod over the 10-mm rod, with a confidence coefficient of 0.90.

9.52 Silicon wafers are scored and then broken into the many small microchips that will he mounted into circuits. Two breaking methods are being compared. Out of 400 microchips broken by method A, 32 are unusable because of faulty breaks. Out of 400 microchips broken by method B, only 28 are unusable. Estimate the difference between proportions of improperly broken microchips for the two breaking methods. Use a confidence coefficient of 0.95. Which method of breaking would you recommend?

9.53 Time-Yankelovich surveys, regularly seen in the news magazine *Time*, report on telephone surveys of approximately 1,000 respondents. In December 1983, 60% of the respondents said that they worry about nuclear war. In a similar survey in June 1983, only 50% said that they worry about nuclear war. (See *Time*, January 2, 1984, p. 51.) The article reporting these figures says that when they are compared "the potential sampling error is plus or minus 4.5%."

a Explain how the 4.5% is obtained and what it means.

b Estimate the true difference in these proportions in a 95% confidence interval.

9.54 An electric circuit contains three resistors, each of a different type. Tests on 10 type I resistors showed a sample mean resistance of 9.1 ohms with a sample standard deviation of 0.2 ohm; tests on 8 type II resistors yielded a sample mean of 14.3 ohms and a sample standard deviation of 0.4 ohm; whereas tests on 12 type III resistors yielded a sample mean of 5.6 ohms and a sample standard deviation of 0.1 ohm. Find a 95% confidence interval for $\mu_I + \mu_{II} + \mu_{III}$, the expected resistance for the circuit. What assumptions are necessary for your answer to be valid?

9.4 Prediction Intervals

Previous sections of this chapter have considered the problem of *estimating* parameters, or population constants. A similar problem involves *predicting* a value for a future observation of a random variable. Given a set of n lifelength measurements on components of a certain type, it may be of interest to form an interval (a prediction interval) in which we think the next observation on the lifelength of a similar component is quite likely to lie.

For independent random variables having a common normal distribution, a prediction interval is easily derived using the t-distribution. Suppose we are to observe n independent, identically distributed normal random variables and then use the information to predict where X_{n+1} might lie. We know from previous discussions that $(\overline{X}_n - X_{n+1})$ is normally distributed with mean zero and variance $(\sigma^2/n) + \sigma^2 = \sigma^2[1 + (1/n)]$. If we use the variables $X_1, \ldots, X_n$ to calculate S^2 as an estimator of σ^2, then

$$T = \frac{\overline{X}_n - X_{n+1}}{S\sqrt{1 + \dfrac{1}{n}}}$$

will have a t-distribution with $(n - 1)$ degrees of freedom. Thus,

$$1 - \alpha = P\left(-t_{\alpha/2} \leq \frac{\overline{X}_n - X_{n+1}}{S\sqrt{1 + \dfrac{1}{n}}} \leq t_{\alpha/2}\right)$$

$$= P\left(\overline{X}_n - t_{\alpha/2}S\sqrt{1 + \dfrac{1}{n}} \leq X_{n+1} \leq \overline{X}_n + t_{\alpha/2}S\sqrt{1 + \dfrac{1}{n}}\right)$$

A Prediction Interval

A prediction interval for X_{n+1}, with coverage probability $(1 - \alpha)$ is given by

$$\overline{x}_n \pm t_{\alpha/2}s\sqrt{1 + \dfrac{1}{n}}$$

Example 9.23

Ten independent observations are taken on bottles corning off a machine designed to fill them to 16 ounces. The $n = 10$ observations show a mean of $\overline{X} = 16.1$ ounces and a standard deviation of $s = 0.01$ ounce. Find a 95% prediction interval for the amount filled (in ounces) in the next bottle to be observed.

Solution Assuming a normal probability model for the amount filled, the interval

$$\bar{x} \pm t_{\alpha/2}s\sqrt{1 + \frac{1}{n}}$$

will provide the answer. With $t_{0.025} = 2.262$, based on 9 degrees of freedom,

$$16.1 \pm (2.261)(0.01)\sqrt{1 + \frac{1}{10}}$$

yields 16.1 ± 0.024, that is, $(16.076, 16.124)$. We are about 95% confident that the next observation will lie between 16.076 and 16.124.

Note that the prediction interval will *not* become arbitrarily small with increasing n. In fact, if n is very large, the term $1/n$ might be ignored, resulting in an interval of the form $\bar{x} \pm t_{\alpha/2}s$. There will be a certain amount of error in predicting the value of a random variable, no matter how much information is available to estimate μ and σ^2.

This type of prediction interval is sometimes used as the basis for control charts when "sampling by variables." That is, if n observations on some important variable are made while a process is in control, then the next observation should lie in the interval $\bar{x}_n \pm t_{\alpha/2}s\sqrt{1 + 1/n}$. If it doesn't, then there may be some reason to suspect that the process has gone out of control. Usually, more than one observation would have to be observed outside the appropriate interval before an investigation into causes of the abnormality would be launched.

Exercises

9.55 In studying the properties of a certain type of resistor, the actual resistances produced were measured on a sample of 15 such resistors. These resistances had a mean of 9.8 ohms and a standard deviation of 0.5 ohm. One resistor of this type is to be used in a circuit. Find a 95% prediction interval for the resistance it will produce. What assumption is necessary for your answer to be valid?

9.56 For three tests with a certain species of freshwater fish, the LC50 measurements for Diazinon, a common insecticide, were 7.8, 1.6, and 1.3 (in parts per million). Assuming normality of the LC50 measurements, construct a 90% prediction interval for the LC50 measurement of the next test with Diazinon.

9.57 It is extremely important for a business firm to be able to predict the amount of downtime for its computer system over the next month. A study of the past five months has shown the downtime to

have a mean of 42 hours and a standard deviation of 3 hours. Predict the downtime for next month in a 95% prediction interval. What assumptions are necessary for your answer to be valid?

9.58 In designing concrete structures, the most important property of the concrete is its compressive strength. Six tested concrete beams showed compressive strengths (in thousand psi) of 3.9, 3.8, 4.4, 4.2, 3.8, and 5.4. Another beam of this type is to be used in a construction project. Predict its compressive strength in a 90% prediction interval. Assume compressive strengths to be approximately normally distributed.

9.59 A particular model of subcompact automobile has been tested for gas mileage 50 times. These mileage figures have a mean of 39.4 mpg and a standard deviation of 2.6 mpg. Predict the gas mileage to be obtained on the next test, with $1 - \alpha = 0.90$.

9.5 Tolerance Intervals

As we saw in Chapter 8, control charts are built around the idea of controlling the center (mean) and variation (variance) of a process variable such as tensile strength of wire or volume of soft drink put into cans. In addition to information on these selected parameters, engineers and technicians often want to determine an interval that contains a certain (high) proportion of the measurements, with a known probability. For example, in the production of resistors it is helpful to have information often in the form "We are 95% confident that 90% of the resistances produced are between 490 and 510 ohms." Such an interval is called a *tolerance interval* for 90% of the population, with confidence coefficient $1 - \alpha = 0.95$. The lower end of the interval (490, in this case) is called the *lower tolerance limit*, and the upper end (510, in this case) is called the *upper tolerance limit*. We still use the term *confidence coefficient* to reflect the confidence we have in a particular method producing an interval that includes the prescribed portion of the population. That is, a method for constructing 95% tolerance intervals for 90% of a population should produce intervals that contain approximately 90% of the population measurements in 95 out of every 100 times it is used.

Tolerance intervals are sometimes specified by a design engineer or production supervisor, in which case measurements must be taken from time to time to see whether the specification is being met. More often, measurements on a process are taken first, and then tolerance limits are calculated to describe how the process is operating. These "natural" tolerance limits are often made available to customers. Suppose the key measurements (resistance, for example) on a process have been collected and studied over a long period of time and are known to have a normal distribution with mean μ and standard deviation σ. We can then easily construct a tolerance interval for 90% of the population, because the interval

$$(\mu - 1.645\sigma, \ \mu + 1.645\sigma)$$

must contain precisely 90% of the population measurements. The confidence coefficient here is 1.0, because we have no doubt that exactly 90% of the measurements fall in the resulting interval. In most cases, however, μ and σ are not known and must be estimated by $\bar{x}$ and s, which are calculated from sample data. The interval

$$(\bar{x} - 1.645s, \ \bar{x} + 1.645s)$$

will no longer cover exactly 90% of the population because of the errors introduced by the estimators. However, it is possible to find a number K such that

$$(\bar{x} - Ks, \ \bar{x} + Ks)$$

has the property that it covers a proportion δ of the population with confidence coefficient $(1 - \alpha)$. Table 19 in the Appendix gives values of K for three values of δ and two values of $(1 - \alpha)$.

Example 9.24

A random sample of 45 resistors was tested with a resulting average resistance produced of 498 ohms. The standard deviation of these resistances was 4 ohms. Find a 95% tolerance interval for 90% of the resistance measurements in the population from which this sample was selected. Assume a normal distribution for this population.

Solution For these data, $n = 45$, $\bar{x} = 498$, and $s = 4$. From Table 19 with $\delta = 0.90$ and $1 - \alpha = 0.95$, we have $K = 2.021$. Thus,

$$(\bar{x} - 2.021s, \bar{x} + 2.021s)$$

i.e., $[498 - 2.021(4), 498 + 2.021(4)]$ i.e., $(489.9, 506.1)$

gives a 95% tolerance interval. We are 95% confident that 90% of the population resistances in the population lie between 489.9 and 506.1 ohms. (In constructing such intervals, the lower tolerance limit should be rounded down and the upper tolerance limit rounded up.)

Notice that, as n increases, the values of K tend toward the corresponding $z_{\alpha/2}$ values. Thus, the tolerance intervals $\bar{x} \pm Ks$ tend to get closer to the intervals $\mu \pm z_{\alpha/2}\sigma$ as n gets large.

The tolerance interval formulation discussed in previous paragraphs requires an assumption of normality for the data on which the interval is based. We now present a simple nonparametric technique for constructing tolerance intervals for any continuous distribution. Suppose a random sample of measurements $X_1, X_2, \ldots, X_n$ is ordered from smallest to largest to produce an ordered set $Y_1, Y_2, \ldots, Y_n$. That is, $Y_1 = \min(X_1, \ldots, X_n)$ and $Y_n = \max(X_1, \ldots, X_n)$. It can then be shown that

$$P[(Y_1, Y_n) \text{ covers at least } \delta \text{ of the population}] = 1 - n\delta^{n-1} + (n - 1)\delta^n$$

Thus, the interval (Y_1, Y_n) is a tolerance interval for a proportion δ of the population with confidence coefficient $1 - n\delta^{n-1} + (n - 1)\delta^n$. For a specified δ, the confidence coefficient is determined by the sample size. We illustrate the use of nonparametric tolerance intervals in the following examples.

Example 9.25

A random sample of $n = 50$ measurements on the lifelength of a certain type of LED display showed the minimum to be $y_1 = 2{,}150$ hours and the maximum to be $y_{50} = 2{,}610$ hours. If (y_1, y_{50}) is used as a tolerance interval for 90% of the population measurements on lifelengths, find the confidence coefficient.

Solution   It is specified that $\delta = 0.90$ and $n = 50$. Thus, the confidence coefficient for an interval of the form (Y_1, Y_{50}) is given by

$$1 - n\delta^{n-1} + (n-1)\delta^n = 1 - 50(0.9)^{49} + 49(0.9)^{50} = 0.97$$

We are 97% confident that the interval $(2{,}150, 2{,}610)$ contains at least 90% of the lifelength measurements for the population under study.

Example 9.26

In the setting of Example 9.25, suppose the experimenter wants to choose the sample size n so that (Y_1, Y_n) is a 90% tolerance interval for 90% of the population measurements. What value should the experimenter choose for n?

Solution   Here the confidence coefficient is 0.90 and $\delta = 0.90$. So we must solve the equation

$$1 - n(0.9)^{n-1} + (n-1)(0.9)^n = 0.90$$

This can be done quickly by trial and error, with a calculator. The value $n = 37$ gives a confidence coefficient of 0.896, and the value $n = 38$ gives a confidence coefficient of 0.905. Therefore, $n = 38$ should be chosen to guarantee a confidence coefficient of at least 0.90. With $n = 38$ observations, (Y_1, Y_n) is a 90.5% tolerance interval for 90% of the population measurements.

Exercises

9.60   A bottle-filling machine is set to dispense 10.0 cc of liquid into each bottle. A random sample of 100 filled bottles gives a mean fill of 10.1 cc and a standard deviation of 0.02 cc. Assuming the amounts dispensed to be normally distributed,
  a Construct a 95% tolerance interval for 90% of the population measurements
  b Construct a 99% tolerance interval for 90% of the population measurements
  c Construct a 95% tolerance interval for 99% of the population measurements

9.61   The average thickness of the plastic coating on electrical wire is an important variable in determining the wearing characteristics of the wire. Ten thickness measurements from randomly selected points along a wire gave the following (in thousandths of an inch):

  8, 7, 10, 5, 12, 9, 8, 7, 9, 10

Construct a 99% tolerance interval for 95% of the population measurements on coating thickness. What assumption is necessary for your answer to be valid?

9.62   A cutoff operation is supposed to cut dowel pins to 1.2 inches in length. A random sample of 60 such pins gave an average length of 1.1 inches and a standard deviation of 0.03 inch. Assuming normality of the pin lengths, find a tolerance interval for 90% of the pin lengths in the population from which this sample was selected. Use a confidence coefficient of 0.95.

9.63   In the setting of Exercise 9.62, construct the tolerance interval if the sample standard deviation is
  a 0.015 inch
  b 0.06 inch
  Compare these answers to that of Exercise 9.62. Does the size of s seem to have a large effect on the length of the tolerance interval?

9.64   In the setting of Exercise 9.62, construct the tolerance interval if the sample size is as follows:
  a $n = 100$
  b $n = 200$
  Does the sample size seem to have a great effect on the length of the tolerance interval, as long as it is "large"?

9.65 For the data of Exercise 9.61 and without the assumption of normality,
  a Find the confidence coefficient if (y_1, y_n) is to be used as a tolerance interval for 95% of the population.
  b Find the confidence coefficient if (y_1, y_n) is used as a tolerance interval for 80% of the population.

9.66 The interval (y_1, y_n) is to be used as a method of constructing tolerance intervals for a proportion δ of a population with confidence coefficient $(1 - \alpha)$. Find the minimum sample sizes needed to ensure that
  a $1 - \alpha = 0.90$ for $\delta = 0.95$
  b $1 - \alpha = 0.90$ for $\delta = 0.80$
  c $1 - \alpha = 0.95$ for $\delta = 0.95$
  d $1 - \alpha = 0.95$ for $\delta = 0.80$

9.6 The Method of Maximum Likelihood

The estimators presented in previous sections of this chapter were justified merely by the fact that they seemed reasonable. For instance, our intuition says that a sample mean should be a good estimator of a population mean, or that a sample proportion should somehow resemble the corresponding population proportion. There are, however, general methods for deriving estimators for unknown parameters of population models. One widely used method is called the *method of maximum likelihood*.

Suppose, for example, that a box contains four balls, of which an unknown number θ are white ($4 - \theta$ are nonwhite). We are to sample two balls at random and count X, the number of white balls in the sample. We know the probability distribution of X to be given by

$$P(X = x) = p(x) = \frac{\binom{\theta}{x}\binom{4 - \theta}{2 - x}}{\binom{4}{2}}$$

Now suppose we observe that $X = 1$. What value of θ will maximize the probability of this event? From the above distribution, we have

$$p(1 \mid \theta = 0) = 0 \qquad\qquad p(1 \mid \theta = 2) = \frac{2}{3}$$

$$p(1 \mid \theta = 1) = \frac{\binom{1}{1}\binom{3}{1}}{\binom{4}{2}} = \frac{3}{6} = \frac{1}{2} \qquad p(1 \mid \theta = 3) = \frac{1}{2}$$

$$p(1 \mid \theta = 4) = 0$$

Hence, $\theta = 2$ maximizes the probability of the observed sample, so we would choose this value, 2, as the maximum likelihood estimate for θ, given that we observed that $X = 1$. You can show for yourself that if $X = 2$. then 4 will be the maximum likelihood estimate of θ.

In general, suppose the random sample $X_1, X_2, \ldots, X_n$ results in observations $x_1, x_2, \ldots, x_n$. In the discrete case, the *likelihood L* of the sample is the product of the marginal probabilities; that is,

$$L(\theta) = p(x_1; \theta)p(x_2; \theta) \cdots p(x_n; \theta)$$

where

$$P(X = x_i) = p(x_i; \theta)$$

with θ being the parameter of interest. In the continuous case, the likelihood of the sample is the product of the marginal density functions; that is,

$$L(\theta) = f(x_1; \theta)f(x_2; \theta) \cdots f(x_n; \theta)$$

In either case, we maximize L as a function of θ to find the maximum likelihood estimator of θ.

If the form of the probability distribution is not too complicated, we can generally use methods of calculus to find the functional form of the maximum likelihood estimator instead of working out specific numerical cases. We illustrate with the following examples.

Example 9.27

Suppose that in a sequence of n independent, identical Bernoulli trials, Y successes are observed. Find the maximum likelihood estimator of p, the probability of successes on any single trial.

Solution Y is really the sum of n independent Bernoulli random variables $X_1, X_2, \ldots, X_n$. Thus, the likelihood is given by

$$L(p) = p^{x_1}(1 - p)^{1-x_1}p^{x_2}(1 - p)^{1-x_2} \cdots p^{x_n}(1 - p)^{1-x_n}$$

$$= p^{\sum_{i=1}^{n} x_i}(1 - p)^{n - \sum_{i=1}^{n} x_i}$$

$$= p^y(1 - p)^{n-y}$$

Because $\ln[L(p)]$ is a monotone function of $L(p)$, both $\ln[L(p)]$ and $L(p)$ will have their maxima at the same value of p. In many cases, it is easier to maximize $\ln[L(p)]$, so we will follow that approach here. Now

$$\ln[L(p)] = y \ln(p) + (n - y) \ln(1 - p)$$

is a continuous function of $p\,(0 < p < 1)$, and thus the maximum value can be found by setting the first derivative equal to zero. We have

$$\frac{\partial \ln L(p)}{\partial p} = \frac{y}{p} - \frac{n - y}{1 - p}$$

Setting

$$\frac{y}{\hat{p}} - \frac{n - y}{1 - \hat{p}} = 0 \qquad \text{we find} \qquad \hat{p} = \frac{y}{n}$$

and hence we take Y/n as the maximum likelihood estimator of p. In this case, as well as in many others, the maximum likelihood estimator agrees with our intuitive estimator given earlier in this chapter.

Example 9.28

Suppose we are to observe n independent lifelength measurements $X_1, \ldots, X_n$ from components known to have lifelengths exhibiting a Weibull model, given by

$$f(x) = \begin{cases} \dfrac{\gamma x^{\gamma-1}}{\theta} e^{-x^{\gamma}/\theta} & x > 0 \\ 0 & \text{elsewhere} \end{cases}$$

Assuming γ is known, find the maximum likelihood estimator of θ.

Solution   In analogy with the discrete case given above, we write the likelihood $L(\theta)$ as the joint density function of $X_1, \ldots, X_n$, or

$$L(\theta) = f(x_1, x_2, \ldots, x_n)$$
$$= f(x_1) f(x_2) \ldots f(x_n)$$
$$= \frac{\gamma x_1^{\gamma-1}}{\theta} e^{-x_1^{\gamma}/\theta} \ldots \frac{\gamma x_n^{\gamma-1}}{\theta} e^{-x_n^{\gamma}/\theta}$$
$$= \left(\frac{\gamma}{\theta}\right)^n (x_1, x_2, \ldots, x_n)^{\gamma-1} e^{-\sum_{i=1}^n x_i^{\gamma}/\theta}$$

Now $\ln[L(\theta)] = n \ln(\gamma) - n \ln(\theta) + (\gamma - 1) \ln(x_1 \ldots x_n) - \dfrac{1}{\theta} \sum_{i=1}^n x_i^{\gamma}$

To find the value of θ that maximizes $L(\theta)$, we take

$$\frac{\partial \ln L(\theta)}{\partial \theta} = -\frac{n}{\theta} + \frac{1}{\theta^2} \sum_{i=1}^n x_i^{\gamma}$$

Setting this derivative equal to zero, we get

$$-\frac{n}{\hat{\theta}} + \frac{1}{\hat{\theta}^2} \sum_{i=1}^n x_i^{\gamma} = 0 \quad \text{i.e.,} \quad \hat{\theta} = \frac{1}{n} \sum_{i=1}^n x_i^{\gamma}$$

Thus, we take $(1/n) \sum_{i=1}^n X_i^{\gamma}$ as the maximum likelihood estimator of θ. The interested reader can check to see that this estimator is unbiased for θ.

The maximum likehood estimators shown in Examples 9.27 and 9.28 are both unbiased, but this is not always the case. For example, if X has a geometric distribution, as given in Chapter 5, then the maximum likelihood estimator of p, the probability of a success on any one trial, is $1/X$. In this case, $E(1/X) \neq p$, so this estimator is not unbiased for p.

Many times the maximum likelihood estimator is simple enough that we can use its probability distribution to find a confidence interval, either exact or approximate, for the parameter in question. We illustrate the construction of an exact confidence interval for a nonnormal case with the exponential distribution. Suppose $X_1, \ldots, X_n$ represents a random sample from a population modeled by the density function

$$f(x) = \begin{cases} \dfrac{1}{\theta} e^{-x/\theta} & x > 0 \\ 0 & \text{elsewhere} \end{cases}$$

Because this exponential density is a special case of the Weibull with $\gamma = 1$, we know from Example 9.28 that the maximum likelihood estimator of θ is $(1/n)\sum_{i=1}^{n}X_i$. We also know that $\sum_{i=1}^{n}X_i$ will have a gamma distribution, but we do not have percentage points of most gamma distributions is standard sets of tables, except for the special case of the χ^2 distribution. Thus, we will try to transform $\sum_{i=1}^{n}X_i$ into something that has a χ^2 distribution.

Let $U_i = 2X_i/\theta$. Then

$$F_U(u) = P(U_i \leq u) = P\left(\frac{2X_i}{\theta} \leq u\right)$$

$$= P\left(X_i \leq \frac{u\theta}{2}\right) = 1 - e^{-(1/\theta)(u\theta/2)}$$

$$= 1 - e^{-u/2}$$

Hence, $\quad f_U(u) = \dfrac{1}{2}e^{-u/2}$

and U_i has an exponential distribution with mean 2, which is equivalent to the $\chi^2(2)$ distribution. It then follows that

$$\sum_{i=1}^{n}U_i = \frac{2}{\theta}\sum_{i=1}^{n}X_i$$

has a χ^2 distribution with $2n$ degrees of freedom. Thus, a confidence interval for θ with confidence coefficient $(1 - \alpha)$ can be formed by finding a $\chi^2_{\alpha/2}(2n)$

and a $\chi^2_{1-\alpha/2}(2n)$ value and writing

$$(1 - \alpha) = P\left[\chi^2_{1-\alpha/2}(2n) \leq \frac{2}{\theta}\sum_{i=1}^{n}X_i \leq \chi^2_{\alpha/2}(2n)\right]$$

$$= P\left[\frac{1}{\chi^2_{\alpha/2}(2n)} \leq \frac{\theta}{2\sum_{i=1}^{n}X_i} \leq \frac{1}{\chi^2_{1-\alpha/2}(2n)}\right]$$

$$= P\left[\frac{2\sum_{i=1}^{n}X_i}{\chi^2_{\alpha/2}(2n)} \leq \theta \leq \frac{2\sum_{i=1}^{n}X_i}{\chi^2_{1-\alpha/2}(2n)}\right]$$

Example 9.29

Consider the first 10 observations (in hundreds of hours) from Table 6.1 on life-lengths of batteries: 0.406, 2.343, 0.538, 5.088, 5.587, 2.563, 0.023, 3.334, 3.491, and 1.267. Using these observations as a realization of a random sample of lifelength measurements from an exponential distribution with mean θ, find a 95% confidence interval for θ.

Solution Using

$$\left[\frac{2\sum_{i=1}^{n}x_i}{\chi^2_{\alpha/2}(2n)}, \frac{2\sum_{i=1}^{n}x_i}{\chi^2_{1-\alpha/2}(2n)}\right]$$

as the confidence interval for θ, we have

$$\sum_{i=1}^{10}x_i = 24.640, \quad \chi^2_{0.975}(20) = 9.591, \text{ and } \chi^2_{0.025}(20) = 34.170$$

Hence, the realization of the confidence interval becomes

$$\left[\frac{2(24.640)}{34.170}, \frac{2(24.640)}{9.591}\right]$$

or (1.442, 5.138). We are about 95% confident that the true mean lifelength is between 144 and 514 hours. The rather wide interval is a reflection of both out relative lack of information in the sample of only $n = 10$ batteries and the large variability in the data.

It is informative to compare the exact intervals for the mean θ from the exponential distribution with what would have been generated if we had falsely assumed the random variables to have a normal distribution and used $\bar{x} \pm t_{\alpha/2}s/\sqrt{n}$ as a confidence interval. Table 9.7 and Figure 9.16 shows one hundred 95% confidence intervals generated by exact methods (using the χ^2 table) for

Table 9.7

Exact Exponential Confidence Intervals ($n = 5$)

Sample	LCL	UCL	Sample	LCL	UCL	Sample	LCL	UCL	Sample	LCL	UCL
1	2.0360	12.8439	26	4.2298	26.6834	51	9.1080	57.4570	76	7.4693	47.1190
*2	1.4945	9.4281	27	5.1712	32.6223	52	4.2073	26.5413	77	8.5458	53.9103
3	4.1570	26.2243	28	8.7703	55.3266	53	5.0834	32.0681	78	2.9840	18.8243
4	3.0777	19.4153	29	2.5468	16.0667	54	5.0498	31.8561	79	9.2678	58.4647
5	5.8994	37.2158	30	5.2786	33.2998	55	3.6093	22.7689	80	4.2078	26.5442
6	6.4954	40.9755	31	7.2621	45.8124	56	5.8484	36.8940	81	8.0267	50.6353
7	2.1970	13.8598	32	2.2776	14.3680	57	5.1935	32.7627	82	9.9686	62.8854
8	2.3772	14.9965	33	4.6354	29.2424	58	4.1803	26.3710	83	2.2581	14.2447
9	5.9072	37.2652	34	3.8028	23.9895	59	4.6671	29.4422	84	9.9971	63.0651
10	7.6724	48.4009	35	1.9877	12.5396	60	5.3942	34.0291	*85	10.2955	64.9478
11	6.6416	41.8977	36	4.6377	29.2569	61	6.0696	38.2899	86	6.0830	38.374
12	4.7809	30.1601	37	7.3677	46.4784	62	3.4519	21.7764	87	3.0037	18.9483
13	3.1717	20.0085	38	3.5592	22.4533	63	5.1959	32.7781	88	2.3759	14.9881
*14	1.5147	9.5555	39	2.9913	18.8707	64	4.7976	30.2655	89	4.0007	25.2376
15	3.7016	23.3513	40	5.7916	36.5356	65	5.7648	36.3669	*90	1.1737	7.4044
16	3.6999	23.3408	41	5.6268	35.4962	66	4.1567	26.2223	91	3.8375	24.2081
17	6.9034	43.5494	42	3.9599	24.9808	67	8.9634	56.5450	92	2.9560	18.6473
18	2.3600	14.8881	43	2.3039	14.5341	68	6.7443	42.5456	93	4.9714	31.3616
19	6.5329	41.2122	44	7.4084	46.7352	69	5.3691	33.8709	94	3.0063	18.9648
20	4.0475	25.5335	45	3.0687	19.3588	70	3.7709	23.7883	95	4.1684	26.2958
21	5.6173	35.4365	46	2.9999	18.9251	71	5.4299	34.2542	96	4.6269	29.1879
22	4.0710	25.6815	47	3.9591	24.9756	72	2.7016	17.0429	97	3.8026	23.9883
23	4.6528	29.3518	48	3.0141	19.0146	73	4.2962	27.1020	98	6.0056	37.8858
24	4.5828	28.9103	49	2.6913	16.9782	74	2.3106	14.5761	*99	1.5153	9.5592
25	3.8090	24.0291	50	4.2829	27.0187	75	7.6502	48.2602	100	7.0408	44.4161

sample of size $n = 5$ from an exponential distribution with $\theta = 10$. Note that five of the intervals (marked with an asterisk) do not include the true θ, as expected. Four miss on the low side, and one misses on the high side.

Table 9.8 repeats the process of generating confidence intervals for θ, with the same samples of size $n = 5$, but now using $\bar{x} \pm t_{\alpha/2} s/\sqrt{n}$ with $\alpha/2 = 0.025$ and 4 degrees of freedom. Note from Figure 9.17 that ten of the intervals do not include the true θ, twice the number expected. Also, the intervals that miss the true

Figure 9.16

Exact exponential confidence intervals ($n = 5$)

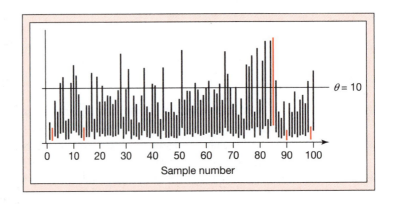

Table 9.8

Confidence Intervals Based on the Normal Distribution (*n* = 5)

Sample	LCL	Mean	UCL	S	Sample	LCL	Mean	UCL	S
*1	1.9750	4.1704	6.3656	1.7682	51	−3.5590	18.6561	40.8713	17.8944
*2	−1.0690	3.0613	7.1916	3.3269	52	−2.9490	8.6179	20.1847	9.3171
3	−1.5500	8.5150	18.5800	8.1074	53	1.6480	10.4124	19.1769	7.0598
4	−1.0190	6.3041	13.6277	5.8991	54	−2.8520	10.3436	23.5391	10.6290
5	4.1830	12.0839	19.9850	6.3644	55	1.6450	7.3930	13.1407	4.6298
6	3.8380	13.3046	22.7712	7.6253	56	2.3440	11.9794	21.6150	7.7615
7	−3.0360	4.5002	12.0368	6.0707	57	−8.5250	10.6380	29.8011	15.4359
*8	0.0100	4.8693	9.7289	3.9144	58	−0.7810	8.5626	17.9059	7.5260
9	1.0040	12.0999	23.1954	8.9374	59	2.3640	9.5598	16.7556	5.7962
10	4.9170	15.7156	26.5146	8.6986	60	−1.3140	11.0491	23.4128	9.9589
11	8.3400	13.6041	18.8683	4.2404	61	1.4650	12.4326	23.3998	8.8341
12	−0.3970	9.7929	19.9832	8.2083	62	−2.9780	7.0707	17.1193	8.0941
13	1.0420	6.4967	11.9513	4.3937	63	−6.0650	10.6429	27.3514	13.4586
*14	−0.2740	3.1026	6.4790	2.7197	64	−4.3780	9.8271	24.0327	11.4426
15	2.5420	7.5821	12.6217	4.0594	65	3.1550	11.8082	20.4612	6.9700
16	3.3980	7.5787	11.7595	3.3678	66	0.2980	8.5143	16.7307	6.6184
17	6.9940	14.1404	21.2868	5.7564	67	−4.6550	18.3600	41.3751	18.5387
*18	1.2440	4.8341	8.4245	2.8920	68	3.1400	13.8144	24.4798	8.5910
19	−6.3830	13.3815	33.1458	15.9202	69	−2.9480	10.9978	24.9433	11.2331
20	1.1220	8.2907	15.4591	5.7742	70	0.3320	7.7240	15.1161	5.9544
21	−3.5930	11.5061	26.6049	12.1621	71	−3.3830	11.1222	25.6275	11.8840
22	−0.4330	8.3387	17.1102	7.0654	72	1.0270	5.5338	10.0406	3.6302
23	−5.9250	9.5305	24.9856	12.4491	73	−1.7200	8.7999	19.3194	8.4735
24	−11.2000	9.3871	29.9740	16.5828	74	−0.7550	4.7328	10.2205	4.4203
25	−4.1660	7.8022	19.7701	9.6402	75	−4.7530	15.6699	36.0924	16.4503
26	1.5640	8.6640	15.7640	5.7191	76	−12.2230	15.2994	42.8221	22.1695
27	3.8180	10.5924	17.3667	5.4567	77	−5.1020	17.5045	40.1115	18.2099
28	−3.4920	17.9644	39.4208	17.2832	78	0.5040	6.1122	11.7204	4.5175
29	−0.6560	5.2168	11.0900	4.7308	79	0.0420	18.9833	37.9249	15.2575
30	1.8370	10.8123	19.7876	7.2296	80	−0.8730	8.6188	18.1102	7.6453
31	2.5810	14.8751	27.1690	9.9027	81	7.2320	16.4411	25.6501	7.4179
32	−1.9410	4.6653	11.2716	5.3214	82	−0.2890	20.4187	41.1267	16.6803
33	−10.7340	9.4949	29.7238	16.2943	*83	−0.6840	4.6252	9.9343	4.2765
34	−2.7250	7.7893	18.3035	8.4692	84	1.1620	20.4771	39.7917	15.5579
*35	−0.5460	4.0716	8.6890	3.7193	85	−13.4710	21.0884	55.6473	27.8373
36	0.8410	9.4996	18.1584	6.9747	86	−0.0860	12.4599	25.0060	10.1059
37	−1.4670	15.0914	31.6496	13.3376	87	−3.6590	6.1524	15.9641	7.9033
38	2.1440	7.2905	12.4370	4.1455	88	−4.6600	4.8666	14.3930	7.6735
39	−2.9910	6.1272	15.2452	7.3446	89	0.9140	8.1946	15.4755	5.8648
40	−4.7140	11.8630	28.4403	13.3530	*90	0.8260	2.4042	3.9824	1.2712
41	4.3560	11.5255	18.6946	5.7747	91	−0.4160	7.8603	16.1361	6.6662
42	−4.4360	8.1112	20.6585	10.1069	92	−0.0930	6.0547	12.2023	4.9519
*43	1.2210	4.7192	8.2178	2.8181	93	2.4510	10.1830	17.9150	6.2281
44	−0.8770	15.1748	31.2261	12.9294	94	−6.6130	6.1578	18.9282	10.2866
45	0.8650	6.2858	11.7065	4.3665	95	−0.4700	8.5382	17.5461	7.2559
46	0.7510	6.1449	11.5392	4.3451	96	−0.0360	9.4772	18.9902	7.6627
47	−1.3250	8.1095	17.5442	7.5997	97	−2.2010	7.7889	17.7785	8.0466
48	−1.1160	6.1740	13.4641	5.8722	98	2.2790	12.3014	22.3234	8.0727
49	−2.8370	5.5128	13.8623	6.7256	*99	1.3830	3.1038	4.8248	1.3862
50	2.5290	8.7729	15.0168	5.0294	100	2.8040	14.4218	31.6476	13.8754

Figure 9.17

Confidence intervals based on the normal distribution ($n = 5$)

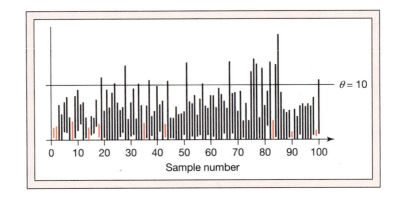

$\theta = 10$ all miss on the low side. That fact, coupled with the fact that the lower bounds are frequently negative even though $\theta > 0$, seems to indicate that the intervals based on normality are too far left compared to the true intervals based on the χ^2 distribution.

The lesson to be learned here is that *one should be careful of inflicting the normality assumption on data*, especially if a more reasonable and mathematically tractable model is available.

9.6.1 Maximum Likelihood Estimators of Certain Functions of Parameters

Before we leave the section on maximum likelihood estimation, we now show how we can carry out the estimation of certain functions of parameters. In general, if we have a maximum likelihood estimator, say Y, for a parameter θ, then any continuous function of θ, say $g(\theta)$, will have a maximum likelihood estimator $g(Y)$.

Suppose measurements $X_1, \ldots, X_n$ again denote independent lifelength measurements from a population modeled by the exponential distribution with mean θ. The *reliability* of each component is defined by

$$R(t) = P(X_i > t)$$

In the exponential case,

$$R(t) = \int_t^\infty \frac{1}{\theta} e^{-x/\theta} \, dx = e^{-t/\theta}$$

Suppose we want to estimate $R(t)$ from $X_1, \ldots, X_n$. We know that the maximum likelihood estimator of θ is $\overline{X}$, and hence the maximum likelihood estimator of $R(t) = e^{-t/\theta}$ is $e^{-t/\overline{x}}$.

Now we can use the results from Chapter 7 to find an approximate confidence interval for $R(t)$ if n is large. Let

$$g(\theta) = e^{-t/\theta}$$

Then $g'(\theta) = (t/\theta^2)e^{-t/\theta}$ and, since $\overline{X}$ is approximately normally distributed with mean θ and standard deviation $\theta/\sqrt{n}$ (since $V(X_i) = \theta^2$), it follows that

$$\sqrt{n}\,\frac{[g(\overline{X}) - g(\theta)]}{|g'(\theta)|\theta} = \frac{\sqrt{n}\,[e^{-t/\overline{x}} - e^{-t/\theta}]}{\left(\dfrac{t}{\theta}\right)e^{-t/\theta}}$$

is approximately a standard normal random variable if n is large. Thus, for large n, $e^{-t/\overline{X}}$ is approximately normally distributed with mean $e^{-t/\theta}$ and standard deviation $(1/\sqrt{n})(t/\theta)e^{-t/\theta}$. Note that this is only a large-sample approximation, as $E(e^{-t/\overline{X}}) \neq e^{-t/\theta}$. This result can be used to form an approximate confidence interval for $R(t) = e^{-t/\theta}$ of the form

$$e^{-t/\overline{x}} \pm z_{\alpha/2}\frac{1}{\sqrt{n}}\left(\frac{t}{\overline{x}}\right)e^{-t/\overline{x}}$$

using $\overline{x}$ as an approximation to θ in the standard deviation.

Example 9.30

Using the 50 observations given in Table 6.1 as a realization of a random sample from an exponential distribution, estimate $R(5) = P(X_i > 5) = e^{5/\theta}$ in an approximate 95% confidence interval.

Solution   For $(1-\alpha) = 0.95$, $z_{\alpha/2} = 1.96$. Using the method indicated above, the interval becomes

$$e^{-t/\overline{x}} \pm (1.96)\frac{1}{\sqrt{n}}\left(\frac{t}{\overline{x}}\right)e^{-t/\overline{x}}$$

or

$$e^{-5/2.267} \pm (1.96)\frac{1}{\sqrt{50}}\left(\frac{5}{2.267}\right)e^{-5/2.267}$$

$$0.110 \pm 0.067 \quad \text{i.e.,} \quad (0.043, 0.177)$$

That is, we are about 95% confident that the probability is between 0.043 and 0.177. For these data, the true value of θ was 2, and $R(5) = e^{-5/2} = 0.082$ is well within this interval. The interval could be narrowed by selecting a larger sample.

Exercises

9.67   If $X_1,\ldots,X_n$ denotes a random sample from a Poisson distribution with mean λ, find the maximum likelihood estimator of λ.

9.68   Because $V(X_i) = \lambda$ in the Poisson case, it follows from the Central Limit Theorem that $\overline{X}$ will be approximately normally distributed with mean λ and variance λ/n, for large n.

a Use the above facts to construct a large-sample confidence interval for λ.

b Suppose that 100 reinforced concrete trusses were examined for cracks. The average number of cracks per truss was observed to be four. Construct an approximate 95% confidence interval for the true mean number of cracks per truss for trusses of this type. What assumptions are necessary for your answer to be valid?

9.69 Suppose $X_1, \ldots, X_n$ denotes a random sample from the normal distribution with mean μ and variance σ^2. Find the maximum lilelihood estimators of μ and σ^2.

9.70 Suppose $X_1, \ldots, X_n$ denotes a random sample from the gamma distribution with a known α but unknown β. Find the maximum likelihood estimator of β.

9.71 The stress resistances for specimens of a certain type of plastic tend to have a gamma distribution with $\alpha = 2$, but β may change with certain changes in the manufacturing process. For eight specimens independently selected from a certain process, the resistances (in psi) were

$$29.2, 28.1, 30.4, 31.7, 28.0, 32.1, 30.1, 29.7$$

Find a 95% confidence interval for β. [*Hint*: Use the methodology outlined for Example 9.28.]

9.72 If X denotes the number of the trial on which the first defective is found in a series of independent quality-control tests, find the maximum likelihood estimator of p, the true probability of observing a defective.

9.73 The absolute errors in the measuring of the diameters of steel rods are uniformly distributed between zero and θ. There such measurements on a standard rod produced errors of 0.02, 0.06, and 0.05 centimeter. What is the maximum likelihood estimate of θ? [*Hint*: This problem cannot be solved by differentiation. Write down the likelihood, $L(\theta)$, for three independent measurements and think carefully about what value of θ produces a maximum. Remember, each measurement must be between zero and θ.]

9.74 The number of improperly soldered connections per microchip in an electronics manufacturing operation follows a binomial distribution with $n = 20$ and p unknown. The cost of correcting these malfunctions, per microchip, is

$$C = 3X + X^2$$

Find the maximum likelihood estimate of $E(C)$ if $\hat{p}$ is available as an estimate of p.

9.7 Summary

We have developed *confidence intervals* for individual means and proportions, linear functions of means and proportions, individual variances, and ratios of variances. In a few cases, we have shown how to calculate confidence intervals for other functions of parameters. We also touched on the notion of a *prediction interval* for a random variable and a *tolerance interval* for a distribution. Most often, these intervals were two-sided, in the sense that they provided both an upper and a lower bound. Occasionally, a one-sided interval is in order.

Most of the estimators used in the construction of confidence intervals were selected on the basis of intuitive appeal. However, the *method of maximum likelihood* provides a technique for finding estimators that are usually close to being unbiased and have small variance. (Most of the intuitive estimators turned out to be maximum likelihood estimators.)

If it is appropriate to treat a population parameter as a random quantity, then *Bayesian* techniques can be used to find estimates. Many of the estimators used in this chapter will be used in a slightly different setting in Chapter 9.

Table 9.9

Summary of Confidence Intervals and Corresponding Conditions

Single Sample Case

μ: Large-sample

$$\overline{X} \pm z_{\alpha/2} \frac{\sigma}{\sqrt{n}}$$

- Random sample
- Large sample

μ: Normal distribution

$$\overline{X} \pm t_{\alpha/2} \frac{s}{\sqrt{n}}$$
$$\text{df} = n - 1$$

- Random sample
- Normal population

p: Large sample

$$\hat{p} \pm z_{\alpha/2} \sqrt{\frac{\hat{p}(1 - \hat{p})}{n}}$$

- Random sample
- Large sample

σ^2: Normal distribution

$$\left(\frac{(n - 1)S^2}{\chi^2_{\alpha/2}(\nu)}, \frac{(n - 1)S^2}{\chi^2_{1-\alpha/2}(\nu)} \right)$$
$$\text{df} = n - 1$$

- Random sample
- Normal population

Multiple Samples Case

$\mu_1 - \mu_2$: Large samples

$$(\overline{X}_1 - \overline{X}_2) \pm z_{\alpha/2} \sqrt{\frac{\sigma_1^2}{n_1} + \frac{\sigma_2^2}{n_2}}$$

- Random samples
- Independent samples
- Large samples

$\theta = a_1\mu_1 + a_2\mu_2 + a_3\mu_3$:
(Linear Function of Means)
Large samples

$$\hat{\theta} \pm z_{\alpha/2} \sqrt{V(\hat{\theta})}$$
$$\hat{\theta} = a_1\overline{X}_1 + a_2\overline{X}_2 + a_3\overline{X}_3$$
$$V(\hat{\theta}) = a_1^2 \left(\frac{\sigma_1^2}{n_1} \right) + a_2^2 \left(\frac{\sigma_2^2}{n_2} \right) + a_3^2 \left(\frac{\sigma_3^2}{n_3} \right)$$

- Random samples
- Independent samples
- Large samples

$\mu_1 - \mu_2$: Normal distribution,
$\sigma_1^2 = \sigma_2^2$

$$(\overline{X}_1 - \overline{X}_2) \pm t_{\alpha/2} S_p \sqrt{\frac{1}{n_1} + \frac{1}{n_2}}$$

$$S_p^2 = \frac{(n_1 - 1)S_1^2 + (n_2 - 1)S_2^2}{n_1 + n_2 - 2}$$
$$\text{df} = n_1 + n_2 - 2$$

- Random samples
- Independent samples
- Normal populations
- Equal population variances

$\theta = a_1\mu_1 + a_2\mu_2 + a_3\mu_3$:
(Linear Function of Means)
Normal distribution,
$\sigma_1^2 = \sigma_2^2 = \sigma_3^2$

$$\hat{\theta} \pm t_{\alpha/2} S_p \sqrt{\frac{a_1^2}{n_1} + \frac{a_2^2}{n_2} + \frac{a_3^2}{n_3}}$$
$$\hat{\theta} = a_1\overline{X}_1 + a_2\overline{X}_2 + a_3\overline{X}_3$$
$$S_p^2 = \frac{\sum_{i=1}^{3}(n_i - 1)S_i^2}{\sum_{i=1}^{3}(n_i - 1)}$$
$$\text{df} = \sum_{i=1}^{3}(n_i - 1)$$

- Random samples
- Independent samples
- Normal populations
- Equal population variances

$\mu_1 - \mu_2$: Normal distribution,
$\sigma_1^2 \neq \sigma_2^2$

$$(\overline{X}_1 - \overline{X}_2) \pm t_{\alpha/2} \sqrt{\frac{S_1^2}{n_1} + \frac{S_2^2}{n_2}}$$
$$\text{df} = \frac{\left(\frac{S_1^2}{n_1} + \frac{S_2^2}{n_2} \right)^2}{\frac{(S_1^2/n_1)^2}{(n_1 - 1)} + \frac{(S_2^2/n_2)^2}{(n_2 - 1)}}$$

- Random samples
- Independent samples
- Normal populations
- Unequal population variances

$p_1 - p_2$: Large samples

$$(\hat{p}_1 - \hat{p}_2) \pm z_{\alpha/2} \sqrt{\frac{\hat{p}_1(1 - \hat{p}_1)}{n_1} + \frac{\hat{p}_2(1 - \hat{p}_2)}{n_2}}$$

- Random samples
- Independent samples
- Large samples

σ_2^2/σ_1^2: Normal distributions

$$\left(\frac{s_2^2}{s_1^2} \frac{1}{F_{\alpha/2}(\nu_2, \nu_1)}, \frac{s_2^2}{s_1^2} F_{\alpha/2}(\nu_1, \nu_2) \right)$$
$$\text{df} = (n_1 - 1, n_2 - 1)$$

- Random samples
- Independent samples
- Normal populations

Supplementary Exercises

9.75 The diameter measurements of an armored electric cable, taken at 10 points along the cable, yield a sample mean of 2.1 centimeters and a sample standard deviation of 0.3 centimeter. Estimate the average diameter of the cable in a confidence interval with a confidence coefficient of 0.90. What assumptions are necessary for your answer to be valid?

9.76 Suppose the sample mean and standard deviation of Exercise 9.75 had come from a sample of 100 measurements. Construct a 90% confidence interval for the average diameter of the cable. What assumption must necessarily be made?

9.77 The Rockwell hardness measure of steel ingots is produced by pressing a diamond point into the steel and measuring the depth of penetration. A sample of 15 Rockwell hardness measurements on specimens of steel gave a sample mean of 65 and a sample variance of 90. Estimate the true mean hardness in a 95% confidence interval, assuming the measurements to come from a normal distribution.

9.78 Twenty specimens of a slightly different steel from that used in Exercise 9.77 were observed and yielded Rockwell hardness measurements with a mean of 72 and a variance of 94. Estimate the difference between the mean hardness for the two varieties of steel in a 95% confidence interval.

9.79 In the fabrication of integrated circuit chips, it is of great importance to form contact windows of precise width. (These contact windows facilitate the interconnections that make up the circuits.) Complementary metal-oxide semiconductor (CMOS) circuits are fabricated by using a photolithography process to form windows. The key steps involve applying a photoresist to the silicon wafer, exposing the photoresist to ultraviolet radiation through a mask, and then dissolving away the photoresist from the exposed areas, revealing the oxide surface. The printed windows are then etched through the oxide layers down to the silicon. Key measurements that help determine how well the integrated circuits may function are the window widths before and after the etching process. (See Phadke et al., *The Bell System Technical Journal*, 62, no. 5, 1983, pp. 1273–1309, for more details.)

Pre-etch window widths for a sample of 10 test locations are as follows (in μm):

2.52, 2.50, 2.66, 2.73, 2.71,
2.67, 2.06, 1.66, 1.78, 2.56

Post-etch window widths for 10 independently selected test locations are

3.21, 2.49, 2.94, 4.38, 4.02,
3.82, 3.30, 2.85, 3.34, 3.91

One important problem is simply to estimate the true average window width for this process.

a Estimate the average pre-etch window width, for the population from which this sample was drawn, in a 95% confidence interval.

b Estimate the average post-etch window width in a 95% confidence interval.

9.80 In the setting of Exercise 9.79, a second important problem is to compare the average window widths before and after etching. Using the data given there, estimate the true difference in average window widths in a 95% confidence interval.

9.81 Window sizes in integrated circuit chips must be of fairly uniform size in order for the circuits to function properly. Thus, the variance of the window width measurements is an important quantity. Refer to the data of Exercise 9.79.

a Estimate the true variance of the post-etch window widths in a 95% confidence interval.

b Estimate the ratio of the variance of post-etch window widths to that of pre-etch window widths in a 95% confidence interval.

c What assumptions are necessary for your answers in parts (a) and (b) to be valid?

9.82 The weights of aluminum grains follow a lognormal distribution, which means that the natural logarithms of the grain weights follow a normal distribution. A sample of 177 aluminum grains has logweights averaging 3.04 with a standard deviation of 0.25. (The original weight measurements were in 10^{-4} gram.) Estimate the true mean logweight of the grains from which this sample was selected, using a confidence coefficient of 0.90. (*Source:* Department of Materials Science, University of Florida.)

9.83 Two types of aluminum powders are blended before a sintering process is begun to form solid aluminum. The adequacy of the blending is gauged by taking numerous small samples from the blend and measuring the weight proportion of one type of powder. For adequate blending, the weight proportions from sample to sample should have small variability. Thus, the variance becomes a key measure of blending adequacy.

Ten samples from an aluminum blend gave the following weight proportions for one type of powder:

0.52, 0.54, 0.49, 0.53, 0.52,
0.56, 0.48, 0.50, 0.52, 0.51

a Estimate the population variance in a 95% confidence interval.

b What assumptions are necessary for your answer in part (a) to be valid? Are the assumptions reasonable for this case?

9.84 It is desired to estimate the proportion of defective items produced by a certain assembly line to within 0.1 with confidence coefficient 0.95. What is the smallest sample size that will guarantee this accuracy no matter where the true proportion of defectives might lie? [*Hint*: Find the value of p that maximizes the variance $p(1 - p)/n$, and choose the value of n that corresponds to it.]

9.85 Suppose that, in a large-sample estimate of $(\mu_1 - \mu_2)$ for two populations with respective variances σ_1^2 and σ_2^2, a total of n observations is to be selected. How should these n observations be allocated to the two populations so that the length of the resulting confidence interval will be minimized?

9.86 Suppose the sample variances given in Exercise 9.35 are good estimates of the population variances. Using the allocation scheme of Exercise 9.85, find the number of measurements to be taken on each coupling agent to estimate the true difference in means to within 1 unit with confidence coefficient 0.95. Compare your answer to that of Exercise 9.36.

9.87 A factory operates with two machines of type A and one machine of type B. The weekly repair costs Y for type A machines are normally distributed with mean μ_1 and variance σ^2. The weekly repair costs X for machines of type B are also normally distributed, but with mean μ_2 and variance $3\sigma^2$. The expected repair cost per week for the factory is then $(2\mu_1 + \mu_2)$. Suppose $Y_1, \ldots, Y_n$ denotes a random sample on costs for type A machines and $X_1, \ldots, X_n$ denotes an independent random sample on costs for type B machines. Use these data to construct a 95% confidence interval for $(2\mu_1 + \mu_2)$.

9.88 In polycrystalline aluminum, the number of grain nucleation sites per unit volume is modeled as having a Poisson distribution with mean λ. Fifty unit-volume test specimens subjected to annealing regime A showed an average of 20 sites per unit volume. Fifty independently selected unit-volume test specimens subjected to annealing regime B showed an average of 23 sites per volume. Find an approximate 95% confidence interval for the difference between the mean site frequencies for the two annealing regimes. Would you say that regime B tends to increase the number of nucleation sites?

9.89 A random sample of n items is selected from the large number of items produced by a certain production line in one day. The number of defectives X is observed. Find the maximum likelihood estimator of the ratio R of the number of defective to nondefective items.

Hypothesis Testing

A quality control inspector randomly selects five bottles filled by a machine set to fill 12 ounces and measures the amounts filled in each bottle. The resulting sample mean is 11.82 ounces per bottle. The dotplot in Figure 10.1 gives a simulated sampling distribution of sample means when the population mean is known to be 12. Considering this dotplot, does this sample mean of 11.82 ounces give sufficient evidence to conclude that on the average the machine is underfilling 12-ounce bottles and needs to be stopped for repairs? What is the likelihood that such an interruption in the filling process is unnecessary?

To answer such questions, we need to use a technique known as *testing of hypotheses.* This is another formal manner of making inferences from sample to population. The methodology allows the weight of sample evidence for or against the experimenter's hypothesis to be assessed probabilistically and is the basis of this scientific method. Again, as in estimation, the sampling distribution of the statistic involved plays a key role in the development of the methodology.

Figure 10.1
Mean amount filled per
bottle (in oz) (*n* = 5)

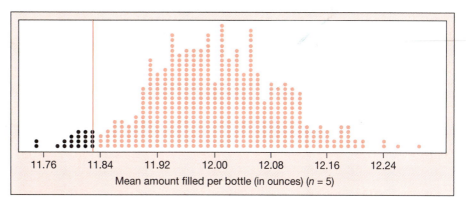

Mean amount filled per bottle (in ounces) (*n* = 5)

Introduction

One method of using sample data to formulate inferences about a population parameter, as seen in Chapter 9, is to produce a confidence interval estimate of the
parameter in question. But it often happens that an experimenter is interested only
in checking a claim, or *hypothesis*, concerning the value of a parameter and is not
really interested in the location or length of the confidence interval itself. For
example, if an electronic component is guaranteed to last at least 200 hours, then
an investigator might be interested in checking the hypothesis that the mean is
really at least 200 hours or it is lower than 200 hours. A confidence interval on
mean is not in itself of great interest, although it can provide a mechanism for
checking, or *testing*, the hypothesis of interest.

Hypothesis testing arises as a natural consequence of the scientific method.
The scientist observes nature, formulates a theory, and then tests the theory against
observations (see Figure 10.2).

In the context of the problem,

- The experimenter theorizes that the parameter of interest takes on a certain
  value or a set of values.
- He or she then takes a sample from the population in question.
- The experimenter compares the observations against the theory.
- If the observations disagree seriously with the theory, then the experimenter
  may reject the theory, or hypothesis.
- If the observations are compatible with the theory, then the experimenter may
  proceed as if the hypothesis were true.

Generally, the process does not stop in either case. New theories are posed, new
experiments are completed, and new comparisons between theory and reality are
made in a continuing cycle.

Note that hypothesis testing requires a decision when the observed sample is
compared with theory.

- How do we decide whether the sample agrees with the hypothesis?
- When do we reject the hypothesis, and when should we not reject it?
- What is the probability that we will make a wrong decision?
- What function of the sample observations should we employ in our decision-
  making process?

The answers to these questions lie in the study of statistical hypothesis testing.

Figure 10.2

Hypothesis testing, a process

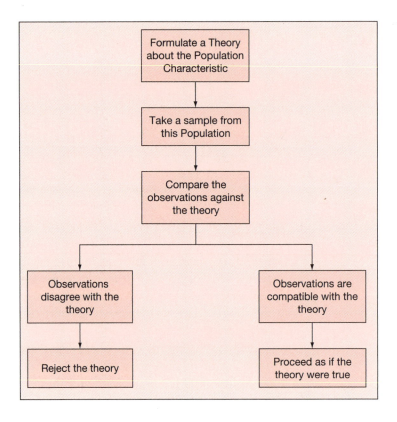

10.1 Terminology of Hypothesis Testing

Consider a filling machine set to fill 12 ounces of juice in bottles. Although each bottle is getting slightly different amounts, the machine is supposed to dispense fill at the rate of 12 ounces per bottle, on the average. The population is all the bottles filled by this machine. If the process-mean deviates significantly from 12 ounces, then the filling process is stopped to make appropriate adjustments. An industrial engineer in charge of monitoring the process suspects that the machine is not working to standard. In other words, it is claimed that the population mean, μ, the true mean amount filled per bottle has taken on some value other than the standard (say, μ_0), which in this case is 12 ounces. Suppose we want to test the validity of this claim; that is, we are interested in determining whether the process mean has deviated from the required 12 ounces per bottle.

Suppose a random sample of size n is taken from such a population with unknown mean μ. It seems intuitively reasonable that our decision about the population mean should be based on the sample mean $\overline{X}$, with extra information provided by the population variance σ^2 or its estimator S^2. For example, if we are making inference about the mean amount filled by the filling machine, then we can make our decision using sample mean with some information about the precision provided by the variance.

A **hypothesis** is a statement about the population parameter or process characteristic.

There are two types of hypotheses, null and alternative hypothesis.

- The null hypothesis (H_0) usually represents the standard operating procedure of a system or known specifications. Initially, this statement is assumed to be true. For example, the standard set of $\mu = 12$ ounces for the filling process is a null hypothesis. It gives a specific value for μ and represents what should happen when the machine is operating according to the specification.
- Often the alternative hypothesis (H_a) is the alternative statement proposed by the researcher, and for that reason it is also known as the *research hypothesis*. The alternative hypothesis specifies those values of the parameter that represent an important change from standard operating procedure or known specifications. It is a contradiction to the null hypothesis. If the sampled evidence suggests the null hypothesis is false, then it will be replaced by the alternative hypothesis. The alternative hypothesis in the above example will be $\mu \neq 12$ ounces; that is, the machine is not filling on the average the amount it is set to dispense and needs to be corrected.

A **null hypothesis** is a statement that specifies a particular value (or values) for the parameter being studied. It is denoted by H_0.

An **alternative hypothesis** is a statement of the change from the null hypothesis that the investigation is designed to check. It is denoted by H_a.

How do we decide which statement is true? Sample observations are collected to determine whether the data support H_0 or H_a.

The **test statistic** (**TS**) is the sample quantity on which the decision to reject or not reject H_0 is based.

The **rejection region** (or **critical region**) is the set of values of the test statistic that leads to the rejection of null hypothesis in favor of an alternative hypothesis.

The complement of the rejection region—in other words, the set of values of the test statistic that fails to lead to the rejection of null hypothesis—is the *non-rejection region*.

The **critical value** (**CV**) is the boundary between the rejection region and the non-rejection region.

Example 10.1

The depth setting on a certain drill press is 2 inches. A small amount of variation in the depths of holes is acceptable, but the average depth of holes that are either too large or too small is not acceptable. Identify the parameter and the pair of hypotheses that are of interest here.

Solution Here we are interested in making an inference about the value of

$$\mu = \text{the mean depth of holes (in inches) drilled by this press}$$

Because the depth setting is on 2 inches, one could hypothesize that the average depth of holes drilled by this machine is 2 inches, that is, $H_0: \mu = 2$ inches. The investigator wants to decide whether the mean depth of holes drilled by this press is either larger or smaller than the setting, that is, $H_a: \mu \neq 2$ inches.

Obviously, the sample mean will serve as an appropriate statistic to estimate μ, but we really want to know whether μ is different from 2 inches. If the null hypothesis is true, then we expect the sample mean to be around 2 inches. But a too large or a too small sample mean compared to 2 inches provides a reason to favor the alternative. Therefore, as shown in Figure 10.3, the rejection region will be in both the tails of the distribution of the test statistic. The notation CV indicates the critical value separating the rejection region from the non-rejection region.

Figure 10.3

Rejection region for a two-tailed test

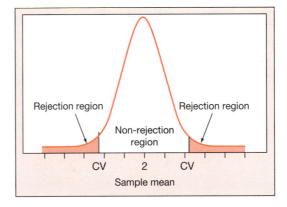

A test with the rejection region in both tails of the distribution of statistic is known as a *two-tailed test*.

Example 10.2

From past experience, it is known that one manufacturing process produces 1% defectives. Recently, some structural changes were made to the design of the process. Any reduction in percent defectives as a result is a welcome change; however, any increase in percent defectives is undesirable and requires immediate attention. Identify the parameter of interest and formulate a pair of hypotheses to be tested.

Solution Here we are interested in making inference about the value of

p = the proportion of defectives produced by the manufacturing process

Because the desirable values of p are at or below 1%, one could hypothesize that the proportion defectives in the population of items produced by this process is at most 1%, that is, $H_0: p \leq 0.01$. However, the investigator is interested in identifying the undesirable increase in the proportion defective. Hence the alternative hypothesis, $H_a: p > 0.01$.

Here sample proportion serves as an appropriate statistic to estimate the population proportion p. But what we are really interested in is whether p is larger than 0.01. If the null hypothesis is true, we expect the sample proportion to be around 0.01. But, if the sample proportion is too large, then we favor the alternative hypothesis. Therefore, as shown in Figure 10.4, the rejection region will be in the right tail of the distribution of the statistic.

Figure 10.4

Rejection region for a right-tailed test

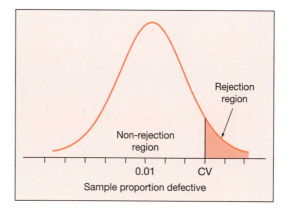

A test with the rejection region in the right tail of the distribution of a statistic is known as a **right-tailed test**.

Example 10.3

A machine is cutting rods with mean diameter 5 mm and variance 0.5 mm^2. Some adjustments are made to the process with the intention of improving the precision of the cutting process. What is the parameter of interest and what null and alternative hypotheses should be tested to check whether the adjustments have been effective in reducing variation in diameters?

Solution Here we are interested in

σ^2 = the variance of diameters of rods cut by this machine

The status quo would be that the adjustments are ineffective and the machine is still cutting rods with a diameter variance at least 0.5 mm^2, that is, $H_0: \sigma^2 \geq 0.5$ mm^2. The contradiction would be that the adjustments are effective and the variance has fallen below 0.5, that is, $H_a: \sigma^2 < 0.5$.

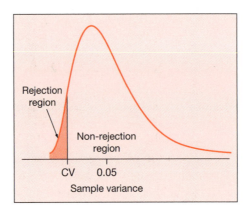

In this example, the sample variance will be an appropriate statistic to estimate the population variance σ^2. But we are really interested in knowing whether σ^2 is smaller than 0.5. If the null hypothesis is true, we expect the sample variance to be around 0.5. But if the sample variance is much smaller than 0.5, then we favor the alternative hypothesis. Therefore, as shown in Figure 10.5, the rejection region will be in the left tail of the distribution of statistic.

> A test with the rejection region in the left tail of the distribution of statistic is known as a **left-tailed test**.

Example 10.4

A new process for drawing wire might lead researchers to believe that, on the average, the tensile strength can be improved by 10 psi over the tensile strength using the standard process. In this context, samples of wire from the new process are tested, and their mean tensile strength is compared to the known mean using the standard process, μ_0. Here the null hypothesis is $H_0: \mu = \mu_0$ and the alternative hypothesis is $H_a: \mu = \mu_0 + 10$, which is the alternative proposed by the researcher.

Type I and Type II Errors   Because the true parameter value is unknown and the decision to "reject null" or "do not reject null" is made based on the sample statistic, we see that there are two ways that errors can be made in the decision process:

- Reject the null hypothesis when it is true, called a *type I error*.
- Fail to reject the null hypothesis when the alternative is true, called a *type II error*.

Type I Error
Rejecting the null hypothesis when it is true.

Type II Error
Failing to reject the null hypothesis when it is false.

Table 10.1

Decisions and Errors in Hypothesis Testing

		Decision Based on the Data	
		Reject H_0	Do Not Reject H_0
Truth	H_0 is True	Type I error	Correct decision
(Unknown)	H_0 is False	Correct decision	Type II error

The possible decision and errors are displayed in Table 10.1.

We denote the probability of a type I error by α, and the probability of a type II error by β. The probability of type I error is also referred to as the *level of significance* or *the size of the rejection region*.

For a one-sided test $H_0: \mu \leq \mu_0$ versus $H_a: \mu > \mu_0$, we reject H_0 if $\overline{X} \geq CV$. Then

$$\alpha = P(\text{type I error}) = P(\text{reject } H_0 \text{ when } \mu = \mu_0)$$

$$= P(\overline{X} > CV \text{ when } \mu = \mu_0)$$

and

$$\beta = P(\text{type II error}) = P(\text{Do not reject } H_0 \text{ when } \mu = \mu_a)$$

$$= P(\overline{X} > CV \text{ when } \mu = \mu_a)$$

Level of Significance

$$\alpha = P(\text{type I error}) = P(\text{reject } H_0 \text{ when } \mu = \mu_0)$$

Probability of Type II error

$$\beta = P(\text{type II error}) = P(\text{Do not reject } H_0 \text{ when } \mu = \mu_a)$$

For a large sample test of a mean (i.e., when sampling distribution of $\overline{X}$ is approximately normal), we can show α and β as in Figure 10.6. In this figure, if μ_0 is the true value of μ under H_0, and μ_a is a specific alternative value of interest, then

- The area to the right of K under the normal curve centered at μ_0 is α. Note that α represents the chance that the null hypothesized value μ_0 will be rejected when in fact μ_0 is the true value of μ.

Figure 10.6

α and β for a statistical test

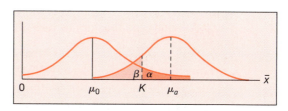

- The area to the left of K under the normal curve centered at μ_a is β. Note that β represents the chance that we will fail to reject the null hypothesized, when in fact μ_a is the true value of μ.

We can see from Figure 10.6 that for a fixed sample size, we cannot lower the probability of both types of errors. If we lower α, then β increases and vice versa. The value of K can be determined by specifying a value for α and using properties of the normal distribution. Using standard deviation as a measure of distance (as seen in Chapter 1 and again in Chapter 6), we can describe how far any point is from the mean of the distribution. Therefore, under H_0 we can express point K as a point that is z_α standard deviations above the mean, that is, $K = \mu_0 + z_\alpha(\sigma/\sqrt{n})$.

Example 10.5

The type I and type II errors for the situation in Example 10.1 can be described as follows:

- Concluding based on the data that the mean depth of holes made by the drill press is not acceptable (too large or too small), when in reality it is 2 inches. This will result in a type I error, which will lead to not using a perfectly good drill press.
- Concluding based on the data that the mean depth of holes made by the drill press is acceptable, when in reality it is too large or too small compared to the acceptable value of 2 inches. This will result in a type II error, which will lead to using a faulty drill to make holes.

Example 10.6

The type I and type II errors for the situation in Example 10.2 can be described as follows:

- Concluding based on the data that the process is producing more than a desirable level of percent defectives, when in reality the process is performing at a desirable level. This will result in a type I error, which will lead to making unnecessary adjustments to a good process.
- Concluding based on the data that the process is performing at the desirable level, when in reality it is not and needs immediate attention. This will result in a type II error, which will lead to failure in making needed process adjustments that could prevent the process from producing an undesirable level of defectives.

Example 10.7

The type I and type II errors for the situation in Example 10.3 can be described as follows:

- Concluding based on the data that the adjustments made to the cutting process are effective in reducing variation, when in reality the process is still performing at the preadjustment level. This will result in a type I error, which will lead to not realizing that the adjusted process is not effective in reducing variation and should be adjusted further.

- Concluding based on the data that the adjustments made to the cutting process have failed, when in reality the new process is functioning with a better precision. This will result in a type II error, which will lead to making further unnecessary adjustments to an already improved process.

Exercises

10.1 The output voltage for a certain electric circuit is specified to be 130. An electrical engineer claims that in his experience these circuits produce less than 130 volts. He decides to take a random sample of independent readings to support his claim. Formulate null and alternative hypotheses to be tested.

10.2 Refer to the situation in Exercise 10.1. Describe type I and type II errors that the engineer could possibly make.

10.3 The Rockwell hardness index for steel is determined by pressing a diamond point into the steel and measuring the depth of penetration. The manufacturer claims that this steel has an average hardness index of at least 64. A competitor decides to check out this claim. Formulate null and alternative hypotheses that the competitor should check in order to test the manufacturer's claim.

10.4 Refer to Exercise 10.3.
   a What decision by the competitor could lead to a type I error?
   b What decision by the competitor could lead to a type II error?

10.5 The hardness of a certain rubber (in degrees Shore) is claimed to be 65 by the manufacturer. One of their customers suspects that the rubber produced by this manufacturer is not as hard as claimed. What null and alternative hypotheses should the customer use to test the manufacturer's claim?

10.6 Refer to Exercise 10.5. Identify type I and type II errors that the customer could possibly make.

10.7 Certain rockets are manufactured with a range of 2,500 meters. It is theorized that the range will be reduced after the rockets are in storage for some time. Develop null and alternative hypothesis to support the theory by rocket scientists of a shorter range of rockets after storage.

10.8 Refer to Exercise 10.7. Describe type I and type II errors that the rocket scientists could possibly make.

10.9 The stress resistance of a certain plastic is specified to be 30 psi. A toy manufacturer uses this plastic. Recently, the toy manufacturer has received several complaints about the quality of toys and suspects lower than claimed stress resistance of the plastic. Formulate null and alternative hypothesis that the toy manufacturer should test in order to confirm his suspicions.

10.10 Refer to Exercise 10.9.
   a Describe type I and type II hypotheses *errors* that the toy manufacturer could possibly make.
   b Describe consequences of making such errors.

10.11 Steel rods with 10 mm diameters are used in the construction of bridges. The steel rods are used only if their yield stress exceeds 490. The state government is reviewing the product of one manufacturer to possibly use in the construction. The product will be tested before issuing a contract. What null and alternative hypotheses should the state government test to determine whether the steel rods meet the requirement?

10.12 Refer to the situation in Exercise 10.11.
   a Describe type I and type II hypotheses that the state government could possibly make.
   b Describe consequences of making such errors.

10.2 # Hypothesis Testing: The Single-Sample Case

In this section, we present tests of hypothesis concerning the mean of a population, the variance of the population, and the probability of success. Some of these tests require large samples and are approximate in nature, whereas others are exact tests for any sample size.

10.2.1   Testing for mean: General distributions case

One of the most common and, at the same time, easiest hypothesis-testing situations to consider is that of testing a population mean when a large random sample of size n from the population of question is available for observation. Denoting the population mean by μ, and the sample mean by $\overline{X}$, we know from Chapter 9 that $\overline{X}$ is a good estimator for μ. Therefore, it seems intuitively reasonable that our decision about μ should be based on $\overline{X}$, with extra information provided by the population variance σ^2 or its estimator S^2.

Suppose it is claimed that μ takes on the specific value μ_0, and we want to test the validity of this claim. How can we tie this testing notion into the notion of a confidence interval established in Chapter 9? Upon reflection, it may seem reasonable that if the resulting confidence interval for μ does *not* contain μ_0, then we should *reject* the hypothesis that $\mu = \mu_0$. On the other hand, if μ_0 is within the confidence limits, then we cannot reject it as a plausible value for μ. Now if μ_0 is not within the interval $\overline{X} \pm z_{\alpha/2}\sigma/\sqrt{n}$, then either

$$\mu_0 < \overline{X} - z_{\alpha/2}\frac{\sigma}{\sqrt{n}} \qquad \text{or} \qquad \mu_0 > \overline{X} + z_{\alpha/2}\frac{\sigma}{\sqrt{n}}$$

i.e.,
$$\frac{\overline{X} - \mu_0}{\sigma/\sqrt{n}} > z_{\alpha/2} \qquad \text{or} \qquad \frac{\overline{X} - \mu_0}{\sigma/\sqrt{n}} < -z_{\alpha/2}$$

Assuming a large sample, so that $\overline{X}$ has approximately a normal distribution, we see that the hypothesized value μ_0 is rejected if

$$|Z| = \left|\frac{\overline{X} - \mu_0}{\sigma/\sqrt{n}}\right| > z_{\alpha/2}$$

for some prescribed α, where Z has approximately a standard normal distribution. In other words, we reject μ_0 as a plausible value for μ if $\overline{X}$ is too many standard deviations (more than $z_{\alpha/2}$) from μ_0. Whereas $(1 - \alpha)$ was called the confidence coefficient in estimation problems, α is referred to as the *significance level* in hypothesis-testing problems.

Example 10.8 ───

The depth setting on a certain drill press is 2 inches. One could then hypothesize that the average depth of all holes drilled by this machine is $\mu = 2$ inches. To check this hypothesis and the accuracy of the depth gauge, a random sample of 100 holes drilled by this machine was measured and found to have a sample mean of 2.005 inches with a standard deviation of 0.03 inch. With $\alpha = 0.05$, can the hypothesis be rejected based on these sample data?

Solution   Here we are interested in testing a hypothesis about $\mu =$ the mean depth of the population of holes drilled by this machine. With $n = 100$, a sufficiently large sample to apply the central limit theorem, it is assumed that $\overline{X}$ is approximately normal in distribution. Thus, the hypothesis $\mu = 2$ is rejected if

$$|Z| = \left| \frac{\bar{X} - \mu_0}{\sigma/\sqrt{n}} \right| > z_{\alpha/2} = 1.96$$

For large n, the sample standard deviation S can be substituted for the unknown population standard deviation σ, and the approximate normality still holds. Thus, the observed test statistic becomes

$$\frac{\bar{x} - \mu_0}{s/\sqrt{n}} = \frac{2.005 - 2.000}{0.03/\sqrt{100}} = 1.67$$

This is actually a t-statistic, but it can be approximated by z for a large n. Because the observed value of $\sqrt{n}(\bar{x} - \mu_0)/s$ is less than 1.96, we cannot reject the hypothesis that 2 is a plausible value of μ. The 95% confidence interval for μ is (1.991, 2.011).

> Note that *not rejecting* the hypothesis that $\mu = 2$ is not the same as *accepting* the hypothesis that $\mu = 2$. When we do not reject the hypothesis $\mu = 2$, we are saying that 2 is a plausible value of μ, but there are other equally plausible values for μ. We cannot conclude that μ is equal to 2 and 2 alone.

Sometimes, a null hypothesis is enlarged to include an interval of possible values. For example, a can for 12 ounces of cola is designed to hold only slightly more than 12 ounces, so it is important that overfilling not be serious. However, underfilling of the cans could have serious implications. Suppose μ represents the mean ounces of cola per can dispensed by a filling machine. Under a standard operating procedure that meets specifications, μ must be 12 or more, so we set H_0 as $\mu \geq 12$. An important change occurs if μ drops below 12, so H_a becomes $\mu < 12$. Under H_0, the vendor is in compliance with the law, and customers are satisfied. Under H_a, the vendor is violating a law, and customers are unhappy. The vendor wants to sample the process regularly to be reasonably sure that a shift from H_0 to H_a has not taken place.

When H_0 is an interval such as $\mu \geq 12$, the equality as $\mu = 12$ still plays a key role. Sample data will be collected, and a value for $\bar{x}$ will be calculated. If $\bar{x}$ is much less than 12, H_0 will be rejected. If the absolute difference between $\bar{x}$ and 12 is large enough to reject H_0, then the differences between $\bar{x}$ and 12.5 or 12.9 will be even larger and will again lead to rejection of H_0. Thus, we need to consider only the boundary point between H_0 and H_a (12 in this case) when calculating an observed value of a test statistic.

For testing $H_0: \mu \geq \mu_0$ versus $H_a: \mu < \mu_0$, we consider only the analogous one-sided confidence interval for μ. Here it is important to establish an upper confidence limit for μ, because we want to know how large μ is likely to be. If μ_0 is larger than the upper confidence limit for μ, then we will reject $H_0: \mu \geq \mu_0$. The one-sided upper limit for μ is $\bar{X} + z_\alpha \sigma / \sqrt{n}$ (see Section 9.2). We reject H_0 in favor of H_a when $\bar{X} + z_\alpha \sigma / \sqrt{n} < \mu_0$ or, equivalently, when

$$\frac{\bar{X} - \mu_0}{\sigma / \sqrt{n}} < -z_\alpha$$

Notice that we will have the same test statistic but the rejection region has changed somewhat.

The corresponding rejection region for testing $H_0: \mu \leq \mu_0$ versus $H_a: \mu > \mu_0$ is given by

$$\frac{\bar{X} - \mu_0}{\sigma / \sqrt{n}} > z_\alpha$$

Example 10.9

A vice-president for a large corporation claims that the average number of service calls on equipment sold by that corporation is no more than 15 per week. To investigate her claim, service records were checked for $n = 36$ randomly selected weeks, with the result $\bar{x} = 17$ and $s^2 = 9$ for the sample data. Does the sample evidence contradict the vice-president's claim at the 5% significance level?

Solution The parameter of interest here is $\mu =$ the mean number of service calls on the equipment sold by the corporation. The vice-president states that, according to her specifications, the mean must be 15 or less. Thus, we are testing

$$H_0: \mu \leq 15 \qquad \text{versus} \qquad H_a: \mu > 15$$

Using as the test statistic

$$Z = \frac{\bar{X} - \mu_0}{\sigma / \sqrt{n}}$$

we reject H_0 for $z > z_{0.05} = 1.645$. Substituting s for σ, we calculate the observed statistic to be

$$\frac{\bar{x} - \mu_0}{s / \sqrt{n}} = \frac{17 - 15}{3 / \sqrt{36}} = 4$$

That is, the observed sample mean is four standard deviations greater than the hypothesized value of $\mu_0 = 15$. With $\alpha = 0.05$, $z_\alpha = 1.645$. Because the test statistic exceeds z_α (see Figure 10.7), we have sufficient evidence to reject the null hypothesis. It does appear that the mean number of service calls exceeds 15.

At this point, the question raised is: What is the likelihood of observing a sample mean of 17 or higher if the vice president's claim is correct?

Figure 10.7

Rejection region for Example 10.9

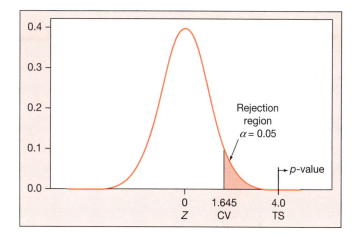

Determining _p_-value The hypothesis-testing procedure outlined above involves specifying a significance level α, finding $z_{\alpha/2}$ (or z_α for one-sided test), calculating a value for the test statistic Z, and rejecting H_0 if $|z| > z_\alpha$ (or the appropriate one-sided adjustment). In this process, α is somewhat arbitrarily determined taking into account the consequences of making such an error. For example, if the stakes are high, a low value of α is selected, such as 0.01 or even 0.005. If the stakes are low, then it is possible to relax and set a value for α, such as 0.10. However, 0.05 is a very commonly used value of α. An alternative to specifying α is to find the _smallest_ significance level at which the observed result would lead to rejection of H_0. This value is called the _p-value_ of the test, or the _attained significance level_. We would then reject H_0 for small _p_-values.

> ### Observed Significance Level or _p_-value
>
> The probability of observing a test statistic value at least as extreme as the one computed from the sample data if the null hypothesis is true.

Example 10.10

Find the _p_-value for situation in Example 10.9.

Solution In Example 10.9, the data provided $z = 4$ as the observed value of the test statistics for testing H_0: $\mu \leq 15$ versus H_a: $\mu > 15$. The _smallest_ significance level that would lead to rejection of H_0 in this case must be such that $z = 4$, that is the area under the standard normal curve to the right of 4 (see Figure 10.7). Thus the _p_-value for this test is 0.0000317, and we would reject H_0 for any significance level greater than or equal to this _p_-value. Note that we did reject H_0 at the 0.05 significance level in Example 10.9 and that _p_-value = 0.0000317 is less than 0.05.

This _p_-value of 0.0000317 indicates that the likelihood of observing a sample mean of 17 or larger when the true mean number of service calls is no more than 15 per week is extremely low.

The p-value is calculated under the assumption that H_0 is true. It can be thought of as the weight of evidence in the sample data regarding whether to reject the null hypothesis. If the p-value is small (close to zero), there is strong evidence against H_0, and it should be rejected. If the p-value is large (say, larger than 0.05), there is weak evidence for rejecting H_0.

Example 10.11

Find the p-value for the situation in Example 10.8.

Solution   In Example 10.8, the data provided TS $= z = 1.67$ as the observed value of the test statistics for testing H_0: $\mu = 2$ versus H_a: $\mu \neq 2$ in. The *smallest* significance level that would lead to rejection of H_0 in this case must be such that $z = 1.67$. Because this is a two-sided test, the area under the standard normal curve to the right of 1.67 is half of the p-value, which implies that the p-value $= 2(0.0475) = 0.095$ (see Figure 10.8). We would reject H_0 for a significance level greater than or equal to this p-value. Note that we did *not* reject H_0 at the 0.05 significance level in Example 10.8 and that 0.05 is less than p-value $= 0.095$.

Figure 10.8

The p-value for Example 10.8

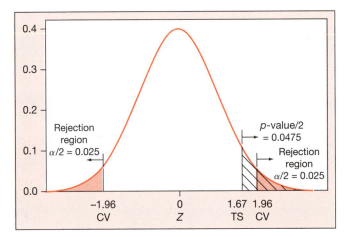

The probability of getting a sample of holes created with this drill with a mean diameter that is at least 1.67 standard deviations away from the depth setting of 2 inches is 0.095, not such a rare event.

In general, use the following four steps to solve any hypothesis testing problem.

1. Identify null and alternative hypothesis to be tested defining parameter(s) of interest.
2. Identify appropriate testing procedure to be used and check underlying conditions.
3. Compute the test statistic value and the p-value (or determine rejection region).
4. Use either the rejection region approach or a p-value approach to make a decision using values computed in step 3, and write the conclusion in the context of the problem.

Example 10.12

Suppose a manufacturer of one type of transistors believes that the mean lifetime (in 100 hours) of transistors produced exceeds 2.0. A random sample of 50 transistors selected from the production recorded lifetimes listed in Table 6.1. Do the sample data provide sufficient evidence to support the manufacturer's belief? Justify your answer using a 5% level of significance.

Solution Here we are interested in testing a hypothesis about the population mean,

$$\mu = \text{the mean lifetime of transistors produced by the manufacturer}$$

The hypotheses of interest are $H_0: \mu \leq 2.0$ and $H_a: \mu > 2.0$. Because a large sample was selected ($n = 50$), we'll use an approximate Z-test for the mean. A Minitab and a JMP output are provided below.

Minitab Printout	JMP Printout
One-Sample Z: Lifetime	Lifetime
Test of mu = 2 vs mu = 2	Test Mean = value
The assumed sigma = 1.932	

Minitab Printout:

One-Sample Z: Lifetime

Test of mu = 2 vs mu = 2
The assumed sigma = 1.932

Variable	N	Mean	StDev	SE Mean
lifetime	50	2.268	1.932	0.273

Variable	95.0% Lower Bound	Z	P
lifetime	1.819	0.98	0.163

JMP Printout:

Lifetime

Test Mean = value

Hypothesized Value	2
Actual Estimate	2.26794
df	49
Std Dev	1.93244
Sigma given	1.932

z Test	
Test Statistic	0.9807
Prob > |z|	0.3268
Prob > z	0.1634
Prob < z	0.8366

From the printout, we see that the sample mean $\bar{x} = 2.268$. Now the question is this: Is the sample mean 2.268 significantly larger than the hypothesized population mean 2.0, or is the difference due to chance variation? Because p-value $= 0.163 > \alpha = 0.05$, we fail to reject the null hypothesis and conclude that the sample data do not support the manufacturer's belief.

Note that the p-value $= P(Z > 0.98) = 0.163$, that is, the probability of observing a sample mean of 2.268 or larger when the true mean lifetime of transistors is 2.0 is equal to 0.163. This p-value indicates that if $\mu \leq 2.0$, then $\bar{X} \geq 2.268$ is not a rare event. When we set $\alpha = 0.05$, we are willing to take a 5% risk of rejecting a true null hypothesis. The p-value $= 0.163$, tells us that if we reject the null hypothesis based on these data, then there is a 16.3% chance that we rejected a true null hypothesis. This chance is much higher than the risk we are willing to take. Therefore, based on these data, we should not reject the null hypothesis.

Also, as seen in Chapter 6, note that the distribution of lifetimes of transistors is very skewed. However, for sample size 50, we get approximate normal distribution for the sample mean, using the central limit theorem.

As discussed earlier, both type I and type II errors cannot be controlled for a fixed-sample-size procedure. Therefore hypothesis test is developed by fixing type I error rate (α). Then the question of interest is: What is the probability of a type II error for a specific alternative?

Example 10.13

In a power-generating plant, pressure in a certain line is supposed to maintain an average of 100 psi over any 4-hour period. If the average pressure exceeds 103 psi for a 4-hour period, serious complications can evolve. During a given 4-hour period, $n = 30$ measurements (assumed random) are to be taken. For testing $H_0: \mu \leq 100$ versus $H_a: \mu = 103$, α is to be 0.01. If $\sigma = 4$ psi for these measurements, calculate the probability of a type II error.

Solution   For testing $H_0: \mu \leq 100$ versus $H_a: \mu = 103$, we reject H_0 if

$$\frac{\overline{X} - \mu_0}{\sigma/\sqrt{n}} > z_{0.01} = 2.33$$

or if

$$\overline{X} = \mu_0 + z_{0.01}\frac{\sigma}{\sqrt{n}} = 100 + 2.33\left(\frac{4}{\sqrt{30}}\right) = 101.7$$

Now if the true mean is really 103, then

$$\beta = P(\overline{X} < 101.7) = P\left(\frac{\overline{X} - \mu_0}{\sigma/\sqrt{n}} < \frac{101.7 - 103}{4/\sqrt{30}}\right)$$

$$= P(Z < -1.78) = 0.0375$$

Under these conditions, the chance of observing a sample mean that is not in the rejection region when the true average pressure is 103 psi is quite small (less than 4%). Thus, the operation can confidently continue if the sample mean of the tests $\bar{x}$ is less than 101.7.

In many practical problems, a specific value for an alternative will not be known, and consequently β cannot be calculated. In those cases, we want to choose an appropriate significance level α and a test statistic that will make β as small as possible. Fortunately, most of the test statistics we use in this text have the property that, for fixed α and fixed sample size, β is nearly as small as possible for any alternative parameter value. In that sense, we are using the best possible test statistic.

Furthermore, we set up our hypothesis so that if the test statistic falls into the rejection region, we reject H_0, knowing that the risk of a type I error is fixed at α. However, if the test statistic is not in the rejection region for H_0, we hedge by stating that the evidence is insufficient to reject H_0. We *do not* affirmatively accept H_0.

The term $(1 - \beta)$ is also referred to as the *power* of the test. The power of the test is defined as the probability of rejecting the null hypothesis when an alternative hypothesis is true. It is often interpreted as the probability of correctly rejecting the false null hypothesis.

The Power of a Statistical Test

Probability of rejecting the null hypothesis when an alternative hypothesis is true.

Because the power of the test is a probability, it takes a value between 0 and 1. Ideally, we would like any statistical test to have power 1, which means never having to make a type II error. However, in view of the fact that the decision is made between null and alternative hypotheses based only on the information about part of the population (i.e., the sample), we cannot totally avoid the possibility of making a type II error. A test with power close to 1 is considered a very good test because it is highly likely to reject the null when it should.

Power measures the ability of a statistical test to detect differences. Larger differences are easier to detect and hence a larger power. Therefore, it is also used to compare different statistical tests and to determine the sample size required to achieve certain predetermined power.

Refer to the situation described in Example 10.13. This test has a

$$\text{Power} = 1 - \beta = 1 - 0.0375 = 0.9625$$

This means that if the true mean pressure is 103 psi, this test will correctly reject the null hypothesis of $\mu \leq 100$, detecting the difference of 3 psi about 96.25% of the time.

The power of a test often increases by increasing type I error rate, decreasing variation, and/or increasing the sample size.

Determining Sample Size   For specified μ_0 and μ_a, we can see from Figure 10.6 that if α is decreased, K will move to the right and β will increase. On the other hand, if α is increased, K will move to the left and β will decrease. However, if the sample size is increased, the sampling distributions will display less variability (they will become narrower and more sharply peaked), and *both* α and β can decrease. We will now show that it is possible to select a sample size to guarantee specified α and β, as long as μ_0 and μ_a are also specified.

For testing H_0: $\mu = \mu_0$ against H_a: $\mu = \mu_a$, where $\mu_a > \mu_0$, we have seen that H_0 is rejected when $\overline{X} > K$, where $K = \mu_0 + z_\alpha(\sigma/\sqrt{n})$. But from Figure 10.6 it is clear that we also have $K = \mu_a + z_\beta(\sigma/\sqrt{n})$. Thus,

$$\mu_0 + z_\alpha(\sigma/\sqrt{n}) = \mu_a - z_\beta(\sigma/\sqrt{n})$$
$$(z_\alpha + z_\beta)(\sigma/\sqrt{n}) = \mu_a - \mu_0$$

or

$$\sqrt{n} = \frac{(z_\alpha + z_\beta)\sigma}{\mu_a - \mu_0}$$

This allows us to find a value of n that will produce a specified α and β for known μ_0 and μ_a.

The sample size for specified α and β when testing $H_0: \mu = \mu_0$ versus $H_a: \mu = \mu_a$ is given by

$$n = \frac{(z_\alpha + z_\beta)^2 \sigma^2}{(\mu_a - \mu_0)^2}$$

Example 10.14

For the power plant of Example 10.13, it is desired to test $H_0: \mu = 100$ versus $H_a: \mu = 103$ with $\alpha = 0.01$ and $\beta = 0.01$. How many observations are needed to ensure this result?

Solution   Because, $\alpha = \beta = 0.01$, $z_\alpha = z_\beta = 2.33$. Also, $\sigma = 4$ for these measurements. We then have

$$n = \frac{(z_\alpha + z_\beta)^2 s^2}{(\mu_\alpha - \mu_0)^2} = \frac{(2.33 + 2.33)^2 (4)^2}{(103 - 100)^2} = 38.6$$

By taking 39 measurements, we can reduce β to 0.01 while also holding α at 0.01.

Testing a Mean: General Distributions Case   (Approximate Z-Test)

$H_0: \mu = \mu_0$

$$TS: Z = \frac{\overline{X} - \mu_0}{\sigma/\sqrt{n}}$$

If unknown, estimate σ using s.

Reject H_0 if p-value $< \alpha$.

Right-tailed test:

$H_a: \mu > \mu_0$
RR: $z > z_\alpha$
p-value: $P(Z > z)$

Left-tailed test:

$H_a: \mu < \mu_0$
RR: $z < -z_\alpha$
p-value: $P(Z < z)$

Two-tailed test:

$H_a: \mu \neq \mu_0$
RR: $z > z_{\alpha/2}$ or $z < -z_{\alpha/2}$
p-value: $2P(Z > |z|)$

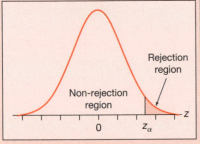

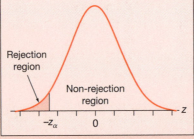

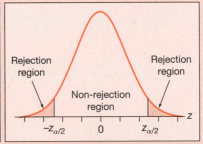

Conditions:

1. A random sample is taken from the population.
2. The sample size is large enough to apply the central limit theorem.

Checking the sample size: There is no fixed value that determines whether a sample size is large or small. To determine how large a sample is needed for the sampling distribution of the sample mean to approximate the normal distribution depends on the shape of the sampled population. If there are no outliers, and the population distribution is not extremely skewed, then we can safely say that $n \geq 40$ is large enough to get approximately normal sampling distribution of $\overline{X}$. Proceed with caution for sample sizes 15–20; use transformation if necessary. Refer to Section 9.2 for more discussion on determining how large a sample is large enough to use normal approximation.

10.2.2 Testing a mean: Normal distribution case

When sample sizes are too small for the central limit theorem to provide a good approximation to the distribution of $\overline{X}$, additional assumptions must be made on the nature of the probabilistic model for the population. Just as in the case of estimation, one common procedure is to assume, whenever it can be justified, that the population measurements fit the normal distribution reasonably well. Then $\overline{X}$, the sample mean for a random sample of size n, will have a normal distribution, and

$$\sqrt{n}(\overline{X} - \mu)/S$$

will have a t-distribution with $(n - 1)$ degrees of freedom. Thus, the hypothesis testing procedure for the population mean will have the same pattern as above, but the test statistic,

$$T = \frac{\overline{X} - \mu}{S/\sqrt{n}}$$

will have a t-distribution, rather than a normal distribution, under $H_0: \mu = \mu_0$.

Testing a Mean: Normal Distribution Case   **(t-Test)**

$H_0: \mu = \mu_0$ TS: $T = \dfrac{\overline{X} - \mu_0}{S/\sqrt{n}}$ $\nu = (n - 1)$ degrees of freedom

Reject H_0 if p-value $< \alpha$

Right-tailed test:	Left-tailed test:	Two-tailed test:
$H_a: \mu > \mu_0$	$H_a: \mu < \mu_0$	$H_a: \mu \neq \mu_0$
RR: $t > t_\alpha(\nu)$	RR: $t < -t_\alpha(\nu)$	RR: $t > t_{\alpha/2}(\nu)$ or $t < -t_{\alpha/2}(\nu)$
p-value: $P(t(\nu) > t)$	p-value: $P(t(\nu) < t)$	p-value: $2P(t(\nu) > |t|)$

Conditions:
1. A random sample is taken from the population.
2. Sampled population is normally distributed, at least approximately.

Checking the normality condition: If the sample size is reasonably large, a dot plot or a histogram can be employed to check whether the shape of the distribution is fairly symmetric, and bell-shaped without any outliers. Also a Q–Q plot or a normal probability plot can be used to check normality of population if sample size is sufficiently large. However, similar to the t-interval for a mean, the t-test for a mean is also very robust to the condition of normality, which means slight variations from normality will not affect the outcome of the test.

Example 10.15

A corporation sets its annual budget for a new plant on the assumption that the average weekly cost for repairs is to be $\mu = 1{,}200$. To see whether this claim is realistic, 10 weekly repair cost figures are obtained from similar plants. The sample is assumed to be random and yields $\bar{x} = 1{,}290$ and $s = 110$. Does this sample indicate that \$1,200 is not a good assumed value for the mean weekly cost of repairs? Use $\alpha = 0.05$. Assume normality of weekly repair costs.

Solution Because the detection of a departure from the assumed average in either direction would be important for budgeting purposes, it is desired to test H_0: $\mu = 1{,}200$ versus H_a: $\mu \neq 1{,}200$, where $\mu =$ the mean weekly repair cost. The observed value of test statistic is

$$t = \frac{\bar{x} - m_0}{s/\sqrt{n}} = \frac{1{,}290 - 1{,}200}{110/\sqrt{10}} = 2.587$$

with $n - 1 = 10 - 1 = 9$ degrees of freedom. From Table 5 in the Appendix, we see that for 9 degrees of freedom,

$$1.833 < t = 2.587 < 2.262$$

Therefore, using the symmetric nature of the t-distribution and the fact that this is a two-tailed test,

$$2(0.010) < p\text{-value} = P(t(9) < -2.587) + P(t(9) > 2.587) < 2(0.025)$$

That is, $0.02 < p\text{-value} < 0.05$. It means that if we reject the null hypothesis, then there is a 2% to 5% chance that we made a wrong decision. Because we are willing to take a 5% chance of rejecting the true null hypothesis and p-value $< \alpha = 0.05$, we reject the null hypothesis. There is reason to suspect that 1,200 is not a good assumed value for the mean weekly repair cost, and perhaps more investigation into these repair costs should be made before the budget is set.

Alternatively using rejection region approach at $\alpha = 0.05$ and $n = 10$, we get $t_{\alpha/2} = t_{0.025} = 2.262$ for $n - 1 = 9$ degrees of freedom. We reject the null hypothesis because $t = 2.587$ is greater than the critical value of 2.262. Therefore, with the rejection region approach we arrive at the same conclusion as using the p-value approach (see Figure 10.9).

Figure 10.9

Rejection region and p-value for Example 10.15

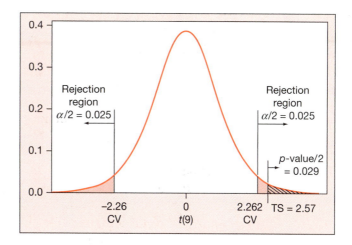

Using technology, it is easy to determine a p-value for a test. For example, using a calculator, we get that the p-value = 0.029, which is less than $\alpha = 0.05$. A Minitab printout is given below for example 10.15. It shows that the p-value = 0.029. Additionally, it also gives a two-sided 95% confidence interval, (1211.3, 1368.7). Because this interval does not enclose the hypothesized value of $\mu_0 = 1200$, we can draw the same conclusion from the confidence interval approach that at 5% level of significance, data indicate the mean weekly repair cost is not $1,200. In fact, $\mu_0 < (1211.3, 1368.7)$ indicates with 95% certainty that the mean weekly repair cost is more than the budgeted amount of $1,200.

One-Sample T

```
Test of mu = 1200 vs not = 1200

 N   Mean StDev SE Mean      95% CI         T     P
10 1290.0 110.0    34.8 (1211.3, 1368.7)  2.59 0.029
```

Example 10.16

Muzzle velocities of eight shells tested with a new gunpowder yield a sample mean of 2,959 feet per second and a standard deviation of 39.4. The manufacturer claims that the new gunpowder produces an average velocity of no less than 3,000 feet per second. Does the sample provide enough evidence to contradict the manufacturer's claim? Use $\alpha = 0.05$.

Solution Here we are interested in testing $H_0: \mu = 3,000$ versus $H_a: \mu < 3,000$ because we want to see whether there is evidence to refute the manufacturer's claim. The parameter of interest is $\mu =$ the mean velocity of the new gunpowder. Assuming that muzzle velocities can be reasonably modeled by a normal probability distribution, we can use a t-test in this situation. The rejection region is $t < -t_{0.05} = -1.895$ for 7 degrees of freedom (see Figure 10.10). The observed test statistic is

$$t = \frac{\bar{x} - m_0}{s/\sqrt{n}} = \frac{2,959 - 3,000}{39.4/\sqrt{8}} = -2.943$$

Figure 10.10

Rejection region and
p-value for Example 10.16

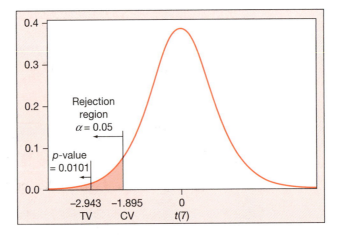

Because $t = -2.943 < -t_a$, we reject the null hypothesis and say that there appears to be good reason to doubt the manufacturer's claim.

It is easy to find the *p*-value using a calculator or a computer program. The TI output below shows that the *p*-value $= P(t(7) < -2.943) = 0.0101$.

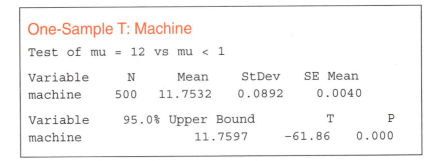

Because the *p*-value $< \alpha$, we reject H_0 and arrive at the same conclusion. This *p*-value tells us that if we reject the null hypothesis based on these data, then there is about 1.01% chance that we made a wrong decision. Because we are willing to take up to a 5% risk of rejecting a true null hypothesis, based on this *p*-value it seems reasonable to reject the null hypothesis.

Example 10.17

Five cans filled by a filling machine (set to fill 12 oz) were selected at random, and the amounts filled in each were measured. The computer output for the collected data is given below:

One-Sample T: Machine

Test of mu = 12 vs mu < 1

Variable	N	Mean	StDev	SE Mean
machine	500	11.7532	0.0892	0.0040

Variable	95.0% Upper Bound	T	P
machine	11.7597	−61.86	0.000

a Does this sample give sufficient evidence to conclude that, on the average, the machine is underfilling 12-ounce cans and should be stopped for adjustments?

b If the manager of the bottling plant decides to stop the machine for adjustments, what is the likelihood that such an interruption in the filling process was unnecessary?

Solution **a** Here we are interested in μ = the mean amount per can filled by the machine. The hypotheses of interest are

H_0: μ = 12 ounces, that is, the filling process is working fine

H_a: μ < 12 ounces, that is, the filling process should be stopped for adjustments.

The Minitab output shows that p-value is almost zero ($p < 0.001$). At such a small p-value, we can reject the null hypothesis and conclude that the filling process should be stopped to make adjustments to the machine setting.

b We are interested in determining the likelihood that such an interruption was unnecessary, that is, the probability of observing such a small sample mean or even smaller when in reality the average amount filled per can is 12 ounces. As we know, this probability is given by the p-value, which is almost 0.

10.2.3   Testing for proportion: Large sample case

Just as in case of interval estimation, the large-sample test for a mean can be easily transformed into a test for a binomial proportion. Recall that if Y has a binomial distribution with mean np, then Y/n is approximately normal for a large n, with mean p and variance $p(1 - p)/n$.

Testing for Proportion (or the Probability of Success) p: Large Sample Case   (Approximate Z-Test)

H_0: $p = p_0$

$$TS: Z = \frac{\hat{p} - p_0}{\sqrt{\dfrac{p_0(1 - p_0)}{n}}}$$

Reject H_0 if p-value $< \alpha$

Right-tailed test:

H_a: $p > p_0$

RR: $z > z_\alpha$

p-value: $P(Z > z)$

Left-tailed test:

H_a: $p < p_0$

RR: $z < -z_\alpha$

p-value: $P(Z < z)$

Two-tailed test:

H_a: $p \neq p_0$

RR: $z > z_{\alpha/2}$ or $z < -z_{\alpha/2}$

p-value: $2P(Z > |z|)$

Conditions:

1. A random sample of size n is taken from the population.
2. The sample size is large enough so that the distribution of $\hat{p}$ is approximately normal.

Checking the normality condition: If $n\hat{p} \geq 10$ and $n(1 - \hat{p}) \geq 10$, then the sample size is large enough to assume that the distribution of $\hat{p}$ is approximately normal.

Example 10.18

A machine in a certain factory must be repaired if it produces more than 10% defectives among the large lot of items it produces in a day. A random sample of 100 items from the day's production contains 15 defectives, and the foreman says that the machine must be repaired. Does the sample evidence support his decision at the 0.01 significance level?

Solution   We want to test $H_0: p \leq 0.10$ (i.e., the machine does not need any repairs) versus the alternative $H_a: p > 0.10$ (i.e., the machine needs to be repaired), where p denotes the proportion of defectives in the population (i.e., all items produced by this machine). The test statistic will be based on Y/n, where Y denotes the number of defectives observed in the sample. We will reject H_0 in favor of H_a, if Y/n is suitably large. From the sample, we can estimate p as $\hat{p} = y/n = 0.15$. Now we know that $\hat{p} = 0.15 > 0.10$, but because we are looking at the sample and not the population, the question that arises is: Is the sample proportion 0.15 significantly higher than hypothesized proportion 0.10, *or* could the difference be due to chance variation, which is part of the sampling process?

Our observed value of test statistic is

$$z = \frac{\hat{p} - p_0}{\sqrt{p_0(1 - p_0)/n}} = \frac{0.15 - 0.10}{\sqrt{0.10(0.90)/100}} = 1.67$$

Figure 10.11

Rejection region and *p*-value for Example 10.18

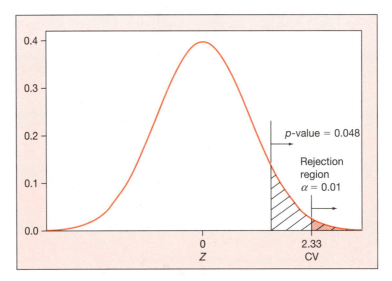

which shows that the observed proportion is 1.67 standard deviations above the hypothesized proportion. Now $z_{0.01} = 2.33$, that is, any proportion more than 2.33 standard deviations above the hypothesized proportion will be considered too large for sampling variation. Because $1.67 < 2.33$, we will not reject the null hypothesis (see Figure 10.11). It is quite likely to get 15% defectives in a sample of size 100 from a process with an overall 0.10 defective rate. The evidence does not support the foreman's decision at 0.01 level of significance.

Alternatively, we can use technology to solve this problem. A TI-89 output and a Minitab output are shown in Figure 10.12. It shows that the p-value $= 0.04779$ ≈ 0.048. When the process has a 10% defective rate, the probability of observing 15% or more defective items in a sample is 0.048. We are able to obtain an accurate p-value using a calculator or a computer. The normal table has its limitations.

Figure 10.12

Hypothesis testing results for Example 10.18 using technology

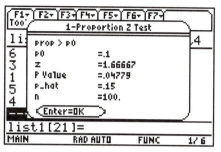

(a) TI-89 output for one proportion z-test

Test and CI for One Proportion

```
Test of p = 0.1 vs p > 0.1

Sample   X   N  Sample p  95% Lower Bound  Z-Value  P-Value
1       15 100  0.150000         0.091267     1.67    0.048

Using the normal approximation.
```

(b) Minitab output for one proportion z-test

Some points to note:

- A smaller α value should be chosen here because the type I error (rejecting H_0 when it is true) has serious consequences, namely, the machine would be shut down unnecessarily.
- Even though the test statistic did not fall in the rejection region, we did not suggest that the null hypothesis be accepted (i.e., the true proportion is 0.10). We simply stated that the sample does not refute the null hypothesis. To be more specific would be to risk the type II error, and we have not computed a probability β of making such an error for any specified alternative proportion, p_a.
- The smallest significance level leading to rejection of H_a in this case is the one producing $z_a = 1.67$, because this is a one-sided test. From Table 4 in the Appendix, $P(Z > 1.67)$; that is, the area to the right of 1.67 is 0.0475, which is the p-value for this test. We would reject H_0 at any significance level at or above this value.
- With the current technological advances, it is possible to obtain an exact p-value using a binomial distribution instead of a normal approximation. As shown in Figure 10.13, the probability of observing 15 or more defectives in a random sample of 100 when the process defective rate is 0.10 (i.e., when the null hypothesis is true) is

$$p\text{-value} = P(X \geq 15 \mid n = 100, p = 0.10) = 0.073$$

Figure 10.13
Exact binomial probability for Example 10.18

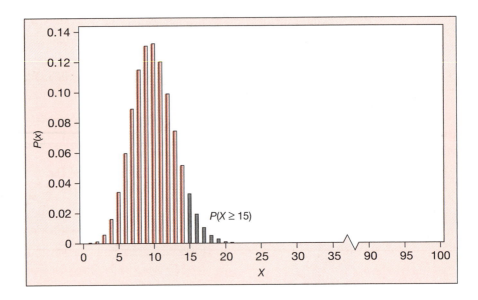

Activity   Is my penny a fair coin?

For this activity, use

- A penny.
- Stirling's sheet given on page 421.

Note that, Stirling's sheet marks "Heads" on one side and "Tails" on the other.

- Start at the top of the sheet and move down the sheet along the curves with each toss.
- Move one step towards side marked "Heads" if heads shows up on that toss, and opposite if tails shows up.
- Continue till you reach the bottom of the sheet. Total 50 tosses are required to reach the bottom.
- Read the percent of heads from the bottom scale. Divide by 100 to get a proportion of heads in 50 tosses.

Test hypothesis $H_0: p = 0.5$ (i.e., my penny is a fair coin) versus $H_a: p \neq 0.5$ (i.e., my penny is biased).

10.2.4   Testing for variance: Normal distribution case

Generally, the variance of the population in the study is unknown. Often it is desired to test a hypothesis about the value of the population variance σ^2. We can do it using the statistic used for constructing the confidence interval for σ^2. Recall that for a random sample of size n from a normal distribution,

$$\frac{(n-1)S^2}{\sigma^2} \text{ has a } \chi^2(n-1) \text{ distribution}$$

Following the same principles as earlier, a procedure for testing hypothesis $H_0: \sigma^2 = \sigma_0^2$ can be developed.

Testing for the Variance σ^2 (Chi-Square Test)

$H_0: \sigma^2 = \sigma_0^2$ $\qquad$ $TS: \chi^2 = \dfrac{(n-1)S^2}{\sigma_0^2}$ $\qquad$ $\nu = (n-1)$ degrees of freedom

Reject H_0 if p-value $< \alpha$

Right-tailed test:	Left-tailed test:	Two-tailed test:
$H_a: \sigma^2 > \sigma_0^2$	$H_a: \sigma^2 < \sigma_0^2$	$H_a: \sigma_1^2 \neq \sigma_0^2$
RR: $\chi^2 > \chi_\alpha^2(\nu)$	RR: $\chi^2 < \chi_{1-\alpha}^2(\nu)$	RR: $\chi^2 > \chi_{\alpha/2}^2(\nu)$ or $\chi^2 < \chi_{1-\alpha/2}^2(u)$
p-value: $P(\chi^2(\nu) > \chi^2)$	p-value: $P(\chi^2(\nu) < \chi^2)$	p-value:

p-value (two-tailed):
$$\begin{cases} 2\,P(\chi^2(\nu) > \chi^2) & \text{if } S^2 > \sigma_0^2 \\ 2\,P(\chi^2(\nu) < \chi^2) & \text{if } S^2 < \sigma_0^2 \end{cases}$$

Conditions:

1. A random sample is taken from the population.
2. Sampled population is normally distributed.

Checking the normality condition: If the sample size is reasonably large, a dot plot or a histogram can be employed to check if the shape of the distribution is fairly symmetric, and bell-shaped without any outliers. Also, a Q–Q plot or a normal probability plot can be used to check normality of population if sample size is sufficiently large. However, unlike the t-test for a mean χ^2 test for variance is *not* robust to the condition of normality, which means variations from normality *will affect* the outcome of the test.

Example 10.19

A machined engine part produced by a certain company is claimed to have a diameter variance no larger than 0.0002 inch. A random sample of ten parts gave a sample variance of 0.0003. Assuming normality of diameter measurements, is there significant evidence to refute the company's claim? Use $\alpha = 0.05$.

Solution Here we are testing $H_0: \sigma^2 \leq 0.0002$ versus $H_a: \sigma^2 > 0.0002$, where σ^2 denotes the variance of machined engine part diameters. At $\alpha = 0.05$ and $n = 10$, we have $\chi_{0.05}^2(9) = 16.919$. Thus, the rejection region is $\chi^2 > 16.919$ (see Figure 10.14). The test statistic is observed to be

$$\chi^2 = \frac{(n-1)s^2}{\sigma_0^2} = \frac{9(0.0003)}{0.0002} = 13.5.$$

Because $\chi^2 = 13.5 < 16.919$, we do not reject H_0, and conclude that we do not have sufficient evidence to refute the company's claim.

Figure 10.14

Rejection region and p-value for Example 10.19

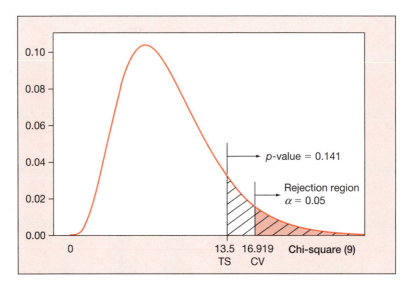

Alternatively, use technology to find the p-value. The Minitab output in Figure 10.15 shows the p-value = 0.141, which is larger than $\alpha = 0.05$, and therefore we arrive at the same conclusion. This p-value indicates that the probability of observing a sample variance of 0.0003 or larger when the population variance is at most 0.0002 is 0.141. In other words, getting a sample with $s^2 = 0.0003$ when $\sigma^2 = 0.0002$ is not a rare event.

Figure 10.15

Minitab printout for Example 10.19

Test and CI for One Standard Deviation

Method

Null hypothesis   Sigma = 0.0141421
Alternative hypothesis   Sigma > 0.0141421

The standard method is only for the normal distribution.
The adjusted method cannot be calculated with summarized data.

Statistics
 N   StDev   Variance
10   0.0173   0.000300

95% One-Sided Confidence Intervals

                    Lower Bound   Lower Bound
Method   for StDev   for Variance
Standard   0.0126   0.000160

Tests
Method   Chi-Square   DF   P-Value
Standard   13.50   9   0.141

Exercises

10.13 The output voltage for a certain electric circuit is specified to be 130. A sample of 40 independent readings on the voltage for this circuit gave a sample mean of 128.6 and a standard deviation of 2.1. Test the hypothesis that the average output voltage is 130 against the alternative that it is less than 130. Use a 5% significance level.

10.14 Refer to Exercise 10.13. If the voltage falls as low as 129, serious consequences may result. For testing $H_0: \mu \geq 130$ versus $H_a: \mu = 129$, find β, the probability of a Type II error, for the rejection region used in Exercise 10.13.

10.15 For testing $H_0: \mu = 130$ versus $H_a: \mu = 129$ with $\sigma = 2.1$, as in Exercise 10.14, find the sample size that will yield $\alpha = 0.05$ and $\beta = 0.01$.

10.16 The Rockwell hardness index for steel is determined by pressing a diamond point into the steel and measuring the depth of penetration. For 50 specimens of a certain type of steel, the Rockwell hardness index averaged 62 with a standard deviation of 8. The manufacturer claims that this steel has an average hardness index of at least 64. Test this claim at the 1% significance level. Find the p-value for this test.

10.17 Steel is sufficiently hard for a certain use as long as the mean Rockwell hardness measure does not drop below 60. Using the rejection region found in Exercise 10.16, find β for the specific alternative $\mu = 60$.

10.18 For testing $H_0: \mu = 64$ versus $H_a: \mu = 60$ with $\sigma = 8$, as in Exercises 10.16 and 10.17, find the sample size required for $\alpha = 0.01$ and $\beta = 0.05$.

10.19 The pH of water coming out of a certain filtration plant is specified to be 7.0. Thirty water samples independently selected from this plant show a mean pH of 6.8 and a standard deviation of 0.9. Is there any reason to doubt that the plant's specification is being maintained? Use $\alpha = 0.05$. Find the p-value for this test.

10.20 A manufacturer of resistors claims that 10% fail to meet the established tolerance limits. A random sample of resistance measurements for 60 such resistors reveals eight to lie outside the tolerance limits. Is there sufficient evidence to refute the manufacturer's claim, at the 5% significance level? Find the p-value for this test.

10.21 For a certain type of electronic surveillance system, the specifications state that the system will function for more than 1,000 hours with probability at least 0.90. Checks on 40 such systems show that 5 failed prior to 1,000 hours of operation. Does this sample provide sufficient

information to conclude that the specification is not being met? Use $\alpha = 0.01$.

10.22 The hardness of a certain rubber (in degrees Shore) is claimed to be 65. Fourteen specimens are tested, resulting in an average hardness measure of 63.1 and a standard deviation of 1.4. Is there sufficient evidence to reject the claim, at the 5% level of significance? What assumption is necessary for your answer to be valid?

10.23 Certain rockets are manufactured with a range of 2,500 meters. It is theorized that the range will be reduced after the rockets are in storage for some time. Six of these rockets are stored for a certain period of time and then tested. The ranges found in the tests are as follows: 2,490, 2,510, 2,360, 2,410, 2,300, and 2,440. Does the range appear to be shorter after storage? Test at the 1% significance level.

10.24 For screened coke, the porosity factor is measured by the difference in weight between dry and soaked coke. A certain supply of coke is claimed to have a porosity factor of 1.5 kilograms. Ten samples are tested, resulting in a mean porosity factor of 1.9 kilograms and a variance of 0.04. Is there sufficient evidence to indicate that the coke is more porous than is claimed? Use $\alpha = 0.05$ and assume the porosity measurements are approximately normally distributed.

10.25 The stress resistance of a certain plastic is specified to be 30 psi. The results from 10 specimens of this plastic show a mean of 27.4 psi and a standard deviation of 1.1 psi. Is there sufficient evidence to doubt the specification at the 5% significance level? What assumption are you making?

10.26 Yield stress measurements on 51 steel rods with 10 mm diameters gave a mean of 485 N/mm^2 and a standard deviation of 17.2. (See Booster, *Journal of Quality Technology*, 1983, for details.) Suppose the manufacturer claims that the mean yield stress for these bars is 490. Does the sample information suggest rejecting the manufacturer's claim, at the 5% significance level?

10.27 The widths of contact windows in certain CMOS circuit chips have a design specification of 3.5 μm (see Phadke et al., *The Bell System Technical Journal*, 1983, for details.) Postetch window widths of test specimens were as follows:

3.21, 2.49, 2.94, 4.38, 4.02,
3.82, 3.30, 2.85, 3.34, 3.91

Can we reject the hypothesis that the design specification is being met, at the 5% significance level? What assumptions are necessary for this test to be valid?

10.28 The variation in window widths for CMOS circuit chips must be controlled at a low level if the circuits are to function properly. Suppose specifications state that $\alpha = 0.30$ for window widths. Can we reject the claim that this specification is being met, using the data of Exercise 10.27? Use $\alpha = 0.05$.

10.29 The Florida Poll of February–March 1984 interviewed 871 adults from around the state. On one question, 53% of the respondents favored strong support of Israel. Would you conclude that a majority of adults in Florida favor strong support of Israel? (*Source: Gainesville Sun*, April 1, 1984.)

10.30 A Yankelovich, Skelly, and White poll reported upon on November 30, 1984, showed that 54% of 2,207 people surveyed thought the U.S. income tax system was too complicated. Can we conclude safely, at the 5% significance level, that the majority of Americans think the income tax system is too complicated?

10.31 Wire used for wrapping concrete pipe should have an ultimate tensile strength of 300,000 pounds, according to a design engineer. Forty tests of such wire used on a certain pipe showed a mean ultimate tensile strength of 295,000 pounds and a standard deviation of 10,000. Is there sufficient evidence to suggest that the wire used on the tested pipe does not meet the design specification at the 10% significance level?

10.32 The break angle in a torsion test of wire used in wrapping concrete pipe is specified to be 40 degrees. Fifty torsion tests of wire wrapping a certain malfunctioning pipe resulted in a mean break angle of 37 degrees and a standard deviation of 6 degrees. Can we say that the wire used on the malfunctioning pipe does not meet the specification at the 5% significance level?

10.33 Soil pH is an important variable in the design of structures that will contact the soil. The pH at a potential construction site was said to average 6.5. Nine test samples of soil from the site gave readings of

    7.3, 6.5, 6.4, 6.1, 6.0, 6.5, 6.2, 5.8, 6.7

Do these readings cast doubt upon the claimed average? (Test at the 5% significance level.)

What assumptions are necessary for this test to be valid?

10.34 The resistances of a certain type of thermistor are specified to be normally distributed with a mean of 10,000 ohms and a standard deviation of 500 ohms, at a temperature of 25°C. Thus, only about 2.3% of these thermistors should produce resistances in excess of 11,000 ohms. The exact measurement of resistances produced is time-consuming, but it is easy to determine whether the resistance is larger than a specified value (say, 11,000). For 100 thermistors tested, 5 showed resistances in excess of 11,000 ohms. Do you think the thermistors from which this sample was selected fail to meet the specifications? Why or why not?

10.35 The dispersion, or variance, of haul times on a construction project is of great importance to the project supervisor, because highly variable haul times cause problems in scheduling jobs. The supervisor of the truck crews states that the range of haul times should not exceed 40 minutes (The range is the difference between the longest and shortest times.) Assuming these haul times to be approximately normally distributed, the project supervisor takes the statement on the range to mean that the standard deviation (σ should be approximately 10 minutes. Fifteen haul times are actually measured and show a mean of 142 minutes and a standard deviation of 12 minutes. Can the claim of ($\sigma = 10$ be refuted at the 5% significance level?

10.36 Aptitude tests should produce scores with a large amount of variation so that an administrator can distinguish between persons with low aptitude and persons with high aptitude. The standard test used by a certain industry has been producing scores with a standard deviation of 5 points. A new test is tried on 20 prospective employees and produces a sample standard deviation of 8 points. Are scores from the new test significantly more variable than scores from the standard? Use $\alpha = 0.05$.

10.37 For the rockets of Exercise 10.23, the variation in ranges is also of importance. New rockets have a standard deviation of range measurements equal to 20 kilometers. Does it appear that storage increases the variability of these ranges? Use $\alpha = 0.05$.

10.3 Hypothesis Testing: The Multiple-Sample Case

Just as we developed confidence interval estimates for linear functions of means when k samples from k different populations are available, we could develop corresponding hypothesis tests. However most tests of population means involve tests of simple differences of the form $(\mu_1 - \mu_2)$. Thus, in this section, we'll consider tests involving equality of two population parameters. The more general case of testing equality of k means will be covered in Chapter 12, using the analysis of variance technique.

10.3.1 Testing the difference between two means: General distributions case

We know from Chapter 9 that in the large-sample case, the estimator $(\bar{X}_1 - \bar{X}_2)$ has an approximate normal distribution. Hence, for testing $H_0: \mu_1 - \mu_2 = D_0$ we can use a test statistic

$$\frac{(\bar{X}_1 - \bar{X}_2) - D_0}{\sqrt{\dfrac{\sigma_1^2}{n_1} + \dfrac{\sigma_2^2}{n_2}}}$$

If σ_1^2 and σ_2^2 are unknown, they can be estimated using s_1^2 and s_2^2, respectively.

Testing the difference between 2 Means ($\mu_1 - \mu_2$): **(Approximate Z-test)**
General Distributions Case

$H_0: \mu_1 - \mu_2 = D_0$

$$TS: z = \frac{(\bar{X}_1 - \bar{X}_2) - D_0}{\sqrt{\dfrac{\sigma_1^2}{n_1} + \dfrac{\sigma_2^2}{n_2}}}$$

Reject H_0 if p-value $< \alpha$.

If σ_1^2 and σ_2^2 are unknown, use s_1^2 and s_2^2 instead.

Right-tailed test:

$H_a: \mu_1 - \mu_2 > D_0$

RR: $z > z_\alpha$

p-value: $P(Z > z)$

Left-tailed test:

$H_a: \mu_1 - \mu_2 < D_0$

RR: $z < -z_\alpha$

p-value: $P(Z < z)$

Two-tailed test:

$H_a: \mu_1 - \mu_2 \neq D_0$

RR: $z > z_{\alpha/2}$ or $z < -z_{\alpha/2}$

p-value: $2 P(Z > |z|)$

> Conditions:
> 1. Random samples are taken from both the populations.
> 2. Both samples are independently taken.
> 3. Both samples are large enough to get an approximate normal distribution for the sample means by applying the central limit theorem.

Checking the sample size condition: Note that both samples must be large enough to use normal approximation. Refer to Section 9.2 for checking the large samples condition.

Example 10.20

A study was conducted to compare the length of time it took men and women to perform a certain assembly-line task. Independent random samples of 50 men and 50 women were employed in an experiment in which each person was timed (in seconds) on identical tasks. The results are summarized here.

	Men	Women
n	50	50
$\bar{x}$	42	38
s^2	18	14

Do the data present sufficient evidence to suggest a difference between the true mean completion times of this task for men and women? Use a 5% significance level.

Solution Let μ_1 be the mean completion time for men and μ_2 be the mean completion time for women. Because we are interested in detecting a difference in either direction, we want to test $H_0: \mu_1 - \mu_2 = 0$ (i.e., no difference) versus $H_a: \mu_1 - \mu_2 \neq 0$ (i.e., difference exists). Because both the samples are large, we can use an approximate Z-test. The test statistic calculated with s_i estimating σ_i is

$$z = \frac{(\bar{X}_1 - \bar{X}_2) - D_0}{\sqrt{\dfrac{s_1^2}{n_1} + \dfrac{s_2^2}{n_2}}} = \frac{(42 - 38) - 0}{\sqrt{\dfrac{18}{50} + \dfrac{14}{50}}} = 5$$

That is, the difference between the sample means is 5 standard deviations from the hypothesized zero difference. Now $p = P(|z| > 5) = 5.7421 \times 10^{-7} < \alpha = 0.05$, we reject H_0. It does appear that the difference in completion times for men and women is real. Such an extremely low p-value indicates that we almost made no mistake in rejecting the null hypothesis (see Figure 10.16). The likelihood of observing such an extreme sample mean, or one even more extreme, when in truth there is no difference between the completion times of men and women, is almost zero.

Alternatively, we arrive at the same conclusion using a rejection region approach with $z_{0.025} = 1.96$. Because $|z| = 5 > 1.96$, reject the null hypothesis and conclude that there is a significant difference in the completion times of men and women.

We can easily get the p-value, using a TI-89 output given in Figure 10.17. It shows that the p-value $= 0.00000057$, which for all practical purposes is almost equal to zero. It shows that if the two population means are equal, then we are not likely to see a difference of 5 or more between the means of two independently drawn samples of size 50 each. So based on the data collected, it makes sense to reject the null hypothesis and conclude that there appears to be a significant difference between the two population means.

Figure 10.16

Rejection region and p-value for Example 10.20

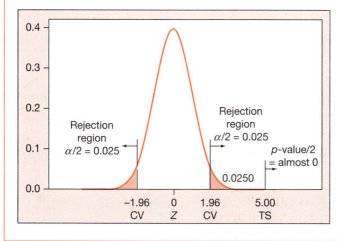

Figure 10.17

TI-89 output for Example 10.20

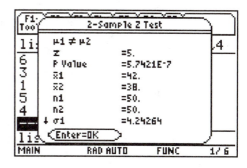

10.3.2 Testing the difference between two means: Normal distributions case

In the previous section, we studied how to compare two population means, using large samples. However, experimenters are not able to take a large number of measurements in all situations. For financial, time-related, or other reasons, often experimenters have to make decisions based on the outcome of small samples. In this section, we are going to describe how to test an hypothesis about two population means with the help of small samples. Although we are interested in comparing two population means, the distribution of the test statistic (and hence the testing procedure) depends on the population variances too. Here we'll consider two situations where the population variances are equal or unequal.

Equal Variances Case　A small-sample case necessitates an assumption about the nature of the probabilistic model for the random variable used. If both populations seem to have normal distributions and if population variances are equal ($\sigma_1^2 = \sigma_2^2 = \sigma^2$), then a t-test can be constructed for testing hypothesis $H_0: \mu_1 - \mu_2 = D_0$, using the fact that test statistic

$$T = \frac{(\bar{X}_1 - \bar{X}_2) - D_0}{S_p\sqrt{\dfrac{1}{n_1} + \dfrac{1}{n_2}}}, \quad \text{where} \quad S_p = \sqrt{\frac{(n_1 - 1)S_1^2 + (n_2 - 1)S_2^2}{(n_1 - 1) + (n_2 - 1)}}$$

has a t-distribution with $(n_1 + n_2 - 2)$ degrees of freedom.

Testing the Difference between 2 Means $(\mu_1 - \mu_2)$:　　(t-Test)
Normal Distributions with Equal Variances

$$H_0: \mu_1 - \mu_2 = D_0 \quad TS: T = \frac{(\bar{X}_1 - \bar{X}_2) - (\mu_1 - \mu_2)}{S_p\sqrt{\dfrac{1}{n_1} + \dfrac{1}{n_2}}}, \quad S_p = \sqrt{\frac{(n_1 - 1)S_1^2 + (n_2 - 1)S_2^2}{(n_1 - 1) + (n_2 - 1)}}$$

Reject H_0 if p-value $< \alpha$ 　　　　　　where　$\nu = n_1 + n_2 - 2$ degrees of freedom

Right-tailed test:	Left-tailed test:	Two-tailed test:
$H_a: \mu_1 - \mu_2 > D_0$	$H_a: \mu_1 - \mu_2 < D_0$	$H_a: \mu_1 - \mu_2 \neq D_0$
RR: $t > t_\alpha(\nu)$	RR: $t < -t_\alpha(\nu)$	RR: $t > t_{\alpha/2}(\nu)$ or $t < -t_{\alpha/2}(\nu)$
p-value: $P(t(\nu) > t)$	p-value: $P(t(\nu) < t)$	p-value: $2\,P(t(\nu) > |t|)$

Conditions:

1. Random samples are taken from both the populations.
2. The samples are independently taken.
3. Both the sampled populations are normally distributed.
4. Both population variances are equal ($\sigma_1^2 = \sigma_2^2$).

Checking the normality condition: Make a dot plot or a histogram for each sample. Check whether the shape of each distribution is fairly symmetric, and bell-shaped without any outliers. In other words, check whether the distribution resembles normal distribution. Alternatively, use a Q–Q plot or a normal probability plot.

Checking the condition of equal variances: Make parallel dotplots and compare the spreads visually.

Example 10.21

The designer of a new sheet-metal stamping machine claims that her new machine can turn out a certain product faster than the machine now in use.

	Standard Machine	New Machine
n	9	9
$\bar{x}$	35.22	31.56
s^2	24.44	20.03

Nine independent trials of stamping the same type of item on each machine gave the accompanying results on times to completion (in seconds). At the 5% significance level, can the designer's claim be substantiated? Assume that the stamping times are normally distributed for both machines.

Solution   We are interested in testing $H_0: \mu_1 - \mu_2 = 0$ versus $H_a: \mu_1 - \mu_2 > 0$ (i.e., $\mu_1 > \mu_2$), where μ_1 is the mean stamping time per item on the standard machine and μ_2 is mean stamping time per item on the new machine. Because the sample variances look to be close enough to assume the equality of the population variances, we can use the pooled variances t-test to compare the mean stamping times of the two methods. The test statistic is calculated as

$$t = \frac{(\bar{X}_1 - \bar{X}_2) - (\mu_1 - \mu_2)}{S_p\sqrt{\dfrac{1}{n_1} + \dfrac{1}{n_2}}} = \frac{(35.22 - 31.56) - 0}{4.72\sqrt{\dfrac{1}{9} + \dfrac{1}{9}}} = 1.65$$

where

$$S_p = \sqrt{\frac{(n_1 - 1)S_1^2 + (n_2 - 1)S_2^2}{(n_1 - 1) + (n_2 - 1)}} = \sqrt{\frac{(9 - 1)24.44 + (9 - 1)20.03}{(9 - 1) + (9 - 1)}}$$
$$= \sqrt{22.24} = 4.72$$

and p-value $= P(t(16) > 1.65) = 0.060$ (Figure 10.18 and Figure 10.19). The p-value of 0.060 indicates that if there is no difference in the mean stamping times

Figure 10.18

Rejection region and p-value for Example 10.21

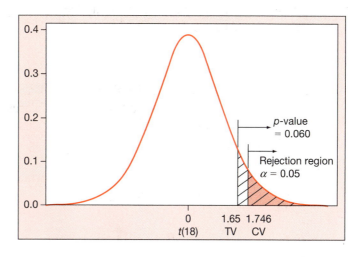

Figure 10.19

Minitab printout for
Example 10.21

```
Two-Sample T-Test and CI

Sample    N      Mean     StDev    SE Mean
1         9     35.22     4.94        1.6
2         9     31.56     4.48        1.5


Difference = mu (1) - mu (2)
Estimate for difference: 3.66
95% lower bound for difference: -0.22
T-Test of difference = 0 (vs >):
T-Value = 1.65   P-Value = 0.060   DF = 16
Both use Pooled StDev = 4.7156
```

of the two machines, then the probability of observing a test statistic value of 1.65 or even larger is 0.060. Thus, there is a higher likelihood that we would be rejecting the null hypothesis wrongfully than the chance that we would like to take of making such a wrong decision (p-value > 0.05). Therefore, do not reject the null hypothesis and conclude that there is not sufficient evidence to substantiate the designer's claim.

Note that the rejection region is $t > 1.746 = t_{0.05}$ for 16 degrees of freedom and the test statistic value 1.65 falls in the non-rejection region.

Example 10.22 ——————————————————————————

A quality control inspector compares the ultimate tensile strength (UTS) measurements for class II and III prestressing wire by taking a sample of five specimens from a roll of each class, for laboratory testing. The sample data (in 1,000 psi) are as follows:

Class II: 253, 261, 258, 255, 256   Class III: 274, 275, 271, 277, 256

Do the true mean UTS measurements for the two classes of wire appear to differ? The tensile strengths are known to be approximately normally distributed.

Solution   Here we are interested in testing $H_0: \mu_{II} = \mu_{III}$ against $H_a: \mu_{II} \neq \mu_{III}$, where μ_i is the true mean UTS measurement for class i ($i = $ II, III) wires. The computer output (Minitab) is shown in Figure 10.20.

For testing the equality of means, the t-statistic is -10.53 with two-sided p-value < 0.001 (essentially zero). Thus, there is a strong evidence to reject the null hypothesis, indicating that the mean UTS measurements differ for class II and class III wire.

Figure 10.20

Minitab printout for Example 10.22

> **Two-Sample T-Test and CI: Class II, Class III**
>
> Two-sample T for Class II vs Class III
>
> | | N | Mean | StDev | SE Mean |
> |---|---|------|-------|---------|
> | Class II | 5 | 256.60 | 3.05 | 1.4 |
> | Class II | 5 | 274.60 | 2.30 | 1.0 |
>
> Difference = mu Class II - mu Class III
> Estimate for difference: -18.00
> 95% CI for difference: (-21.94, -14.06)
> T-Test of difference = 0 (vs not =): T-Value = -10.53
> P-Value = 0.000 DF = 8
> Both use Pooled StDev = 2.70

Unequal Variances Case If one suspects that the variances in two normal populations under study are *not* equal, then the *t*-test based on pooling the sample variances should not be used. What should be used in its place? Unfortunately, the answer to that question is not easy because there is no exact way to solve this problem. From Chapter 8, we know that statistic

$$T = \frac{(\bar{X}_1 - \bar{X}_2) - D_0}{\sqrt{\dfrac{s_1^2}{n_1} + \dfrac{s_2^2}{n_2}}}$$

has approximately *t*-distribution under $H_0: \mu_1 - \mu_2 = D_0$, with the following degrees of freedom,

$$\mathrm{df} = \frac{\left[\left(s_1^2/n_1\right) + \left(s_2^2/n_2\right)\right]^2}{\left[\left(s_1^2/n_1\right)^2/\left(n_1 - 1\right)\right] + \left[\left(s_2^2/n_2\right)^2/\left(n_2 - 1\right)\right]}$$

where s_1^2 and s_2^2 are the sample variances. It is recommended that these computations be done by computer or calculator.

Testing the Difference between 2 Means $(\mu_1 - \mu_2)$: (Approximate *t*-Test)
Unequal Variances Case

$$H_0: \mu_1 - \mu_2 = D_0 \qquad \text{TS: } T = \frac{(\bar{X}_1 - \bar{X}_2) - D_0}{\sqrt{\dfrac{S_1^2}{n_1} + \dfrac{S_2^2}{n_2}}}$$

$$v = \frac{\left[\left(S_1^2/n_1\right) + \left(S_2^2/n_2\right)\right]^2}{\left[\left(S_1^2/n_1\right)^2/\left(n_1 - 1\right)\right] + \left[\left(S_2^2/n_2\right)^2/\left(n_2 - 1\right)\right]} \text{ degrees of freedom}$$

Right-tailed test:	Left-tailed test:	Two-tailed test:
$H_a: \mu_1 - \mu_2 > D_0$	$H_a: \mu_1 - \mu_2 < D_0$	$H_a: \mu_1 - \mu_2 \neq D_0$
RR: $t > t_\alpha(v)$	RR: $t < -t_\alpha(v)$	RR: $t > -t_{\alpha/2}(v)$ or $t < -t_{\alpha/2}(v)$
p-value: $P(t(v) > t)$	p-value: $P(t(v) < t)$	p-value: $2P(t(v) > |t|)$

Conditions:

1. Random samples are taken from both the populations.
2. The samples are independently taken.
3. Both the sampled populations are normally distributed.

Checking the normality condition: Make a dot plot or a histogram for each sample. Check whether the shape of each distribution is fairly symmetric, and bell-shaped without any outliers. In other words, check whether the distribution resembles normal distribution. Alternatively, use a Q–Q plot or a normal probability plot if sample sizes are sufficiently large.

Checking the condition of unequal variances: Make parallel dotplots and compare the spreads visually. In most cases, if the ratio of sample variances is larger than 3, then it is safe to assume that the population variances are different.

Example 10.23

The prestressing wire on each of two concrete pipes manufactured at different times was compared for torsion properties. Ten specimens randomly selected from each pipe were twisted in a laboratory apparatus until they broke. The number of revolutions until complete failure was recorded. The results are as follows, with C1 and C2 denoting the two concrete pipes:

C1: 5.83, 5.66, 4.75, 3.00, 3.37, 3.63, 4.00, 4.63, 4.25, 4.13
C2: 3.38, 2.81, 7.00, 1.50, 5.88, 5.25, 4.08, 7.63, 4.50, 4.88

Is there any evidence to suggest that the true mean revolutions to failure differ for the wire on the two pipes?

Solution   Careful study of the datasets and dotplots in Figure 10.21 shows that C1 results appear to be less variable than the C2 results.

Figure 10.21

Number of revolutions until complete failure

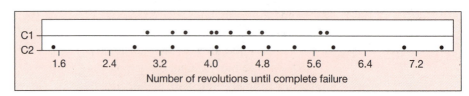

Number of revolutions until complete failure

It doesn't seem reasonable here to assume that these two sets of measurements come from populations with the same variance. Thus, we'll use the approximate *t*-test outlined above. The Minitab printout is shown in Figure 10.22. Note that the unequal variances case is a default option in most statistical software. It is also a recommended procedure unless we have good evidence that variances are nearly equal.

The observed *t*-statistic is −0.55 with 13 degrees of freedom. The *p*-value of 0.589 suggests we cannot reject H_0. Therefore, conclude that there is no evidence to suggest the mean torsion differs for two types of pipes. The *p*-value indicates the probability of observing a difference of 0.366 or extreme in the sample means when the mean number of revolutions for two concrete pipes are the same is 0.589. Not such a rare event.

Figure 10.22

Minitab printout for Example 10.23

```
Two-Sample T-Test and CI: C1, C2

Two-sample T for C1 vs C2

          N     Mean    StDev    SE Mean
C1       10    4.325    0.920      0.29
C2       10    4.69     1.87       0.59


Difference = mu (C1) - mu (C2)
Estimate for difference: -0.366
95% CI for difference: (-1.792, 1.060)
T-Test of difference = 0 (vs not =):
T-Value = -0.55   P-Value = 0.589 DF = 13
```

Activity Does a longer wing length result in a longer mean flight time?

The *response* to be measured is the flight time of the helicopter. It is defined as the time for which the helicopter remains in flight.

Refer to data collected in Table 9.6. Compute the mean and the standard deviation for each sample. Test the hypothesis $H_0: \mu_5 = \mu_6$ against $H_a: \mu_5 < \mu_6$, where μ is the mean flight time.

10.3.3 Testing the difference between the means for paired samples

The two-sample *t*-test given above works only in the case of *independent* samples. On many occasions, however, two samples will arise in a dependent fashion. A commonly occurring situation is that in which repeated observations are taken on the *same* sampling unit, such as counting the number of accidents in various plants both before and after a safety awareness program is instituted. The counts in one plant may be independent of the counts in another, but the two counts (before and after) within any one plant will be dependent. Thus, we must develop a mechanism for analyzing measurements that occur in pairs.

Let $(X_1, Y_1), \ldots, (X_n, Y_n)$ denote a random sample of paired observations. That is, (X_i, Y_i) denotes two measurements taken in the same sampling unit, such as counts of accidents within the ith plant taken before and after the safety awareness program is put into effect, or lifetimes of two components within the same system. Suppose the hypothesis of interest concerns the difference between $E(X_i)$ and $E(Y_i)$. The two-sample tests developed earlier cannot be used for this purpose because of the dependence between X_i and Y_i. However, $E(X_i) - E(Y_i) = E(X_i - Y_i)$. Thus the comparison of $E(X_i)$ and $E(Y_i)$ can be made by looking at the mean of the differences $(X_i - Y_i)$. Letting $D_i = X_i - Y_i$, the hypothesis $E(X_i) - E(Y_i) = 0$ is equivalent to the hypothesis $E(D_i) = 0$.

To construct a test statistic with a known distribution, we assume $D_1, \ldots, D_n$ is a random sample of *differences*, each possessing a normal distribution with mean μ_D and variance σ_D^2. To test $H_0: \mu_D = 0$, employ the test statistic

$$T = \frac{\overline{D} - 0}{S_D/\sqrt{n}}$$

where

$$\overline{D} = \frac{1}{n}\sum_{i=1}^{n} D_i \quad \text{and} \quad S_D^2 = \frac{1}{n-1}\sum_{i=1}^{n}(D_i - \overline{D})^2$$

If the differences have the normal distribution, the test statistic T will have a t-distribution with $(n - 1)$ degrees of freedom when the null hypothesis is true (see Chapter 8 for the sampling distribution of T). Thus, a two-sample problem is reduced to a one-sample problem.

Testing the Difference between Means μ_d for Paired Samples (*t*-Test)

$H_0: \mu_d = \mu_{d0}$

TS: $T = \dfrac{\bar{d} - \mu_{d0}}{S_d/\sqrt{n}}$

$v = (n - 1)$ degrees of freedom

Right-tailed test:	Left-tailed test:	Two-tailed test:
$H_a: \mu > \mu_{d0}$	$H_a: \mu < \mu_{d0}$	$H_a: \mu \neq \mu_{d0}$
RR: $t > t_\alpha(v)$	RR: $t < -t_\alpha(v)$	RR: $t > -t_{\alpha/2}(v)$ or $t < -t_{\alpha/2}(v)$
p-value: $P(t(v) > t)$	p-value: $P(t(v) < t)$	p-value: $2P(t(v) > |t|)$

Conditions:

1. A random sample of differences is taken from the population of differences.
2. Sampled population of differences is normally distributed.

Remember to check the assumption of normality of the *differences* and not the individual samples.

Example 10.24

Two methods of determining percentage of iron in ore samples are to be compared by subjecting 12 ore samples to each method. The results of the experiment are listed in table below.

Figure 10.23

Rejection region and p-value for Example 10.24

Ore Sample	Method A	Method B	d_i
1	38.25	38.25	0.00
2	31.68	31.71	−0.03
3	26.24	26.25	−0.01
4	41.29	41.33	−0.04
5	44.81	44.80	0.01
6	46.37	46.39	−0.02
7	35.42	35.46	−0.04
8	38.41	38.42	−0.01
9	42.68	42.70	−0.02
10	46.71	46.76	−0.05
11	29.20	29.18	0.02
12	30.76	30.79	−0.03

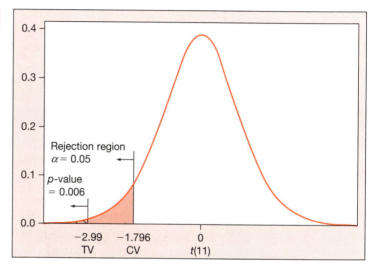

Do the data provide evidence that on the average method B measures a higher average percentage of iron than method A? Use $\alpha = 0.05$.

Solution Let $\mu_D =$ the mean difference in the percentage of iron from two different methods. By taking difference of (Method A − Method B), we want to test $H_0: \mu_D \geq 0$ versus $H_a: \mu_D < 0$. If method B has mean higher iron percentage, then μ_D will be negative. For the column of differences we get the mean and standard deviation of differences: $\bar{d} = -0.01833$, and $s_d = 0.02125$. It follows that

$$t = \frac{\bar{d} - 0}{s_d/\sqrt{n}} = \frac{-0.01833}{0.02125/\sqrt{12}} = 2.99.$$

The Minitab printout in Figure 10.24 gives p-value $= P(t(11) < -2.99) = 0.006$, which is less than $\alpha = 0.05$ (also see Figure 10.23). Hence, we reject the null hypothesis that $\mu_D \geq 0$ and accept the alternative $\mu_D < 0$. At $\alpha = 0.05$, we conclude that on the average method B has a higher percentage of iron than method A.

Because $\alpha = 0.05$ and $n - 1 = 11$ degrees of freedom, the rejection region consists of those values of t smaller than $-t_{0.05} = -1.796$. Hence, we reject the null hypothesis that $\mu_D \geq 0$, accept the alternative $\mu_D < 0$, and arrive at the same conclusion.

Figure 10.24

Minitab printout for Example 10.24

Paired T-Test and CI: Method A, Method B

Paired T for Method A — Method B

	N	Mean	StDev	SE Mean
Method A	12	37.65	7.00	2.02
Method B	12	37.67	7.00	2.02
Difference	12	-0.01833	0.02125	0.00613

95% upper bound for mean difference: -0.00732

T-Test of mean difference = 0 (vs < 0):

T-Value = -2.99   P-Value = 0.006

Example 10.25

Concrete-filled tube columns can provide excellent seismic resistant structural properties such as high strength, high conductivity, and large energy absorption capacity. Hu, Asle, Huang, Wu, and Wu (*Journal of Structural Engineering*, 2003) proposed proper material constitutive models for concrete-filled tube columns and verified their failure strength by the nonlinear finite element program ABAQUS against experimental data. Table below gives failure strength (kN) as measured by experimental data and analysis.

Failure Strength (kN)											
Experiment	2727.6	2902.6	2463.1	3744.5	3855.3	3807.3	3610.1	3596.3	3457.0	3507.1	3184.2	3104.7
Analysis	2741.0	2885.0	2476.0	3745.0	3862.0	3812.0	3609.0	3568.0	3465.0	3541.0	3213.0	3092.0

Does the mean failure strength measured by experimental data differ significantly from the analytically determined failure strength? Use $\alpha = 0.05$.

Solution Let μ_D be the mean difference between the experimental analytical strength. We are interested in testing

$$H_0: \mu_D = 0 \quad \text{against} \quad H_a: \mu_D \neq 0$$

The computer output from JMP is given in Figure 10.25. The histogram and the boxplot show that the distribution of differences is fairly symmetric with no outliers. The assumption of normality of the population of differences seems reasonable. Because the *p*-value = 0.4445 is larger than $\alpha = 0.05$, do not reject the null hypothesis and conclude that the mean difference between the experimental and analytical measurements is not significant. Note that in this situation, we do want the analytical and experimental measurements to be similar.

Figure 10.25

JMP output for Example 10.25

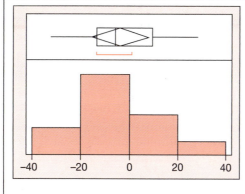

Test Mean 5 Value

Hypothesized Value	0
Actual Estimate	-4.1
df	11
Std Dev	17.907
	t Test
Test Statistic	-0.7931
Prob > |t|	0.4445
Prob > t	0.7778
Prob < t	0.2222

10.3.4   Testing the ratio of variances: Normal distributions case

We can compare the variances of two normal populations by looking at the ratio of sample variances. For testing $H_0: \sigma_1^2 = \sigma_2^2$ against $H_a: \sigma_1^2 \neq \sigma_2^2$, we can calculate ratio S_1^2/S_2^2 and reject the null hypothesis if this statistic is either very large or very small. The precise rejection region can be found by observing that $F = S_1^2/S_2^2$ has an F-distribution when $H_0: \sigma_1^2 = \sigma_2^2$ is true. Thus, we reject the null hypothesis for $F > F_{a/2}(\nu_1, \nu_2)$ or $F < F_{1-a/2}(\nu_1, \nu_2)$. A one-sided test can be constructed similarly.

Testing the Ratio of Variances   (*F*-Test)

$H_0: \sigma_1^2 = \sigma_2^2$ $TS: F = \dfrac{S_1^2}{S_2^2}$ $\nu_1 = n_1 - 1$ and $\nu_2 = n_2 - 1$ degrees of freedom

Reject H_0 if p-value $< \alpha$

Right-tailed test:

$H_a: \sigma_1^2 > \sigma_2^2$

RR: $F > F_a(\nu_1, \nu_2)$

p-value: $P(F(\nu_1, \nu_2) > F)$

Left-tailed test:

$H_a: \sigma_1^2 < \sigma_2^2$

RR: $F < F_{1-a}(\nu_1, \nu_2)$

p-value: $P(F(\nu_1, \nu_2) < F)$

Two-tailed test:

$H_a: \sigma_1^2 \neq \sigma_2^2$

RR: $F > F_{a/2}(\nu_1, \nu_2)$ or

     $F < F_{1-a/2}(\nu_1, \nu_2)$

p-value:

$2P(F(\nu_1, \nu_2) > F)$ if $S_1^2 > S_2^2$

$2P(F(\nu_1, \nu_2) < F)$ if $S_1^2 < S_2^2$

Conditions:

1. Random samples are taken from both the populations.
2. Both samples are independently taken.
3. Both the sampled populations are normally distributed.

As discussed in Chapter 9, for constructing a confidence interval for the ratio of population variances, this testing of hypothesis procedure is also not robust to the condition of normality of both populations. Therefore, it should be used only if the population distributions are known to be normal.

Example 10.26

Suppose the diameter variance of the machined engine parts in Example 10.17 is to be compared with the diameter variance of a similar part manufactured by a competitor. The variances of random samples are as follows:

Manufacturer	$n_1 = 10$	$S_1^2 = 0.0003$
Competitor	$n_2 = 20$	$S_2^2 = 0.0001$

Is there evidence to conclude that $\sigma_2^2 < \sigma_1^2$ at the 5% significance level? Assume normality for both populations.

Solution We are interested in testing $H_0: \sigma_1^2 \leq \sigma_2^2$ versus $H_a: \sigma_1^2 > \sigma_2^2$. The outcome of the statistic $F = S_1^2/S_2^2$, which has an F-distribution if $\sigma_1^2 = \sigma_2^2$, is calculated as

$$F = \frac{S_1^2}{S_2^2} = \frac{0.0003}{0.0001} = 3$$

We reject H_0 because $F = 3 > F_{0.05}(9,19) = 2.42$. It looks like the competitor's engine parts have less variation in the diameters.

The p-value for the test can be easily obtained using technology. For example, TI-89 gives p-value $= 0.02096 < \alpha = 0.05$. Therefore, we arrive at the same conclusion.

10.3.5 Robustness of Statistical Procedures

A word on the *robustness* of statistical tests of hypotheses to underlying conditions is in order. Tests based on t, χ^2 and F-distributions assume that the sample measurements come from normally distributed populations. What if the populations are not normal? Will these tests still work? As seen earlier, the ability of a test to perform satisfactorily when the conditions are not met is called *robustness*. In that sense, the t-tests are quite robust; that is, they perform well even when the normality conditions are violated by a considerable amount.

On the other hand, tests for variances based on χ^2 and F-distributions are not very robust to the condition of normality of populations. The worst offender is the F-test for equality of variances. If the population distributions are slightly skewed, this test will reject the null hypothesis of equal variances too often. What is thought, for example, to be a 0.05-level test may, in fact, be a 0.10-level test. The test is actually testing the null hypothesis of equal variances from normal distributions against the alternative of unequal variances from normal distributions *or* distributions that are skewed. So, be careful when you interpret the outcome of an F-test on variances; rejection of the null hypothesis may come about for a variety of reasons.

Exercises

10.38 Two designs for a laboratory are to be compared with respect to the average amount of light produced on table surfaces. Forty independent measurements (in footcandles) are taken in each laboratory, with the following results:

Design I	Design II
$n_1 = 40$	$n_2 = 40$
$\bar{x}_1 = 28.9$	$\bar{x}_2 = 32.6$
$s_1^2 = 15.1$	$s_2^2 = 15.8$

Is there sufficient evidence to suggest that the designs differ with respect to the average amount of light produced? Use $\alpha = 0.05$.

10.39 Shear-strength measurements derived from unconfined compression tests for two types of soils gave the following results (measurements in tons per square foot):

Soil Type I	Soil Type II
$n_1 = 30$	$n_2 = 40$
$\bar{x}_1 = 1.65$	$\bar{x}_2 = 1.43$
$s_1 = 0.26$	$s_2 = 0.22$

Do the soils appear to differ with respect to average shear strength, at the 1% significance level?

10.40 A study was conducted by the Florida Game and Fish Commission to assess the amounts of chemical residues found in the brain tissue of brown pelicans. For DDT, random sample of $n_1 = 10$ juveniles and $n_2 = 13$ nestlings gave the following results (measurements in parts per million):

Juveniles	Nestlings
$n_1 = 10$	$n_2 = 13$
$\bar{y}_1 = 0.041$	$\bar{y}_2 = 0.026$
$s_1 = 0.017$	$s_2 = 0.016$

Test the hypothesis that there is no difference between mean amounts of DDT found in juveniles and nestlings versus the alternative that the juveniles have a larger mean amount. Use $\alpha = 0.05$. (This test has important implications regarding the build-up of DDT over time.)

10.41 The strength of concrete depends, to some extent, on the method used for drying. Two different drying methods showed the following results for independently tested specimens (measurements in psi):

Method I	Method II
$n_1 = 7$	$n_2 = 10$
$\bar{x}_1 = 3.250$	$\bar{x}_2 = 3.240$
$s_1 = 210$	$s_2 = 190$

Do the methods appear to produce concrete with different mean strengths? Use $\alpha = 0.05$. What assumptions are necessary in order for your answer to be valid?

10.42 The retention of nitrogen in the soil is an important consideration in cultivation practices, including the cultivation of forests. Two methods for preparing plots for planting pine trees after clear-cutting were compared on the basis of the percentage of labeled nitrogen recovered. Method A leaves much of the forest floor intact, while method B removes most of the organic material. It is clear that method B will produce a much lower recovery of nitrogen from the forest floor. The question of interest is whether or not method B will cause more nitrogen to be retained in the microbial biomass, as a compensation for having less organic material available. Percentage of nitrogen recovered in the microbial biomass was measured on six text plots for each method. Method A plots showed a mean of 12 and a standard deviation of 1. Method B plots showed a mean of 15 and a standard deviation of 2. At the 10% significance level, should we say that the mean percentage of recovered nitrogen is larger for method B? (See Vitousek and Matson, *Science*, July 6, 1984, for more details.)

10.43 In the presence of reverberation, native as well as nonnative speakers of English have some trouble recognizing consonants. Random samples of 10 natives and 10 nonnatives were given the Modified Rhyme Test, and the percentages of correct responses were recorded. (See Nabelek and Donahue, *Journal of the Acoustical Society of America*, 1984.) The data are as follows:

Natives:	93, 85, 89, 81, 88, 88, 89, 85, 85, 87
Nonnatives:	76, 84, 78, 73, 78, 76, 70, 82, 79, 77

Is there sufficient evidence to say that nonnative speakers of English have a smaller mean percentage of correct responses at the 5% significance level?

10.44 Biofeedback monitoring devices and techniques in the control of physiologic functions help astronauts control stress. One experiment related to this topic is discussed by G. Rotondo, et al., *Acta Astronautica*, 1983. Six subjects were placed in a stressful situation (using video games), followed by a period of biofeedback and adaptation. Another group of six subjects was placed under the same stress and then simply told to relax. The first group had an average heart rate of 70.4 and a standard deviation of 15.3. The second group had an average heart rate of 74.9 and a standard deviation of 16.0. At the 10% significance level, can we say that the average heart rate with biofeedback is lower than that without biofeedback? What assumptions are necessary for your answer to be valid?

10.45 The t-test for comparing means assumes that the population variances are equal. Is that a valid assumption based on the data of Exercise 10.44? (Test the equality of population variances at the 10% significance level.)

10.46 Data from the U.S. Department of Interior–Geological Survey give flow rates for a small river in North Florida. It is of interest to compare flow rates for March and April, two relatively dry months. Thirty-one measurements for March showed a mean flow rate of 6.85 cubic feet per second and a standard deviation of 1.2. Thirty measurements for April showed a mean of 7.47 and a standard deviation of 2.3. Is there sufficient evidence to say that these two months have different average flow rates, at the 5% significance level?

10.47 An important measure of the performance of a machine is the mean time between failures (MTBF). A certain printer attached to a word processor was observed for a period of time during which 10 failures were observed. The times between failures averaged 98 working hours, with a standard deviation of 6 hours. A modified version of this printer was observed for eight failures, with the times between failures averaging 94 working hours, with a standard deviation of 8 hours. Can we say, at the 1% significance level, that the modified version of the printer has a smaller MTBF? What assumptions are made in this test? Do you think the assumptions are reasonable for the situation?

10.48 It is claimed that tensile-strength measurements for a 12 mm–diameter steel rod should, on the average, be at least 8 units (N/mm^2) higher than the tensile-strength measurements for a 10 mm–diameter steel rod. Independent samples of 50 measurements each for the two sizes of rods gave the following results:

10-mm Rod	12-mm Rod
$\bar{x}_1 = 545$	$\bar{x}_2 = 555$
$s_1 = 24$	$s_2 = 18$

Is the claim justified at the 5% significance level?

10.49 Alloying is said to reduce the resistance in a standard type of electrical wire. Ten measurements on a standard wire yielded a mean resistance of 0.19 ohm and a standard deviation of 0.03. Ten independent measurements on an alloyed wire yielded a mean resistance of 0.11 ohm and a standard deviation of 0.02. Does alloying seem to reduce the mean resistance in the wire? Test at the 10% level of significance.

10.50 Two different machines, A and B, used for torsion tests of steel wire were tested on 12 pairs of different types of wire, with one member of each pair tested on each machine. The results (break angle measurements) were as follows:

					Wire Type							
	1	2	3	4	5	6	7	8	9	10	11	12
Machine A	32	35	38	28	40	42	36	29	33	37	22	42
Machine B	30	34	39	26	37	42	35	30	30	32	20	41

a Is there evidence at the 5% significance level to suggest that machines A and B give different average readings?

b Is there evidence at the 5% significance level to suggest that machine B gives a lower average reading than machine A?

10.51 Six rockets, nominally with a range of 2,500 meters, were stored for some time and then tested. The ranges found in the tests were 2,490, 2,510, 2,360, 2,410, 2,300, and 2,440. Another group of six rockets, of the same type, was stored for the same length of time but in a different manner. The ranges for these six were 2,410, 2,500, 2,360, 2,290, 2,310, and 2,340. Do the two storage methods produce significantly different mean ranges? Use $\alpha = 0.05$ and assume range measurements to be approximately normally distributed with the same variance for each manner of storage.

10.52 The average depth of bedrock at two possible construction sites is to be compared by driving five piles at random locations within each site. The results, with depths in feet, are as follows:

Site A	Site B
$n_1 = 5$	$n_2 = 5$
$\bar{x}_1 = 142$	$\bar{x}_2 = 134$
$s_1 = 14$	$s_2 = 12$

Do the average depths of bedrock differ for the two sites at the 10% significance level? What assumptions are you making?

10.53 Gasoline mileage is to be compared for two automobiles, A and B, by testing each automobile on five brands of gasoline. Each car used one tank of each brand, with the following results (in miles per gallon):

Brand	Auto A	Auto B
1	28.3	29.2
2	27.4	28.4
3	29.1	28.2
4	28.7	28.0
5	29.4	29.6

Is there evidence to suggest a difference between true average mileage figures for the two automobiles? Use a 5% significance level.

10.54 The two drying methods for concrete were used on seven different mixes, with each mix of concrete subjected to each drying method. The resulting strength-test measurements (in psi) are given below:

Mix	Method I	Method II
A	3,160	3,170
B	3,240	3,220
C	3,190	3,160
D	3,520	3,530
E	3,480	3,440
F	3,220	3,210
G	3,120	3,120

Is there evidence of a difference between average strengths for the two drying methods at the 10% significance level?

10.55 Two procedures for sintering copper are to be compared by testing each procedure on six different types of powder. The measurement of interest is the porosity (volume percentage due to voids) of each test specimen. The results of the tests are as follows:

Powder	Procedure I	Procedure II
1	21	23
2	27	26
3	18	21
4	22	24
5	26	25
6	19	16

Is there evidence of a difference between true average porosity measurements for the two procedures? Use $\alpha = 0.05$.

10.56 Refer to Exercise 10.41. Do the variances of the strength measurements differ for the two drying methods? Test at the 10% significance level.

10.57 Refer to Exercise 10.51. Is there sufficient evidence to say that the variances among range measurements differ for the two storage methods? Use $\alpha = 0.10$.

10.58 Refer to Exercise 10.52. Does site A have significantly more variation among depth measurements than site B? Use $\alpha = 0.05$.

10.59 It is well-known that the linear systems arising from spatial discretizations of fluid mechanics problems grow rapidly as the size of the problem increases. The solution process involves the inversion of a Laplacian in each spatial direction. The larger number of iterations involved lead to longer solution times. Zsaki et al. (*International Journal for Numerical Methods in Fluids*, 2003) studied how to effectively solve the vector Laplacian system by applying finite-element tearing and interconnecting (FETI) method. In this method, the total number of inner iterations changes depending on the mesh size used. They performed numerical tests for different mesh sizes with and without the reconjugation method and the inner FTEI iterations were recorded for each case.

Mesh size	Inner FETI (w/o)	Inner FETI (with Reconj)
1/8	217	95
1/12	252	107
1/16	269	109
1/20	271	113
1/28	291	115
1/32	291	118
1/38	291	118
1/40	293	119
1/48	311	119

Is the inner FETI significantly higher without reconjugation than that with reconjugation? Justify your answer using a 5% level of significance.

10.60 The Tennessee Board of Architectural and Engineering Examiners requested the Tennessee Society of Professional Engineers and the Consulting Engineers of Tennessee to look into various issues related to the professional registration of engineers. In response to this request, a Blue Ribbon Committee was formed, which sent a survey to engineering deans during 1999 and 2000. The results were reported in a manuscript by Madhavan and Malasri (*Journal of Professional Issues in Engineering Education and Practice*, 2003). They reported the following:

a Of 789 engineering faculty in Texas, 486 were registered, while of 96 faculty in Arkansas, 62 were registered. Is there a significant difference in the percentage of registered engineering faculty in Texas and Arkansas? Use $\alpha = 0.05$.

b Of 162 engineering faculty in Nebraska, 65 were registered, while of 189 faculty in Colorado, 21 were registered. Is there a significant difference in the percentage of registered engineering faculty in Nebraska and Colorado? Use $\alpha = 0.05$.

10.61 Bennett et al. (*Journal of Professional Issues in Engineering Education and Practice*, 2002) studied conditions of existing bridges in Tennessee that were built during 1950–56 using prestressed concrete segments. They recorded number of spans, length, and average daily traffic among other characteristics. The average daily traffic (number of vehicles) data for bridges with short (1–3 spans) and long (4 + spans) bridges were reported as follows:

1–3 spans	4 + spans
1,680	1,640
1,650	1,640
410	1,300
3,700	
730	
630	
400	
400	

Is there a significant difference in the mean daily traffic on bridges with different number of spans? Use $\alpha = 0.05$.

10.62 Chitizadeh and Varghani (*IEE Proceedings Communications*, 2003) proposed a new crossbar switch with input queue. Crossbar switches are used for connecting multiple inputs to multiple outputs in a matrix manner. They are used in information processing applications such as packet switching as well as in sorting machines. Chitizadeh and Varghani studied variation in throughput with respect to switch size. The results from a simulation study for a full load and the theoretical throughput (in % of total input) are listed in listed in the table below for switch sizes 2 to 16.

Switch Size	Theory	Simulation	Switch Size	Theory	Simulation
2	100	98	10	60	51
3	100	96	11	57	47
4	100	93	12	53	42
5	93	84	13	50	41
6	82	75	14	48	40
7	76	70	15	42	38
8	73	60	16	41	35
9	63	55			

a Is there a significant difference between the simulated and theoretical mean throughput? Justify using a 5% level of significance.

b Plot the theoretical throughput against the simulated throughout. Are there any systematic differences? What would you call these? Make suggestions for adjusting the theoretical throughput using this information.

10.63 In material sciences, measuring temperatures accurately is of great importance. Optical pyrometry is an established method used for measuring temperatures. These pyrometric temperature measurements are affected by selection of operating wavelength and atmospheric conditions. To solve some of the measurement problems, Rohner and Newmann (*Transactions of the ASME*, 2003) developed a new system based on UV-B (ultraviolet) wavelength range for pyrometry and tested on a blackbody without solar irradiation. They recorded the percentage of deviation of the UV temperatures from the blackbody temperatures. Two sets of data were collected; one based on the calibration at 1400°C and the other at 1500°C. The data are listed in table below.

Black Body Temperature (°C)	Percentage of Deviation	
	At 1400°C Calibration	At 1500°C Calibration
1300	0.90	1.15
1320	0.80	1.05
1340	0.40	0.65
1360	0.50	0.75
1380	0.30	0.55
1400	0.00	0.30
1420	−0.05	0.25
1440	−0.05	0.20
1460	−0.15	0.15
1480	−0.25	0.05
1500	−0.40	−0.05

Is the mean percent deviation significantly higher at 1500°C? Use $\alpha = 0.05$.

10.64 Pfeifer, Schubert, Liauw, and Emig (*Trans IChemE*, 2003) studied dynamic behavior of microreactors for methanol steam reforming, which is used in fuel cell applications. They measured temperatures on the top cover plate of reactor type RS2 at a set point of 310°C for two different electric controls (grouping and single control) of the heat cartridges. Temperatures measured at 11 specific thermocouple holes are

listed in table at the right. Is the mean temperature for grouping significantly higher than the
mean temperature for single control? Justify at a
5% error rate.

Number of Thermocouple Holes	Temperature (°C)	
	Grouping	**Single Control**
1	285	277
2	285	280
3	287	283
4	289	285
5	290	289
6	293	290
7	293	290
8	293	290
9	292	291
10	291	291
11	291	291

10.4 χ^2 Tests on Frequency Data

Earlier we saw how to construct tests of a hypothesis concerning a binomial parameter p in the large-sample case. The test was based on statistic Y, the number (or
frequency) of successes in n trials of an experiment. The observation of interest
was a frequency count rather than a continuous measurement such as a lifetime,
reaction time, velocity, and so on. Now we take a detailed look at three types of
situations in which hypothesis-testing problems arise for frequency, or count data.
All the results in this section are approximations that work well only when samples are reasonably large. Later we'll discuss what we mean by "large" samples.

10.4.1   Testing parameters of the multinomial distribution

The first of the three situations involves testing a hypothesis concerning the parameters of a multinomial distribution. (See Chapter 5 for a description of this distribution
and some of its properties.) Suppose n observations (trials) come from a multinomial
distribution with k possible outcomes per trial. Let X_i, $i = 1, \ldots, k$, denote the number of trials resulting in outcome i, with p_i denoting the probability that any one trial
will result in outcome i. Recall that $E(X_i) = np_i$. Suppose we want to test the hypothesis that the p_i's have specified values, that is, $H_0: p_1 = p_{10}, p = p_{20}, \ldots, p_k = p_{k0}$.
The alternative will be the most general one, which simply states "at least one equality fails to hold." To test the validity of this hypothesis, we can compare the observed
count in cell i, X_i, with what we would expect that count to be if H_0 were true, namely,
$E(X_i) = np_{i0}$. So the test statistic is based on $X_i - E(X_i)$, $i = 1, \ldots, k$. We don't
know if these differences are large or small unless we can standardize them in some
way. It turns out that a good statistic to use squares these differences and divides them
by $E(X_i)$, resulting in the test statistic

$$X^2 = \sum_{i=1}^{k} \frac{[X_i - E(X_i)]^2}{E(X_i)}$$

In repeated sampling from a multinomial distribution for which H_0 is true, X^2 will have approximately a $\chi^2(k-1)$ distribution. The null hypothesis is rejected for large values of X^2 ($X^2 > \chi_\alpha^2(k-1)$), because X^2 will be large if there are large discrepancies between X_i and $E(X_i)$.

Testing Parameters of the Multinomial Distribution (Chi-Square Test)

H_0: $p_i = p_{i0}$ ($i = 1, 2, \ldots, k$), where $P(\text{Outcome } i) = p_i$.
H_a: At least one probability differs from the specified value.

$$TS: X^2 = \sum_{i=1}^{k} \frac{(X_i - np_{i0})^2}{np_{i0}} \quad \text{with} \quad \nu = (k-1) \text{ degrees of freedom}$$

Reject the null hypothesis if

- $X^2 > \chi^2(\nu)$ (using the rejection region approach).
- p-value $< \alpha$, where p-value $= P(\chi^2(\nu) > X^2)$ (using the p-value approach).

Conditions:

1. A random sample of size n is taken.
2. The sample size n is large enough to get an approximate chi-square distribution for the test statistic.

Checking the Condition of Large n If for all expected counts $np_{i0} \geq 5$, then we have a large enough sample to use chi-square approximation. If some cells have $np_{i0} < 5$, then such cells can be combined with some other adjoining cells in some logical manner to satisfy the requirement of expected counts.

Example 10.27

The ratio of the number of items produced in a factory by three shifts—first, second, and third—is 4:2:1, due primarily to the decreased number of employees on the later shifts. This means that 4/7 of the items produced come from the first shift, 2/7 from the second, and 1/7 from the third. It is hypothesized that the number of defectives produced should follow this same ratio. A sample of 50 defective items was tracked back to the shift that produced them, with the following results:

	Shift 1	Shift 2	Shift 3
No. of defectives	20	16	14

Test the hypothesis indicated above with $\alpha = 0.05$.

Solution   We are interested in testing $H_0: p_1 = 4/7, p_2 = 2/7, p_3 = 1/7$ against H_a: The proportions differ from those indicated in the null hypothesis. Under H_0, the expected number of defectives from the three shifts, respectively, are

$$E(X_1) = np_1 = 50(4/7) = 28.57, \quad E(X_2) = np_2 = 50(2/7) = 14.29,$$
and
$$E(X_3) = np_3 = 50(1/7) = 7.14$$

Because all the expected counts are at least 5, we can compute the χ^2-test statistic as

$$\sum_{i=1}^{3} \frac{[X_i - E(X_i)]^2}{E(X_i)} = \frac{[20 - 28.57]^2}{28.57} + \frac{[16 - 14.29]^2}{14.29} + \frac{[14 - 7.14]^2}{7.14} = 9.367$$

Now $\chi^2_{0.05}(2) = 5.991$, and because our observed χ^2 is larger than this critical χ^2 value, we reject H_0. In conclusion, the ratio of the number of defective items produced in this factory by three shifts differs from 4:2:1.

This χ^2 test is always a two-tailed test, and although the observed 14 is approximately twice the expected 7.14, the observed 20 is less than the expected 28.57, and both observed cells are out of line. This test does not indicate which discrepancy is more serious.

10.4.2   Testing equality among binomial parameters

The second situation in which a X^2 statistic can be used on frequency data is in the testing of equality among binomial parameters for k separate populations. In this case, k independent random samples are selected that result in k binomially distributed random variables $Y_1, \ldots, Y_k$, where Y_i is based on n_i trials with success probability p_i on each trial. The problem is to test the null hypothesis $H_0: p_1 = p_2 = \cdots = p_k$ against the alternative of at least one inequality. There are now $2k$ cells to consider, as outlined in Table 10.2.

Table 10.2

Cells for k Binomial Observations

	Observation					
	1	2	3	$\ldots$	k	Total
Successes	y_1	y_2	y_3		y_k	y
Failures	$n_1 - y_1$	$n_2 - y_2$	$n_3 - y_3$		$n_k - y_k$	$n - y$
Total	n_1	n_2	n_3		n_k	n

$$n = \sum_{i=1}^{k} n_i \quad \text{and} \quad y = \sum_{i=1}^{k} y_i$$

As in the case of testing multinomial parameters, we should construct the test statistic by comparing the observed cell frequencies with the expected cell frequencies. However, the null hypothesis does not specify values for $p_1, p_2, \ldots, p_k$, and hence the expected cell frequencies must be estimated. We now discuss the expected cell frequencies and their estimators.

Because each Y_i has a binomial distribution, we know from Chapter 5 that

$$E(Y_i) = n_i p_i \quad \text{and} \quad E(n_i - Y_i) = n_i - n_i p_i = n_i(1 - p_i)$$

Under the null hypothesis that $p_1 = p_2 = \cdots = p_k = p$, we estimate the common value of p by pooling the data from all k samples. The minimum-variance unbiased estimator of p is

$$\frac{1}{n}\sum_{i=1}^{k} Y_i = \frac{Y}{n} = \frac{\text{total number of successes}}{\text{total sample size}}$$

The estimators of the expected cell frequencies are then taken to be

and
$$\hat{E}(Y_i) = n_i\left(\frac{Y}{n}\right)$$

$$\hat{E}(n_i - Y_i) = n_i\left(1 - \frac{Y}{n}\right) = n_i\left(\frac{n - Y}{n}\right)$$

In each case, note that the estimate of an expected cell frequency is found by the following rule:

$$\text{estimated expected cell frequency} = \frac{(\text{column total})(\text{row total})}{\text{overall total}}$$

To test $H_0\colon p_1 = \cdots = p_k = p$, we make use of the statistic

$$X^2 = \sum_{i=1}^{k}\left\{\frac{[Y_i - \hat{E}(Y_i)]^2}{\hat{E}(Y_i)} + \frac{[(n_i - Y_i) - \hat{E}(n_i - y_i)]^2}{\hat{E}(n_i - Y_i)}\right\}$$

This statistic has approximately a $\chi^2(k - 1)$ distribution, as long as the sample sizes are reasonably large.

Testing Equality among Binomial Parameters   **(Chi-Square Test)**

$H_0\colon p_1 = p_2 = \cdots = p_k$.
H_a: At least one probability differs from the rest.

$$\text{TS: } X^2 = \sum_{i=1}^{k}\left[\frac{(Y_i - n_i Y/n)^2}{n_i Y/n} + \frac{((n - Y_i) - n_i(n - Y)/n)^2}{n_i(n - Y)/n}\right] \quad \text{with} \quad \nu = (k - 1)$$

degrees of freedom

Reject the null hypothesis if

- $X^2 > \chi^2(\nu)$ (using the rejection region approach).
- p-value $< \alpha$, where p-value $= P(\chi^2(\nu) > X^2)$ (using the p-value approach).

Conditions:

1. A random sample of size n is taken.
2. The sample size n is large enough to get an approximate chi-square distribution for the test statistic.

Checking the Condition of Large n If all expected counts, $n_i Y/n \geq 5$, and $n_i(n - Y)/n \geq 5$, then we have a large enough sample to use a chi-square approximation. If some cells have expected counts < 5, then such cells can be combined with some other adjoining cells in some logical manner to satisfy the requirement of expected counts.

Example 10.28

A chemical company is experimenting with four different mixtures of a chemical designed to kill a certain species of insect. Independent random samples of 200 insects are subjected to one of the four chemicals, and the number of insects dead after 1 hour of exposure is counted. The results are summarized in table below.

Observed Counts	Mix 1	Mix 2	Mix 3	Mix 4	Total
Dead	124	147	141	142	**564**
Not dead	76	53	59	48	**236**
Total	**200**	**200**	**200**	**200**	**800**

We want to test the hypothesis that the rate of kill is the same for all four mixtures. Test at the 5% level of significance.

Solution   We are testing $H_0: p_1 = p_2 = p_3 = p_4 = p$, where p_i denotes the probability that an insect subjected to chemical i dies in the indicated length of time. Now the estimated cell frequencies under H_0 are found by

$$\frac{(n_1)(y)}{n} = \frac{(\text{Column 1 total})(\text{Row 1 total})}{n} = \frac{(200)(564)}{800} = 141$$

$$\frac{(n_1)(n - y)}{n} = \frac{(\text{Column 1 total})(\text{Row 2 total})}{n} = \frac{(200)(236)}{800} = 59$$

and so on. Note that in this case all estimated frequencies in any one row are equal. The estimated expected frequencies are given in table below:

Estimated Expected Counts	Mix 1	Mix 2	Mix 3	Mix 4	Total
Dead	141	141	141	141	**564**
Not dead	59	59	59	59	**236**
Total	**200**	**200**	**200**	**200**	**800**

Because all the expected frequencies are at least 5, we can proceed with the χ^2 test. Calculating the observed χ^2 value, we get

$$\chi^2 = \frac{(124 - 141)^2}{141} + \cdots + \frac{(48 - 59)^2}{59} = 10.72$$

Because $\chi^2_{0.05}(3) = 7.815 < 10.72$, we reject the hypothesis of equal kill rates for the four chemicals and conclude that at least two mixtures have different kill rates.

10.4.3 Test of independence

The third type of experimental situation that makes use of a χ^2 test again involves the multinomial distribution. However, this time the cells arise from a double classification scheme. For example, employees could be classified according to gender and marital status, which results in four cells, married male, married female, unmarried male, and unmarried female.

	Male	Female
Married		
Unmarried		

A random sample of n employees would then give values of random frequencies to the four cells. In these two-way classifications, a common question to ask is, Does the row criterion depend on the column criterion? In other words, is the row criterion contingent upon the column criterion?

In this context, the two-way tables like the one just illustrated are referred to as *contingency tables*. We might want to know whether marital status depends on the gender of the employee or whether these two classification criterion seem to be independent of each other.

The null and the alternative hypotheses in these two-way tables are as follows:

- H_0: The row criterion and the column criterion are independent of each other.
- H_a: The row and the column criteria are not independent of each other.

Looking at the following 2×2 tables, we find

	B_1	B_2
A_1	5	15
A_2	10	30

(a)

	B_1	B_2
A_1	5	30
A_2	10	15

(b)

- An array such as the one in table (a) would not refute independence, because within each column the chance of being in row 1 is about 1 out of 3. That is, the chance of being in row 1 does not depend on which column you happen to be in.

- An array like the one in table (b) would support the alternative hypothesis of dependence, because the chance of being in row 1 changes from column 1 (1 in 3) to column 2 (2 in 3). That is the chance of being in row 1 depends on which column you happen to be in.

In general, the cell frequencies and cell probabilities can be labeled as

X_{ij} = the observed frequency of cell in the row i and column j.
p_{ij} = the probability of being in the cell in the row i and column j.

Under the null hypothesis of independence between rows and columns, $p_{ij} = p_{i.}p_{.j}$, where $p_{i.}$ denotes the probability of being in row i and $p_{.j}$ denotes the probability of being in column j. Also, $E(X_{ij}) = np_{ij} = np_{i.}p_{.j}$. Now $p_{i.}$ and $p_{.j}$ must be estimated and the best estimators are $p_{i.} = X_{i.}/n$ and $p_{.j} = X_{.j}/n$ (see Table 10.3).

Table 10.3

A $r \times c$ Contingency Table

		Column				
		1	$\cdots$	**j**	$\cdots$	**c**	**Row Total**
Row	1	X_{11}	$\cdots$	X_{1j}	$\cdots$	X_{1c}	X_{11}
	$\vdots$	$\vdots$		$\vdots$		$\vdots$	$\vdots$
	i	X_{i1}	$\cdots$	X_{ij}	$\cdots$	X_{ic}	$X_{i.}$
	$\vdots$	$\vdots$		$\vdots$		$\vdots$	$\vdots$
	r	X_{r1}	$\cdots$	X_{rj}	$\cdots$	X_{rc}	$X_{r.}$
Column Total		$X_{.1}$	$\cdots$	$X_{.j}$	$\cdots$	$X_{.c}$	n

Thus, the best estimator of an expected cell frequency is

$$\widehat{E(X_{ij})} = n\hat{p}_{i.}\hat{p}_{.j} = n\left(\frac{X_{i.}}{n}\right)\left(\frac{X_{.j}}{n}\right) = \frac{X_{i.}X_{.j}}{n}$$

Once again, we see that the estimate of an expected cell frequency is given by

$$\frac{x_{i.}x_{.j}}{n} = \frac{(\text{row } i \text{ total})(\text{column } j \text{ total})}{n}$$

The test statistic for testing the independence hypothesis is

$$X^2 = \sum_{j=1}^{c}\sum_{i=1}^{r}\frac{\left(X_{ij} - E(X_{ij})\right)^2}{E(X_{ij})}$$

which has approximately a χ^2 distribution with $(r-1)(c-1)$ degrees of freedom, where r is the number of rows and c is the number of columns in the contingency table.

Test of Independence   (Chi-Square Test)

H_0: Two factors of interest are independent.
   Or, there is no association between two factors of interest.

H_a: Two factors of interest are not independent.
   Or, there is an association between two factors of interest.

$$TS: X^2 = \sum_{j=1}^{c}\sum_{i=1}^{r} \frac{\left(X_{ij} - E(X_{ij})\right)^2}{E(X_{ij})}$$

where $E(X_{ij}) = \dfrac{X_{i.}X_{.j}}{n}$ and $\nu = (r - 1)(c - 1)$ degrees of freedom.

Reject the null hypothesis if

- $X^2 > \chi^2(\nu)$ (using the rejection region approach).
- p-value $< \alpha$, where p-value $= P(\chi^2(\nu) > X^2)$ (using the p-value approach).

Conditions:

1. A random sample of size n is taken.
2. The sample size n is large enough to get an approximate χ^2 distribution for the test statistic.

Checking the Condition of Large n If for all expected counts, $E_{ij} \geq 5$ ($i = 1, 2, \ldots, r$, and $j = 1, 2, \ldots, c$), then we have a large enough sample to use chi-square approximation. If some cells have $E_{ij} < 5$, then the corresponding rows and/or columns can be combined with some other rows and/or columns in some logical manner to satisfy the requirement of expected counts.

Example 10.29

A sample of 200 machined parts is selected from the 1-week production of a machine shop that employs three machinists. The parts are inspected to determine whether they are defective, and they are categorized according to which machine did the work. The observed counts are given in Table 10.4.

   Is the defective/nondefective classification independent of machinist classification? Make a decision at the 1% significance level.

Table 10.4

Observed (Expected) Cell Counts for Example 10.29

	Machinist A	Machinist B	Machinist C	Total
Defective	10 (9.92)	8 (10.88)	14 (11.2)	32
Nondefective	52 (52.08)	60 (57.12)	56 (58.8)	168
Total	**62**	**68**	**70**	**200**

Solution Here we are interested in testing the following pair of hypotheses:

H_0: The defective/nondefective classification is independent of machinist classification

H_a: The defective/nondefective classification depends on machinist classification

Here we'll use a χ^2 test of independence. First, we must find the estimated expected cell frequencies, as follows:

$$\hat{E}(X_{11}) = \frac{x_{1.}x_{.1}}{n} = \frac{(32)(62)}{200} = 9.92 \quad \cdots \quad \hat{E}(X_{23}) = \frac{x_{2.}x_{.3}}{n} = \frac{(168)(70)}{200} = 58.8$$

The expected cell frequencies are shown in parenthesis in Table 10.4. All expected cell frequencies are greater than 5. Now the test statistic is

$$\chi^2 = \frac{(10 - 9.22)^2}{9.22} + \cdots + \frac{(56 - 58.8)^2}{58.8} = 1.74$$

The degrees of freedom are $\nu = (r-1)(c-1) = (2-1)(3-1) = 2$. From Table 6 in the Appendix (or calculator output in Figure 10.26), we get the critical value $\chi^2_{0.01}(2) = 9.21$, and p-value $= P(\chi^2(2) > 1.74) = 0.4186$. Because $1.74 < 9.21$ or p-value $> \alpha = 0.01$, we cannot reject the null hypothesis of independence between the machinist and defective/nondefective classification. There is not sufficient evidence to indicate difference in the rate of defectives produced by three machinists.

Figure 10.26

TI-89 output for Example 10.29

Exercises

10.65 Chinowsky (*Journal of Professional Issues in Engineering Education and Practice*, 2002) researched the education background of design and construction executives. Each executive was classified by degree (B.S., M.S., Ph.D.) and type of degree (engineering, business, other). At the 5% level of significance, is there a significant association between the degree and the type of degree among the executives?

	Engineering	Business	Other
BS	172	28	50
M.S.	55	56	14
Ph.D.	7	1	7

10.66 Two types of defects, A and B, are frequently seen in the output of a certain manufacturing process. Each item can be classified into one of the four classes AB, $A\bar{B}$, $\bar{A}B$, $\bar{A}\bar{B}$, where $\bar{A}$ denotes the absence of the type-A defect. For 100 inspected items, the following frequencies were observed:

$$AB \ 48, \ A\bar{B} \ 18, \ \bar{A}B \ 21, \ \bar{A}\bar{B} \ 13$$

Test the hypothesis that the four categories, in the order listed, occur in the ratio 5:2:2:1. (Use $\alpha = 0.05$.)

10.67 Vehicles can turn right, turn left, or continue straight ahead at a certain intersection. It is hypothesized that half the vehicles entering this intersection will continue straight ahead. Of the other half, equal proportions will turn right and left. Fifty vehicles were observed to have the following behavior:

	Straight	Left Turn	Right Turn
Frequency	28	12	10

Test the stated hypothesis at the 10% level of significance.

10.68 A manufacturer of stereo amplifiers has three assembly lines. We want to test the hypothesis that the three lines do not differ with respect to the number of defectives produced. Independent samples of 30 amplifiers each are selected from the output of the lines, and the number of defectives is observed. The data are as follows:

	Line		
	I	II	III
Number of Defectives	6	5	9
Sample Size	30	30	30

Conduct a test of the hypothesis given above at the 5% significance level.

10.69 Two inspectors are asked to rate independent samples of textiles from the same loom. Inspector A reports that 18 out of 25 samples fall in the top category, while inspector B reports that 20 out of 25 samples merit the top category. Do the inspectors appear to differ in their assessments? Use $\alpha = 0.05$.

10.70 Two chemicals, A and B, are designed to protect pine trees from a certain disease. One hundred trees are sprayed with chemical A, and 100 are sprayed with chemical B. All trees are subjected to the disease, with the following results:

	A	B
Infected	20	16
Sample size	100	100

Do the chemicals appear to differ in their ability to protect the trees? Test at the 1% significance level.

10.71 A study of the relationship between athletic involvement and academic achievement for college students sampled randomly 852 students. The selected students were categorized according to amount of athletic involvement and grade-point averages at graduation. The results are listed in table below.

		Athletic Involvement		
		None	1–3 Semesters	4 or More Semesters
GPA	Below Mean	290	94	42
	Above Mean	238	125	63
		528	219	105

a Do final grade-point averages appear to be independent of athletic involvement? (Use $\alpha = 0.05$.)

b For students with four or more semesters of athletic involvement, is the proportion with GPAs above the mean significantly different, at the 0.05 level, from the proportion with GPAs below the mean?

10.72 Refer to Exercise 10.66. Test the hypothesis that the type A defects occur independently of the type B defects. Use $\alpha = 0.05$.

10.73 A new sick-leave policy is being introduced into a firm. A sample of employee opinions showed the following breakdowns by sex and opinion:

	Favor	Oppose	Undecided
Male	31	44	6
Female	42	36	8

Does the reaction to the new policy appear to be related to sex? Test at the 5% significance level.

10.74 A sample of 150 people is observed using one of four entrances to a commercial building. The data are as follows:

Entrance	1	2	3	4
No. of People	42	36	31	41

a Test the hypothesis that all four entrances are used equally often. Use $\alpha = 0.05$.

b Entrances 1 and 2 are on a subway level, while entrances 3 and 4 are on ground level. Test the hypothesis that subway and ground-level entrances are used equally often, use $\alpha = 0.05$.

10.75 The use of aerosol dispensers has been somewhat controversial because of their possible effects on the ozone. A random sample of 300 women in 2003 found that 35% feared the aerosol effects on ozone, whereas only 29% of a sample of 300 stated this fear in 2007. Is there a significant difference between the true proportions for these two years, at the 1% level of significance?

10.76 The study referenced in Exercise 10.75 was actually a 5-year study, with a random sample of 300 respondents per year. The percents of respondents stating that the aerosol dispensers did not spray properly were as follows:

2003	2004	2005	2006	2007
73%	62%	65%	69%	64%

Are there significant differences among these percentages at the 1% level of significance?

10.77 Industrial workers who are required to wear respirators were checked periodically to see that the equipment fit properly. From past records, it was expected that 1.2% of those checked would fail (i.e., their equipment would not fit property). The data from one series of checks were as follows:

Time After First Measurement (months)	Number of Workers Measured	Observed Failures	Expected Failures (1.2% of col. 2)
0–6	28	1	0.34
6–12	92	3	1.10
12–18	270	1	3.24
18–24	702	9	8.42
24–30	649	10	7.79
30–36	134	0	1.61
>36	93	0	1.12
Total	**1,968**	**24**	**23.6**

Can the asumption of the constant 1.2% failure rate be rejected at the 5% significance level?

10.78 In a study of 1,000 major U.S. firms in 2001, 54% stated that quality of life was a major factor in locating corporate headquarters. A similar study of 1,000 firms in 2006 showed 55% making this statement. Is this a significant change at the 5% significance level?

10.5   Goodness of Fit Tests

Thus far, the main emphasis has been on choosing a probabilistic model for a measurement, or a set of measurements, taken on some natural phenomenon. We have discussed the basic probabilistic manipulations one can make as well as checking the condition of normality when estimating certain unknown parameters or testing hypotheses concerning them. The rigorous way of checking to see whether the data we observe do, in fact, agree with the underlying probabilistic model assumed for these data is called *goodness of fit* and is discussed in this section. As an example, two widely applicable techniques, the χ^2 test for goodness of fit for discrete distributions and Kolmogorov-Smirnov test for continuous distributions will be presented in this section.

10.5.1 χ^2 test

Let Y denote a discrete random variable that can take on values $y_1, y_2, \ldots$. Under the null hypothesis H_0, we assume a probabilistic model for Y. That is, we assume $P(Y = y_i) = p_i$, where p_i may be completely specified or may be a function of other unknown parameters. For example, we could assume a Poisson model for Y, and thus

$$P(Y = y_i) = p_i = \lambda^{y_i} e^{-\lambda}/y_i!$$

In addition, we could assume that λ is specified at λ_0, or we could assume that nothing is known about λ in which case λ would have to be estimated from the sample data. The alternative hypothesis is the general one that the model in H_0 does not fit.

- Define null and alternative hypotheses.
  H_0: $P(Y = y_i \mid \theta) = p_i$ against H_a: Model in H_0 does not fit.
  The parameter(s) θ may or may not be known.
- Take a sample consisting of n independently observed values of Y.
- For this sample, let F_i = number of times $(Y = y_i)$ is observed = frequency of y_i.
- Under H_0: $P(Y = y_i) = p_i$, calculate the expected frequency as $E(F_i) = np_i$.
- If θ are unknown, estimate them from sample data. Then $\hat{E}(F_i)$ is also estimated and is referred to as $\hat{E}(F_i)$, the estimated expected frequency.
- If the frequency counts F_i are quite small, then group consecutive values of Y together and add the corresponding frequencies and probabilities. As a rule of thumb, values of Y should be grouped so that the expected count is at least 5 in every cell.
- After this grouping is completed, let k denote the number of cells obtained. Compare the observed values of F_i with what is expected under the null hypothesis. The test statistic will be based on the differences $[F_i - E(F_i)]$ as in previously discussed χ^2 tests. The test statistic for the null hypothesis that $P(Y = y_i) = p_i$, $i = 1, 2, \ldots$, is given by

$$X^2 = \sum_{i=1}^{k} \frac{[F_i - \hat{E}(F_i)]^2}{\hat{E}(F_i)}$$

- For large n, this test statistic has approximately χ^2 distribution with degrees of freedom given by $[(k - 1) - \text{the number of parameters estimated}]$.

Example 10.30

The number of accidents per week Y in a certain factory was checked for a random sample of 50 weeks. The following results were recorded.

Y	0	1	2	3 or more
Frequency	32	12	6	0

Is it reasonable to conclude that Y follows a Poisson distribution? Use $\alpha = 0.05$.

Solution   We are interested in testing the following pair of hypotheses:

H_0: Y follows a Poisson distribution, that is, $P(y_i) = \lambda^{y_i} e^{-\lambda}/y_i$ ($y_i = 0, 1, 2, \ldots$).
H_a: Y does not follow a Poisson distribution.

Because λ is unknown, it must be estimated, and the best estimate is $\hat{\lambda} = \overline{Y}$. For the given data, $\overline{y} = [0(32) + 1(12) + 2(6)]/50 = 0.48$. Because the last cell has zero frequency, we'll combine last two categories into one, giving us a total of three cells, as seen in table on the next page.

Y_i	$\hat{p}_i = P(Y = Y_i)$	$\hat{E}(F_i) = np_i$	F_i
0	$e^{-\hat{\lambda}} = e^{-0.48} = 0.6788$	$50(0.6788) = 30.94$	32
1	$\hat{\lambda}e^{-\hat{\lambda}} = (0.48)e^{-0.48} = 0.2970$	$50(0.2970) = 14.85$	12
2 or more	$1 - e^{-\hat{\lambda}} - \hat{\lambda}e^{-\hat{\lambda}} = 1 - 0.6788 - 0.2970 = 0.0242$	$50(0.0242) = 4.21$	6

The observed χ^2 value is then

$$\chi^2 = \frac{(32 - 30.94)^2}{30.94} + \frac{(12 - 14.85)^2}{14.85} + \frac{(6 - 4.21)^2}{4.21} = 1.34$$

The test statistic has 1 degree of freedom, because $k = 3$ and one parameter is estimated. Now $\chi^2_{0.05}(1) = 3.841 > 1.34$ *or* p-value $= P(\chi^2(1) > 1.34) = 0.247 < \alpha = 0.05$; hence, we do not have sufficient evidence to reject H_0. The data do appear to fit the Poisson model reasonably well.

10.5.2   Kolmogorov-Smirnov test

It is possible to construct χ^2 tests for goodness of fit to continuous distributions, but the procedure is a little more subjective because the continuous random variable does not provide natural cells into which the data can be grouped. For the continuous case, the Kolmogorov-Smirnov (K-S) statistic, which compares the empirical distribution function of a random sample with a hypothesized theoretical distribution function, is described here.

- Define the null and alternative hypotheses.
  H_0: A continuous random variable Y has a distribution function given by $F(y)$.
  H_a: $F(y)$ is not the true distribution function for Y.
- Take a random sample of size n, say, $y_1, y_2, \ldots, y_n$.
- Reorder the observed values from smallest to largest, and denote the ordered y_i's by $y_{(1)} \leq y_{(2)} \leq \cdots \leq y_{(n)}$. That is, if $y_1 = 7, y_2 = 9$, and $y_3 = 3$, then $y_{(1)} = 3$, $y_{(2)} = 7$, and $y_{(3)} = 9$.
- Determine the empirical distribution function as follows:

  $F_n(y) =$ fraction of the sample less than or equal to y

  $$= \begin{cases} \dfrac{(i-1)}{n} & \text{if } y_{(i-1)} \leq y \leq y_{(i)} \quad i = 1, 2, \ldots, n \\ 1 & \text{if } y \geq y_{(n)} \end{cases}$$

  where we let $y_0 = -\infty$.
- Estimate all unknown parameters from the data.
- If the null hypothesis is true, then $F(y)$ should be close to $F_n(y)$. The test statistic must measure the closeness of $F(y)$ to $F_n(y)$ over the whole range of y-values. See Figure 10.27 for a typical plot of $F(y)$ and $F_n(y)$.

Figure 10.27

A plot of $F_n(y)$ and $F(y)$

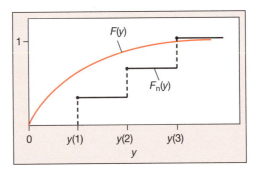

- Compute the K-S statistic D based on the maximum distance between $F(y)$ and $F_n(y)$ as

$$D = \max_{y} [F(y) - F_n(y)]$$

- The null hypothesis is rejected if D is too large. Because $F(y)$ and $F_n(y)$ are nondecreasing and $F_n(y)$ is constant between sample observations, the maximum deviation between $F(y)$ and $F_n(y)$ will occur either at one of the observation points $y_1, y_2, \ldots, y_n$ or immediately to the left of one of these points. Because $D = \max(D^+, D^-)$, it is necessary to check only

$$D^+ = \max_{1 \le i \le n} \left[\frac{i}{n} - F(y_i)\right] \quad \text{and} \quad D^- = \max_{1 \le i \le n} \left[F(y_i) - \frac{i-1}{n}\right]$$

- The critical values for this test are given in Table XX of Appendix.

Example 10.31

Consider the first 10 observations of Table 6.1 to be a random sample from a continuous distribution. Is it reasonable to assume that these data are from an exponential distribution with mean 2? Use the $\alpha = 0.05$ significance level.

Solution   We are interested in testing the following pair of hypotheses.

H_0: $F(y)$ is exponential with $\theta = 2$, i.e. $F(y_i) = 1 - e^{-y_i/2}$.
H_a: $F(y)$ is not exponential with $\theta = 2$.

Order 10 observations ($y_{(i)}$) and for each ordered observation find the value of $F(y_i)$.

Because $D^+ = 0.0866$ and $D^- = 0.2901$, $D = \max(D^+, D^-) = 0.2901$. From Table XX of the Appendix, the critical value for a two-sided test with $n = 10$ and $\alpha = 0.05$ is 0.409. Because $0.409 > D = 0.2901$, at the $\alpha = 0.05$ significance level, we do not reject the null hypothesis. Notice, we are saying that we cannot reject the exponential model with $\theta = 2$ as a plausible model for these data. This does not imply that the exponential model is the *best* model for these data, because several other possible models would not be rejected either.

i	$y_{(i)}$	$F(y_i)$	i/n	$(i-1)/n$	$i/n - F(y_i)$	$F(y_i) - (i-1)/n$
1	0.023	0.0114	0.1	0.0	0.0886	0.0114
2	0.406	0.1837	0.2	0.1	0.0163	0.0837
3	0.538	0.2359	0.3	0.2	0.0641	0.0359
4	1.267	0.4693	0.4	0.3	−0.0693	0.1930
5	2.343	0.6901	0.5	0.4	−0.1901	0.2901
6	2.563	0.7224	0.6	0.5	−0.1224	0.2224
7	3.334	0.8112	0.7	0.6	−0.1112	0.2112
8	3.491	0.8254	0.8	0.7	−0.0254	0.1254
9	5.088	0.9214	0.9	0.8	−0.0214	0.1214
10	5.587	0.9388	1.0	0.9	0.0612	0.0388
					$D^+ = 0.0866$	$D^- = 0.2901$

- Samples of a size less than 20 do not allow for discrimination among distributions; that is, many different distributions may all appear to fit equally well.
- There are numerous goodness-of-fit statistics that could be considered in addition to Kolmogorov-Smirnov, such as *chi-square, Cramer-von Mises, Anderson-Darling, Shapiro-Wilk W, likelihood ratio (LR)*, and *Watson* statistics. Some goodness-of-fit statistics are distribution-specific. Although there are some differences in their mathematical development, the general idea behind them is the same.
- With technological development, it is much easier to use some software to assess fit of a distribution. Note that different software provide output for different goodness-of-fit techniques.

It is much easier to fit a distribution and assess its fit using computer programs. See the following printout from JMP for the data in Example 10.31. The JMP uses the *likelihood* ratio test to assess goodness of exponential fit, which gives *p*-value ≈ 1.0. Therefore, at any reasonable level of significance we would not reject the null hypothesis that these data might be from the exponential distribution with $\theta = 2$.

Figure 10.28

Exponential probability plot for data in Example 10.31

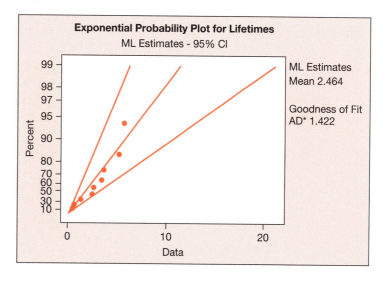

Alternatively, the fit can be assessed visually using probability plots. See Figure 10.28 for an exponential probability plot generated using Minitab. It gives the *Anderson-Darling* goodness-of-fit statistic. The middle straight line represents the exponential model with $\theta = 2$. The closely scattered data points around this line indicate adequacy of fit for this model.

10.6 Using Computer Programs to Fit Distributions

A graphical check of normality introduced in Chapter 6 is derived from the fact that if X is normally distributed with mean μ and standard deviation σ, then $Z = (X - \mu)/\sigma$ has a standard normal distribution, or $X = \mu + \sigma Z$. We can use this result to check the assumption of normality.

- Let $x_{(1)} \leq x_{(2)} \leq \cdots \leq x_{(n)}$ represent ordered sample observations.
- Then $x_{(i)}$ can be thought as the sample percentile $i/(n + 1)$. Compute the corresponding standard normal percentile known as a z-score, $z_{(i)}$. It is also known as the *normal score*.
- Plot $x_{(i)}$ against $z_{(i)}$, $i = 1, 2, \ldots, n$.
- This plot will be close to a straight line if the data come from a normal distribution. The plot will deviate substantially from a straight line if the data do not come from a normal distribution.

Many computer packages calculate normal scores corresponding to sample data, so this plot is easy to produce. However different packages might use different approximations for calculating percentiles, resulting in slightly different plots.

Example 10.32

Soil–water flux measurements (in centimeters/day) were taken at 20 experimental plots in a field. The soil–water flux was measured in a draining soil profile in which steady-state soil–water flow conditions had been established. Theory and empirical evidence have suggested that the measurements should fit a lognormal distribution. The data are recorded (y_i) as logarithms of the actual measurements. Test the hypothesis that the y_i's are from a normal distribution. Use $\alpha = 0.05$.

0.3780	0.5090	0.6230	0.6860	0.7350
0.7520	0.7580	0.8690	0.8890	0.8890
0.8990	0.9370	0.9820	1.0220	1.0370
1.0880	1.1230	1.2060	1.3340	1.4230

Solution   Here we are interested in testing the following pair of hypotheses:

H_0: The data came from a normal population.
H_a: The data did not come from a normal population.

Figure 10.29

Normal probability plot for soil-water flux data

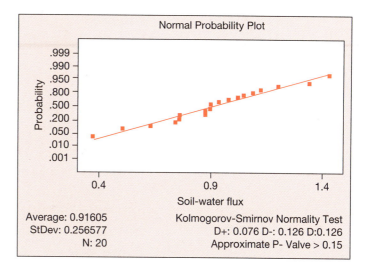

Under the null hypothesis, the random variable Y has a normal distribution with mean μ and variance σ^2, where μ and σ^2 are unknown and must be estimated from the data. Once again this can be easily achieved using some computer program.

The points in Figure 10.29 do fall rather close to the straight line, substantiating the claim that the data come from a normal distribution. The straight line in the plot is the theoretical normal distribution line. Also note that the approximate p-value > 0.15 for the K-S test. Thus, the null hypothesis of normal distribution for the data is not rejected.

The plot also gives the sample estimates of mean and standard deviation. It is possible to estimate the mean and standard deviation from the plot.

- The soil–water flux measurement corresponding to the 50th percentile (i.e., probability = 0.50) gives the mean to be approximately equal to 0.91.
- The range of the data is $R = 1.4 - 0.4 = 1.0$. It gives the standard deviation to be approximately equal to $R/4 = 0.25$.

Example 10.33

Taraman (1974) reported the failure times of 24 machine tools.

$$70, 29, 60, 28, 64, 32, 44, 24, 35, 31, 38, 35,$$
$$52, 23, 40, 28, 46, 33, 46, 27, 37, 34, 41, 28$$

Are these data likely to be from exponential or lognormal population? Use $\alpha = 0.05$.

Solution The JMP printout for fitting exponential and lognormal distributions is shown in Figure 10.30 along with the outcome of the K-S test.

For exponential fit, p-value $= 0.01 < \alpha = 0.05$. Thus, reject the null hypothesis that the data fit an exponential distribution and conclude that the data are not from an exponential population. The curve and the histogram indicate the same.

For the lognormal fit, p-value $= 0.15 > \alpha = 0.05$. Thus, we fail to reject the null hypothesis that the data are from a lognormal distribution. It seems the lognormal distribution is a reasonable fit to these data. The curve and the histogram confirm it.

Figure 10.30

JMP printout for Example 10.33

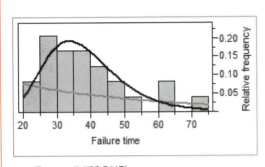

——— Exponential(38.5417)
——— LogNormal(3.60633,0.29452)

Fitted Exponential

▼ Parameter Estimates

Type	Parameter	Estimate	Lower 95%	Upper 95%
Scale	σ	38.541667	26.486944	59.167265

▼ Goodness-of-Fit Test

Kolmogorov's D

D		Prob>D
0.449406	<	0.0100*

Note: Ho = The data is from the Exponential distribution. Small p-values reject Ho.

Fitted LogNormal

▼ Parameter Estimates

Type	Parameter	Estimate	Lower 95%	Upper 95%
Scale	μ	3.6063323	3.4836266	3.729038
Shape	σ	0.2945204	0.2274193	0.4024748

▼ Goodness-of-Fit Test

Kolmogorov's D

D		Prob>D
0.110384	>	0.1500

Note: Ho = The data is from the LogNormal distribution. Small p-values reject Ho.

Exercises

10.79 For fabric coming off a certain loom, the number of defects per square yard is counted on 50 sample specimens, each 1 square yard in size. The results are as follows:

Number of Defects	Frequency of Observation
0	0
1	3
2	5
3	10
4	14
5	8
6 or more	10

Test the hypothesis that the data come from a Poisson distribution. Use $\alpha = 0.05$.

10.80 The data in the following table show the frequency counts for 400 observations on the number of bacterial colonies within the field of a microscope, using samples of milk film. (Bliss and Owens, *Biometrics*, 1953.) Test the hypothesis that the data come from a Poisson distribution. Use $\alpha = 0.05$.

Number of Colonies per Field	Frequency of Observations
0	56
1	104
2	80
3	62
4	42
5	27
6	9
7	9
8	5
9	3
10	2
11	0
19	1

10.81 The number of accidents experienced by machinists in a certain industry was observed for a certain period of time with the following results. (Bliss and Fisher, *Biometrics*, 1953.)

Accidents per Machinist	0	1	2	3	4	5	6	7	8
Frequency (number of machinists)	296	74	26	8	4	4	1	0	1

At the 5% level of significance, test the hypothesis that the data came from a Poisson distribution.

10.82 The following data are observed LC50 values on copper in a certain species of fish (measurements in parts per million):

0.075, 0.10, 0.23, 0.46, 0.10, 0.15, 1.30, 0.29, 0.31, 0.32, 0.33, 0.54, 0.85, 1.90, 9.00

It has been hypothesized that the natural logarithms of these data fit the normal distribution. Check this claim at the 5% significance level. Also, make a normal scores plot if a computer is available.

10.83 The time (in seconds) between vehicle arrivals at a certain intersection was measured for a certain time period, with the following results:

9.0, 10.1, 10.2, 9.3, 9.5, 9.8, 14.2, 16.1, 8.9, 10.5, 10.0, 18.1, 10.6, 16.8, 13.6, 11.1

a Test the hypothesis that these data come from an exponential distribution. Use $\alpha = 0.05$.
b Test the hypothesis that these data come from an exponential distribution with a mean of 12 seconds. Use $\alpha = 0.05$.

10.84 The weights of copper grains are hypothesized to follow a lognormal distribution, which implies that the logarithms of the weight measurements should follow a normal distribution. Twenty weight measurements, in 10^{-4} gram, are as follows:

2.0, 3.0, 3.1, 4.3, 4.4, 4.8, 4.9, 5.1, 5.4, 5.7, 6.1, 6.6, 7.3, 7.6, 8.3, 9.1, 11.2, 14.4, 16.7, 19.8

Test the hypothesis stated above at the 5% significance level.

10.85 Fatigue life, in hours, (A.C. Cohen, et al., *Journal of Quality Technology*, 16, no. 3, 1984, p. 165.) for 10 bearings of a certain type was as follows:

152.7, 172.0, 172.5, 173.3, 193.0, 204.7, 216.5, 234.9, 262.6, 422.6

Test the hypothesis, at the 5% significance level, that these data follow a Weibull distribution with $\gamma = 2$.

10.86 The times to failure of eight turbine blades in jet engines, in 10^3 hours, were as follows (Pratt & Whitney Aircraft, PWA 3001, 1967.):

3.2, 4.7, 1.8, 2.4, 3.9, 2.8, 4.4, 3.6

Do the data appear to fit a Weibull distribution with $\gamma = 3$? Use $\alpha = 0.025$.

10.87 Records were kept on the time of successive failures of the air conditioning system of a Boeing 720 jet airplane. If the air conditioning system has a constant failure rate, then the intervals between successive failure must have an exponential distribution. The observed intervals, in hours, between successive failures are as follows (F. Proschan, *Technometrics,* 1963.):

23, 261, 87, 7, 120, 14, 62, 47, 225, 71, 246, 21, 42, 20, 5, 12, 120, 11, 3, 14, 71, 11, 14, 11, 16, 90, 1, 16, 52, 95

Do these data seem to follow an exponential distribution at the 5% significance level? If possible, construct a normal scores plot for these data. Does the plot follow a straight line?

10.7 Acceptance Sampling

From Chapter 1 to this point, emphasis has continually been focused on the use of statistical techniques for improving the quality of processes (or total quality management). The old-style quality-control plans consisted of items being inspected as they left the factory, with the good ones being sold and the bad ones being discarded or reworked. That philosophy has given way to a more cost-efficient philosophy of making things right in the first place, and statistical techniques like the tools introduced in Chapter 1 and the control charts of Chapter 8 reflect the modern view.

As a result of this shift in emphasis, acceptance sampling plans (an old-style technique for inspecting lots) have fallen into disfavor. However, many constructive uses of acceptance sampling plans still exist in applications that are "process-centered" rather than "product-centered." Suppose worker A solders wires in circuit boards that are then passed on to worker B for installation into a system. It might be worthwhile to sample soldered connections occasionally, pulling them apart to assess the strength of the soldering, so that the quality of this operation in maintained. (Obviously, every connection cannot be tested due to both the destructive nature of the test and time constraints.) A sampling plan must be designed to efficiently measure the defect rate for the soldering process. Here, worker A is the "producer." Notice that the use of sampling within the process is far different from the old scheme of manufacturing the systems first and then testing a sample of the finished products.

10.7.1 Acceptance Sampling by Attributes

This section discusses the problem of making a decision with regard to the proportion of defective items in a finite lot based on the number of defectives observed in a sample from that lot. Inspection in which a unit of product is classified simply as defective or nondefective is called *inspection by attributes*.

Lots of items, either raw materials or finished products, are sold by a producer to a consumer with some guarantee as to quality. In this section, quality will be determined by the proportion p of defective items in the lot. To check on the quality characteristics of a lot, the consumer will sample some of the items, test them, and observe the number of defectives. Generally, some defectives are allowable, because defective-free lots may be too expensive for the consumer to purchase. But if the number of defectives is too large, the consumer will reject the lot and return it to the producer. In the decision to accept or reject the lot, sampling is generally used, because the cost of inspecting the entire lot may be too high or the inspection process may be destructive.

Before sampling inspection takes place for a lot, the consumer must have in mind a proportion p_0 of defectives that will be acceptable. Thus, if the true proportion of defectives p is no greater than p_0, the consumer will want to accept the lot, and if p is greater than p_0, the consumer will want to reject the lot. This maximum proportion of defectives satisfactory to the consumer is called the *acceptable quality level* (AQL).

Note that we now have a hypothesis-testing problem in which we are testing $H_0: p \leq p_0$ versus $H_a: p > P_0$. A random sample of n items will be selected from the lot and the number of defectives Y observed. The decision concerning rejection or nonrejection of H_0 (i.e., rejecting or accepting the lot) will be based on the observed value of Y.

The probability of a type I error α in this problem is the probability that the lot will be rejected by the consumer when, in fact, the proportion of defectives is satisfactory to the consumer. This is referred to as the *producer's risk*. The value of α calculated for $p = p_0$ is the upper limit of the proportion of good lots rejected by the sampling plan being considered. The probability of a type II error β calculated for some proportion of defectives p_1, where $p_1 > p_0$, represents the probability that an unsatisfactory lot will be accepted by the consumer. This is referred to as the *consumer's risk*. The value of β calculated for $p = p_1$ is the upper limit of the proportion of bad lots accepted by the sampling plan, for all $p \geq p_1$.

The null hypothesis $H_0: p \leq p_0$ will be rejected if the observed value of Y is larger than some constant a. Because the lot is accepted if $Y \leq a$, the constant a is called the *acceptance number*. In this context, the significance level is given by

$$\alpha = P(\text{reject } H_0 \text{ when } p = p_0) = P(Y > a \text{ when } p = p_0)$$

and

$$1 - \alpha = P(\text{not reject } H_0 \text{ when } p = p_0)$$
$$= P(\text{accepting the lot when } p = p_0) = P(A)$$

Because p_0 may not be known precisely, it is of interest to see how $P(A)$ behaves as a function of the true p, for a given n and a. A plot of these probabilities is called an *operating characteristic curve*, abbreviated OC curve. We illustrate its calculation with the following numerical example.

Example 10.34

One sampling inspection plan calls for $n = 10$, and $a = 1$, while another calls for $n = 25$ and $a = 3$. Plot the operating characteristic curves for both plans. If the plant using these plans can operate efficiently on 30% defective raw materials, considering the price, but cannot operate efficiently if the proportion gets close to 40%, which plan should be chosen?

Solution We must calculate $P(A) = P(Y \leq a)$ for various values of p in order to plot the OC curves. Using Table 2 in the Appendix, we have

$n = 10, a = 1$		$n = 25, a = 3$	
$p = 0$	$p(A) = 1$	$p = 0$	$p(A) = 1$
$p = 0.1$	$p(A) = 0.736$	$p = 0.1$	$p(A) = 0.764$
$p = 0.2$	$p(A) = 0.376$	$p = 0.2$	$p(A) = 0.234$
$p = 0.3$	$p(A) = 0.149$	$p = 0.3$	$p(A) = 0.033$
$p = 0.4$	$p(A) = 0.046$	$p = 0.4$	$p(A) = 0.002$
$p = 0.6$	$p(A) = 0.002$		

These values are plotted in Figure 10.31. Note that the OC curve for the ($n = 10$, $a = 1$) plan drops slowly and does not get appreciably small until p is in the neighborhood of 0.4. For p around 0.3, this plan still has a fairly high probability of accepting the lot. The other curve ($n = 25$, $a = 3$) drops much more rapidly and falls to a small $P(A)$ at $p = 0.3$. The latter plan would be better for the plant, if it can afford to sample $n = 25$ items out of each lot before making a decision.

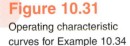

Figure 10.31

Operating characteristic curves for Example 10.34

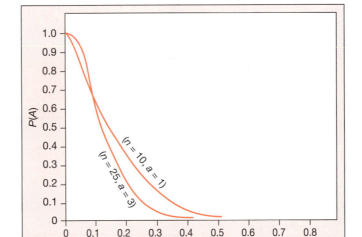

Numerous handbooks give operating characteristic curves and related properties for a variety of sampling plans for inspecting by attributes; one of the most widely used sets of plans is Military Standard 105D (MIL-STD-105D).

Even though the probability of acceptance, as calculated in Example 10.34, generally does not depend on the lot size as long as the lot is large compared to the sample size, MIL-STD-105D is constructed so that sample size increases with lot size. The practical reasons for the increasing sample sizes are that small random samples are sometimes difficult to obtain from large lots and that discrimination between good and bad lots may be more important for large lots. The adjustments in sample sizes for large lots are based on empirical evidence rather than on strict probabilistic considerations.

MIL-STD-105D contains plans for three sampling levels, labeled I, II, and III. Generally, sampling or inspection level II is used. Level I gives slightly smaller samples and less discrimination, whereas level III gives slightly larger samples and more discrimination. Table 10 in the Appendix gives the code letters for various lot of batch sizes and inspection levels I, II, and III. This code letter is used in entering Table 11, 12, or 13 to find the sample size and acceptance number to be used for specified values of the AQL.

In addition to having three inspection levels, MIL-STD-105D also has sampling plans for normal, tightened, and reduced inspection. These plans are given in Tables 11, 12, and 13, respectively. Normal inspection is designed to protect the producer against the rejection of lots with percent defectives less than the AQL. Tightened inspection is designed to protect the consumer from accepting lots with percent defectives greater than the AQL. Reduced inspection is introduced as an economical plan to be used if the quality history of the lots is good. When sampling a series of lots, rules for switching among normal, tightened, and reduced sampling are given in MIL-STD-105D. When MIL-STD-105D is used on a single lot, the OC curves for various plans, given in the handbook, should be studied carefully before a decision is made on the optimal plan for a particular problem.

Example 10.35

A lot of size 1,000 is to be sampled with normal inspection at level II. If the AQL of interest is 10%, find the appropriate sample size and acceptance number from MIL-STD-105D.

Solution   Entering Table 10 in the Appendix for a lot of size 1,000, we see that the code letter for level II is J. Because normal inspection is to be used, we enter Table 11 at row J. The sample size is 80, and, moving over to the column headed 10% AQL, we see that the acceptance number is 14. Thus, we sample 80 items and accept the lot if the number of defectives is less than or equal to 14. This plan will have a high probability of accepting any lot with a true percentage of defectives below 10%.

10.7.2   Acceptance Sampling by Variables

In acceptance sampling by attributes, we merely record, for each sampled item, whether or not the item is defective. The decision to accept or reject the lot then depends on the number of defectives observed in the sample. However, if the characteristic under study involves a measurement on each sampled item, we could base the decision to accept or reject on the measurements themselves, rather than simply on the number of items having measurements that do not meet a certain standard.

For example, suppose a lot of manufactured steel rods is checked for quality, with the quality measurement being the diameter. Rods are not allowed to exceed 10 centimeters in diameter, and a rod with a diameter in excess of 10 centimeters is declared defective for this special case. In attribute sampling, the decision to accept or reject the lot is based on the number of defective rods observed in a sample. But the decision could be made instead on the basis of some function on the actual diameter observations themselves, say, the mean of these observations. If the sample mean is close to or in excess of 10 centimeters, we would expect the lot to contain a high percentage of defectives, whereas if the sample mean is much smaller than 10 centimeters, we would expect the lot to contain relatively few defectives.

Using actual measurements of the quality characteristic as a basis for acceptance or rejection of a lot is called *acceptance sampling by variables*. Sampling by variables often has advantages over sampling by attributes, since the variables method may contain more information for a fixed sample size than the attribute method. Also, when looking at the actual measurements obtained by the variables method, the experimenter may gain some insights into the degree of nonconformance and may quickly be able to suggest methods for improving the product.

For acceptance sampling by variables, let

$$U = \text{upper specification limit} \quad \text{and} \quad L = \text{lower specification limit}$$

If X denotes the quality measurement being considered, the U is the maximum allowable value of X on an acceptable item. (U is equal to 10 centimeters in the rod-diameter example.) Similarly, L is the minimum allowable value for X on an acceptable item. Some problems may specify U, other L, and still others both U and L.

As in acceptance sampling by attributes, an AQL is specified for a lot, and a sampling plan is then determined that will give high probability of acceptance to lots with percent defectives lower than the AQL and low probability of acceptance to lots with percent defectives higher than the AQL. In practice, an *acceptability constant k* is determined such that a lot is accepted if

$$\frac{U - \overline{X}}{S} \geq k \quad \text{or} \quad \frac{\overline{X} - L}{S} \geq k$$

where $\overline{X}$ is the sample mean and S the sample standard deviation of the quality measurements.

As in the case of sampling by attributed, handbooks of sampling plans are available for acceptance sampling by variables. One of the most widely used of these handbooks is Military Standard 414 (MIL-STD-414). Tables 14 and 15 of the Appendix will help illustrate how MIL-STD-414 is used. Table 14A shows which tables AQL value to use if the problem calls for an AQL not specifically tabled. Table 14B gives the sample size code letter to use with any possible lot size. Sample sizes are tied to lot sizes, and different levels of sampling are used as in MIL-STD-105D. Inspection level IV is recommended for general use.

Table 15 gives the sample size and value of k for a fixed code letter and AQL. Both normal and tightened inspection procedures can be found in this table. The use of these tables is illustrated in Example 10.36.

Example 10.36

The tensile strengths of wires in a certain lot of size 400 are specified to exceed 5 kilograms. We want to set up an acceptance sample plan with AQL = 1.4%. With inspection level IV and normal inspection, find the appropriate plan from MIL-STD-414.

Solution Looking first at Table 14A, we see that a desired AQL of 1.4% would correspond to a tabled value of 1.5%. Finding that lot size of 400 for inspection level IV in Table 14B yhields a code letter of I.

Looking at Table 15 in the row labeled I, we find a sample size of 25. For that same row, the column headed with an AQL of 1.5% gives $k = 1.72$. Thus, we are to sample 25 wires from the lot and accept the lot if

$$\frac{\overline{X} - 5}{S} \geq 1.72$$

where $\overline{X}$ and S are the mean and standard deviation for the 25 tensile strength measurements.

MIL-STD-414 also gives a table for reduced inspection, much like that in MIL-STD-105D. The handbook provides OC curves for the sampling plans tabled there and, in addition, gives sampling plans based on sample ranges rather than standard deviations.

The theory related to the development of the sampling plans for inspection by variables is more difficult than that for inspection by attributes and cannot be developed fully here. However, we outline the basic approach.

Suppose again that X is the quality measurement under study and U is the upper specification limit. The probability that an item is defective is then given by $P(X > U)$. (This probability also represents the proportion of defective items in the lot.) Now we want to accept only lots for which $P(X > U)$ is small, say, less than or equal to a constant M, But $P(X > U) \leq M$ is equivalent to

$$P\left(\frac{X - \mu}{\sigma} > \frac{U - \mu}{\sigma}\right) \leq M$$

where $\mu = E(X)$ and $\sigma^2 = V(X)$. If, in addition, X has a normal distribution, then the latter probability statement is equivalent to

$$P\left(z > \frac{U - \mu}{\sigma}\right) \leq M$$

where Z has a standard normal distribution. It follows that $P(X > U)$ will be "small" if and only if $(U - \mu)/\sigma$ is "large." Thus,

$$p(X > U) \leq M \quad \text{is equivalent to} \quad \frac{U - \mu}{\sigma} \geq k*$$

for some constant $k*$. Because μ and σ are unknown, the actual acceptance criterion used in practice becomes

$$\frac{U - \overline{X}}{S} \geq k$$

where k is chosen so that $P(X > U) \leq M$.

For normally distributed X, $P(X > U)$ can be estimated by making use of the sample information provided by $\overline{X}$ and S, and these estimates are tabled in MIL-STD-414. These estimated are used in the construction of OC curves. Also, tables are given that show the values of M, in addition to the value of k, for sampling plans designed for specified AQLs.

Exercises

10.88 Construct operating characteristic curves for the following sampling plans, with sampling by attributes.
  a $n = 10$, $a = 2$
  b $n = 10$, $a = 4$
  c $n = 20$, $a = 2$
  d $n = 20$, $a = 4$

10.89 Suppose a lot of size 3,000 is to be sampled by attributes. An AQL of 4% is specified for acceptable lots. Find the appropriate level II sampling plan under
  a Normal inspection
  b Tightened inspection
  c Reduced inspection

10.90 Fuses of a certain type must not allow the current passing through them to exceed 20 amps. A lot of 1,200 of these fuses is to be checked for quality by a lot-acceptance plan with inspection by variables. The desired AQL is 2%. Find the appropriate sampling plan for level IV sampling and
  a Normal inspection
  b Tightened inspection

10.91 Refer to Exercise 10.90. If, under normal inspection, the sample mean turned out to be 19.2 amps and the sample standard deviation 0.3 amp, what would you conclude?

10.92 For a lot of 250 one-gallon cans filled with a certain industrial chemical, each gallon is specified to have a percentage of alcohol in excess of L. If the AQL is specified to be 10%, find an appropriate lot-acceptance sampling plan (level IV inspection) under
  a Normal inspection
  b Tightened inspection

10.8 Summary

Hypothesis testing parallels the components of the scientific method. A new treatment comes onto the scene and is claimed to be better than the standard treatment with respect to some parameter, such as proportion of successful outcomes. An experiment is run to collect data on the new treatment. A decision is then made as to whether the data support the claim. The parameter value for the standard is the *null hypothesis*, and the nature of the change in parameter claimed for the new treatment is the *alternative* or *research hypothesis*. Guides to which decision is most appropriate are provided by the probabilities of *type I and type II errors*. For the simple cases that we have considered, hypothesis-testing problems parallel estimation problems in that similar statistics are used. Some exceptions were noted, such as the X^2 tests on frequency data and goodness-of-fit tests. These latter tests are not motivated by confidence intervals.

How does one choose appropriate test statistics? We did not discuss a general method of choosing a test statistic as we did for choosing an estimator (the maximum likelihood principle). General methods of choosing test statistics with good properties do, however, exist. The reader interested in the theory of hypothesis testing should consult a text on mathematical statistics. The tests presented in this chapter have some intuitive appeal as well as good theoretical properties.

Supplementary Exercises

10.93 An important quality characteristic of automobile batteries is their weight, since that characteristic is sensitive to the amount of lead in the battery plates. A certain brand of heavy-duty battery has a weight specification of 69 pounds. Forty batteries selected from recent production have an average weight of 64.3 pounds and a standard deviation of 1.2 pounds. Would you suspect that something has gone wrong with the production process? Why or why not?

10.94 Quality of automobile batteries can also be measured by cutting them apart and measuring plate thicknesses. A certain brand of batteries is specified to have positive plate thicknesses of 120 thousandths of an inch and negative plate thicknesses of 100 thousandths of an inch. A sample of batteries from recent production was cut apart, and plate thicknesses were measured. Sixteen positive plate thickness measurements averaged 111.6, with a standard deviation of 2.5. Nine negative plate thickness measurements averaged 99.44, with a standard deviation of 3.7. Do the positive plates appear to meet the thickness specification? Do the negative plates appear to meet the thickness specification?

10.95 The mean breaking strength of cotton threads must be at least 215 grams for the thread to be used in a certain garment. A random sample of 50 measurements on a certain thread gave a mean breaking strength of 210 grams and a standard deviation of 18 grams. Should this thread be used on the garments?

10.96   Prospective employees of an engineering firm are told that engineers in the firm work at least 45 hours per week, on the average. A random sample of 40 engineers in the firm showed that, for a particular week, they averaged 44 hours of work with a standard deviation of 3 hours. Are the prospective employees being told the truth?

10.97   Company A claims that no more than 8% of the resistors it produced fail to meet the tolerance specifications. Tests on 100 randomly selected resistors yielded 12 that failed to meet the specifications. Is the company's claim valid? Test at the 10% significance level.

10.98   An approved standard states that the average LC50 for DDT should be ten parts per million for a certain species of fish. Twelve experiments produced LC50 measurements of

    16, 5, 21, 19, 10, 5, 8, 2, 7, 2, 4, 9

Do the data cast doubt on the standard at the 5% significance level?

10.99   A production process is supposed to be producing 10-ohm resistors. Fifteen randomly selected resistors showed a sample mean of 9.8 ohms and a sample standard deviation of 0.5 ohm. Are the specifications of the process being met? Should a two-tailed test be used here?

10.100   For the resistors of Exercise 10.99, the claim is made that the standard deviation of the resistors produced will not exceed 0.4 ohm. Can the claim be refuted at the 10% significance level?

10.101   The abrasive resistance of rubber is increased by adding a silica filler and a coupling agent to chemically bond the filler to the rubber polymer chains. Fifty specimens of rubber made with a type I coupling agent gave a mean resistance measurement of 92, with the variance of the measurements being 20. Forth specimens of rubber made with a type II coupling agent gave a mean of 98 and a variance of 30 on resistance measurements. Is there sufficient evidence to say that the mean resistance to abrasion differs for the two coupling agents at the 5% level of significance?

10.102   Two different types of coating for pipes are to be compared with respect to their ability to aid in resistance to corrosion. The amount of corrosion on a pipe specimen is quantified by measuring the maximum pit depth. For coating A, 35 specimens showed an average maximum pit depth of 0.18 cm. The standard deviation of these maximum pit depths was 0.02 cm. For coating B, the maximum pit depths for 30 specimens had a mean of 0.21 cm and a standard deviation of 0.03 cm. Do the mean pit depths appear to differ for the two types of coating? Use $\alpha = 0.05$.

10.103   Bacteria in water samples are sometimes difficult to count, but their presence can easily be detected by culturing. In 50 independently selected water samples from a lake, 43 contained certain harmful bacteria. After adding a chemical to the lake water, another 50 water samples showed only 22 with the harmful bacteria. Does the addition of the chemical significantly reduce the proportion of samples containing the harmful bacteria? $\alpha = 0.025$.

10.104   A large firm made up of several companies has instituted a new quality-control inspection policy. Among 30 artisans sampled from Company A, only 5 objected to the new policy. Among 35 artisans sampled from Company B, 10 objected to the policy. Is there a significant difference, at the 5% significance level, between the proportions voicing *no* objections to the new policy?

10.105   *Research Quarterly*, May 1979, reports on a study of impulses applied to the ball by tennis rackets of various construction. Three measurements on ball impulses were taken on each type of racket. For a Classic (wood) racket, the mean was 2.41 and the standard deviation was 0.02. For a Yamaha (graphite) racket, the mean was 2.22 and the standard deviation was 0.07. Is there evidence of a significant difference between the mean impulses for the two rackets at the 5% significance level?

10.106   For the impulse measurements of Exercise 10.105, is there sufficient evidence to say that the graphite racket gives more variable results that the wood racket? Use $\alpha = 0.05$.

10.107   An interesting and practical use of the χ^2 test comes about in the testing for segregation of species of plants or animals. Suppose that two species of plants, A and B, are growing on a test plot. To assess whether or not the species tend to segregate, n plants are randomly sampled from the plot, and the species of each sampled plant *and* the species of its *nearest* neighbor are recorded. The recorded. The data are then arranged in a table as follows:

		Nearest Neighbour	
		A	B
Sample Plant	A	a	b
	B	c	d
			n

If *a* and *d* are large relative to *b* and *c*, we would be inclined to say that the species tend to segregate. (Most of A's neighbors and of type A, and most of B's neighbors are of type B). If *b* and *c* are large compared to *a* and *d*, we would say that the species tend to be overly mixed. In either of these cases (segregation or overmixing), a χ^2 test should yield a large value, and the hypothesis of random mixing would be rejected. For each of the following cases, test the hypothesis of random mixing (or, equivalently, the hypothesis that the species of a sampled plants is independent of the species of its nearest neighbor). Use $\alpha = 0.05$ in each case.

a $a = 20, b = 4, c = 8, d = 18$
b $a = 4, b = 20, c = 18, d = 8$
c $a = 20, b = 4, c = 18, d = 8$

10.108 "Love is not blind." An article on this topic in the *Gainesville Sun*, June 22, 1992, states that 72 blindfolded people tried to distinguish their partner from two decoys of similar age by feeling their foreheads. And 58% of them were correct! Is this as amazing a result as the article suggests, or could the result have been achieved by guessing? What is an appropriate null hypothesis here? What is an appropriate alternative hypothesis?

10.109 Is knowledge of right-of-way laws at four-legged intersections associated with similar knowledge for T intersections? Montgomery and Carstens (*Journal of Transportation Engineering*, May 1987) show the results of a questionnaire designed to investigate this question (among others). How would you answer the question posed using data given here?

		Response to T-Intersection Question		
		Correct	Incorrect	
Response to Four-Legged-Intersection Question	Correct	141	145	286
	Incorrect	13	188	201
		154	133	487

10.110 Chinowsky (*Journal of Professional Issues in Engineering Education and Practice*, 2002) researched the education background of design and construction executives. Of 111 construction executives 70 had a Bachelor of Science degree in Engineering. Of 139 design executives, 102 had a Bachelor of Science degree in Engineering. At the 5% level of significance, is there a significant difference in the percentage of executives with a Bachelor of Science degree in Engineering among the construction and design executives? Justify your answer.

10.111 Scruggs and Iwan (*Journal of Structural Engineering*, 2003) studied Implementation of a Brushless DC Machine (BDC) as a force actuator, for use in suppressing vibrations in civil structures. The data for semiactive BDC and magnetorheological (MR) fluid dampers are given in table below. The interstory drifts (x_i in cm), and absolute accelerations (a_i in m/s^2) were recorded.

	Semiactive	
	BDC	**MR**
d_i (cm)	0.103	0.114
	0.088	0.090
	0.060	0.101
x_i (cm)	0.103	0.114
	0.152	0.185
	0.198	0.212
a_i (m/s^2)	2.50	6.96
	2.53	7.39
	4.17	7.03

a At the 5% level of significance, is there a significant difference in the mean distances (d_i) measured for BDC and MR semiactive dampers?
b At the 5% level of significance, is there a significant difference in the mean inventory drifts measured for BDC and MR semiactive dampers?
c At the 5% level of significance, is there a significant difference in the mean absolute accelerations measured for BDC and MR semiactive dampers?

10.112 Seim et al. (*Journal of Structural Engineering*, 2003) Studied the post-strengthening of concrete slabs with externally bonded carbon fiber reinforced polymers. The moment capacity of the slab specimens calculated according to the design guidelines at a reduced 65% level and that reached in the tests is listed in table below.

Slab	Moment Capacity Expected for $\epsilon_{L,k} = 0.65\%$ M_{design} (kN m)	Experimental Maximum Moment M_{exp} (kN m)
1	113	121
2	108	124
3	116	126
4	113	123
5	113	114
6	108	116
7	80.4	72.5
8	54.7	45.9
9	80.4	69.6
10	80.4	65.8

a Is there a significant difference in the experimental and analytical mean moment? Justify your answer using the 5% level of significance.

b Suppose the first six pairs of measurement correspond to slabs of group A and the last four to slabs of group B. Now how will you analyze these data to determine whether the difference in mean moment (analytical and experimental) is statistically significant?

10.113 It is a common practice to use composite beams with formed steel deck as the floor system in steel frame structures. Park, Kim, and Yang (*Journal of Structural Engineering*, 2003) performed a series of tests on composite beams with web opening. The applied loads at the first yield and at the first occurrence of transverse cracks are given in table below. Is the mean load at the first yield significantly higher than the mean yield at the appearance of the first traverse crack?

First Yield	Transverse
45.2	37.0
65.1	27.9
56.2	36.7
52.1	32.7
47.1	35.3
58.8	47.0
49.0	49.0

10.114 A lack of knowledge of a job or profession often results in it being dismissed as a career option. Does knowledge lead to the decision to be an engineer? To investigate this question Hamill and Hodgkinson (*Proceedings of ICE*, 2003) surveyed 566 11- to 16-year-olds, and classified them by their level of knowledge about engineering and the likelihood of becoming an engineer (see table below). Are the subject knowledge and decision to become an engineer independent? Justify using the 5% error rate.

Likelihood of Becoming an Engineer	Level of Knowledge about Engineering				
	Nothing at All	Not Much	A Little	Quite a Lot	A Lot
Very unlikely	179	132	53	8	4
Fairly unlikely	5	44	32	15	5
Fairly likely	3	4	16	16	5
Very likely	3	3	9	20	10

10.115 A horizontal directional drilling training was conducted for the California Department of Transportation engineers and safety inspectors. A pretest was administered prior to the training and a final exam was given at the end of the training. Ariaratnam, Najafi, and Morones (*Journal of Professional Issues in Engineering Education and Practice*, 2002) reported the results, as shown in table below.

Area	n	Pretest Results		Final Test Results	
		Mean Percentage	Standard Deviation	Mean Percentage	Standard Deviation
San Bernardino	53	53.05	9.66	74.39	8.34
Orange County	65	56.97	8.39	76.28	7.31

a Are the mean pretest scores significantly different for the two areas, San Bernardino and Orange County? Justify at the 5% error rate.

b Are the mean final test scores significantly different for the two areas, San Bernardino and Orange County? Justify at the 5% error rate.

c We are interested in determining whether there is a significant improvement in the mean scores from the pretest to the final test. Do we have sufficient information to answer this question? If yes, show your work. If no, justify why not.

d We are not sure how the engineers were selected for this training. Will it have any effect on your conclusions? Describe.

e Suppose the participants constitute all the engineers of the California Department of transportation. Will your answer to parts (a) and (b) change? Justify your answer.

Inference for Regression Parameters

Does the life of machine tools depend upon the cutting speed and the depth of cuts? How long is a machine tool expected to last when used to cut depths of 0.03 inch at a speed of 450 (see Figure 11.1)?

To answer such questions, we need to estimate the function relating the life of machine tools to its cutting speed and the depth of cuts. We also need to estimate slope and intercept, and test for the significance of those estimates. Following up on the work of Chapter 2, we now proceed to build models in which the mean of one variable can be written as a function of two or more independent variables. For example, the average amount of energy required to heat a house depends not only on the air temperature but also on the size of the house, the amount of insulation, and the type of heating unit, among other things. We will use multiple regression models for estimating means and for predicting future values of key variables under study.

Figure 11.1
The relationship between machine life and cutting speed and depth

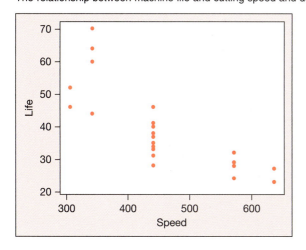 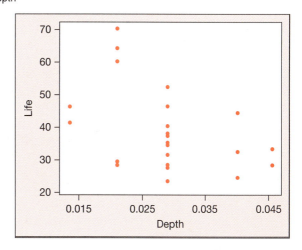

Introduction

In previous chapters, we estimated and tested hypotheses concerning a population mean by making use of measurements on a single variable from that population. Frequently, however, the mean of one variable is dependent upon one or more related variables. For example, the average amount of energy required to heat houses of a certain size depends on the air temperature during the days of the study. In this chapter, we learn to construct confidence intervals for and to test hypotheses concerning the slope and intercept. Extending the same idea, we learn to develop models describing the mean of one variable as a function of two or more variables.

11.1 Regression Models with One Predictor Variable

There are different types of relationships among different variables. Some variables have an exact relationship between them. The model that hypothesizes an exact relationship and does not allow for the random error in prediction is known as a *deterministic model*. Many scientific relations, and laws of physics are deterministic relations. For example, the relation between the Celsius scale (C) and Fahrenheit scale (F) of temperatures is a deterministic relation given by $F = 32 + (9/5)C$. For a given temperature measured on a Celsius scale, there is exactly one value on a Fahrenheit scale; that is, all the data points will fall perfectly on the line described earlier and no deviation from the line will be allowed (see Figure 11.2).

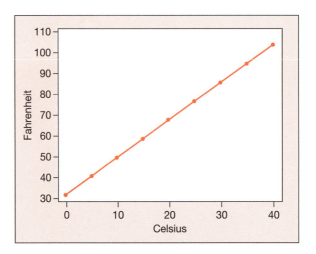

Figure 11.2
A deterministic model

The primary usefulness of deterministic models is in the description of physical laws. For example, Ohm's law describes the relationship between current and resistance in a deterministic manner. Newton's laws of motion are other examples of deterministic models. However, it should be noted that, in all these examples, the laws hold precisely only under ideal conditions. Laboratory experiments rarely reproduce these laws exactly. Usually, random error will be introduced by the experiment, so the laws will provide only approximations to reality.

In reality, most observed relationships are not deterministic. In Chapter 2, we developed a model to predict peak power load as a function of daily maximum temperature. Does the exact relationship exist between peak power load and maximum temperature? We see that it is not possible for several reasons. Peak load depends on factors besides the maximum temperature, such as the capacity of the plant, the number of customers the plant services, and the geographical location of the plant. However, even if all these factors were included in a model, it is still unlikely that we would be able to predict the peak load *exactly*. There will almost certainly be some variation in peak power load due strictly to *random phenomenon* that cannot be modeled or explained. If we believe there will be unexplained variation in peak power load, then we need to use a *probabilistic model* that accounts for this *random error*. The probabilistic model includes both a deterministic component and a random error component as

$$Y = \text{deterministic component} + \text{random error}$$

where Y is the random variable to be predicted. We will always assume that the mean value of the random error equals zero. This is equivalent to assuming that the mean value of Y, $E(Y)$, equals the deterministic component of the models.

In the previous two chapters, we discussed the simplest form of a probabilistic model. We learned how to make inference about the mean of Y, $E(Y)$, when $E(Y) = \mu$, where μ is a constant. However, we realized that this relation did not imply that Y would equal μ exactly, but instead that Y would equal to μ plus or minus a random error.

Figure 11.3

The probabilistic model $Y = \mu + \varepsilon$

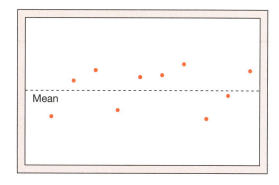

In particular, if we assume Y is normally distributed with mean μ and variance σ^2, then we may write the probabilistic model as $Y = \mu + \varepsilon$, where the random component (ε) is normally distributed with mean zero and variance σ^2. Some possible observations generated by this model are shown in Figure 11.3.

In this chapter, we discuss how to generalize the model to allow $E(Y)$ to be a function of other variables. Let us hypothesize that the mean of Y is a straight-line function of x. As seen in Chapter 2, a *linear regression model* or *linear regression equation* gives a straight line probabilistic relationship between two variables as follows:

$$Y = \beta_0 + \beta_1 x + \varepsilon$$

where

- Y is the *response variable*.
- x is the *explanatory variable*.
- β_0 is the *Y-intercept*. It is the value of y, for $x = 0$.
- β_1 is the *slope* of the line. It gives the amount of change in y for every unit of change in the value of x.
- ε is the random error component.

For a given x, the difference between the observed measurement and the expected value based on the modeled relation gives the error in estimation or residual, that is, $e_i = \hat{\varepsilon}_i = y_i - \hat{y}_i$ (Figure 11.4).

One method for modeling relation is the *least-squares* approach. This approach chooses the values of Y-intercept and slope that minimize the sum of squared errors (SSE). A line that minimizes the SSE is referred to as the *least-squares regression line*. It is also known as the *line of best fit*.

Suppose we decide that Y can be modeled as a function of x using $Y = \beta_0 + \beta_1 x + \varepsilon$. We want to estimate β_0 and β_1 from sampled data (or measurements taken). Using the estimated Y-intercept and slope, the estimated regression line is $\hat{y} = \hat{\beta}_0 + \hat{\beta}_1 x$. We are interested in values of $\hat{\beta}_0$ and $\hat{\beta}_1$ such that SSE is minimized, where

$$SSE = \sum_{i=1}^{n}(y_i - \hat{y}_i)^2 = \sum_{i=1}^{n}(y_i - \hat{\beta}_0 - \hat{\beta}_1 x_i)^2$$

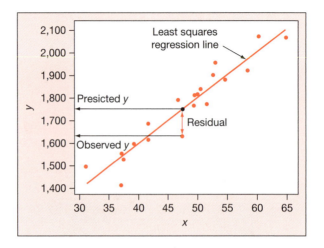

Figure 11.4

The least-squares regression line and residuals

Now differentiate SSE with respect to $\hat{\beta}_0$ and $\hat{\beta}_1$, set the results equal to zero, and solve for $\hat{\beta}_0$ and $\hat{\beta}_1$.

$$\frac{\partial(SSE)}{\partial\hat{\beta}_0} = -2\sum_{i=1}^{n}(y_i - \hat{\beta}_0 - \hat{\beta}_1 x_i) = 0$$

$$\frac{\partial(SSE)}{\partial\hat{\beta}_1} = -2\sum_{i=1}^{n}x_i(y_i - \hat{\beta}_0 - \hat{\beta}_1 x_i) = 0$$

The solution to these two equations is

$$\hat{\beta}_1 = \frac{\sum_{i=1}^{n}(x_i - \bar{x})(y_i - \bar{y})}{\sum_{i=1}^{n}(x_i - \bar{x})^2} = \frac{SS_{xy}}{SS_{xx}} \quad \text{and} \quad \hat{\beta}_0 = \bar{y} - \hat{\beta}_1\bar{x}$$

A line estimated using least-squares estimates of the Y-intercept and the slope has the smallest sum of squared errors. Any other line will have a larger error sum of squares than this line. It means, on the average, this line is likely to give estimates closer to the actual values than any other line. The line of best fit will always pass through the point $(\bar{X}, \bar{Y})$. The sum of squared errors for the line of best fit is computed as

$$SSE = \sum e_i^2 = \sum(y_i - \hat{y}_i)^2 = SS_{yy} - \hat{\beta}_1 SS_{xy}$$

Example 11.1

In their article, Mulekar and Kimball (2004, Stats) used 16 years of data to study several hurricane characteristics. DeMaria et al. (2001, **http://www.bbsr.edu/rpi/research/demaria/demaria4.html**) collected data on North Atlantic tropical cyclones from 1988–2003. This dataset, known as the Extended Best Track (EBT) dataset, contains observations of hurricane characteristics for all storms at 6-hour intervals from the tropical depression to the dissipation stage. The minimum sea-level pressure (psmin), measured in hecto-Pascals (hPa), and maximum low-level wind speed (vmax), measured in meters per second (m/s), are two

Figure 11.5

JMP output for hurricane data

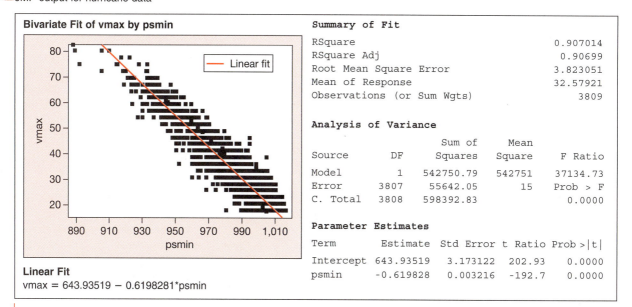

Bivariate Fit of vmax by psmin

Linear Fit

vmax = 643.93519 − 0.6198281*psmin

Summary of Fit

RSquare	0.907014
RSquare Adj	0.90699
Root Mean Square Error	3.823051
Mean of Response	32.57921
Observations (or Sum Wgts)	3809

Analysis of Variance

Source	DF	Sum of Squares	Mean Square	F Ratio
Model	1	542750.79	542751	37134.73
Error	3807	55642.05	15	Prob > F
C. Total	3808	598392.83		0.0000

Parameter Estimates

Term	Estimate	Std Error	t Ratio	Prob >|t|
Intercept	643.93519	3.173122	202.93	0.0000
psmin	−0.619828	0.003216	−192.7	0.0000

characteristics used to describe the intensity of hurricanes. Are these two characteristics related to each other? If yes, estimate the relation between vmax and psmin.

Solution The regression output using JMP for the hurricane data is given in Figure 11.5.

- The scatterplot shows a very strong negative linear relation between vmax and psmin.
- The coefficient of determination, $R^2 = 0.907$, indicates that about 90.7% variation in minimum surface pressure can be explained using the relation between vmax and psmin. In other words, about 9.3% variation in minimum surface pressure can be attributed to other factors.
- The relation between maximum low-level wind speed and minimum surface pressure is estimated by

$$\text{vmax} = 643.94 - 0.62\ \text{psmin}$$

- The estimated Y-intercept is $\hat{\beta}_0 = 643.94$. Technically, it means that when there is no minimum surface pressure (psmin = 0), on the average maximum low-level winds of speed 643.94 m/s are observed. However, we should be very careful about this interpretation, because no data are available for psmin around zero, and we do not know whether the relation described by the scatterplot in Figure 11.5 holds true for psmin < 888 (the lowest recorded value). In fact, a careful examination of the scatterplot for low values of psmin shows slight change in direction (cannot confirm due to lack of data), which indicates that this linear relation may not hold true for psmin values closer to zero.
- The estimated slope of relation is $\hat{\beta}_1 = 0.62$; that is, for every one hPa increase in the minimum surface pressure, the maximum low level wind speed drops on the average by 0.62 m/s.

- As described earlier, these estimates of the Y-intercept and slope were obtained by minimizing the error sum of squares (SSE):

$$\min(SSE) = \sum_{i=}^{3809} (y_i - \hat{y}_i)^2 = \sum_{i=}^{3809} (y_i - \hat{\beta}_0 - \hat{\beta}_1 x_i)^2 = 55{,}642.05$$

Any other relation between psmin and vmax will result in an SSE larger than 55,642.05. Overall, this relation will give a smaller error. The estimated standard deviation, $\sigma = 3.82$ tells that on the average the maximum low-level winds predicted using this linear model will differ by 3.82 m/s from the observed value.

Exercises

11.1 Use the method of least squares to fit a straight line to the following six data points:

x	1	2	3	4	5	6
y	1	2	2	3	5	5

a What are the least-squares estimates of β_0 and β_1?
b Plot the data points and graph the least-squares line. Does the line pass through the data points?

11.2 Use the method of least squares to fit a straight line to the following data points:

x	−2	−1	0	1	2
y	4	3	3	1	−1

a What are the least-squares estimates of β_0 and β_1?
b Plot the data points and graph the least-squares line. Does the line pass through the data points?

11.3 The elongation of a steel cable is assumed to be linearly related to the amount of force applied. Five identical specimens of cable gave the following results when varying forces were applied:

Force (x)	1.0	1.5	2.0	2.5	3.0
Elongation (y)	3.0	3.8	5.4	6.9	8.4

Use the method of least squares to fit the line

$$Y = \beta_0 + \beta_1 x + \varepsilon$$

11.4 A company wants to model the relationship between its sales and the sales for the industry as a whole. For the following data, fit a straight line by the method of least squares.

Year	Company Sales y ($ million)	Industry Sales x ($ million)
2002	0.5	10
2003	1.0	12
2004	1.0	13
2005	1.4	15
2006	1.3	14
2007	1.6	15

11.5 It is thought that abrasion loss in certain steel specimens should be a linear function of the Rockwell hardness measure. A sample of eight specimens gave the following results:

Rochwell Hardness (x)	60	62	63	67	70	74	79	81
Abrasion Loss (y)	251	245	246	233	221	202	188	17

Fit a straight line to these measurements.

11.6 The Organization of Petroleum Exporting Countries (OPEC), a cartel of crude-oil suppliers, controls the crude oil prices and production. Do they affect the gasoline prices paid by motorists? Table on the next page gives the average annual prices of crude oil and unleaded gasoline in the United States (U.S. Census Bureau, *Statistical Abstracts of the United States,* 2009). The price of crude oil is in dollars per barrel (a refiner acquisition cost) and the price of unleaded gasoline is in dollars per gallon (paid by motorists at the pump) for the years 1976–2007.

a Use these data to calculate the least-squares line that describes the relationship between the price of a gallon of gas and the price of a barrel of crude oil.
b Plot your least-squares line on a scatterplot of the data. Does your least-squares line appear to be an appropriate characterization of the relationship between y and x? Explain.

Year	Crude Oil	Unleaded Gasoline	Year	Crude Oil	Unleaded Gasoline
1976	10.89	0.61	1991	19.06	1.14
1977	11.96	0.66	1992	18.43	1.13
1978	12.46	0.67	1993	16.41	1.11
1979	17.72	0.90	1994	15.59	1.11
1980	28.07	1.25	1995	17.23	1.15
1981	35.24	1.38	1996	20.71	1.23
1982	31.87	1.30	1997	19.04	1.23
1983	28.99	1.24	1998	12.52	1.06
1984	28.63	1.21	1999	17.51	1.17
1985	26.75	1.20	2000	28.26	1.51
1986	14.55	0.93	2001	22.95	1.46
1987	17.90	0.95	2002	24.10	1.36
1988	14.67	0.95	2003	28.53	1.59
1989	17.97	1.02	2004	36.98	1.88
1990	22.22	1.16			

c If the price of crude oil fell to $20 per barrel, to what level (approximately) would the price of regular gasoline fall? Justify your response.

11.7 A study is made of the number of parts assembled as a function of the time spend on the job. Twelve employees are divided into three groups, assigned to three time intervals, with the following results:

Time x	Number of Parts Assembled y
10 minutes	27, 32, 26, 34
15 minutes	35, 30, 42, 47
20 minutes	45, 50, 52, 49

Fit the model $Y = \beta_0 + \beta_1 x + \varepsilon$ by the method of least squares.

11.8 Laboratory experiments designed to measure LC50 values for the effect of certain toxicants on fish are basically run by two different methods. One method has water continuously flowing through laboratory tanks, and the other has static water conditions. For purposes of establishing criteria for toxicants, the Environmental Protection Agency (EPA) wants to adjust all results to the flow-through condition. Thus, a model is needed to relate the two methods. Observations on certain toxicants examined under both static and flow-through conditions yielded the following measurements (in parts per million):

Toxicant	LC50 Flow-Through y	LC50 Static x
1	23.00	39.00
2	22.30	37.50
3	9.40	22.20
4	9.70	17.50
5	0.15	0.64
6	0.28	0.45
7	0.75	2.62
8	0.51	2.36
9	28.00	32.00
10	0.39	0.77

Fit the model $Y = \beta_0 + \beta_1 x + \varepsilon$ by the method of least squares.

11.2 The Probability Distribution of the Random Error Component

In Section 11.1, we hypothesized the form of $E(Y)$ to be a linear function of x and learned to use the sample data to estimate unknown parameters in the model. The hypothesized model relating Y to x is $Y = \beta_0 + \beta_1 x + \varepsilon$.

Recall that to make inferences about the mean of a population using a small sample, we used a t-statistic and assumed that the data were randomly obtained from a normally distributed population. Similarly, when making inferences about the parameters of a regression model, we make assumptions about the distribution of the error component.

The conditions about the probability distribution of ε are described in Figure 11.6. We can summarize these assumptions as follows:

1. The error component ε is normally distributed, at least approximately.
2. The mean of ε is zero.
3. The variance of ε is σ^2, a constant for all values of x.
4. The errors associated with different observations are random and independent.

Note that this is the same as saying $Y_1, Y_2 \cdots Y_n$ are random samples from normally distributed populations, with Y_i having mean $\beta_0 + \beta_1 x_i$ and variance σ^2 free of X.

Various techniques exist for testing the validity of the conditions, and there are some alternative techniques to be used when the assumptions appear to be invalid. In Chapter 2, we learned to assess adequacy of a linear model by checking conditions (3) and (4). Condition (1) about the normality of distribution is required for the inferences about the parameters of a regression model, especially when the sample size is small. Condition (2) related to the fact that we are proposing a simple linear model as the mean value of Y, or $E(Y)$.

Variation among observations is measured using their standard deviation. However, now we are interested in the variation in observations around the line, that is, how far they deviate from the line of best fit. Note that the probability distribution of ε would be completely specified if the variance σ^2 were known. To estimate σ^2, we use the SSE for the least-squares model. The estimate s^2 of σ^2 is calculated by dividing SSE by the number of degrees of freedom associated with the error component. We use 2 degrees of freedom to estimate Y-intercept and slope of the straight-line model, leaving $(n - 2)$ degrees of freedom for the error variance estimation. Thus, the unbiased estimator of σ^2 is

$$s^2 = \frac{SSE}{n - 2} \quad \text{where} \quad SSE = \sum e_i^2 = \sum (y_i - \hat{y}_i)^2 = SS_{yy} - \hat{\beta}_1 SS_{xy}$$

Figure 11.6

The probability distribution of ε

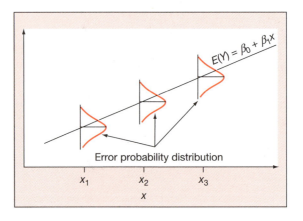

Example 11.2

In the peak power load example (Chapter 2), we calculated $SS_{yy} = 19,262.6$, $SS_{xy} = 2,556$, and $\hat{\beta}_1 = 6.7175$. Then we calculate the SSE for the least-squares line as

$$SSE = SS_{yy} - \hat{\beta}_1 SS_{xy} = 19,263.6 - 6.7175(2,556) = 2,093.43$$

Because there were 10 data points used, we have $n - 2 = 8$ degrees of freedom for estimating σ^2. Thus,

$$s^2 = \frac{SSE}{n - 2} = \frac{2,093.43}{8} = 261.68$$

is the estimated variance, and $s = \sqrt{261.68} = 16.18$ is the estimated standard deviation of ε. It shows, on the average, how far the observed power load values are from the predicted value for a given temperature.

We can get an intuitive feeling for the distribution of the random errors by recalling that about 95% of the observations from a normal distribution lie within two standard deviations of the mean. Thus, we expect approximately 95% of the observations to fall within $2s$ of the estimated mean $\hat{y} = \hat{\beta}_0 + \hat{\beta}_1 x$. For the peak power load example, note that all 10 observations fall within $2s = 2(16.18) = 32.36$ of the least-squares regression line.

In Chapter 2, we learned to use the technique called *residual analysis* to check assumptions. We can easily review the residual plot to check the assumption of normality of random errors. We expect to see about 68% of errors within one standard deviation, about 95% within two standard deviations, and almost all within three standard deviations of the mean (which is expected to be 0).

Exercises

11.9 Calculate SSE and s^2 for the data of Exercise 11.1.

11.10 Calculate SSE and s^2 for the data of Exercise 11.5. Interpret s^2 value in context of problem.

11.11 Calculate SSE and s^2 for the data of Exercise 11.6. Interpret s^2 value in context of problem.

11.12 Calculate SSE and s^2 for the data of Exercise 11.7. Interpret s^2 value in context of problem.

11.13 Calculate SSE and s^2 for the data of Exercise 11.8. Interpret s^2 value in context of problem.

11.14 J. Matis and T. Wehrly (*Biometrics*, 35, no. 1, March 1979) report the data given in table below on the proportion of green sunfish that survive a fixed level of thermal pollution for varying lengths of time:

a Fit the linear model $Y = \beta_0 + \beta_1 x + \varepsilon$ by the method of least squares.

b Plot the data points and graph the line found in (a). Does the line fit through the points?

c Compute SSE and s^2. Interpret variance in context.

Proportion of Survivors (y)	1.00	0.95	0.95	0.90	0.85	0.70	0.65	0.60	0.55	0.40
Scaled Time (x)	0.10	0.15	0.20	0.25	0.30	0.35	0.40	0.45	0.50	0.55

11.3   Making Inferences about Slope β_1

Different pairs of samples (same size) from the same population result in different estimates of slope. Therefore, we need to take into account the sampling distribution of $\hat{\beta}_1$ for making an inference about $\hat{\beta}_1$.

> ### Sampling Distribution of $\hat{\beta}_1$
>
> If we assume that the error components are independent normal random variables with mean zero and constant variance σ^2, the sampling distribution of the least-squares estimator $\hat{\beta}_1$ of the slope will be normal, with mean β_1 (the true slope) and standard deviation $\sigma_{\hat{\beta}_1} = \sigma/\sqrt{SS_{xx}}$.

Because σ, the standard deviation of ε, is usually unknown, the statistic $(\hat{\beta}_1 - \beta_1)/s_{\hat{\beta}_1}$ has a student's t-distribution with $(n - 2)$ degrees of freedom, where $s_{\hat{\beta}_1} = s/\sqrt{SS_{xx}}$ is the estimated standard deviation of the sampling distribution of $\hat{\beta}_1$. Note that the statistic has the usual form of a t-statistic, a normally distributed estimator divided by an estimate of its standard deviation.

11.3.1   Estimating slope using a confidence interval

One way to make inference about the slope β_1 is to estimate it by using a confidence interval, and the other way is to test a claim about its value.

> ### 100 $(1 - \alpha)$ Percent Confidence Interval for Slope $\hat{\beta}_1$
>
> $$\hat{\beta}_1 \pm t_{\alpha/2}(n - 2)s_{\hat{\beta}_1}$$
>
> Conditions:
>
> 1. The error component ε is normally distributed.
> 2. The mean of ε is zero.
> 3. The variance of ε is σ^2, a constant for all values of x.
> 4. The errors associated with different observations are independent.

Checking Conditions The conditions can be checked using the residual plot (see Chapter 2 for discussion). The condition of normality can be checked using a histogram or a dotplot. It can also be checked using the Q–Q plot or the normal probability plots (discussed in Chapter 6).

Example 11.3

Refer to the peak power load example from Chapter 2. Construct a 95% confidence interval for the slope β_1.

Solution   The estimated slope for relation between peak power load and temperature is 6.7175. It measures the average increase in the peak power load with respect to a 1-degree increase in temperature. A 95% confidence interval for the slope β_1 is

$$\hat{\beta}_1 \pm t_{\alpha/2}(n-2)s_{\hat{\beta}_1}, \quad \text{i.e.,} \quad 6.7175 \pm 2.306\left(\frac{16.18}{\sqrt{380.5}}\right),$$

$$\text{i.e.,} \quad 6.7175 \pm 1.913, \quad \text{i.e.,} \quad (4.80, 8.63)$$

Thus, we are 95% confident that the true slope of the peak power load model is between 4.80 and 8.63. It implies that the true slope is positive and the maximum daily temperature is a good predictor for the peak power load.

11.3.2   Testing a hypothesis about slope

Another way to make an inference about the slope β_1 is to test a hypothesis about the specific value of the slope. Refer back to the peak power load example. Suppose that the daily peak power load is *completely unrelated* to the maximum temperature on a given day. This implies that the mean $E(Y) = \beta_0 + \beta_1 x$ does not change as x changes. In the straight-line model, this means that the true slope β_1 is equal to zero (see Figure 11.7).

Note that if $\beta_1 = 0$, the model simply reduces to $Y = \beta_0 + \varepsilon$, the constant mean model. Therefore, to test the null hypothesis that x contributes no information regarding the prediction of Y against the alternative hypothesis that these variables are linearly related with a nonzero slope, we test

$$H_0: \beta_1 = 0 \quad \text{versus} \quad H_a: \beta_1 \neq 0$$

Figure 11.7

Graph of the model
$Y = \beta_{\hat{0}} + \varepsilon$

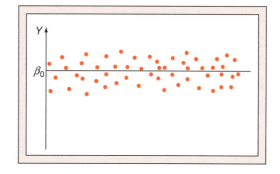

The appropriate test statistic is found by considering the sampling distribution of $\hat{\beta}_1$, the least-squares estimator of slope β_1. The t-statistic defined earlier is used to test hypothesis about the slope.

- What conclusion can be drawn if the data do not support the rejection of null hypothesis? We know from the previous discussion (Chapter 10) on the philosophy of hypothesis testing that such a test statistic value does not lead us to accept the null hypothesis; that is, we do not conclude that $\beta_1 = 0$. Two possibilities arise at this point. One possibility is that there is really no relation between $E(Y)$ and x. Another possibility is that additional data might indicate that β_1 differs from zero, or a more complex relationship may exist between Y and x, requiring the fitting of a model other than the simple linear model.
- If the data support the alternative hypothesis, we conclude that x contributes information for the prediction of Y using the simple linear model (although the relationship between $E(Y)$ and x could be more complex than a straight-line model).

Thus, to some extent this is a test of the adequacy of the hypothesized model.

Testing the Slope of a Straight-Line Model (t-test)

$H_0: \beta_1 = \beta_{10}$ $TS: T = \dfrac{\hat{\beta}_1 - \beta_{10}}{S/\sqrt{SS_{xx}}}$ $\nu = (n - 2)$ degrees of freedom

Reject H_0 if p-value $< \alpha$

Right-tailed test:	Left-tailed test:	Two-tailed test:
$H_a: \beta_1 > \beta_{10}$	$H_a: \beta_1 < \beta_{10}$	$H_a: \beta_1 \neq \beta_{10}$
RR: $t > t_\alpha(\nu)$	RR: $t < -t_\alpha(\nu)$	RR: $t > t_{\alpha/2}(\nu)$ or $t < -t_{\alpha/2}(\nu)$
p-value: $P(t(\nu) > t)$	p-value: $P(t(\nu) < t)$	p-value: $2\,P(t(\nu) > |t|)$

Conditions:

1. The error component ε is normally distributed.
2. The mean of ε is zero.
3. The variance of ε is σ^2, a constant for all values of x.
4. The errors associated with different observations are independent.

Checking Conditions The conditions can be checked using the residual plot (see Chapter 2 for discussion). The assumption of normality can be checked using a histogram or a dotplot. It can also be checked using the Q–Q plot or the normal probability plots (discussed in Chapter 6).

Example 11.4

Refer to the peak power load example (Chapter 2). Suppose we are interested in testing the adequacy of the hypothesized linear relationship between the peak power load and the maximum daily temperature (Figure 11.8). To test $H_0: \beta_1 = 0$

Figure 11.8

Simple linear model fitted to
power load and temperature data

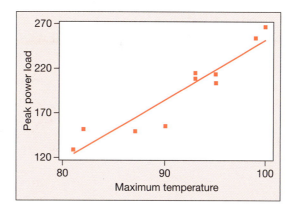

against $H_a: \beta_1 \neq 0$, we choose $\alpha = 0.05$. Because we have 10 pairs of measurements, the degrees of freedom for error are 8, and $t_{0.025}(8) = 2.306$. Thus, the rejection region is $t < -2.306$ or $t > 2.306$. We previously calculated $\hat{\beta}_1 = 6.7175$, $s = 16.18$, and $SS_{xx} = 380.5$. Thus,

$$t = \frac{\hat{\beta}_1 - 0}{16.18/\sqrt{380.5}} = 8.10$$

Because this test statistic value falls in the upper-tail rejection region ($t = 8.10 > 2.306$), we reject the null hypothesis and conclude that the slope β_1 is not equal to zero. The sample evidence indicates that the peak power load tends to increase as a day's maximum temperature increases.

Alternatively, we could use some statistical program to test this hypothesis. Figure 11.9 shows outputs from Minitab and JMP. Both the outputs show that $p < 0.0001$ for testing for the slope. Therefore, we reject the null hypothesis of zero slope and arrive at the same conclusion.

Figure 11.9

Minitab and JMP outputs for power load data

Minitab output

The regression equation is
load = -420 + 6.72 temperature

Predictor	Coef	SE Coef	T	P
Constant	-419.85	76.06	-5.52	0.001
temperat	6.7175	0.8294	8.10	0.000

S = 16.18   R-Sq = 89.1%   R-Sq(adj) = 87.8%

(handwritten: reject H_0, $< \alpha/2$, < -2.306, > 2.306, good correlation, β_0, β_1)

JMP printout

Linear Fit
Load = -419.8491 + 6.717477 Temperature

Summary of Fit

RSquare	0.8913
RSquare Adj	0.8777
Root Mean Square Error	16.177
Mean of Response	194.8
Observations (or Sum Wgts)	10

Parameter Estimates

Term	Estimate	Std Error	t Ratio	Prob >|t|
Intercept	-419.8491	76.0577	-5.52	0.0006
Temperature	6.7175	0.8293	8.10	<.0001

Example 11.5

Refer to Example 2.7 of overtolerance (OT) and wall reduction (WR) from Chapter 2. The Minitab printout of fitting a line to this data is as follows:

```
The regression equation is
WR = 3.23 + 215 OT

Predictor        Coef        SE Coef           T           P
Constant       3.2310         0.2725       11.86       0.000
OT            215.30          32.41         6.64       0.000

S = 0.9553        R-Sq = 65.7%         R-Sq(adj) = 64.2%
```

It shows that the estimated error variance is $s = 0.9553$. Suppose we are interested in testing

$$H_0: \beta_1 = 0 \text{ (OT contributes no information for the prediction of}$$
$$\text{WR using a simple linear model.)}$$

$$H_a: \beta_1 \neq 0 \text{ (OT contributes information for the prediction of}$$
$$\text{WR using a simple linear model.)}$$

The t-test results from the printout show that the test statistic value is equal to 6.64 and the p-value for the t-test is less than 0.0001. At any reasonable significance level, we would reject the null hypothesis and conclude that the overtolerance provides information for the prediction of wall reduction. The positive slope indicates that the amount of wall reduction in fact increases with the overtolerance at the rate of 215 units.

Example 11.6

To assess the appropriateness of the web crippling design rules in the North American Specification (NAS) and the Australian/New Zealand Standard (AS/NZS) for channel sections, Young and Hancock (*Journal of Structural Engineering*, 2003) conducted a series of tests on cold-formed steel channels under end loading (ETF). Table 11.1 shows the web crippling test strength (P_{Exp}) and the design strengths calculated using the NAS specification (P_{NAS}) and the AS/NZS Standard ($P_{AS/NZS}$) for end loading.

 Is there any association between the test strength determined analytically and that using NAS and AS/NZS standards? Is there any association between the test strength determined by NAS and AS/NZS standards?

 The scatterplot matrix (Figure 11.10) shows that there is a very strong positive association between the test strengths determined analytically and using NAS and

Table 11.1

Test and Design Strengths

P_{EXP}	P_{NAS}	$P_{AS/NZS}$
26.5	25.1	35.7
23.1	20.9	33.6
30.1	25.5	35.0
25.4	21.3	33.2
34.5	25.4	35.7
27.8	20.9	33.1
42.6	38.5	52.5
41.6	31.6	48.7
63.0	63.8	80.8
53.8	52.6	75.3
63.5	61.8	78.3
52.1	51.2	73.2

Figure 11.10

Correlation matrix for test strengths

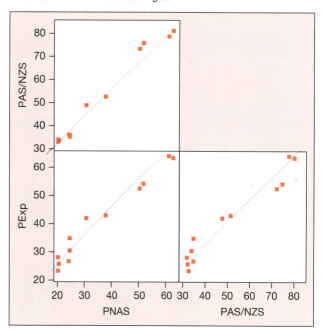

AS/NZS standards. The table of correlation coefficients below also confirms our observation from the scatterplot. Such a near perfect association shows a strong agreement between two different standards used, as well as between the analytical and experimental results.

Correlation Coefficient	Exp	NAS	AS/NZS
Exp	1.0000	0.9826	0.9744
NAS	0.9826	1.0000	0.9892
AS/NZS	0.9744	0.9892	1.0000

Let us take a look at the design strengths calculated using two different standards (NAS and AS/NZS). It is not enough to just have a strong association between the analytical test results given by two standards, we are interested in determining if two standards give similar results. In other words, we are interested in testing a hypothesis that the slope of linear association between the results by two standards is equal to 1.

H_0: $\beta_1 = 1$ (Both standards result in similar strengths.)

H_a: $\beta_1 \neq 1$ (One standard gives consistently higher strengths than the other.)

We have $n = 12$ pairs of measurements. With 10 degrees of freedom, we will reject the null hypothesis at 5% significance level, if $|t| > t_{0.025}(10) = 2.228$. From the following JMP printout (Figure 11.11), we get $\hat{\beta}_1 = 1.2002$ and $s = 3.638$. Also, remember that sum of squares for X is the numerator of the variance of X. Thus, using $s_x = 16.46$ for NAS, we get $SS_{xx} = (n - 1)s_x^2 = (12 - 1)16.46^2 = 2980.2476$. Now compute the test statistic as

$$t = \frac{\hat{\beta}_1 - \beta_{10}}{s/\sqrt{SS_{xx}}} = \frac{1.2002 - 1.0}{3.0638/\sqrt{2980.2476}} = 3.567$$

Alternatively, we could read $s_{\beta_1} = 0.056118$ from the printout and compute the test statistic value as $(1.2002 - 1.0)/0.056118 = 3.567$. Because $t = 3.567 > 2.228$, we reject the null hypothesis and conclude that the slope of the linear relation is not equal to 1. In fact, a close inspection of the scatterplot in Figure 11.11 shows that the AS/NZS standard consistently estimated strength higher than the NAS standards. Also, the discrepancy is increasing with the test strengths.

Let us check whether the assumptions for the t-test are met. The residual plot and the normal quantile plot generated using JMP are given in Figure 11.12. The residual plot does not show any patterns or trends, indicating that the assumption of independent residuals is reasonable. Also, the fairly equal spread indicates that the assumption of constant variance of residuals is reasonable. From the quantile plot, we see that the residuals do not show a very good linear pattern; however, it is reasonable to assume a nearly normal distribution for residuals. This conclusion is also supported by the outcome of the Shapiro-Wilk W test. At any reasonable α, there is no strong evidence against the null hypothesis of normality of residuals ($p = 0.1734$).

Figure 11.11

A JMP printout of a simple linear fit to strength data

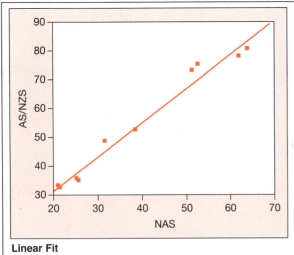

Summary of Fit

RSquare	0.978606
RSquare Adj	0.976467
Root Mean Square Error	3.063766
Mean of Response	51.25833
Observations (or Sum Wgts)	12

Parameter Estimates

Term	Estimate	Std Error	t Ratio	Prob >|t|
Intercept	7.3897863	2.233681	3.31	0.0079
NAS	1.2002338	0.056118	21.39	<.0001

NAS Moments

Mean	36.55
Std Dev	16.460946
Std Err Mean	4.7518657
N	12

Linear Fit

AS/NZS = 7.3897863 + 1.2002338 NAS

Figure 11.12

JMP printout for the residual and the normal quantile plot for a linear fit in Figure 11.11

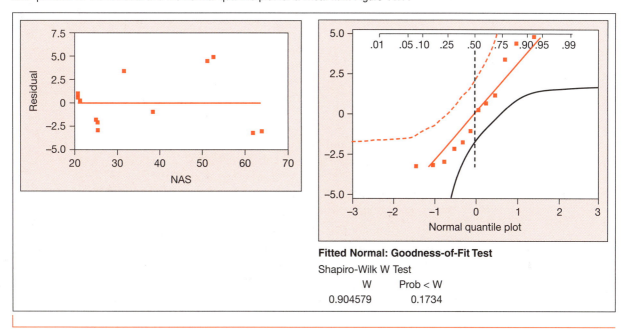

Fitted Normal: Goodness-of-Fit Test

Shapiro-Wilk W Test

W	Prob < W
0.904579	0.1734

Example 11.7

Taraman (1974) reported the failure times (tool life in min) of 24 machine tools and their corresponding cutting speeds (in feet per min [fpm]). The experiment was conducted using workpiece materials of SAE 1018 cold-rolled steel, 4 inches in diameter and 2 feet long. Is cutting speed a good predictor for the tool life? The tool life and the cutting speeds are listed in table below.

Speed	Life	Speed	Life	Speed	Life	Speed	Life
340	70	340	44	305	52	305	46
570	29	570	24	635	23	635	27
340	60	440	35	440	40	440	37
570	28	440	31	440	28	440	34
340	64	440	38	440	46	440	41
570	32	440	35	440	33	440	28

Solution   The JMP printout for linear regression is given in Figure 11.13. The scatterplot indicates that the straight-line model can be used to describe the tool life as a function of cutting speed. The best line of fit is given by

$$\text{Tool Life} = 83.686361 - 0.1003215 \text{ Cutting Speed}$$

The coefficient of determination, $R^2 = 0.606$ tells us that about 61% variation in tool lifelength is attributable to the cutting speed. There is still about a 39% variation in tool life that is not explained by this linear model used. The remaining variation may be attributable to some other factors not considered here, but affecting the tool life. Because the p-value < 0.001, we reject H_0: $\beta_1 = 0$ and conclude that cutting speed contributes significant amount of information toward the prediction of tool life.

Figure 11.13

JMP printout for machine tool life data

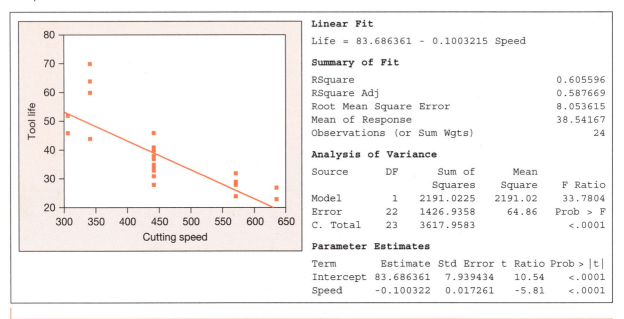

Linear Fit

Life = 83.686361 - 0.1003215 Speed

Summary of Fit

RSquare	0.605596
RSquare Adj	0.587669
Root Mean Square Error	8.053615
Mean of Response	38.54167
Observations (or Sum Wgts)	24

Analysis of Variance

Source	DF	Sum of Squares	Mean Square	F Ratio
Model	1	2191.0225	2191.02	33.7804
Error	22	1426.9358	64.86	Prob > F
C. Total	23	3617.9583		<.0001

Parameter Estimates

Term	Estimate	Std Error	t Ratio	Prob > |t|
Intercept	83.686361	7.939434	10.54	<.0001
Speed	-0.100322	0.017261	-5.81	<.0001

11.3.3 Connection between inference for slope and correlation coefficient

Note that the Pearson's correlation coefficient r is computed using the same quantities used in fitting the least-squares line. In Chapter 2, we learned that slope and correlation coefficient are related $r = \hat{\beta}_1 s_x/s_y$, where

$$r = \frac{SS_{xy}}{\sqrt{SS_{xx}\, SS_{yy}}} \quad \text{and} \quad \hat{\beta}_1 = \frac{SS_{xy}}{SS_{xx}}$$

Because both r and $\hat{\beta}_1$ provide information about the utility of the model, it is not surprising that there is a similarity in their computational formulas: particularly note that SS_{xy} appears in the numerators of both the expressions. Because terms in the denominators are always positive, r and $\hat{\beta}_1$ will always be of the same sign (either both positive or both negative). Positive values of r imply that y increases as x increases, and negative values of r imply that y decreases as x increases. As shown below, we can use the sample information to test the null hypothesis that the normal random variables X and Y are uncorrelated (i.e., $\rho = 0$), where ρ is the population correlation coefficient. However, note that this test cannot be used for testing the hypothesis of $\rho = \rho_0 \neq 0$.

Testing the Strength of Linear Association (t test)

$H_0: \rho = 0$ 	 TS: $T = r\sqrt{\dfrac{n-2}{1-r^2}}$ 	 $\nu = (n-2)$ degrees of freedom

Reject H_0 if p-value $< \alpha$

Right-tailed test:

$H_a: \rho > 0$

RR: $t > t_\alpha(\nu)$

p-value: $P(t(\nu) > t)$

Left-tailed test:

$H_a: \rho < 0$

RR: $t < -t_\alpha(\nu)$

p-value: $P(t(\nu) < t)$

Two-tailed test:

$H_a: \rho \neq 0$

RR: $t > t_{\alpha/2}(\nu)$ 	 or 	 $t < -t_{\alpha/2}(\nu)$

p-value: $2P(t(\nu) > |t|)$

Conditions:

1. The pairs of observations are obtained independently.
2. The pairs of observations are taken from a bivariate normal population.

Note that testing hypothesis $H_0: \rho = 0$ (random variables X and Y are uncorrelated) is equivalent to testing hypothesis $H_0: \beta_1 = 0$ (the slope of the straight-line model is 0).

For testing $\beta_1 = 0$, 	 $t = \dfrac{\hat{\beta}_1}{s/\sqrt{SS_{xx}}}.$

For testing $\rho = 0$, 	 $t = r\sqrt{\dfrac{n-2}{1-r^2}}.$

However,

$$r\sqrt{\frac{n-2}{1-r^2}} = \sqrt{n-2}\left[\frac{SS_{xy}}{\sqrt{SS_{xx}SS_{yy}}}\right]\frac{1}{\sqrt{1 - SS_{xy}^2/(SS_{xx}SS_{yy})}}$$

$$= \sqrt{n-2}\left(\frac{SS_{xy}}{SS_{xx}}\right)\left(\frac{\sqrt{SS_{xx}}}{\sqrt{SS_{yy} - SS_{xy}^2/SS_{xx}}}\right) = \frac{\hat{\beta}_1\sqrt{SS_{xx}}}{s}$$

The only real difference between the least-squares slope estimate $\hat{\beta}_1$ and the correlation coefficient r is the measurement scale, because r is unitless, whereas $\hat{\beta}_1$ depends on the units of X and Y. The information they provide about the utility of the least-squares model to some extent is redundant. However, the test for the correlation requires the additional assumption that the variables X and Y have a bivariate normal distribution, whereas the test of slope does not require that x be a random variable. For example, we could use the straight-line model to describe the relationship between the yield of a chemical process Y and the temperature at which the reaction takes place x, even if the temperature is controlled and therefore nonrandom. However, the correlation coefficient would have no meaning in this case, because the bivariate normal distribution requires that both variables be marginally normal.

Exercises

11.15 Refer to Exercise 11.1. Is there sufficient evidence to say that the slope of the line is significantly different from zero? Use $\alpha = 0.05$.

11.16 Refer to Exercise 11.2. Estimate β_1 in a confidence interval with confidence coefficient 0.90.

11.17 Refer to Exercise 11.3. Estimate the amount of elongation per unit increase in force. Use a confidence coefficient of 0.95.

11.18 Refer to Exercise 11.4. Do industry sales appear to contribute any information to the prediction of company sales? Test at the 10% level of significance.

11.19 Refer to Exercise 11.5. Does Rockwell hardness appear to be linearly related to abrasion loss? use $\alpha = 0.05$.

11.20 In Exercise 11.6, the price of a gallon of gasoline y was modeled as a function of the price of a barrel of crude oil x using the following equation: $Y = \beta_0 + \beta_1 x + \varepsilon$. Estimates of β_0 and β_1 were found to be 25.883 and 3.115, respectively.

  a Estimate the variance and standard deviation of ε.

  b Suppose the price of a barrel of crude oil is $15. Estimate the mean and standard deviation of the price of a gallon of gas under these circumstances.

  c Repeat part (b) for a crude oil price of $30.

  d What assumptions about ε did you make in answering parts (b) and (c)?

11.21 Refer to Exercise 11.7. Estimate the increase in expected number of parts assembled per 5-minute increase in time spent on the job. Use a 95% confidence coefficient.

11.22 Refer to Exercise 11.8. Does it appear that flow-through LC50 values are linearly related to static LC50 values? Use $\alpha = 0.05$.

11.23 It is well known that large bodies of water have a mitigating effect on the temperature of the surrounding land masses. On a cold night in central Florida, temperatures (in degrees Fahrenheit) were recorded at equal distances along a transect running in the downwind direction from a large lake. The data are in table below.

  a Fit the linear model $Y = \beta_0 + \beta_1 x + \varepsilon$ by the method of least squares.

  b Find SSE and s^2.

  c Does it appear that temperatures decrease significantly as distance from the lake increases? Use $\alpha = 0.05$.

Site (x)	1	2	3	4	5	6	7	8	9	11
Temperature (y)	37.00	36.25	35.41	34.92	34.52	34.45	34.40	34.00	33.62	33.90

11.4 Using the Linear Model for Estimation and Prediction

If we are satisfied that a useful model has been found to describe the relationship between peak power load and maximum daily temperature, we are ready to accomplish the original objective in constructing the model, that is, using it for prediction, estimation, and so on. The most common uses of a probabilistic model for making inferences can be divided into two categories.

  1. The use of the model for estimating the mean value of Y, $E(Y)$, for a specific value of x. For our peak power load example, we might want to estimate the mean peak power load for days during which the maximum temperature is 90°F.

  2. The use of the model to predict a particular value of Y for a given x. That is, we might want to predict the peak power load for a particular day during which the maximum temperature will be 90°F.

In the first case, we are attempting to estimate the mean value of Y over a very large number of days. In the second case, we are trying to predict the Y value for a single day. Which of these uses of the model, estimating the mean value of Y or predicting an individual value of Y (for the same value of x), can be accomplished with the greater accuracy?

Before answering this question, we first consider the problem of choosing an estimator (or predictor) of the mean (or individual) Y value. We use the least-squares model

$$\hat{y} = \hat{\beta}_0 - \hat{\beta}_1 x$$

both to estimate the mean value of Y and to predict a particular value of Y for a given value of x. In the power load example, we found

$$\hat{y} = -419.85 + 6.7175x$$

So the estimated mean peak power load for all days when $x = 90$ (maximum temperature is 90°F) is

$$\hat{y} = -419.85 + 6.7175(90) = 184.72$$

The identical value is used to predict the Y value when $x = 90$. That is, both the estimated mean and the predicted value of Y are $y = 184.72$ when $x = 90$, as shown in Figure 11.14.

The difference between these two uses of the probabilistic model lies in the relative precision of the estimate and the prediction. The precisions are measured by the expected squared distances between the estimator/predictor y and the quantity being estimated or predicted. In the case of estimation, we are trying to estimate

$$E(Y) = \beta_0 + \beta_1 x_p$$

where x_p is the particular value of x at which the estimate is being made. The estimator is

$$\hat{y} = \hat{\beta}_0 + \hat{\beta}_1 x_p$$

Figure 11.14

Estimated mean value and predicted individual value of peak power load for temperature $x = 90$°F

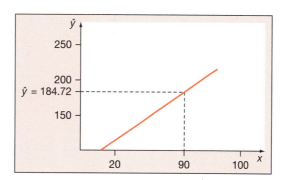

the sources of error in estimating $E(Y)$ are the estimators $\hat{\beta}_0$ and $\hat{\beta}_1$. The variance can be shown to be

$$\sigma_{\hat{Y}}^2 = E\{[\hat{Y} - E(\hat{Y})]^2\} = \sigma^2\left[\frac{1}{n} + \frac{(x_p - \bar{x})^2}{SS_{xx}}\right]$$

In the case of predicting a particular y value, we are trying to predict

$$Y_p = \beta_0 + \beta_1 x_p + \varepsilon_p$$

where ε_p is the error associated with the particular Y value Y_p. Thus, the sources of prediction error are the estimator $\hat{\beta}_0$ and $\hat{\beta}_1$ and the error ε_p. The variance associated with ε_p is σ^2, and it can be shown that this is the additional term in the expected squared prediction error:

$$\sigma^2(y_p - \hat{y}) = E[(\hat{Y} - Y_p)^2] = \sigma^2 + \sigma_{\hat{y}}^2$$

$$= \sigma^2\left[1 + \frac{1}{n} + \frac{(x_p - \bar{x})^2}{SS_{xx}}\right]$$

The sampling errors associated with the estimator and predictor are summarized in Figures 11.15 and 11.16. They graphically depict the difference between the error of estimating a mean value of Y and the error of predicting a future value of Y for $x = x_p$. Note that the error of estimation is just the vertical difference between the least-squares line and the true line of means, $E(Y) = \beta_0 + \beta_1 x$. However, the error in predicting some future value of Y is the sum of two errors: the error of estimating the mean of Y, $E(Y)$, shown in Figure 11.15, plus the random error ε_p that is a component of the value Y_p to be predicted. Note that both errors will be smallest when $x_p = \bar{x}$. The farther x_p lies from the mean, the larger the errors of estimation and prediction.

Figure 11.15

Error in estimating the mean value of Y for a given value of x

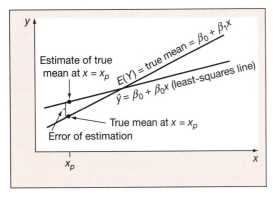

Figure 11.16

Error in predicting a future value of Y for a given value of x

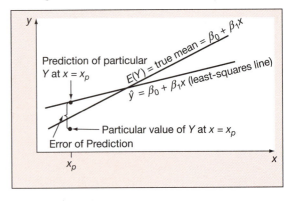

Sampling Errors for the Estimator of the Mean of Y and the Predictor of an Individual Y

1. The standard deviation of the sampling distribution of the estimator $\hat{Y}$ of the mean value of Y at a fixed x is

$$\sigma_{\hat{Y}} = \sigma \sqrt{\frac{1}{n} + \frac{(x - \bar{x})^2}{SS_{xx}}}$$

where σ is the standard deviation of the random error ε.

2. The standard deviation of the prediction error for the predictor $\hat{Y}$ of an individual Y-value at a fixed $\bar{x}$ is

$$\sigma_{(Y-\hat{Y})} = \sigma \sqrt{1 + \frac{1}{n} + \frac{(x - \bar{x})^2}{SS_{xx}}}$$

where σ is the standard deviation of the random error ε.

The true value of σ will rarely be known. Thus, we estimate σ by s and calculate the estimation and prediction intervals as shown next.

A 100$(1 - \alpha)$ Percent Confidence Interval for the Mean Value of Y at a Fixed x

$$\hat{Y} \pm t_{\alpha/2}(n - 2)(\text{estimated standard deviation of } \hat{Y})$$

or

$$\hat{Y} \pm t_{\alpha/2}(n - 2)s \sqrt{\frac{1}{n} + \frac{(x - \bar{x})^2}{SS_{xx}}}$$

A 100$(1 - \alpha)$ Percent Prediction Interval for Individual Y at a Fixed x

$$\hat{Y} \pm t_{\alpha/2}(n - 2)[\text{estimated standard deviation of } (Y - \hat{Y})]$$

or

$$\hat{Y} \pm t_{\alpha/2}(n - 2)s \sqrt{1 + \frac{1}{n} + \frac{(x - \bar{x})^2}{SS_{xx}}}$$

Note that, the term *prediction interval* is used when the interval is intended to enclose the value of a random variable. The term *confidence interval* is reserved for estimation of population parameters (such as the mean).

Example 11.8

Refer to the peak power load example (Chapter 2).

a Find a 95% confidence interval for the mean peak power load when the maximum daily temperature is 90°F.

b Predict the peak power load for a day during which the maximum temperature is 90°F.

Solution

a A 95% confidence interval for estimating the mean peak power load at a high temperature of 90°F is given by

$$\hat{y} \pm t_{\alpha/2}(n-2)s\sqrt{\frac{1}{n} + \frac{(x - \bar{x})^2}{SS_{xx}}}$$

We have previously calculated $\hat{y} = 184.72$, $s = 16.18$, $\bar{x} = 91.5$, and $SS_{xx} = 380.5$. With $n = 10$ and $\alpha = 0.05$, $t_{0.025}(8) = 2.306$, and thus the 95% confidence interval is

$$\hat{y} \pm t_{\alpha/2}(n-2)s\sqrt{\frac{1}{n} + \frac{(x - \bar{x})^2}{SS_{xx}}}$$

$$184.72 \pm (2.306)(16.18)\sqrt{\frac{1}{10} + \frac{(90 - 91.5)^2}{380.5}} \quad \text{i.e.,} \quad 184.72 \pm 12.14$$

Thus, we can be 95% confident that the *mean* peak power load is between 172.58 and 196.86 megawatts for days with a maximum temperature of 90°F.

b Using the same values we find

$$\hat{y} \pm t_{\alpha/2}(n-2)s\sqrt{1 + \frac{1}{n} + \frac{(x - \bar{x})^2}{SS_{xx}}}$$

$$184.72 \pm (2.306)(16.18)\sqrt{1 + \frac{1}{10} + \frac{(90 - 91.5)^2}{380.5}} \quad \text{i.e.,} \quad 184.72 \pm 39.24$$

Thus, we are 95% confident that the peak power load will be between 145.48 and 223.96 megawatts on a particular day when the maximum temperature is 90°F.

A comparison of the confidence interval for the mean peak power load and the prediction interval for a single day's peak power load for a maximum temperature of 90°F ($x = 90$) is illustrated in Figure 11.17. It is important to note that the prediction interval for an individual value of Y will always be wider than the corresponding confidence interval for the mean value of Y.

Figure 11.17

A 95% confidence interval for mean peak power load and a prediction interval for peak power load when $x = 90$

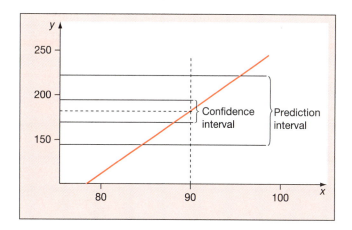

Figure 11.18

The danger of using a model to predict outside the range of sample values of x

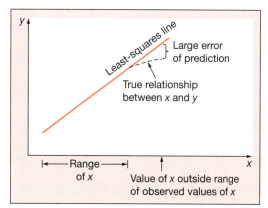

Be careful not to use the least-squares prediction equation to estimate the mean value of Y or to predict a particular value of Y for values of x that fall outside the range of the values of x contained in your sample data. The model might provide a good fit to the data over the range of x values contained in the sample but a very poor fit for values of x outside this region (see Figure 11.18). Failure to heed this warning may lead to errors of estimation and prediction that are much larger than expected.

Example 11.9

Refer to the example of overtolerance and wall reduction in Chapter 2.

a Estimate the mean wall reduction for OT $= 0.0020$ using a 95% confidence level.

b Predict the amount of wall reduction for OT $= 0.0020$ on one test block, using a 95% confidence level.

Solution Let us use the JMP printout given in Figure 11.19 in computing intervals. The inside pair of curves gives the confidence intervals and the outside pair gives the prediction intervals.

Using the least-squares estimate of the linear relation between WR and OT, we get,

$$\hat{y} = 3.231042 + 215.3032(0.0020) = 3.66$$

Figure 11.19

Linear regression printout for overtolerance data

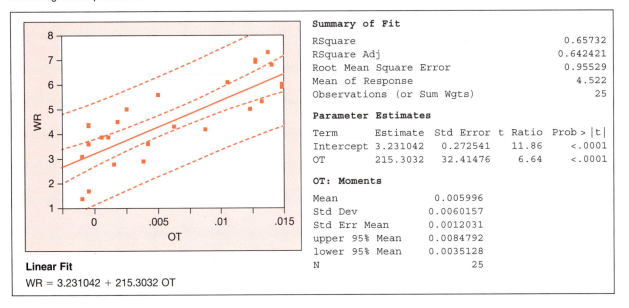

Linear Fit

WR = 3.231042 + 215.3032 OT

With $n = 25$ pairs of observations, the error degrees of freedom are 23. Thus, $t_{0.025}(23) = 2.06866$. The printout gives, $s = 0.95529$, $\bar{x} = 0.005996$, and $SS_{xx} = (25 - 1)0.0060157^2 = 0.0008685$.

a The 95% confidence interval for estimating the mean wall reduction at OT = 0.0020 is

$$\hat{y} \pm t_{\alpha/2}(n - 2)s\sqrt{\frac{1}{n} + \frac{(x - \bar{x})^2}{SS_{xx}}}$$

i.e.,

$$3.66 \pm 2.069(0.955)\sqrt{\frac{1}{25} + \frac{(0.002 - 0.006)^2}{0.00087}}$$

i.e.,

$$3.66 \pm 0.4775, \quad \text{i.e.,} \quad (3.183, 4.138)$$

Thus, we are 95% confident that the mean wall reduction is between 3.18 and 4.14 for an overtolerance of 0.0020.

b The 95% prediction interval for predicting the wall reduction for a test block with OT = 0.0020 is

$$\hat{y} \pm t_{\alpha/2}(n - 2)s\sqrt{1 + \frac{1}{n} + \frac{(x - \bar{x})^2}{SS_{xx}}}$$

i.e.,

$$3.66 \pm 2.069(0.955)\sqrt{1 + \frac{1}{25} + \frac{(0.002 - 0.006)^2}{0.00087}}$$

i.e.,

$$3.66 \pm 2.033, \quad \text{i.e.,} \quad (1.627, 5.693)$$

Thus, we are 95% confident that the wall reduction will be between 1.627 and 5693 for the test block with an overtolerance of 0.0020.

It is easy to get computer programs to compute the confidence intervals and prediction intervals for the range of the data and plot them on the scatterplot, known as confidence *bands* and *prediction bands*, respectively. The JMP printout (Figure 11.19) shows such confidence and prediction bands. The inner pair of curves corresponds to the confidence bands, and the outer pair corresponds to the prediction bands. Notice that the prediction bands are wider than the confidence bands due to additional error term involved in prediction intervals. Also, note that the bands are narrowest at $\bar{x}$ and become wider on both sides as x moves away from $\bar{x}$.

Exercises

11.24 Refer to Exercise 11.1.
  a Estimate the mean value of Y for $x = 2$, using a 95% confidence interval.
  b Find a 95% prediction interval for a response Y at $x = 2$.

11.25 Refer to Exercise 11.3. Estimate the expected elongation for a force of 1.8, using a 90% confidence interval.

11.26 Refer to Exercise 11.4. If in 1988 the industry sales figure is 16, find a 95% prediction interval for the company sales figure. Do you see any possible difficulties with the solution to this problem?

11.27 Refer to Exercise 11.5. Estimate, in a 95% confidence interval, the mean abrasion loss for steel specimens with a Rockwell hardness measure of 75. What assumptions are necessary for your answer to be valid?

11.28 Refer again to Exercise 11.5. Predict, in a 95% prediction interval, the abrasion loss for a particular steel specimen with a Rockwell hardness measure of 75. What assumptions are necessary for your answer to be valid?

11.29 Refer to Exercise 11.7. Estimate, in a 90% confidence interval, the expected number of parts assembled by employees working for 12 minutes.

11.30 Refer to Exercise 11.8. Find a 95% prediction interval for a flow-through measurement corresponding to a static measurement of 21.0.

11.31 Refer to Exercise 11.3. If no force is applied, the expected elongation should be zero (that is, $\beta_0 = 0$). Construct a test of $H_0: \beta_0 = 0$ versus $H_a: \beta_0 \neq 0$, using the data of Exercise 11.3. Use $\alpha = 0.05$.

11.32 In planning for an initial orientation meeting with new electrical engineering majors, the chairman of the Electrical Engineering Department wants to emphasize the importance of doing well in the major courses in getting better-paying jobs after graduation. To support this point, the chairman plans to show that there is a strong positive correlation between starting salaries for recent electrical engineering graduates and their grade-point averages in the major courses. Records for seven of last year's electrical engineering graduates are selected at random and given in the following table:

Grade-Paint Average in Major Courses x	Starting Salary y ($ thousand)
2.58	66.5
3.27	68.6
3.85	69.5
3.50	69.2
3.33	68.5
2.89	66.6
2.23	65.6

  a Find the least-squares prediction equation.
  b Plot the data and graph the line as a check on your calculations. Do the data provide sufficient evidence that grade-point average provides information for predicting starting salary?
  c Find a 95% prediction interval for the starting salary of a graduate whose grade-point average is 3.2.
  d What is the mean starting salary for graduates with grade-point averages equal to 3.0? Use a 95% confidence interval.

11.33 Lawton (*Journal of Pressure Vessel Technology*, 2003) describes experiments conducted to determine temperature and heat transfer at the bore surface of the barrel at or near the commencement of rifling and to support the assessment of new ammunition for the 155 mm gun. The wear rate of propellant (m/s) and the maximum bore temperature (°C) were as shown in shown in table below.

Firing	Temp	Rate	Firing	Temp	Rate
1	975	0.14	14	815	0.045
2	960	0.14	15	790	0.040
3	975	0.13	16	895	0.044
4	965	0.12	17	950	0.044
5	960	0.11	18	925	0.042
6	955	0.081	19	885	0.040
7	950	0.078	20	895	0.034
8	850	0.067	21	815	0.030
9	855	0.065	22	805	0.029
10	820	0.055	23	825	0.027
11	830	0.055	24	820	0.026
12	895	0.056	25	725	0.028
13	925	0.055	26	675	0.020

a Develop a model to predict the wear rate as a function of the maximum bore temperature.

b Develop confidence and prediction bonds.

c Estimate the mean wear rate for the maximum bore temperature 900°C. Use a 5% error rate.

d Predict the wear rate for the maximum bore temperature 900°C. Use a 5% error rate.

11.34 The yield pressure of an autofrettaged cannon is controlled by the applied and residual stresses in the cannon and by its material strength. Unanticipated yielding of cannon (or any other pressure vessel) can be a safety hazard. Underwood. Monk, Audino, and Parker (*Journal of Pressure Vessel Technology*, 2003) studied the effect of applied pressure (measured in MPa) on the permanent outer diameter (OD) hoop strain, through laboratory tests. The laboratory recorded permanent OD strain (%) and applied pressure (measured in MPa) as shown in Table 11.2.

a Develop a model to estimate OD strain as a function of applied pressure for each of the three yield strengths. Use transformation if necessary.

b Assess the fit of each model.

c Estimate the OD strain for 700 MPa applied pressure at a yield strength of 1022 MPa, as well as one with a 1105 MPa.

Table 11.2

Applied Pressure and Outer Diameter Strain Data

Yield Strength					
1022 MPa		1105 MPa		1177 MPa	
Pressure	OD Strain	Pressure	OD Strain	Pressure	OD Strain
635	0.010	690	0.010	745	0.012
650	0.012	705	0.012	760	0.015
665	0.019	720	0.018	775	0.019
675	0.027	735	0.026	785	0.026
690	0.044	745	0.048	800	0.034
705	0.090	760	0.058	815	0.053
		775	0.092	830	0.080

11.5 Multiple Regression Analysis

Most practical applications of regression analysis utilize models that are more complex than the simple straight-line model. For example, a realistic model for a power plant's peak power load would include more than just the daily high temperature. Factors such as humidity, day of the week, and season are a few of the many variables that might be related to peak load. Thus, we would want to incorporate these and other potentially important independent variables into the model to make more accurate predictions.

Probabilistic models that include terms involving x^2, x^3 (or higher-order terms), or more than one independent variable are called *multiple regression models*. The general form of these models is

$$Y = \beta_0 + \beta_1 x_1 + \beta_2 x_2 + \cdots + \beta_k x_k + \varepsilon$$

The dependent variable Y is now written as a function of k independent variables $x_1, x_2, \ldots, x_k$. The random error term is added to allow for deviation between the deterministic part of the model, $\beta_0 + \beta_1 x_1 + \cdots + \beta_k x_k$, and the value of the dependent variable Y. The random error component makes the model probabilistic rather than deterministic. The value of the coefficient β_i determines the contribution of the independent variable x_i, and β_0 is the Y-intercept. The coefficients $\beta_0, \beta_1, \ldots, \beta_k$ will usually be unknown, since they represent population parameters.

Polynomial Regression   At first glance, it might appear that the regression model shown above would not allow for anything other that straight-line relationships between Y and the independent variables, but this is not the case. Actually, $x_1, x_2, \ldots, x_k$ can be functions of variables as long as the functions do not contain unknown parameters. For example, the yield Y of a chemical process could be a function of the independent variables

$$x_1 = \text{temperature at which the reaction takes place}$$
$$x_2 = (\text{temperature})^2 = x_1^2$$
$$x_3 = \text{pressure at which the reaction takes place}$$

We might hypothesize the model

$$E(Y) = \beta_0 + \beta_1 x_1 + \beta_2 x_1^2 + \beta_3 x_3$$

Although this model contains a second-order (or "quadratic") term x_1^2 and therefore is curvilinear, the model is still *linear in the unknown parameters* β_0, β_1, β_2 and β_3. Multiple regression models that are linear in the unknown parameters (the β's) are called *linear models*, even though they may contain nonlinear independent variables. The kth degree polynomial regression model is given by

$$Y = \beta_0 + \beta_1 x + \beta_2 x^2 + \cdots + \beta_k x^k + \varepsilon$$

It is a special case of the general additive multiple regression model with $x_1 = x, x_2 = x^2, \ldots, x_k = x^k$. The quadratic regression model (i.e., the model involving terms of up to degree 2) is a commonly used polynomial regression model. It models the mean of Y values, $E(Y)$, using a parabolic curve.

Example 11.10

Figure 11.20 shows the electricity usage as a function of area of a house. The electricity usage increases with the area; however, the rate of change is not constant across all the values of area. In fact, as the area increases the rate at which electricity usage increases seems to have decreased, which makes sense because the larger house does not necessarily mean a larger family and more usage. The estimated polynomial of degree 2 is

$$\text{usage} = -1{,}216.14 + 2.39893 \text{ size} - 0.00045 \text{ (size)}^2$$

For a house with 2,000 square feet area, the predicted electricity usage is

$$\text{usage} = -1{,}216.14 + 2.39893(2{,}000) - 0.00045(2{,}000)^2 = 1{,}781.72$$

Figure 11.20
Quadratic regression model

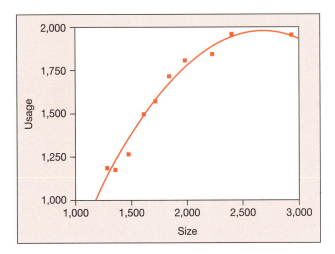

Qualitative Predictor Variables: So far, we have developed models using only quantitative variables as the predictor variables. However, sometimes the response variable is dependent on qualitative variables too. We use a *dummy variable*, taking values 0 or 1 to indicate the relevant category of the qualitative variable as follows:

$$\text{Dummy variable} = \begin{cases} 1 & \text{if the catagorical variable takes a specific value(s)} \\ 0 & \text{otherwise} \end{cases}$$

Example 11.11

In addition to the size of the house, the selling price of a house might also depend on whether the house is a brick house or a stucco house. But the style of the house (brick or stucco) is a qualitative variable, and we need to recode it before we can use it in the model. Suppose we define a dummy variable "style" as follows:

$$\text{Style} = \begin{cases} 1 & \text{if the house is a brick house} \\ 0 & \text{otherwise} \end{cases}$$

Then use model

$$\text{Price} = \beta_0 + \beta_1 \text{ Area} + \beta_2 \text{ Style} + \varepsilon$$

The mean selling price of the house depends on whether it is a brick or stucco house.

For a brick house: $\qquad\qquad E(\text{Price}) = \beta_0 + \beta_1 \text{ Area} + \beta_2$

For a stucco house: $\qquad\qquad E(\text{Price}) = \beta_0 + \beta_1 \text{ Area}$

The coefficient β_2 is the difference in the mean prices of same-sized brick and stucco houses. If $\beta_2 > 0$ then on the average the brick houses cost more than the stucco houses. Figure 11.21 shows the relation between the mean selling prices of houses in Gainesville, Florida, as a function of the area of the house and the style (brick or stucco), estimated using data collected from sales recorded over a 1-month period.

Figure 11.21

Regression function for model with one dummy variable

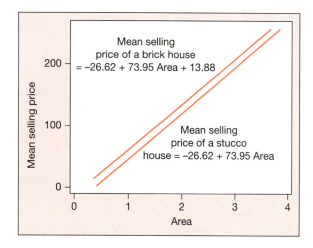

For a qualitative variable with m categories, we need $(m - 1)$ dummy variables to account for all the levels of the variable. Because only two styles of houses (brick and stucco) were sampled, only one dummy variable is needed to describe it.

Example 11.12

Refer to Example 11.11 scenario. Suppose houses with four different styles—brick, stucco, wooden siding, and aluminum siding—were sampled. To take into account four ($m = 4$) categories, we need three dummy variables as follows:

$$\text{Style-I} = \begin{cases} 1 & \text{if the house has brick} \\ 0 & \text{otherwise} \end{cases}$$

$$\text{Style-II} = \begin{cases} 1 & \text{if the house has stucco} \\ 0 & \text{otherwise} \end{cases}$$

$$\text{Style-III} = \begin{cases} 1 & \text{if the house has wooden siding} \\ 0 & \text{otherwise} \end{cases}$$

We can use these three dummy variables to represent four styles of the house as follows:

Style of house	Style-I	Style-II	Style-III
Brick	1	0	0
Stucco	0	1	0
Wooden siding	0	0	1
Aluminum siding	0	0	0

Then the model becomes,

$$\text{Price} = \beta_0 + \beta_1 \text{ Area} + \beta_2 \text{ Style-I} + \beta_3 \text{ Style-III} + \beta_4 \text{ Style-III} + \varepsilon.$$

It gives the mean selling price of the house as a function of the style of house.

For a brick house: $E(\text{Price}) = \beta_0 + \beta_1 \text{ Area} + \beta_2$
For a stucco house: $E(\text{Price}) = \beta_0 + \beta_1 \text{ Area} + \beta_3$
For a wooden siding house: $E(\text{Price}) = \beta_0 + \beta_1 \text{ Area} + \beta_4$
For an aluminum siding house: $E(\text{Price}) = \beta_0 + \beta_1 \text{ Area}$

- Here β_2 is the difference in the mean prices of same-sized brick and aluminum siding houses. If $\beta_2 > 0$, then on the average the brick houses cost more than the aluminum siding houses.
- Here β_3 is the difference in the mean prices of same-sized stucco and aluminum siding houses. If $\beta_3 > 0$, then on the average the stucco houses cost more than the aluminum siding houses.
- Here β_4 is the difference in the mean prices of same-sized wooden siding and aluminum siding houses. If $\beta_4 > 0$, then on the average the wooden siding houses cost more than the aluminum siding houses.

Interaction between Variables   When more than one independent variable is used in the model, some variables might react with each other and exercise a different effect on the response. Then we say that there is an *interaction* between the two variables. The interaction term is included in the model by adding a product term of the two variables as follows:

$$y = \beta_0 + \beta_1 x_1 + \beta_2 x_2 + \beta_3 x_1 x_2 + \varepsilon$$

With more variables in the model, more than one interaction term can be included in the model. Interactions among more than two variables can also be studied. For example, an interaction model with three explanatory variables is

$$Y = \beta_0 + \beta_1 x_1 + \beta_2 x_2 + \beta_3 x_3 + \beta_4 x_1 x_2 + \beta_5 x_1 x_3 + \beta_6 x_2 x_3 + \beta_7 x_1 x_2 x_3 + \varepsilon$$

Example 11.13

Refer to the Example 11.12 scenario. There might not be a fixed difference in the mean selling prices of the same-sized brick and stucco houses over the range of area studied. The difference in the mean selling prices of same-sized brick and stucco houses might decrease with the increase in area of the house. In other words, for larger houses, the difference in selling prices of brick and stucco houses is smaller than that for the smaller houses. Then we say that there is an *interaction* between area and style. The interaction model used for the selling price is written as

$$\text{Price} = \beta_0 + \beta_1 \text{ Area} + \beta_2 \text{ Style} + \beta_3 \text{ Area} \times \text{Style} + \varepsilon$$

The mean selling price of the house depends on whether it is a brick or stucco house.

For a brick house: $\quad E(\text{Price}) = \beta_0 + (\beta_1 + \beta_3) \text{ Area} + \beta_2$

For a stucco house: $\quad E(\text{Price}) = \beta_0 + \beta_1 \text{ Area}$

Figure 11.22 shows the prediction equations with interaction between the area and the style of the house. Note that the lines are not parallel to each other. On the other hand, the model without interaction (Figure 11.21) shows two lines parallel to each other.

Figure 11.22

Regression function for model with interaction

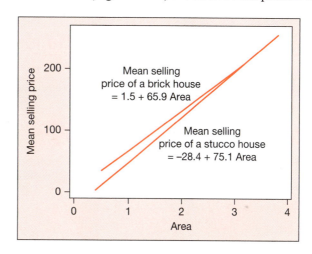

Nonlinear Models Not all relations are linear in nature. Many relations are nonlinear, but some nonlinear relationships can be transformed into linear relationships. For example, the model $Y = \beta_0 e^{(\beta_1 x_1 + \beta_2 x_2)} \varepsilon$ describes a nonlinear relation. By taking logarithms of both sides we get, $\ln(Y) = \ln(\beta_0) + \beta_1 x_1 + \beta_2 x_2 + \ln(\varepsilon)$, which is a linear multiple regression model. An appropriate transformation can be selected based on the theoretical relationships among variables or from the nature of relationships described by different plots.

For example, Kim, Koo, Choi, and Kim (*Transactions of the ASME*, 2003) investigated the validity of the present ASME code in evaluating the integrity of a reactor pressure vessel under pressurized thermal stock conditions. Figure 11.23 shows the elastic hoop stress (in MPa) distribution after 700 seconds. It is clear that the relationship is not linear. The scatterplot also shows the logarithmic transformation of the relationship overlaid on it, which may be a reasonable model to consider. The transformed fit is

$$\ln(\text{Hoop Stress}) = 5.93225 - 0.92334x/a$$

For more information on transformation of data, refer to Chapter 2, Section 8.

Figure 11.23

Nonlinear relationship and logarithmic transformation

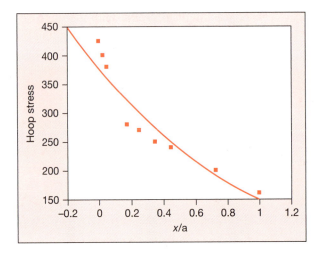

11.5.1   Fitting the model: The least-squares approach

The model used for fitting multiple regression models is identical to that used for fitting the simple straight-line model—the method of least squares. That is, we choose the estimated model

$$\hat{y} = \hat{\beta}_0 + \hat{\beta}_1 x_1 + \cdots + \hat{\beta}_k x_k$$

that minimizes

$$SSE = \sum_{i=1}^{n} (y_i - \hat{y}_i)^2$$

As in the case of simple linear model, the sample estimates $\hat{\beta}_0, \hat{\beta}_1, \ldots, \hat{\beta}_k$ are obtained as a solution of a set of simultaneous linear equations.

The primary difference between fitting the simple and multiple regression models is computational difficulty. The $(k + 1)$ simultaneous linear equations that must be solved to find the $(k + 1)$ estimated coefficients $\hat{\beta}_0, \hat{\beta}_1, \ldots, \hat{\beta}_k$ are difficult to solve (sometimes nearly impossible) with a pocket or desk calculator. Consequently, we resort to the use of computers. Many computer packages are available for fitting multiple regression models using the method of least squares. We present outputs from some of the popular computer packages instead of presenting the tedious calculations required to fit the models. Most statistical programs have similar regression outputs (see Section 11.10 on Reading Computer Printouts).

Example 11.14

Recall the peak power load data from Chapter 2 (repeated on the next page). We previously used a straight-line model to describe the peak power load and daily high temperature relationship. But note a slight upward curvature in the trend. The peak power load seems to be increasing at a faster rate for higher daily temperatures. Now suppose we want to hypothesize a curvilinear relationship

$$Y = \beta_0 + \beta_1 x + \beta_2 x^2 + \varepsilon$$

Daily High Temperature x (°F)	Peak Power Load y (megawatts)
95	214
82	152
90	156
81	129
99	254
100	266
93	210
95	204
93	213
87	150

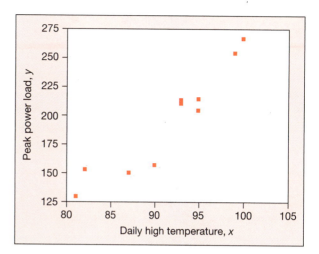

The curvature present in this relationship provides some support for the inclusion of the term $\beta_2 x^2$ in the model. Part of the Minitab output for the peak power load data is reproduced in Figure 11.24. The least-square estimates of the β parameters appear in the column labeled "Coef." We can see that $\hat{\beta}_0 = 1{,}784.2$, $\hat{\beta}_1, = -42.39$, and $\hat{\beta}_2 = 0.27$. Therefore, the equation that minimizes the SSE for the data is

$$\hat{y} = 1{,}784.2 - 42.39x + 0.27x^2$$

The SSE for this best line of fit is equal to 1,175.1.

Figure 11.24

Computer printout for power load data

Regression Analysis: load versus temperature, temp*2

```
The regression equation is
load = 1784 - 42.4 temperature + 0.272 temp*2

Predictor        Coef    SE Coef        T          P
Constant       1784.2      944.1     1.89      0.101
temperat       -42.39      21.00    -2.02      0.083
temp*2         0.2722     0.1163     2.34      0.052

S = 12.96   R-Sq = 93.9%   R-Sq(adj) = 92.2%

Analysis of Variance

Source           DF        SS        MS        F        P
Regression        2   18088.5    9044.3    53.88    0.000
Residual Error    7    1175.1     167.9
Total             9   19263.6
```

Note that the graph of the multiple regression model (Figure 11.25) provides a good fit to the data. Lack of pattern in the residual plot in Figure 11.26 also confirms it. However, before we can more formally measure the utility of the model, we need to estimate the variance of the error component ε.

Figure 11.25

Plot of curvilinear model

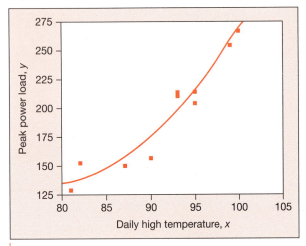

Figure 11.26

Residual plot for the model in Figure 11.25

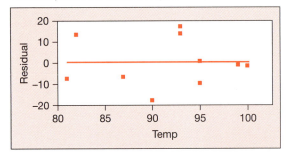

11.5.2   Estimation of error variance σ^2

The specification of the probability distribution of the random error component ε of the multiple regression model follows the same general outline as for the straight-line model. We assume that

- The errors ε are normally distributed with mean zero and constant variance σ^2 for any set of values for the independent variables $x_1, x_2, \ldots, x_k$.
- The errors are assumed to be independent.

Given these assumptions, the remaining task in specifying the probability distribution of ε is to estimate σ^2.

Example 11.15

In the quadratic model (Example 11.14) describing peak power load as a function of daily high temperature, we found a minimum SSE = 1,175.1. Now we want to use this quantity to estimate the variance of ε. Recall that the estimator for the straight-line model was $s^2 = SSE/(n-2)$, and note that the denominator is n minus the number of estimated β parameters, which is $(n-2)$ in the straight-line model. Because we must estimate one more parameter β_2 for the quadratic model $Y = \beta_0 + \beta_1 x + \beta_2 x^2 + \varepsilon$, the estimate of σ^2 is $s^2 = SSE/(n-3)$. That is, the denominator becomes $(n-3)$ because there are now three β parameters in the model.

The numerical estimate for our example is

$$s^2 = \frac{SSE}{10-3} = \frac{1,175.1}{7} = 167.87$$

where s^2 is called the mean square for error, or MSE. This estimate of σ^2 is shown in Figure 11.24 in the column labeled MS and the row labeled Error.

For the general multiple regression model

$$Y = \beta_0 + \beta_1 x_1 + \beta_2 x_2 + \cdots + \beta_k x_k + \varepsilon$$

we must estimate the $(k + 1)$ parameters, $\beta_0, \beta_1, \beta_2, \ldots, \beta_k$. Thus, the estimator of σ^2 is the SSE divided by the quantity $[n - (\text{number of estimated } \beta \text{ parameters})]$.

Estimator of σ^2 for Multiple Regression Model with k Independent Variables

$$\text{MSE} = \frac{\text{SSE}}{n - (\text{number of estimated } \beta \text{ parameters})} = \frac{\text{SSE}}{n - (k + 1)}$$

We use the estimator of σ^2 both to check the adequacy of the model (Sections 10.4 and 10.5) and to provide a measure of the reliability of predictors and estimates when the model is used for those purposes (Section 10.8). Thus, you can see that the estimation of σ^2 plays an important part in the development of a regression model.

The estimator of σ^2 is used

- To check the adequacy of the model (Sections 11.5.3 and 11.5.4)
- To provide a measure of the reliability of predictors and estimates when the model is used for those purposes (Section 11.9)

11.6 Inference in Multiple Regression

In the multiple regression analysis, the response variable is described as a function of more than one predictor variable. Therefore, there are several types of inferences that can be made using this model. In the simple linear model studied earlier, the test for the slope (t-test) is equivalent to the test for the utility of the model (F-test). However, in the multiple regression they differ on the account of having more than one slope parameter.

11.6.1 A test of model adequacy

To find a statistic that measures how well a multiple regression model fits a set of data, we use the multiple regression equivalent of R^2, the coefficient of determination for the straight-line model (Chapter 2). Thus, we define the *multiple coefficient of determination R^2* as

$$R^2 = 1 - \frac{\displaystyle\sum_{i=1}^{n}(y_i - \hat{y}_i)^2}{\displaystyle\sum_{i=1}^{n}(y_i - \bar{y})^2} = 1 - \frac{\text{SSE}}{\text{SS}_{yy}}$$

where $\hat{y}_i$ is the predicted value of Y_i using the underlying model. Just for the simple linear model, R^2 represents the fraction of the sample variation of the y values (measured by SS_{yy}) that is explained by the least-squares prediction equation. Thus, $R^2 = 0$ implies a complete lack of fit of the model to the data, and $R^2 = 1$ implies a perfect fit, with the model passing through every data point. In general, $0 \leq R^2 \leq 1$, and the larger the value of R^2, the better the model fits the data.

For the peak power load data, the value of $R^2 = 0.939$ is indicated in Figure 11.26. It implies that using the independent variable daily high temperature in a quadratic model results in 93.9% reduction in the total *sample* variation (measured by SS_{yy}) of peak power load Y. Thus, R^2 is a sample statistic that tells us how well the model fits the data and thereby represents the measure of the adequacy of the model.

The fact that R^2 is a sample statistic implies that it can be used to make inference about the utility of the entire model for predicting the population of y values at each setting of the independent variables.

Testing the Utility of a Multiple Regression Model: The Global *F*-Test

H_0: $\beta_1 = \beta_2 = \cdots = \beta_k = 0$.

H_a: At least one of the β parameters is nonzero.

$$F = \frac{R^2/k}{(1 - R^2)/[n - (k + 1)]}$$

RR: $F > F_\alpha(k, n - [k + 1])$

Using the p-value approach: Reject H_0 if p value $< \alpha$, where the p-value $= P(F(k, n - [k + 1]) > F)$.

Conditions:

1. The error component ε is normally distributed.
2. The mean of ε is zero.
3. The variance of ε is σ^2, a constant for all values of x.
4. The errors associated with different observations are independent.

The tail values of the F-distribution are given in Tables 7 and 8 in the Appendix. The F-test statistic becomes large as the coefficient of determination R^2 becomes large. To determine how large F must be before we can conclude at a given significance level that the model is useful for predicting Y, we set up the rejection region as $F > F_\alpha(k, n - [k + 1])$.

Example 11.16

For the electrical usage example, the test of H_0: $\beta_1 = \beta_2 = 0$ against H_a: at least one of the coefficients is nonzero, would test the global utility of the model. Now $n = 10, k = 2$ gives $n - (k + 1) = 7$ degrees of freedom. At $\alpha = 0.05$, we reject H_0: $\beta_1 = \beta_2 = 0$ if $F > F_{0.05}(2, 7) = 4.74$. That is the same as saying reject H_0 if the p-value < 0.05, where the p-value $= P(F(2, 7) > F)$. From the computer printout in Figure 11.24 we find that the computed F is 53.88. Because this value greatly exceeds the tabulated value of 4.74, we conclude that at least one of the model coefficient β_1 and β_2 is nonzero. Therefore, the global F-test indicates that the quadratic model $Y = \beta_0 + \beta_1 x + \beta_2 x^2 + \varepsilon$ is useful for predicting peak power load. The printout shows that the p-value $= P(F(2, 7) > 53.88) \approx 0.000$, that is, p-value $< 0.001 < \alpha$. Therefore, we arrive at the same conclusion using the p-value.

Before placing full confidence in the value of R^2 and the F-test, however, we must examine the residuals to see whether there is anything unusual in the data. In the case of fitting y as a function of a single x, the residuals were plotted against x. In the multiple regression problem, there may be more than one x against which the residuals need to be plotted, and any of these plots could be informative. To standardize procedure a bit, it is customary in such cases to plot residuals against the predicted values, $\hat{y}$. This plot should look similar to the plot of residuals against x when there is only a single x to consider.

Two residual plots for the peak power load example are shown below, one with predicted values and one with x-values on the horizontal axis. Note that both plots show fairly random patterns, which helps confirm the adequacy of the model. Also, the patterns in the two plots are similar to each other.

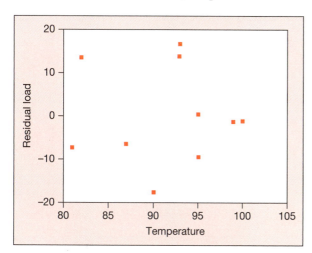
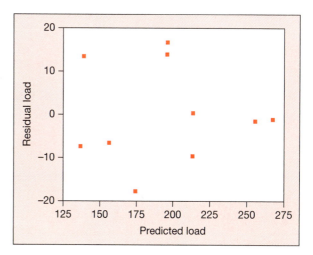

Example 11.17

A study is conducted to determine the effects of company size and the presence or absence of a safety program on the number of work hours lost due to work-related accidents. A total of 40 companies are selected for the study, 20 randomly chosen from companies with no active safety program, and the other 20 from companies that have enacted active safety programs. Each company is monitored for a 1-year period. The collected data are shown in table below, and the following model is proposed:

$$E(Y) = \beta_0 + \beta_1 x_1 + \beta_2 x_2$$

where
Y = lost work hours over the 1-year study period
x_1 = number of employees
$x_2 = \begin{cases} 1 & \text{if an active safety program is used} \\ 0 & \text{if no sctive safety program is used} \end{cases}$

x_1	x_2	y	x_1	x_2	y	x_1	x_2	y	x_1	x_2	y
6,490	0	121	1,986	0	23	3,077	1	44	9630	1	86
7,244	0	169	9,653	0	177	6,600	1	73	2905	1	40
7,943	0	172	9,429	0	178	2,732	1	8	6308	1	44
6,478	0	116	2,782	0	65	7,014	1	90	1908	1	36
3,138	0	53	8,444	0	146	8,321	1	71	8542	1	78
8,747	0	177	6,316	0	129	2,422	1	37	4750	1	47
2,020	0	31	2,363	0	40	9,581	1	111	6056	1	56
4,090	0	94	7,915	0	167	9,326	1	89	7052	1	75
3,230	0	72	6,928	0	115	6,818	1	72	7794	1	46
8,786	0	171	5,526	0	123	4,831	1	35	1701	1	6

a Test the utility of the model by testing H_0: $\beta_1 = \beta_2 = 0$.
b Interpret the least-squares coefficients $\hat{\beta}_1$ and $\hat{\beta}_2$.

Solution   Given below is the computer printout for the least-squares fit of the model.

$$Y = \beta_0 + \beta_1 x_1 + \beta_2 x_2 + \varepsilon$$

```
Regression Analysis: y versus x1, x2

The regression equation is
y = 31.7 + 0.0143 x1 - 58.2 x2

Predictor         Coef      SE Coef          T          P
Constant        31.670        8.560       3.70      0.001
x1            0.014272     0.001222      11.68      0.000
x2             -58.223        6.316      -9.22      0.000

S = 19.9678     R-Sq = 85.9%     R-Sq(adj) = 85.2%

Analysis of Variance

Source              DF          SS         MS          F          P
Regression           2       90058      45029     112.94      0.000
Residual Error      37       14752        399
Total               39      104811
```

a We want to use this information to test

$H_0: \beta_1 = \beta_2 = 0$.

H_a: At least one β is nonzero, i.e., the model is useful for predicting Y.

The rejection region for this test at $\alpha = 0.05$ is $F > F_{0.05}(2, 37) = 3.25$. Because $R^2 = 0.86$, the test statistic value is

$$F = \frac{R^2/k}{(1 - R^2)/[n - (k + 1)]} = \frac{0.86/2}{(1 - 0.86)/37} = 113.6$$

From the printout the p-value $= P(F(2, 37) > 113.6) < 0.001$. Thus, we can be very confident that the model contributes information for the prediction of lost work hours, because $F = 113.6 > 3.25$ or the p-value $< \alpha = 0.05$. The F-value is also given in the printout, so we do not have to compute it. Using the printout value also helps avoid rounding errors; note the difference between the calculated and the printout F-values.

b The estimated values of β_1 and β_2 are shown on the printout. The least-squares model is

$$\hat{y} = 31.67 + 0.0143x_1 - 58.22x_2$$

The coefficient $\hat{\beta}_1 = 0.0143$ represents the estimated slope of the number of lost hours (in thousands) versus the number of employees. In other words, we estimate that on average, each employee loses 14.3 work hours (0.0143 thousand hours) annually due to accidents. Note that this slope applies both to plants with safety programs and to plants without safety programs.

The coefficient $\hat{\beta}_2 = -58.2$ is *not* a slope because it is the coefficient of the dummy variable x_2. Instead, $\hat{\beta}_2$ represents the estimated mean change in lost hours between plants with no safety programs ($x_2 = 0$) and plants with safety programs ($x_2 = 1$), assuming both plants have the same number of employees. Our estimate indicates that an average of 58,200 fewer work hours are lost by plants with safety programs than by plants without them.

Recall that the variable x_2 in Example 11.17 is called the *dummy* or *indicator variable*. Dummy variables are used to represent categorical or qualitative independent variables, such as the presence or absence of a safety program. Note that the coefficient of the dummy variable represents the expected difference in lost work hours between companies with and without safety programs, assuming the companies have the same number of employees, for example:

• The expected number of lost work hours for a company with 500 employees and no safety program is (according to the model)

$$E(Y) = \beta_0 + \beta_1(500) + \beta_2(0) = \beta_0 + \beta_1(500)$$

- The expected number of lost work hours for a company with 500 employees and an active safety program is (according to the model)

$$E(Y) = \beta_0 + \beta_1(500) + \beta_2(1) = \beta_0 + \beta_1(500) + \beta_2$$

As can be seen here, the difference in means is β_2. Thus, the use of the dummy variable allows us to assess the effects of a qualitative variable on the mean response.

We should be careful not to be too excited about this very large F-value, because we have concluded only that the model contributes *some* information about Y. The width of the prediction intervals and confidence intervals are better measures of the amount of information the model provides about Y. Note on the computer printout that the estimated standard deviation of this model is 19.97. Because most (usually 90% or more) of the values of Y will fall within two standard deviations of the predicted value, we have the notion that the model predictions will usually be accurate to within about 40,000 hours. We will make this notion more precise in Section 11.6.4, but the standard deviation helps us obtain a preliminary idea of the amount of information the model contains about Y.

11.6.2   Estimating and testing hypotheses about individual β parameters

After determining that the model is useful by testing and rejecting $H_0: \beta_1 = \beta_2 = \cdots = \beta_k = 0$, we might be interested in making inferences about particular β parameters that have practical significance. For example, testing

$$H_0: \beta_j = 0 \quad (j = 1, 2, \ldots k)$$

A test of this hypothesis can be performed using a student's t-test. The t-test utilizes a test statistic analogous to that used to make inferences about the slope of the simple line model (Section 11.3). The t-statistic is formed by dividing the sample estimate $\hat{\beta}_j$ of the population coefficient β_j by $s_{\hat{\beta}_j}$, the estimated standard deviation of the repeated sampling distribution of $\hat{\beta}_2$:

Test statistic: $t = \hat{\beta}_j / s_{\hat{\beta}_j}$

Most computer packages list the estimated standard deviation $s_{\hat{\beta}_j}$ and the calculated t-values for each of the estimated model coefficients β_j. The rejection region for the test is found in exactly the same way as the rejection regions for the t-tests for the population means, using the error degrees of freedom. The error degrees of freedom is $(n - [k + 1])$.

Example 11.18

In the peak power load example, we fit the model $E(Y) = \beta_0 + \beta_1 x + \beta_2 x^2$, where Y is the peak power load and x is the daily maximum temperature. A test of particular interest would be

$$H_0: \beta_2 = 0 \text{ (No quadratic relationship exists.)}$$

$$H_a: \beta_2 > 0 \text{ (The peak power load increases at an increasing rate as the daily maximum temperature increases.)}$$

With the error degrees of freedom $(n - 3) = 7$, the rejection region (see Figure 11.27) for the one-tailed test with $\alpha = 0.05$ is

Rejection region: $t > t_\alpha(n - 3)$, i.e., $t > 1.895$

Figure 11.27

Rejection region for test of $H_0: \beta_2 = 0$

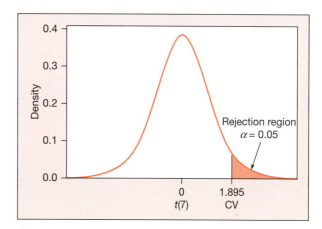

The computer printout for the peak power load example is given on the next page. It shows that $\hat{\beta}_2 = 0.2721$, $s_{\hat{\beta}_2} = 0.1163$. Thus, the test statistic value is $t = \hat{\beta}_2/s_{\hat{\beta}_2} = 0.2721/0.1163 = 2.34$ as shown in the printout. The p-value for this test is $P(t(7) > 2.34) = 0.0519/2 = 0.0259$ (note that the computer printout gave the p-value for the two-tailed test, whereas we are interested in the one-tailed test of hypothesis). Because the p-value $< \alpha = 0.05$, we reject the null hypothesis and conclude that the quadratic term $\beta_2 x^2$ contributes to the prediction of peak power load. There is evidence that the peak power load increases at an increasing rate as the daily maximum temperature increases.

We can also form a confidence interval for the parameter β_2 as follows:

$$\hat{\beta}_2 \pm t_{\alpha/2}(n - 3)s_{\hat{\beta}_2}, \quad \text{i.e.,} \quad 0.272 \pm 1.895(0.116), \quad \text{i.e.,} \quad (0.052, 0.492)$$

This interval gives an estimate of the rate of curvature in mean peak power load as the daily temperature increases. Note that all values in the interval are positive, confirming the conclusion of our test.

```
Response Load Whole Model
Summary of Fit

RSquare                              0.939001
RSquare Adj                          0.921572
Root Mean Square Error              12.95633
Mean of Response                        194.8
Observations (or Sum Wgts)                 10

Analysis of Variance

Source       DF   Sum of Squares   Mean Square    F Ratio
Model         2        18088.535       9044.27    53.8777
Error         7         1175.065        167.87   Prob > F
C. Total      9        19263.600                  <.0001

Parameter Estimates

Term             Estimate   Std Error   t Ratio   Prob >|t|
Intercept       1784.1883     944.123      1.89      0.1007
Temperature     -42.38624    21.00079     -2.02      0.0833
temp2           0.2721606     0.11634      2.34      0.0519
```

Test of an Individual Coefficient in the Multiple Regression Model   (*t*-test)

$H_0: \beta_j = 0$ $TS: T = \dfrac{\hat{\beta}_j}{s_{\hat{\beta}_j}}$ $\nu = (n - [k + 1])$ degrees of freedom

Reject H_0 if *p*-value $< \alpha$ n = number of observations
k = number of β parameters in the model, excluding β_0

Right-tailed test

$H_a: \beta_j > 0$

RR: $t > t_\alpha(\nu)$

p-value: $P(t(\nu) > t)$

Left-tailed test

$H_a: \beta_j < 0$

RR: $t < -t_\alpha(\nu)$

p-value: $P(t(\nu) < t)$

Two-tailed test

$H_a: \beta_j \neq 0$

RR: $t > t_{\alpha/2}(\nu)$ or $t < -t_{\alpha/2}(\nu)$

p-value: $2P(t(\nu) > |t|)$

Conditions:

1. The error component ε is normally distributed.
2. The mean of ε is zero.
3. The variance of ε is σ^2, a constant for all values of x.
4. The errors associated with different observations are independent.

If all the tests of model utility indicate that the model is useful for predicting Y, can we conclude that the best prediction model has been found? Unfortunately, we cannot. There is no way of knowing (without further analysis) whether the addition of other independent variables will improve the utility of the model.

Example 11.19

Refer to Example 11.17, in which we modeled a plant's lost work hours as a function of number of employees and the presence or absence of a safety program. Suppose a safety engineer believes that safety programs tend to help larger companies even more than smaller companies. Thus, instead of a relationship like that shown in Figure 11.28a, in which the rate of increase in mean lost work hours per employee is the same for companies with and without safety programs, the engineer believes that the relationship is like that shown in Figure 11.28b. Note that although the mean number of lost work hours is smaller for companies with safety programs no matter how many employees the company has, the magnitude of the difference becomes greater as the number of employees increases. When the slope of the relationship between $E(Y)$ and one independent variable (x_1) depends on the value of a second independent variable (x_2), as is the case here, we say that x_1 and x_2 interact. The model for mean lost work hours that includes interaction is written

$$E(Y) = \beta_0 + \beta_1 x_1 + \beta_2 x_2 + \beta_3 x_1 x_2$$

Note that the increase in mean lost work hours $E(Y)$ for each one-person increase in number of employees x_1 is no longer given by the constant β_1 but is now given by $\beta_1 + \beta_3 x_2$. That is, the amount that $E(Y)$ increases for each 1-unit increase in x_1 is dependent on whether the company has a safety program. Thus, the two variables x_1, and x_2 interact to affect Y.

Figure 11.28

Examples of No-Interaction and Interaction Models

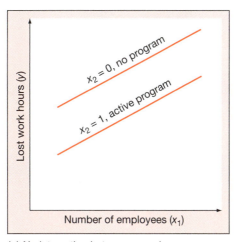

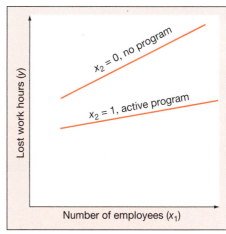

(a) No interaction between x_1 and x_2

(b) interaction between x_1 and x_2

The 40 data points listed in Example 11.17 were used to fit the model with interaction. A portion of the computer printout is shown on the next page. Test the hypothesis that the effect of safety programs on mean lost work hours is greater for larger companies.

```
The regression equation is
y = -0.74 + 0.019 x1 + 5.4 x2 - 0.0107 x1*x2

Predictor          Coef         Stdev      t-ratio          P
Constant         -0.739         7.927        -0.09      0.926
x1              0.019696      0.001219        16.16      0.000
x2                 5.37         11.08         0.48      0.631
x1*x2          -0.010737      0.001715        -6.26      0.000

S = 14.01              R-Sq = 93.3%          R-Sq(adj) = 92.7%


Analysis of Variance

SOURCE          DF          SS          MS           F          P
Regression       3        97749       32583      166.11      0.000
Error           36         7062         196
Total           39       104811
```

Solution The interaction model is

$$E(Y) = \beta_0 + \beta_1 x_1 + \beta_2 x_2 + \beta_3 x_1 x_2$$

so the slope of $E(Y)$ versus x_1 is β_1 for companies with no safety program and $\beta_1 + \beta_3$ for those with safety programs. Our research hypothesis is that the slope is smaller for companies with safety programs, so we will test

$$H_0: \beta_3 = 0 \quad \text{versus} \quad H_a: \beta_3 < 0$$

Test statistic:
$$t = \frac{\hat{\beta}_3}{s_{\hat{\beta}_3}} = -6.26$$

Rejection region: for $\alpha = 0.05$,

$$t < -t_{0.05}[n - (k + 1)] \quad \text{or} \quad t < -t_{0.05}(36) = -1.645$$

The printout above indicates that t-value corresponding to β_3 is -6.26, and the corresponding p-value $= P(t(36) < -6.26) < 0.001 \approx 0$. Because $t = -6.26 < -1.645$, or the p-value is almost zero, we reject the null hypothesis. The safety engineer can conclude that the larger the company, the greater the benefits of the safety program in terms of reducing the mean number of lost work hours.

In comparing the results for Examples 11.17 and 11.19, we see that the model with interaction fits the data better than the model without the interaction term. The R^2 value is greater for the model with interaction, and the interaction coefficient is significantly different from zero. How else might we compare how well these two models fit the data? Thinking back for a moment to Chapter 2, we might want to look at residuals, as we did for the simple straight-line model.

Figure 11.29 shows the residual plot for the model *without* interaction and the histogram for these residuals. Figure 11.30 shows the residual plot and the histogram of residuals for the model *with* interaction. Note that the plot in Figure 11.30a seems to show more "randomness" in the points and less spread

Figure 11.29

Model without interaction (Example 11.17)

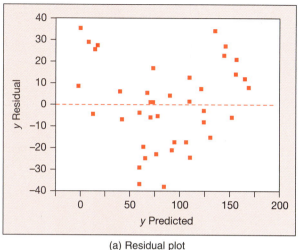

(a) Residual plot

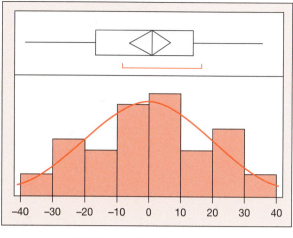

(b) Histogram of residuals

Figure 11.30

Model with interaction (Example 11.19)

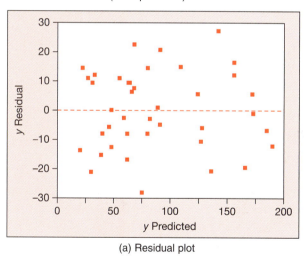

(a) Residual plot

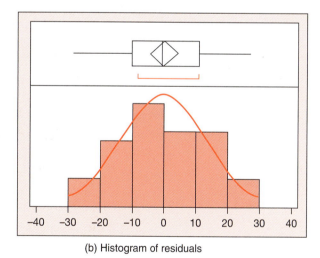

(b) Histogram of residuals

along the vertical axis than the one in Figure 11.29a. The histogram in Figure 11.30b is more compact and more symmetrically mound-shaped than the histogram in Figure 11.29b. All of these features indicate that the interaction model fits the data better than the model without interaction.

11.6.3 Using a multiple regression model for estimation and prediction

In Section 11.4 we discussed the use of the least-squares line in these situations:

- For estimating the mean value Y, $E(Y)$, for some value of x, say x_p
- For predicting value of Y to be observed in the future when $x = x_p$

Recall that the least-squares line yielded the same value for both the estimate of $E(Y)$ and the prediction of some future value of Y. That is, both were the results of substituting x_p into the prediction equation $\hat{y}_p = \hat{\beta}_0 + \hat{\beta}_1 x_p$ and calculating $\hat{y}_p$. However, the confidence interval for the mean $E(Y)$ was narrower than the prediction interval for Y because of the additional uncertainty attributable to the random error ε when predicting the future value of Y. These same concepts carry over to the multiple regression model.

Example 11.20

Refer to the example of peak power load (Example 11.14). Suppose we want to estimate the mean peak power load for a given daily high temperature $x_p = 90°F$. Assuming the quadratic model represents the true relationship between the peak power load and maximum temperature, we want to estimate

$$E(Y) = \beta_0 + \beta_1 x_p + \beta_2 x_p^2 = \beta_0 + \beta_1(90) + \beta_2(90)^2$$

Substituting into the least-squares prediction equation, the estimate of $E(Y)$ is

$$\hat{y} = \hat{\beta}_0 + \hat{\beta}_1(90) + \hat{\beta}_2(90)^2$$
$$= 1{,}784.188 - 42.3862(90) + 0.27216(90)^2 = 173.93$$

Figure 11.31

Confidence and prediction intervals for $x = 90$

x	ESTIMATED MEAN VALUE	LOWER 95%CL FOR MEAN	UPPER 95%CL FOR MEAN
90	173.9279	159.1462	188.7096

(a) Estimated mean value and confidence interval for $x = 90$

x	PREDICTED VALUE	LOWER 95%CL INDIVIDUAL	UPPER 95%CL INDIVIDUAL
90	173.9279	139.9113	207.9445

(b) Predicted value and prediction interval for $x = 90$

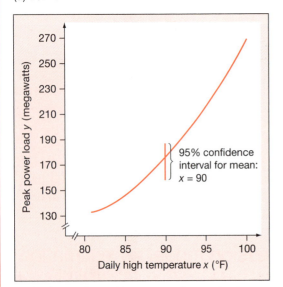

(c) Confidence interval for mean peak power load

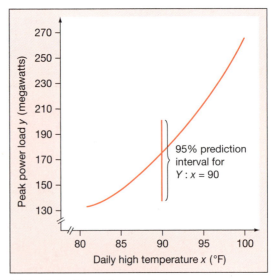

(d) Prediction interval for electrical usage

To form a confidence interval for the mean, we need to know the standard deviation of the sampling distribution for the estimator $\hat{y}$. For multiple regression models, the form of the standard deviation is rather complex. However, computer programs allow us to obtain the confidence intervals for the mean value of Y for a given combination of values of independent variables. Figure 11.31a shows the mean value and corresponding 95% confidence interval for $x = 90$. Note that $\hat{y} = 173.93$, which agrees with our earlier calculation. The 95% confidence interval for the true mean of Y is shown to be 159.15 to 188.71 (see Figures 11.31a and 11.31c).

If we were interested in predicting the electrical usage for a particular day on which the high temperature is 90°F, $\hat{y} = 173.93$ would be used as the predicted value. The printout in Figure 11.31b gives the prediction interval for $x = 90$ to be 139.91 to 207.94 (also see Figure 11.31d). Note that the prediction interval for $x = 90$ is wider than the confidence interval for the mean value.

Figure 11.32

Confidence and prediction bands for power load data

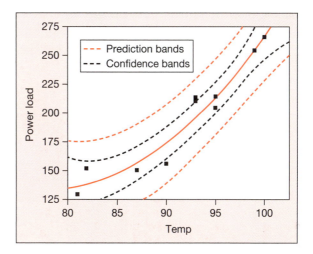

Unfortunately, not all computer packages have the capability to produce confidence intervals for means and prediction intervals for particular Y values. However, most packages will generate confidence and prediction bands, as shown in Figure 11.32. The middle thick curve is the fitted model. The two inner curves are the confidence bands for means, and the outermost curves correspond to the prediction bands for particular Y values.

Example 11.21

Refer to the example of tool life (Example 11.7). Along with the tool failure times and cutting speed, Taraman (1974) also reported corresponding feed rate (in inches per revolution [ipr] and depth of cut (in inches) for 24 tools. The tool life data are listed in table on the next page.

a Does the tool life depend on the cutting speed, feed rate, and depth of cut?
b Obtain the least-squares equation that can be used to predict tool life using speed, feed rate, and depth of cut.
c Estimate the mean life of a machine tool that is being used to cut depths of 0.03 inch at a speed of 450 fpm with a feed rate of 0.01 ipr.

Tool	Speed	Feed	Depth	Tool Life	Tool	Speed	Feed	Depth	Tool Life
1	340	0.00630	0.02100	70.0	13	305	0.00905	0.02900	52.0
2	570	0.00630	0.02100	29.0	14	635	0.00905	0.02900	23.0
3	340	0.01410	0.02100	60.0	15	440	0.00472	0.02900	40.0
4	570	0.01416	0.02100	28.0	16	440	0.01732	0.02900	28.0
5	340	0.00630	0.02100	64.0	17	440	0.00905	0.01350	46.0
6	570	0.00630	0.04000	32.0	18	440	0.00905	0.04550	33.0
7	340	0.01416	0.04000	44.0	19	305	0.00905	0.02900	46.0
8	570	0.01416	0.04000	24.0	20	635	0.00905	0.02900	27.0
9	440	0.00905	0.02900	35.0	21	440	0.00472	0.02900	37.0
10	440	0.00905	0.02900	31.0	22	440	0.01732	0.02900	34.0
11	440	0.00905	0.02900	38.0	23	440	0.00905	0.01350	41.0
12	440	0.00905	0.02900	35.0	24	440	0.00905	0.04550	28.0

Solution   **a**   Here Tool Life is a response variable, whereas, speed, feed, and depth of cut are predictor variables. First, make scatterplots of life of tool versus the predictor variables and compute correlation coefficients, as shown in Figure 11.33.

Figure 11.33

Scatterplots and correlation coefficients for tool life data

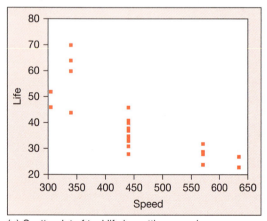

(a) Scatterplot of tool life by cutting speed

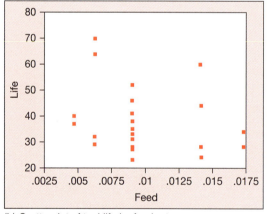

(b) Scatterplot of tool life by feed rate

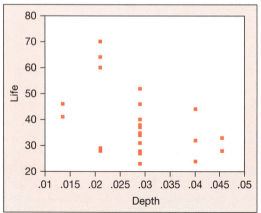

(c) Scatterplot of tool life by depth of cut

Correlations: Speed, Feed, Depth, Life

	Speed	Feed	Depth
Feed	-0.003		
	0.988		
Depth	0.110	0.099	
	0.607	0.645	
Life	-0.778	-0.221	-0.425
	0.000	0.300	0.038

Cell Contents: Pearson correlation
                                P-Value

(d) A correlation matrix for tool life, cutting speed, feed rate, and depth of cut

The scatterplots and the correlation table, both indicate the following:

- There is a fairly strong and statistically significant (at $\alpha = 0.05$) negative relation between the tool life and the cutting speed. Cutting at a higher speed reduces the tool life.
- There is a weak but statistically significant (at $\alpha = 0.05$) negative relation between the tool life and the depth of cut. Making deeper cuts reduces the tool life.
- There is a very weak negative relation between the tool life and the feed rate; the relation is not significant (at $\alpha = 0.05$).

```
Response Life

Summary of Fit
RSquare                                    0.758391
RSquare Adj                                 0.72215
Root Mean Square Error                      6.61109
Mean of Response                           38.54167
Observations (or Sum Wgts)                       24

Analysis of Variance
Source         DF   Sum of Squares  Mean Square   F Ratio
Model           3        2743.8281      914.609   20.9262
Error          20         874.1302       43.707  Prob > F
C. Total       23        3617.9583                <.0001

Parameter Estimates
Term        Estimate      Std Error    t Ratio  Prob >|t|
Intercept   101.76536      8.333102      12.21    <.0001
Speed        -0.095777     0.014258      -6.72    <.0001
Feed       -667.9724     386.2308        -1.73    0.0991
Depth      -472.3043     161.8143        -2.92    0.0085
```

b The JMP printout for the multiple regression is given below. It gives the least-squares equation

$$\text{Tool life} = 101.7654 - 0.0958 \text{ Speed} - 667.9724 \text{ Feed rate}$$
$$- 472.3043 \text{ Depth of cut}$$

Note that all the estimated coefficients are negative; that is, as the speed, feed rate, and depth increase, the tool life decreases. The $R^2 = 0.758$, up from 0.61 for the simple linear model using speed only. Using this multiple regression model, about 76% of variation in tool life can be explained, an increase of about 15%.

Because the p-value < 0.0001 for the F-test (model utility test), reject the null hypothesis of all the slope parameters are equal to zero. We conclude that above described multiple regression model using speed, feed, and depth contributes to the prediction of tool life.

The individual t-tests of $\beta_i = 0$ resulted in $p < 0.001$ for speed and $p = 0.0085$ for depth, both of them below any reasonable level of significance. Therefore, we can say that both speed and depth contribute significant amount of information for prediction of tool life. Note that the p-value for feed is equal to 0.0991, which is between 5% and 10% level of significance. In the presence of speed and depth, the feed rate provides some useful information for predicting tool life.

c A tool that is used to cut depths of 0.03 inch at a speed of 450 fpm with a feed rate of 0.01 ipr is expected to last on the average

$$\text{Tool life} = 101.7654 - 0.0958\,(450) - 667.9724\,(0.01)$$
$$- 472.3043\,(0.03) = 37.81 \text{ min}$$

Using JMP, we can obtain a 95% confidence interval for the mean tool life as (34.97, 40.66). It tells us that we are 95% confident that the mean life of such a tool used to cut depths of 0.03 inch at a speed of 450 fpm with a feed rate of 0.01 ipr will be between 34.97 min and 40.66 min.

Exercises

11.35 Suppose that you fit the model

$$Y = \beta_0 + \beta_1 x_1 + \beta_2 x_2 + \beta_3 x_1 x_2 + \beta_4 x_1^2 + \beta_5 x_2^2 + \varepsilon$$

to $n = 30$ data points and that

$$\text{SSE} = 0.37 \quad \text{and} \quad R^2 = 0.89$$

a Do the values of SSE and R^2 suggest that the model provides a good fit to the data? Explain.

b Is the model of any use in predicting y? Test the null hypothesis that $E(Y) = \beta_0$, that is,

$$H_0\colon \beta_1 = \beta_2 = \cdots = \beta_3 = 0$$

against the alternative hypothesis

$$H_a\colon \text{At least one of the parameters}$$
$$\beta_1, \beta_2, \ldots, \beta_5 \text{ is nonzero.}$$

Use $\alpha = 0.05$.

11.36 In hopes of increasing the company's share of the fine-food market, researchers for a meat-processing firm that prepares meats for exclusive restaurants are working to improve the quality of its hickory-smoked hams. One of their studies concerns the effect of time spent in the smokehouse on the flavor of the ham. Hams that were in the smokehouse for varying amount of time were each subjected to a taste test by a panel of ten food experts. The following model was thought by the researchers to be appropriate:

$$Y = \beta_0 + \beta_1 t + \beta_2 t^2 + \varepsilon$$

where

$Y = $ mean of the taste scores for the ten experts
$t = $ time in the smokehouse (hours)

Assume that the least-squares model estimated using a sample of 20 hams is

$$\hat{y} = 20.3 + 5.2t - 0.0025t^2$$

and that $s_{\beta_2} = 0.0011$. The coefficient of determination is $R^2 = 0.79$.

a Is there evidence to indicate that the overall model is useful? Test at $\alpha = 0.05$.

b Is there evidence to indicate that the quadratic terms is important in this model? Test at $\alpha = 0.05$.

11.37 Because the coefficient of determination R^2 never decreases when a new independent variable is added to the model, it is tempting to include many variables in a model to force R^2 to be near 1. However, doing so reduces the degrees of freedom available for estimating σ^2, which adversely affects our ability to make reliable inferences. As an example, suppose you want to use 18 economic indicators to predict next year's GNP. You fit the model

$$Y = \beta_0 + \beta_1 x_1 + \beta_2 x_2 + \cdots + \beta_{17} x_{17} + \beta_{18} x_{18} + \varepsilon$$

where $Y = $ GNP and $x_1, x_2, \ldots, x_{18}$, are indicators. Only 20 years of data ($n = 20$) are used to fit the model, and you obtain $R^2 = 0.95$. Test to see whether this impressive-looking R^2 value is large enough for you to infer that this model is useful, that is, that at least one term in the model is important for predicting GNP. Use $\alpha = 0.05$.

11.38 A utility company of a major city reported the average utility bills listed in the table below for a standard-size home during the last year.

Month	Average Monthly Temperature x (°F)	Average Utility Bill y (dollars)
January	38	99
February	45	91
March	49	78
April	57	61
May	69	55
June	78	63
July	84	80
August	89	95
September	79	65
October	64	56
November	54	74
December	41	93

a Plot the points in a scatterplot.

b Use the methods of Chapter 9 to fit the model

$$Y = \beta_0 + \beta_1 x + \varepsilon$$

What do you conclude about the utility of this model?

c Hypothesize another model that might better describe the relationship between the average utility bill and average temperature. If you have access to a computer package, fit the model and test its utility.

11.39 To run a manufacturing operation efficiently, it is necessary to know how long it takes employees to manufacture the product. Without such information, the cost of making the product cannot be determined. Furthermore, management would not be able to establish an effective incentive plan for its employees, because it would not know how to set work standards (Chase and Aquilano, *Production and Operations Man-agement*, 1979). Estimates of production time are frequently obtained by using time studies. The data in the following table were obtained from a recent time study of a sample of 15 employees on an automobile assembly line. The recorded values are y = time to complete task (minutes) and x = months of experience.

a The computer printout for fitting the model $y = \beta_0 + \beta_1 x + \beta_2 x^2 + \varepsilon$ is shown below. Find the least-squares prediction equation.

Time to Complete Task y	Months of Experience x	Time to Complete Task y	Months of Experience x
10	24	17	3
20	1	18	1
15	10	16	7
11	15	16	9
11	17	17	7
19	3	18	5
11	20	10	20
13	9		

```
Predictor      Coef      Stdev   t-ratio       p
Constant    20.0911     0.7247     27.72   0.000
x           -0.6705     0.1547     -4.33   0.001
x*x        0.009535   0.006326      1.51   0.158

s = 1.091   r-sq = 91.6%   R-sq(adj) = 90.2%


Analysis of Variance

SOURCE       DF       SS       MS       F       P
Regression    2  156.119   78.060   65.59   0.000
Error        12   14.281    1.190
Total        14  170.000
```

b Plot the fitted equation on a scatterplot of the data. Is there sufficient evidence to support the inclusion of the quadratic term in the model? Explain.

c Test the null hypothesis that $\beta_2 = 0$ against the alternative hypothesis that $\beta_2 \neq 0$. Use $\alpha = 0.01$. Does the quadratic term make an important contribution to the model?

d Your conclusion in part (c) should have been to drop the quadratic term from the model. Do so, and fit the "reduced model" $y = \beta_0 + \beta_1 x + \varepsilon$ to the data.

e Define β_1, in the context of this exercise. Find a 90% confidence interval for β_1, in the reduced model of part (d).

11.40 A researcher wanted to investigate the effects of several factors on production-line supervisors' attitudes toward handicapped workers. A study was conducted involving 40 randomly selected supervisors. The response Y, a supervisor's attitude toward handicapped workers, was measured with a standardized attitude scale. Independent variables used in the study were

$$x_1 = \begin{cases} 1 & \text{if the supervisor is female} \\ 0 & \text{if the supervisor is male} \end{cases}$$

x_2 = number of years of experiences in a supervisory job

The researcher fit the model

$$Y = \beta_0 + \beta_1 x_1 + \beta_2 x_2 + \beta_3 x_2^2 + \varepsilon$$

to the data with the following results:

$$\hat{y} = 50 + 5x_1 + 5x_2 - 0.1x_2^2$$
$$s_{\hat{\beta}_3} = 0.03$$

a Do these data provide sufficient evidence to indicate that the quadratic term, years of experience x_2^2, is useful for predicting attitude score? Use $\alpha = 0.05$.

b Sketch the predicted attitude score $\hat{y}$ as a function of the number of years of experience x_2 for male supervisors ($x_1 = 0$). Next, substitute ($x_1 = 1$) into the least-squares equation and obtain a plot of the prediction equation for female supervisors. [*Note:* For both males and females, plotting $\hat{y}$ for $x_2 = 0, 2, 4, 6, 8$, and 10 will produce a good picture of the prediction equations. The vertical distance between the males' and females' prediction curves is the same for all values of x_2.]

11.41 To project personnel needs for the Christmas shopping season, a department store wants to project sales for the season. The sales for the previous Christmas season are an indication of what to expect for the current season. However, the projection should also reflect the current economic environment by taking into consideration sales for a more recent period. The following model might be appropriate:

$$Y = \beta_0 + \beta_1 x_1 + \beta_2 x_2 + \varepsilon$$

where

$$x_1 = \text{previous Christmas sales}$$
$$x_2 = \text{sales for August of current year}$$
$$y = \text{sales for upcoming Christmas}$$

(All units are in thousands of dollars.) Data for ten previous years were used to fit the prediction equation, and the following values were calculated:

$$\hat{\beta}_1 = 0.62 \qquad s_{\hat{\beta}_1} = 0.273$$
$$\hat{\beta}_2 = 0.55 \qquad s_{\hat{\beta}_2} = 0.181$$

Use these results to determine whether there is evidence to indicate that the mean sales this Christmas are related to this year's August sales in the proposed model.

11.42 Suppose you fit the second-order model

$$Y = \beta_0 + \beta_1 x + \beta_2 x^2 + \varepsilon$$

to $n = 30$ data points. Your estimate of β_2, is $\hat{\beta}_2 = 0.35$, and the standard error of the estimate is $s_{\hat{\beta}_2} = 0.13$.

a Test the null hypothesis that the mean value of Y is related to x by the (first-order) linear model

$$E(Y) = \beta_0 + \beta_1 x$$

Test $H_0: \beta_2 = 0$ against the alternative hypothesis ($H_a: \beta_2 \neq 0$) that the true relationship is given by the quadratic model (a second-order linear model)

$$E(Y) = \beta_0 + \beta_1 x + \beta_2 x^2$$

Use $\alpha = 0.05$.

b Suppose you wanted only to determine whether the quadratic curve opens upward, that is, whether the slope of the curve increases as x increases. Give the test statistic and the rejection region for the test for $\alpha = 0.05$. Do the data support the theory that the slope of the curve increases as x increases? Explain.

c What is the value of the F statistic for testing the null hypothesis that $\beta_2 = 0$?

d Could the F statistic in part (c) be used to conduct the tests in parts (a) and (b)? Explain.

11.43 How is the number of degrees of freedom available for estimating σ^2 (the variance of ε) related to the number of independent variables in a regression model?

11.44 An employer has found that factory workers who are with the company longer tend to invest more in a company investment program per year than workers who have less time with the company. The following model is believed to be adequate in modeling the relationship of annual amount invested Y to years working for the company x:

$$Y = \beta_0 + \beta_1 x + \beta_2 x^2 + \varepsilon$$

The employer checks the records for a sample of 50 factory employees for a previous year and fits the above model to get $\hat{\beta}_2 = 0.0015$ and $s_{\hat{\beta}_2} = 0.00712$. The basic shape of a quadratic model depends upon whether $\beta_2 < 0$ or $\beta_2 > 0$. Test to see whether the employer can conclude that $\beta_2 > 0$. Use $\alpha = 0.05$.

11.45 Automobile accidents result in a tragic loss of life and, in addition, represent a serious dollar loss to the nation's economy. Shown in the Table 11.3 are the number of highway deaths (to the nearest hundred) and the number of licensed vehicles (in hundreds of thousands) for the years 1950–1979. (The years are coded 1–30 for convenience.) During the years 1974–1979 (years 25–30 in the table), the nationwide 55-mile-per-hour speed limit was in effect.

a Write a second-order model relating the number y of highway deaths for a year to the number x_1 of licensed vehicles.

Table 11.3

Automobile Accident Data

Year	Deaths (y)	Number of Vehicles (x_1)	Year	Deaths (y)	Number of Vehicles (x_1)
1	34.8	49.2	16	49.1	91.8
2	37.0	51.9	17	53.0	95.9
3	37.8	53.3	18	52.9	98.9
4	38.0	56.3	19	54.9	103.1
5	35.6	58.6	20	55.8	107.4
6	38.4	62.8	21	54.6	111.2
7	39.6	65.2	22	54.3	116.3
8	38.7	67.6	23	56.3	122.3
9	37.0	68.8	24	55.5	129.8
10	37.9	72.1	25	46.4	134.9
11	38.1	74.5	26	45.9	137.9
12	38.1	76.4	27	47.0	143.5
13	40.8	79.7	28	49.5	148.8
14	43.6	83.5	29	51.5	153.6
15	47.7	87.3	30	51.9	159.4

Source: U.S. Department of Transportation.

b The computer printout for fitting the model to the data is shown below. Is there sufficient evidence to indicate that the model provides information for the prediction of the number of annual highway deaths? Test using $\alpha = 0.05$.

c Give the p-value for the test of part (b) and interpret it.

d Does the second-order term contribute information for the prediction of y? Test by using $\alpha = 0.05$.

e Give the p-value for the test of part (d) and interpret it.

```
Predictor      Coef     Stdev  t-ratio      p
Constant     -1.408     6.895    -0.20  0.840
x1           0.8455    0.1462     5.78  0.000
x1*x1    -0.0033220 0.0007093    -4.68  0.000

s = 3.706  R-sq = 76.7%  R-sq(adj) = 75.0%


Analysis of Variance

SOURCE       DF       SS       MS      F      P
Regression    2  1222.16   611.08  44.50  0.000
Error        27   370.79    13.73
Total        29  1592.95
```

f Plot the residuals from the fitted model against x_1. Do you think this is the best model to use here?

11.46 If producers (providers) of goods (services) are able to reduce the unit cost of their goods (services) by increasing the scale of their operation, they are the beneficiaries of an economic force known as *economies of scale*. Economies of scale cause a firm's long-run average costs to decline (Ferguson and Maurice, *Econom-ics Analysis*, 1971). The question of whether economies of scale, diseconomies of scale, or neither (i.e., constant economies of scale) exist in the U.S. motor-freight common carrier industry has been debated for years. In an effort to settle the debate within a specific subsection of the trucking industry, T. Sugrue, M. Ledford, and W. Glaskowsky (*Transportation Journal*, 1982, pp. 27–41) used regression analysis to model the relationship between each of a number of profitability/cost measures and the size of the operation. In one case, they modeled expense per vehicle mile Y_1 as a function of the firm's total revenue X. In another case, they modeled expense per ton mile Y_2 as a function of X. Data were collected from 264 firms, and the following least-squares results were obtained:

Dependent Variable	$\hat{\beta}_0$	$\hat{\beta}_1$	r	F
Expense per vehicle mile	2.279	−0.00000069	−0.00783	1.616
Expense per ton mile	0.1680	−0.000000066	−0.0902	2.148

a Investigate the usefulness of the two models estimated by Sugrue, Ledford, and Glaskowsky. Use $\alpha = 0.05$. Draw appropriate conclusions in the context of the problem.

b Are the observed significance levels of the hypothesis tests you conducted in part (a) greater than 0.10 or les than 0.10? Explain.

c What do your hypothesis tests of part (a) suggest about economies of scale in the subsection of the trucking industry investigated by Sugrue, Ledford, and Glaskowsky—the long-haul, heavy-load, intercity general-freight common carrier section of the industry? Explain.

11.47 Krishna and Baten (*Trans Ichem E*, 2003) Examined the ability of computational fluid dynamics to model the complex two-phase hydrodynamics of sieve trays. A total of 14 experiments were conducted in which the superficial gas velocity (m/s) and the clear liquid height (m) were recorded in table below.

Expt.	Velocity	Height	Expt.	Velocity	Height
1	0.50	0.070	8	0.71	0.063
2	0.48	0.069	9	0.75	0.062
3	0.48	0.068	10	0.76	0.031
4	0.53	0.066	11	0.80	0.060
5	0.57	0.065	12	0.86	0.059
6	0.62	0.065	13	0.91	0.060
7	0.66	0.063	14	0.91	0.059

Develop an appropriate model to estimate clear liquid height as a function of the superficial gas velocity.

11.48 Lawton (*Journal of Pressure Vessel Technology*, 2003) described experiments conducted to assess new ammunition for 155 mm gun. Two new barrels were used, one fired charge N, a normal propellant without any wear-reducing additive, whereas the other fired charge M, modified by the addition of wear-reducing additive. To study the influence of rate of fire (rounds/min) on the number of rounds to cook-off when firing charges, the number of rounds fired before the barrel reached its cook-off temperature and the firing rate were recorded for charges N and M as shown in table below.

Charge N		Charge M	
Rate of fire	Number of Rounds to Cook-off	Rate of Fire	Number of Rounds to Cook-off
0.60	160	0.9	160
0.65	115	1.0	145
0.70	90	1.5	115
0.80	80	2.0	100
1.00	70	3.0	92
1.50	61	5.0	85
2.00	58	10.0	70
3.00	54		
5.00	48		
10.00	35		

a Develop appropriate model to predict the number of cook-off rounds as a function of the rate of fire for ammunition N.

b Develop appropriate model to predict the number of cook-off rounds as a function of the rate of fire for ammunition M.

c Assess the fit of each model.

d Compare both the models to determine whether the additive is effective in extending the number of cook-off rounds in 155 mm gun.

11.7   Model Building: A Test for a Portion of a Model

In Sections 11.6.1 and 11.6.2, we discussed these tests:

- A test of model adequacy (an F-test) that involves testing all the parameters in a multiple regression model (H_0: $\beta_1 = \beta_2 = \cdots = \beta_k = 0$) using the coefficient of determination R^2
- A t-test for individual model parameters, H_0: $\beta_i = 0$

But the question that arises is, How do we know which variables to use in the model? We now develop a test of sets of β parameters representing a *portion* of the model, a technique that is useful in constructing a multiple regression model.

Suppose we are interested in developing a model to estimate Y. The pool of possible independent variables associated with Y that can be used to develop the model contains variables $X_1, X_2, \ldots, X_k$. For example, to estimate the selling price of a house, data on variables such as area, age, style, number of rooms, number of bedrooms, number of floors, location of house, and so on may be available for houses in a given geographical area. However, all variables in the pool might not provide a significant amount of information for prediction of sale price. We are interested in building a model using variables providing the most significant information about the response variable. Looking at the association between the response variable (Y) and all the predictor variables in the available pool ($X_1, X_2, \ldots, X_k$), we can select the predictor variable that is associated most strongly with the response variable to enter into the model.

Example 11.22

The following table gives data collected on houses sold during a 1-month period in Gainesville, Florida. The data are given on the following variables:

$$y = \text{the selling price of the house (in thousands of dollars)}$$
$$x_1 = \text{area of the house (in thousand square feet)}$$
$$x_2 = \text{number of bedrooms in the house}$$
$$x_3 = \text{number of bathrooms in the house}$$
$$x_4 = \text{style of the house (1 = brick, and 0 = stucco)}$$

y	x_1	x_2	x_3	x_4	y	x_1	x_2	x_3	x_4	y	x_1	x_2	x_3	x_4	y	x_1	x_2	x_3	x_4
48.5	1.10	3	1	0	76.0	1.46	3	2	0	99.9	1.51	3	2	1	117.9	1.99	4	2	0
55.0	1.01	3	2	0	72.9	1.56	4	2	0	95.5	1.54	3	2	1	110.0	1.55	3	2	0
68.0	1.45	3	2	0	73.0	1.22	3	2	0	98.5	1.51	3	2	0	115.0	1.67	3	2	0
137.0	2.40	3	3	1	70.0	1.40	2	2	0	100.1	1.85	3	2	0	124.0	2.40	4	2	0
309.4	3.30	4	3	0	76.0	1.15	2	2	0	99.9	1.62	4	2	1	129.9	1.79	4	2	1
17.5	0.40	1	1	0	69.0	1.74	3	2	0	101.9	1.40	3	2	1	124.0	1.89	3	2	0
19.6	1.28	3	1	0	75.5	1.62	3	2	0	101.9	1.92	4	2	0	128.0	1.88	3	2	1
24.5	0.74	3	1	0	76.0	1.66	3	2	0	102.3	1.42	3	2	1	132.4	2.00	4	2	1
34.8	0.78	2	1	0	81.8	1.33	3	2	0	110.8	1.56	3	2	1	139.3	2.05	4	2	1
32.0	0.97	3	1	0	84.5	1.34	3	2	0	105.0	1.43	3	2	1	139.3	2.00	4	2	1
28.0	0.84	3	1	0	83.5	1.40	3	2	0	98.9	2.00	3	2	0	139.7	2.03	3	2	1
49.9	1.08	2	2	0	86.0	1.15	2	2	1	106.3	1.45	3	2	1	142.0	2.12	3	3	0
59.9	0.99	2	1	0	86.9	1.58	3	2	1	106.5	1.65	3	2	0	141.3	2.08	4	2	1
61.5	1.01	3	2	0	86.9	1.58	3	2	1	116.0	1.72	4	2	1	147.5	2.19	4	2	0
60.0	1.34	3	2	0	86.9	1.58	3	2	1	108.0	1.79	4	2	1	142.5	2.40	4	2	0
65.9	1.22	3	1	0	87.9	1.71	3	2	0	107.5	1.85	3	2	0	148.0	2.40	5	2	0
67.9	1.28	3	2	0	88.1	2.10	3	2	0	109.9	2.06	4	2	1	149.0	3.05	4	2	0
68.9	1.29	3	2	0	85.9	1.27	3	2	0	110.0	1.76	4	2	0	150.0	2.04	3	3	0
69.9	1.52	3	2	0	89.5	1.34	3	2	0	120.0	1.62	3	2	1	172.9	2.25	4	2	1
70.5	1.25	3	2	0	87.4	1.25	3	2	0	115.0	1.80	4	2	1	190.0	2.57	4	3	1
72.9	1.28	3	2	0	87.9	1.68	3	2	0	113.4	1.98	3	2	0	280.0	3.85	4	3	0
72.5	1.28	3	1	0	88.0	1.55	3	2	0	114.9	1.56	3	2	0					
72.0	1.36	3	2	0	90.0	1.55	3	2	0	115.0	2.19	3	2	0					
71.0	1.20	3	2	0	96.0	1.36	3	2	1	115.0	2.07	4	2	0					

Figure 11.34

Printout of stepwise selection

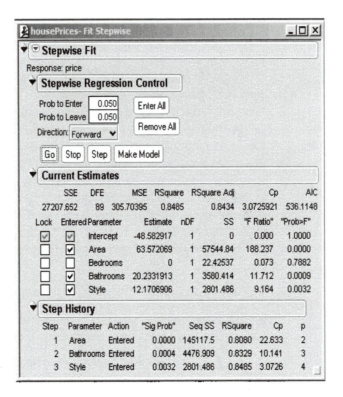

We are interested in developing a model to predict the selling price of a house. The pool of predictor variables contains four variables. Let's use a stepwise procedure to select predictors to be included in the model from this pool of variables.

Figure 11.34 gives the JMP printout of a computing intensive procedure for variable selection known as *stepwise selection procedure*. It shows that stepwise procedure selected only three out of four possible predictor variables, at $\alpha = 0.05$. The selected variables are area, number of bathrooms, and style of the house.

Figure 11.35 shows that all three selected variables are contributing to the prediction of selling prices. The $R^2 = 0.8485$; that is, about 85% of variation in selling prices is attributable to the area, number of bathrooms, and style of the house. What would have happened if we had also included the number of bedrooms in the model? Let's fit the full model and find out.

Figure 11.36 gives the JMP output for the model that includes all four predictor variables. It shows that in the presence of the area, number of bathrooms, and style, the number of bedrooms does not have any significant amount of information to contribute for the prediction of selling price. In fact, if we compare the R^2 values for two models, we notice that the improvement is negligible (0.8485 to 0.8486), indicating it is not worth adding number of bathrooms to the model. In practical situations, we'll have one less variable to track in the prediction process.

Figure 11.35

Multiple regression model for selling price of houses

```
Response price
Summary of Fit

RSquare                          0.848513
RSquare Adj                      0.843406
Root Mean Square Error           17.48439
Mean of Response                 99.54409
Observations (or Sum Wgts)             93

Analysis of Variance

                  Sum of      Mean
Source     DF    Squares     Square    F Ratio
Model       3   152395.94    50798.6   166.1694
Error      89    27207.65      305.7   Prob > F
C. Total   92   179603.59               <.0001

Parameter Estimates

Term        Estimate Std Error t Ratio Prob >|t|
Intercept    -48.583     8.798   -5.52    <.0001
Area          63.572     4.634   13.72    <.0001
Bathrooms     20.233     5.912    3.42   0.0009
Style         12.170     4.020    3.03   0.0032
```

Figure 11.36

Full model for selling prices

```
Response price
Summary of Fit

RSquare                          0.848638
RSquare Adj                      0.841758
Root Mean Square Error           17.57621
Mean of Response                 99.54409
Observations (or Sum Wgts)             93

Analysis of Variance

                  Sum of      Mean
Source     DF    Squares     Square    F Ratio
Model       4   152418.36    38104.6   123.3466
Error      88    27185.23      308.9   Prob > F
C. Total   92   179603.59               <.0001

Parameter Estimates

Term        Estimate Std Error t Ratio Prob >|t|
Intercept    -46.021    12.984   -3.54   0.0006
Area          64.621     6.071   10.64    <.0001
Bedrooms      -1.1436     4.245   -0.27   0.7882
Bathrooms     19.863     6.100    3.26   0.0016
Style         12.413     4.139    3.00   0.0035
```

Such variable selection methods use *a test for a portion of a model*. Suppose we have k predictors, $X_1, X_2, \ldots, X_k$, available for developing a model to estimate Y. A model using all k available variables is known as a *complete model*. A model based on a subset of g ($g < k$) variables is known as a *reduced model*. The complete and reduced models are given as follows:

Complete model: $E(Y) = \beta_0 + \beta_1 x_1 + \cdots + \beta_g x_g + \beta_{g+1} x_{g+1} + \cdots + \beta_k x_k$
Reduced model: $E(Y) = \beta_0 + \beta_1 x_1 + \cdots + \beta_g x_g$

We want to use some data to test which of these two models is more appropriate. This can be done by testing the hypothesis that the extra β parameters in the complete model are equal to zero, that is,

$H_0: \beta_{g+1} = \cdots = \beta_k = 0$.
H_a: At least one of the β parameters is nonzero.

Follow these steps for testing this hypothesis.

- Use the least-squares method to fit the complete model. Calculate the corresponding Error Sum of Squares, SSE_C.
- Use the least-squares method to fit the reduced model. Calculate the corresponding Error Sum of Squares, SSE_R.
- Calculate the difference $(SSE_R - SSE_C)$. If the added terms in the model contribute to the prediction of y, then SSE_C should be much smaller than SSE_R, and the difference $(SSE_R - SSE_C)$ will be large. In other words, the larger the difference, the greater the weight of the evidence that the variables contribute to the prediction of y.

The Error Sum of Squares will always decrease when a new term is added to the model. The question is whether this decrease is large enough to conclude that it is due to more than just an increase in the number of terms and to chance. We use an F-test to determine whether the decrease in SSE is due to chance.

- Calculate an F-statistic as follows:

$$F = \frac{\text{Drop in SSE/number of additional } \beta \text{ parameters}}{s^2 \text{ for complete model}}$$

$$= \frac{(SSE_R - SSE_C)/(k - g)}{SSE_C/[n - (k + 1)]}$$

- When the assumptions about the error term ε are satisfied and the null hypothesis is true, this F-statistic has an F-distribution with $\nu_1 = (k - g)$ and $\nu_2 = n - (k + 1)$ degrees of freedom. Notice that ν_1 is the number of β parameters being tested, and ν_2 is the number of degrees of freedom associated with error for the complete model.
- If the added terms *do* contribute to the model (i.e., H_a is true), then we expect the F-statistic to be large. Thus, we reject H_0 when the F-statistic exceeds F_α for a preselected α.

Testing a Portion of a Model $\hfill$ (*F*-test)

Complete model:

$$E(Y) = \beta_0 + \beta_1 x_1 + \cdots + \beta_g x_g + \beta_{g+1} x_{g+1} + \cdots + \beta_k x_k$$

Reduced model:

$$E(Y) = \beta_0 + \beta_1 x_1 + \cdots + \beta_g x_g$$

$H_0: \beta_{g+1} = \cdots = \beta_k = 0.$
$H_a:$ At least one of the β parameters is nonzero.

$$F = \frac{(SSE_R - SSE_C)/(k - g)}{SSE_C/[n - (k + 1)]}$$

$RR: F > F_\alpha(k - g, n - [k + 1])$

Using the p-value approach: Reject H_0 if p value $< \alpha$,

$$p\text{-value} = P(F(k - g, n - [k + 1]) > F)$$

SSE_C = Error SS for the complete model
SSE_R = Error SS for the reduced model
$k - g$ = number of added β parameters
n = number of observations
$k + 1$ = number of β parameters in the complete model including

Conditions:

1. The error component ε is normally distributed.
2. The mean of ε is zero.
3. The variance of ε is σ^2, a constant for all values of x.
4. The errors associated with different observations are independent.

The F-test can be used to determine whether *any* set of terms should be included in a model by testing the null hypothesis that the members of a particular set of β parameters simultaneously equal zero.

Example 11.23

Refer to the example of tool life data.

a In Example 11.7, we developed a least-squares equation describing tool life as a function of speed. In Example 11.21, we developed a multiple regression equation to describe tool life as a function of speed, depth, and feed rate. Are two additional predictors (depth and feed rate) contributing significantly to the prediction of tool life?

b If there is any *interaction* between these factors, we would like to develop a model to estimate tool life that includes the interaction terms.

Solution   a In this situation, we can define the complete and reduced models as follows:

Complete model: $E(\text{life}) = \beta_0 + \beta_1(\text{speed}) + \beta_2(\text{feed}) + \beta_3(\text{depth})$
Reduced model: $E(\text{life}) = \beta_0 + \beta_1(\text{speed})$

To determine whether two additional predictors contribute to the prediction of tool life, we test the hypothesis H_0: $\beta_2 = \beta_3 = 0$. The JMP printouts for the reduced and complete models are given in Figure 11.37. Because $k - g = 2$, and $n - (k + 1) = 24 - (3 + 1) = 20$, at the 5% level of significance the rejection region is given by $F > F_{0.05}(2, 20) = 3.49$. From Figure 11.37a we get $SSE_R = 1426.9358$, and from Figure 11.37b we get, $SSE_C = 874.1302$. Then

$$F = \frac{(SSE_R - SSE_C)/(k - g)}{SSE_C/[n - (k + 1)]} = \frac{(1426.9358 - 874.1302)/2}{874.1302/20} = 6.324$$

which is much higher than the critical value 3.49, leading to the rejection of the null hypothesis. Thus, we conclude that at least one of the β parameters is nonzero and is contributing to the prediction of tool life.

Figure 11.37

JMP printout for reduced and complete models for tool life data

Linear Fit

Life = 83.686361 - 0.1003215 Speed

Summary of Fit

RSquare	0.605596
RSquare Adj	0.587669
Root Mean Square Error	8.053615
Mean of Response	38.54167
Observations (or Sum Wgts)	24

Analysis of Variance

Source	DF	Sum of Squares	Mean Square	F Ratio
Model	1	2191.0225	2191.02	33.7804
Error	22	1426.9358	64.86	Prob > F
C. Total	23	3617.9583		<.0001

Parameter Estimates

Term	Estimate	Std Error	t Ratio	Prob >|t|
Intercept	83.686361	7.939434	10.54	<.0001
Speed	-0.100322	0.017261	-5.81	<.0001

(a) Simple linear regression model (reduced)

Response Life

Summary of Fit

RSquare	0.758391
RSquare Adj	0.72215
Root Mean Square Error	6.61109
Mean of Response	38.54167
Observations (or Sum Wgts)	24

Analysis of Variance

Source	DF	Sum of Squares	Mean Square	F Ratio
Model	3	2743.8281	914.609	20.9262
Error	20	874.1302	43.707	Prob > F
C. Total	23	3617.9583		<.0001

Parameter Estimates

Term	Estimate	Std Error	t Ratio	Prob >|t|
Intercept	101.76536	8.333102	12.21	<.0001
Speed	-0.095777	0.014258	-6.72	<.0001
Feed	-667.9724	386.2308	-1.73	0.0991
Depth	-472.3043	161.8143	-2.92	0.0085

(b) Multiple regression model (complete)

b. Let us consider the following complete model that includes all the interaction terms, and the reduced model that included only the *main effects*.

Complete model:

$$E(\text{life}) = \beta_0 + \beta_1(\text{speed}) + \beta_2(\text{feed}) + \beta_3(\text{depth}) + \beta_4(\text{speed*feed})$$
$$+ \beta_5(\text{feed*depth}) + \beta_6(\text{speed*depth}) + + \beta_7(\text{speed*feed*depth})$$

Reduced model:

$$E(\text{life}) = \beta_0 + \beta_1(\text{speed}) + \beta_2(\text{feed}) + \beta_3(\text{depth})$$

To determine whether the interaction terms contribute to the prediction of life of tools, we test the hypothesis

$H_0: \beta_4 = \beta_5 = \beta_6 = \beta_7 = 0$
$H_a:$ At least one of the β parameters is nonzero.

Because $k - g = 4$, and $n - (k + 1) = 24 - (7 + 1) = 16$, at the 5% level of significance the rejection region is given by $F > F_{0.05}(4, 16) = 3.01$. The following JMP printouts show that $SSE_C = 483.1569$ (Figure 11.38a), and $SSE_R = 874.1302$ (Figure 11.38b). Then

$$F = \frac{(SSE_R - SSE_C)/(k - g)}{SSE_C/[n - (k + 1)]} = \frac{(874.1302 - 483.1569)/4}{483.1569/16} = 3.24$$

Because $F = 3.24 > 3.01$, we reject the null hypothesis and conclude that not all the added β parameters are zero. It is possible that not all the

Figure 11.38

Tool life models with and without interaction terms

Response Life

Summary of Fit

RSquare	0.866456
RSquare Adj	0.80803
Root Mean Square Error	5.495207
Mean of Response	38.54167
Observations (or Sum Wgts)	24

Analysis of Variance

Source	DF	Sum of Squares	Mean Square	F Ratio
Model	7	3134.8015	447.829	14.8301
Error	16	483.1569	30.197	Prob > F
C. Total	23	3617.9583		<.0001

Parameter Estimates

Term	Estimate	Std Error	t Ratio	Prob >|t|
Intercept	315.40	75.85	4.16	<.0001
Speed	-0.54	0.16	-3.34	0.004
Feed	-13993.00	6886.00	-2.03	0.059
Depth	-8562.00	2749.00	-3.12	0.007
Speed*Feed	26.34	14.58	1.81	0.090
Feed*Depth	516692.00	239253.00	2.16	0.046
Speed*Depth	16.60	5.71	2.91	0.010
Speed*Feed*Depth	-1015.726	494.982	-2.05	0.0569

(a) Tool life model with interaction terms

Response Life

Summary of Fit

RSquare	0.758391
RSquare Adj	0.72215
Root Mean Square Error	6.61109
Mean of Response	38.54167
Observations (or Sum Wgts)	24

Analysis of Variance

Source	DF	Sum of Squares	Mean Square	F Ratio
Model	3	2743.8281	914.609	20.9262
Error	20	874.1302	43.707	Prob > F
C. Total	23	3617.9583		<.0001

Parameter Estimates

Term	Estimate	Std Error	t Ratio	Prob >|t|
Intercept	101.765	8.333	12.21	<.0001
Speed	-0.096	0.014	-6.72	<.0001
Feed	-667.972	386.231	-1.73	0.0991
Depth	-472.304	161.814	-2.92	0.0085

(b) Tool life model without interaction terms

interaction terms have a significant contribution toward prediction of life of tools, but at least one of the interaction terms is significant. The equation of best fit is

$$\text{Life} = 315 - 0.539\,\text{Speed} - 13{,}993\,\text{Feed} - 8{,}562\,\text{Depth}$$
$$+ 26.3\,\text{Speed*Feed} + 16.6\,\text{Speed*Depth} + 51{,}6692\,\text{Feed*Depth}$$
$$- 1{,}016\,\text{Speed*Feed*Depth}$$

Note that the interaction terms Speed*Feed and Speed*Feed*Depth are not significant at the 5% level of significance. Can we remove them from the present model? It will be interesting to see how the model without terms Speed*Feed and Speed*Feed*Depth fits the data and how the outcome is affected.

The following example will demonstrate both the need for a test for the portion of a regression model and the flexibility of multiple regression models through an application of dummy variables.

Example 11.24

Suppose a construction firm wishes to compare the performance of its three sales engineers, using the mean profit per sales dollar. The sales engineers bid on the jobs in two states, so the true mean profit per sales dollar is to be considered a function of two factors: sales engineer and state.

		State	
		S_1	S_2
Sales Engineer	E_1	μ_{11}	μ_{12}
	E_2	μ_{21}	μ_{22}
	E_3	μ_{31}	μ_{32}

μ_{ij} = the mean profit per sales dollar by the ith engineer from the job in the jth state, ($i = 1, 2, 3$ and $j = 1, 2$)

Because both sales engineer and state are *qualitative* factors, dummy variables will be used to represent them in the multiple regression model. To represent the state effect in the model, define

$$x_1 = \begin{cases} 1 & \text{if state} = S_2 \\ 0 & \text{if state} = S_1 \end{cases}$$

To represent the sales engineer effect in the model, define

$$x_2 = \begin{cases} 1 & \text{if sales engineer} = E_2 \\ 0 & \text{if sales engineer} = E_1 \text{ or } E_3 \end{cases} \quad x_3 = \begin{cases} 1 & \text{if sales engineer} = E_3 \\ 0 & \text{if sales engineer} = E_1 \text{ or } E_2 \end{cases}$$

Note that two dummy variables are defined to represent three sales engineers. Using Y = profit per sales dollar, the model is

$$E(Y) = \overbrace{\beta_0 + \beta_1 x_1}^{\text{State}} + \overbrace{\beta_2 x_2 + \beta_3 x_3}^{\text{Sales Engineer}} + \overbrace{\beta_4 x_1 x_2 + \beta_5 x_1 x_3}^{\text{State} \times \text{Engineer Intraction}}$$

Table 11.4

Correspondence between Means μ_{ij} and Model Parameters β

	State	
Sales Engineer	S_1 (i.e., $x_1 = 0$)	S_2 (i.e., $x_1 = 1$)
E_1 (i.e., $x_2 = 0$, $x_3 = 0$)	$\mu_{11} = E(Y\|E_1, S_1)$ $= \beta_0$	$\mu_{12} = E(Y\|E_1, S_2)$ $= \beta_0 + \beta_1$
E_2 (i.e., $x_2 = 1$, $x_3 = 0$)	$\mu_{21} = E(Y\|E_2, S_1)$ $= \beta_0 + \beta_2$	$\mu_{22} = E(Y\|E_2, S_2)$ $= \beta_0 + \beta_1 + \beta_2 + \beta_4$
E_3 (i.e., $x_2 = 0$, $x_3 = 1$)	$\mu_{31} = E(Y\|E_3, S_1)$ $= \beta_0 + \beta_3$	$\mu_{32} = E(Y\|E_3, S_2)$ $= \beta_0 + \beta_1 + \beta_3 + \beta_5$

Note that there are six parameters in the model: $\beta_0, \beta_1, \beta_2, \beta_3, \beta_4,$ and β_5, one corresponding to each mean, as shown in Table 11.4. Any other definitions of dummy variables will generate a different correspondence between means μ_{ij} and β parameters of the model. Note that one fewer dummy variables than the number of factor levels (one dummy variable for two states, and two dummy variables for three sales engineers) will yield an exact correspondence between the number of means and the number of model parameters.

Additive Effects Now suppose the construction firm wants to test whether the relative performance of the three sales engineers is the same in both states (see Figure 11.39a), In this case, although the level of profit may shift between states, the shift is the same for all three sales engineers. Therefore, the *relative* position of the engineers' mean profits is the same. This phenomenon is referred to as *additivity* of state and sales engineer effects and results in simplification of the model. The interaction terms $\beta_4 x_1 x_2$ and $\beta_5 x_1 x_3$ are now unnecessary, because only β_1 is needed to represent the additive effect of state (see Table 11.5). Note that the difference between the state means for each sales engineer is β_1. The description of this relationship requires only four model parameters.

$$E(Y) = \beta_0 + \beta_1 x_1 + \beta_2 x_2 + \beta_3 x_3$$

Figure 11.39

Model with and without interaction

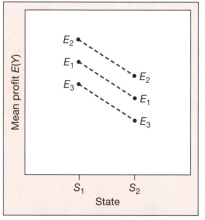

(a) Model without interaction: Effects are additive

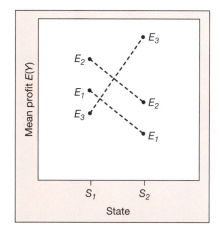

(b) Model with interaction: Effects are non-additive

Table 11.5

Representation of Mean Profit in the Additive Model

		State	
		S_1	S_2
Sales Engineer	E_1	$\mu_{11} = E(Y \mid x_1 = 0, x_2 = 0, x_3 = 0)$ $= \beta_0$	$\mu_{12} = E(Y \mid x_1 = 1, x_2 = 0, x_3 = 0)$ $= \beta_0 + \beta_1$
	E_2	$\mu_{21} = E(Y \mid x_1 = 0, x_2 = 1, x_3 = 0)$ $= \beta_0 + \beta_2$	$\mu_{22} = E(Y \mid x_1 = 1, x_2 = 1, x_3 = 0)$ $= \beta_0 + \beta_1 + \beta_2$
	E_3	$\mu_{31} = E(Y \mid x_1 = 0, x_2 = 0, x_3 = 1)$ $= \beta_0 + \beta_3$	$\mu_{32} = E(Y \mid x_1 = 1, x_2 = 0, x_3 = 1)$ $= \beta_0 + \beta_1 + \beta_3$

Non-Additive Effects If the sales engineers perform differently in the two states, their mean profits will not maintain the same relative positions. An example of this *non-additive* relationship between state and sales engineer is shown in Figure 11.39b. The description of this more complex relationship requires all six model parameters.

$$E(Y) = \beta_0 + \beta_1 x_1 + \beta_2 x_2 + \beta_3 x_3 + \beta_4 x_1 x_2 + \beta_5 x_1 x_3$$

Thus, we can see that the difference between an additive and a non-additive relationship is the interaction terms $\beta_4 x_1 x_2$ and $\beta_5 x_1 x_3$. If β_4 and β_5 are zero, the effects are additive; otherwise, the state/sales engineer effects are non-additive.

Let us call this interaction model, the *complete model*, and the reduced model the *main effects model*. We can determine which of these two models is more appropriate by testing that the β parameters in the interaction terms equal zero:

$$H_0: \beta_4 = \beta_5 = 0$$
$$H_a: \text{At least one intraction } \beta \text{ parameters is nonzero}$$

Example 11.25

The profit per sales dollar Y for six combinations of sales engineers and states is shown in table below. Note that the number of construction jobs per combination varies from one for levels (E_1, S_2) to three for levels (E_1, S_1). A total of 12 jobs are sampled.

		State	
		S_1	S_2
Sales	E_1	0.065, 0.073, 0.068	0.036
Engineer	E_2	0.078, 0.082	0.050, 0.043
	E_3	0.048, 0.046	0.061, 0.062

a Assume the interaction between engineer (E) and state (S) is negligible. Fit the model for expected profit per sales dollar with the interaction terms omitted.
b Fit the complete model for the expected profit per sales dollar, allowing for the fact that interactions might occur.
c Test the hypothesis that the interaction terms do not contribute to the model.

Solution   **a** The computer printout for the main effects model

$$E(Y) = \beta_0 + \overbrace{\beta_1 x_1}^{\substack{S \\ \text{Main Effect}}} + \overbrace{\beta_2 x_2 + \beta_3 x_3}^{\substack{E \\ \text{Main Effect}}}$$

is given in Figure 11.40a. The least-squares prediction equation is

$$\hat{y} = 0.0645 - 0.0158\, x_1 + 0.0067\, x_2 - 0.0023\, x_3$$

Figure 11.40
JMP printouts for profit models with and without interactions

<table>
<tr><td colspan="2">

Response sales

Summary of Fit

RSquare	0.362032
RSquare Adj	0.122794
Root Mean Square Error	0.01375
Mean of Response	0.059333
Observations (or Sum Wgts)	12

Analysis of Variance

Source	DF	Sum of Squares	Mean Square	F Ratio
Model	3	0.00085826	0.000286	1.5133
Error	8	0.00151241	0.000189	Prob > F
C. Total	11	0.00237067		0.2838

Parameter Estimates

Term	Estimate	Std Error	t Ratio	Prob >|t|
Intercept	0.0644545	0.00718	8.98	<.0001
x1	-0.015818	0.008291	-1.91	0.0928
x2	0.0067045	0.009941	0.67	0.5190
x3	-0.002295	0.009941	-0.23	0.8232

</td><td colspan="2">

Response sales

Summary of Fit

RSquare	0.971457
RSquare Adj	0.947671
Root Mean Square Error	0.003358
Mean of Response	0.059333
Observations (or Sum Wgts)	12

Analysis of Variance

Source	DF	Sum of Squares	Mean Square	F Ratio
Model	5	0.00230300	0.000461	40.8414
Error	6	0.00006767	0.000011	Prob > F
C. Total	11	0.00237067		0.0001

Parameter Estimates

Term	Estimate	Std Error	t Ratio	Prob >|t|
Intercept	0.0686667	0.001939	35.42	<.0001
x1	-0.032667	0.003878	-8.42	0.0002
x2	0.0113333	0.003066	3.70	0.0101
x3	-0.021667	0.003066	-7.07	0.0004
x1*x2	-0.000833	0.00513	-0.16	0.8763
x1*x3	0.0471667	0.00513	9.19	<.0001

</td></tr>
<tr><td colspan="2">(a) Main effects model</td><td colspan="2">(b) Complete model including interactions</td></tr>
</table>

b The computer printout for the complete model

$$E(Y) = \beta_0 + \beta_1 x_1 + \beta_2 x_2 + \beta_3 x_3 + \beta_4 x_1 x_2 + \beta_5 x_1 x_3$$

is given in Figure 11.40b. The least-squares prediction equation is

$$\hat{y} = 0.0687 - 0.0327 x_1 + 0.0113 x_2 - 0.0217 x_3 - 0.0008 x_1 x_2 + 0.0472 x_1 x_3$$

c We are interested in testing the hypothesis

$H_0: \beta_4 = \beta_5 = 0$ (The interaction terms do not contribute to the model.)
H_a: At least one of interaction β parameters is nonzero.

From the printouts in Figure 11.40, we get

   Main effects model: $SSE_R = 0.00151241$
   Interaction model: $SSE_C = 0.00006767$

Because $k - g = 2$, and $n - (k + 1) = 12 - 6 = 6$, at the 5% level of significance the rejection region is given by $F > F_{0.05}(2, 6) = 5.14$. Then the test statistic is

$$F = \frac{(SSE_R - SSE_C)/(k - g)}{SSE_C/[n - (k + 1)]} = \frac{(0.00151241 - 0.00006767)/2}{0.00006767/6} = 64.04$$

Because the calculated $F = 64.04$ greatly exceeds 5.14, we are quite confident in concluding that at least one of the interaction terms contributes to the prediction of Y, profit per sales dollar. Further investigation is needed to decide which interaction terms should be retained in the prediction model. For example, it looks like the interaction term X_1*X_2 is not significant. Also, notice that R^2 increased from 36% to 97% by including the interaction terms.

Exercises

11.49 Suppose you fit the regression model

$$Y = \beta_0 + \beta_1 x_1 + \beta_2 x_2 + \beta_3 x_3 + \beta_4 x_4 + \varepsilon$$

to $n = 25$ data points and you want to test the null hypothesis $\beta_1 = \beta_2 = 0$.
  a Explain how you would find the quantities necessary for the F-statistic.
  b How many degrees of freedom would be associated with F?

11.50 An insurance company is experimenting with three different training programs, A, B, and C, for its salespeople. The following main effects model is proposed:

$$E(Y) = \beta_0 + \beta_1 x_1 + \beta_2 x_2 + \beta_3 x_3$$

where

  Y = monthly sales (in thousands of dollars)
  x_1 = number of months of experience
  $x_2 = \begin{cases} 1 & \text{if training program B was used} \\ 0 & \text{otherwise} \end{cases}$
  $x_3 = \begin{cases} 1 & \text{if training program C was used} \\ 0 & \text{otherwise} \end{cases}$

Training program A is the base level.
  a What hypothesis would you test to determine whether the mean monthly sales differ for salespeople trained by the three programs?
  b After experimenting with 50 salespeople over a 5-year period, the complete model is fit, with the result

$$\hat{y} = 10 + 0.5x_1 + 1.2x_2 - 0.4x_3$$
$$SSE = 140.5$$

Then the reduced model $E(Y) = \beta_0 + \beta_1 x_1$ is fit to the same data, with the result

$$\hat{y} = 11.4 + 0.4x_1 \quad SSE = 183.2$$

Test the hypothesis you formulated in part (a). Use $\alpha = 0.05$.

11.51 A large research and development company rates the performance of each of the members of its technical staff once a year. Each person is rated on a scale of zero to 100 by his or her immediate supervisor, and this merit rating is used to help determine the size of the person's pay raise for the coming year. The company's personnel department is interested in developing a regression model to help them forecast the merit rating that an applicant for a technical position will receive after he or she has been with the company three years. The company proposes to use the following model to forecast the merit ratings of applicants who have just completed their graduate studies and have no prior related job experience:

$$E(Y) = \beta_0 + \beta_1 x_1 + \beta_2 x_2 + \beta_3 x_1 x_2 + \beta_4 x_1^2 + \beta_5 x_2^2$$

where

  Y = applicant's merit rating after three years
  x_1 = applicant's grade-point average (GPA) in graduate school
  x_2 = applicant's verbal score on the Graduate Record Examination (percentile)

A random sample of $n = 40$ employees who have been with the company more than there years was selected. Each employee's merit rating after three years, his or her graduate-school GPA, and the percentile in which the verbal Graduate Record Exam score fell were recorded. The above model was fit to these data and a portion of the resulting computer shown below.

SOURCE	DF	SUM OF SQUARES	MEAN SQUARE
MODEL	5	4911.56	982.31
ERROR	34	1830.44	53.84
TOTAL	39	6742.00	R-SQUARE
			0.729

The reduced model $E(Y) = \beta_0 + \beta_1 x_1 + \beta_2 x_2$ was fit to the same data, and the resulting computer printout is partially reproduced below.

SOURCE	DF	SUM OF SQUARES	MEAN SQUARE
MODEL	2	3544.84	1772.42
ERROR	37	3197.16	86.41
TOTAL	39	6742.00	R-SQUARE
			0.526

a Identify the null and alternative hypotheses for a test to determine whether the complete model contributed information for the prediction of Y.

b Identify the null and alternative hypotheses for a test to determine whether a second-order model contributes more information than a first-order model for the prediction of Y.

c Conduct the hypothesis test you described in part (a). Test using $\alpha = 0.05$. Draw the appropriate conclusions in the context of the problem.

d Conduct the hypothesis test you described in part (b). Test using $\alpha = 0.05$. Draw the appropriate conclusions in the context of the problem.

11.52 In an attempt to reduce the number of work hours lost due to accidents, a company tested each of three safety programs, A, B, and C, at three of the company's nine factories. The proposed complete model is

$$E(Y) = \beta_0 + \beta_1 x_1 + \beta_2 x_2 + \beta_3 x_3$$

where

Y = total work hours lost due to accidents for a one-year period beginning six months after the plan is instituted

x_1 = total work hours lost due to accidents during the year before the plan was instituted

$x_2 = \begin{cases} 1 & \text{if program B is in effect} \\ 0 & \text{otherwise} \end{cases}$

$x_3 = \begin{cases} 1 & \text{if program C is in effect} \\ 0 & \text{otherwise} \end{cases}$

After the programs have been in effect for 18 months, the complete model is fit to the $n = 9$ data points, with the result

$$\hat{y} = -2.1 + 0.88 x_1 - 150 x_2 + 35 x_3$$

$$\text{SSE} = 1{,}527.27$$

Then the reduced model $E(Y) = \beta_0 + \beta_1 x_1$ is fit, with the result

$$\hat{y} = 15.3 + 0.84 x_1 \quad \text{SSE} = 3{,}113.14$$

Test to see whether the mean work hours lost differ for the three programs. Use $\alpha = 0.05$.

11.53 The following model was proposed for testing salary discrimination against women in a state university system:

$$E(Y) = \beta_0 + \beta_1 x_1 + \beta_2 x_2 + \beta_3 x_1 x_2 + \beta_4 x_2^2$$

where

Y = annual salary (in thousands of dollars)

$x_1 = \begin{cases} 1 & \text{if female} \\ 0 & \text{if male} \end{cases}$

x_2 = experience (years)

Below is a portion of the computer printout that results from fitting this model to a sample of 200 faculty members in the university system:

SOURCE	DF	SUM OF SQUARES	MEAN SQUARE
MODEL	4	2351.70	587.92
ERROR	195	783.90	4.02
TOTAL	195	3135.60	R-SQUARE
			0.7500

Do these data provide sufficient evidence to support the claim that the mean salary of faculty members is dependent on sex? Use $\alpha = 0.05$.

11.8 Other Regression Models

Besides the regression models discussed so far, there are other situations where regression techniques are used for model building. Here we present three such examples.

- *Response surface* method is used to obtain the combination of values of predictor variables that result in the optimal value for the response variable.
- *Modeling time trend* is used when data are collected over time; in other words, successive points in time when data are collected become one of the predictor variables.
- *Logistic regression* is used to model a categorical variable as a function of numeric and other categorical variables.

11.8.1 Response surface method

Many companies manufacture products that are at least partially chemically produced (e.g., steel, paint, gasoline). In many instances, the quality of the finished product is a function of factors such as the temperature and pressure at which the chemical reactions take place. Then the company might be interested in determining values of temperature and pressure at which the quality of product is maximized.

Example 11.26

Suppose a manufacturer wanted to model the quality Y of a product as a function of the temperature x_1 and the pressure x_2 at which it is produced. Four inspectors independently assign a quality score between zero and 100 to each product, and then the quality y is calculated by averaging the four scores. An experiment is conducted by varying temperature between 80 and 100°F and varying pressure between 50 and 60 pounds per square inch. The resulting data ($n = 27$) are given in table below.

x_1(°F)	x_2 (pounds per square inch)	y	x_1(°F)	x_2 (pounds per square inch)	y	x_1(°F)	x_2 (pounds per square inch)	y
80	50	50.8	90	50	63.4	100	50	46.6
80	50	50.7	90	50	61.6	100	50	49.1
80	50	49.4	90	50	63.4	100	50	46.4
80	55	93.7	90	55	93.8	100	55	69.8
80	55	90.9	90	55	92.1	100	55	72.5
80	55	90.9	90	55	97.4	100	55	73.2
80	60	74.5	90	60	70.9	100	60	38.7
80	60	73.0	90	60	68.8	100	60	42.5
80	60	71.2	90	60	71.3	100	60	41.4

Step 1 The first step is to hypothesize a model relating product quality to the temperature and pressure at which the product was manufactured. A model that will

allow us to find the setting of temperature and pressure that maximizes quality is the equation for a *paraboloid*. The visualization of the paraboloid appropriate for this application is an inverted bowl-shaped surface, and the corresponding mathematical model for the mean quality at any temperature/pressure setting is

$$E(Y) = \beta_0 + \beta_1 x_1 + \beta_2 x_2 + \beta_3 x_1^2 + \beta_4 x_2^2 + \beta_5 x_1 x_2$$

This model is also referred to as a *complete second-order model* because it contains all first- and second-order terms in x_1 and x_2.[†] Note that a model with only first-order terms (a plane) would have no curvature, so the mean quality could not reach a maximum within the experimental region of temperature/pressure even if the data indicated the probable existence of such a value. The inclusion of second-order terms allows curvature in the three-dimensional response surface traced by mean quality, so a maximum mean quality can be reached if the experimental data indicate that one exists.

Step 2 Next we use the least-squares technique to estimate the model coefficients $\beta_0, \beta_1, \ldots, \beta_5$ of the paraboloid, using this data. The least-squares model (see Figure 11.41) is

$$\hat{y} = -5{,}127.90 + 31.10 x_1 + 139.75 x_2 - 0.133 x_1^2 - 1.14 x_2^2 - 0.146 x_1 x_2$$

A three-dimensional graph of this model is shown in Figure 11.43.

Figure 11.41

Printout for product quality example

Figure 11.42

Residual plot for product quality example

Response quality

```
Summary of Fit

RSquare                          0.993006
RSquare Adj                      0.991341
Root Mean Square Error           1.678696
Mean of Response                 66.96296
Observations (or Sum Wgts)             27
```

```
Analysis of Variance

Source     DF  Sum of Squares  Mean Square   F Ratio
Model       5       8402.2645      1680.45  596.3239
Error      21         59.1784         2.82  Prob > F
C. Total   26       8461.4430                <.0001
```

```
Parameter Estimates

Term         Estimate  Std Error  t Ratio  Prob > |t|
Intercept  -5127.899     110.296   -46.49      <.0001
temp        31.096389    1.344413    23.13      <.0001
press      139.74722     3.140054    44.50      <.0001
temp2       -0.133389    0.006853   -19.46      <.0001
press2      -1.144222    0.027413   -41.74      <.0001
temp*press    -0.1455    0.009692   -15.01      <.0001
```

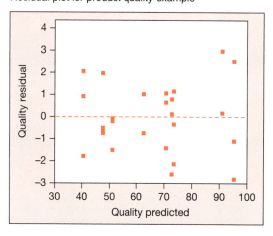

[†]Recall from analytic geometry that the order of a term is the sum of the exponents of the variables in the term. Thus, $\beta_5 x_1 x_2$, is a second-order term, as is $\beta_3 x_1^2$. The term $\beta_1 x_1^2 x_2$ is a third-order term.

Figure 11.43

Plot of second-order least-squares model for the product quality example

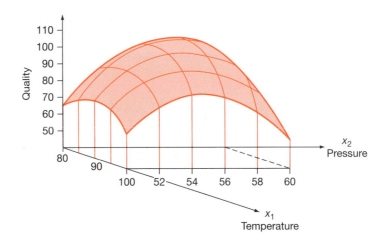

Step 3 The next step is to specify the probability distribution of ε, the random error component. We assume that ε is normally distributed, with a mean of zero and a constant variance σ^2. Furthermore, we assume that the errors are independent. The estimate of the variance σ^2 is given in the printout (Figure 11.41) as

$$s^2 = \text{MSE} = \frac{\text{SSE}}{n - (k + 1)} = \frac{59.1784}{27 - (5 + 1)} = 2.818$$

Step 4 Next we want to evaluate the adequacy of the model. First, note that $R^2 = 0.993$. This implies that 99.3% of the variation in y, observed quality ratings for the 27 experiments, is accounted for by the model. The statistical significance of this can be tested:

$$H_0: \beta_1 = \beta_2 = \beta_3 = \beta_4 = \beta_5 = 0$$
$$H_a: \text{At least one model coefficient is nonzero.}$$

Test statistic:
$$F = \frac{R^2/k}{(1 - R^2)/[n - (k + 1)]}$$

Rejection region: For $\alpha = 0.05$, $\quad F > F_\alpha[k, n - (k + 1)] = F_{0.05}(5, 21)$

Since $F = 596.32 > 2.57$ or p-value $= 0.0001$ (see Figure 11.41), we conclude that the model does contribute information about product quality.

Now we examine residuals for unusual patterns. Figure 11.42 shows the residuals from the second-order model plotted against the predicted values. The pattern appears to be fairly random, but the predicted values do seem to cluster into three groups. Can you think of a reason for this clustering?

Step 5 The culmination of the modeling effort is to use the model for estimation and/or prediction. In this example, suppose the manufacturer is interested in estimating the mean product quality for the setting of temperature and pressure at which the estimated model reaches a maximum. To find this setting, we solve the equations

$$\frac{\partial \hat{y}}{\partial x_1} = \hat{\beta}_1 + 2\hat{\beta}_3 x_1 + \hat{\beta}_5 x_2 = 0$$

$$\frac{\partial \hat{y}}{\partial x_2} = \hat{\beta}_2 + 2\hat{\beta}_4 x_2 + \hat{\beta}_5 x_1 = 0$$

for x_1 and x_2, obtaining $x_1 = 86.25°$ and $x_2 = 55.58$ pounds per square inch. The fact that $\partial^2 \hat{y}/\partial x_1^2 = 2\hat{\beta}_3 < 0$ and $\partial^2 \hat{y}/\partial x_2^2 = 2\hat{\beta}_4 < 0$ ensures that this setting of x_1 and x_2 corresponds to a maximum value of $\hat{y}$.

To obtain an estimated mean quality for this temperature–pressure combination, we use the least-squares model:

$$\hat{y} = \hat{\beta}_0 + \hat{\beta}_1(86.25) + \hat{\beta}_2(55.58) + \hat{\beta}_3(86.25)^2 + \hat{\beta}_4(55.58)^2 + \hat{\beta}_5(86.25)(55.58)$$

The estimated mean value is given in a partial reproduction of the regression printout for this example. The estimated mean quality is 96.9, and a 95% confidence interval is 95.5 to 98.3. Thus, we are confident that the mean quality rating will be between 95.5 and 98.3 when the product is manufactured at 86.25°F and 55.58 pounds per square inch.

X1	X2	PREDICTED VALUE	LOWER 95% CL FOR MEAN	UPPER 95% CL FOR MEAN
86.25	55.58	96.87401860	95.46463236	98.28340483

Activity

Goal: To determine the optimal wing length and body width that will maximize the flight time of helicopters.

The *response* to be measured is flight time of the helicopter. It is defined as the time for which the helicopter remains in flight.

1. Use the template of paper helicopter from Chapter 3. Cut along the solid lines and fold along the dotted lines.
2. Prepare helicopters of different wing lengths ranging from 4.5 inches to 7.5 inches and different body widths ranging from 2 inches to 6 inches.
3. Release the helicopter from above your head while standing alone on a step stool or a ladder. Start the timer at the drop of the helicopter. Stop the timer when the helicopter hits the floor.
4. Measure the time in flight (in sec) for different runs of these helicopters flown in random order.
5. Use the response surface technique to find the optimal wing length and body width.

11.8.2   Modeling a time trend

As seen in Chapter 2, often data are collected periodically over time, and we are interested in developing a model that describes response variable as a function of time.

Example 11.27 ——————————————————————————————————

Table below gives data on the annul fuel consumption (FC, billion gallons) for vans. The classification "vans" includes vans, pickups, and SUVs. The U.S. Federal Highway Commission collected this data for years 1970–1999.

Year	FC-SUV	Year	FC-SUV	Year	FC-SUV	Year	FC-SUV
1970	12.3	1984	25.6	1990	35.6	1996	47.4
1975	19.1	1985	27.4	1991	38.2	1997	49.4
1980	23.8	1986	29.1	1992	40.9	1998	50.5
1981	23.7	1987	30.6	1993	42.9	1999	52.8
1982	22.7	1988	32.7	1994	44.1		
1983	23.9	1989	33.3	1995	45.6		

Consumers, car manufacturers, and environmental scientists carefully watch fuel consumption of vehicles. A time series plot (Figure 11.44) shows an increasing trend, that is, an evidence that the fuel consumption of vans has increased over the years. What would be a reasonable model to explain the pattern in fuel consumption of vans over the years? Figure 11.44 suggests a linear trend in fuel consumption.

The least-squares fit of a simple-linear model is given in Figure 11.45. This linear model produces a high R^2 (96%) and a significant slope ($p < 0.0001$), but the residual plot (Figure 11.45) suggests that the model does not adequately describe the trend involved. There is a definite pattern of decreasing residuals followed by increasing residuals. Perhaps there is a curvature and it can be accounted for by fitting a logarithmic model (log[FC] versus time). See the output of fitting a logarithmic model in Figure 11.46.

Figure 11.44

Fuel consumption over the years

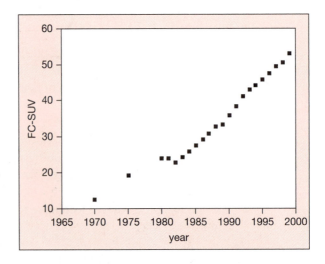

Figure 11.45

JMP printout for linear fit

```
Linear Fit

FC-SUV = -2890.312 + 1.4710977 year

Summary of Fit

RSquare                         0.960641
RSquare Adj                     0.958673
Root Mean Square Error          2.30936
Mean of Response                34.16364
Observations (or Sum Wgts)            22

Analysis of Variance

                   Sum of      Mean
Source     DF      Squares     Square     F Ratio
Model       1    2603.3481    2603.35    488.1452
Error      20     106.6628       5.33    Prob > F
C. Total   21    2710.0109               <.0001

Parameter Estimates

Term        Estimate Std Error t Ratio Prob >|t|
Intercept -2890.312   132.366  -21.84     <.0001
year       1.4710977  0.066584   22.09     <.0001
```

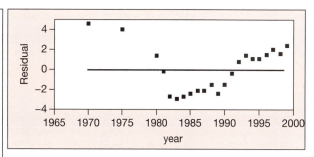

Residual plot of the simple linear model

Figure 11.46

JMP printout for logarithmic fit

```
Transformed Fit Log

Log(FC-SUV) = -92.81396 + 0.0484343 year

Summary of Fit

RSquare                         0.984406
RSquare Adj                     0.983627
Root Mean Square Error          0.047277
Mean of Response                3.471232
Observations (or Sum Wgts)            22

Analysis of Variance

                   Sum of      Mean
Source     DF      Squares     Square     F Ratio
Model       1    2.8219891    2.82199     1262.58
Error      20    0.0447019    0.00224    Prob > F
C. Total   21    2.8666911               0.0001

Parameter Estimates

Term        Estimate Std Error t Ratio Prob>|t|
Intercept -92.81396   2.709773  -34.25    <.0001
year       0.0484343  0.001363   35.53    <.0001
```

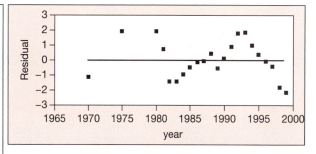

Residual plot for logarithmic model

Fitting log(FC) as a function of year produces a slightly better fit ($R^2 = 98\%$), however, the residual plot (Figure 11.46) still shows a pattern. Even this model does not adequately describe the pattern in fuel consumption over time.

A careful look at Figure 11.44 suggests that two different lines might fit the data better. It seems that the trend in fuel consumption prior to 1983 differs from that afterward. So, we need to fit different lines for fuel consumption up to 1983 and fuel consumption after 1983. This can be accomplished by adding an indicator variable to the model to keep track of the year. Let's define it as

$$\text{Period} = \begin{cases} 0 & \text{if year} \leq 1982 \\ 1 & \text{if year} \geq 1983 \end{cases}$$

Because slopes will be different for the two lines, we must include an interaction term in the model. Modeling fuel consumption as a function of year, period, and year∗period we get the output in Figure 11.47 output.

The fit has improved a little ($R^2 = 99.6\%$), and all the terms in this interaction model are significant. Moreover, the residual plot (Figure 11.47) shows randomly scattered residuals without the two-line pattern of Figure 11.45. The Shapiro-Wilk W test (Figure 11.47) for testing the null hypothesis that residuals are normally distributed resulted in the p-value of 0.3033. So, normality of residuals may be a reasonable assumption.

Figure 11.47

JMP printout for model with interaction term

Response FC-SUV

Summary of Fit

RSquare	0.995726
RSquare Adj	0.995014
Root Mean Square Error	0.80214
Mean of Response	34.16364
Observations (or Sum Wgts)	22

Analysis of Variance

Source	DF	Sum of Squares	Mean Square	F Ratio
Model	3	2698.4292	899.476	1397.943
Error	18	11.5817	0.643	Prob > F
C. Total	21	2710.0109		<.0001

Parameter Estimates

Term	Estimate	Std Error	t Ratio	Prob > |t|
Intercept	-1825.179	157.6883	-11.57	<.0001
year	0.9332016	0.079737	11.70	<.0001
period	-1785.288	176.4004	-10.12	<.0001
year*period	0.8993965	0.089079	10.10	<.0001

Residual FC-SUV: Fitted Normal

Goodness-of-Fit Test

Shapiro-Wilk W Test

W	Prob < W
0.949250	0.3033

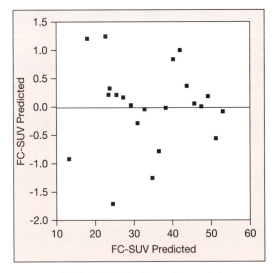

Residual plot for interaction model

As seen from this example, fitting models to data can be a complex task involving a variety of models. Often, there is no single linear model that adequately explains all patterns in the data. Nevertheless, the linear models are very useful for explaining the key features of the data.

11.8.3 Logistic regression

Regression analysis is useful for relating qualitative responses to the quantitative predictor variables. We discussed the use of indicator functions to include categorical variables as the possible predictors. However, in all the examples discussed so far the response variable was always a quantitative variable. In many real life situations, we are interested in relating a categorical variable with a quantitative variable. For example, outcome of strength test (passes or does not pass) as a function of amount of pressure applied, or ability of workers to complete the task (completes on time or not) as a function of experience. In other words, we are interested in predicting the outcome of a binary response variable using some numerical (and categorical) variable(s). In such situations, we use logistic regression to model the relationship between a binary response variable and a numerical predictor variable.

Consider a simple linear model, $Y = \beta_0 + \beta_1 x + \varepsilon$, where Y is a binary response variable (i.e., taking on the values of either 0 or 1). Suppose $P(Y_i = 1) = p_i$, then $P(Y_i = 0) = (1 - p_i)$, and the expected value of Y is $E(Y_i) = 1p_i + 0(1 - p_i) = p_i$. However, the assumption $E(\varepsilon) = 0$ gives $E(Y_i) = \beta_0 + \beta_1 x_i$. Equating two expressions, we get

$$E(Y_i) = \beta_0 + \beta_1 x_i = p_i$$

In other words, the mean response for a given x is the probability that $Y_i = 1$ because $E(Y_i)$ is a probability and probability is always between 0 and 1. This puts a restriction on the expected value of Y, which might not be satisfied by many relations in practice. Unfortunately, errors from such a model do not satisfy assumptions required by the linear regression model.

- Residual (error) is defined as $\varepsilon_i = Y_i - [\beta_0 + \beta_1 x_i]$ and takes only two values. If $Y_i = 0$, then $\varepsilon_i = -[\beta_0 + \beta_1 x_i]$, and if $Y_i = 1$, then $\varepsilon_i = 1 - [\beta_0 + \beta_1 x_i]$. Obviously, the assumption of normally distributed errors is not appropriate.
- Because the response variable is binary, the error variance is given by

$$\sigma^2 = p_i(1 - p_i) = (\beta_0 + \beta_1 x_i)(1 - \beta_0 - \beta_1 x_i)$$

Clearly the error variance depends on the value of the predictor variable, and the assumption of constant error variance for all values of X is unreasonable.

These constraints rule out the possibility of using a linear regression to model the relation between a binary response variable and a numerical predictor variable.

A nonlinear relation, particularly S-shaped, is often used to describe such a relationship. It is also known as *sigmoidal* and is given by

$$p = E(Y) = \frac{\exp(\beta_0 + \beta_1 x)}{1 + \exp(\beta_0 + \beta_1 x)}$$

Then the ln(odds) can be modeled as a linear function of the predictor variable:

$$\ln(\text{odds}) = \ln\left(\frac{p}{1 - p}\right) = \hat{\beta}_0 + \hat{\beta}_1 x$$

The model fitted gives

$$\ln\left(\frac{p}{1 - p}\right) = \hat{\beta}_0 + \hat{\beta}_1 x$$

Now it is easy to see that $\hat{\beta}_1$ is the difference between estimated ln(odds) for two different values of X. The parameters of this model are estimated using a maximum likelihood method. Because no closed-form solution exists for the values of $\hat{\beta}_0$ and $\hat{\beta}_1$, computer-intensive numerical search methods are used to find estimates. Therefore, we'll simply show how to interpret and use results given by standard statistical software. The prediction equation is given by

$$p_i = P(Y_i = 1|x) = \frac{\exp(\hat{\beta}_0 + \hat{\beta}_1 x)}{1 + \exp(\hat{\beta}_0 + \hat{\beta}_1 x)}$$

The estimated regression coefficient $\hat{\beta}_1$ is related to the odds ratio as follows:

$$\beta_1 = \ln[\text{odds}(x + 1)] - \ln[\text{odds}(x)] = \ln\left[\frac{\text{odds}(x + 1)}{\text{odds}(x)}\right]$$

Thus, the regression coefficient $\hat{\beta}_1$ can be interpreted as the increase in odds of occurrence of an event for every unit increase in the value of the predictor. If more than one predictor variable is used (numerical or categorical) a multiple logistic regression model can be developed as

$$E(Y) = \frac{\exp(\beta_0 + \beta_1 x_1 + \beta_2 x_2 + \cdots + \beta_k x_k)}{1 + \exp(\beta_0 + \beta_1 x_1 + \beta_2 x_2 + \cdots + \beta_k x_k)}$$

where $X_1, X_2, \ldots, X_k$ are k predictor variables.

Example 11.28

A manufacturer of suspension cables conducts strength tests on one type of cables in the laboratory. The strength test involves applying a specific amount of pressure to the cable and recording the outcome. The outcome of the strength test is recorded as "Pass" or "Fail," where "Pass" indicates the cable did not break. In this test, increasing the amount of pressure cannot be applied to determine the breaking point because the first application of pressure weakens this type of cable, increasing the likelihood of breakage. The experiment was conducted by selecting a sample of

25 cable pieces, applying the strength test at different levels of pressure, and recording the outcomes. The outcome is "Fail" if the cable breaks at the level of applied pressure; otherwise, it is "Pass." The data collected are recorded in table below.

Pressure	Result	Pressure	Result	Pressure	Result	Pressure	Result	Pressure	Result
3943	Pass	3900	Pass	4143	Fail	4035	Pass	3848	Pass
4163	Fail	3942	Pass	4146	Fail	3914	Pass	4007	Fail
3812	Pass	3732	Pass	3962	Fail	3853	Pass	4077	Fail
3888	Pass	4480	Fail	4335	Fail	4030	Fail	3907	Pass
3926	Fail	3940	Fail	3822	Pass	3967	Pass	4251	Fail

In this case, we are interested in modeling the outcome of the strength test as a function of pressure applied. The response variable test result is a binary variable taking values "Pass" or "Fail," and the explanatory variable, pressure applied, is numeric. Therefore, we cannot use regular regression technique to estimate the relation between the outcome and pressure. But we can use a logistic model to describe the relation between the likelihood of failing the test and the pressure applied. We will fit the model

$$E(\text{Test result}) = \frac{\exp(\beta_0 + \beta_1 \text{ Pressure})}{1 + \exp(\beta_0 + \beta_1 \text{ Pressure})}$$

The dotplot in Figure 11.48 shows that higher the pressure, the more likely the cable is to break. JMP output for a logistic fit for probability of failing the strength test is given in Figure 11.49. The whole model test gives the p-value < 0.0001, so we conclude that the model fits the data well. Specifically, we can look at the chi-square test for pressure. Here we are testing the null hypothesis of $H_0: \beta_1 = 0$ against the alternative $H_a: \beta_1 \neq 0$. The p-value for the test is 0.0233, which is lower than $\alpha = 0.05$. Therefore, we reject the null hypothesis and conclude that the pressure applied is a good predictor of the outcome of the test.

The predictor equation is given by

$$P(\text{Cable fails the test at given pressure})$$
$$= \frac{\exp(-98.91 + 0.2489 \text{ Pressure})}{1 + \exp(-98.91 + 0.02489 \text{ Pressure})}$$

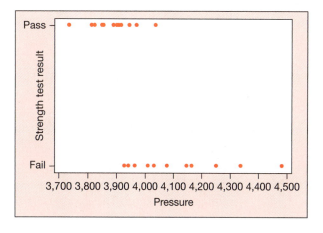

Figure 11.49

JMP printout for logistic model

Logistic Fit of Test by Pressure

Whole Model Test

Model	-LogLikelihood	DF	ChiSquare	Prob > ChiSq
Difference	9.248571	1	18.49714	<.0001
Full	8.060103			
Reduced	17.308674			

RSquare (U)   0.5343
Observations (or Sum Wgts)   25

Converged by Gradient
Parameter Estimates

Term	Estimate	Std Error	ChiSquare	Prob > Chisq
Intercept	-98.91055	43.394065	5.20	0.0226
Pressure	0.02489308	0.0109705	5.15	0.0233

For log odds of Fail/Pass

Note that the estimate of slope is positive, indicating the probability of cable failure will increase as pressure applied to the cable increases.

Now $e^{\hat{\beta}_1} = e^{0.02489} = 1.025$ is the odds ratio for pressure. The likelihood of failure increases by about 2.5% for every unit increase in pressure. Figure 11.50 shows the fit of the sigmoidal model to the data collected.

We can use the predictor equation to predict the likelihood of the test result for a given pressure. For example, if pressure = 4,050 were to be applied, then

$$P(\text{Cable fails the test at Pressure 4,250})$$

$$= \frac{\exp[-98.91 + 0.02489\,(4050)]}{1 + \exp[-98.91 + 002489\,(4050)]} = 0.6545$$

So there is a 65% chance that the cable might break if 4,050 pounds per square inch pressure were applied.

Figure 11.50

Sigmoidal model for cable strength data

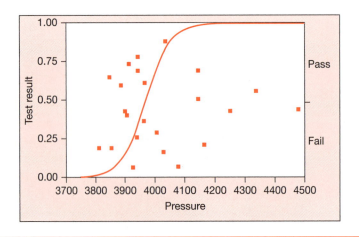

11.9 Checking Conditions and Some Pitfalls

In the simple linear regression analysis, when we use least-squares technique to estimate the line of best fit, we assume that the errors are independent and they are distributed with mean zero. When applying any inferential procedure, we also make a distributional assumption about errors. We assume that the errors are normally distributed with a constant variance σ^2 for all values of x. From a given set of data, we never know for certain that the assumptions are satisfied.

- How far can we deviate from the assumptions and still expect a regression analysis to yield reliable results?
- How can we detect departures (if they exist) from the assumptions?
- What can we do about departures from assumptions?

We have already discussed these issues in context of simple linear regression. Now we'll discuss them in the setting of a multiple regression.

11.9.1 Checking conditions

Remember the multiple regression model for Y is

$$Y = \beta_0 + \beta_1 x_1 + \beta_2 x_2 + \cdots + \beta_k x_k + \varepsilon$$

for a given set of values of $x_1, x_2, \ldots, x_k$ and a random error ε. We make certain assumptions about the random error. The properties of the sampling distributions of the estimators $\hat{\beta}_0, \hat{\beta}_1, \ldots, \hat{\beta}_k$, and ultimately the inferential procedures used, depend on the assumptions concerning the probability distributions of the random errors. Altogether, to make any inferences in multiple regression, we make four assumptions:

1. For any given set of values $x_1, x_2, \ldots, x_k$, the random error ε has a normal probability distribution.
2. For any given set of values of $x_1, x_2, \ldots, x_k$, the mean value of the random error $E(\varepsilon) = $ zero.
3. For any given set of values $x_1, x_2, \ldots, x_k$, the variance of random error ε is equal to σ^2.
4. The random errors are independent (in a probabilistic sense).

It is unlikely that the assumptions are satisfied exactly for many practical situations. If departures from the assumptions are not too great, experience has shown that a least-squares regression analysis procedure produces estimates (predictions and statistical test results) that possess, for all practical purposes, the properties specified in this chapter. Although the random errors will rarely satisfy these assumptions exactly, the reliability of regression analysis will hold approximately for many types of data encountered in practice.

1. For any given set of values $x_1, x_2, \ldots, x_k$, the random error ε has a normal probability distribution.

   The normality condition can be checked easily by making a histogram or dotplot of the residuals. A visual inspection of characteristics (such as symmetry, bell-shape, etc.) can determine whether the condition of normality is reasonable. This is a fairly robust condition; in other words, a slight variation from normality does not affect the reliability of results. However, large deviations from normality are a serious problem. This approach of checking normality might not work well for small samples with large variation. Alternatively, we can use normal probability plots or Q–Q plots, described in Chapter 6, to check this condition. Departures from the assumption of normality usually lead to lower power for the inference procedures. Sometimes the problem of departure from normality can be corrected by using appropriate normalizing transformation of data.

   For example, Figure 11.51 gives a normal quantile plot for the residuals for a linear relation between crude oil and unleaded gasoline prices (data in Exercise 11.6). The residuals pretty much follow a straight-line pattern, except maybe for one measurement that is slightly away from the line of normal distribution. There is no reason to decide that the residuals do not follow normal distribution.

2. The mean value of the random error for any given set of values of $x_1, x_2, \ldots, x_k$ is $E(\varepsilon) = 0$.

   One consequence of this assumption is that the mean $E(Y)$ for a specific set of values of $x_1, x_2, \ldots x_k$ is $E(Y) = \beta_0 + \beta_1 x_1 + \beta_2 x_2 + \cdots + \beta_k x_k$. That is,

$$
Y \;=\; \underbrace{E(Y)}_{\substack{\text{Mean of } y \text{ for specific} \\ \text{values of } x_1, x_2, \ldots x_k}} \;+\; \underbrace{\varepsilon}_{\text{Random error}}
$$

   The second consequence of this assumption is that the least-squares estimators $\hat{\beta}_0, \hat{\beta}_1, \ldots, \hat{\beta}_k$ will be unbiased regardless of the remaining assumptions made about the random errors and their probability distributions.

   Figure 11.52 gives the residual plot for the linear model fitted to the crude oil unleaded gasoline prices Exercise 11.6. It shows that all the residuals are distributed around zero, making the average residual equal to zero.

3. For any given set of values $x_1, x_2, \ldots, x_k$, the variance of random error ε is equal to σ^2. If the variance of the random error ε changes from one set of predictor variable values to another, then we say that a condition called *heteroscedasticity* has occurred. An easy method of detecting nonhomogeneous variances of random errors is a visual inspection of residual plot for the changing spread of residuals. We can sometimes transform the data so that the error variance is stabilized and the standard least-squares methodology becomes appropriate. Some common transformations are as follows:

   a If the error variance is proportional to the mean of y for a given set of x-values, $E(Y \mid x_1, x_2, \ldots, x_k)$, then the errors are likely to be Poisson distributed. In this case, use a square-root transformation to stabilize the error variance (develop a model of $\sqrt{y}$ as a function of $x_1, x_2, \ldots, x_k$).

Figure 11.51

Normal quantile plot

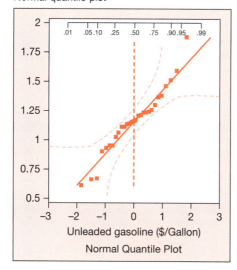

Unleaded gasoline ($/Gallon)

Normal Quantile Plot

Figure 11.52

Residual plot for crude oil unleaded gasoline prices

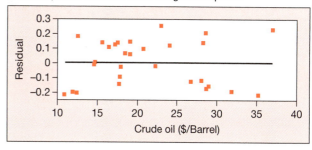

b If the error standard deviation is proportional to the mean of y for a given set of x-values, $E(Y \mid x_1, x_2, \ldots, x_k)$, then use a logarithmic transformation to stabilize the error variance (develop a model of $\log(y)$ as a function of $x_1, x_2, \ldots, x_k$).

c If the error standard deviation is proportional to the square of the mean of y for a given set of x values, $E(Y \mid x_1, x_2, \ldots, x_k)$, then use a reciprocal transformation to stabilize the error variance (develop a model of $1/y$ as a function of $x_1, x_2, \ldots, x_k$).

Empirical Selection of Transformation In those experiments where data are collected for fixed x-values with multiple replications, one can select the variance-stabilizing transformation empirically. Suppose the variance of error is not same for different values of x_i, that is $\sigma_i \propto \mu_i^\lambda$, where $\mu_i = E(Y \mid x_i)$. It is same as saying $\sigma_i = \theta \mu_i^\lambda$, where θ is a constant of proportionality. We are interested in estimating required transformation λ. Taking logarithms, we get a linear relation $\log(\sigma_i) = \log(\theta) + \lambda \log(\mu_i)$. Because σ_i and μ_i are unknown, we use their estimates, s_i and $\bar{y}_i$ respectively. Plot s_i versus $\bar{y}_i$, fit a straight-line model, and estimate λ from the slope of this linear equation. Of course, this method is useful for a large data set with replicated responses for fixed values of predictor variables.

Example 11.29

Let us refer to the example of tool life. Figure 11.53 shows a residual plot for tool life as a function of speed. It looks like the variation in residuals is decreasing (slightly) with the increasing lifetime. Now plot $\ln(s)$ versus $\ln(\bar{y})$ for fixed values of speed (see Figure 11.54). The linear model fit gives estimated relation to be

$$\text{Log(Std Dev[Life])} = -3.001527 + 1.257362 * \text{Log(Mean[Life])}$$

This relation gives the estimate $\hat{\lambda} = 1.26$ that can be used for variance-stabilizing transformation.

Figure 11.53

Residual plot for tool life versus speed

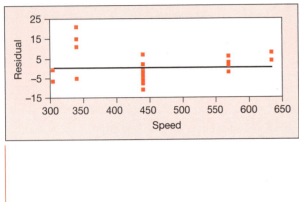

Figure 11.54

Scatterplot of $\ln(s)$ vs. $\ln(\bar{y})$

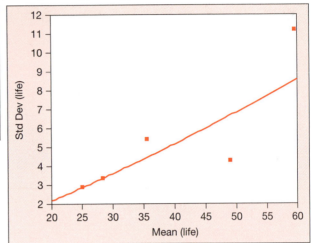

4. The random errors are independent (in a probabilistic sense).

If the observations are likely to be correlated (as in the case of data collected over time), we must check for correlation between random errors. Use of improper randomization technique can potentially result in correlated errors. The problem of correlated errors is a very serious problem. It is also difficult to correct. Plotting the residuals in time order of data collection is useful for detecting correlation between residuals. The runs of positive or negative residuals indicate correlated errors, that is, violation of condition of independent errors. If the skills of the experimenter improve or the capabilities of the testing equipment decrease over time, it will show up in the residual time plot as increased variation in residuals over time.

Example 11.30

Let us look at two examples. Figure 11.55 shows the time-ordered residuals for monthly bill data Exercise 11.37 where bills were recorded for the months January through December. Figure 11.56 shows a time-ordered plot of residuals for the example of crude oil and unleaded gasoline prices data Exercise 11.6. The prices were reported for the years 1976–2004. Clearly, there is no pattern among the residuals for monthly bill data, and the condition of independent errors is reasonable. On the other hand, there is a clear increasing trend among the residuals of unleaded gasoline prices data indicating the errors are positively correlated, making the condition of independent errors unreasonable.

Figure 11.55

Time-ordered residuals for monthly bill data

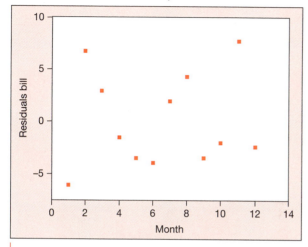

Figure 11.56

Time-ordered residuals for unleaded gasoline data

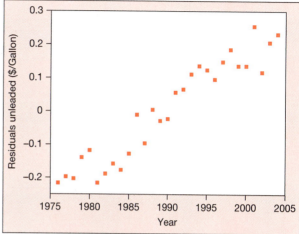

Frequently, the data are observational; that is, we simply observe an experimental unit and record values for $y, x_1, x_2, \ldots, x_k$. Do these data violate the condition that the values $x_1, x_2, \ldots, x_k$ are fixed? For this particular case, we can assume that $x_1, x_2, \ldots, x_k$ are measured without error, and the mean value $E(Y)$ can be viewed as a conditional mean. That is, $E(Y)$ is the mean value of Y, given that the x variables assume a specific set of values. With this modification, the least-squares analysis is applicable to observational data.

11.9.2   Some pitfalls

There are several problems you should be aware of when constructing a prediction model for some response Y. We discuss a few of the most important in this section.

Problem 1: Parameter Estimability Suppose you want to fit a model relating a firm's monthly profit Y to the advertising expenditure x. We propose the first-order model

$$E(Y) = \beta_0 + \beta_1 x$$

Now suppose we have three months of data, and the firm spent $1,000 on advertising during each month. The data are shown in Figure 11.57. You can see the problem: The parameters of the line cannot be estimated when all the data are concentrated at a single x-value. Recall that it takes two points (x-values) to fit a straight line. Thus, the parameters are not estimable when only one x-value is observed. A similar problem would occur if we attempted to fit the second-order model

$$E(Y) = \beta_0 + \beta_1 x + \beta_2 x^2$$

Figure 11.57

Profit and advertising expenditure data for three months

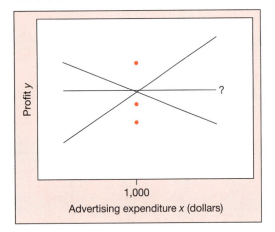

Figure 11.58

Only two x-values observed—the second-order model is not estimable

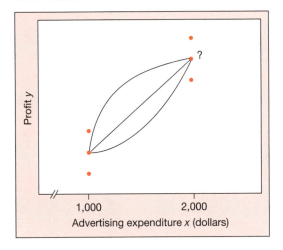

to a set of data for which only *one* or *two* different x values were observed (see Figure 11.58). At least three different x values must be observed before a second-order model can be fit to a set of data (that is, before all three parameters are estimable). In general, the number of levels of x must be at least one more than the order of the polynomial in x that you want to fit.

The independent variables will almost always be observed at a sufficient number of levels to permit estimation of the model parameters. However, when the computer program you are using suddenly refuses to fit a model, the problem is probably inestimable parameters.

Problem 2: Multicollinearity Often two or more of the independent variables used in the model for $E(Y)$ will contribute redundant information because they are correlated with one another. For example, suppose we want to construct a model to predict the gasoline mileage rating of a truck as a function of its load x_1 and the horsepower x_2 of its engine. In general, you would expect heavy loads to require greater horsepower and to result in lower mileage ratings. Thus, although both x_1 and x_2 contribute information for the prediction of mileage rating, some of the information is overlapping because x_1 and x_2 are correlated.

If the model

$$E(Y) = \beta_0 + \beta_1 x_1 + \beta_2 x_2$$

were fit to a set of data, we might find the t-values for both $\hat{\beta}_1$ and $\hat{\beta}_2$ (the least-squares estimates) to be nonsignificant. However, the F-test for H_0: $\beta_1 = \beta_2 = 0$ would probably be highly significant. The tests might seem to be contradictory, but really they are not. The t-tests indicate that the contribution of one variable, say $x_1 =$ load, is not significant after the effect of $x_2 =$ horsepower has been discounted (because x_2 is also in the model). The significant F-test, on the other hand, tells us that at least one of the two variables is making a contribution to the prediction of y (i.e., β_1, β_2, or both differ from zero). In fact, both are probably contributing, but the contribution of one overlaps that of the other. When highly correlated independent variables are present in a regression model, the results can be confusing. The researcher might want to include only one of the variables in the final model.

Problem 3: Prediction Outside the Experimental Region By the late 1960s, many research economists had developed highly technical models to relate the state of the economy to various economic indices and other independent variables. Many of these models were multiple regression models where, for example, the dependent variable Y might be next year's growth in GNP and the independent variables might include this year's rate of inflation, this year's Consumer Price Index, and other factors. In other words, the model might be constructed to predict the state of next year's economy using this year's knowledge.

Unfortunately, these models were almost unanimously unsuccessful in predicting the recession that occurred in the early 1970s. What went wrong? One of the problems was that the regression models were used to predict Y for values of the independent variables that were outside the region in which the model had been developed. For example, the inflation rate in the late 1960s, when the models were developed, ranged from 6% to 8%. When the double-digit inflation of the early 1970s became a reality, some researchers attempted to use the same models to predict future growth in GNP. As you can see in Figure 11.59, the model can be very accurate for predicting Y when x is in the range of experimentation, but using the model outside that range is a dangerous practice.

Figure 11.59

Using a regression model outside the experimental region

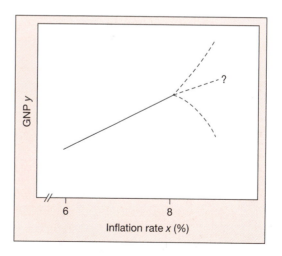

11.10 Reading Computer Printouts

The computations for a regression analysis are quite complicated and cumbersome. However a number of statistical computer packages are available that can lighten the computation load. Most packages produce printouts that contain common information. To illustrate, the standard computer printouts for a regression analysis of the peak power load data using JMP and Minitab are given in Figure 11.60 and Figure 11.61 respectively. Suppose we are fitting the interaction model $(y = \beta_0 + \beta_1 x_1 + \beta_2 x_2 + \beta_3 x_1 x_2 + \varepsilon)$ to the accident data (Example 11.17).

The components of the printout corresponding to the following terms used in this chapter are identified in the printouts using these respective numbers

1. The prediction equation
2. The column of estimates of regression coefficients $(\hat{\beta}_0, \hat{\beta}_1, \hat{\beta}_2, \hat{\beta}_3)$
3. The column of estimated standard error of regression coefficients $(S_{\hat{\beta}_i})$
4. The column of test statistic values (t-statistic $t = \hat{\beta}_i / s_{\hat{\beta}_i}$ for testing $H_0: \beta_i = 0$)
5. The observed significance level (p-value) for t-test in (4)

Figure 11.60

JMP printout for accident data

JMP Printout

Response y
Whole Model

Summary of Fit

RSquare	0.932624	(8)
RSquare Adj	0.92701	
Root Mean Square Error	14.00564	(11)
Mean of Response	87.075	
Observations (or Sum Wgts)	40	

Analysis of Variance

Source	DF	Sum of Squares	Mean Square	F Ratio
Model	3 (14)	97749.09	32583.0	(6) 166.1062
Error	(9) 36 (13)	7061.68	(10) 196.2	Prob > F
C. Total	39 (12)	104810.77		(7) <.0001

Parameter Estimates

	(2) Estimate	(3) Std Error	(4) t Ratio	(5) Prob >|t|
Term				
Intercept	31.052863	6.005189	5.17	<.0001
x1	0.0143271	0.000857	16.71	<.0001
x2	-58.217	4.429921	-13.14	<.0001
(x1-5921.9)* (x2-0.5)	-0.010737	0.001715	-6.26	<.0001

Figure 11.61

Minitab printout for accident data

Minitab Printout

Regression Analysis: y versus x1, x2, x1*x2

The regression equation is
y = -0.74 + 0.0197 x1 + 5.4 x2 - 0.0107 x1*x2 (1)

	(2)	(3)	(4)	(5)
Predictor	Coef	SE Coef	T	P
Constant	-0.739	7.927	-0.09	0.926
x1	0.019696	0.001219	16.16	0.000
x2	5.37	11.08	0.48	0.631
x1*x2	-0.010737	0.001715	-6.26	0.000

(11) S = 14.01   (8) R-Sq = 93.3%   R-Sq(adj) = 92.7%

Analysis of Variance

Source	DF	SS	MS	F	P
Regression	3	(14) 97749	32583	(6) 166.11	(7) 0.000
Residual Error	(9) 36	(13) 7062	(10) 196		
Total	39	(12) 104811			

6. The F-statistic for testing utility of the model
7. The observed significance level for the F-test in (6)
8. R^2 = Coefficient of determination = $1 - SSE/SS_{yy}$
9. The degrees of freedom for SSE
10. s^2 = Error variance = MSE
11. $s = \sqrt{s^2}$ = Standard deviation of error
12. SS_{yy} = Total sum of squares
13. SSE = Error sum of squares
14. $SSR = SS_{yy} - SSE$ = Regression sum of squares

You can also compute the value of the F-statistic directly from the mean square entries given in the Analysis of Variance table. Thus, an equivalent form of the F-statistic for testing $H_0: \beta_1 = \beta_2 = \cdots \beta_k = 0$ is

$$F = \frac{\text{Mean Square for Regression}}{\text{Mean Square for Error (or residuals)}} = \frac{\text{Mean Square for Regression}}{s^2}$$

These quantities are given in the printout in the column labeled "Mean Square or MS." Thus, for the accident data (see Figure 11.61) we get $F = 32{,}583/196 = 166$, a value that agrees with that given in the printout. The logic behind this test and other tests of hypotheses concerning sets of β parameters was discussed earlier in this chapter.

11.11 Summary

We have discussed inference techniques for simple linear regression and some of the methodology of *multiple regression analysis*, a technique for modeling a dependent variable Y as a function of several independent variables $x_1, x_2, \ldots, x_k$. The steps we follow in constructing and using regression models are:

Step 1. The form of the probabilistic model is hypothesized.
Step 2. The model coefficients are estimated by using least squares.
Step 3. The probability distribution of ε is specified and σ^2 is estimated.
Step 4. The adequacy of the model is checked.
Step 5. If the model is deemed useful, it may be used to make estimates and to predict values of Y to be observed in the future.

We stress that this chapter is not intended to be a complete coverage of regression analysis. Whole texts have been devoted to this topic. However, we have presented the core necessary for a basic understanding of regression.

Supplementary Exercises

11.54 A reactor pressure vessel is usually clad in stainless steel to prevent corrosion and radiation embrittlement. Thus, the occurrence of subclad cracks is higher than that of surface cracks. Kim et al. (*Transactions of the ASME*, 2003) performed 3-D elastic-plastic finite element analysis to investigate the validity of present ASME code in evaluating the integrity of a vessel under pressurized thermal shock conditions. The hoop stress (in MPa) data recorded for different values of x/a are given in the following table.

x/a	Hoop Stress
0	425
0.05	400
0.08	380
0.175	280
0.25	270
0.35	250
0.45	240
0.725	200
1.00	160

a Plot the hoop stress versus x/a values. Describe the relation between the two variables.
b Estimate the relation between the hoop stress and x/a values, using the least-squares method.
c Does it appear that the hoop stress changes significantly with x/a values. Use $\alpha = 0.05$.

11.55 Table 11.6 gives the national air-quality data for the years 1980–2000 (National Air Quality and Emissions Report, 2003). All values are measured in ppm.

a Plot the air quality data versus year. Describe any trends observed in the air quality measurements over the years data were collected.
b Make a scatterplot matrix of all the air quality variables. Comment on the nature of relations among these variables.
c Estimate the strongest relation. Assess the fit of the model using $\alpha = 0.05$.

11.56 Stay and Barocas (*International Journal for Numerical Methods in Fluids*, 2003) demonstrated a localized reduction technique called "stitching" for hydrodynamic problems for which known analytical solution is available. They discussed the computational advantage of the coupled Stokes-Raynolds, using a fully-flooded deformable-roll coating example. The difference between the peak pressure calculated with

Table 11.6

Air Quality Data for Years 1981–2000

Year	CO	Pb	NO$_2$	Ozone	SO$_2$
1981	8.4	0.58	0.024	0.126	0.0102
1982	8.1	0.58	0.023	0.125	0.0095
1983	7.9	0.47	0.023	0.137	0.0093
1984	7.8	0.45	0.023	0.125	0.0095
1985	7.1	0.28	0.023	0.123	0.0090
1986	7.2	0.18	0.023	0.118	0.0088
1987	6.7	0.13	0.023	0.125	0.0086
1988	6.4	0.12	0.023	0.136	0.0087
1989	6.4	0.10	0.023	0.116	0.0085
1990	5.9	0.08	0.022	0.114	0.0079
1991	5.6	0.08	0.019	0.111	0.0081
1992	5.3	0.07	0.019	0.105	0.0076
1993	5.0	0.06	0.019	0.107	0.0074
1994	5.1	0.05	0.020	0.106	0.0072
1995	4.6	0.05	0.019	0.112	0.0057
1996	4.3	0.05	0.019	0.105	0.0057
1997	4.1	0.04	0.018	0.104	0.0056
1998	3.9	0.04	0.018	0.110	0.0055
1999	3.7	0.04	0.018	0.107	0.0053
2000	3.4	0.04	0.017	0.100	0.0051

GOMA (a full-Newton finite element program for free and moving boundary problems) and that calculated with coupled Stokes-Raynolds code (error in pressure) along with the stitch length (in cm), is recorded in table below. Also given are the CPU solve-times (in seconds) for the coupled Stokes-Raynolds code.

Stitch Length	Error	CPU Time	Stitch Length	Error	CPU Time
0.1	0.01	140	0.8	0.25	85
0.2	0.11	130	0.9	0.24	81
0.3	0.15	118	1.0	0.21	73
0.4	0.17	102	1.1	0.20	65
0.5	0.20	105	1.2	0.18	55
0.6	0.24	100	1.3	0.18	58
0.7	0.25	85	1.4	0.15	50

a Plot the error in pressure against the stitch length. Describe the nature of the relation.

b Plot the error in pressure against the CPU time. Describe the nature of the relation.

c Estimate both the relations using the least-squares method.

d Assess the fit of each model using $\alpha = 0.05$.

11.57 The use of the variable-orifice damper has blossomed in Japan. These dampers provide a mechanically simple and low-power consumer means to protect structures against natural disaster by reducing structural vibrations. Spencer and Nagarajaiah (*Journal of Structural Engineering*, 2003) give the following data for buildings recently completed or currently under construction.

Stories	Height (m)	No. of Semi-active Dampers
11	56.0	42
31	140.5	72
25	119.9	38
28	136.6	60
38	172.0	88
54	241.4	356
19	98.9	27
30	104.9	66
23	100.4	28

a Plot the number of semi-active dampers against the height of the tower. Describe the nature of the relation.

b Estimate the relation using the least-squares method.

c Does the number of dampers used increase significantly with the height of the tower? Use $\alpha = 0.05$.

11.58 Moles et al. (*Chemical Engineering Research and Design*, 2003) studied how computation times of the stochastic methods are reduced using parallelization (parallel implementation of different global optimization methods).

Speed-Up (y)	No. of Processors (x)
1.00	1
1.95	2
2.60	3
4.30	4
4.80	6

a Make a scatterplot. Speed-up is measured as percent increase in computation times. Describe the nature of the relation between speed-up and the number of processors.

b Estimate the relation between speed-up and number of processors using the least-squares method.

c Does the speed-up increase significantly with the number of processors used? Use $\alpha = 0.05$.

11.59 Wang and Reinhardt (*Journal of Pressure Vessel Technology*, 2003) studied the assessment of a steam generator tube. They analyzed a finite element model with a percentage degraded area (PDA) in the range of 10% to 90% and recorded the finite element limit load. The following table gives the PDA and normalized finite element limit loads.

PDA	Limit Load
10	0.80
20	0.67
30	0.52
40	0.53
50	0.45
60	0.32
70	0.27
80	0.19
90	0.10

a Plot the limit load against the PDA. Describe the nature of the relation.

b Estimate the relation using the least-squares method.

c Does the load limit decrease significantly with the PDA? Use $\alpha = 0.05$.

11.60 Carrasco and Uribeotxeberria (*IEE Proceedings Communications*, 2003) recorded the total number of operations per input pattern required to implement two different algorithms, (SRK and RTRL) for different number of receiving antenna elements. They are reported in the table below.

No. of Antennas	2	3	4	5	6	7	8
SRK	80	156	256	380	528	700	896
RTRL	260	340	420	500	580	660	740

a Make a scatterplot of the number of operations versus the number of antennas used for each algorithm used.

b Comment on the nature of the relationships observed.

c Estimate the nature of the relation, using the appropriate model for each type of algorithm.

d Test for slope at a 5% error rate.

11.61 In the 1960s, promotion of underground utilities as an alternative to traditional overhead distribution began. The costs for two types of utilities differ, and the power company has to provide both estimates to the potential developer of the property. Stanford (*Journal of Energy Engineering*, 1993) reported data collected on 47 randomly selected subdivisions company files for which the costs were determined previously (Table 11.7). Both types of costs are determined for each project for two years as the material costs and various other input values change from year to year. The characteristics measured were as follows:

C1: Number of lots in the project
C2: Total front-line footage of lots
C3: Number of corner lots
C4: Total length of streets
C5: Number of cul-de-sacs
C6: Total length of curved streets within the project
C7: Number of accesses to the project
C8: Number of intersections within the project

a Develop a model to estimate the year 1 cost of developing underground utilities.
b Develop a model to estimate the year 1 cost of developing overhead utilities.
c Develop a model to estimate the year 2 cost of developing underground utilities.
d Develop a model to estimate the year 2 cost of developing overhead utilities.

11.62 Olajossy, Gawdzik, Budner, and Dula (*Trans IchemE*, 2003) conducted an experiment of plant output and recorded the ratio of recycled methane to feed methane. They also recorded the product purity (vol. %) and efficiency of methane recovery (%). The data are given in the table below.

Plant Capacity	Recycle Ratio	Product Purity	Efficiency of Methane Recovery
47.5	2.10	97.5	86.5
48.0	2.05	98.5	87.5
49.5	2.05	97.5	87.5
51.0	1.95	97.0	88.5
53.0	1.85	96.5	89.5
54.0	1.75	96.0	90.5
57.0	1.65	95.0	91.5
59.0	1.60	92.5	92.0
62.0	1.45	91.5	92.0
67.0	1.25	88.5	91.0
72.0	1.10	86.5	93.0

a Make scatterplots of the recycled ratio, product purity, and efficiency of recovery versus the plant capacity.
b Describe the nature of each relation in context.

c Estimate each relation using an appropriate model.
d Because all three variables provide outcomes of the same experiments, comment on their behavior collectively in the context of the problem.

11.63 Rohner and Newmann (*Transactions of the ASME*, 2003) recorded the percentage of deviation of UV temperatures from black body temperatures. Two sets of data were collected, one based on the calibration at 1400°C and the other at 1500°C. The data are given in the table below.

Black Body Temperature (°C)	Percentage of Deviation	
	1400°C	1500°C
1300	0.90	1.15
1320	0.80	1.05
1340	0.40	0.65
1360	0.50	0.75
1380	0.30	0.55
1400	0.00	0.30
1420	−0.05	0.25
1440	−0.05	0.20
1460	−0.15	0.15
1480	−0.25	0.05
1500	−0.40	−0.05

a Make a scatterplot of the percent deviation versus black body temperature for each calibration.
b Fit appropriate model to the data for each calibration level and assess the fit of the model.
c Determine the temperature at which the percent deviation is zero.

11.64 Spath and Amos (*Journal of Solar Energy Engineering*, 2003) studied the process of hydrogen production via thermal decomposition of methane, using a solar reactor. The table on page 664 shows the hydrogen production rate, Y (in mol/s) at different solar irradiance, X (in w/m^2) for 70% conversion for the three different heliostat sizes.
a Make a scatterplot for each heliostat size.
b Comment on the nature of the relation between hydrogen production rate and solar irradiance.
c Develop models to predict the hydrogen production.
d Estimate the mean production at solar irradiance of 500 w/m^2.

11.65 Brittorf and Kresta (*Trans IchemE*, 2003) conducted experiments for solid-liquid mixing in stirred tanks and studied cloud height for solid

Table 11.7

Underground Utilities Data

Sub-division	Year 2 Cost Under-ground	Year 2 Cost Overhead	Year 1 Cost Under-ground	Year 1 Cost Overhead	C1	C2/C1	C3	C4	C5	C6	C7	C8
1	26,233	23,937	24,614	22,882	55	44	2	1,280	1	540	1	1
2	51,606	36,984	48,702	35,453	59	121	16	5,320	0	600	6	8
3	44,393	38,085	42,173	36,643	77	86	18	4,390	1	630	5	6
4	16,611	12,528	15,742	12,047	17	179	3	2,100	0	2,100	4	1
5	16,411	12,085	15,577	11,539	25	94	1	1,240	0	90	2	1
6	37,200	34,233	35,230	33,055	46	122	10	3,060	2	1,740	3	5
7	45,724	41,481	43,649	39,686	64	129	6	4,770	0	1,300	3	3
8	10,971	9,545	10,393	9,200	16	131	0	1,050	1	1,050	1	0
9	5,650	3,800	5,318	5,378	11	43	0	360	1	0	1	0
10	33,351	27,837	31,695	26,737	52	113	8	3,330	1	2,200	3	4
11	34,323	25,864	32,632	24,762	43	101	7	3,380	2	2,800	3	4
12	7,419	6,659	6,953	6,265	15	51	0	340	1	0	1	0
13	10,657	8,386	10,054	8,063	16	123	0	1,000	0	470	2	0
14	10,925	7,918	10,318	7,564	13	132	2	960	0	150	3	1
15	15,612	12,465	14,830	11,942	23	91	2	1,320	1	580	3	2
16	5,344	4,561	5,065	4,402	6	134	1	530	1	300	2	1
17	31,968	27,701	30,296	26,562	43	111	7	3,350	0	0	6	4
18	8,980	7,080	8,485	6,889	17	69	0	540	1	250	1	0
19	36,243	31,262	34,404	29,880	48	115	9	3,870	1	700	3	4
20	12,689	10,712	12,086	10,383	20	118	0	1,250	1	500	1	0
21	24,908	20,340	23,607	19,473	36	113	6	2,400	1	520	2	3
22	7,929	6,208	7,476	5,949	9	107	0	960	0	0	1	0
23	14,633	10,051	13,892	9,686	21	71	2	1,120	0	140	3	1
24	16,272	12,542	15,324	11,976	20	132	8	2,120	0	400	6	1
25	9,200	6,588	7,883	6,484	7	145	0	450	1	450	1	0
26	22,638	18,062	21,616	17,470	35	123	2	2,250	1	0	1	1
27	28,118	21,163	26,602	20,286	39	102	7	2,320	2	1,000	3	3
28	26,894	24,111	25,412	23,083	41	91	7	2,400	0	200	4	2
29	32,313	27,040	30,663	25,989	54	83	8	3,150	1	580	3	4
30	62,727	48,240	59,185	46,378	80	113	14	5,700	2	3,150	3	6
31	55,651	49,864	52,482	47,857	76	111	10	5,220	2	2,100	6	6
32	5,397	42,230	5,084	3,996	8	89	0	500	1	0	1	0
33	15,610	14,625	14,750	13,957	15	152	1	1,850	1	1,600	3	1
34	30,920	23,243	29,287	22,489	54	88	4	2,460	0	0	4	1
35	69,279	56,139	65,729	54,016	121	76	18	5,850	2	950	5	8
36	27,032	22,218	25,610	21,339	38	96	5	2,700	1	900	5	3
37	26,150	20,185	24,819	19,400	29	103	6	2,730	1	250	2	3
38	32,894	28,175	31,294	27,070	49	118	11	1,670	0	1,100	5	5
39	3,294	2,832	3,091	2,723	5	52	0	100	1	0	1	0
40	65,894	60,466	62,442	57,632	89	107	12	5,770	5	5,750	2	5
41	26,297	18,009	24,933	17,118	32	122	4	3,150	0	260	2	2
42	38,036	30,010	35,746	28,669	84	42	6	1,890	3	600	2	3
43	12,191	8,470	11,539	7,983	27	29	1	510	0	510	3	1
44	12,387	10,387	11,683	9,726	28	28	2	240	1	240	1	0
45	13,334	9,010	12,614	8,456	30	29	0	396	1	0	1	0
46	36,530	26,031	34,635	24,405	80	31	2	1,419	0	288	3	0
47	42,668	30,712	40,355	28,973	95	25	0	2,109	0	400	4	3

| Heliostat Size | | | | | |
| 2,188 m^2 | | 4,375 m^2 | | 8,750 m^2 | |
Solar Irradiance	Hydrogen Production	Solar Irradiance	Hydrogen Production	Solar Irradiance	Hydrogen Production
50	0	25	0	20	0.0
200	2	100	2	100	4.0
400	4	200	4	200	7.5
600	6	400	7.5	300	10.5
800	7.5	600	10.5	400	13.5
1,000	9.0	800	13.5	500	16.5
1,025	9.5	900	14.5		
		1,000	16.0		
		1,025	16.5		

suspensions. They recorded the off-bottom clearance and impeller diameter, and measured the core velocity in water. Develop a model to predict the core velocity using data given below.

Off-Bottom Clearance	Impeller Diameter	Core Velocity
0.19	0.15	0.33
0.19	0.19	0.32
0.19	0.28	0.29
0.19	0.38	0.27
0.19	0.57	0.18
0.33	0.13	0.32
0.33	0.20	0.30
0.33	0.26	0.28
0.33	0.33	0.27
0.33	0.50	0.17
0.50	0.16	0.44
0.50	0.34	0.33
0.43	0.21	0.32
0.43	0.43	0.22
0.32	0.21	0.25
0.32	0.32	0.23
0.50	0.25	0.23
0.50	0.34	0.21
0.58	0.19	0.29
0.58	0.29	0.25
0.58	0.39	0.22

11.66 After a regression model is fit to a set of data, a confidence interval for the mean value of Y at a given setting of the independent variables will *always* be narrower than the corresponding prediction interval for a particular value of Y at the same setting of the independent variables. Why?

11.67 Before a particular job is accepted, a computer at a major university estimates the cost of running the job to see whether the user's account contains enough money to cover the cost. As part of the job submission, the user must specify estimated values for two variables: central processing unit (CPU) time and number of lines printed. Although the CPU time required and the number of lines printed do not account for the complete cost of the run, it is though that knowledge of their values should allow a good prediction of job cost. The following model is proposed to explain the relationship of CPU time and lines printed to job cost:

$$E(Y) = \beta_0 + \beta_1 x_1 + \beta_2 x_2 + \beta_3 x_1 x_2$$

where

$$Y = \text{job cost (in dollars)}$$
$$x_1 = \text{number of lines printed}$$
$$x_2 = \text{CPU time}$$

Records from 20 previous runs were used to fit this model. The SAS printout is shown in Figure 11.62

a Identify the least-squares model that was fit to the data.

b What are the values of SSE and s^2 (estimate of σ^2) for the data?

c Explain what is meant by the statement "This value of SSE [see part (b)] is minimum."

11.68 Refer to Exercise 11.67 and the portion of the SAS printout shown in Figure 11.62

a Is there evidence that the model is useful (as a whole) for predicting job cost? Test at $\alpha = 0.05$.

b Is there evidence that the variables x_1 and x_2 interact to affect Y? Test at $\alpha = 0.01$.

c What assumptions are necessary for the tests conducted in parts (a) and (b) to be valid?

11.69 Refer to Exercise 11.67 and the portion of the SAS printout shown. Use a 95% confidence interval to estimate the mean cost of computer jobs that require 42 seconds of CPU time and print 2,000 lines.

Figure 11.62

SAS output for Exercise 11.67

SOURCE	DF	SUM OF SQUARES	MEAN SQUARE	F VALUE	PR > F
MODEL	3	43.25090461	14.41696820	84.96	0.0001
ERROR	16	2.71515039	0.16969690		
CORRECTED TOTAL	19	45.9660550		R-SQUARE	STD DEV
				0.940931	0.41194283

PARAMETER		ESTIMATE	T FOR H0: PARAMETER = 0	PR > T	STD ERROR OF ESTIMATE
INTERCEPT		0.04564705	0.22	0.8318	0.21082636
X1		0.00078505	5.80	0.0001	0.00013537
X2		0.23737262	7.50	0.0001	0.03163301
X1*X2		-0.00003809	-.299	0.0086	0.0001273

X1	X2	PREDICTED VALUE	LOWER 95% CL FOR MEAN	UPPER 95% CL FOR MEAN
2000	42	8.38574865	7.32284845	9.44864885

11.70 Several states now require all high school seniors to pass an achievement test before they can graduate. On the test, the seniors must demonstrate their familiarity with basic verbal and mathematical skills. Suppose the educational testing company that creates and administers these exams wants to model the score Y on one of their exams as a function of the student's IQ x_1 and socioeconomic status (SES). The SES is a categorical (or *qualitative*) variable with three levels: low, medium, and high.

$$x_2 = \begin{cases} 1 & \text{if SES is medium} \\ 0 & \text{if SES is low or high} \end{cases}$$

$$x_3 = \begin{cases} 1 & \text{if SES is high} \\ 0 & \text{if SES is low or medium} \end{cases}$$

Data were collected for a random sample of 60 seniors who had taken the test, and the model

$$E(Y) = \beta_0 + \beta_1 x_1 + \beta_2 x_2 + \beta_3 x_3$$

was fit to these data, with the results shown in the SAS printout in Figure 11.63.
a Identify the least-squares equation.
b Interpret the value of R^2 and test to determine whether the data provide sufficient evidence to indicate that this model is useful for predicting achievement test scores.
c Sketch the relationship between predicted achievement test score and IQ for the three levels of SES. [*Note:* Three graphs of $\hat{y}$ versus x_1 must be drawn: the first for the low-SES model ($x_2 = x_3 = 0$), the second for the medium-SES model ($x_2 = 1$, $x_3 = 0$), and the third for the high-SES model ($x_2 = 0$, $x_3 = 1$). Note that increase in predicted achievement test score per unit increase in IQ is the same for all three levels of SES; that is, all three lines are parallel.]

11.71 Refer to Exercise 11.70. The same data are now used to fit the model

$$E(Y) = \beta_0 + \beta_1 x_1 + \beta_2 x_2 + \beta_3 x_3 + \beta_4 x_1 x_2 + \beta_5 x_1 x_3$$

Thus, we now add the interaction between IQ and SES to the model. The SAS printout for this model is shown in Figure 11.64.
a Identify the least-squares prediction equation.
b Interpret the value of R^2 and test to determine whether the data provide sufficient evidence to indicate that this model is useful for predicting achievement test scores.
c Sketch the relationship between predicted achievement test score and IQ for the three levels of SES.
d Test to determine whether these is evidence that the mean increase in achievement test score per unit increase in IQ differs for the three levels of SES.

11.72 The EPA wants to model the gas mileage rating Y of automobiles as a function of their engine size x. A quadratic model

$$Y = \beta_0 + \beta_1 x + \beta_2 x^2$$

Figure 11.63
SAS output for Exercise 11.70

SOURCE	DF	SUM OF SQUARES	MEAN SQUARE	F VALUE	PR > F
MODEL	3	12268.56439492	4089.52146497	188.33	0.0001
ERROR	56	1216.01893841	21.71462390		STD DEV
CORRECTED TOTAL	59	13484.58333333		R-SQUARE	4.65989527
				0.909822	

PARAMETER	ESTIMATE	T FOR H0: PARAMETER = 0	PR > T	STD ERROR OF ESTIMATE
INTERCEPT	-13.06166081	-3.21	0.0022	4.07101383
X1	0.74193946	17.56	0.0001	0.04224805
X2	18.60320572	12.49	0.0001	1.48895324
X3	13.40965415	8.97	0.0001	1.49417069

Figure 11.64
SAS output for Exercise 11.71

SOURCE	DF	SUM OF SQUARES	MEAN SQUARE	F VALUE	PR > F
MODEL	5	12515.10021009	2503.02004202	139.42	0.0001
ERROR	54	969.48312324	17.95339117		
CORRECTED TOTAL	59	13484.58333333		R-SQUARE	STD DEV
				0.928104	4.23714422

PARAMETER	ESTIMATE	T FOR H0: PARAMETER = 0	PR > T	STD ERROR OF ESTIMATE
INTERCEPT	0.60129643	0.11	0.9096	5.26818519
X1	0.59526252	10.70	0.0001	0.05563379
X2	-3.72536406	-0.37	0.7115	10.01967496
X3	-16.23196444	-1.90	0.0631	8.55429931
X1*X2	0.23492147	2.29	0.0260	0.10263908
X1*X3	0.30807756	3.53	0.0009	0.08739554

is proposed. A sample of 50 engines of varying sizes is selected, and the miles per gallon rating of each is determined. The least-squares model is

$$\hat{y} = 51.3 - 10.1x + 0.15x^2$$

The size x of the engine is measured in hundreds of cubic inches. Also, $s_{\hat{\beta}_2} = 0.0037$ and $R^2 = 0.93$.

a Sketch this model between $x = 1$ and $x = 4$.

b Is there evidence that the quadratic term in the model is contributing to the prediction of the miles per gallon rating Y? Use $\alpha = 0.05$.

c Use the model to estimate the mean miles per gallon rating for all cars with 350-cubic-inch engines ($x = 3.5$).

d Suppose a 95% confidence interval for the quantity estimated in part (c) is (17.2, 18.4). Interpret this interval.

e Suppose you purchase an automobile with a 350-cubic-inch engine and determine that the miles per gallon rating is 14.7. Is the fact that this value lies outside the confidence interval given in part (d) surprising? Explain.

11.73 To increase the motivation and productivity of workers, an electronics manufacturer decides to experiment with a new pay incentive structure at one of two plants. The experimental plan will be tried at plant A for six months, while workers at plant B will remain on the original pay plan. To evaluate the effectiveness of the new plan, the average assembly time for part of an electronic system was measured for employees at both plants at the beginning and end of the six-month period. Suppose the following model was proposed:

$$Y = \beta_0 + \beta_1 x_1 + \beta_2 x_2 + \varepsilon$$

where

Y = assembly time (hours) at end of six-month period

x_1 = assembly time (hours) at beginning of six-month period

$$x_2 = \begin{cases} 1 & \text{if plant A (dummy variable)} \\ 0 & \text{if plant B} \end{cases}$$

A sample of $n = 42$ observations yielded

$$\hat{y} = 0.11 + 0.98x_1 - 0.53x_2$$

where

$$s_{\hat{\beta}_1} = 0.231 \qquad s_{\hat{\beta}_2} = 0.48$$

Test to see whether, after allowing for the effect of initial assembly time, plant A had a lower mean assembly time than plant B. Use $\alpha = 0.01$. [*Note:* When the (0, 1) coding is used to define a dummy variable, the coefficient of the variable represents the difference between the mean response at the two levels represented by the variable. Thus, the coefficient β_2 is the difference in mean assembly time between plant A and plant B at the end of the six-month period, and $\hat{\beta}_2$ is the sample estimator of that difference.]

11.74 One fact that must be considered in developing a shipping system that is beneficial to both the customer and the seller is time of delivery. A manufacturer of farm equipment can ship its products by either rail or truck. Quadratic models are thought to be adequate in relating time of delivery to distance traveled for both modes of transportation. Consequently, it has been suggested that the following model be fit:

$$E(Y) = \beta_0 + \beta_1 x_1 + \beta_2 x_2 + \beta_3 x_1 x_2 + \beta_4 x_2^2$$

where

Y = shipping time

$$x_1 = \begin{cases} 1 & \text{if rail} \\ 0 & \text{if truck} \end{cases}$$

x_2 = distance to be shipped

a What hypothesis would you test to determine whether the data indicate that the quadratic distance term is useful in the model, that is, whether curvature is present in the relationship between mean delivery time and distance?

b What hypothesis would you test to determine whether there is a difference in mean delivery time by rail and by truck?

11.75 Refer to Exercise 11.74. Suppose the proposed second-order model is fit to a total of 50 observations on delivery time. The sum of squared errors is SSE = 226.12. Then the reduced model

$$E(Y) = \beta_0 + \beta_2 x_2 + \beta_4 x_2^2$$

is fit to the same data, and SSE = 259.34. Test to see whether the data indicate that the mean delivery time differs for rail and truck deliveries.

11.76 *Operations management* is concerned with planning and controlling those organizational functions and systems that produce goods and services (Schroeder, *Operations Management: Decision Making in the Operations Function,* McGraw-Hill, 1981). One concern of the operations manager of a production process is the level of productivity of the process. An operations manager at a large manufacturing plant is interested in predicting the level of productivity of assembly line A next year (i.e., the number of units that will be produced by the assembly line next year). To do so, she has decided to use regression analysis to model the level of productivity Y as a function of time x. The number of units produced by assembly line A was determined for each of the past 15 years ($x = 1, 2, \ldots, 15$). The model

$$E(Y) = \beta_0 + \beta_1 x + \beta_2 x^2$$

was fit to these data, using Minitab. The results shown in the printout in below were obtained.

```
THE REGRESSION EQUATION IS
Y = -1187 - 1333 X1 - 45.6 X2

                          ST. DEV.    T-RATIO=
     COLUMN  COEFFICIENT  OF COEF.   COEF./S.D.
       --        -1187       446       -2.66
X1     C2         1333       128       10.38
X2     C3       -45.59      7.80       -5.84


THE ST. DEV. OF Y ABOUT REGRESSION LINE IS
S = 501


WITH (1523)  = 12 DEGRESS OF FREEDOM
R-SQUARED  = 97.3 PERCENT
R-SQUARED  = 96.9 PERCENT. ADJUSTED FOR D.F.


ANALYSIS OF VARIANCE

DUE TO        DF           SS  MS = SS/DF
REGRESSION     2      110578719    55289359
RESIDUAL      12        3013365      251114
TOTAL         14      113592083
```

a Identify the least-squares prediction equation.

b Find R^2 and interpret its value in the context of this problem.

c Is there sufficient evidence to indicate that the model is useful for predicting the productivity of assembly line A? Test using $\alpha = 0.05$.

d Test the null hypothesis $H_0: \beta_2 = 0$ against the alternative hypothesis $H_a: \beta_2 \neq 0$ by using $\alpha = 0.05$. Interpret the results of your test in the context of this problem.

e Which (if any) of the assumptions we make about ε in regression analysis are likely to be violated in this problem? Explain.

11.77 Many companies must accurately estimate their costs before a job is begun in order to acquire a contract and make a profit. For example, a heating and plumbing contractor may base cost estimate for new homes on the total area of the house, the number of baths in the plans, and whether central air conditioning is to be installed.

a Write a first-order model relating the mean cost of material and labor $E(Y)$ to the area, number of baths, and central air conditioning variables.

b Write a complete second-order model for the mean cost as a function of the same three variables as in (a).

c How could you test the research hypothesis that the second-order terms are useful for predicting mean cost?

11.78 Refer to Exercise 11.77. The contractor samples 25 recent jobs and fits both the complete second-order model part (b) and the reduced main effects model in part (a) so that a test can be conducted to determine whether the additional complexity of the second-order model is necessary. The resulting SSE and R^2 are given in the table below.

a Is there sufficient evidence to conclude that the second-order terms are important for predicting the mean cost?

	SSE	R^2
First-order	8.548	0.950
Second-order	6.133	0.964

b Suppose the contractor decides to use the main effects model to predict costs. Use the global F-test to determine whether the main effects model is useful for predicting costs.

11.79 A company that services two brands of microcomputers would like to be able to predict the amount of time it takes a service person to perform preventive maintenance on each brand. It believes the following predictive model is appropriate:

$$Y = \beta_0 + \beta_1 x_1 + \beta_2 x_2 + \varepsilon$$

where

$$Y = \text{maintenance time}$$

$$x_1 = \begin{cases} 1 & \text{if brand A} \\ 0 & \text{if brand I} \end{cases}$$

$$x_2 = \text{service person's number of months of experience in preventive maintenance}$$

Ten different service people were randomly selected, and each was randomly assigned to perform preventive maintenance on either a brand A or brand I microcomputer. The data in table below were obtained.

Maintenance Time (hours)	Brand	Experience (months)
2.0	1	2
1.8	1	4
0.8	0	12
1.1	1	12
1.0	0	8
1.5	0	2
1.7	1	6
1.2	0	5
1.4	1	9
1.2	0	7

a Fit the model to the data.

b Investigate whether the overall model is useful. Test using $\alpha = 0.05$.

c Find R^2 for the fitted model. Does the value of R^2 support your findings in part (b)? Explain.

d Find a 90% confidence interval for β_2. Interpret your result in the context of this exercise.

e Use the fitted model to predict how long it will take a person with six months of experience to service a brand I microcomputer.

f How long would it take the person referred to in part (e) to service 10 brand I microcomputers? List any assumptions you made in reaching your prediction.

g Find a 95% prediction interval for the time required to perform preventive maintenance on a brand A microcomputer by a person with four months of experience.

11.80 Many colleges and universities develop regression models for predicting the grade-point average (GPA), y, of incoming freshmen. This predicted GPA can then be used to help make admission decisions. Although most models use many independent variables to predict GPA, we will illustrate by choosing two variables:

$$x_1 = \text{verbal score on college entrance examination (percentile)}$$

$$x_2 = \text{mathematics score on college entrance examination (percentile)}$$

The data in table above are obtained for a random sample of 40 freshmen at one college.

a Fit the first-order model (no quadratic and no interaction terms)

$$Y = \beta_0 + \beta_1 x_1 + \beta_2 x_2 + \varepsilon$$

Interpret the value of R^2, and test whether these data indicate that the terms in the model are useful for predicting freshman GPA. Use $\alpha = 0.05$.

b Sketch the relationship between predicted GPA $\hat{y}$ and verbal score x_1 for the following mathematics scores: $x_2 = 60, 75,$ and 90.

11.81 Refer to Exercise 11.80. Now fit the following second-order model to the data:

$$Y = \beta_0 + \beta_1 x_1 + \beta_2 x_2 + \beta_3 x_1^2 + \beta_4 x_2^2 \\ + \beta_5 x_1 x_2 + \varepsilon$$

a Interpret the value of R^2, and test whether the data indicate that this model is useful for predicting freshman GPA. Use $\alpha = 0.05$.

b Sketch the relationship between predicted GPA $\hat{y}$ and the verbal score x_1 for the following mathematics scores: $x_2 = 60, 75,$ and 90. Compare these graphs with those for the first-order model in Exercise 11.80.

c Test to see whether the interaction term $\beta_5 x_1 x_2$ is important for the prediction of GPA. Use $\alpha = 0.10$. Note that this term permits the distance between three mathematics score curves for GPA versus verbal score to change as the verbal score changes.

11.82 Plastics made under different environmental conditions are known to have differing strengths. A scientist would like to know which combination of temperature and pressure yields a plastic with a high breaking strength. A small preliminary experiment was run at two pressure levels and two temperature levels. The following model was proposed:

$$E(Y) = \beta_0 + \beta_1 x_1 + \beta_2 x_2 + \beta_3 x_1 x_2$$

where

Y = breaking strength (pounds)
x_1 = temperature (°F)
x_2 = pressure (pounds per square inch)

A sample of $n = 16$ observations yielded

$$\hat{y} = 226.8 + 4.9 x_1 + 1.2 x_2 - 0.7 x_1 x_2$$

with

$$s_{\hat{\beta}_1} = 1.11 \quad s_{\hat{\beta}_2} = 0.27 \quad s_{\hat{\beta}_3} = 0.34$$

Do the data in table below indicate that there is an interaction between temperature and pressure? Test by using $\alpha = 0.05$.

x_1	x_2	Y	x_1	x_2	Y
81	87	3.49	79	75	3.45
68	99	2.89	81	62	2.76
57	86	2.73	50	69	1.90
100	49	1.54	72	70	3.01
54	83	2.56	54	52	1.48
82	86	3.43	65	79	2.98
75	74	3.59	56	78	2.58
58	98	2.86	98	67	2.73
55	54	1.46	97	80	3.27
49	81	2.11	77	90	3.47
64	76	2.69	49	54	1.30
66	59	2.16	39	81	1.22
80	61	2.60	87	69	3.23
100	85	3.30	70	95	3.82
83	76	3.75	57	89	2.93
64	66	2.70	74	67	2.83
83	72	3.15	87	93	3.84
93	54	2.28	90	65	3.01
74	59	2.92	81	76	3.33
51	75	2.48	84	69	3.06

11.83 Air pollution regulations for power plants are often written so that the maximum amount of pollutant that can be emitted is increased as the plant's output increases. Suppose the data in the table below are collected over a period of time:

Output x (megawatts)	Sulfur Dioxide Emission Y (parts per million)
525	143
452	110
626	173
573	161
422	105
712	240
600	165
555	140
675	210

a Plot the data in a scatterplot.

b Use the least-squares method to fit a straight line relating sulfur dioxide emission to output. Plot the least-squares line on the scatterplot.

c Use a computer program to fit a second-order (quadratic) model relating sulfur dioxide emission to output.

d Conduct a test to determine whether the quadratic model provides a better description of the relationship between sulfur dioxide and output than the linear model.

e Use the quadratic model to estimate the mean sulfur dioxide emission when the power output is 500 megawatts. Use a 95% confidence interval.

11.84 The recent expansion of U.S. grain exports has intensified the importance of the linkage between the domestic grain transportation system and international transportation. As a first step in evaluating the economies of this interface, Martin and Clement (*Transportation Journal,* 1982) used multiple regression to estimate ocean transport rates for grain shipped from the Lower Columbia River international ports. These ports include Portland, Oregon; and Vancouver, Longview, and Kalama, Washington. Rates per long ton Y were modeled as a function of the following independent variables:

x_1 = shipment size in long tons
x_2 = distance to destination port in miles
x_3 = bunker fuel price in dollars per barrel
$x_4 = \begin{cases} 1 & \text{if American flagship} \\ 0 & \text{if foreign flagship} \end{cases}$
x_5 = size of the port as measured by the U.S. Defense Mapping Agency's Standards
x_6 = quantity of grain exported from the region during the year of interest

The method of least squares was used to fit the model to 140 observations from the period 1978 through 1980. The following results were obtained:

i	$\hat{\beta}_1$	*t*-statistic for testing $\beta_i = 0$
0	−18.469	−2.76
1	−0.367	−5.62
2	6.434	3.64
3	−0.269	−2.25
4	1.799	12.96
5	50.292	19.14
6	2.275	1.17
7	−0.018	−2.69

$$R^2 = 0.8979 \quad F = 130.665$$

a Using $\alpha = 0.01$, test
$H_0: \beta_1 = \beta_2 = \beta_3 = \beta_4 = \beta_5 = \beta_6 = \beta_7 = 0$.
Interpret the results of your test in the context of the problem.

b Binkley and Harrer (*American Journal of Agricultural Economics,* 1979) estimated a similar rate function by using multiple regression, but they used different independent variables. The coefficient of determination for their model was 0.46. Compare the explanatory power of the Binkley and Harrer model with that of Martin and Clement.

c According to the least-squares model, do freight charges increase with distance? Do they increase at an increasing rate? Explain.

11.85 An economist has proposed the following model to describe the relationship between the number of items produced per day (output) and the number of hours of labor expended per day (input) in a particular production process:

$$Y = \beta_0 + \beta_1 x + \beta_2 x^2 + \varepsilon$$

where

Y = number of items produced per day
x = number of hours of labor per day

A portion of the computer printout that results from fitting this model to a sample of 25 weeks of production data is shown below. Test the hypothesis that, as the amount of input increases, the amount of output also increases, but at a decreasing rate. Do the data provide sufficient evidence to indicate that the *rate* of increase in output per unit increase of input decreases as the input increases? Test by using $\alpha = 0.05$.

```
THE REGRESSION EQUATION IS
Y = -6.17 + 2.04 X1 - 0.0323 X2


                                  ST. DEV.  T-RATIO =
          COLUMN   COEFFICIENT  OF COEF. COEF./S.D.
            --        -6.173      1.666      -3.71
X1        C2           2.036      0.185      11.02
X2        C3         -0.03231    0.00489     -6.60


THE ST.DEV. OF Y ABOUT REGRESSION LINE IS
S = 1.243
WITH  (25-3) = 22   DEGREES OF FREEDOM
R-SQUARED = 95.5 PERCENT
R-SQUARED = 95.1 PERCENT, ADJUSTED FOR D.F.


ANALYSIS OF VARIANCE

     DUE TO      DF          SS     MS = SS/DF
REGRESSION       2      718.168      359.084
RESIDUAL        22       33.992        1.545
TOTAL           24      752.160
```

11.86 An operations engineer is interested in modeling $E(Y)$, the expected length of time per month (in hours) that a machine will be shut down for repairs, as a function of the type of machine (001 or 002) and the age of the machine (in years). He has proposed the following model:

$$E(Y) = \beta_0 + \beta_1 x_1 + \beta_2 x_1^2 + \beta_3 x_2$$

where
x_1 = age of machine

$$x_1 = \begin{cases} 1 & \text{if machine type 001} \\ 0 & \text{if machine type 002} \end{cases}$$

Data were obtained on $n = 20$ machine breakdowns and were used to estimate the parameters of the above model. A portion of the regression analysis computer printout is shown in figure below.

```
SOURCE    DF    SUM OF SQUARES    MEAN SQUARE
MODEL      3         2396.364        798.788
ERROR     16          128.586          8.037
TOTAL     19         2524.950       R-SQUARE
                                       0.949
```

The reduced model $E(Y) = \beta_0 + \beta_1 x_1 + \beta_2 x_2$ was fit to the same data. The regression analysis computer printout is partially reproduced in figure below.

```
SOURCE    DF    SUM OF SQUARES    MEAN SQUARE
MODEL      2          2342.42        1171.21
ERROR     17           182.53          10.74
TOTAL     19          2524.95       R-SQUARE
                                       0.928
```

Do these data provide sufficient evidence to conclude that the second-order (x_1^2) terms in the model proposed by the operations engineer is necessary? Test by using $\alpha = 0.05$.

11.87 Refer to Exercise 11.86. The data that were used to fit the operations engineer's complete and reduced models are displayed in table below.
a Use these data to test the null hypothesis that $\beta_1 = \beta_2 = 0$ in the complete model. Test by using $\alpha = 0.10$.
b Carefully interpret the results of the test in the context of the problem.

Downtime per Month (hours)	Machine Type	Machine Age (years)
10	001	1.0
20	001	2.0
30	001	2.7
40	001	4.1
9	001	1.2
25	001	2.5
19	001	1.9
41	001	5.0
22	001	2.1
12	001	1.1
10	002	2.0
20	002	4.0
30	002	5.0
44	002	8.0
9	002	2.4
25	002	5.1
20	002	3.5
42	002	7.0
20	002	4.0
13	002	2.1

11.88 Refer to Exercise 11.45, where we presented data on the number of highway deaths and the number of licensed vehicles on the road for the years 1950–1979. We mentioned that the number of deaths Y may also have been affected by the existence of the national 55-mile-per-hour speed limit during the years 1974–1979 (i.e., years 25–30). Define the dummy variable

$$x_2 = \begin{cases} 1 & \text{if 55-mile-per-hour speed limit} \\ & \text{was in effect} \\ 0 & \text{if not} \end{cases}$$

a Introduce the variable x_2 into the second-order model of Exercise 11.45 to account for the presence or absence of the 55-mile-per-hour speed limit in a given year. Include terms involving the interaction between x_2 and x_1.
b Refer to your model for part (a). Sketch on a single piece of graph paper your visualization of the two response curves, the second-order curves relating Y to x_1 before and after the imposition of the 55-mile-per-hour speed limit.
c Suppose that x_1 and x_2 do not interact. How would that fact affect the graphs of the two response curves of part (b)?
d Suppose that x_1 and x_2 interact. How would this fact affect the graphs of the two response curves?

11.89 In Exercise 11.45, we fit a second-order model to data relating the number Y of U.S. highway deaths per year to the number x_1 of licensed vehicles on the road. In Exercise 11.88, we added a qualitative variable x_2 to account for the presence or absence of the 55-mile-per-hour national speed limit. The accompanying SAS computer printout (Figure 11.86) gives the results of fitting the model

$$E(Y) = \beta_0 + \beta_1 x_1 + \beta_2 x_1^2 + \beta_3 x_2$$
$$+ \beta_4 x_1 x_2 + \beta_5 x_1^2 x_2$$

to the data. Use this printout (Figure 11.65) and the printout for Exercise 11.45 to determine whether the data provide sufficient evidence to indicate that the qualitative variable (speed limit) contributes information for the prediction of the annual number of highway deaths. Test by using $\alpha = 0.05$. Discuss the practical implications of your test results.

11.90 *Productivity* has been defined as the relationship between inputs and outputs of a productive system. In order to manage a system's productivity, it is necessary to measure it. Productivity is typically measured by dividing a measure of system output by a measure of the inputs to the system. Some examples of productivity measures are sales/salespeople, yards of carpet laid/number of carpet layers, shipments./[(direct labor) + (indirect labor) + (materials)]. Notice that productivity can be improved either by producing greater output with the same inputs or by producing the same out-put with fewer inputs. In manufacturing operations, productivity ratios like the third example above generally vary with the volume of output produced. The production data in Table 11.8 were collected for a random sample of months for the three regional plants of a particular manufacturing firm. Each plant manufactures the same product.

a Construct a scatterplot for these data. Plot the North-plant data using dots, the South-plant data using small circles, and the West-plant data using small triangles.

b Visually fit each plant's response curve to the scatterplot.

c Based on the results of part (b), propose a second-order regression model that could be used to estimate the relationship between productivity and volume for the three plants.

d Fit the model you proposed in part (c) to the data.

e Do the data provide sufficient evidence to conclude that the productivity response curves for the three plants differ? Test by using $\alpha = 0.05$.

f Do the data provide sufficient evidence to conclude that your second-order model contributed more information for the prediction of productivity than does a first-order model? Test by using $\alpha = 0.05$.

g Next month, 890 units are scheduled to be produced at the West plant. Use the model you developed in part (d) to predict next month's productivity ratio at the West plant.

11.91 The U.S. Federal Highway Administration publishes data on reaction distance for drivers and

Figure 11.65

SAS output for Exercise 11.89

```
DEPENDENT VARIABLE: DEATHS
```

SOURCE	DF	SUM OF SQUARES	MEAN SQUARE	F VALUE
MODEL	5	1428.02852498	285.60570500	41.56
ERROR	24	164.91847502	6.87160313	PR > F
CORRECTED TOTAL	29	1592.94700000		0.0001

R-SQUARE	C.V.	ROOT MSE	DEATHS MEAN
0.896470	5.7752	2.62137428	45.39000000

PARAMETER	ESTIMATE	T FOR H0: PARAMETER = 0	PR > T	STD ERROR OF ESTIMATE
INTERCEPT	16.94054815	2.21	0.0372	7.68054270
VEHICLES	0.34942808	1.89	0.0715	0.18528901
VEHICLES*VEHICLES	-0.00017135	-0.16	0.8724	0.00105538
LIMIT	41.38712279	0.11	0.9155	385.80676597
VEHICLES*LIMIT	-0.75118159	-0.14	0.8878	5.26806364
VEHICLE*VEHICLE* LIMIT	0.00245917	0.14	0.8921	0.01794470

Table 11.8

North Plant		South Plant		West Plant	
Productivity Ratio	No. of Units Produced	Productivity Ratio	No. of Units Produced	Productivity Ratio	No. of Units Produced
1.30	1,000	1.43	1,015	1.61	501
0.90	400	1.50	925	0.74	140
1.21	650	0.91	150	1.19	303
0.75	200	0.99	222	1.88	930
1.32	850	1.33	545	1.72	776
1.29	600	1.15	402	1.39	400
1.18	756	1.51	709	1.86	810
1.10	500	1.01	176	0.99	220
1.26	925	1.24	392	0.79	160
0.93	300	1.49	699	1.59	626
0.81	258	1.37	800	1.82	640
1.12	590	1.39	660	0.91	190

braking distance for a typical automobile traveling on dry pavement. Some of the data appear in the table below. Total stopping distance is the sum of reaction distance and braking distance. Find simple linear regression models for estimating each average

a Reaction distance from speed

b Braking distance from speed

c Total stopping distance from speed

Use your answers to fill in a line of the table for an automobile traveling 55 mph.

Speed (mph)	Reaction Distance (feet)	Breaking Distance (feet)
20	22	20
30	33	40
40	44	72
50	55	118
60	66	182
70	77	266

11.92 In building a model to study automobile fuel consumption, Biggs and Akcelik (*Journal of Transportation Engineering*, 1987) began by looking at the relationship between idle fuel consumption and engine capacity. Suppose the data are as shown in table below.

Idle Fuel Consumption (mL/s)	Engine Size (L)
0.18	1.2
0.21	1.2
0.17	1.2
0.31	1.8
0.34	1.8
0.29	1.8
0.42	2.5
0.39	2.5
0.45	2.5
0.52	3.4
0.61	3.4
0.44	3.4
0.62	4.2
0.65	4.2
0.59	4.2

What model do you think best explains the relationship between idle fuel consumption and engine size?

11.93 One important feature of the plates used in high-performance thin-layer chromatography is the water uptake capability. Six plates were modified by increasing the amount of cyanosilane (CN). Water uptake measurements (Rf) were then recorded for three component mixtures

(BaP, BaA, Phe) used with each plate. Six measurements were made on each mixture/plate combination. The data are in shown in table below. Fit a model relating % CN to mean Rf for each compound. (The models could differ.) Do the values of s suggest any complication?

Plate No.	Type of Plate	Compound	Mean Rf	s
1	0% CN	BaP	0.17	0.021
		BaA	0.25	0.26
		Phe	0.36	0.033
2	1% CN	BaP	0.21	0.014
		BaA	0.30	0.023
		Phe	0.44	0.046
3	5% CN	BaP	0.23	0.021
		BaA	0.31	0.016
		Phe	0.41	0.025
4	10%CN	BaP	0.36	0.028
		BaA	0.45	0.026
		Phe	0.59	0.027
5	20% CN	BaP	0.46	0.45
		BaA	0.58	0.026
		Phe	0.73	0.027
6	30% CN	BaP	0.46	0.069
		BaA	0.56	0.70
		Phe	0.66	0.095

Source: A. Colsmjo and M. Ericsson, *Journal of High Resolution Chromatography*, vol. 10, April 1987, pp. 177–180.

11.94 Low temperatures on winter nights in Florida have caused extensive damage to the state's agriculture. Thus, it is important to be able to predict freezes quite accurately. In an attempt to use satellite data for this purpose, a study was done to compare the surface temperatures detected by the infrared sensors of HCMM and GOES-East satellites. GOES temperature data compared to HCMM sample means are given in the table below. Find an appropriate model relating the mean HCMM reading to the GOES temperature. What would happen to the R^2 value if the actual HCMM data points were used rather than their means?

GOES Temp. (K)	HCMM Temp. (K)			
	Mean	S.D.	Range Min.	Max.
275.5	274.5	1.01	272.8	281.2
275.5	275.2	1.25	273.2	281.6
277.5	277.6	2.72	273.6	284.2
276.5	277.3	3.06	273.2	283.9
277.5	277.8	3.53	273.6	283.9
281.0	278.9	0.80	276.9	280.8
277.5	277.9	3.56	273.6	283.5
278.0	279.2	3.49	273.2	283.9
280.0	280.6	3.01	277.3	287.5
285.0	290.9	3.28	281.6	294.2
278.0	276.9	2.07	274.5	283.1
284.5	284.4	2.78	277.3	287.2
284.5	291.5	2.75	282.0	294.2
283.5	285.5	1.87	277.3	287.2
280.0	279.8	2.36	278.9	287.5

Source: E. Chen and H. Allen, *Remote Sensing of the Environment*, vol. 21, 1987, pp. 341–353.

Analysis of Variance

Does the percent yield of BON acid change depending on the reaction temperature and reaction hour? Is there significant interaction between the reaction temperature and the reaction pressure?

To answer such questions, we need to study how to analyze outcomes of experiments with two or more factors and the interaction effects of different factors (see Figure 12.1). Statistical thinking and statistical components are essential components of process improvement. Many different ideas and tools covered so far are pulled together in designed experiments. Here we revisit the notions of data collection, estimation, hypothesis testing, and regression analysis for the purpose of comparing treatments and making decisions as to which treatment is optimal. Carefully planned and executed experiments are the key to increased knowledge of any process under study and to improved quality.

Figure 12.1

Means plot for yield of BON acid

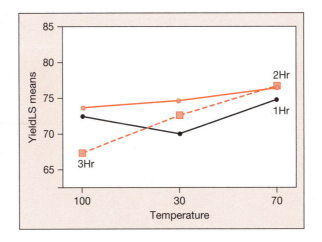

12.1 A Review of Designed Experiments

The goal of a statistical investigation is the improvement of a process. In Chapter 3, we studied basics of designing experiments. A *designed experiment* is a plan for data collection and analysis directed at making a decision about whether one treatment is better than another. We may employ estimation procedures or hypothesis testing procedures in the analysis, but the goal is always a decision about the relative effectiveness of treatments. A designed experiment should follow the scientific method, the model for problem solving, and the key elements outlined in Chapter 3.

Experimentation is scientific. An investigation should involve the following steps:

- Begin with a conjecture (question or hypothesis) formulated from prior knowledge.
- Collect new data relevant to the conjecture.
- Based on the collected data, decide whether the conjecture appears to be true.
- If the conjecture does appear to hold, refine it and check, using additional data.
- If the conjecture does not appear to hold, modify and check it again.

Thus, experimentation and the development of knowledge are part of an ongoing process that moves back and forth from conjecture to verification (data) in a continuous cycle (see Figure 12.2). Experimentation is problem solving. Hence, the procedure of stating the question or problem clearly, collecting and analyzing data, interpreting the data to make decisions, verifying the decisions, and planning the next action provides the pragmatic steps to follow in any experimental investigation.

Experimentation is organized and carried out according to a detailed plan.

- A question or conjecture is clearly stated.
- Treatment and control variables are defined.
- Lurking variables (if any) are identified.
- Randomization is employed to assign treatments to experimental units.
- Unbiased methods of measurement are selected for data collection.
- The data are collected and analyzed using appropriate statistical techniques.

Figure 12.2
A schematic diagram
designed experiments

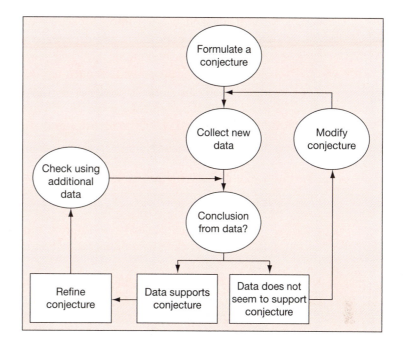

Example 12.1

Computer chips (semiconductors) are manufactured in a round wafer (about 3 inches in diameter) that holds many chips. The wafers are scored and broken apart to retrieve the chips for actual use in circuits. This process is rather crude and causes much waste. The question of investigation, then, is this: How can the breaking process be improved to gain a higher yield of usable chips?

Suppose an engineer suggests a new method (say, method B) to compete with the standard (say, method A). These two methods form the treatments to be compared in an experiment; after all, we cannot just take the word of the developer that the new method is better! There may be many background (lurking) variables, including the day or time the wafer was made, the temperature of the ovens, the chemical composition of the oxides, the work crew who made the wafers, and so on, affecting the outcome of the process. Accordingly, we might design different experiments, and thereby analyze the data collected differently. Let us look at three possible scenarios.

- **Completely randomized design:** Initially, the experimenter decides to control none of the lurking variables directly, but rather to attempt to control all of them indirectly by randomization. A number of wafers are selected at random to be broken by method A, and others are randomly selected to be broken by method B. Hopefully, the various characteristics of the wafers will be somewhat balanced between the two methods by this randomization. This simple design of experiment is called a *completely randomized design*. It works well for simulations in which direct controls are either not necessary or impossible to arrange.

- **Randomized block design:** On thinking more carefully about this scenario, the experimenter decides that the day of the week on which the wafer was made and broken into chips may be important because of differing environmental conditions and different workers. To account for this possibility, the experimental design is changed so that both method A and method B are used on wafers each day of a 5-day workweek. Specifically, the methods are used throughout the day on a sample of wafers from each day's production, and the order in which the two methods are used is randomized as well, with the restriction that the same number of observations must be taken on each method each day. The resulting design is called a *randomized block design*. The days form the blocks in which an equal number of measurements are taken on each treatment. The results allow possible differences to show up even if there is great variation in the outcome measures from day to day. In the randomized block design, the differences between days can also be measured, but this difference is not the main focus. The blocks are used as a device to gain more information about the treatment differences.

• **Factorial design:** The number of usable chips per wafer turns out to be much the same from day to day for either treatment, so blocking can be eliminated from future experiments of this type. But another consideration now comes into play. The process under discussion is actually a two-stage process involving scoring the wafer so that the chips are defined and then breaking apart the silicon wafer, hopefully along the score lines. The new method, method B, involves both deeper score lines and a different approach to breaking. Because there are now two types of score lines and two types of breaking, a slightly different arrangement of "treatments" is in order. If each score-line depth/breaking method combination is used, four treatments result (shallow score lines and breaking method A, deep score lines and breaking method A, shallow score line and breaking method B, and deep score line and breaking method B). Such a design is called a *factorial design* or a factorial arrangement of treatments. Once four treatments are defined, the measurements on those treatments could actually be taken by using a completely randomized design.

These three designs are the main focus of this chapter. We now proceed with the discussion of how to analyze data from such designed experiments by methods that come under the general topic of *analysis of variance*. These methods extend some of the techniques of Chapters 9 and 10 and make use of regression analysis (Chapters 2 and 11) as a computational tool.

12.2   Analysis of Variance (ANOVA) Technique

Recall the two sample *t*-test (pooled variance) studied in Chapter 10. It is designed to test the equality of two population means using two independent random samples, one from each population under study, for example, comparing the mean stamping times of two machines or comparing mean tensile strengths of class II and class III prestressing wires. However, often we are interested in comparing more than two population means, for instance, comparing the mean stamping times of five machines. Suppose four sheet metals are randomly selected from each machine in the study and the stamping times are recorded. Here five stamping machines are referred to as the *treatments* being studied. Each sheet metal used in the experiment is called an *experimental unit*. The stamping time measured for each test run is a *response*. This is a *designed experiment* and the collected data are typically recorded as in Table 12.1.

Here y_{ij} is the time recorded for the jth sheet from the ith machine ($i = 1, 2, 3, 4, 5$ and $j = 1, 2, 3, 4$). Now, using the sample means, $\bar{y}_1,\dots,\bar{y}_5$, we would like to decide whether different machines have significantly different mean stamping times. This is an example of a *completely randomized design, balanced case having one observation per experimental unit*. Suppose we used a two-sample t-test, studied in Chapter 10, on each possible pair of machines; there would be 10 such paired comparisons. Let's use $\alpha = 0.05$ for each of these tests. Then, using this procedure, what is the probability of finding *at least* one significant pairwise mean difference when in fact *none* exists? This probability, known as *experimentwise error rate*, is denoted by α_E. It is shown to be approximately equal to $1 - (1 - \alpha)^p$, where p denotes the number of separate tests to be made. In this experiment with five stamping machines, the experimentwise error rate is $\alpha_E \approx 1 - (1 - 0.05)^{10} \approx 40\%$.

As we saw here, all possible pairwise comparisons magnify the chances of deciding the five means are unequal when in fact they are all equal. Also, the error rate increases as the number of population means to be compared increases. Therefore, using a two-sample t-test to compare all possible pairs of means is not an appropriate technique when the experiment involves more than two treatments. For comparing more than two treatment means, analysis of variance, referred to as ANOVA, which protects against the excessive error rate, is recommended.

The analysis of variance technique, as the name suggests, involves splitting variation in responses into different parts, each attributable to a different source. For example, part of the variation in stamping times might be due to the use of different machines, and part due to a combination of some other factors. Let us look at another example.

Table 12.1

Data Layout for a Completely Randomized Experiment

Machines (Treatments)				
Machine 1	**Machine 2**	**Machine 3**	**Machine 4**	**Machine 5**
y_{11}	y_{21}	y_{31}	y_{41}	y_{51}
y_{12}	y_{22}	y_{32}	y_{42}	y_{52}
y_{13}	y_{23}	y_{33}	y_{43}	y_{53}
y_{14}	y_{24}	y_{34}	y_{44}	y_{54}
$\bar{y}_1$	$\bar{y}_2$	$\bar{y}_3$	$\bar{y}_4$	$\bar{y}_5$

In a technical report in 1976, the Federal Highway Administration reported results of a traffic engineering study. This study investigated traffic delays at intersections with signals on suburban streets. Three types of signals were utilized in the study: (1) pretimed, (2) semi-actuated, and (3) fully actuated. Pretimed signals operate on a fixed-time cycle regardless of the traffic pattern. An intersection equipped with any device to detect approaching traffic is known as actuated. A semi-actuated signal has devices to detect traffic approaching from minor streets, but the major street is pretimed. A fully actuated signal functions completely on the basis of the traffic pattern detected. Five intersections were randomly assigned for each type of signal. The measure of traffic delay used in the study was the average stopped time per vehicle at each of the intersections (seconds per vehicle). The data collected are given in Table 12.2.

Let's plot the data and compare the stop times visually. The line in the middle represents the overall sample mean (mean of all 15 measurements), and the short lines within each signal type are the sample means for each signal type. In general, we see from the dotplot (Figure 12.3) that the stop times at fully actuated signals are lower than those at the pretimed and semi-actuated signals. Overall, the stop times vary from signal to signal. But, also notice the following:

- Variation in the stop times for signals of the same type
- Variation in the mean stop times for the three signal types

The mean of all the observations without any consideration to the signal type is called the *overall sample mean*. Here overall sample mean is about 23.5 seconds. Variation among all stop times without taking into account the signal type is called *total variation*. Note that not all observations are the same. What makes them different?

Here we can see that signal type is one factor affecting the stop times. Even if there were no effect of signal type on the stop times, it would be unlikely that the three sample means would be identical. The variation in the mean stop times of the

Table 12.2

Stopping Times at Signals

Pretimed	Semi-actuated	Fully Actuated
36.6	17.5	15.0
39.2	20.6	10.4
30.4	18.7	18.9
37.1	25.7	10.5
34.1	22.0	15.2
35.48	20.9	14.0

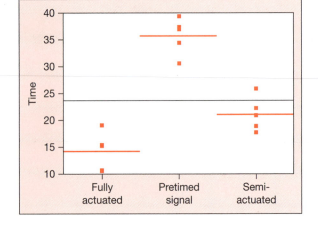

Figure 12.3

Dotplot of stopping times at the three types of signals

three samples due to the use of different signal types is called *between-treatment variation*. It is also known as *explained variation* because it can be explained using differential effects of the signal types.

Then why are stop times for the same signal type different? Although the same type of signal was installed in the similar type of intersection (signal types randomly assigned to intersections), some other effects may go uncontrolled, such as differences in the traffic patterns or weather conditions. The variation in stop times for different locations with the same signal type is called *within-treatment variation*. It is also known as *unexplained variation* because no single cause is obvious for this variation.

So, the total variation is composed of two parts, one that can be explained using treatments and the other that is unexplained. We can write this result as

$$\text{Total variation} = \text{Between-treatment variation} + \text{Within-treatment variation}$$

Both sources of variation reflect sampling variation. Additionally, the between-group variation reflects variation due to differential treatment effects. Now that we have identified sources of variation in the stop times, how can we decide how much of the variation is due to each source? In other words, how do we find a number that will describe the amount of variation? The total variation, between-treatment variation, and within-treatment variation are quantified using *total sum of squares*, *treatment sum of squares*, and *error (or residual) sum of squares*, respectively.

The populations of interest in the experimental design problems are generally referred to as treatments. The notation used for the k population (or k treatment) problem is summarized in Table 12.3.

The total sample size for the experiment is $n = n_1 + n_2 + \cdots + n_k$, the overall total of observations is $T_0 = \sum y_{ij} = T_1 + T_2 + \cdots + T_k$, and the overall sample mean is $\bar{y} = T_0/n$. If we have the same number of observations for all the

Table 12.3

Notation Used for the *k* Population Problem

	Populations (treatments)			
	1	**2**	**...**	**k**
Mean	μ_1	μ_2	...	μ_k
Variance	σ_1^2	σ_2^2	...	σ_k^2
	Independent Random Samples			
Sample Size	n_1	n_2	...	n_k
Sample Totals	T_1	T_2	...	T_k
Sample Means	$\bar{y}_1$	$\bar{y}_2$	...	$\bar{y}_3$
Sample Variances	s_1^2	s_2^2	...	s_k^2

treatments, then the design is said to be *balanced*. Balanced designs are easier to evaluate and more powerful in detecting differences.

The *total sum of squares* (TSS) measures the amount of variation among all the responses without any reference to their dependence on the factors involved, or group affiliation. It is computed by adding squared distances between each response and the overall mean.

$$TSS = \sum_{i=1}^{k} \sum_{j=1}^{n_i} (y_{ij} - \bar{y})^2$$

Total n measurements contribute to the total sum of squares, but their mean is the overall sample mean. Thus in reality, only $(n - 1)$ independent deviations from the overall mean contribute to the Total sum of squares. Thus, $(n - 1)$ degrees of freedom are associated with the total sum of squares.

The *between-group sum of squares*, also known as *sum of squares for treatment* (SST) measures the explained variation, that is, the amount of variation in responses that can be explained using their dependence on the factors involved. It is computed as the weighted total of the squared distances between the mean response for each group and the grand mean.

$$SST = \sum_{i=1}^{k} n_i (\bar{y}_i - \bar{y})^2$$

Similarly, with k treatments, there are n deviations of sample means from the overall mean contributing to the treatment sum of squares. However, the mean of sample means is equal to the overall mean. Therefore, there are only $(k - 1)$ independent contributing deviations. Thus, $(k - 1)$ degrees of freedom are associated with the treatment sum of squares.

The *error sum of squares* or *within-group sum of squares* (SSE) measures the unexplained variation in responses. It is defined as the variation among responses from experimental units in the same group (or receiving the same treatment). It is computed as the total of the squared distances between the responses and the corresponding group means.

$$SSE = \sum_{i=1}^{k} \sum_{j=1}^{n_i} (y_{ij} - \bar{y}_i)^2$$

Within each group, there are n_i measurements, that is, n_i contributing deviations from the group mean to the error sum of squares (see Figure 12.4). However, within each group, the mean of all observations equal the group mean. Hence, only $n_i - 1$ independent deviations from each group are contributing to the error sum of squares. Therefore, $\sum (n_i - 1) = (n - k)$ degrees of freedom are associated with the error sum of squares. Note that this is an extension of the numerator for pooled s^2.

Figure 12.4

Random samples from *k* populations

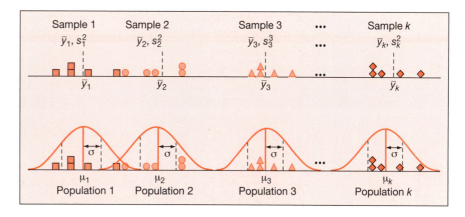

Because the total variation is composed of two sources of variation, we get

$$
\underbrace{\sum_{i=1}^{k}\sum_{j=1}^{n_i}(y_{ij}-\bar{y})^2}_{\text{Total Sum of Squares}} \;=\; \underbrace{\sum_{i=1}^{k}n_i(\bar{y}_i-\bar{y})^2}_{\text{Sum of Squares for Treatment}} \;+\; \underbrace{\sum_{i=1}^{k}\sum_{j=1}^{n_i}(y_{ij}-\bar{y}_i)^2}_{\text{Sum of Squares for Error}}
$$

and

$$
\underbrace{(n-1)}_{\substack{\text{Degrees of Freedom for Total}\\\text{Sum of Squares}}} \;=\; \underbrace{(k-1)}_{\substack{\text{Degrees of Freedom for Sum}\\\text{of Squares for Treatment}}} \;+\; \underbrace{(n-k)}_{\substack{\text{Degrees of Freedom for Sum}\\\text{of Squares for Error}}}
$$

However, the number of groups and the number of observations per group are different. Therefore, we cannot compare the treatment sum of squares and error sum of squares directly. Instead, we need to determine the mean variation due to each source of variation and compare.

- The *between-group mean squares* also known as *mean squares for treatments* (MST) measures the mean variation among the group means or the mean variation in responses due to factor. It is computed by dividing the between-group sum of squares by the degrees of freedom.
- The *within-group (error) mean squares* (MSE) measures the mean chance variation. It is computed by dividing the error sum of squares by the degrees of freedom. The MSE is also an unbiased estimator of common population variance σ^2, regardless of whether the null hypothesis is true or false.

Here the research problem of interest is, Are the mean stop times at the intersections the same regardless of the type of signal, or do they differ by the type of signal?

To answer this question, let us look at the dotplot of the data (Figure 12.3). The stopping times seem to be different for the three signal types, but are the differences significant or they are just chance variations? The statistical question of interest is, Do the sample means differ just by chance, or are there really differences among the population means?

We can use sample means to answer this question. First, let us define μ_i = The mean stopping time at signal-type i (i = 1, 2, 3). Then we are interested in testing the hypotheses:

H_0: $\mu_1 = \mu_2 = \mu_3$, i.e., all three signal types have the same mean stop times.
H_a: At least two signal types have different mean stop times.

In general, the ANOVA technique is used to test the following hypotheses about k population means:

H_0: $\mu_1 = \mu_2 = \ldots = \mu_k$ i.e., all the populations have same mean.
H_a: At least two populations have different means.

The test statistic used for this test is the F-ratio. The F-ratio is a summary statistic computed as a ratio of the mean variation due to factor to the mean variation due to chance. It is computed as F-ratio = MST/MSE. We have seen in Chapter 8 that under certain conditions the sampling distribution of the ratio of variances is an F-distribution. If all the population (treatment) probability distributions are normal and the population variances are equal (i.e., $\sigma_1^2 = \sigma_2^2 = \ldots = \sigma_k^2$), then the F-ratio has an F-distribution. All this information is then put together in a table known as the ANOVA table as follows:

ANOVA Table

Sources of Variation	Degrees of freedom	Sum of Squares	Mean Squares	F-ratio	p-value
Treatment	Treatment df	SST	MST	F = MST/MSE	P(F > F-ratio)
Error	Error df	SSE	MSE		
Total	Total df	TSS			

How are these measures of variation used to make a decision about the equality of population means? Note that F-ratio is computed as

$$F\text{-ratio} = \frac{\text{MST}}{\text{MSE}} = \frac{\sum_{i=1}^{k} n_i(\bar{y}_i - \bar{y})^2/(k-1)}{\sum_{i=1}^{k} \sum_{j=1}^{n_i} (y_{ij} - \bar{y})^2/(n-k)}$$

By inspecting this ratio, we can see that large deviations among the sample means will cause the numerator to be large and thus may provide evidence for the rejection of the null hypothesis. The denominator is an average within-sample variance and is not affected by differences among the sample means. When the population means for all treatments are the same (i.e., $\mu_1 = \mu_2 = \cdots = \mu_k$), treatment mean squares will also estimate the average chance variation, and the F-ratio will be close to 1. If the F-ratio is considerably larger than 1, then the average variation among treatment means is a lot more than the average chance variation.

Note that rejection of the null hypothesis does not imply all the treatment means are different, nor does it tell which treatment means are different. It simply tells us that at least two treatment means are significantly different at the predetermined significance level.

This technique of ANOVA can be extended to the analysis of data collected from experiments involving more than one factor, such as a randomized block design or a factorial experiment. For example, in the experiment of the three signal types, the total variation in stopping times is split into two parts, variation due to differences in signal types, and the chance variation. Now suppose it is suspected that the traffic pattern varies by the time of the day and might have some effect on stop times. Therefore, five different time frames are selected and in each time frame data are collected from all three signal types, as seen in Table 12.4:

This design is known as a *randomized block design* with time of day as blocks. Now the total variation in stop times can be split into three different sources: (1) variation due to type of signal (treatment), (2) variation due to time of the day (block), and (3) chance variation (error). If we define y_{ij} as the mean stop time per vehicle during the jth time frame at the ith signal type, then the composition of total variation can be given as follows:

$$\underbrace{\sum_{i=1}^{3}\sum_{j=1}^{5}(y_{ij}-\bar{y})^2}_{\substack{\text{Total Sum of}\\ \text{Squares}}} = \underbrace{\sum_{i=1}^{3}5(\bar{y}_{i.}-\bar{y})^2}_{\substack{\text{Sum of Squares for}\\ \text{Signal Type}}} + \underbrace{\sum_{j=1}^{5}3(\bar{y}_{.j}-\bar{y})^2}_{\substack{\text{Sum of Squares for}\\ \text{Time Frame}}} + \underbrace{\sum_{i=1}^{3}\sum_{j=1}^{5}(y_{ij}-\bar{y}_{.i}-\bar{y}_{.j}+\bar{y})^2}_{\substack{\text{Sum of Sqares}\\ \text{for Error}}}$$

and

$$\underbrace{(15-1)}_{\substack{\text{Degrees of Freedom for}\\ \text{Total Sum of Squares}}} = \underbrace{(3-1)}_{\substack{\text{Degrees of Freedom}\\ \text{for Signal Type}}} + \underbrace{(5-1)}_{\substack{\text{Degrees of Freedom}\\ \text{for Time Frame}}} + \underbrace{(15-3-5+1)}_{\substack{\text{Degrees of Freedom}\\ \text{for Error}}}$$

In such a case, we might be interested in testing two different hypotheses, one comparing the mean stop times for different signal types, and the other comparing the mean stop times for different time frames.

In summary, analysis of variance is a technique that can be used to analyze data collected from different kinds of experiments involving different number of factors. In Chapter 11, we saw that the ANOVA technique can also be used to check the utility of the model in regression analysis.

Table 12.4

Signal Data for a Block Design

Time of Day	Pretimed	Semi-actuated	Fully Actuated
7 AM–9 AM	36.6	17.5	15.0
10 AM–12 PM	39.2	20.6	10.4
1 PM–3 PM	30.4	18.7	18.9
4 PM–6 PM	37.1	25.7	10.5
7 PM–9 PM	34.1	22.0	15.2

Exercises

12.1 Complete the following ANOVA table for a completely randomized experiment:

Sources of Variation	Degrees of Freedom	Sum of Squares	Mean Squares	F-ratio
Treatment	_____	1,580.35	_____	_____
Error	20	_____	_____	
Total	23	1,728.35		

  a How many treatments are in this experiment?
  b What is the total number of observations taken?
  c If the same number of observations were taken for each treatment, determine the number of observations taken per treatment.

12.2 It has been hypothesized that treatment (after casting) of a plastic used in optic lenses will improve wear. Four different treatments are to be tested. To determine whether any differences in mean wear exist among treatments, 28 castings from a single formulation of the plastic were made, and 7 castings were randomly assigned to each treatment. Wear was determined by measuring the increase in "haze" after 200 cycles of abrasion (with better wear indicated by small increases). Determine the degrees of freedom for each term in the accompanying ANOVA table.

Sources of Variation	Degrees of Freedom
Treatment	_____
Error	_____
Total	_____

12.3 A partially completed ANOVA table for a completely randomized design is shown here:

Sources of Variation	Degrees of Freedom	Sum of Squares	Mean Squares	F-ratio
Treatment	4	24.7	_____	_____
Error	_____	_____	_____	
Total	34	62.4		

  a Complete the ANOVA table.
  b How many treatments are involved in the experiment?

12.4 The concentration of a catalyst used in producing grouted sand is thought to affect its strength. An experiment designed to investigate the effects of three different concentrations of the catalyst utilized five specimens of grout per concentration. The strength of a grouted sand was determined by placing the test specimen in a press and applying pressure until the specimen broke. Determine the degrees of freedom for each term in the accompanying ANOVA table.

Sources of Variation	Degrees of Freedom
Treatment	_____
Error	_____
Total	_____

12.3 Analysis of Variance for the Completely Randomized Design

Earlier, we discussed an example of three signal types in which the intersections were randomly selected to receive one of three types of signals in the study. Example 10.21 reported results from a random sample of nine measurements on stamping times for a standard machine and an independently selected random sample of nine measurements from a new machine. These are examples of a completely randomized design.

A **completely randomized design** (CRD) is a plan for collecting data in which an independent random sample is selected from each population of interest or each treatment is randomly assigned to experimental units.

Table 12.5

Data Layout for a Completely Randomized Experiment

	Treatments			
	1	**2**		**k**
	(μ_1, σ_1^2)	(μ_2, σ_2^2)	...	(μ_k, σ_k^2)
	y_{11}	y_{21}	...	y_{51}
	y_{12}	y_{22}	...	y_{52}
	$\vdots$	$\vdots$	$\vdots$	$\vdots$
	y_{1n_1}	y_{2n_2}	...	y_{5n_k}
Sample Size	n_1	n_2	...	n_k	n
Sample Totals	T_1	T_2	...	T_k	T_0
Sample Means	$\bar{y}_1$	$\bar{y}_2$	...	$\bar{y}_3$	$\bar{y}$
Sample Variances	s_1^2	s_2^2	...	s_k^2	

In a completely randomized design, the experimental units are assigned to treatments at random, or samples are taken independently and randomly from each treatment (population). The general layout of the data collected from a completely randomized design with k treatments and the basic statistic computed from it are shown in Table 12.5:

As discussed in Section 12.2, total variation is split into different parts due to different sources. In a CRD, the total variation is split into two parts, namely, variation between treatments and residual or error. Compute the sum of squares and the ANOVA table as follows:

$$\text{TSS} = \sum_{i=1}^{k} \sum_{j=1}^{n_i} (y_{ij} - \bar{y})^2 \quad \text{SST} = \sum_{i=1}^{k} n_i (\bar{y}_i - \bar{y})^2 \quad \text{SSE} = \sum_{i=1}^{k} \sum_{j=1}^{n_i} (y_{ij} - \bar{y}_{i.})^2$$

ANOVA Table

Sources of Variation	Degrees of Freedom	Sum of Squares	Mean Squares	F-ratio	p-value
Treatment	$(k-1)$	SST	MST = SST/$(k-1)$	$F = \dfrac{\text{MST}}{\text{MSE}}$	P(F > F-ratio)
Error	$(n-k)$	SSE	MSE = SSE/$(n-k)$		
Total	$(n-1)$	TSS			

The hypothesis test for comparing k treatment means of a CRD is as follows:

> ### Test to Compare k Treatment Means for a Completely Randomized Design
>
> $H_0: \mu_1 = \mu_2 = \cdots = \mu_k$ that is, all the populations have the same mean.
>
> H_a: At least two populations have different means.
>
> Test Statistic: $F = \dfrac{\text{MST}}{\text{MSE}}$
>
> Rejection Region: $F > F_\alpha(k - 1, n - k)$
>
> #### Conditions:
> 1. Each sample is taken randomly and independently from the respective population, or treatments are assigned at random to experimental units.
> 2. Each population has a normal probability distribution approximately.
> 3. All the population variances are equal (i.e., $\sigma_1^2 = \sigma_2^2 = \cdots = \sigma_k^2 = \sigma^2$), or nearly so.

Checking Conditions

- The conditions of random and independent samples must be satisfied for valid conclusions from the test. However, the conditions of normality of populations and equal variances are fairly robust.
- The F-test works reasonably well if the populations are only approximately normal (symmetric, mound-shaped without outliers), with similar-sized spreads. The assumption of normality can be checked using Q–Q plots if a sufficient number of observations are available per treatment.
- The assumption of homogeneity of variances, more serious than that of normality, can be checked by using a parallel dotplot and assessing the spread, or by using one of the tests of homogeneity of variances (such as Levene's test or Bartlett's test). Most statistical programs provide results for such tests easily. Another easier way to roughly check the condition of equality of variances is to compute a ratio of the largest sample standard to the smallest sample standard deviation. If this ratio is at the most 3, then it is reasonable to conclude that there is no evidence of heterogeneity of variances.

The MSE estimates the common variance of all the populations; in other words, $\hat{\sigma}^2 = \text{MSE}$, regardless of whether the null hypothesis is true or not.

Example 12.2

Refer to the example of the three signal types. The Federal Highway Administration collected data on mean stop times from three types of signals on suburban streets: (1) pretimed, (2) semi-actuated, and (3) fully actuated. Five intersections were randomly selected for installing each type of signal. The data collected (also reported in Table 12.2) and the summary statistics are shown on next page.

Pretimed	Semi-actuated	Fully Actuated
36.6	17.5	15.0
39.2	20.6	10.4
30.4	18.7	18.9
37.1	25.7	10.5
	22.0	15.2
$n_1 = 5$	$n_2 = 5$	$n_3 = 5$
$T_1 = 177.4$	$T_2 = 104.5$	$T_3 = 70.0$
$\bar{y}_{1.} = 23.48$	$\bar{y}_{2.} = 20.9$	$\bar{y}_{3.} = 14.0$
$S_1 = 3.37$	$S_2 = 3.19$	$S_3 = 3.59$

There are a total of 15 observations for the three signal types, and $\bar{y} = (177.4 + 104.5 + 70.0)/3 = 23.46$.

$$\text{TSS} = \sum(y_{ij} - \bar{y})^2 = (36.6 - 23.46)^2 + \cdots + (15.2 - 23.46)^2 = 1{,}340.46$$

with $(15 - 1) = 14$ degrees of freedom

$$\text{SST} = \sum n_i(\bar{y}_{i.} - \bar{y})^2 = 5(23.48 - 23.46)^2 + 5(20.9 - 23.46)^2$$
$$+ 5(14.0 - 23.46)^2 = 1{,}202.63$$

with $(3 - 1) = 2$ degrees of freedom

$$\text{SSE} = \sum(y_{ij} - \bar{y}_{i.})^2 = (36.6 - 23.48)^2 + \cdots + (37.1 - 23.48)^2$$
$$+ \cdots + (15.2 - 14.0)^2 = 137.83$$

with $(14 - 2) = 12$ degrees of freedom.

However, it is easier to compute SSE as the difference between TSS and SST (SSE = TSS – SST). Therefore,

and
$$\text{MST} = \frac{\text{SST}}{\text{df}} = \frac{1{,}202.63}{2} = 601.314,$$

$$\text{MSE} = \frac{\text{SSE}}{\text{df}} = \frac{137.83}{12} = 11.486$$

It gives us the test statistic value F-ratio $= 601.314/11.486 = 52.35$. Now we can put all this information into an ANOVA table:

Sources of Variation	Degrees of Freedom	Sum of Squares	Mean Squares	*F*-ratio
Signals	2	1,202.63	601.314	52.35
Error	12	137.83	11.486	
Total	14	1,340.46		

We are interested in testing the hypothesis:

H_0: $\mu_1 = \mu_2 = \mu_3$, i.e., the mean stopping times at the three types of signals are the same.

H_1: The mean stopping times for at least two types of signals are different.

where

μ_1 = The true mean stopping time at the intersection with a pretimed signal
μ_2 = The true mean stopping time at the intersection with a semi-actuated signal
μ_3 = The true mean stopping time at the intersection with a fully actuated signal

The dotplot of Figure 12.2 shows a fairly similar spread of data for all three signals, indicating that the assumption of equal variances is reasonable. The common variance is estimated to be 11.486 (see MSE from the ANOVA table). Because the data collected are mean stop time per vehicle computed from all vehicles stopped at the intersection over a 2-hour time period, using the central limit theorem (see Chapter 8), it is reasonable to assume that the data are from approximately normal populations.

At a 5% significance level, $F_{0.05}(2,12) = 3.885 < F\text{-ratio} = 52.35$ (see Figure 12.5). Therefore, we reject the null hypothesis of equal means and conclude that at least two types of signals have different mean stop times.

Figure 12.5

Rejection region for *F*-test

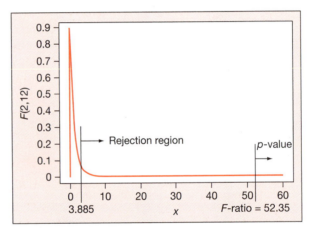

Instead of using the *F*-ratio, which also needs an *F*-table, we can use the *p*-value approach to make a decision. Most statistical software and calculators (such as the TI-89) easily provide the *F*-statistic with the corresponding *p*-value.

Example 12.3

Over the past two decades, computers and computer trainings have become an integral part of curricula of schools and universities. As a result, younger business professionals tend to be more comfortable with computers than their more senior counterparts. Gary Dickson (1984), Professor of Management Information Systems at the University of Minnesota, investigated the computer literacy of middle managers with 10 years or more of management experience. Nineteen managers from the Twin City area were randomly sampled and asked to complete a questionnaire. A higher score on the questionnaire indicates more computer knowledge. Prior to completing the questionnaire, the managers were asked to describe their knowledge and experience with computers. Using this information, the managers were classified as possessing a low (A), medium (B), or high (C) level of technical computer expertise. The data collected are given in Table 12.7. Is there sufficient evidence to conclude that the mean score differs for the three groups of managers?

Level of Expertise	Score on Test	Total	n	Mean	Std Dev
A	82, 114, 90, 80, 88, 93, 80, 105	732	8	91.500	12.3056
B	128, 90, 130, 110, 133, 130, 104	825	7	117.857	16.6175
C	156, 128, 151, 140	575	4	143.750	12.4466

Solution The dotplot in JMP output below shows the variation in sample mean scores for managers in the three different groups. The statistical question of interest here is, Is variation among the sample means scores attributable to chance variation or are the population mean scores for the three groups different? So, we are interested in testing $H_0: \mu_A = \mu_B = \mu_C$ against H_a: At least two group means are different, where μ_i = the population mean score for group $i (i = A, B, C)$. The JMP printout for ANOVA is given below.

The dotplot shows that the three samples each display about the same amount of variability. Also, the ratio of the largest to the smallest standard deviation is only 1.35, so there is no reason to conclude that the condition of homogeneity of population variances is violated. However, the sample sizes are too small to show normality (or lack of it) in a definitive way. The p-value for the F-test is < 0.0001, which means at any reasonable significance level we would reject the null hypothesis of equal population mean scores for three groups of managers and conclude that at least two groups differ in their mean scores.

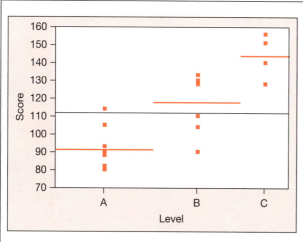

One-way Anova

Summary of Fit

Rsquare	0.70582
Adj Rsquare	0.669047
Root Mean Square Error	14.10143
Mean of Response	112.2105
Observations (or Sum Wgts)	19

Analysis of Variance

Source	DF	Sum of Squares	Mean Square	F Ratio	Prob > F
Level	2	7633.551	3816.78	19.1942	<.0001
Error	16	3181.607	198.85		
C. Total	18	10815.158			

A graphical approach is an easy way of checking the condition of equal variances. Alternatively, we could conduct a test for equality of more than two variances. There are several tests described in the literature and available from different statistical software, but most of them require larger sample sizes than in the above two examples.

If the assumption of equal variances required by the analysis of variance procedure cannot be met, sometimes data can be transformed to stabilize variance. Variance-stabilizing transformations are described in Chapter 11 in the context of regression situations.

Example 12.4

Porosity of metal, an important determinant of strength and other properties, can be measured by looking at cross sections of the metal under a microscope. The *pore* (void) is dark, the metal is light, and the boundary between these two phases is often clearly delineated. If the grid is laid on the microscopic field, the number of intersections between the grid lines and the pore boundaries is proportional to the length of that boundary per unit area (and hence is proportional to the pore surface area per unit volume of metal). Such porosity counts were made for samples of antimony, Linde copper, and electrolytic copper (see table below) from the experiments conducted by the Materials Science Department at the University of Florida.

Do the mean counts show significant differences?

Antimony					Linde Copper					Electrolytic Copper				
10	9	9	9	10	14	12	15	14	10	42	46	44	39	50
11	14	11	8	11	17	16	11	13	14	34	42	40	36	37
7	6	9	7	8	15	11	16	12	6	46	42	43	50	32
10	12	14	9	8	13	20	17	10	16	41	37	49	28	34
8	9	8	7	10	10	13	12	17	13	34	38	43	39	42
$\bar{y}_{AN} = 9.36$					$\bar{y}_{LC} = 13.48$					$\bar{y}_{EC} = 40.32$				
$s^2_{AN} = 4.07$					$s^2_{LC} = 9.01$					$s^2_{EC} = 31.58$				

Solution   The boxplots in Figure 12.6a show the original data. Obviously, the variation is not similar across the three samples, so ANOVA will not be an appropriate technique to compare population means. In fact, the variances are nearly proportional to the means (see Table 12.6), which indicates that a square root transformation will help stabilize variance (refer to Chapter 11 for variance-stabilizing transformations). The boxplots in Figure 12.6b show the data transformed using the square root transformation. Observe that the square root transformation does make the variability about the same for all three samples (confirmed by Table 12.6).

From looking at the data, it is obvious that the electrolytic copper has a much higher count. What is not obvious is that the mean count for Linde copper is also significantly higher than the mean count for antimony.

Figure 12.6

Boxplots for data and transformed data

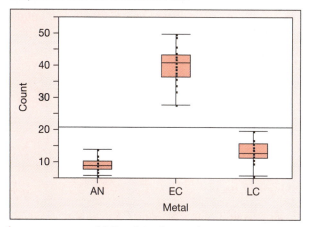

(a) Boxplots of count data

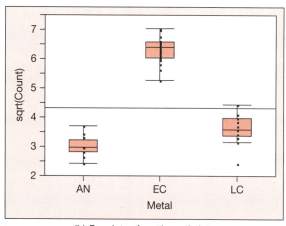

(b) Boxplots of sqrt(count) data

Table 12.6

Mean and Standard Deviation of Counts and Transformed Counts

	Count			**√Count**		
	AN	**LC**	**EC**	**AN**	**LC**	**EC**
Mean	9.36	13.48	40.32	3.043	3.648	6.335
Std Dev	2.018	3.002	5.618	0.323	0.425	0.447
n	25	25	25	25	25	25

Figure 12.7

JMP output for ANOVA on transformed scores

One-way Analysis of sqrt(Count) by Metal

One-way Anova

```
Summary of Fit
RSquare                          0.929572
Adj Rsquare                      0.927616
Root Mean Square Error           0.401898
Mean of Response                 4.341846
Observations (or Sum Wgts)             75
```

Analysis of Variance

Source	DF	Sum of Squares	Mean Square	F Ratio	Prob > F
metal	2	153.49828	76.7491	475.1627	<.0001
Error	72	11.62957	0.1615		
C.Total	74	165.12785			

Now let us use ANOVA on the $\sqrt{\text{Counts}}$ to test H_0: $\sqrt{\mu_{\text{AN}}} = \sqrt{\mu_{\text{LC}}} = \sqrt{\mu_{\text{EC}}}$. The JMP printout in Figure 12.7 shows that the p-value < 0.0001. At such a small p-value for the F-test, we reject the null hypothesis of equality of mean square root count for three metals, and conclude that at least two mean square root counts are different.

It means that the mean counts for at least two metals are different too; in other words, the hypothesis H_0: $\mu_{AN} = \mu_{LC} = \mu_{EC}$ is rejected too. So the mean counts differ for the tree metals; however, the analysis must be done on the transformed data because of the lack of homogeneity of variances.

Activity

Goal: To compare mean flight times of helicopters with wing lengths of 4.5, 5.5, and 6.5 inches.

The *response* to be measured is flight time of the helicopter. It is defined as the time for which the helicopter remains in flight.

1. Prepare three helicopters with wing lengths of 4.5, 5.5, and 6.5 inches, respectively using template from Chapter 3.
2. Ten runs for each wing length are planned.
3. Using the random number table, randomize the order of runs of the three wing lengths. Record the order in the column labeled "Wing Length" in Table 12.7.
4. Release the helicopter from above your head while standing alone on a step stool or a ladder. Start the timer at the drop of the helicopter. Stop the timer when the helicopter hits the floor.
5. Measure the time in flight (in sec) for the helicopters in the predetermined order. Record the flight times.
6. Reorganize the data in Table 12.8.
7. Analyze the data from this completely randomized experiment, using the ANOVA technique.

Table 12.7

Data Recording Sheet with Random Order

Flight Number	Wing Length	Flight Time	Flight Number	Wing Length	Flight Time	Flight Number	Wing Length	Flight Time
1	_____	_____	11	_____	_____	21	_____	_____
2	_____	_____	12	_____	_____	22	_____	_____
3	_____	_____	13	_____	_____	23	_____	_____
4	_____	_____	14	_____	_____	24	_____	_____
5	_____	_____	15	_____	_____	25	_____	_____
6	_____	_____	16	_____	_____	26	_____	_____
7	_____	_____	17	_____	_____	27	_____	_____
8	_____	_____	18	_____	_____	28	_____	_____
9	_____	_____	19	_____	_____	29	_____	_____
10	_____	_____	20	_____	_____	30	_____	_____

Table 12.8

Reorganized Data Recording Sheet

Wing Length 4.5 inches		Wing Length 5.5 inches		Wing Length 6.5 inches	
Run	**Time**	**Run**	**Time**	**Run**	**Time**
1	_____	1	_____	1	_____
2	_____	2	_____	2	_____
3	_____	3	_____	3	_____
4	_____	4	_____	4	_____
5	_____	5	_____	5	_____
6	_____	6	_____	6	_____
7	_____	7	_____	7	_____
8	_____	8	_____	8	_____
9	_____	9	_____	9	_____
10	_____	10	_____	10	_____

Exercises

12.5 Independent random samples were selected from three normally distributed populations with common (but unknown) variance σ^2. The data are shown in table below:

Sample 1	Sample 2	Sample 3
3.1	5.4	1.1
4.3	3.6	0.2
1.2	4.0	3.0
	2.9	

a Compute the appropriate sums of squares and mean squares, and fill in the appropriate entries in the analysis of variance table shown below:

Source	df	SS	MS	F - Ratio
Treatments	____	____	____	____
Error	____	____	____	
Total	____	____		

b Test the hypothesis that the population means are equal (that is, $\mu_1 = \mu_2 = \mu_3$) against the alternative hypothesis that at least one mean is different from the other two. Test by using $\alpha = 0.05$.

12.6 A partially completed ANOVA table for a completely randomized design is shown below:
  a Complete the ANOVA table.
  b How many treatments are involved in the experiment?

c Do the data provide sufficient evidence to indicate a difference among the population means? Test by using $\alpha = 0.01$.

Source	df	SS	MS	F - Ratio
Treatments	4	24.7	____	____
Error	____		____	
Total	34	62.4		

12.7 Some varieties of nematodes (round worms that live in the soil and frequently are so small they are invisible to the naked eye) feed upon the roots of lawn grasses and other plants. This pest, which is particularly troublesome in warm climates, can be treated by the application of nematicides. Data collected on the percentage of kill for nematodes for four particular rates of application (dosages given in pounds of active ingredient per acre) are listed in table below:

Rate of Application			
2	**3**	**5**	**7**
86	87	94	90
82	93	99	85
76	89	97	86
			91

Do the data provide sufficient evidence to indicate a difference in the mean percentage of kill for the four different rates of application of nematicide? Use $\alpha = 0.05$.

12.8 It has been hypothesized that treatment (after casting) of a plastic used in optic lenses will improve wear. Four different treatments are to be tested. To determine whether any differences in mean wear exist among treatments, 28 castings from a single formulation of the plastic were made, and seven castings were randomly assigned to each treatment. Wear was determined by measuring the increase in "haze" after 200 cycles of abrasion (with better wear indicated by small increases). The data are listed in table below:

Treatment			
A	B	C	D
9.16	11.95	11.47	11.35
13.29	15.15	9.54	8.73
12.07	14.75	11.26	10.00
11.97	14.79	13.66	9.75
13.31	15.48	11.11	11.71
12.32	13.47	15.05	12.45
11.78	13.06	14.86	12.38

a Is there evidence of a difference in mean wear among the four treatments? Use $\alpha = 0.05$.
b Estimate the mean difference in haze increase between treatments B and C, using a 99% confidence interval. [*Hint*: Use the two-sample t-statistic.]
c Find a 90% confidence interval for the mean wear for lenses receiving treatment A. [*Hint*: Use the one-sample t-statistic.]

12.9 The application of *management by objectives* (MBO), a method of performance appraisal, was the object of a study by Shetty and Carlisle of Utah State University ("Organizational Correlates of a Management by Objectives Program," *Academy of Management Journal*, 1974). The study dealt with the reactions of a university faculty to an MBO program. One hundred nine faculty members were asked to comment on whether they thought the MBO program was successful in improving their performance within their respective departments and the university. Each response was assigned a score from 1 (significant improvement) to 5 (significant decrease). The following table shows the sample sizes, sample totals, mean scores, and sums of squares of deviations *within* each sample for samples of scores corresponding to the four academic ranks. Assume that the four samples can be viewed as independent random samples of scores selected from among the four academic ranks.
a Perform an analysis of variance for the data.
b Arrange the results in an analysis of variance table.

	Academic Rank			
	Instructor	Assistant Professor	Associate Professor	Professor
Sample Size	15	41	29	24
Sample Total	42.960	145.222	92.249	73.224
Sample Mean	2.864	3.542	3.181	3.051
Within-sample Sum of Squared Deviations	2.0859	14.0186	7.9247	5.6812

c Do the data provide sufficient evidence to conclude that there is a difference in mean scores among the four academic ranks? Test by using $\alpha = 0.05$.
d Find a 95% confidence interval for the difference in mean scores between instructors and professors.
e Do the data provide sufficient evidence to indicate a difference in mean scores between non-tenured faculty members (instructors and assistant professors) and tenured faculty members (Associate Professors and Professors)?

12.10 The concentration of a catalyst used in producing grouted sand is thought to affect its strength. An experiment designed to investigate the effects of three different concentrations of the catalyst utilized five specimens of grout per concentration. The strength of a grouted sand was determined by placing the test specimen in a press and applying pressure until the specimen broke. The pressures required to break the specimens, expressed in pounds per square inch, are shown in table below:

Concentration of Catalyst		
35%	40%	45%
5.9	6.8	9.9
8.1	7.9	9.0
5.6	8.4	8.6
6.3	9.3	7.9
7.7	8.2	8.7

Do the data provide sufficient evidence to indicate a difference in mean strength of the grouted sand among the three concentrations of catalyst? Test by using $\alpha = 0.05$.

12.11 Several companies are experimenting with the concept of paying production workers (generally paid by the hour) on a salary basis. It is believed that absenteeism and tardiness will increase under this plan, yet some companies feel that the working environment and overall productivity will improve. Fifty production workers under the salary plan are monitored at Company A, and fifty workers under the hourly plan are monitored at Company B. The number of work hours missed due to tardiness or absenteeism over a one-year period is recorded for each worker. The results are partially summarized in the following table:

Source	df	SS	MS	F-ratio
Company	——	3,237.2	——	——
Error	——	16,167.7	——	
Total	99			

a Fill in the missing information in the table.

b Is there evidence at the $\alpha = 0.05$ level of significance that the mean number of hours missed differs for employees of the two companies?

12.12 One of the selling points of golf balls is their durability. An independent testing laboratory is commissioned to compare the durability of three different brands of golf balls. Balls of each type will be put into a machine that hits the balls with the same force that a golfer uses on the course. The number of hits required until the outer covering cracks is recorded for each ball, with the results given the table below. Ten balls from each manufacturer are randomly selected for testing.

	Brand	
A	**B**	**C**
310	261	233
235	219	289
279	263	301
306	247	264
237	288	273
284	197	208
259	207	245
273	221	271
219	244	298
301	228	276

Is there evidence that the mean durabilities of the three brands differ? Use $\alpha = 0.05$.

12.13 Eight independent observations on percent copper content were taken on each of four castings of bronze. The sample means for each casting are given in the table below.

Casting	1	2	3	4
Means	80	81	86	90

Is there sufficient evidence to say that there are differences among the mean percentages of copper for the four castings? Use $\alpha = 0.01$. For these data, SSE = 700.

12.14 Three thermometers are used regularly in a certain laboratory. To check the relative accuracies of the thermometers, they are randomly and independently placed in a cell kept at zero degrees Celsius. Each thermometer is placed in the cell four times, with the results (readings in degrees Celsius) listed below:

	Thermometer	
1	**2**	**3**
0.10	− 0.20	0.90
0.90	0.80	0.20
− 0.80	− 0.30	0.30
− 0.20	0.60	− 0.30

Are there significant differences among the means for the three thermometers? Use $\alpha = 0.05$.

12.15 Casts of aluminum were subjected to four standard heat treatments and their tensile strengths were measured. Five measurements were taken on each treatment, with the results (in 1,000 psi), shown below.

	Treatment		
A	**B**	**C**	**D**
35	41	42	31
31	40	49	32
40	43	45	30
36	39	47	32
32	45	48	34

Perform an analysis of variance. Do the mean tensile strengths differ from treatment to treatment, at the 5% significance level?

12.16 How does flextime, which allows workers to set their individual work schedules, affect worker job satisfaction? Researchers recently conducted a study to compare a measure of job satisfaction for workers, using three types of work scheduling: flextime, staggered starting hours, and fixed hours. Workers in each group worked according to their specified work-scheduling system for four months. Although each worker filled out job satisfaction questionnaires both before and after the four-month test period, we will examine only the post–test-period scores. The sample sizes, means, and standard deviations of the scores for the three groups are shown in the table below:

a Assume that the data were collected according to a completely randomized design. Use the information in the table to calculate the treatment totals and SST.

b Use the values of the sample standard deviation to calculate the sum of squares of deviations *within* each of the three samples. Then calculate SSE, the sum of these quantities.

c Construct an analysis of variance table for the data.

d Do the data provide sufficient evidence to indicate differences in mean job satisfaction scores among the three groups? Test by using $\alpha = 0.05$.

e Find a 90% confidence interval for the difference in mean job satisfaction scores between workers on flextime and workers on fixed schedules.

f Do the data provide sufficient evidence to indicate a difference in mean scores between workers on flextime and workers using staggered starting hours? Test by using $\alpha = 0.05$.

	Group		
	Flextime	**Staggered**	**Fixed**
Sample Size	27	59	24
Mean	35.22	31.05	28.71
Standard Deviation	10.22	7.22	9.28

12.4   Relation of ANOVA for CRD with a *t*-Test and Regression

12.4.1   Equivalence between a *t*-test and an *F*-test for CRD with two treatments

The completely randomized design is an extension of the two-sample *t*-test, designed to test the hypothesis of equality of two population means, assuming the populations have the same variance. When comparing two population means using independent random samples, performing a two-sided *t*-test is equivalent to doing an *F*-test of ANOVA. For comparison of two treatment means, ANOVA gives the same results as the *t*-test. But, ANOVA is recommended for comparison of more than two treatment means because of the compounding errors if multiple *t*-tests are used.

Remember, in a two-sample *t*-test $H_0: \mu_1 = \mu_2$ versus $H_a: \mu_1 \neq \mu_2$, the test statistic is calculated as

$$t = \frac{(\bar{y}_1 - \bar{y}_2)}{s_p\sqrt{\dfrac{1}{n_1} + \dfrac{1}{n_2}}} \quad \text{where} \quad s_p^2 = \frac{(n_1 - 1)\, s_1^2 + (n_2 - 1)\, s_2^2}{(n_1 - 1) + (n_2 - 1)}$$

is the "pooled" sample variance used to estimate the common population variance. Remember from Chapter 6 that $t^2 = F$. As we can easily show,

$$t^2 = \frac{(\bar{y}_1 - \bar{y}_2)^2}{s_p^2\left(\dfrac{1}{n_1} + \dfrac{1}{n_2}\right)} = \frac{\displaystyle\sum_{i=1}^{2} n_i(\bar{y}_i - \bar{y})^2}{s_p^2} = \frac{\text{MST}}{\text{MSE}} = F\text{-ratio}$$

Although the test statistic t has a t-distribution with $(n_1 + n_2 - 2)$ degrees of freedom, the test statistic F-ratio has an F-distribution with $\nu_1 = 1$ and $\nu_2 = n_1 + n_2 - 2$ degrees of freedom. Thus, we can test the same hypothesis of equal means for two populations using a t-test or an F-test of ANOVA.

Example 12.5

Suppose a company receives a supply of class II prestressing wire from two different suppliers. A quality control inspector tests randomly selected samples to check their ultimate tensile strengths (UTS). The UTS data (in 1,000 psi) collected from laboratory testing of class II wires from both the suppliers are shown as follows:

Supplier 1:	253, 261, 258, 255, 256
Supplier 2:	264, 265, 261, 257, 256

Do the true mean UTS measurements for the wire from two suppliers appear to differ?

Solution Because we are interested in testing hypothesis $H_0: \mu_1 = \mu_2$ versus $H_a: \mu_1 \neq \mu_2$, where μ_i is the true mean UTS for the wire from supplier i $(i = 1, 2)$, we can use either a t-test or an F-test. The JMP printouts for both the tests are given below:

T Test

Assuming equal variances

	Difference	t Test	DF	Prob > |t|
Estimate	-4.0000	-1.768	8	0.1151
Std Error	2.2627			
Lower 95%	-9.2179			
Upper 95%	1.2179			

One-way Analysis of UTS By Supplier

One-way Anova

Analysis of Variance

Source	DF	Sum of Squares	Mean Square	F Ratio	Prob > F
Supplier	1	40.00000	40.0000	3.1250	0.1151
Error	8	102.40000	12.8000		
C. Total	9	142.40000			

Figure 12.8

Association between *t*-test and *F*-test

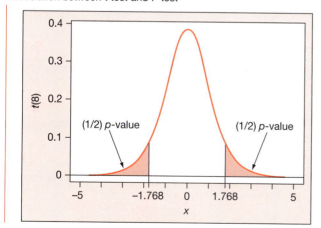

(a) *p*-value for *t*-test

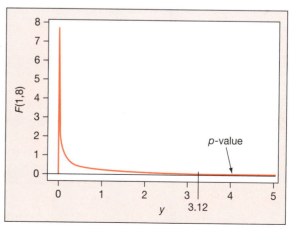

(b) *p*-value for *F*-test

As we can see from the printout, test statistic values for t- and F-tests respectively are $t = -1.768$, and F-ratio = 3.125. We can easily see that $t^2 = (-1.768)^2 = 3.125 = F$-ratio. Note that the p-value for both the tests is equal. Because p-value = 0.1151 is larger than $\alpha = 0.05$, we fail to reject the null (using either test) and conclude that there is no evidence of significant difference between the mean UTI for wires provided by the two suppliers. Figure 12.8a and 12.8b, respectively, show the p-values for the t- and F-tests.

In the ANOVA, the same idea is extended to more than two populations. For testing

$$H_0: \mu_1 = \mu_2 = \cdots = \mu_k \quad \text{versus} \quad H_a: \text{At least two treatment means differ}$$

the "pooled" variance from k independent random samples is computed as

$$s_p^2 = \frac{(n_1 - 1)\,s_1^2 + (n_2 - 1)\,s_2^2 + \cdots + (n_k - 1)\,s_k^2}{(n_1 - 1) + (n_2 - 1) + \cdots + (n_k - 1)} = \frac{\sum_{i=1}^{k} \sum_{j=1}^{n_i} (y_{ij} - \bar{y}_i)^2}{(n - k)} = \text{MSE}$$

Then the test statistic is constructed as

$$\frac{\sum_{i=1}^{k} n_i(\bar{y}_i - \bar{y})^2/(k - 1)}{s_p^2} = \frac{\text{MST}}{\text{MSE}} = F\text{-ratio}$$

which follows an F-distribution with $(k - 1)$ and $(n - k)$ degrees of freedom. By inspecting this ratio, we can see that large deviations among the sample means will cause the numerator to be large and thus may provide evidence for the rejection of the null hypothesis. The denominator is an average within-sample variance and is not affected by differences among the sample means.

Example 12.6

Refer to the example of the three signal types. The mean and standard deviation for the three samples are listed below:

Level	Number	Mean	Std Dev
Fully Actuated	5	14.00	3.59
Pretimed	5	35.48	3.37
Semi-actuated	5	20.90	3.19

Because the assumption of equal variances for the three populations is reasonable, let us estimate the common variance, using a pooled sample variance as follows:

$$s_p^2 = \frac{(5 - 1)3.59^2 + (5 - 1)3.37^2 + (5 - 1)3.19^2}{(5 - 1) + (5 - 1) + (5 - 1)} = 11.47$$

Note that $\hat{\sigma}^2 = s_p^2 = \text{MSE}$ (from the ANOVA table).

12.4.2 ANOVA and regression analysis

We used the ANOVA technique in the regression analysis (Chapter 11) too. What is the connection between the regression analysis and the analysis of variance? The analysis of variance, as presented in Section 12.3, can be produced using the regression techniques described in Chapters 2 and 11. This approach can be used if the computer program has multiple regression techniques available.

Example 12.7

Consider Example 12.3 involving managers with three different expertise levels. This experiment involves three treatments, A, B, and C. Let the response variable Y = the test score of one manager. Then we can model Y as

$$Y = \beta_0 + \beta_1 x_1 + \beta_2 x_2 + \varepsilon$$

where ε is the random error with $E(\varepsilon) = 0$ and $V(\varepsilon) = \sigma^2$. The indicator variables x_1 and x_2 are defined as follows:

$$x_1 = \begin{cases} 1 & \text{if the manager is in group B} \\ 0 & \text{otherwise} \end{cases}$$

and

$$x_2 = \begin{cases} 1 & \text{if the manager is in group C} \\ 0 & \text{otherwise} \end{cases}$$

For a response Y_A from treatment A, $x_1 = 0$ and $x_2 = 0$; hence,

$$E(Y_A) = \beta_0 = \mu_A$$

For a response Y_B from treatment B, $x_1 = 1$ and $x_2 = 0$; hence,

$$E(Y_B) = \beta_0 + \beta_1 = \mu_B.$$

It follows that $\quad \beta_1 = \mu_B - \mu_A$
In like manner, $\quad E(Y_C) = \beta_0 + \beta_2 = \mu_C$
and thus, $\quad \beta_2 = \mu_C - \mu_A$

The null hypothesis of interest, $H_0: \mu_A = \mu_B = \mu_C$ (in ANOVA), is now equivalent to $H_0: \beta_1 = \beta_2 = 0$ (in regression).

For fitting the model $Y = \beta_0 + \beta_1 x_1 + \beta_2 x_2 + \varepsilon$ by the method of least squares, arrange the data as shown in the table on next page. The method of least squares will give

$$\hat{\beta}_0 = \bar{y}_A, \quad \hat{\beta}_1 = \bar{y}_B - \bar{y}_A, \quad \text{and} \quad \hat{\beta}_2 = \bar{y}_C - \bar{y}_A$$

	y	x_1	x_2
Treatment A	82	0	0
	114	0	0
	90	0	0
	80	0	0
	88	0	0
	93	0	0
	80	0	0
	105	0	0
Treatment B	128	1	0
	90	1	0
	130	1	0
	110	1	0
	133	1	0
	130	1	0
	104	1	0
Treatment C	156	0	1
	128	0	1
	151	0	1
	140	0	1

The error sum of squares for this complete model (see Figure 12.9) is $SSE_2 = 3,181.61$. Under the null hypothesis $H_0: \beta_1 = \beta_2 = 0$, the reduced model becomes $Y = \beta_0 + \varepsilon$. The error sum of squares, when this model is fit by the method of least squares, denoted by SSE_1, is computed to be $SSE_1 = TSS = 10,815.2$.

Using the F-test discussed in Chapter 11 for testing $H_0: \beta_1 = \beta_2 = 0$, we get

$$F_{Reg} = \frac{(SSE_1 - SSE_2)/2}{SSE_2/(n-3)} = \frac{(TSS - SSE_2)/2}{SSE_2/(n-3)} = \frac{SST/2}{SSE_2/(n-3)} = \frac{MST}{MSE} = F_{ANOVA}$$

Thus, the F-test arising from the regression formulation is equivalent to the F-test of ANOVA given in Section 12.3. All the analysis of variance problems can be formulated in terms of regression problem. From both the printouts (Figures 12.10a and 12.10b), note that

$$\hat{\beta}_0 = \bar{y}_A = 91.5, \quad \hat{\beta}_1 = \bar{y}_B - \bar{y}_A = 117.85 - 91.5 = 26.35,$$

$$\text{and} \quad \hat{\beta}_2 = \bar{y}_C - \bar{y}_A = 143.75 - 91.5 = 52.25$$

Figure 12.10a shows the quantile plot for residuals. Although the linear fit is not very good, there is no evidence to reject the null hypothesis of normal distribution. Figure 12.10b shows the residual plot. The points are distributed around zero with about the same variability within each treatment group. Thus, the conditions underlying the analysis of variance (or linear regression) seem to be met satisfactorily.

Figure 12.9

JMP printouts for regression and comparison of means approaches

```
Response Score

Whole Model

Summary of Fit
RSquare                              0.70582
RSquare Adj                         0.669047
Root Mean Square Error              14.10143
Mean of Response                    112.2105
Observations (or Sum Wgts)                19


Analysis of Variance
Source      DF    Sum of      Mean
                  Squares     Square      F Ratio
Model        2    7633.551    3816.78     19.1942
Error       16    3181.607     198.85     Prob > F
C. Total    18   10815.158                 <.0001


Parameter Estimates
Term        Estimate   Std Error   t Ratio   Prob>|t|
Intercept       91.5     4.98561     18.35    <.0001
x1         26.357143     7.298186     3.61    0.0023
x2             52.25     8.63533      6.05    <.0001
```

(a) Regression approach

```
Response Score

Whole Model

Summary of Fit
RSquare                              0.70582
RSquare Adj                         0.669047
Root Mean Square Error              14.10143
Mean of Response                    112.2105
Observations (or Sum Wgts)                19


Analysis of Variance
Source      DF    Sum of      Mean
                  Squares     Square      F Ratio
Model        2    7633.551    3816.78     19.1942
Error       16    3181.607     198.85     Prob > F
C. Total    18   10815.158                 <.0001


Least Squares Means Table
Level    Least Sq Mean    Std Error       Mean
A            91.50000     4.9856099     91.500
B           117.85714     5.3298411    117.857
C           143.75000     7.0507171    143.750
```

(b) Comparison of means approach

Figure 12.10

Checking assumptions for residuals

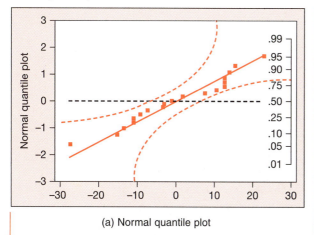

(a) Normal quantile plot

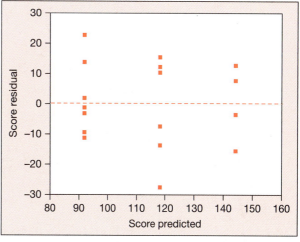

(b) Residual plot

Exercises

12.17 Refer to Exercise 12.5. Answer part (b) by writing the appropriate linear model and using regression techniques.

12.18 Refer to Exercise 12.7. Test the assumption of equal percentages of kill for the four rates of application by writing a linear model and using regression techniques.

12.19 Refer to Exercise 12.14. Write a linear model for this experiment. Test for significant differences among the means for the three thermometers by using regression techniques.

12.5 Estimation for Completely Randomized Design

Confidence intervals for treatment means and differences between treatment means can be produced by the methods introduced in Chapter 10. Recall that we are assuming all treatment populations to be normally distributed with a common variance. Thus, the confidence intervals based on the t-distribution can be employed. The common population variance σ^2 is estimated by the pooled sample variance $s^2 = \text{MSE}$.

Because we have k means in an analysis of variance problem, we might want to construct a number of confidence intervals based on the same set of experimental data. For example, we might want to construct confidence intervals for all k means individually or for all possible differences between pairs of means. If such multiple intervals are to be used, we must use extreme care in selecting the confidence coefficient for the individual intervals so that the experimentwise error rate remains small. For example, suppose we construct two confidence intervals,

$$\overline{Y}_A \pm t_{\alpha/2} \frac{s}{\sqrt{n}} \quad \text{and} \quad \overline{Y}_B \pm t_{\alpha/2} \frac{s}{\sqrt{n}}$$

Now

$$P(\overline{Y}_A \pm t_{\alpha/2} s/\sqrt{n} \text{ includes } \mu_A) = 1 - \alpha,$$

and

$$P(\overline{Y}_B \pm t_{\alpha/2} s/\sqrt{n} \text{ includes } \mu_B) = 1 - \alpha.$$

But

$$P(\overline{Y}_A \pm t_{\alpha/2} s/\sqrt{n} \text{ includes } \mu_A \text{ and } \overline{Y}_B \pm t_{\alpha/2} s/\sqrt{n} \text{ includes } \mu_B) < 1 - \alpha$$

so the simultaneous coverage probability is *less than* $(1 - \alpha)$, the confidence level that we used on each interval. There are several methods of conducting multiple comparisons (such as Bonferroni, Tukey's, etc.) so that the confidence level is controlled at $(1 - \alpha)$. We'll only describe the Bonferroni method in detail and use computer printouts of others.

Method If c intervals are to be constructed on one set of experimental data, one method of keeping the simultaneous coverage probability (or confidence coefficient) at a value of *at least* $(1 - \alpha)$ is to make the individual confidence coefficients as close as possible to $1 - (\alpha/c)$. This technique for multiple confidence intervals is outlined below:

Confidence Intervals for Means in the Completely Randomized Design: Bonferroni Method

Suppose c intervals are to be constructed from one set of data. Single treatment mean μ_i:

$$\bar{y}_i \pm t_{\alpha/2c} \frac{s}{\sqrt{n}}$$

Difference between two treatment means:

$$(\bar{y}_i - \bar{y}_j) \pm t_{\alpha/2c}\, s\sqrt{\frac{1}{n_i} + \frac{1}{n_j}}$$

Note that $s = \sqrt{\text{MSE}}$ and that all t-values depend on $(n - k)$ degrees of freedom.

Example 12.8

Refer to Example 12.3. Construct confidence intervals for all three possible differences between pairs of treatment means so that the simultaneous confidence coefficient is 0.95.

Solution Remember, we rejected the null hypothesis of equal mean scores for three groups. Because there are three intervals to construct ($c = 3$), each interval should be of the form

$$(\bar{y}_i - \bar{y}_j) \pm t_{0.05/2(3)}\, s\sqrt{\frac{1}{n_i} + \frac{1}{n_j}}$$

Now with $n - k = 16$ degrees of freedom, $t_{0.05/2(3)} = t_{0.00833} = 2.67323$ (using a TI-89). However, t-tables do not provide t-values for area 0.0833. Thus, using Table 5 in the Appendix, we can use $t_{0.00833} \approx t_{0.01} = 2.583$ (approximate). Then the three intervals are constructed as follows:

$$\mu_A - \mu_B: \quad (\bar{y}_A - \bar{y}_B) \pm t_{0.00833}\, s\sqrt{\frac{1}{n_A} + \frac{1}{n_B}}$$

$$= (91.5 - 117.9) \pm 2.67323(14.1014)\sqrt{\frac{1}{8} + \frac{1}{7}} = -26.4 \pm 19.51$$

$$\mu_A - \mu_C: \quad (\bar{y}_A - \bar{y}_C) \pm t_{0.00833}\, s \sqrt{\frac{1}{n_A} + \frac{1}{n_C}}$$

$$= (91.5 - 143.8) \pm 2.67323(14.1014)\sqrt{\frac{1}{8} + \frac{1}{4}} = -52.3 \pm 23.08$$

$$\mu_B - \mu_C: \quad (\bar{y}_B - \bar{y}_C) \pm t_{0.00833}\, s \sqrt{\frac{1}{n_B} + \frac{1}{n_C}}$$

$$= (117.9 - 143.8) \pm 2.67323(14.1014)\sqrt{\frac{1}{7} + \frac{1}{4}} = -25.9 \pm 23.63$$

In interpreting these results, we would be inclined to conclude that all pairs of means differ, because none of these three intervals contains zero. Furthermore, inspection of the means leads to the inference that $\mu_A < \mu_B < \mu_C$. The questionnaire appears to be a valid predictor of the level of manager's technical knowledge of computers. Our combined confidence level for all three intervals is at least 0.95.

Most statistical programs provide some multiple comparisons procedure(s). To get Bonferroni intervals, one can always use a procedure that controls individual error rate (such as Fisher's or Student's t) by setting the confidence level at $\alpha/2c$, where c is the number of intervals to be constructed. In practice, we are more likely to use the procedures provided by the software used than to compute intervals by hand. Figure 12.11 gives Bonferroni comparisons for Example 12.3 data about managers and expertise. The mean score is significantly different for all three managers.

Figure 12.11

JMP output for Bonferroni comparisons

Comparisons for each pair using Student's t

```
            t                Alpha
       2.67205              0.0167

Abs(Dif)-LSD               C           B           A
C                     -26.644       2.276      29.176
B                       2.276     -20.141       6.856
A                      29.176       6.856     -18.840

Positive values show pairs of means that are significantly different.

Level                                        Mean
C           A                             143.75000
B                    B                    117.85714
A                              C           91.50000

Levels not connected by same letter are significantly different.

Level      - Level       Difference      Lower CL     Upper CL
C              A          52.25000        29.17593     75.32407
B              A          26.35714         6.85599     45.85829
C              B          25.89286         2.27579     49.50992
```

Example 12.9

Refer to Example 12.2 of the three signal types. Remember, we rejected the null hypothesis of equal mean stop times for three types of signals. Obviously, now the question arises: Which signal type has the lowest mean stop time?

a Construct 95% simultaneous confidence intervals for the pairwise difference in mean stop times of the different signal types, and determine which signal type has the lowest mean stop time.
b Construct a 95% confidence interval for the mean stop time of the best signal.

Solution
a The JMP output for the 95% simultaneous confidence intervals, that is, Bonferroni intervals by setting error rate of 0.0167 for each comparison, is given below. It shows 95% intervals for pairs of means as follows:

$\mu_{Pre} - \mu_{FA}$	$\mu_{SA} - \mu_{FA}$	$\mu_{Pre} - \mu_{SA}$
(15.52, 27.44)	(0.94, 12.86)	(8.62, 20.54)

Note that all three intervals do not contain zero, indicating the mean stop times are different for all three signal types. Also, the column on means tells us that $\hat{\mu}_{FA} < \hat{\mu}_{SA} < \hat{\mu}_{PA}$. Therefore, we conclude that the fully actuated signals, which have the lowest mean stop time, are the best.

b From the printout, we get, $s = \sqrt{MSE} = 3.389$. Using degrees of freedom for error, 12, we get $t_{0.025}(12) = 2.179$. Thus, using the method described in Chapter 9 of constructing a confidence interval, we get

$$\bar{y}_{FA} \pm \frac{s}{\sqrt{n_{FA}}} \Rightarrow 14.0 \pm \frac{3.389}{\sqrt{5}} \Rightarrow 14.0 \pm 1.52 \Rightarrow (12.48, 15.52)$$

Comparisons for each pair using Student's t

t	Alpha
2.77840	0.0167

Abs(Dif)-LSD	Pretimed	Semi-actuated	Fully Actuated
Pretimed	-5.955	8.625	15.525
Semi-actuated	8.625	-5.955	0.945
Fully Actuated	15.525	0.945	-5.955

Positive values show pairs of means that are significantly different.

Level			Mean
Pretimed	A		35.480000
Semi-actuated		B	20.900000
Fully Actuated		C	14.000000

Levels not connected by same letter are significantly different.

Level	-Level	Difference	Lower CL	Upper CL
Pretimed	Fully Actuated	21.48000	15.52472	27.43528
Pretimed	Semi-actuated	14.58000	8.62472	20.53528
Semi-actuated	Fully Actuated	6.90000	0.94472	12.85528

Thus, we are 95% confident that the mean stop time for fully actuated signals is between 12.48 and 15.52 seconds.

Exercises

12.20 Refer to Exercise 12.5.

  a Find a 90% confidence interval for $(\mu_2 - \mu_3)$. Interpret the interval.

  b What would happen to the width of the confidence interval in part (a) if you quadrupled the number of observations in the two samples?

  c Find a 95% confidence interval for μ_2.

  d Approximately how many observations would be required if you wanted to be able to estimate a population mean correct to within 0.4 with probability equal to 0.95?

12.21 Refer to Exercise 12.6.

  a Suppose $\bar{y}_1 = 3.7$ and $\bar{y}_2 = 4.1$. Do the data provide sufficient evidence to indicate a difference between μ_1 and μ_2? Assume that there are seven observations for each treatment. Test by using $\alpha = 0.10$.

  b Refer to part (a). Find a 90% confidence interval for $(\mu_1 - \mu_2)$.

  c Refer to part (a). Find a 90% confidence interval for μ_1.

12.22 Refer to Exercise 12.7.

  a Estimate the true difference in mean percentage of kill between rate 2 and rate 5. Use a 90% confidence interval.

  b Construct confidence intervals for the six possible differences between treatment means, with simultaneous confidence coefficient close to 0.90.

12.23 Refer to Exercise 12.10.

  a Find a 95% confidence interval for the difference in mean strength for specimens produced with a 35% concentration of catalyst versus those containing a 45% concentration of catalyst.

  b Construct confidence intervals for the three treatment means, with simultaneous confidence coefficient approximately 0.90.

12.24 Refer to Exercise 12.13. Casting 1 is a standard, and castings 2, 3, and 4 involve slight modifications to the process. Compare the standard with each of the modified processes by producing three confidence intervals with simultaneous confidence coefficient approximately 0.90.

12.6   Analysis of Variance for the Randomized Block Design

12.6.1   ANOVA for the RBD

In the completely randomized design, only one source of variation, the treatment-to-treatment variation, is identified, specifically considered in the design, and analyzed after the data are collected. That is, observations are taken from each of k treatments, and then an inferential procedure is used to determine how the treatment means may differ.

However, in practical situations, the responses are subject to other sources of variation besides treatments. Suppose, for example, that an engineer is studying the gas mileage resulting from four brands of gasoline. He randomly assigns three cars (I, II, and III) to each of four brands of gasoline (A, B, C, and D) to be compared. Then the resulting design would be a completely randomized design with three measurements per treatment, as shown in Figure 12.12. Two undesirable conditions might result from this experiment:

Figure 12.12

A CRD for gasoline mileage study

Brand of Gasoline			
A	B	C	D
—	—	—	—
—	—	—	—
—	—	—	—

Figure 12.13

A RBD for gasoline mileage study

Car		
I	II	III
B	A	D
A	C	B
C	B	A
D	D	C

- The auto-to-auto variation would add to the variance of the measurements within each sample and hence inflate the MSE. The *F*-test could then turn out to be insignificant even when the differences do exist among the true treatment means.
- All of the A responses could turn out to be from automobile I, and all of the B responses from automobile II. In that case, if the mean for A and the mean for B were significantly different, we wouldn't know how to interpret the results. Are the brands really different, or are the automobiles different with respect to gas mileage? The effect of brand of gasoline and the effect of automobile will be confounded.

A randomized block design will help avoid both of these difficulties. If more than one automobile is used in the study, then automobiles would form another important source of variation. To control for this additional source of variation, it is important to run each brand of gasoline at least once in each automobile. In this type of experiment, the brands of gasoline are the *treatments* and the automobiles are called the *blocks*. If the order of running the brands in each automobile is randomized, the resulting design is called a *randomized block design (RBD)*. One possible randomization that might result in the structure of design is given in Figure 12.13. In this resulting design, for auto I, brand B was run first, followed in order by brands A, C, D, and so on.

As studied in Chapter 3, *block* is a group of homogeneous experimental units, that is, experimental units similar in characteristics. The blocking helps the experimenter control a source of variation in response so that any true treatment differences are likely to show up in the analysis.

A **randomized block design** is a plan for collecting data in which each of *k* treatments is measured once in each of *b* blocks. The order of the treatments within the blocks is randomized.

After the measurements are completed in a randomized block design, they can be rearranged into a two-way table, as follows:

		Treatments				Totals
		1	**2**	**...**	**k**	
Blocks	**1**	y_{11}	y_{12}	...	y_{1k}	B_1
	2	y_{21}	y_{22}	...	y_{2k}	B_2
	$\vdots$	$\vdots$	$\vdots$	$\vdots$	$\vdots$	$\vdots$
	b	y_{b1}	y_{b2}	...	y_{bk}	B_b
Totals		T_1	T_2	...	T_k	$T_0 = \sum y_{ij}$

Note that there are k treatments and b blocks. We indicate treatment totals by T_i and block totals by B_j. Because each treatment occurs once and only once in each block, the total sample size $n = bk$. The overall sample total $T_0 = \sum y_{ij} = \sum T_i = \sum B_j$, and the overall sample mean is $\bar{y} = T_0/n$. Let $\bar{y}_{i.} = T_i/b$ be the ith treatment mean$(i = 1, 2, \ldots, k)$, and $\bar{y}_{.j} = B_j/k$ be the jth block mean$(i = 1, 2, \ldots, b)$. Note from the data table that the total sum of squares (TSS) can now be partitioned into a treatment sum of squares (SST), a block sum of squares (SSB), and an error sum of squares (SSE). The sum of squares terms are computed as follows and the ANOVA table is completed.

$$\text{TSS} = \sum_{i=1}^{t}\sum_{j=1}^{b}(y_{ij} - \bar{y})^2 \quad \text{SST} = \sum_{i=1}^{t}b(\bar{y}_{i.} - \bar{y})^2$$

$$\text{SSB} = \sum_{i=1}^{b}t(\bar{y}_{.j} - \bar{y})^2 \quad \text{SSE} = \sum_{i=1}^{t}\sum_{j=1}^{b}(\bar{y}_{ij} - \bar{y}_{i.} - \bar{y}_{.j} + \bar{y})^2$$

ANOVA Table for RBD

Sources of Variation	Degrees of Freedom	Sum of Squares	Mean Squares	F-ratio	p-value
Treatments	$(k-1)$	SST	MST = SST/$(k-1)$	F_T = MST/MSE	$P(F > F_T)$
Blocks	$(b-1)$	SSB	MSB = SSB/$(b-1)$	F_B = MSB/MSE	$P(F > F_B)$
Error	$(n-k)$	SSE	MSE = SSE/$(n-k)$		
Total	$(n-1)$	TSS			

A test of the null hypothesis that the k treatment means $\mu_1, \mu_2, \ldots, \mu_k$ are equal is an F-test constructed as a ratio of MST to MSE. In an analogous fashion, we could construct a test of the hypothesis that the block means are equal. The resulting F-test for this case would be the ratio of MSB to MSE.

Test to Compare k Treatment Means for a Randomized Block Design

$H_0: \mu_1 = \mu_2 = \cdots = \mu_k$
H_a: At least two treatment means differ.

Test Statistic: $F_T = \dfrac{\text{MST}}{\text{MSE}}$

Rejection region: $F_T > F_\alpha(k - 1, (k - 1)(b - 1))$

Is Blocking Effective?

Test to Compare b Block Means for a Randomized Block Design

H_0: All b block means are equal (i.e., blocking is not effective).
H_a: At least two block means differ (i.e., blocking is effective).

> Test Statistic: $F_B = \dfrac{\text{MSB}}{\text{MSE}}$
>
> Rejection region: $F_B > F_\alpha(b - 1, (k - 1)(b - 1))$
>
> ## Conditions:
>
> 1. Treatments are randomly assigned to units within each block.
> 2. The measurements for each treatment/block population are approximately normally distributed.
> 3. The variances of probability distributions are equal.

Example 12.10

Four chemical treatments for fabric are to be compared with regard to their ability to resist stains. Two different types of fabric are available for the experiment, so it is decided to apply each chemical to a sample of each type of fabric. The result is a randomized block design with four treatments and two blocks. The measurements (resistance scores) are listed below:

Resistance Scores		Block (Fabric)		Means
		1	**2**	
Treatment (Chemical)	1	5	9	7.0
	2	3	8	5.5
	3	8	13	10.5
	4	4	6	5.0
Means		5	9	7.0

a Is there evidence of significant differences among the mean resistance scores for the chemicals? Use $\alpha = 0.05$.

b Is blocking effective (i.e., is there evidence of significant differences among the mean resistance scores for the two fabric types)? Use $\alpha = 0.05$.

Solution For these data, the sum of squares is computed as follows and the ANOVA table is completed.

$$\text{TSS} = \sum_{i=1}^{4} \sum_{j=1}^{2} (y_{ij} - \bar{y})^2 = (5 - 7)^2 + \cdots + (6 - 7)^2 = 72$$

$$\text{SST} = \sum_{i=1}^{4} b(\bar{y}_{i.} - \bar{y})^2 = 2(7 - 7)^2 + \cdots + 2(5 - 7)^2 = 37$$

$$\text{SSB} = \sum_{i=1}^{2} t(\bar{y}_{.j} - \bar{y})^2 = 4(5 - 7)^2 + 4(9 - 7)^2 = 32$$

$$\text{SSE} = \sum_{i=1}^{4} \sum_{j=1}^{2} (\bar{y}_{ij} - \bar{y}_{i.} - \bar{y}_{.j} + \bar{y})^2$$

$$= (5 - 7 - 5 + 7)^2 + \cdots + (6 - 5 - 9 + 7)^2 = 3$$

ANOVA Table for RBD

Sources of Variation	Degrees of Freedom	Sum of Squares	Mean Squares	F-ratio
Treatments	3	37	$37/3 = 12.33$	$F_T = 12.33/1 = 12.33$
Blocks	1	32	$32/1 = 32$	$F_B = 32/1 = 32$
Error	3	3	$3/3 = 1$	
Total	7	72		

a Here we are interested in testing $H_0: \mu_1 = \mu_2 = \mu_3 = \mu_4$, where μ_i is the mean resistance score for the ith chemical ($i = 1, 2, 3, 4$). Because $F_{0.05}(3, 3) = 9.28$ is less than our observed F-ratio for treatments, 12.33, we reject the null hypothesis and conclude that there is significant evidence that the mean resistance differs for at least two treatments.

b Here we are interested in testing $H_0: \mu_1 = \mu_2$, where μ_j is the mean resistance score for the jth fabric ($j = 1, 2$). We also see from the ANOVA table that the F-ratio for blocks is 32. Because $F_{0.05}(1, 3) = 10.13 < 32$, we reject the null hypothesis and conclude that there is evidence of significant difference between the block (fabric) means. That is, the fabrics seem to differ with respect to their ability to resist stains when treated with these chemicals. The blocking is effective; in other words, the decision to "block out" the variability between fabrics appears to have been a good one.

Figure 12.14

Minitab and JMP outputs for RBD for Example 12.10

Minitab Printout

Two-way ANOVA: resistance versus chemical, fabric

```
Analysis of Variance for resistance
Source     DF     SS      MS      F        P
chemical    3   37.00   12.33   12.33   0.034
fabric      1   32.00   32.00   32.00   0.011
Error       3    3.00    1.00
Total       7   72.00
```

JMP Printout

Response Resistance

Whole Model

Summary of Fit

RSquare	0.958333
RSquare Adj	0.902778
Root Mean Square Error	1
Mean of Response	7
Observations (or Sum Wgts)	8

Analysis of Variance

Source	DF	Sum of Squares	Mean Square	F Ratio
Model	4	69.000000	17.2500	17.2500
Error	3	3.000000	1.0000	Prob > F
C.Total	7	72.000000		0.0207

Effect Tests

Source	Nparm	DF	Sum of Squares	F Ratio	Prob > F
Chemical	3	3	37.000000	12.3333	0.0341
Fabric	1	1	32.000000	32.0000	0.0109

The Minitab and JMP printouts for this analysis are shown in Figure 12.14. The computer printouts show that,

a. For testing equality of means for chemicals, the p-value $= 0.034$. Because the p-value $= 0.034 < \alpha = 0.05$, reject the null hypothesis and conclude that at least two chemicals have different mean resistance scores.

b. For testing equality of means for fabrics, the p-value $= 0.011$. Because the p-value $= 0.011 < \alpha = 0.05$, reject the null hypothesis and conclude that the two types of fabrics used have different mean resistance scores; that is, the blocking is effective.

Activity

Goal: To compare mean flight times of helicopters with wing lengths of 4.5, 5.5, and 6.5 inches while controlling for the effect of the dropper.

The *response* to be measured is flight time of the helicopter. It is defined as the time for which the helicopter remains in flight. It is suspected that there is a dropper effect on flight times. So, use the dropper as a blocking factor. If a group has four students, rotate the role of dropper among the students, thereby using the dropper as a blocking factor.

1. Prepare three helicopters with wing lengths of 4.5, 5.5, and 6.5 inches, respectively using template from Chapter 3.
2. Using the random number table, for each dropper randomize the order of runs of the three wing lengths.
3. Have the student release the helicopter from above his or her head while standing alone on a step stool or a ladder. Start the timer at the drop of the helicopter. Stop the timer when the helicopter hits the floor.
4. Measure the time in flight (in sec) for the helicopters in predetermined order. Record the flight times in Table 12.9.
5. Reorganize the data in Table 12.10.
6. Analyze using ANOVA for the randomized block design.

Table 12.9

Data Recording Sheet for Paper Helicopter Experiment

Dropper 1			Dropper 2			Dropper 3			Dropper 4		
Flight Number	Wing Length	Flight Time	Flight Number	Wing Length	Flight Time	Flight Number	Wing Length	Flight Time	Flight Number	Wing Length	Flight Time
1	___	___	1	___	___	1	___	___	1	___	___
2	___	___	2	___	___	2	___	___	2	___	___
3	___	___	3	___	___	3	___	___	3	___	___

Table 12.10

Data Table for Paper Helicopter Experiment

	Dropper 1	Dropper 2	Dropper 3	Dropper 4
Wing length 4.5 in	_____	_____	_____	_____
Wing length 5.5 in	_____	_____	_____	_____
Wing length 6.5 in	_____	_____	_____	_____

Exercises

12.25 A randomized block design was conducted to compare the mean responses for three treatments, A, B, and C, in four blocks. The data are shown below.

		Block 1	Block 2	Block 3	Block 4
	A	3	6	1	2
Treatment	B	5	7	4	6
	C	2	3	2	2

a Compute the appropriate sums of squares and mean squares and fill in the entries in the analysis of variance table shown below:

Source	df	SS	MS	F Ratio
Treatment				
Block				
Error				
Total				

b Do the data provide sufficient evidence to indicate a difference among treatment means? Test using $\alpha = 0.05$.

c Do the data provide sufficient evidence to indicate that blocking was effective in reducing the experimental error? Test using $\alpha = 0.05$.

12.26 The analysis of variance for a randomized block design produced the ANOVA table entries shown below:

Source	df	SS	MS	F Ratio
Treatment	3	27.1	_____	_____
Block	5	_____	14.90	_____
Error	_____	33.4		
Total	_____			

a Complete the ANOVA table.

b Do the data provide sufficient evidence to indicate a difference among the treatment means? Test by using $\alpha = 0.01$.

c Do the data provide sufficient evidence to indicate that blocking was a useful design strategy to employ for this experiment? Explain.

12.27 An evaluation of diffusion boding of zircaloy components is performed. The main objective is to determine which of three elements—nickel, iron, or copper—is the best bonding agent. A series of zircaloy components is bonded, using each of the possible bonding agents. Since there is a great deal of variation among components machined from different ingots, a randomized block design is used, blocking on the ingots. A pair of components from each ingot is bonded together, using each of the three agents, and the pressure (in units of 1,000 pounds per square inch) required to separate the bonded components is measured. The data are listed below:

Ingot	Nickel	Iron	Copper
		Bonding Agent	
1	67.0	71.9	72.2
2	67.5	68.8	66.4
3	76.0	82.6	74.5
4	72.7	78.1	67.3
5	73.1	74.2	73.2
6	65.8	70.8	68.7
7	75.6	84.9	69.0

Is there evidence of a difference in pressure required to separate the components among the three bonding agents? Use $\alpha = 0.05$.

12.28 A construction firm employs three cost estimators. Usually, only one estimator works on each potential job, but it is advantageous to the company if the estimators are consistent enough so that it does not matter which of the three estimators is assigned to a particular job. To check the consistency of the estimators, several jobs are selected and all three estimators are asked to make estimates. The estimates for each job (in thousands of dollars) by each estimator are given below.

	Estimator		
Job	**A**	**B**	**C**
1	27.3	26.5	28.2
2	66.7	67.3	65.9
3	104.8	102.1	100.8
4	87.6	85.6	86.5
5	54.5	55.6	55.9
6	58.7	59.2	60.1

a Do these estimates provide sufficient evidence that the means for the estimators differ? Use $\alpha = 0.05$.

b Present the complete ANOVA summary table for this experiment.

12.29 A power plant that uses water from the surrounding bay for cooling its condensors is required by the EPA to determine whether discharging its heated water into the bay has a detrimental effect on the flora (plant life) in the water. The EPA requests that the power plant make its investigation at three strategically chosen locations, called *stations*. Stations 1 and 2 are located near the plant's discharge tubes, while Station 3 is located farther out in the bay. During one randomly selected day in each of four months, a diver sent down to each of the stations randomly samples a square-meter area of the bottom and counts the number of blades of the different types of grasses present. The result are listed the table below for one important grass type:

	Station		
Month	**1**	**2**	**3**
May	28	31	53
June	25	22	61
July	37	30	56
August	20	26	48

a Is there sufficient evidence to indicate that the mean number of blades found per square meter per month differs for the three stations? Use $\alpha = 0.05$.

b Is there sufficient evidence to indicate that the mean number of blades found per square meter differs across the four months? Use $\alpha = 0.05$.

12.30 The Bell Telephone Company's long-distance phone charges may appear to be exorbitant when compared with the charges of some of its competitors, but this is because a comparison of charges between competing companies is often analogous to comparing apples and eggs. Bell's charges for individuals are on a per-call basis. In contrast, its competitors often charge a monthly minimum long-distance fee, reduce the charges as the usage rises, or do both. Shown in Table 12.11 is a sample of long-distance charges from Orlando, Florida, to 12 cities for three

Table 12.11

Long-Distance Charges of Phone Calls from Orlando, Florida

From Orlando to	Time	Length of Call (minutes)	Company		
			1	**2**	**3**
New York	Day	2	$0.77	$0.79	$0.66
Chicago	Evening	3	0.69	0.71	0.59
Los Angeles	Day	2	0.87	0.88	0.66
Atlanta	Evening	1	0.22	0.23	0.20
Boston	Day	3	1.15	1.19	0.99
Phoenix	Day	5	1.92	1.98	1.65
West Palm Beach	Evening	2	0.49	0.42	0.40
Miami	Day	3	1.12	1.05	0.99
Denver	Day	10	3.85	3.96	3.30
Houston	Evening	1	0.22	0.23	0.20
Tampa	Day	3	1.06	1.00	0.99
Jacksonville	Day	3	1.06	1.00	0.99

non-Bell companies offering long-distance service. The data were reported in an advertisement in the *Orlando Sentinel*, March 19, 1984. A note in fine print below the advertisement states that the rates are based on "30 hours of usage" for each of the servicing companies. The data in Table 12.11 are pertinent for companies making phone calls to large cities. Therefore, assume that the cities receiving the calls were randomly selected from among all large cities in the United States.

a What type of design was used for the data collection?

b Perform an analysis of variance for the data. Present the results in an ANOVA table.

c Do the data provide sufficient evidence to indicate differences in mean charges among the three companies? Test by using $\alpha = 0.05$.

d Company 3 placed the advertisement, so it might be more relevant to compare the charges for companies 1 and 2. Do the data provide sufficient evidence to indicate a difference in mean charges for these two companies? Test by using $\alpha = 0.05$.

12.31 From time to time, one branch office of a company must make shipments to a certain branch office in another state. There are three package delivery services between the two cities where the branch offices are located. Since the price structures for the three delivery services are quite similar, the company wants to compare the delivery times. The company plans to make several different types of shipments to its branch office. To compare the carriers, each shipment will be sent in triplicate, one with each carrier. The results listed in table below are the delivery times in hours.

	Carrier		
Shipment	**I**	**II**	**III**
1	15.2	16.9	17.1
2	14.3	16.4	16.1
3	14.7	15.9	15.7
4	15.1	16.7	17.0
5	14.0	15.6	15.5

Is there evidence of a difference in mean delivery times among the three carriers? Use $\alpha = 0.05$.

12.32 Due to increased energy shortages and costs, utility companies are stressing ways in which home and apartment utility bills can be cut. One utility company reached an agreement with the owner of a new apartment complex to conduct a

test of energy-saving plans for apartments. The tests were to be conducted before the apartments were rented. Four apartments were chosen that were identical in size, amount of shade, and direction faced. Four plans were to be tested, one on each apartment. The thermostat was set at 75°F in each apartment, and the monthly utility bill was recorded for each of the three summer months. The results are listed in table below.

	Treatment			
Month	**1**	**2**	**3**	**4**
June	$74.44	$68.75	$71.34	$65.47
July	89.96	73.47	83.62	72.33
August	82.00	71.23	79.98	70.87

Treatment 1: no insulation
Treatment 2: insulation in walls and ceilings
Treatment 3: no insulation; awnings for windows
Treatment 4: insulation and awnings for windows

a Is there evidence that the mean monthly utility bills differ for the four treatments? Use $\alpha = 0.01$.

b Is there evidence that blocking is important, that is, that the mean utility bills differ for the three months? Use $\alpha = 0.05$.

12.33 A chemist runs an experiment to study the effects of four treatments on the glass transition temperature of a particular polymer compound. Raw material used to make this polymer is bought in small batches. The material is thought to be fairly uniform within a batch but variable between batches. Therefore, each treatment was run on samples from each batch, with the results shown in table below. (showing temperature in k):

	Treatment			
Batch	**I**	**II**	**III**	**IV**
1	576	584	562	543
2	515	563	522	536
3	562	555	550	530

a Do the data provide sufficient evidence to indicate a difference in mean temperature among the four treatments? Use $\alpha = 0.05$.

b Is there sufficient evidence to indicate a difference in mean temperature among the three batches? Use $\alpha = 0.05$.

c If the experiment is conducted again in the future, would you recommend any changes in its design?

Table 12.12

Number of Strikes Per Year for Different Industries

Year	Food and Kindred Products	Primary Metal Industry	Electrical Equipment and Supplies	Fabricated Metal Products	Chemicals and Allied Products
1976	227	197	204	309	129
1977	221	239	199	354	111
1978	171	187	190	360	113
1979	178	202	195	352	143
1980	155	175	140	280	89
1981	109	114	106	203	60

Source: *Statistical Abstract of the United States*, 1989.

12.34 Table 12.12 lists the number of strikes that occurred per year in five U.S. manufacturing industries over the period from 1976 to 1981. Only work stoppages that continued for at least one day and involved six or more workers were counted.

  a Perform an analysis of variance and determine whether there is sufficient evidence to conclude that the mean number of strikes per year differs among the five industries. Test by using $\alpha = 0.05$.

  b Construct the appropriate ANOVA table.

12.35 Hagihara and Miyazaki (*Transactions of the ASME*, 2003) studied bifurcation buckling of conical roof shells often used on oil storage tanks. The shells of 4(2) 12 mm thickness were used in the experiment. The number of circumferential waves at bifurcation buckling was reported, as shown in table below, for the radius of 15 m, for four different pulse widths (0, 20, 100, 320 ms).

  a Determine whether the mean number of waves differ by shell thickness. Use $\alpha = 0.05$.

  b State the underlying assumptions.

  c Check whether the assumptions are satisfied.

Pulse Width (ms)	Thickness 4	6	8	10	12
0	40	35	33	31	30
20	13	10	9	9	8
100	64	53	46	42	38
320	65	53	47	42	39

12.6.2 Relation between a paired *t*-test and an *F*-test for RBD

In Section 12.4, we discussed the association between the *t*-test (equal variances case) and the *F*-test for ANOVA. The ANOVA for a CRD is a generalization of pooled variances of the *t*-test. Similarly, ANOVA for a RBD is a generalization of a paired *t*-test (refer to Chapter 10). For a randomized block design with two treatments and *b*-blocks, a two-sided paired *t*-test of equal means and an *F*-test of ANOVA give the same results. Just remember that $t^2 = F$. However, when more than two treatment means are to be compared using a RBD, use of ANOVA is recommended to control experimentwise error rate.

Example 12.11

Ermolova (*IEE Proceedings Communications,* 2003) studied the performance of conventional and modified communication systems using simulations. For different output backoff (OBO) values, the solid-state power amplifier (SSPA) model was used and the estimates of the gain were recorded as follows for the case when channel noise is negligible:

OBO (dB)	2.93	3.452	3.859	4.412	4.825
Modified system	1.776	1.637	1.554	1.460	1.405
Conventional system	1.901	1.730	1.629	1.517	1.451

Does the mean gain using a modified system differ significantly from that obtained using a conventional system?

Solution   Here we have a matched pairs design with two treatments, namely the modified and the conventional system. For each OBO level, two measurements were taken, one on a conventional system and one on a modified system. If we define the mean difference in the gains by the two systems as $\mu_D = \mu_M - \mu_C$, then we are interested in testing the hypothesis $H_0: \mu_D = 0$, that is, that there is no difference in the mean gain by modified and conventional systems, against the alternative $H_a: \mu_D \neq 0$, that is, that the mean gain by the modified system differs from the mean gain by the conventional system.

The JMP output (Figure 12.15) shows that the test statistic value is $t = 5.67$, with *p*-value = 0.0048, which is less than $\alpha = 0.05$. Hence, we reject the null

Figure 12.15

Results of *t* and *F* tests

Matched Pairs

Difference: Conventional-Modified

```
Conventional    1.6456      t-Ratio   5.671048
Modified        1.5664          DF           4
Mean
Difference      0.0792   Prob > |t|     0.0048
Std Error       0.01397     Prob > t     0.0024
Upper95%        0.11797     Prob < t     0.9976
Lower95%        0.04043
N                    5
Correlation     0.99997
```

Response Gain

Summary of Fit

RSquare	0.991489
RSquare Adj	0.98085
Root Mean Square Error	0.022082
Mean of Response	1.606
Observations (or Sum Wgts)	10

Analysis of Variance

Source	DF	Sum of Squares	Mean Square	F Ratio
Model	5	0.22720760	0.045442	93.1943
Error	4	0.00195040	0.000488	Prob > F
C. Total	9	0.22915800		0.0003

Effect Tests

Source	Nparm	DF	Sum of Squares	F Ratio	Prob > F
system	1	1	0.01568160	32.1608	0.0048
OBO	4	4	0.21152600	108.4526	0.0002

(a) Results of the matched pairs design

(b) Results of the RBD

hypothesis and conclude that the mean gain by the modified system significantly differs from that by the conventional system.

We can consider each OBO level as a different block in which each of two treatments are run. Five OBO levels form five different blocks. As a result, we have a randomized block design with two treatments and five blocks. Let us look at the JMP printout for the RBD (Figure 12.17b). It shows that for testing the hypothesis $H_0: \mu_M = \mu_C$, the F-ratio = 32.16 with p-value = 0.0048. Note that $t^2 = (5.67)^2 = 32.15 = F$-ratio for the system, and the p-value is the same for both the tests, hence they result in the same conclusion.

Although both the tests result in the same conclusion about the treatments, the RBD design provides more information about the factors in the experiment. In addition to treatment information, the RBD also provides information about the effectiveness of blocking, which is not available from the paired t-test.

12.6.3 ANOVA for RBD and regression analysis

Section 12.4.1 contains a discussion of the regression approach to the analysis of a completely randomized design. A similar model can be written for the randomized block design.

Example 12.12

Refer to the randomized block design of Example 12.10. Let Y be a response variable, and then we can write

$$Y = \beta_0 + \beta_1 x_1 + \beta_2 x_2 + \beta_3 x_3 + \beta_4 x_4 + \varepsilon$$

where

$$x_1 = \begin{cases} 1 & \text{if the response is from block 2} \\ 0 & \text{otherwise} \end{cases}$$

$$x_2 = \begin{cases} 1 & \text{if the response is from treatment 2} \\ 0 & \text{otherwise} \end{cases}$$

$$x_3 = \begin{cases} 1 & \text{if the response is from treatment 3} \\ 0 & \text{otherwise} \end{cases}$$

$$x_4 = \begin{cases} 1 & \text{if the response is from treatment 4} \\ 0 & \text{otherwise} \end{cases}$$

and ε is the random error term. The error sum of squares for this model is denoted by SSE_2. Testing the null hypothesis that the four treatment means are equal is now equivalent to testing $H_0: \beta_2 = \beta_3 = \beta_4 = 0$. The reduced model is then

$$Y = \beta_0 + \beta_1 x_1 + \epsilon$$

which, when fit by the method of least squares, will produce an error sum of squares denoted by SSE_1. The F-test for $H_0: \beta_2 = \beta_3 = \beta_4 = 0$ versus the alternative hypothesis, that at least one β_i, $i = 2, 3, 4$, is different from zero then has the form

$$F = \frac{(SSE_1 - SSE_2)/(k - 1)}{SSE_2/(b - 1)(k - 1)} = \frac{SST/(k - 1)}{SSE_2/(b - 1)(k - 1)} = \frac{MST}{MSE}$$

The test is equivalent to the F-test for the equality of treatment means for a RBD given in Section 12.6.1. The regression printouts for the two models (complete and reduced) are shown in Figure 12.16. The sum of squared errors (SSE) for the complete and reduced model are highlighted in Figure 12.16.

Complete model: $SSE_2 = 3$ Reduced model: $SSE_1 = 40$

Then

$$F = \frac{(40 - 3)/3}{3/3} = 12.33$$

Figure 12.16

Printouts for complete and reduced models

Response Resistance

Whole Model

Summary of Fit

RSquare	0.958333
RSquare Adj	0.902778
Root Mean Square Error	1
Mean of Response	7
Observations (or Sum Wgts)	8

Analysis of Variance

Source	DF	Sum of Squares	Mean Square	F Ratio
Model	4	69.000000	17.2500	17.2500
Error	3	3.000000	1.0000	Prob > F
C.Total	7	72.000000		0.0207

Parameter Estimates

Term	Estimate	Std Error	t Ratio	Prob>|t|
Intercept	5	0.790569	6.32	0.0080
x1	4	0.707107	5.66	0.0109
x2	-1.5	1	-1.50	0.2306
x3	3.5	1	3.50	0.0395
x4	-2	1	-2.00	0.1393

(a) Complete model

Response Resistance

Whole Model

Summary of Fit

RSquare	0.444444
RSquare Adj	0.351852
Root Mean Square Error	2.581989
Mean of Response	7
Observations (or Sum Wgts)	8

Analysis of Variance

Source	DF	Sum of Squares	Mean Square	F Ratio
Model	1	32.000000	32.0000	4.8000
Error	6	40.000000	6.6667	Prob > F
C.Total	7	72.000000		0.0710

Parameter Estimates

Term	Estimate	Std Error	t Ratio	Prob>|t|
Intercept	5	1.290994	3.87	0.0082
x1	4	1.825742	2.19	0.0710

(b) Reduced model

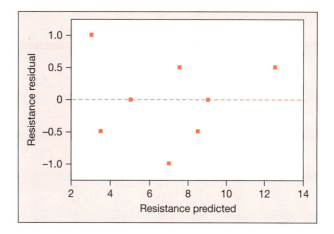

which is exactly the same as F calculated in Example 12.10. The regression approach leads us to the same conclusion as did the ANOVA approach: There is evidence that the mean stain resistances of at least four chemicals differ.

The residual plot above shows residuals distributed randomly around zero, without specific pattern. It indicates the selected model fits data well.

12.6.4 Bonferroni method for estimation for RBD

As in the case of the completely randomized design, we might want to estimate the mean for a particular treatment, or the difference between the means of two treatments, after conducting the initial F-test. Because we will generally be interested in c intervals simultaneously, we will want to set the confidence coefficient of each individual interval at $(1 - \alpha/c)$. This will guarantee that the probability of all c intervals simultaneously covering the parameters being estimated is at least $1 - \alpha$. A summary of the estimation procedure is given here:

Confidence Intervals for Means in the Randomized Block Design

Suppose c intervals are to be constructed from one set of data.

The single treatment mean μ_i:

$$\bar{y}_i \pm t_{\alpha/2c}\frac{s}{\sqrt{b}}$$

The difference between the two treatment means $\mu_i - \mu_j$:

$$(\bar{y}_i - \bar{y}_j) \pm t_{\alpha/2c}s\sqrt{\frac{1}{b} + \frac{1}{b}}$$

Note that $s = \sqrt{\text{MSE}}$ and that all t-values depend on $(k - 1)(t - 1)$ degrees of freedom, where k is the number of treatments and b the number of blocks.

Example 12.13

Refer to the four chemical treatments and two (fabrics) blocks of Example 12.10. We are interested in estimating simultaneously all possible differences with a simultaneous confidence coefficient of at least 0.90, approximately.

Solution   Because there are four treatment means (μ_1, μ_2, μ_3 and μ_4), there will be $c = 6$ differences of the form $\mu_i - \mu_j$. Because the simultaneous confidence level is to be $\alpha = 0.90$, $\alpha/2c = 0.10/2(6) = 0.00833$. Using $(k - 1)(b - 1) = 3(1) = 3$ degrees of freedom, $t_{0.00833} = 4.85737$ (using the TI-89's inverse t-function). If using Table 5 of the Appendix, we need to approximate 0.00833 with 0.01 and get $t_{0.00833} \approx t_{0.01} = 4.541$. For the given data, the sample means are

$$\bar{y}_1 = 7.0, \quad \bar{y}_2 = 5.5, \quad \bar{y}_3 = 10.5, \quad \text{and} \quad \bar{y}_4 = 5.0$$

Also, $s = \sqrt{\text{MSE}} = 1$; thus, any interval of the form $(\bar{y}_i - \bar{y}_j) \pm t_{\alpha/2c}s\sqrt{\dfrac{1}{b} + \dfrac{1}{b}}$ becomes

$$(\bar{y}_i - \bar{y}_j) \pm 4.85737(1)\sqrt{\frac{1}{2} + \frac{1}{2}}, \text{ i.e., } (\bar{y}_i - \bar{y}_j) \pm 4.85737 \text{ (exact using } t_{0.00833})$$

$$(\bar{y}_i - \bar{y}_j) \pm 4.541(1)\sqrt{\frac{1}{2} + \frac{1}{2}}, \text{ i.e., } (\bar{y}_i - \bar{y}_j) \pm 4.541 \text{ (approximate using } t_{0.01})$$

The six confidence intervals are shown in table below. The sample data would suggest that μ_2 and μ_3 are significantly different and that μ_3 and μ_4 are significantly different, because these intervals do not contain zero.
These confidence intervals show that

- The mean resistance score for chemical 3 is significantly higher than that for chemical 2.
- The mean resistance score for chemical 3 is significantly higher than that for chemical 4.
- All other differences in mean resistance scores are not significant.

Parameter	Interval (exact)	Interval (approximate)
$\mu_1 - \mu_2$	1.5 ± 4.85737	1.5 ± 4.541
$\mu_1 - \mu_3$	-3.5 ± 4.85737	-3.5 ± 4.541
$\mu_1 - \mu_4$	2.0 ± 4.85737	2.0 ± 4.541
$\mu_2 - \mu_3$	-5.0 ± 4.85737	-5.0 ± 4.541
$\mu_2 - \mu_4$	0.5 ± 4.85737	0.5 ± 4.541
$\mu_3 - \mu_4$	5.5 ± 4.85737	5.5 ± 4.541

Exercises

12.36 Refer to Exercise 12.25. Answer parts (b) and (c) by fitting a linear model to the data and using regression techniques.

12.37 Refer to Exercise 12.29. Answer part (a) by fitting a linear model to the data and using regression techniques.

12.38 Suppose that three automobile engine designs are being compared to determine differences in mean time between breakdowns.

a Show how three test automobiles and three different test drivers could be used in a 3×3 Latin square design aimed at comparing the engine designs.

b Write the linear model for the design in part (a).

c What null hypothesis would you test to determine whether the mean time between breakdowns differs for the engine designs?

12.39 Refer to Exercise 12.27

a Form a 95% confidence interval to estimate the true mean difference in pressure between nickel and iron. Interpret this interval.

b Form confidence intervals on the three possible differences between the means for the bonding agents, using a simultaneous confidence coefficient of approximately 0.90.

12.40 Refer to Exercise 12.28. Use a 90% confidence interval to estimate the true difference between the mean responses given by estimators B and C.

12.41 Refer to Exercise 12.29. What are the significant differences among the station means? (Use a simultaneous confidence coefficient of 0.90.)

12.42 Refer to Exercise 12.30. Compare the three telephone companies' mean rates by using a simultaneous confidence coefficient of approximately 0.90.

12.43 Refer to Exercise 12.31.

a Use a 99% confidence interval to estimate the difference between the mean delivery times for carriers I and II.

b What assumptions are necessary for the validity of the procedures you used in part (a)?

12.7 The Factorial Experiment

The response of interest Y in many experimental situations is related to one or more variables that can be controlled by the experimenter. For example,

- The yield Y in an experimental study of the process of manufacturing a chemical may be related to the temperature x_1 and pressure x_2 at which the experiment is run. The variables, temperature and pressure, can be controlled by the experimenter.
- In a study of heat loss through ceiling of houses, the amount of heat loss Y will be related to the thickness of insulation x_1 and temperature differential x_2 between the inside and outside of the house. Again, the thickness of insulation and the temperature differential are controlled by the experimenter as he or she designs the study.
- The strength Y of concrete may be related to the amount of aggregate x_1, the mixing time x_2, and the drying time x_3.

As discussed earlier, the variables that are thought to affect the response of interest and are controlled by the experimenter are called *factors*. The various settings of these factors in an experiment are called *levels*. A factor/level combination then defines a *treatment*. In the chemical yield example, the temperatures of interest might be 90°C, 100°C, and 110°C. These are the three experimental levels of the factor "temperature." The pressure settings (levels) of interest might be 400 and 450 psi.

Setting temperature at 90°C and pressure at 450 psi would define a particular treatment, and observations on yield could then be obtained for this combination of levels. Note that three levels of temperature and two levels of pressure would result in $2 \times 3 = 6$ different treatments, if all factor/level combinations are to be used.

An experiment in which treatments are defined by specified factor/level combinations is defined as a *factorial experiment*. In this section, we assume that r observations on the response of interest are realized from each treatment, with each observation being independently selected. That is, the factorial experiment is run in a completely randomized design with r observations per treatment. The objective of the analysis is to decide whether, and to what extent, the various factors affect the response variable. Does the yield of chemical increase as temperature increases? Does the strength of concrete decrease as less and less aggregate is used? These are the types of questions that we'll be able to answer using the results of this section.

12.7.1 Analysis of variance for the factorial experiment

In the production of a certain industrial chemical, the yield Y depends on the cooking time x_1 and the cooling time x_2. Two cooking times and two cooling times are of interest. Only the relative magnitudes of the levels are important in the analysis of factorial experiments, and thus we can call the two levels of cooking time 0 and 1, that is, $x_1 = 0$ or 1. Similarly, we can call the levels of cooling times 0 and 1 as well; in other words, x_2 can take on the values 0 or 1. For convenience, we will refer to cooking time as "factor A," and cooling time as "factor B." Then we have four different treatment combinations in this experiment, and we are interested in comparing their effects on response. Schematically, we can show the arrangement of r observations for each of four treatments (see Figure 12.17).
Four treatment combinations:

1. $x_1 = 0$, and $x_2 = 0$
2. $x_1 = 0$, and $x_2 = 1$
3. $x_1 = 1$, and $x_2 = 0$
4. $x_1 = 1$, and $x_2 = 1$

Here the total $4r$ runs (r runs per treatment combination) of the experiment are conducted in random order. Because the two factors whose effect on response is being studied are applied together to the experimental units, they might interact

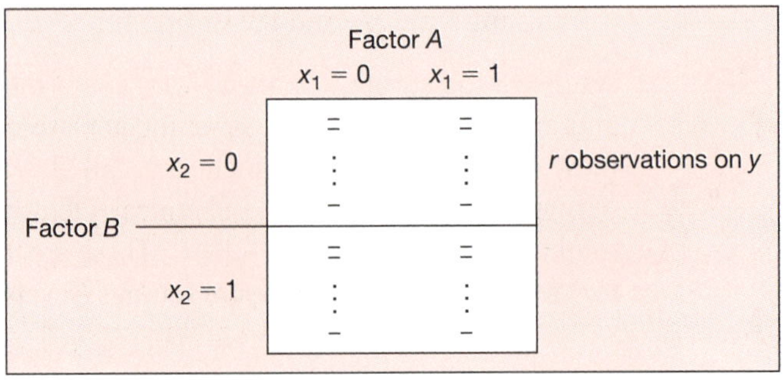

Figure 12.17

A 2×2 factorial experiment with r observations per treatment together to affect the response. If the rate of change of expected response $E(Y)$ as x_1 goes from 0 to 1 is the same for each value of x_2, then we say that there is *no interaction* between factor A and factor B. However, if the rate of change of expected response as x_1 goes from 0 to 1 is different for different values of x_2, then we say that there is *interaction* between factor A and factor B. The same idea can be extended to more than two levels of factors.

Consider an experiment with two factors (A and B), with number of levels a and b, respectively. Table 12.13 shows a typical data table for a $a \times b$ factorial experiment with r replications.

We can see from the data table that this is essentially a completely randomized experiment with $a \times b$ treatment combinations. So, the total sum of squares is divided into treatment sum of squares and error sum of squares. The treatment sum of squares can be further partitioned into a sum of squares for A, a sum of squares for B, and an interaction sum of squares. Therefore, for a two factor factorial experiment in a CRD,

$$\text{Total SS} = \overbrace{\text{SS(A)} + \text{SS(B)} + \text{SS(AB)}}^{\text{SST}} + \text{SSE}$$

Table 12.13

A General Layout of a Two-Factor

Factorial with *r* Replications

Factor *A*

		1	2	3	$\cdots$	a	Totals
	1	— — T_{11}	— — T_{12}	— — T_{13}		— — T_{3a}	B_1
	2	— — T_{21}	— — T_{22}	— — T_{23}		— — T_{2a}	B_2
Factor *B*	3	— — T_{31}	— — T_{32}	— — T_{33}		— — T_{3a}	B_3
	$\vdots$						
	b	— — T_{b1}	— — T_{b2}	— — T_{b3}		— — T_{ba}	B_b
	Totals	A_1	A_2	A_3		A_a	$\sum y$

T_{ij} = Total of observations in row i and column j

Table 12.14

ANOVA Table for a Two-Factor Factorial Experiment in a CRD

Source	df	SS	MS	F-ratio
Treatments	$ab - 1$	SST	$MST = \dfrac{SST}{ab - 1}$	
Factor A	$a - 1$	SS(A)	$MS(A) = \dfrac{SS(A)}{a - 1}$	MS(A)/MSE
Factor B	$b - 1$	SS(B)	$MS(B) = \dfrac{SS(B)}{b - 1}$	MS(B)/MSE
$A \times B$ Interaction	$(a - 1)(b - 1)$	SS($A \times B$)	$MS(A \times B) = \dfrac{SS(A \times B)}{(a - 1)(b - 1)}$	MS($A \times B$)/MSE
Error	$ab(r - 1)$	SSE	$MSE = \dfrac{SSE}{ab(r - 1)}$	
Total	$n - 1$	TSS		

Note: The assumption is made that all factor combination have independent normal distributions with equal variances.

Let y_{ijk} be the response measured on experimental unit of kth replication receiving ith level of factor A and jth level of factor B. Suppose $\bar{y}_{i..}$ is the mean response for ith level of factor A, $\bar{y}_{.j.}$ is the mean response for jth level of factor B, and $\bar{y}_{ij.}$ is the mean response of all replications for the treatment combination (i, j). Then, using $n = abr$ = total number of observations, these sums of squares can be computed as follows and the ANOVA table can be completed (see Table 12.14).

$$\text{TSS} = \sum_{i=1}^{a} \sum_{j=1}^{b} \sum_{k=1}^{r} (y_{ijk} - \bar{y})^2 \qquad \text{SST} = r \sum_{i=1}^{a} \sum_{j=1}^{b} (\bar{y}_{ij.} - \bar{y})^2$$

$$\text{SS}(A) = br \sum_{i=1}^{a} (\bar{y}_{i..} - \bar{y})^2 \qquad \text{SS}(B) = ar \sum_{j=1}^{a} (\bar{y}_{.j.} - \bar{y})^2$$

$$\text{SS}(AB) = r \sum_{i=1}^{a} (\bar{y}_{ij.} - \bar{y}_{i..} - \bar{y}_{.j.} + \bar{y})^2 \qquad \text{SSE} = \sum_{i=1}^{a} \sum_{j=1}^{b} \sum_{k=1}^{r} (y_{ijk} - \bar{y}_{ij.})^2$$

Here we are interested in testing three different hypotheses, one about the interaction effect and two about the main effects of two factors. First, test the hypothesis

H_0: There is no interaction between factors A and B.
H_a: Factors A and B interact to affect response.

If you reject this hypothesis, thus establishing that there is evidence of interaction, then *do not* proceed with tests for the main effects. Significant interaction indicates there are some differences among the treatment means. At this point, it may be best simply to estimate the individual treatment means or differences between means by the methods given in Section 12.5.

If the interaction hypothesis is not rejected, then it is appropriate to test the following two hypothesis about the main effects of two factors:

H_0: There is no difference in the mean response at all levels of factor A.
H_a: The mean response differs for at least two levels of factor A.

and

H_0: There is no difference in the mean response at all levels of factor B.
H_a: The mean response differs for at least two levels of factor B.

For these tests, it is assumed that all factor combinations have independent normal distributions with equal variances.

Example 12.14

For the chemical experiment with two good cooking times (factor A) and two cooling times (factor B), the yields are as given in the table below with $r = 2$ observations per treatment. The combinations of cooking and cooling times were used in random order.

		Factor A		Mean
		$x_1 = 0$	$x_1 = 1$	
Factor B	$x_2 = 0$	9, 8 $\bar{y}_{00.} = 8.5$	5, 6 $\bar{y}_{10.} = 5.5$	$\bar{y}_{.0.} = 7.5$
	$x_2 = 1$	8, 7 $\bar{y}_{01.} = 7.5$	3, 4 $\bar{y}_{11.} = 3.5$	$\bar{y}_{.1.} = 5.5$
	Mean	$\bar{y}_{0..} = 8.0$	$\bar{y}_{1..} = 4.5$	$\bar{y} = 6.25$

Do mean yields differ significantly depending on the cooking times and cooling times?

Solution Using the formulas listed above, compute the sum of squares and complete the ANOVA table.

$$\text{TSS} = \sum_{i=1}^{2}\sum_{j=1}^{2}\sum_{k=1}^{2}(y_{ijk} - \bar{y})^2 = (9 - 6.25)^2 + \cdots + (4 - 6.25)^2 = 31.5$$

$$\text{SST} = r\sum_{i=1}^{a}\sum_{j=1}^{b}(\bar{y}_{ij.} - \bar{y})^2 = 2(8.5 - 6.25)^2 + \cdots + 2(3.5 - 6.25)^2 = 29.5$$

$$\text{SS}(A) = br\sum_{i=1}^{a}(\bar{y}_{i..} - \bar{y})^2 = 2(2)[(8 - 6.25)^2 + (4.5 - 6.25)^2] = 24.5$$

$$\text{SS}(B) = ar\sum_{j=1}^{a}(\bar{y}_{.j.} - \bar{y})^2 = 2(2)[(7.5 - 6.25)^2 + (5.5 - 6.25)^2] = 4.5$$

$$\text{SS}(AB) = r\sum_{i=1}^{a}(\bar{y}_{ij.} - \bar{y}_{i..} - \bar{y}_{.j.} + \bar{y})^2$$
$$= 2[(8.5 - 8.0 - 7.5 + 6.25)^2 + \cdots + (3.5 - 4.5 - 5.5 + 6.25)^2]$$
$$= 0.5$$

$$\text{SSE} = \sum_{i=1}^{a}\sum_{j=1}^{b}\sum_{k=1}^{r}(y_{ijk} - \bar{y}_{ij.})^2 = (9 - 8.5)^2 + \cdots + (3 - 3.5)^2 = 2.0$$

ANOVA for Yield Data

Source	df	SS	MS	F-ratio
Treatments	3	29.5		
A	1	24.5	24.5	$24.5/0.5 = 49.0$
B	1	4.5	4.5	$4.5/0.5 = 9.0$
$A \times B$	1	0.5	0.5	$0.5/0.5 = 1.0$
Error	1	2.0	0.5	
Total	**4**	**31.5**		

First, let us test for interaction effect. Because $F_{0.05}(1, 4) = 7.71$ and the F-ratio $= 1.0$ for the interaction, we conclude that the interaction effect is not significant. Figure 12.18 shows the *interaction plot*, also known as the *means plot*. In this plot, the means of all combinations of levels of the two factors are plotted. It shows that at both levels of factor A, the mean cooking time decreased as we changed factor B from level 0 to level 1. The almost parallel lines indicate that the rate of decrease is similar for both values of A; in other words, there is no significant interaction effect.

We can now proceed with the "main effect" test for factors A and B. The F-ratio of 49.0 for factor A is highly significant. Thus, there is a significant difference between the mean response at $x_1 = 0$ and that at $x_1 = 1$. Looking at the data, we see that the yield falls off as cooking time goes from the low level to the high level.

The F-ratio of 9.0 for factor B is also significant. The mean yield also seems to decrease as cooling time is changed from the low to the high level.

Figure 12.18

The means (or interaction) plot

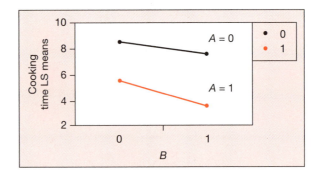

Example 12.15

Suppose a chemical experiment like the one of Example 12.14 (two factors each at two levels) gave the results shown in the following table. Perform an analysis of variance. If the interaction is significant, construct confidence intervals for the six possible differences between treatment means.

	Factor A		Mean
	$x_1 = 0$	$x_1 = 1$	
$x_2 = 0$	9, 8 $\bar{y}_{00.} = 8.5$	5, 6 $\bar{y}_{10.} = 5.5$	$\bar{y}_{.0.} = 7.5$
$x_2 = 1$	3, 4 $\bar{y}_{11.} = 3.5$	8, 7 $\bar{y}_{01.} = 7.5$	$\bar{y}_{.1.} = 5.5$
Mean	$\bar{y}_{0..} = 6.0$	$\bar{y}_{1..} = 6.5$	$\bar{y} = 6.25$

Solution The data involve the same responses as in Example 12.14, but the observations in the lower left and right cells are interchanged. Thus, we still have

$$\text{TSS} = \sum_{i=1}^{2}\sum_{j=1}^{2}\sum_{k=1}^{2}(y_{ijk} - \bar{y})^2 = (9 - 6.25)^2 + \cdots + (4 - 6.25)^2 = 31.5$$

$$\text{SST} = r\sum_{i=1}^{a}\sum_{j=1}^{b}(\bar{y}_{ij.} - \bar{y})^2 = 2(8.5 - 6.25)^2 + \cdots + 2(6.5 - 6.25)^2 = 29.5$$

$$\text{SSE} = \sum_{i=1}^{a}\sum_{j=1}^{b}\sum_{k=1}^{r}(y_{ijk} - \bar{y}_{ij.})^2 = (9 - 8.5)^2 + \cdots + (3 - 3.5)^2 = 2.0$$

However, now

$$\text{SS}(A) = br\sum_{i=1}^{a}(\bar{y}_{i..} - \bar{y})^2 = 2(2)[(6 - 6.25)^2 + (6.5 - 6.25)^2] = 0.5$$

$$\text{SS}(B) = ar\sum_{j=1}^{a}(\bar{y}_{.j.} - \bar{y})^2 = 2(2)[(7.5 - 6.25)^2 + (5.5 - 6.25)^2] = 4.5$$

$$\begin{aligned}
\text{SS}(AB) &= r\sum_{j=1}^{a}(\bar{y}_{ij.} - \bar{y}_{i..} - \bar{y}_{.j.} + \bar{y})^2 \\
&= 2[(8.5 - 6.0 - 7.5 + 6.25)^2 + \cdots + (7.5 - 6.5 - 5.5 + 6.25)^2 \\
&= 0.5
\end{aligned}$$

Now the ANOVA table is as follows:

ANOVA for Yield Data

Source	df	SS	MS	F-ratio
Treatments	3	29.5		
A	1	0.5	24.5	0.5/0.5 = 1.0
B	1	4.5	4.5	4.5/0.5 = 9.0
A × B	1	0.5	24.5	24.5/0.5 = 49.0
Error	1	2.0	0.5	
Total	**4**	**31.5**		

At the 5% level of significance, the interaction term is highly significant, a fact also confirmed by the interaction plot of crossing lines (Figure 12.19). It indicates that when factor $A = 0$, the mean cooking time decreased when the factor B level changed from 0 to 1, but it increased for $A = 1$, showing totally opposite trends and hence interaction between factors A and B. Therefore, we will not make any "main effect" tests but instead will place confidence intervals on all possible differences between treatment means. We will identify treatment combinations using levels of two factors. Following the format of Section 12.5, we can construct intervals of the form.

Figure 12.19

The means (or interaction) plot

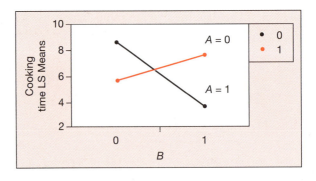

$$(\bar{y}_{i'} - \bar{y}_{i'}) \pm t_{\alpha/2c}\, s \sqrt{\frac{1}{n_i} + \frac{1}{n_{i'}}}$$

With four treatment combinations, we get $c = 4$ pairs. Also, $n_i = n_i = 2$ and $s = \sqrt{\text{MSE}} = \sqrt{0.5} = 0.707$ is based on four degrees of freedom. To construct 95% confidence simultaneous intervals, we use $1 - \alpha/2c = 0.05/2(6) = 0.004167$. Using a TI-89, we get $t_{0.004167} = 4.8509$ (using Table 5 of the Appendix, we can approximate this value as $t_{0.004167} \approx t_{0.005} = 4.604$). Thus, all six intervals will be of the form

$$(\bar{y}_i - \bar{y}_{i'}) \pm 4.8509(0.707)\sqrt{\frac{1}{2} + \frac{1}{2}}, \quad \text{i.e.,} \quad (\bar{y}_i - \bar{y}_{i'}) \pm 3.43$$

Because the sample means are $\bar{y}_{00} = 8.5, \bar{y}_{01} = 3.5, \bar{y}_{10} = 5.5$, and $\bar{y}_{11} = 7.5$, the confidence intervals would then be as listed in the following table.

Parameter	Interval	Parameter	Interval
$\mu_{00} - \mu_{01}$	5.0 ± 3.43	$\mu_{01} - \mu_{10}$	-2.0 ± 3.43
$\mu_{00} - \mu_{10}$	3.0 ± 3.43	$\mu_{01} - \mu_{11}$	-4.0 ± 3.43
$\mu_{00} - \mu_{11}$	1.0 ± 3.43	$\mu_{10} - \mu_{11}$	-2.0 ± 3.43

Therefore, the significant differences are between μ_{00} and μ_{01}, and μ_{01} and μ_{11}. In practical terms, the yield is reduced significantly as x_1 goes from 0 to 1 while x_2 remains at 1. Also, the yield is increased significantly as x_2 goes from 0 to 1 while x_1 remains at 0. Because the difference between μ_{00} and μ_{11} is not significant, the best choices for maximizing yield appear to be either ($x_1 = 0$ and $x_2 = 0$) or ($x_1 = 1$ and $x_2 = 1$).

Activity

Goal: To compare mean flight times of helicopters with wing lengths of 4.5, 5.5, and 6.5 inches and body widths of 1 and 2 inches.

The *response* to be measured is flight time of the helicopter. It is defined as the time for which the helicopter remains in flight. We are interested in the effect

Treatment Number	Wing Length	Body Width
1	4.5 inches	1 inch
2	4.5 inches	2 inches
3	5.5 inches	1 inch
4	5.5 inches	2 inches
5	6.5 inches	1 inch
6	6.5 inches	2 inches

of two factors, namely wing length and body width, on flight times. So, use all combinations of these two factors as treatments, as listed in table above:

Thus, we have a 2×3 factorial experiment. Let's do five replications of each treatment combination. Then we'll have a total of 30 measurements.

If a group has four students, rotate the role of dropper among students; in other words, use the dropper as a blocking factor.

1. Prepare six helicopters with wing-length and body-width combinations listed in the above table.
2. Using the random number table, randomize the order of five runs each of six treatments.
3. Have a student release the helicopter from above his/her head while standing alone on a step stool or a ladder. Start the timer at the drop of the helicopter. Stop the timer when the helicopter hits the floor.
4. Measure the time in flight (in sec) for the helicopters in a predetermined order. Record the flight times in Table 12.15.
5. Reorganize the data as shown in Table 12.16:
6. Analyze using ANOVA for the factorial experiment.
7. Prepare a length*width interaction plot. Is the interaction effect significant?

Table 12.15

Data Recording Sheet

Flight Number	Treatment Number	Flight Time	Flight Number	Treatment Number	Flight Time	Flight Number	Treatment Number	Flight Time
1	———	———	11	———	———	21	———	———
2	———	———	12	———	———	22	———	———
3	———	———	13	———	———	23	———	———
4	———	———	14	———	———	24	———	———
5	———	———	15	———	———	25	———	———
6	———	———	16	———	———	26	———	———
7	———	———	17	———	———	27	———	———
8	———	———	18	———	———	28	———	———
9	———	———	19	———	———	29	———	———
10	———	———	20	———	———	30	———	———

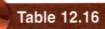

Table 12.16

Data Table

	Wing Length 4.5 in.	Wing Length 5.5 in.	Wing Length 6.5 in.
Body width 1 in.			
Body width 2 in.			

Factorial experiment and regression analysis: The analysis of any factorial experiment could be accomplished by writing a linear regression model for response Y as a function of x_1 and x_2, and then employing the theory of Chapters 10 and 11. Because Y supposedly depends on x_1 and x_2, we could start with the simple model

$$E(Y) = \beta_0 + \beta_1 x_1 + \beta_2 x_2$$

In Figure 12.20, we see that this model implies that the rate of change of $E(Y)$ as x_1 goes from 0 to 1 is the same (β_1) for each value of x_2. This is referred to as a no-interaction case.

It is quite likely that the rate of change of $E(Y)$ as x_1 goes from 0 to 1 will be different for different values of x_2. To account for this *interaction*, we write the model as

$$E(Y) = \beta_0 + \beta_1 x_1 + \beta_2 x_2 + \beta_3 x_1 x_2$$

The term $\beta_3 x_1 x_2$ is referred to as the interaction term. Now

$$\text{if } x_2 = 0, \text{ we have } E(Y) = \beta_0 + \beta_1 x_1$$

$$\text{if } x_2 = 1, \text{ we have } E(Y) = (\beta_0 + \beta_2) + (\beta_1 + \beta_3)x_1$$

Figure 12.20

Means plot for a 2 × 2 factorial experiment with no interaction

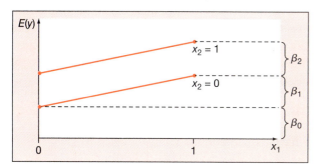

Figure 12.21

Means plot for a 2 × 2 factorial experiment with interaction

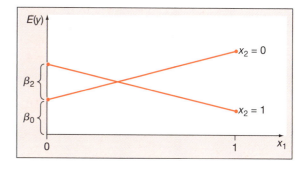

The slope (rate of change) when $x_2 = 0$ is β_1, but the slope when $x_2 = 1$ changes to $\beta_1 + \beta_3$. Figure 12.21 depicts the possible interaction case.

One way to proceed with an analysis of this 2 × 2 factorial experiment is as follows. First, test the hypothesis $H_0: \beta_3 = 0$. If this hypothesis is rejected, thus establishing the evidence of the presence of the interaction effect, *do not* proceed with tests on β_1 and β_2. The significance of β_3 is enough to establish the fact that there are some differences among the treatment means. It may then be best simply to estimate the individual treatment means, or differences between means, by the methods given in Section 12.5.

If the hypothesis $H_0: \beta_3 = 0$ is not rejected, it is then appropriate to test the hypotheses $H_0: \beta_2 = 0$ and $H_0: \beta_1 = 0$. These are often referred to as tests of "main effects." The test of $\beta_2 = 0$ actually compares the observed mean of all observations at $x_2 = 1$ with the corresponding mean at $x_2 = 0$, regardless of the value of x_1. That is, each level of B is averaged over all levels of A, and then the levels of B are compared. This is a reasonable procedure when there is no evidence of interaction, because the change in response from the low to high level of B is essentially the same for each level of A. Similarly, the test of $\beta_1 = 0$ compares the mean response at the high level of factor A ($x_1 = 1$) with that at the low level of A ($x_1 = 0$), with the means computed over all levels of factor B.

Using the method of fitting complete and reduced models (Chapter 11), we complete the analysis of fitting the models given and calculating SSE for each:

1. SSE_1: $Y = \beta_0 + \beta_1 x_1 + \beta_2 x_2 + \beta_3 x_1 x_2 + \varepsilon$
2. SSE_2: $Y = \beta_0 + \beta_1 x_1 + \beta_2 x_2 + \varepsilon$
3. SSE_3: $Y = \beta_0 + \beta_1 x_1 + \varepsilon$
4. SSE_4: $Y = \beta_0 + \varepsilon$

Now SS $(A \times B) = SSE_2 - SSE_1$, the sum of squares for the interaction; also, $SS(B) = SSE_3 - SSE_2$, the sum of squares due to factor B; and $SS(A) = SSE_4 - SSE_3$, the sum of squares due to factor A. This approach works well if a computer is available for fitting the indicated models.

12.7.2   Fitting higher-order models

Suppose the chemical example under discussion had three cooking times (levels of factor A) and two cooling times (levels of factor B). If the three cooking times were equally spaced (such as 20 minutes, 30 minutes, and 40 minutes), then the levels of A could be coded as $x_1 = -1$, $x_1 = 0$, and $x_1 = 1$. Again, only the *relative* magnitudes of the levels are important in the analysis. Because there are now three levels of factor A, a quadratic term in x_1 can be added to the model. Thus, the complete model would have the form

$$E(Y) = \beta_0 + \beta_1 x_1 + \beta_2 x_1^2 + \beta_3 x_2 + \beta_4 x_1 x_2 + \beta_5 x_1^2 x_2$$

Note that if $x_2 = 0$ we simply have a quadratic model in x_1, given by

$$E(Y) = \beta_0 + \beta_1 x_1 + \beta_2 x_1^2$$

If $x_2 = 1$, we have another quadratic model in x_1, but the coefficients have changed:

$$E(Y) = (\beta_0 + \beta_3) + (\beta_1 + \beta_4)x_1 + (\beta_2 + \beta_5)x_1^2$$

If $\beta_4 = \beta_5 = 0$, the two curves will have the same shape, but if either β_4 or β_5 differs from zero, the curves will differ in shape. Thus β_4 and β_5 are both components of interaction (see Figure 12.22.)

In the analysis, we first test $H_0: \beta_4 = \beta_5 = 0$ (no interaction). If we reject this hypothesis, then we do no further F-tests, but we may compare treatment means. If we do not reject the hypothesis of no interaction, we proceed with tests of $H_0: \beta_3 = 0$ (no main effect for factor B) and $H_0: \beta_1 = \beta_2 = 0$ (no main effect for factor A). The test of $H_0: \beta_3 = 0$ merely compares the mean responses at the high and low levels of factor B. The test of $H_0: \beta_1 = \beta_2 = 0$ looks for both linear and quadratic trends among the three means for the levels of factor A, averaged across the levels of factor B.

Figure 12.22

$E(Y)$ for a 3×2 Factorial Experiment

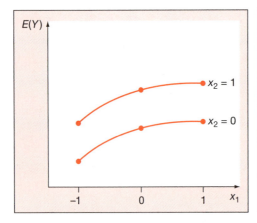
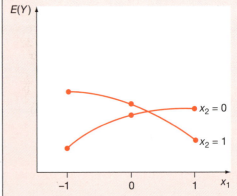

Example 12.16

In the manufacture of a certain beverage, an important measurement is the percentage of impurities present in the final product. The following data show the percentage of impurities present in samples taken from products manufactured at three different temperatures (factor A) and two sterilization times (factor B). The three levels of A were actually 75°, 100°, and 125°C. The two levels of B were actually 15 minutes and 20 minutes. The data are recorded in the table below where the last number in each cell is the cell total.

		Factor A			
		75°C	**100°C**	**125°C**	
Factor B	15 min.	14.05	10.55	7.55	
		14.93	9.48	6.59	63.15
		28.98	20.03	14.14	
	20 min.	16.56	13.63	9.23	
		15.85	11.75	8.78	75.80
		32.41	25.38	18.01	
		61.39	45.41	32.15	138.95

a Perform an analysis of variance.

b Perform an analysis of variance using the regression approach.

Solution

a With $a = 3$, $b = 2$, and $r = 2$, we get the following ANOVA table. Because $F_{0.05}(2, 5) = 5.14$ and $F_{0.05}(1, 6) = 5.99$, it is clear that the interaction is not significant and that the main effects for both factor A and factor B are highly significant. The means for factor A tend to decrease as the temperature goes from low to high, and this decreasing trend is of approximately the same degree for both the low and high levels of factor B.

Source	df	SS	MS	F Ratio
Treatments	5	121.02		
A	2	107.18	53.59	90.83
B	1	13.34	13.34	22.59
$A \times B$	2	0.50	0.25	0.42
Error	6	3.54	0.59	
Total	**11**	**124.56**		

b For the regression approach, we begin by fitting the complete model, with interaction, as shown in Figure 12.23. These models use a $(-1, 0, +1)$ coding for A and a $(-1, +1)$ coding for B. Let us consider the following two models to test for significance of interaction term, that is, $H_0: \beta_4 = \beta_5 = 0$.

Complete model: $E(Y) = \beta_0 + \beta_1 x_1 + \beta_2 x_1^2 + \beta_3 x_2 + \beta_4 x_1 x_2 + \beta_5 x_1^2 x_2$
Reduced model: $E(Y) = \beta_0 + \beta_1 x_1 + \beta_2 x_1^2 + \beta_3 x_2$

The JMP outputs for these two models are shown in Figure 12.23 and Figure 12.24, respectively. From Figure 12.23, we get $SSE_1 = 3.541$ and from Figure 12.24 we get $SSE_2 = 4.047$. Thus, the F-statistic becomes

$$\frac{(4.047 - 3.54)/2}{3.541/6} = \frac{0.25}{0.59} = 0.42$$

To test for significance of the main effect for factor B, reduce the model even further by letting the coefficient of x_2 equal zero. See Figure 12.25 for the Minitab output of this reduced model for Factor B.

Now $SSE_3 = 17.382$. Thus,

$$SS(B) = SSE_3 - SSE_2 = 17.382 - 4.047 = 13.335$$

Figure 12.23

JMP output for the model with interaction (complete model)

```
Summary of Fit

RSquare                          0.971573
RSquare Adj                      0.947884
Root Mean Square Error           0.768218
Mean of Response                 11.57917
Observations (or Sum Wgts)             12

Analysis of Variance

                    Sum of     Mean
Source     DF      Squares    Square   F Ratio
Model       5    121.02154   24.2043   41.0132
Error       6      3.54095    0.5902   Prob > F
C. Total   11    124.56249              0.0001

Parameter Estimates

Term       Estimate  Std Error  t Ratio  Prob>|t|
Intercept  11.3525    0.384109    29.56   <.0001
x1         -3.655     0.271606   -13.46   <.0001
x2          1.3375    0.384109     3.48   0.0131
x1*x2       0.055     0.271606     0.20   0.8462
x1**2       0.34      0.470435     0.72   0.4970
x1**2*x2   -0.425     0.470435    -0.90   0.4011
```

Figure 12.24

JMP output for the model without interaction (reduced model)

```
Summary of Fit

RSquare                          0.967512
RSquare Adj                      0.955329
Root Mean Square Error           0.711233
Mean of Response                 11.57917
Observations (or Sum Wgts)             12

Analysis of Variance

                    Sum of     Mean
Source     DF      Squares    Square   F Ratio
Model       3    120.51567   40.1719   79.4143
Error       8      4.04682    0.5059   Prob > F
C. Total   11    124.56249             <.0001

Parameter Estimates

Term        Estimate   Std Error  t Ratio  Prob >|t|
Intercept  11.3525     0.355616     31.92   <.0001
x1         -3.655      0.251459    -14.54   <.0001
x2          1.0541667  0.205315      5.13   0.0009
x1**2       0.34       0.435539      0.78   0.4575
```

Figure 12.25

Reduced model to test for the effect of factor B

```
Summary of Fit

RSquare                          0.860455
RSquare Adj                      0.829445
Root Mean Square Error           1.389725
Mean of Response                 11.57917
Observations (or Sum Wgts)             12

Analysis of Variance

                    Sum of     Mean
Source     DF      Squares    Square   F Ratio
Model       2    107.18047   53.5902   27.7478
Error       9     17.38203    1.9313   Prob > F
C. Total   11    124.56249              0.0001

Parameter Estimates

Term       Estimate  Std Error  t Ratio  Prob >|t|
Intercept  11.3525    0.694863    16.34   <.0001
x1         -3.655     0.491342    -7.44   <.0001
x1**2      0.34       0.851029     0.40   0.6988
```

Also, $SSE_4 = SST = 124.563$. Thus, $SS(A) = SSE_4 - SSE_3 = 124.563 - 17.382 = 107.81$ Note that $SS(B)$ and $SS(A)$, as calculated from the regression models, agree with the calculations from the analysis of variance table.

Figure 12.26

Normal scores plot of residuals from the main effects model

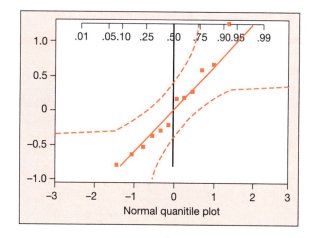

Because the regression model allows easy checking of residuals, we plot these residuals (from the main effects model, with no interaction) against their normal scores, as shown in Figure 12.26. The nearly straight-line plot indicates that the normality assumption is reasonable. The residuals are also plotted against the three levels of A, the two levels of B and the predicted values ($\hat{y}$) in Figure 12.27.

Figure 12.27

Residual plots for the main effects model

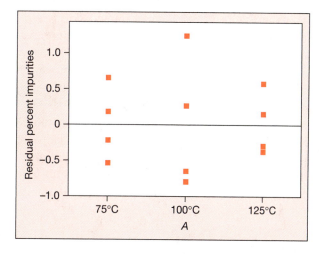

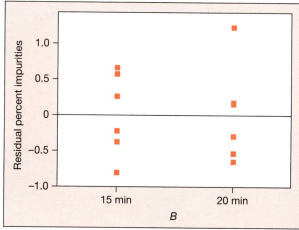

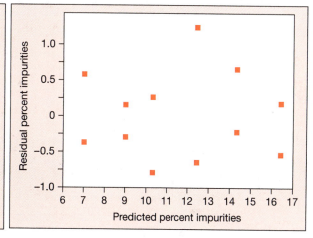

Note that variation is greatest at the middle level of A and the high level of B, which might suggest that the process is more difficult to control at those levels. The plot of residuals against predicted values shows a fairly random scatter, as should be produced by a model that fits well.

Two-level factorial experiments (2×2 or, in general, $k \times 2$ if there are k factors of interest) are often used as screening experiments in preliminary investigations. If a process depends on many factors, two-level experiments are run to determine which factors seem to be affecting the response most dramatically. A higher-level experiment can be run subsequently to find optimal settings for the levels of those factors. Because two-level factorial experiments are so common, we present more details related to their analysis, focusing on graphical techniques.

12.7.3   Graphical methods of analysis

Going back to the data of Example 12.14 for a 2×2 factorial experiment, we first observe that the numerical values of x_1 and x_2 have no effect on the analysis. Thus, we will simply refer to the low levels as '$-$' and the high levels as '$+$'. We then have two observations on yield at A^-B^- (low cooking time, low cooling time), two at A^-B^+, and so on. A convenient way to display the data is on back-to-back dotplots, as in Figure 12.28. We can immediately see a large shift from high to low values, as we move from A^- to A^+, and a less dramatic shift downward as we move from B^- to B^+.

In a similar way, interaction terms can be plotted. The upper left and lower right cells of the data table make positive ($+$) contributions to the interaction, and the other two cells make negative ($-$) contributions. Thus, the interaction dotplot looks like Figure 12.29. Although the $(AB)^+$ values are more spread out, there is little shift of location as we move from $(AB)^-$ to $(AB)^+$. Thus, there is little interaction effect.

The shifts seen in Figures 12.28 and 12.29 can be quantified more precisely in terms of averages. The average of all eight yields is 6.25, while the average of A^+ yields is 4.5 and the average of A^- yields is 8.0. Similar averages can be calculated

Figure 12.28

Dotplots for factors A and B

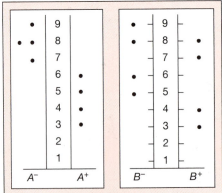

Figure 12.29

Dotplot for interaction

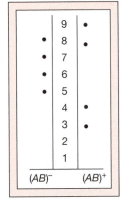

for B^+ and B^-. Figure 12.30 shows the overall average as the center line and the factor-level (cell) averages plotted relative to this center. This is a convenient way to show all averages on one plot.

Notice that the distance between A^+ and the overall mean is the same as the distance between A^- and the overall mean, with a similar result holding for factor B. This distance,

$$\frac{1}{2}(A^+ \text{ average } - A^- \text{ average}) = -1.75$$

is called the *main effect* due to A. It tells us that, on the average, the high cooking time lowers the yield by 1.75 units compared to the experimentwide average yield. Similarly, the main effect for B, -0.75, tells us that the high cooling time lowers the average yield by 0.75 units.

The *interaction effect* is measured by

$$\frac{1}{2}[(AB)^+ \text{average } - (AB)^- \text{ average}] = -0.25$$

and turns out to be small compared to the main effects in this case.

A simple and elegant graphical interpretation of main effects and interaction is developed in Figure 12.31. First we plot the A^- and A^+ averages and connect them with a dashed line. Since we can think of A^- as being at -1 and A^+ as being at $+1$ on the horizontal axis, these points are two units apart, This makes the slope of the dashed line equal to the main effect for A, -1.75 in this case.

We then plot the A^-B^- and A^+B^- averages and connect them with the line labeled B$^-$. Repeating the procedure for A^-B^+ and A^+B^- gives the B^+ line. The slope of the B^+ line is

$$\frac{1}{2}(A^+B^+ \text{ average } - A^-B^+ \text{ average}) = \frac{1}{2}(3.5 - 7.5) = -2.0$$

Figure 12.30

Plot of factor-level means

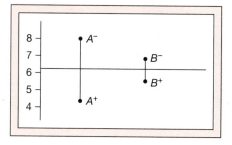

Figure 12.31

Plot of main effect and interaction lines

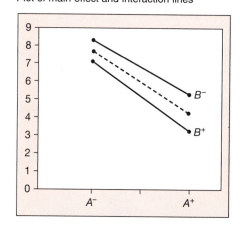

The difference between the slopes of the dashed line (main effect factor A) and the A^+ line is

$$[-2.0 - (-1.75)] = -0.25$$

which is the interaction effect. Whereas a main effect measures the difference between an overall average and a factor-level average, an interaction effect measures the difference between the slope of a main effect line and the slope for that factor plotted at a fixed level of a second factor. To see what happens to these lines for a situation with a large interaction, make a similar plot for the data in Example 12.16.

Example 12.17

An experiment was conducted to compare the strengths of two brands of facial tissues, the famous brands X and Y. The two brands (two levels of factor A) were each tested under both dry and damp conditions (the two levels of factor B). Thus, the four treatments tested are

A^-B^-	(brand X, dry)
A^+B^-	(brand Y, dry)
A^-B^+	(brand X, damp)
A^+B^+	(brand Y, damp)

Four tissues were tested under each treatment. The strength measurement was produced as follows, A tissue was stretched over the opening of a plastic cup and held by a rubber band. Then a marble was dropped onto the stretched tissue. The lowest height (in inches) from which a marble that went through the tissue was dropped is the recorded strength measurement. The data are shown below. (One of the authors actually conducted the experiment!)

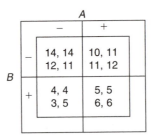

The 14 in the upper left corner of data table means that a tissue of brand X that was dry broke first from a marble dropped from 14 inches above its surface. Analyze the data for significant main effects and interaction.

Solution   Since these are two-digit measurements, stem-and-leaf plots can be used in place of dotplots. The main effect and interaction plots are shown in Figure 12.32.

Figure 12.32

The main effect and interaction plots

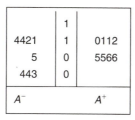

These plots show little shift as we move from A^- to A^+, although A^- values are more spread out than A^+ values. A dramatic shift is seen as we move from B^- to B^+, indicating a large main effect for moisture content. There is some slight shift to larger values as we move from $(AB)^-$ to $(AB)^+$, signaling a possible interaction.

The main effect for A (brand) is

$$\frac{1}{2}(A^+ \text{ average} - A^- \text{ average}) = \frac{1}{2}(8.250 - 8.375) = -0.0625$$

which indicates little difference in strength between brands.

The main effect for B (dry versus damp) is

$$\frac{1}{2}(B^+ \text{ average} - B^- \text{ average}) = \frac{1}{2}(4.750 - 11.875) = -3.5625$$

a rather large value in comparison to the sizes of the measurements. It should not be surprising that damp tissues have less strength than dry tissues. We do not have to experiment very carefully to see this result!

The interaction effect is

$$\frac{1}{2}[(AB)^+ \text{ average} - (AB)^- \text{ average}] = \frac{1}{2}(9.125 - 7.500) = 0.8125$$

This is much larger than the main effect for A and indicates that the interaction is obscuring the effect of brands. A more careful look at the data suggests that brand X tissues tend to be stronger than brand Y tissues when dry, but brand Y tissues are slightly stronger when damp. When do you want tissues to be strong? Do you think you could state a preference for either brand?

The main effect plot here could be misleading because of the interaction, so we show only the interaction plot in Figure 12.33. Note that the two solid lines have obviously different slopes, indicating some interaction.

The analysis of variance shown in Figure 12.34 for these data shows a large interaction mean square and a very large mean square due to factor B, just as we expected. Both of these effects are highly significant.

The analyses of factorial experiments with more than two factors proceed along similar lines.

Figure 12.33

Interaction plot for Example 12.16

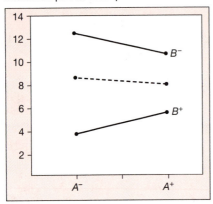

Figure 12.34

Minitab output of Analysis of variance

```
Analysis of Variance

Source          DF           SS            MS

A                1        0.063         0.063
B                1      203.062       203.062

Interaction      1       10.563        10.563

Error           12       11.750         0.979

Total           15      225.437
```

Example 12.18

Fishing boats stay out at sea for days at a time. The fish caught are frozen immediately on the boat, using one of three standard methods (A, B, C). A fishers' association is interested in comparing three different defrosting methods (I, II, III) in terms of preserving the quality of fish. Fish frozen for 1 day and for 8 days were used in the experiment, as duration of freezing might have an effect on the quality of defrosted fish. The quality of fish is recorded as an index computed from the results of several tests. Fish frozen by one of the three methods was defrosted using one of the three methods after being kept frozen for 1 or 8 days. Two samples of each combination were run, and the quality of the defrosted fish was recorded. The data are given in table below.

Does the quality of defrosted fish differ by freezing method, defrosting method, and duration? Do these factors interact to affect the quality?

Defrosting Method		1 Day			8 Days		
		I	**II**	**III**	**I**	**II**	**III**
Freezing Method	**A**	73	70	65	65	68	75
		74	69	67	67	74	73
	B	75	71	69	70	69	68
		76	72	69	72	81	61
	C	74	70	70	72	70	70
		74	65	65	69	80	74

Solution   This is a 2 × 3 × 3 factorial experiment in completely randomized design with two replications. Therefore, there are a total of 18 treatment combinations. There are three factors in the experiment and their levels:

- Duration (2 levels): 1 day and 8 days
- Freezing method (3 levels): A, B, and C
- Defrosting Method (3 levels): I, II, and III

Figure 12.35

JMP output for frozen fish data

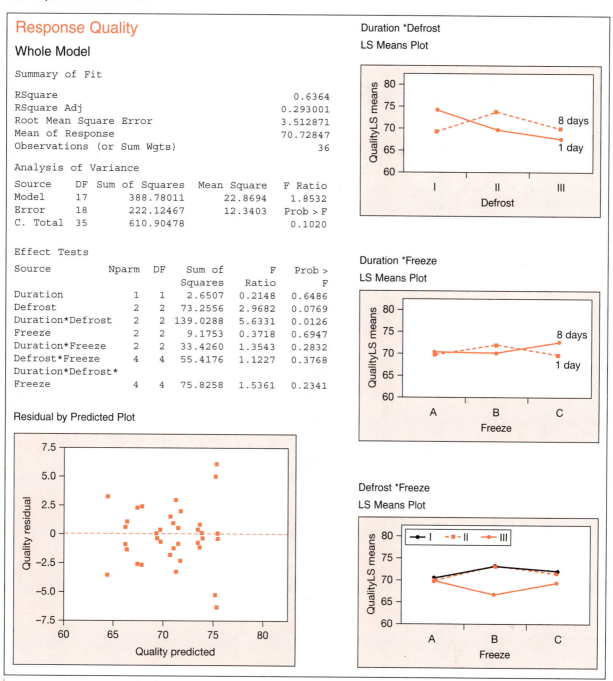

With three factors, we get three 2-factor interactions and one 3-factor interaction. The JMP output for this experiment is given in Figure 12.35.

Start with testing the highest order interaction and work your way down to main effects. For testing,

H_0: There is no 3-factor interaction Duration*Defrost*Freeze effect on the quality.

H_a: 3-factor interaction Duration*Defrost*Freeze affects quality.

The p-value $= 0.2341$ is higher than any reasonable level of significance. Therefore, fail to reject the null hypothesis and conclude that there is no evidence of the presence of a 3-factor interaction.

Now we can test for three 2-factor interactions, Duration*Defrost ($p = 0.0126$), Duration*Freeze ($p = 0.2832$), and Defrost*Freeze ($p = 0.3768$). Their p-values lead to the conclusion that only the interaction Duration*Defrost is significant. Of the three main effects (Duration, Defrost, and Freeze), we can only test for Freeze and it is not significant ($p = 0.6947$). We can construct simultaneous confidence intervals for all other possible differences in means. There are 6 means for combinations of duration and defrosting method ($\mu_{1,I}, \mu_{1,II}, \mu_{1,III}, \mu_{8,I}, \mu_{8,II}, \mu_{8,III}$) resulting in 15 pairs of means.

From the means plots in Figure 12.35, it seems defrosting method I preserves the quality of fish frozen for 1 day better than the quality of fish frozen for 8 days. On the other hand, methods II and III preserve the quality of fish frozen for 8 days better than the quality of fish frozen for 1 day.

Example 12.19

Process and Design Development (1969) reported results of a chemical experiment. The chemical process was run at three different temperatures (30°C, 70°C, and 100°C), three different pressures (30, 70, and 100 psi) for three different reaction times (1, 2, and 3 hours). The resulting percentage yield of BON acid from the process was recorded. The data is presented in the table below.

		1 hour			2 hours			3 hours		
	Temperature	**30°**	**70°**	**100°**	**30°**	**70°**	**100°**	**30°**	**70°**	**100°**
Pressure	**30 psi**	68.5	72.8	72.5	74.5	72.0	75.5	70.5	69.5	65.0
	70 psi	73.0	80.1	72.5	75.0	81.5	70.0	72.5	84.5	66.5
	100 psi	68.7	72.0	73.1	74.6	76.0	76.0	74.7	76.0	70.5

Does the yield differ significantly by temperature, pressure, and reaction time? Do these factors interact to affect the yield?

Solution   This is a $3 \times 3 \times 3$ factorial experiment in a completely randomized design with one replication. Therefore, there are a total of 27 treatment combinations. The three factors in the experiment and their levels are as follows:

- Reaction time (3 levels): 1 hour, 2 hours, and 3 hours
- Temperature (3 levels): 30°C, 70°C, and 100°C
- Pressure (3 levels): 30, 70 and 100 psi

With the three factors, we get three 2-factor interactions and one 3-factor interaction. However, without replication we will not be able to estimate error, and thereby we won't be able to perform any statistical tests. Suppose we assume that there is no highest order interaction, namely time*temperature*pressure. Then the SS(Time*Temp*Press) provides an estimate of σ^2. The JMP printout of this experiment is given below.

We can start with 2-factor interactions. Suppose we test for interaction Temp*Press first.

H_0: There is no Temp*Press interaction.

H_a: Temperature and pressure interact to affect yield.

The output gave the p-value for the F-test as $0.001 < \alpha = 0.05$. Therefore, reject the null hypothesis and conclude that the interaction Temp*Press is significant. In other words, different temperatures have different effects on the yield when the reaction is run at different pressure levels.

Similarly we can test for the other two 2-factor interactions, namely, Time*Temp and Time*Press. The p-values for these interactions are 0.0098 and 0.0559, respectively. At the 5% level of significance, we conclude that the interaction Time*Temp is significant; however, the interaction Time*Press is not significant. The interaction plots (Figure 12.36) also indicate the same.

All three main effects Time ($p = 0.0096$), Temp ($p = 0.0005$), and Press ($p = 0.0023$) are statistically significant. However, in the presence of significant interactions, we need to exercise caution in interpreting the significance of main effects.

Response Yield

Whole Model

Summary of Fit

RSquare	0.959918
RSquare Adj	0.869732
Root Mean Square Error	1.550119
Mean of Response	73.25926
Observations (or Sum Wgts)	27

Analysis of Variance

Source	DF	Sum of Squares	Mean Square	F Ratio
Model	18	460.36222	25.5757	10.6438
Error	8	19.22296	2.4029	Prob > F
C. Total	26	479.58519		0.0010

Effect Tests

Source	Nparm	DF	Sum of Squares	F Ratio	Prob > F
Time	2	2	42.1118	8.7628	0.0096
Temp	2	2	110.7318	23.0416	0.0005
Time*Temp	4	4	67.7614	7.0501	0.0098
Pressure	2	2	68.1363	14.1781	0.0023
Time*Pressure	4	4	35.1837	3.6606	0.0559
Temp*Pressure	4	4	136.4370	14.1952	0.0010

Figure 12.36
Interaction plots for
BON acid data

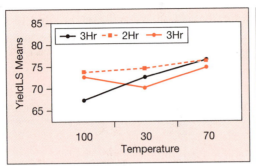

(a) Time * Temperature Means plot

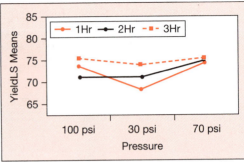

(b) Time * Pressure Means plot

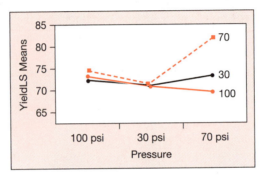

(c) Temperature * Pressure Means plot

Exercises

12.44 In pressure sintering of alumina., two important variables controlled by the experimenter are the pressure and time of sintering. An experiment involving two pressures and two times, with three specimens tested on each pressure/time, combination, showed the densities (in v/cc) listed in table below.

Time	Pressure					
	100 psi			200 psi		
10 min.	3.6	3.5	3.3	3.8	3.9	3.8
20 min.	3.4	3.7	3.7	4.1	3.9	4.2

a Perform an analysis of variance, constructing appropriate tests for interaction and main effects. Use $\alpha = 0.05$.

b Estimate the difference between the true densities for specimens sintered for 20 minutes and those sintered for 10 minutes. Use a 95% confidence coefficient.

12.45 The yield percentage Y of a chemical process depends on the temperature at which the

process is run and the length of time the process is active. For two levels of temperature and two lengths of time, the yields are listed in table below (with two observations per treatment):

Time	Temperature	
	Low	High
Low	24	28
	45	30
High	26	23
	28	22

a Perform an analysis of variance, testing first for interaction. Use $\alpha = 0.05$.

b If interaction is significant, construct confidence intervals on the six possible differences between treatment means, using a simultaneous confidence coefficient of approximately 0.90.

12.46 The yield percentage Y of a certain precipitate depends on the concentration of the reactant and the rate of addition of diammonium hydrogen phosphate. Experiments were run at three different concentration levels and three addition rates. The yield percentages, with two observations per treatment, are listed in table below.

Addition Rate	Concentration of Reactant		
	−1	0	1
−1	90.1	92.4	96.4
	90.3	91.8	96.8
0	91.2	94.3	98.2
	92.3	93.9	97.6
1	92.4	96.1	99.0
	92.5	95.8	98.9

a Perform an analysis of variance, constructing all appropriate tests at the 5% significance level.

b Estimate the average yield percentage, in a 95% confidence interval, for the treatment at the middle level of both concentration and addition rate.

12.47 How do women compare with men in their ability to perform laborious tasks that require strength? Some information on this question is provided in a study by Phillips and Pepper of the fire-fighting ability of men and women. ("Shipboard Fire-Fighting Performance of Females and Males," *Human Factors*, 1982). Phillips and Pepper conducted a 2 × 2 factorial experiment to investigate the effect of the factor sex (male or female) and the factor weight (light or heavy) on the length of time required for a person to perform a particular fire-fighting task. Eight persons were selected for each of the 2 × 2 = 4 sex/weight categories of the 2 × 2 factorial experiment, and the length of time needed to complete the task was recorded for each of the 32 persons. The means and standard deviations of the four samples are shown in table below.

	Light		Heavy	
	Mean	Standard Deviation	Mean	Standard Deviation
Female	18.30	6.81	14.50	2.93
Male	13.00	5.04	12.25	5.70

a Calculate the total of the $n = 8$ time measurements for each of the four categories of the 2 × 2 factorial experiment.

b Use the result of part (a) to calculate the sums of squares for sex, weight, and the sex/weight interaction.

c Calculate each sample variance. Then calculate the sum of squares of deviations *within* each sample for each of the four samples.

d Calculate SSE. [*Hint:* SSE is the pooled sum of squares of the deviations calculated in part (c).]

e Now that you know SS(Sex), SS(Weight), SS(Sex × Weight), and SSE, find SS (Total).

f Summarize the calculations in an analysis of variance table.

g Explain the practical significance of the presence (or absence) of sex/weight interaction. Do the data provide evidence of a sex/weight interaction?

h Do the data provide sufficient evidence to indicate a difference in time required to complete the task between light men and light women? Test by using $\alpha = 0.05$.

i Do the data provide sufficient evidence to indicate a difference in time to complete the task between heavy men and heavy women? Test by using $\alpha = 0.05$.

12.48 Refer to Exercise 12.47. Phillips and Pepper (1982) give data on another 2 × 2 factorial experiment utilizing 20 males and 20 females. The experiment involved the same treatments, with ten persons assigned to each sex/weight category. The response measured for each person was the pulling force the person was able to exert on the starter cord of P-250 fire pump. The means and standard deviations of the four samples (corresponding to the 2 × 2 = 4 categories of the experiment) are shown in table below.

	Light		Heavy	
	Mean	Standard Deviation	Mean	Standard Deviation
Female	46.26	14.23	62.72	13.97
Male	88.07	8.32	86.29	12.45

a Use the procedures outlined in Exercise 11.42 to perform an analysis of variance for the experiment. Display your results in an ANOVA table.

b Explain the practical significance of the presence (or absence) of sex/weight interaction. Do the data provide sufficient evidence of a sex/weight interaction?

c Do the data provide sufficient evidence to indicate a difference in force exerted between light men and light women? Test by using $\alpha = 0.05$.

d Do the data provide sufficient, evidence to indicate a difference in force exerted between heavy men and heavy women? Test by using $\alpha = 0.05$.

12.49 In analyzing coal samples for ash content, two types of crucibles and three temperatures were used in a complete factorial arrangement, with as listed in table below.

		Crucible			
		Steel		**Silica**	
Temperature	825	8.7	7.2	9.3	9.1
	875	9.4	9.6	9.7	9.8
	925	10.1	10.2	10.4	10.7

(Two independent observations were taken on each treatment.) Note that "crucible" is not a quantitative factor, but it can still be considered to have two levels even though the levels cannot be ordered as to high and low.

Perform an analysis of variance, conducting appropriate tests at the 5% significance level. Does there appear to be a difference between the two types of crucible with respect to average ash content?

12.50 Four different types of heads were tested on each of two sealing machines. Four independent measurements of strain were then made on each head/machine combination. The data are given table below.

		Head		
	1	**2**	**3**	**4**
Machine A	3	2	6	3
	0	1	5	0
	4	1	8	1
	1	3	8	1
Machine B	6	7	2	4
	8	6	0	7
	6	3	1	6
	5	4	2	7

a Perform an analysis of variance, making all appropriate tests at the 5% level of significance.
b If you were to use machine A, which head type would you recommend for maximum strain resistance in the seal? Why?

12.51 An experiment was conducted to determine the effects of two alloying elements (carbon and manganese) on the ductility of specimens of metal. Two specimens from each treatment were measured, and the data (in work required to break a specimen of standard dimension) are listed in table below.

		Carbon	
		0.2%	**0.5%**
Manganese	0.5%	34.5	38.2
		37.5	39.4
	1.0%	36.4	42.8
		37.1	43.4

a Perform an analysis of variance, conducting all appropriate tests at the 5% significance level.
b Which treatment would you recommend to maximize average breaking strength? Why?

12.52 Eisinger et al. (*Transactions of the ASME*, 2003) conducted laboratory scale model tests of one simulated stack liner exposed to airflow at ambient atmospheric conditions. The tests were conducted at three different flow velocities (v m/s) to obtain baseline data and data with two and three guide vanes inserted into the 90-degree bends of the pipe. Two replications of each combination were tested, and the maximum acoustic pressure (P Pa) and the pressure drop through the bends (Δp Pa) were recorded as shown in table below.

v m/s	No Vanes		2 Guide Vanes		3 Guide Vanes	
	P	**Δp**	**P**	**Δp**	**P**	**Δp**
21.5	0.408	202	0.185	130	0.173	130
π	0.538	199	0.189	132	0.185	132
23.5	0.653	230	0.303	160	0.292	160
	0.853	238	0.360	161	0.399	161
24.5	0.860	250	0.387	170	0.343	170
	1.100	254	0.530	172	0.443	172

a For measuring the maximum acoustic pressure, is there any interaction effect between the flow velocity and the number of guide vanes inserted?
For measuring the pressure drop, is there any interaction effect between the flow velocity and the number of guide vanes inserted?
b Does the mean maximum acoustic pressure differ significantly by the flow velocity?
Does the mean pressure drop differ significantly by the flow velocity?
c Does the mean maximum acoustic pressure differ significantly by the number of guidelines used?
Does the mean pressure drop differ significantly by the number of guidelines used?

12.8   Summary

Before measurements are actually made in any experimental investigation, careful attention should be paid to the *design* of the experiment. If the experimental objective is the comparison of *k* treatment means and no other major source of variation is present then a *completely randomized design* will be adequate. If a second source of variation is present, it can often be controlled through the use of a *randomized block design*. These two designs serve as an introduction to the topic of design of experiments, but several more complex designs, such as the *Latin square*, can be considered.

The analysis of variance for either design considered in this chapter can be carried out either through a linear model (regression) approach or through the use of direct calculation formulas. The reader would benefit by trying both approaches on some examples to gain familiarity with the concepts and techniques.

The *factorial experiment* arises in situations in which treatments are defined by various combinations of factor levels. Notice that the factorial arrangement defines the treatments of interest but is *not* in itself a design. Factorial experiments can be run in completely randomized, randomized block, or other designs, but we considered their use only in the completely randomized design in this chapter. Many more topics dealing with factorial experiments can be found in texts that focus on experimental design and analysis of variance.

Supplementary Exercises

12.53 Three methods have been devised to reduce the time spent in transferring materials from one location to another. With no previous information available on the effectiveness of these three approaches, a study is performed. Each approach is tried several times, and the amount of time to completion (in hours) is recorded in the table below:

Method		
A	B	C
8.2	7.9	7.1
7.1	8.1	7.4
7.8	8.3	6.9
8.9	8.5	6.8
8.8	7.6	
	8.5	

a What type of experimental design was used?

b Is there evidence that the mean time to completion of the task differs for the three methods? Use $\alpha = 0.01$.

c Form a 95% confidante interval for the mean time to completion for method B.

12.54 One important consideration in determining which location is best for a new retail business is the amount of traffic that passes the location each business day. Counters are placed at each of four locations on the five weekdays, and the number of cars passing each location is recorded in the table below:

Day	Location			
	I	II	III	IV
1	453	482	444	395
2	500	605	505	490
3	392	400	383	390
4	441	450	429	405
5	427	431	440	430

a What type of design does this represent?

b Is there evidence of a difference in the mean number of cars per day at the four locations?

c Estimate the difference between the mean numbers of cars that pass locations I and III each weekday.

12.55 Mileage tests were performed to compare three different brands of unleaded gas. Four different automobiles were used in the experiment, and each brand of gas was used in each car until the mileage was determined. The results, in miles per gallon, are shown in the table below:

	Automobile			
Brand	1	2	3	4
A	20.2	18.7	19.7	17.9
B	19.7	19.0	20.3	19.0
C	18.3	18.5	17.9	21.1

a Is there evidence of a difference in the mean mileage rating among the three brands of gasoline? Use $\alpha = 0.05$.

b Construct the ANQVA summary table for this experiment.

c Is there evidence of a difference in the mean mileage for the four models? Thai is, is blocking important in this type of experiment? Use $\alpha = 0.05$.

d Form a 99% confidence interval for the difference between the mileage ratings of brands B and C.

e Form confidence intervals for the three possible differences between the means for the brands, with a simultaneous confidence coefficient of approximately 0.90.

12.56 The following table shows the partially completed analysis of variance for a two-factor factorial experiment.

Source	df	SS	MS	F ratio
A	3	2.6		
B	5	9.2		
AB			3.1	
Error		18.7		
Total	47			

a Complete the analysis of variance table.

b Give the number of levels for each factor and the number of observations per factor-level combination.

c Do the data provide sufficient evidence to indicate an interaction between the factors? Test by using $\alpha = 0.05$.

d State the practical implications of your test results in part (c).

12.57 The data shown in the table below: are for a 4×3 factorial experiment with two observations per factor-level combination.

		Level of B		
		1	2	3
Level of A	1	2	5	1
		4	6	3
	2	5	2	10
		4	2	9
	3	7	1	5
		10	0	3
	4	8	12	7
		7	11	4

a Perform an analysis of variance for the data and display the results in an analysis of variance table.

b Do the data provide sufficient information to indicate an interaction between the factors? Test by using $\alpha = 0.05$.

c Suppose the objective of the experiment is to select the factor-level combination with the largest mean. Based on the data and using a simultaneous confidence coefficient of approximately 0.90, which pairs of means appear to differ?

12.58 The set of activities and decisions through which a firm moves from its initial awareness of an innovative industrial procedure to its final adoption or rejection of the innovation is referred to as the *industrial adoption process*. The process can be described as having five stages: (1) awareness, (2) interest (additional information requested), (3) evaluation (advantages and disadvantages compared), (4) trial (innovation tested), and (5) adoption. As part of a study of the industrial adoption process, S. Ozanne and R. Churchill (1971) hypothesized that firms use a greater number of informational inputs to the process (e.g., visits by salespersons) in the later stages than in the earlier stages. In particular, they tested the hypotheses that a greater number of informational inputs are used in the interest stage than in the awareness stage and that a greater number are used in the evaluation stage than in the interest stage. Ozanne and Churchill collected the information given in Table 12.17 on the number of informational inputs used by a sample of 37 industrial firms that had recently adopted a particular new automatic machine tool.

Table 12.17

Number of Informational Inputs

Company	Number of Information Sources Used			Company	Number of Information Sources Used		
	Awareness Stage	Interest Stage	Evaluation Stage		Awareness Stage	Interest Stage	Evaluation Stage
1	2	2	3	20	1	1	1
2	1	1	2	21	1	2	1
3	3	2	3	22	1	4	3
4	2	1	2	23	2	3	3
5	3	2	4	24	1	1	1
6	3	4	6	25	3	1	4
7	1	1	2	26	1	1	5
8	1	3	4	27	1	3	1
9	2	3	3	28	1	1	1
10	3	2	4	29	3	2	2
11	1	2	4	30	4	2	3
12	3	3	3	31	1	2	3
13	4	4	4	32	1	3	2
14	4	2	7	33	1	2	2
15	3	2	2	34	2	3	2
16	4	5	4	35	1	2	4
17	2	1	1	36	2	2	3
18	1	1	3	37	2	3	2
19	1	3	2				

Source: S. Qzanne and R. Churchill, *Marketing Models, Behavioral Science Applications*, Intex Educational Publishers, 1971, pp. 249–265.

a Do the data provide, sufficient evidence to indicate that differences exist in the mean number of informational inputs of the three stages of the industrial adoption process studied by Ozanne and Churchill? Test by using $\alpha = 0.05$.

b Find a 90% confidence interval for the difference in the mean number of informational inputs between the evaluation and awareness stages.

12.59 England has experimented with different 40-hour workweeks to maximize, production and minimize expenses. A factory tested a 5-day week (8 hours per day), a 4-day week (10 hours per day), and a $3\frac{1}{2}$ day week (12 hours per day), with the weekly production results shown in the table below (in thousands of dollars worth of items produced).

8-Hour Day	10-Hour Day	12-HourDay
87	75	95
96	82	76
75	90	87
90	80	82
72	73	65
86		

a What type of experimental design was employed here?

b Construct an ANOVA summary table for this experiment.

c Is there evidence of a difference in the mean productivities for the three lengths of work-days?

d Form a 90% confidence interval for the mean weekly productivity when 12-hour workdays are used.

12.60 To compare the preferences of technicians for three brands of calculators, each of three technicians was required to perform an identical series of calculations on each of the three calculators, A, B, and C. To avoid the possibility of fatigue, a suitable time period separated each set of calculations, and the calculators were used in random order by each technician. A preference rating, based on a 0–100 scale, was recorded for each machine/technician combination. These data are shown in table below.

	Calculator Brand		
Technician	A	B	C
1	85	90	95
2	70	70	75
3	65	60	80

a Do the data provide sufficient evidence to indicate a difference in technician preference among the three brands? Use $\alpha = 0.05$.

b Why did the experimenter have each technician test all three calculators? Why not randomly assign three different technicians to each calculator?

12.61 Sixteen workers were randomly selected to participate in an experiment to determine the effects of work scheduling and method of payment on attitude toward the job. Two types of scheduling were employed, the standard 8 to 5 workday and a modification whereby the worker was permitted to start the day at either 7 A.M. or 8 A.M. and to vary the starting time as desired; in addition, the worker was allowed to choose, on a daily basis, either a $\frac{1}{2}$-hour or 1-hour lunch period. The two methods of payment were a standard hourly rate and a reduced hourly rate with an added piece rate based on the worker's production. Four workers were randomly assigned to each of the four scheduling/payment combinations, and each completed an attitude test after one month on the job. The test scores are shown in table below.

		Payment	
Scheduling	Hourly Rate	Hourly and Piece Rate
8 to 5	54, 68	89, 75
	55, 63	71, 83
Worker-modified Schedule	79, 65	83, 94
	62, 74	91, 86

a Construct an analysis of variance table for the data.

b Do the data provide sufficient information to indicate a factor interaction? Test by using $\alpha = 0.05$. Explain the practical implications of your test.

c Do the data indicate that any of the scheduling/payment combinations produce a mean attitude score that is clearly higher than the other three? Test by using a simultaneous Type I error rate of approximately $\alpha = 0.05$, and interpret your results.

12.62 An experiment was conducted to compare the yields of orange juice for six different juice extractors. Because of a possibility of a variation in the amount of juice per orange from one truckload of oranges to another, equal weights of oranges from a single truckload were assigned to each extractor, and this process was repeated for 15 loads. The amount of juice recorded for each extractor and each truckload produced the following sums of squares:

Source	df	SS	MS	F-ratio
Extractor	——	84.71	——	——
Truckload	——	159.29	——	——
Error	——	94.33	——	
Total	——	339.33		

a Complete the ANOVA table.

b Do the data provide sufficient evidence to indicate a difference in mean amount of juice extracted by the six extractors? Use $\alpha = 0.05$.

12.63 A farmer wants to determine the effect of five different concentrations of lime on the pH (acidity) of the soil on a farm. Fifteen soil samples are to be used in the experiment, five from each of the three different locations. The five soil samples from each location are then randomly assigned to the five concentrations of lime, and one week after the lime is applied the pH of the soil is measured. The data are shown in the table below.

	Lime Concentration				
Location	0	1	2	3	4
I	3.2	3.6	3.9	4.0	4.1
II	3.6	3.7	4.2	4.3	4.3
III	3.5	2.9	4.0	3.9	4.2

a What type of experimental design was used here?

b Do the data provide sufficient evidence to indicate that the five concentrations of lime have different mean soil pH levels? Use $\alpha = 0.05$.

c Is there evidence of a difference in soil pH levels among locations? Test by using $\alpha = 0.05$.

12.64 In the hope of attracting more riders, a city transit company plans to have express bus service from a suburban terminal to the downtown business district. These buses should save travel time. The city decides to perform a study of the effect of four different plans (such as a special bus lane and traffic signal progression) on the travel time for the buses. Travel times (in minutes) are measured for several weekdays during a morning rush-hour trip while each plan is in effect. The results are recorded in table below.

	Plan		
1	**2**	**3**	**4**
27	25	34	30
25	28	29	33
29	30	32	31
26	27	31	
	24	36	

a What type of experimental design was employed?

b Is there evidence of a difference in the mean travel times for the four plans? Use $\alpha = 0.01$.

c Form a 95% confidence interval for the difference between plan 1 (express lane) and plan 3 (a control: no special travel arrangements).

12.65 Five sheets of writing paper are randomly selected from each of three batches produced by a certain company. A measure of brightness is obtained for each sheet, with the results listed in table below.

	Batch	
1	**2**	**3**
28	34	27
32	36	25
25	32	29
27	38	31
26	39	21

a Do the mean brightness measurements seem to differ among the three batches? Use $\alpha = 0.05$.

b If you were to select the batch with the largest mean brightness, which would you select? Why?

12.66 In order to provide its clients with comparative information on two large suburban residential communities, a realtor wants to know the average home value in each community. Eight homes are selected at random within each community and are appraised by the realtor. The appraisals are given in table below (in thousands of dollars). Can you conclude that the average home value is different in the two communities? You have three ways of analyzing this problem [parts (a) through (c)].

Community	
A	**B**
43.5	73.5
49.5	62.0
38.0	47.5
66.5	36.5
57.5	44.5
32.0	56.0
67.5	68.0
71.5	63.5

a Use the two-sample t-statistic to test $H_0: \mu_A = \mu_B$.

b Consider the regression model

$$y = \beta_0 + \beta_1 x + \varepsilon$$

where

$$x = \begin{cases} 1 & \text{if community } B \\ 0 & \text{if community } A \end{cases}$$

y = appraised price

Since $\beta_1 = \mu_B - \mu_A$, testing $H_0: \beta_1 = 0$ is equivalent to testing $H_0: \mu_A = \mu_B$. Use the partial reproduction of the regression printout shown in Figure 12.37 to test $H_0: \beta_1 = 0$. Use $\alpha = 0.05$.

c Use ANOVA method to test $H_0: \mu_A = \mu_B$ Use $\alpha = 0.05$.

d Using the results of the three tests in parts (a) through (c), verify that the tests are equivalent (for this special case, $k = 2$) for the completely randomized design in terms of the test-statistic value and rejection region. For the three methods used, what are the advantages and disadvantages (limitations) of using each method in analyzing results, for this type of experimental design?

12.67 Drying stresses in wood produce acoustic emissions (AE), and the rate of acoustic emission can be monitored as a check on drying conditions. The data shown in Table 12.18 are the results of measuring peak valued of acoustic emission over 10-second intervals for wood specimens with

Figure 12.37

Minitab output for Exercise 12.66

```
SOURCE              DIF    SUM OF SQUARES    MEAN SQUARE    F VALUE        PR > F
MODEL                1        40.64062500    40.64062500       0.21        0.6501
ERROR               14      2648.71875000   189.19419643
CORRECTED TOTAL     15      2689.35937500

                                                          R-SQUARE       ROOT MSE
                                                          0.015112    13.75478813


PARAMETER          ESTIMATE           T FOR H0:      PR >|T|    STD ERROR OF
                                  PARAMETER = 0                    ESTIMATE
INTERCEPT       53.25000000              10.95       0.0001      4.86305198
X                3.18750000               0.46       0.6501      6.87739406
```

Table 12.18

Acoustic Emission Rates

Test No.	Diameter (mm)	Thickness (mm)	Dry-Bulb Temp. (°C)	RH (%)	Peak Values of AE Event Count Rates (1/10 s)
1	180	20	40	40	780
2	180	20	50	50	695
3	180	20	60	30	1,140
4	180	25	40	40	270
5	180	25	50	50	340
6	180	25	60	30	490
7	180	30	40	50	160
8	180	30	50	30	270
9	180	30	60	40	185
10	125	20	40	50	620
11	125	20	50	30	1,395
12	125	20	60	40	1,470
13	125	25	40	30	645
14	125	25	50	40	1,020
15	125	25	60	50	755
16	125	30	40	30	545
17	125	30	50	40	635
18	125	30	60	50	335

Source: M. Naguchi, S. Kitayama, K. Satiyoshi, and J. Umetso, *Forest Products Journal*, 37(1), 1987, pp. 28–34.

controlled diameter, thickness, temperature, and relative humidity (RH). Which factors appear to affect the AE rates?

12.68 In a study of wood-laminating adhesives by Kreibich and Hemingway (*Forest Products Journal*, 37(2), 1987, pp. 43–46), tannin is added to the usual phenol-resorcinol-formaldehyde (PRF) resin and tested for strength. For four mixtures of resin and tannin and two types of test (test A, dry shear test; test B, two-hour boil test), the resulting shear strengths of the wood specimens (in psi) were as listed in table below. What can you conclude about the effects of the tannin mixtures on the strength of the laminate?

		Test	
		A	**B**
	50/50	1,960	1,460
		2,267	140
	40/60	1,450	1,620
PRF/Tannin Ratio		1,683	2.287
	30/70	2,080	760
		1,750	130
	23/77	1,493	650
		1,850	90

12.69 Acid precipitation studies in southeastern Arizona involved sampling rain from two sites (Bisbee and the Tucson Research Ranch of the National Audubon Society) and sampling air from smelter plumes. The data for five elements, in terms of their ratios to copper, are given in the Table 12.19. One objective of the study is to see if the plume samples differ from the rain samples at the oilier sites. What can you say about this aspect of the study?

12.70 Lumber moisture content (MC) must he monitored during drying in order to maximize the lumber quality and minimize the drying coats. Four types of

Table 12.19

Summary Statistics for Metals in Precipitation and Plume Samples

Sample	Statistic	Mass Ratio (g/g)				
		As/Cu	Cd/Cu	Pb/Cu	Sb/Cu	Zn/Cu
Bisbee (8/84–10/84)	Mean	0.46	0.068	1.03	0.073	1.64
	S.E.	0.15	0.017	0.16	0.019	0.31
	Min.	0.14	0.029	0.31	0043	0.58
	Max.	1.33	0.19	0.26	0.108	3.79
	n	7	10	13	3	13
TRR (8/84–10/84)	Mean	0.48	0.087	0.94	0.078	4.2
	S.E.	0.12	0.024	0.17	0.018	1.1
	Min.	0.13	0.019	0.26	0.037	0.8
	Max.	1.3	0.30	2.1	0.16	12.2
	n	11	11	11	7	11
TRR (8/84–9/85)	Mean	0.56	0.074	0.82	0.10	9.7
	S.E.	0.067	0.011	0.07	0 016	1.6
	Min.	0.086	0.014	0.08	0.020	0.8
	Max.	1.33	0.30	2.1	0.17	55
	n	31	31	49	11	49
Smelter Plumes	Mean	0.72	0.077	0.90	0.054	1.7
	S.E.	0.37	0.022	0.23	0.034	0.67
	Min.	0.20	0.01	0.44	0.014	0.11
	Max.	2.2	0.15	1.6	0.19	3.8
	n	5	5	4	5	5

Source: C. Blanchard and M. Stromberg, *Atmospheric Environment*, vol. 21, no. 11, 1987, pp. 2375–2381.

Table 12.20

Variation in Average MC between Three Repeat Runs (CV = coefficient of variation = SD/$\bar{x}$)

	Resistance	Power-Loss	High-Frequency	Low-Frequency
20% MC Level	$\bar{X} = 17.8$ $SD = 0.3$ $CV = 1.7\%$	$\bar{X} = 14.2$ $SD = 0.2$ $CV = 1.4\%$	$\bar{X} = 16.8$ $SD = 0.1$ $CV = 0.06\%$	$\bar{X} = 19.4$ $SD = 0.1$ $CV = 0.3\%$
15% MC Level	$\bar{X} = 13.5$ $SD = 0.2$ $CV = 1.1\%$	$\bar{X} = 11.9$ $SD = 0.1$ $CV = 0.9\%$	$\bar{X} = 13.7$ $SD = 0.1$ $CV = 0.8\%$	$\bar{X} = 16.9$ $SD = 0.1$ $CV = 0.5\%$
10% MC Level	$\bar{X} = 11.4$ $SD = 0.1$ $CV = 0.9\%$	$\bar{X} = 10.1$ $SD = 0.1$ $CV = 1.3\%$	$\bar{X} = 11.3$ $SD = 0.1$ $CV = 0.6\%$	$\bar{X} = 15.1$ $SD = 0.1$ $CV = 0.5\%$

moisture meters were studied on three levels of moisture content in lumber by Breiner, Arganbright, and Pong (*Forest Products Journal*, 37(4), 1987, pp. 9–16). Of the four meters used, two were handheld (resistance and power-loss types) and two were in-line (high- and low-frequency). From

Table 12.21

Results of Maximum Force Tests: Mean (interquartile range)

Test	High-Proficiency		Medium-Proficiency	
	Popular	Classical	Popular	Classical
C major scale	72 (53–83)	63(51–72)	51(43–59)	44(38–57)
G_5 at 95 dB				
Maximum force	85(63–97)	73(59–33)	54(46–66)	46(28–56)
Normal force	24(13–23)	27(20–35)	28(22–35)	26(l8–29)
Minimum force	22(18–30)	23(16–27)	24(20–28)	23(15–30)

the data given in Table 12.20, can you determine whether any significant differences exist among the meters for the various MC levels?

12.71 Mouthpiece forces produced while playing the trumpet were studied by Barbenel, Kerry, and Davies (*Journal of Biomechanics*, 1988). Twenty-five high-proficiency players and 35 medium-proficiency players produced results summarized in Table 12.21.

Given only mean values and interquartile ranges, how would you describe the differences, if any, between high-proficiency and medium-proficiency players playing a C major scale? Do there appear to be differences in mean forces between popular and classical play of C major scales? Describe the main differences, if any, among maximum, normal, and minimum forces measured during the playing of G_5 at 95 dB.

12.72 Aerosol particles in the atmosphere influence the transmission of solar radiation by absorption and scattering Absorption causes direct heating of the atmosphere, and scattering causes some radiation to be reflected to space and diffuses radiation reaching the ground. Thus, accurate measurements of absorption and scattering coefficients are quite important. The data given in Table 12.22 come from a study by Kilsby and Smith (*Atmospheric Environment*, 1987). Estimate the difference between aircraft and ground mean scattering coefficient (with a 95% confidence coefficient), assuming the samples are of size 3. Repeat, assuming the samples are of size 20. Perform similar estimations for the absorption coefficients.

12.73 The water quality beneath five retention/recharge basins for an urban area was investigated by the EPA. Data on salinity of the water for samples taken at varying depths from each basin are shown in table below. How would you suggest comparing the mean salinity for the five basins, in light of the

Basin Name	Depth of Sample, m	Salinity		
		Mean	Standard Deviation	No. of Observations
F	2.74	0.123	0.063	10
	8.53	0.099	0.016	3
	15.8	0.112	0.035	15
	20.7	0.153	0.075	18
G	2.74	0.119	NC	1
	14.0	0.541	NC	1
	23.8	0.744	0.090	16
M	3.64	0.294	NC	1
	11.3	0.150	0.035	7
	20.7	0.148	NC	1
	26.2	0.109	0.034	12
EE	3.66	0.085	0.042	13
	7.01	0.158	0.054	12
	10.1	0.175	0.035	14
MM	3.66	0.050	0.012	5
	8.53	0.054	0.015	8
	17.7	0.061	0.020	8

Source: H. Nightingale, *Water Resources Bulletin*, vol. 23, no. 2, 1987, pp. 197–205.

Table 12.22

Comparison of Ground and Aircraft Measurements around Noon on July 27, 1984 (standard deviations given in parentheses)

Instrument	Quantity	Aircraft Measurement	Ground Measurement	Ratio Aircraft/Ground
Nephelometer	Scattering coefficient at 0.55 μm (10^{-4}m)	1.2($\pm$0.2)	0.8($\pm$0.2)	1.5($\pm$0.5)
Filter sample	Absorption coefficient at 0.55 μm (10^{-6}/m)	15.5($\pm$2.5)	8.3($\pm$1.3)	1.9($\pm$0.6)
FSSP	Particle concentration $0.25 < r < 3.75$ μm (cm^{-3})	3.5($\pm$1.0)	2.6($\pm$1.0)	1.3($\pm$0.8)

fact that measurement depths are not the same? Carry out a statistical analysis of these means.

12.74 How do back and arm strengths of men and women change as the strength test moves from static to dynamic situations? This question was investigated by Kumar, Chaffin, and Redfern (*Journal of Biomechanics*. 1988). The dynamic conditions required movement at 20, 60, and 100 cm/sec. Ten male and ten female young adult volunteers without a history of back pain participated in the study. From the data given in table below, analyze the differences between male and female strengths, the differences between back and arm strengths, and the pattern of strengths as one moves from a static strength test through three levels of dynamic strength tests.

12.75 Instantaneous flood discharge rates (in m^3/sec) were recorded over many years for 11 rivers in Connecticut. The sample means and standard deviations are reported in Table 12.23. Noting that the data are highly variable, how would you compare the mean discharge rates for these rivers? Perform an analysis to locate significant differences among these means, if possible.

12.76 Does channelization have an effect on stream flow? This question is to be answered from data on annual stream flow and annual precipitation for two watersheds in Florida. (See Shirmohammadi, Sheridan, and Knisel, *Water Resources Bulletin*, 1987) The data arc summarized in Table 12.24. Does there seem to be a significant difference in mean stream flow before

		Task	Mean Peak Strength
Males	Back	Static	726
		Slow	672
		Medium	639
		Fast	597
	Arm	Static	521
		Slow	399
		Medium	332
		Fast	275
Females	Back	Static	503
		Slow	487
		Medium	432
		Fast	436
	Arm	Static	296
		Slow	266
		Medium	221
		Fast	192

and after channelization, even with differences in precipitation present?

12.77 Scruggs and Iwan (*Journal of Structural Engineering*, 2003) studied implementation of a brushless DC machine as a force actuator for use in suppressing vibrations in civil structures. The data for no device, inactive device, semi-active brushless DC, semi-active magnetorheological (MR), and semi-active passive fluid dampers are given in table below. The performance of devices was measured using distances (d_i in cm), interstory drifts (x_i in cm), and absolute accelerations (a_i in m/s2).

	No Device	Inactive	Semi-active BDC	Semi-active MR	Semi-active Passive
d_i (cm)	0.542	0.404	0.103	0.114	0.107
	0.313	0.223	0.088	0.090	0.125
	0.199	0.140	0.060	0.101	0.089
x_i (cm)	0.542	0.404	0.103	0.114	0.107
	0.826	0.509	0.152	0.185	0.191
	0.961	0.624	0.198	0.212	0.277
a_i (m/s^2)	8.660	5.610	2.50	6.960	2.960
	10.51	7.670	2.530	7.390	3.830
	13.87	9.770	4.170	7.030	6.180

a At the 5% level of significance, compare mean distances for the three semi-active dampers.

b At the 5% level of significance, compare mean interstory drifts for the three semi-active dampers.

c At the 5% level of significance, compare mean absolute acceleration for the three semi-active dampers.

d At the 5% level of significance, compare mean distances for all five groups.

e At the 5% level of significance, compare mean interstory drifts for all five groups.

Table 12.23

Mean Discharge Rates of Rivers

River	Location	Drainage Area (sq. km)	Years of Record	Record	$\bar{x}$	s_x
Scantic	Broad Brook	178	1930–1982	53	42.0	51.3
Burlington	Burlington	92	1937–1982	46	9.7	7.3
Connecticut	Hartford	3,500	1905–1982	78	980.1	761.5
North Branch Park	Hartford	125	1956–1982	27	45.6	50.4
Salmon	East Hampton	575	1930–1981	52	107.4	93.6
Eightmile	North Plain	320	1938–1982	45	30.4	26.0
East Branch Eightmile	North Lyme	197	1937–1982	46	24.9	23.3
Quinnipiac	Wallingford	205	1932–1982	51	62.3	37.9
Blackberry	Canaan	215	1943–1982	30	57.8	67.8
Tenmile	Gaylordsville	430	1921–1982	52	103.5	81.3
Still	Lanesville	495	1935–1982	48	45.7	37.9

Source: D. Jain and V. Singh, Water Resources Bulletin, vol. 23 no. 1, 1987, pp. 59–71.

Table 12.24

Mean Stream Flow before and after Channelization

Watershed	Area (km^2)	Status	Record Period	Annual Precipitation (mm) Mean	Annual Precipitation (mm) S.D.	Annual Stream Flow (mm) Mean	Annual Stream Flow (mm) S.D.
W3	40.7	Before channelization	1956–1963	1,259	230.6	315.7	200.9
		After channelization	1966–1981	1,205	166.0	279.5	177.7
W2	255.6	Before channelization	1956–1961	1,308	272.9	442.6	256.3
		After channelization	1969–1972	1,301	216.1	396.6	216.6

f At the 5% level of significance, compare mean absolute acceleration for all five groups.

g Looking at the interstory drifts, is it better to have brushless DC dampers than none at all? Justify your answer.

12.78 Wet air oxidation is a technology used for treating wastewaters that are either too diluted to incinerate or too concentrated or toxic to biologically degrade. Oliviero et al. (*Transactions of IChemE*, 2003) recorded the concentrations of liquid phase reaction intermediates and time profiles. Table below gives carbon concentrations in various intermediates during the catalytic oxidation of aniline at 200°C over time (in min).

Time (min)	Carbon Concentration (mmol L^{-1})		
	Group A	Group B	Group C
20	0.80	2.00	3.6
40	2.00	2.05	4.5
60	3.00	1.80	6.0
80	3.50	1.95	7.1
100	3.25	1.85	7.0
120	3.25	2.00	6.3
140	3.10	2.10	5.9
160	3.15	2.00	4.8
180	2.70	2.00	4.8

a Determine whether the mean concentrations of the three groups differ significantly at the 5% level of significance.

b Is blocking by time effective in this experiment?

12.79 Triano et al. (*Transactions of the ASME*, 2003) conducted uniaxial tension/compression tests on a servo-hydraulic test machine. Tests were run for four different materials at total strains of 1%, 2%, 3%, and 4%. The resulting Bauschinger modulus reduction (BMR) was recorded. The data is given in table below.

Total Strain	Material			
	A 723-1130	A723-1330	PH13-8-ST-1355	PH13-8-1380
1%	0.917	0.925	0.953	0.948
2%	0.863	0.874	0.852	0.876
3%	0.813	0.819	0.842	0.835
4%	0.798	0.808	0.827	0.788

a Analyze the data, assuming there is no interaction between the total strain and material used. State the hypotheses that can be tested clearly.

b Suppose no randomization was used for these test runs. Then discuss its effect on the results.

12.80 Zhao (*Journal of Pressure Vessel Technology*, 2003) performed fatigue tests on a servo-hydraulic machine at a temperature of 240°C, under total strain control, using a symmetrical push–pull triangular wave mode. Six total strain amplitudes were used and seven randomly assigned specimens were tested at each total strain amplitude level. The resulting stress amplitude (σ/MPa), Young's modulus (E/MPa), virtual stress amplitude (S/MPa), and fatigue life (N/cycles) were recorded, as shown in Table 12.25.

a Compare mean stress amplitude for the six strain ranges, using the 5% error rate.

b Compare mean Young's modulus for the six strain ranges, using the 5% error rate.

c Compare mean virtual stress amplitude for the six strain ranges, using the 5% error rate.

d Compare mean fatigue life for the six strain ranges, using the 5% error rate.

Table 12.25

Results of Fatigue Tests on a Servo-Hydraulic Machine

Strain Range	Stress Amplitude	Young's Modulus	Virtual Stress Amplitude	Fatigue Life N/Cycles
0.0130	329.37	137552	894.09	4000
	335.71	169422	1101.24	3500
	422.05	299535	1946.98	560
	330.28	220183	1431.19	1500
	337.85	164325	1068.11	3200
	333.04	184254	1198.65	2000
	385.50	192171	1249.11	1200
0.0100	373.39	241208	1206.04	3000
	334.85	244234	1221.17	2400
	419.69	358094	1790.47	760
	392.95	194530	972.65	5900
	404.73	354716	1773.58	1200
	338.76	242400	1212.00	3300
	352.83	207852	1039.26	4600
0.0086	316.10	191470	823.32	6850
	343.78	266916	1147.74	6700
	302.63	223253	959.99	5300
	367.24	328928	1414.39	1650
	339.44	224091	963.59	4600
	287.35	158214	680.32	8600
	382.83	295128	1269.05	1350
0.0068	301.07	158165	537.76	17200
	361.29	251241	854.22	4800
	425.92	338703	1151.59	1600
	376.35	259641	882.78	2300
	337.44	261882	890.40	5000
	386.85	269112	914.98	5800
	371.37	263197	894.87	4000
0.0056	366.87	249739	699.27	6000
	325.60	187557	525.16	2500
	298.74	235468	659.31	8600
	282.19	247643	693.40	10500
	366.63	228714	640.40	3000
	333.16	217964	610.30	25500
	353.08	287525	805.07	7100
0.0046	296.94	321357	739.12	15000
	359.32	437130	1005.40	4500
	300.98	310770	714.77	13000
	268.74	210857	484.97	46000
	335.49	256683	590.37	8300
	335.60	255791	588.32	17000
	313.26	281830	648.21	15900

Appendix Tables

Table 1

Random Numbers

Column Row	1	2	3	4	5	6	7	8	9	10	11	12	13	14
1	10480	15011	01536	02011	81647	91646	69179	14194	62590	36207	20969	99570	91291	90700
2	22368	46753	25595	85393	30995	89198	27982	53402	93965	34095	52666	19174	39615	99505
3	24130	48360	22527	97265	76393	64809	15179	24830	49340	32081	30680	19655	63348	58629
4	42167	93093	06243	61680	07856	16376	39440	53537	71341	57004	00849	74917	97758	16379
5	37570	39975	81837	16656	06121	91782	60468	81305	49684	60672	14110	06927	01263	54613
6	77912	06907	11008	42751	27756	53498	18602	70659	90655	15053	21916	81825	44394	42880
7	99562	72905	56420	69994	98872	31016	71194	18738	44013	48840	63213	21069	10634	12952
8	96301	91977	05463	07972	18876	20922	94595	56869	69014	60045	18425	84903	42508	32307
9	89579	14342	63661	10281	17453	18103	57740	84378	25331	12566	58678	44947	05585	56941
10	85475	36875	53342	53988	53060	59533	38867	62300	08158	17983	16439	11458	18593	64952
11	28918	69578	88231	33276	70997	79936	56865	05859	90106	31595	01547	85590	91610	78188
12	63553	40961	48235	03427	49626	69445	18663	72695	52180	20847	12234	90511	33703	90322
13	09429	93969	52636	92737	88974	33488	36320	17617	30015	08272	84115	27156	30613	74952
14	10365	61129	87529	85689	48237	52267	67689	93394	01511	26358	85104	20285	29975	89868
15	07119	97336	71048	08178	77233	13916	47564	81056	97735	85977	29372	74461	28551	90707
16	51085	12765	51821	51259	77452	16308	60756	92144	49442	53900	70960	63990	75601	40719
17	02368	21382	52404	60268	89368	19885	55322	44819	01188	65255	64835	44919	05944	55157
18	01011	54092	33362	94904	31273	04146	18594	29852	71585	85030	51132	01915	92747	64951
19	52162	53916	46369	58586	23216	14513	83149	98736	23495	64350	94738	17752	35156	35749
20	07056	97628	33787	09998	42698	06691	76988	13602	51851	46104	88916	19509	25625	58104
21	48663	91245	85828	14346	09172	30168	90229	04734	59193	22178	30421	61666	99904	32812
22	54164	58492	22421	74103	47070	25306	76468	26384	58151	06646	21524	15227	96909	44592
23	32639	32363	05597	24200	13363	38005	94342	28728	35806	06912	17012	64161	18296	22851
24	29334	27001	87637	87308	58731	00256	45834	15398	46557	41135	10367	07684	36188	18510
25	02488	33062	28834	07351	19731	92420	60952	61280	50001	67658	32586	86679	50720	94953
26	81525	72295	04839	96423	24878	82651	66566	14778	76797	14780	13300	87074	79666	95725
27	29676	20591	68086	26432	46901	20849	89768	81536	86645	12659	92259	57102	80428	25280
28	00742	57392	39064	66432	84673	40027	32832	61362	98947	96067	64760	64584	96096	98253
29	05366	04213	25669	26422	44407	44048	37937	63904	45766	66134	75470	66520	34693	90449
30	91921	26418	64117	94305	26766	25940	39972	22209	71500	64568	91402	42416	07844	69618
31	00582	04711	87917	77341	42206	35126	74087	99547	81817	42607	43808	76655	62028	76630
32	00725	69884	62797	56170	86324	88072	76222	36086	84637	93161	76038	65855	77919	88006
33	69011	65795	95876	55293	18988	27354	26575	08625	40801	59920	29841	80150	12777	48501
34	25976	57948	29888	88604	67917	48708	18912	82271	65424	69774	33611	54262	85963	03547

Column Row	1	2	3	4	5	6	7	8	9	10	11	12	13	14
35	09763	83473	73577	12908	30883	18317	28290	35797	05998	41688	34952	37888	38917	88050
36	91576	42595	27958	30134	04024	86385	29880	99730	55536	84855	29080	09250	79656	73211
37	17955	56349	90999	49127	20044	59931	06115	20542	18059	02008	73708	83517	36103	42791
38	46503	18584	18845	49618	02304	51038	20655	58727	28168	15475	56942	53389	20562	87338
39	92157	89634	94824	78171	84610	82834	09922	25417	44137	48413	25555	21246	35509	20468
40	14577	62765	35605	81263	39667	47358	56873	56307	61607	49518	89656	20103	77490	18062
41	98427	07523	33362	64270	01638	92477	66969	98420	04880	45585	46565	04102	46880	45709
42	34914	63976	88720	82765	34476	17032	87589	40836	32427	70002	70663	88863	77775	69348
43	70060	28277	39475	46473	23219	53416	94970	25832	69975	94884	19661	72828	00102	66794
44	53976	54914	06990	67245	68350	82948	11398	42878	80287	88267	47363	46634	06541	97809
45	76072	29515	40980	07391	58745	25774	22987	80059	39911	96189	41151	14222	60697	59583
46	90725	52210	83974	29992	65831	38857	50490	83765	55657	14361	31720	57375	56228	41546
47	64364	67412	33339	31926	14883	24413	59744	92351	97473	89286	35931	04110	23726	51900
48	08962	00358	31662	25388	61642	34072	81249	35648	56891	69352	48373	45578	78547	81788
49	95012	68379	93526	70765	10592	04542	76463	54328	02349	17247	28865	14777	62730	92277
50	15664	10493	20492	38391	91132	21999	59516	81652	27195	48223	46751	22923	32261	85653
51	16408	81899	04153	53381	79401	21438	83035	92350	36693	31238	59649	91754	72772	02338
52	18629	81953	05520	91962	04739	13092	97662	24822	94730	06496	35090	04822	86774	98289
53	73115	35101	47498	87637	99016	71060	88824	71013	18735	20286	23153	72924	35165	43040
54	57491	16703	23167	49323	45021	33132	12544	41035	80780	45393	44812	12515	98931	91202
55	30405	83946	23792	14422	15059	45799	22716	19792	09983	74353	68668	30429	70735	25499
56	16631	35006	85900	98275	32388	52390	16815	69298	82732	38480	73817	32523	41961	44437
57	96773	20206	42559	78985	05300	22164	24369	54224	35083	19687	11052	91491	60383	19746
58	38935	64202	14349	82674	66523	44133	00697	35552	35970	19124	63318	29686	03387	59846
59	31624	76384	17403	53363	44167	64486	64758	75366	76554	31601	12614	33072	60332	92325
60	78919	19474	23632	27889	47914	02584	37680	20801	72152	39339	34806	08930	85001	87820
61	03931	33309	57047	74211	63445	17361	62825	39908	05607	91284	68833	25570	38818	46920
62	74426	33278	43972	10119	89917	15665	52872	73823	73144	89662	88970	74492	51805	99378
63	09066	00903	20795	95452	92648	45454	09552	88815	16553	51125	79375	97596	16296	66092
64	42238	12426	87025	14267	20979	04508	64535	31355	86064	29472	47689	05974	52468	16834
65	16153	08002	26504	41744	81959	65642	74240	56302	00033	67107	77510	70625	28725	34191
66	21457	40742	29820	96783	29400	21840	15035	34537	33310	06116	95240	15957	16572	06004
67	21581	57802	02050	89728	17937	37621	47075	42080	97403	48626	68995	43805	33386	21597
68	55612	78095	83197	33732	05810	24813	86902	60397	16489	03264	88525	42786	05269	92532
69	44657	66999	99324	51281	84463	60563	79312	93454	68876	25471	93911	25650	12682	73572
70	91340	84979	46949	81973	37949	61023	43997	15263	80644	43942	89203	71795	99533	50501
71	91227	21199	31935	27022	84067	05462	35216	14486	29891	68607	41867	14951	91696	85065
72	50001	38140	66321	19924	72163	09538	12151	06878	91903	18749	34405	56087	82790	70925
73	65390	05224	72958	28609	81406	39147	25549	48542	42627	45233	57202	94617	23772	07896
74	27504	96131	83944	41575	10573	08619	64482	73923	36152	05184	94142	25299	84387	34925
75	31769	94851	39117	89632	00959	16487	65536	49071	39782	17095	02330	74301	00275	48280

Table 1

(Continued)

Column Row	1	2	3	4	5	6	7	8	9	10	11	12	13	14
76	11508	70225	51111	38351	19444	66499	71945	05422	13442	78675	84081	66938	93654	59894
77	37449	30362	06694	54690	04052	53115	62757	95348	78662	11163	81651	50245	34971	52924
78	46515	70331	85922	38329	57015	15765	97161	17869	45349	61796	66345	81073	49106	79860
79	30986	81223	42416	58353	21532	30502	32305	86482	05174	07901	54339	58861	74818	46942
80	63798	64995	46583	09785	44160	78128	83991	42865	92520	83531	80377	35909	81250	54238
81	82486	84846	99254	67632	43218	50076	21361	64816	51202	88124	41870	52689	51275	83556
82	21885	32906	92431	09060	64297	51674	64126	62570	26123	05155	59194	52799	28225	85762
83	60336	98782	07408	53458	13564	59089	26445	29789	85205	41001	12535	12133	14645	23541
84	43937	46891	24010	25560	86355	33941	25786	54990	71899	15475	95434	98227	21824	19585
85	97656	63175	89303	16275	07100	92063	21942	18611	47348	20203	18534	03862	78095	50136
86	03299	01221	05418	38982	55758	92237	26759	86367	21216	98442	08303	56613	91511	75928
87	79626	06486	03574	17668	07785	76020	79924	25651	83325	88428	85076	72811	22717	50585
88	85636	68335	47539	03129	65651	11977	02510	26113	99447	68645	34327	15152	55230	93448
89	18039	14367	61337	06177	12143	46609	32989	74014	64708	00533	35398	58408	13261	47908
90	08362	15656	60627	36478	65648	16764	53412	09013	07832	41574	17639	82163	60859	75567
91	79556	29068	04142	16268	15387	12856	66227	38358	22478	73373	88732	09443	82558	05250
92	92608	82674	27072	32534	17075	27698	98204	63863	11951	34648	88022	56148	34925	57031
93	23982	25835	40055	67006	12293	02753	14827	23235	35071	99704	37543	11601	35503	85171
94	09915	96306	05908	97901	28395	14186	00821	80703	70426	75647	76310	88717	37890	40129
95	59037	33300	26695	62247	69927	76123	50842	43834	86654	70959	79725	93872	28117	19233
96	42488	78077	69882	61657	34136	79180	97526	43092	04098	73571	80799	76536	71255	64239
97	46764	86273	63003	93017	31204	36692	40202	35275	57306	55543	53203	18098	47625	88684
98	03237	45430	55417	63282	90816	17349	88298	90183	36600	78406	06216	95787	42579	90730
99	86591	81482	52667	61582	14972	90053	89534	76036	49199	43716	97548	04379	46370	28672
100	38534	01715	94964	87288	65680	43772	39560	12918	86537	62738	19636	51132	25739	56947

Source: Abridged from W. H. Beyer, ed., *CRC Standard Mathematical Tables*, 24th ed. (Cleveland, The Chemical Rubber Company), 1976. Reproduced by permission of the publisher.

Table 2

Binomial Probabilities

Tabulated values are $\sum_{x=0}^{k} p(x)$. (Computations are rounded at the third decimal place.)

(a) $n = 5$

k \ p	0.01	0.05	0.10	0.20	0.30	0.40	0.50	0.60	0.70	0.80	0.90	0.95	0.99
0	0.951	0.774	0.590	0.328	0.168	0.078	0.031	0.010	0.002	0.000	0.000	0.000	0.000
1	0.999	0.977	0.919	0.737	0.528	0.337	0.188	0.087	0.031	0.007	0.000	0.000	0.000
2	1.000	0.999	0.991	0.942	0.837	0.683	0.500	0.317	0.163	0.058	0.009	0.001	0.000
3	1.000	1.000	1.000	0.993	0.969	0.913	0.812	0.663	0.472	0.263	0.081	0.023	0.001
4	1.000	1.000	1.000	1.000	0.998	0.990	0.969	0.922	0.832	0.672	0.410	0.226	0.049

(b) $n = 10$

k \ p	0.01	0.05	0.10	0.20	0.30	0.40	0.50	0.60	0.70	0.80	0.90	0.95	0.99
0	0.904	0.599	0.349	0.107	0.028	0.006	0.001	0.000	0.000	0.000	0.000	0.000	0.000
1	0.996	0.914	0.736	0.376	0.149	0.046	0.011	0.002	0.000	0.000	0.000	0.000	0.000
2	1.000	0.988	0.930	0.678	0.383	0.167	0.055	0.012	0.002	0.000	0.000	0.000	0.000
3	1.000	0.999	0.987	0.879	0.650	0.382	0.172	0.055	0.011	0.001	0.000	0.000	0.000
4	1.000	1.000	0.998	0.967	0.850	0.633	0.377	0.166	0.047	0.006	0.000	0.000	0.000
5	1.000	1.000	1.000	0.994	0.953	0.834	0.623	0.367	0.150	0.033	0.002	0.000	0.000
6	1.000	1.000	1.000	0.999	0.989	0.945	0.828	0.618	0.350	0.121	0.013	0.001	0.000
7	1.000	1.000	1.000	1.000	0.998	0.988	0.945	0.833	0.617	0.322	0.070	0.012	0.000
8	1.000	1.000	1.000	1.000	1.000	0.998	0.989	0.954	0.851	0.624	0.264	0.086	0.004
9	1.000	1.000	1.000	1.000	1.000	1.000	0.999	0.994	0.972	0.893	0.651	0.401	0.096

(c) $n = 15$

k \ p	0.01	0.05	0.10	0.20	0.30	0040	0.50	0.60	0.70	0.80	0.90	0.95	0.99
0	0.860	0.463	0.206	0.035	0.005	0.000	0.000	0.000	0.000	0.000	0.000	0.000	0.000
1	0.990	0.829	0.549	0.167	0.035	0.005	0.000	0.000	0.000	0.000	0.000	0.000	0.000
2	1.000	0.964	0.816	0.398	0.127	0.027	0.004	0.000	0.000	0.000	0.000	0.000	0.000
3	1.000	0.995	0.944	0.648	0.297	0.091	0.018	0.002	0.000	0.000	0.000	0.000	0.000
4	1.000	0.999	0.987	0.836	0.515	0.217	0.059	0.009	0.001	0.000	0.000	0.000	0.000
5	1.000	1.000	0.998	0.939	0.722	0.403	0.151	0.034	0.004	0.000	0.000	0.000	0.000
6	1.000	1.000	1.000	0.982	0.869	0.610	0.304	0.095	0.015	0.001	0.000	0.000	0.000
7	1.000	1.000	1.000	0.996	0.950	0.787	0.500	0.213	0.050	0.004	0.000	0.000	0.000
8	1.000	1.000	1.000	0.999	0.985	0.905	0.696	0.390	0.131	0.018	0.000	0.000	0.000
9	1.000	1.000	1.000	1.000	0.996	0.966	0.849	0.597	0.278	0.061	0.002	0.000	0.000
10	1.000	1.000	1.000	1.000	0.999	0.991	0.941	0.783	0.485	0.164	0.013	0.001	0.000
11	1.000	1.000	1.000	1.000	1.000	0.998	0.982	0.909	0.703	0.352	0.056	0.005	0.000
12	1.000	1.000	1.000	1.000	1.000	1.000	0.996	0.973	0.873	0.602	0.184	0.036	0.000
13	1.000	1.000	1.000	1.000	1.000	1.000	1.000	0.995	0.965	0.833	0.451	0.171	0.010
14	1.000	1.000	1.000	1.000	1.000	1.000	1.000	1.000	0.995	0.965	0.794	0.537	0.140

Table 2

(Continued)

(d) $n = 20$

k \ p	0.01	0.05	0.10	0.20	0.30	0.40	0.50	0.60	0.70	0.80	0.90	0.95	0.99
0	0.818	0.358	0.122	0.002	0.001	0.000	0.000	0.000	0.000	0.000	0.000	0.000	0.000
1	0.983	0.736	0.392	0.069	0.008	0.001	0.000	0.000	0.000	0.000	0.000	0.000	0.000
2	0.999	0.925	0.677	0.206	0.035	0.004	0.000	0.000	0.000	0.000	0.000	0.000	0.000
3	1.000	0.984	0.867	0.411	0.107	0.016	0.001	0.000	0.000	0.000	0.000	0.000	0.000
4	1.000	0.997	0.957	0.630	0.238	0.051	0.006	0.000	0.000	0.000	0.000	0.000	0.000
5	1.000	1.000	0.989	0.804	0.416	0.126	0.021	0.002	0.000	0.000	0.000	0.000	0.000
6	1.000	1.000	0.998	0.913	0.608	0.250	0.058	0.006	0.000	0.000	0.000	0.000	0.000
7	1.000	1.000	1.000	0.968	0.772	0.416	0.132	0.021	0.001	0.000	0.000	0.000	0.000
8	1.000	1.000	1.000	0.990	0.887	0.596	0.252	0.057	0.005	0.000	0.000	0.000	0.000
9	1.000	1.000	1.000	0.997	0.952	0.755	0.412	0.128	0.017	0.001	0.000	0.000	0.000
10	1.000	1.000	1.000	0.999	0.983	0.872	0.588	0.245	0.048	0.003	0.000	0.000	0.000
11	1.000	1.000	1.000	1.000	0.995	0.943	0.748	0.404	0.113	0.010	0.000	0.000	0.000
12	1.000	1.000	1.000	1.000	0.999	0.979	0.868	0.584	0.228	0.032	0.000	0.000	0.000
13	1.000	1.000	1.000	1.000	1.000	0.994	0.942	0.750	0.392	0.087	0.002	0.000	0.000
14	1.000	1.000	1.000	1.000	1.000	0.998	0.979	0.874	0.584	0.196	0.011	0.000	0.000
15	1.000	1.000	1.000	1.000	1.000	1.000	0.994	0.949	0.762	0.370	0.043	0.003	0.000
16	1.000	1.000	1.000	1.000	1.000	1.000	0.999	0.984	0.893	0.589	0.133	0.016	0.000
17	1.000	1.000	1.000	1.000	1.000	1.000	1.000	0.996	0.965	0.794	0.323	0.075	0.001
18	1.000	1.000	1.000	1.000	1.000	1.000	1.000	0.999	0.992	0.931	0.608	0.264	0.017
19	1.000	1.000	1.000	1.000	1.000	1.000	1.000	1.000	0.999	0.988	0.878	0.642	0.182

(e) $n = 25$

k \ p	0.01	0.05	0.10	0.20	0.30	0.40	0.50	0.60	0.70	0.80	0.90	0.95	0.99
0	0.778	0.277	0.072	0.004	0.000	0.000	0.000	0.000	0.000	0.000	0.000	0.000	0.000
1	0.974	0.642	0.271	0.027	0.002	0.000	0.000	0.000	0.000	0.000	0.000	0.000	0.000
2	0.998	0.873	0.537	0.098	0.009	0.000	0.000	0.000	0.000	0.000	0.000	0.000	0.000
3	1.000	0.966	0.764	0.234	0.033	0.002	0.000	0.000	0.000	0.000	0.000	0.000	0.000
4	1.000	0.993	0.902	0.421	0.090	0.009	0.000	0.000	0.000	0.000	0.000	0.000	0.000
5	1.000	0.999	0.967	0.617	0.193	0.029	0.002	0.000	0.000	0.000	0.000	0.000	0.000
6	1.000	1.000	0.991	0.780	0.341	0.074	0.007	0.000	0.000	0.000	0.000	0.000	0.000
7	1.000	1.000	0.998	0.891	0.512	0.154	0.022	0.001	0.000	0.000	0.000	0.000	0.000
8	1.000	1.000	1.000	0.953	0.677	0.274	0.054	0.004	0.000	0.000	0.000	0.000	0.000
9	1.000	1.000	1.000	0.983	0.811	0.425	0.115	0.013	0.000	0.000	0.000	0.000	0.000
10	1.000	1.000	1.000	0.994	0.902	0.586	0.212	0.034	0.002	0.000	0.000	0.000	0.000
11	1.000	1.000	1.000	0.998	0.956	0.732	0.345	0.078	0.006	0.000	0.000	0.000	0.000
12	1.000	1.000	1.000	1.000	0.983	0.846	0.500	0.154	0.017	0.000	0.000	0.000	0.000
13	1.000	1.000	1.000	1.000	0.994	0.922	0.655	0.268	0.044	0.002	0.000	0.000	0.000
14	1.000	1.000	1.000	1.000	0.998	0.966	0.788	0.414	0.098	0.006	0.000	0.000	0.000
15	1.000	1.000	1.000	1.000	1.000	0.987	0.885	0.575	0.189	0.017	0.000	0.000	0.000
16	1.000	1.000	1.000	1.000	1.000	0.996	0.946	0.726	0.323	0.047	0.000	0.000	0.000
17	1.000	1.000	1.000	1.000	1.000	0.999	0.978	0.846	0.488	0.109	0.002	0.000	0.000
18	1.000	1.000	1.000	1.000	1.000	1.000	0.993	0.926	0.659	0.220	0.009	0.000	0.000
19	1.000	1.000	1.000	1.000	1.000	1.000	0.998	0.971	0.807	0.383	0.033	0.001	0.000
20	1.000	1.000	1.000	1.000	1.000	1.000	1.000	0.991	0.910	0.579	0.098	0.007	0.000
21	1.000	1.000	1.000	1.000	1.000	1.000	1.000	0.998	0.967	0.766	0.236	0.034	0.000
22	1.000	1.000	1.000	1.000	1.000	1.000	1.000	1.000	0.991	0.902	0.463	0.127	0.002
23	1.000	1.000	1.000	1.000	1.000	1.000	1.000	1.000	0.998	0.973	0.729	0.358	0.026
24	1.000	1.000	1.000	1.000	1.000	1.000	1.000	1.000	1.000	0.996	0.928	0.723	0.222

Table 3

Poisson Distribution Function

$$F(x, \lambda) = \sum_{k=0}^{x} e^{-\lambda} \frac{\lambda^k}{k!}$$

λ \ x	0	1	2	3	4	5	6	7	8	9
0.02	0.980	1.000								
0.04	0.961	0.999	1.000							
0.06	0.942	0.998	1.000							
0.08	0.923	0.997	1.000							
0.10	0.905	0.995	1.000							
0.15	0.861	0.990	0.999	1.000						
0.20	0.819	0.982	0.999	1.000						
0.25	0.779	0.974	0.998	1.000						
0.30	0.741	0.963	0.996	1.000						
0.35	0.705	0.951	0.994	1.000						
0.40	0.670	0.938	0.992	0.999	1.000					
0.45	0.638	0.925	0.989	0.999	1.000					
0.50	0.607	0.910	0.986	0.998	1.000					
0.55	0.577	0.894	0.982	0.998	1.000					
0.60	0.549	0.878	0.977	0.997	1.000					
0.65	0.522	0.861	0.972	0.996	0.999	1.000				
0.70	0.497	0.844	0.966	0.994	0.999	1.000				
0.75	0.475	0.827	0.959	0.993	0.999	1.000				
0.80	0.449	0.809	0.953	0.991	0.999	1.000				
0.85	0.427	0.791	0.945	0.989	0.998	1.000				
0.90	0.407	0.772	0.937	0.987	0.998	1.000				
0.95	0.387	0.754	0.929	0.981	0.997	1.000				
1.00	0.368	0.736	0.920	0.981	0.996	0.999	1.000			
1.1	0.333	0.699	0.900	0.974	0.995	0.999	1.000			
1.2	0.301	0.663	0.879	0.966	0.992	0.998	1.000			
1.3	0.273	0.627	0.857	0.957	0.989	0.998	1.000			
1.4	0.247	0.592	0.833	0.946	0.986	0.997	0.999	1.000		
1.5	0.223	0.558	0.809	0.934	0.981	0.996	0.999	1.000		
1.6	0.202	0.525	0.783	0.921	0.976	0.994	0.999	1.000		
1.7	0.183	0.493	0.757	0.907	0.970	0.992	0.998	1.000		
1.8	0.165	0.463	0.731	0.891	0.964	0.990	0.997	0.999	1.000	
1.9	0.150	0.434	0.704	0.875	0.956	0.987	0.997	0.999	1.000	
2.0	0.135	0.406	0.677	0.857	0.947	0.983	0.995	0.999	1.000	
2.2	0.111	0.355	0.623	0.819	0.928	0.975	0.993	0.998	1.000	
2.4	0.091	0.308	0.570	0.779	0.904	0.964	0.988	0.997	0.999	1.000
2.6	0.074	0.267	0.518	0.736	0.877	0.951	0.983	0.995	0.999	1.000

Source: Reprinted by permission from E. C. Molina, *Poisson's Exponential Binomial Limit* (Princeton, N.J., D. Van Nostrand Company, Inc.), 1947.

Table 3

(Continued)

λ \ x	0	1	2	3	4	5	6	7	8	9
2.8	0.061	0.231	0.469	0.692	0.848	0.935	0.976	0.992	0.998	0.999
3.0	0.050	0.199	0.423	0.647	0.815	0.916	0.966	0.988	0.996	0.999
3.2	0.041	0.171	0.380	0.603	0.781	0.895	0.955	0.983	0.994	0.998
3.4	0.033	0.147	0.340	0.558	0.744	0.871	0.942	0.977	0.992	0.997
3.6	0.027	0.126	0.303	0.515	0.706	0.844	0.927	0.969	0.988	0.996
3.8	0.022	0.107	0.269	0.473	0.668	0.816	0.909	0.960	0.984	0.994
4.0	0.018	0.092	0.238	0.433	0.629	0.785	0.889	0.949	0.979	0.992
4.2	0.015	0.078	0.210	0.395	0.590	0.753	0.867	0.936	0.972	0.989
4.4	0.012	0.066	0.185	0.359	0.551	0.720	0.844	0.921	0.964	0.985
4.6	0.010	0.056	0.163	0.326	0.513	0.686	0.818	0.905	0.955	0.980
4.8	0.008	0.048	0.143	0.294	0.476	0.651	0.791	0.887	0.944	0.975
5.0	0.007	0.040	0.125	0.265	0.440	0.616	0.762	0.867	0.932	0.968
5.2	0.006	0.034	0.109	0.238	0.406	0.581	0.732	0.845	0.918	0.960
5.4	0.005	0.029	0.095	0.213	0.373	0.546	0.702	0.822	0.903	0.951
5.6	0.004	0.024	0.082	0.191	0.342	0.512	0.670	0.797	0.886	0.941
5.8	0.003	0.021	0.072	0.170	0.313	0.478	0.638	0.771	0.867	0.929
6.0	0.002	0.017	0.062	0.151	0.285	0.446	0.606	0.744	0.847	0.916
6.2	0.002	0.015	0.054	0.134	0.259	0.414	0.574	0.716	0.826	0.902
6.4	0.002	0.012	0.046	0.119	0.235	0.384	0.542	0.687	0.803	0.886
6.6	0.001	0.010	0.040	0.105	0.213	0.355	0.511	0.658	0.780	0.869
6.8	0.001	0.009	0.034	0.093	0.192	0.327	0.480	0.628	0.755	0.850
7.0	0.001	0.007	0.030	0.082	0.173	0.301	0.450	0.599	0.729	0.830

λ \ x	10	11	12	13	14	15	16	17	18	19
2.8	1.000									
3.0	1.000									
3.2	1.000									
3.4	0.999	1.000								
3.6	0.999	1.000								
3.8	0.998	0.999	1.000							
4.0	0.997	0.999	1.000							
4.2	0.996	0.999	1.000							
4.4	0.994	0.998	0.999	1.000						
4.6	0.992	0.997	0.999	1.000						
4.8	0.990	0.996	0.999	1.000						
5.0	0.986	0.995	0.998	0.999	1.000					
5.2	0.982	0.993	0.997	0.999	1.000					
5.4	0.977	0.990	0.996	0.999	1.000					
5.6	0.972	0.988	0.995	0.998	0.999	1.000				
5.8	0.965	0.984	0.993	0.997	0.999	1.000				
6.0	0.957	0.980	0.991	0.996	0.999	0.999	1.000			
6.2	0.949	0.975	0.989	0.995	0.998	0.999	1.000			
6.4	0.939	0.969	0.986	0.994	0.997	0.999	1.000			
6.6	0.927	0.963	0.982	0.992	0.997	0.999	0.999	1.000		
6.8	0.915	0.955	0.978	0.990	0.996	0.998	0.999	1.000		
7.0	0.901	0.947	0.973	0.987	0.994	0.998	0.999	1.000		

Table 3

(Continued)

λ \ x	0	1	2	3	4	5	6	7	8	9
7.2	0.001	0.006	0.025	0.072	0.156	0.276	0.420	0.569	0.703	0.810
7.4	0.001	0.005	0.022	0.063	0.140	0.253	0.392	0.539	0.676	0.78S
7.6	0.001	0.004	0.019	0.055	0.125	0.231	0.365	0.510	0.648	0.765
7.8	0.000	0.004	0.016	0.048	0.112	0.210	0.338	0.481	0.620	0.741
8.0	0.000	0.003	0.014	0.042	0.100	0.191	0.313	0.453	0.593	0.717
8.5	0.000	0.002	0.009	0.030	0.074	0.150	0.256	0.386	0.523	0.653
9.0	0.000	0.001	0.006	0.021	0.055	0.116	0.207	0.324	0.456	0.587
9.5	0.000	0.001	0.004	0.015	0.040	0.089	0.165	0.269	0.392	0.522
10.0	0.000	0.000	0.003	0.010	0.029	0.067	0.130	0.220	0.333	0.458
10.5	0.000	0.000	0.002	0.007	0.021	0.050	0.102	0.179	0.279	0.397
11.0	0.000	0.000	0.001	0.005	0.015	0.038	0.079	0.143	0.232	0.341
11.5	0.000	0.000	0.001	0.003	0.011	0.028	0.060	0.114	0.191	0.289
12.0	0.000	0.000	0.001	0.002	0.008	0.020	0.046	0.090	0.155	0.242
12.5	0.000	0.000	0.000	0.002	0.005	0.015	0.035	0.070	0.125	0.201

λ \ x	10	11	12	13	14	15	16	17	18	19
7.2	0.887	0.937	0.967	0.984	0.993	0.997	0.999	0.999	1.000	
7.4	0.871	0.926	0.961	0.980	0.991	0.996	0.998	0.999	1.000	
7.6	0.854	0.915	0.954	0.976	0.989	0.995	0.998	0.999	1.000	
7.8	0.835	0.902	0.945	0.971	0.986	0.993	0.997	0.999	1.000	
8.0	0.816	0.888	0.936	0.966	0.983	0.992	0.996	0.998	0.999	1.000
8.5	0.763	0.849	0.909	0.949	0.973	0.986	0.993	0.997	0.999	0.999
9.0	0.706	0.803	0.876	0.926	0.959	0.978	0.989	0.995	0.998	0.999
9.5	0.645	0.752	0.836	0.898	0.940	0.967	0.982	0.991	0.996	0.998
10.0	0.583	0.697	0.792	0.864	0.917	0.951	0.973	0.986	0.993	0.997
10.5	0.521	0.639	0.742	0.825	0.888	0.932	0.960	0.978	0.988	0.994
11.0	0.460	0.579	0.689	0.781	0.854	0.907	0.944	0.968	0.982	0.991
11.5	0.402	0.520	0.633	0.733	0.815	0.878	0.924	0.954	0.974	0.986
12.0	0.347	0.462	0.576	0.682	0.772	0.844	0.899	0.937	0.963	0.979
12.5	0.297	0.406	0.519	0.628	0.725	0.806	0.869	0.016	0.948	0.969

λ \ x	20	21	22	23	24	25	26	27	28	29
8.5	1.000									
9.0	1.000									
9.5	0.999	1.000								
10.0	0.998	0.999	1.000							
10.5	0.997	0.999	0.999	1.000						
11.0	0.995	0.998	0.999	1.000						
11.5	0.992	0.996	0.998	0.999	1.000					
12.0	0.988	0.994	0.997	0.999	0.999	1.000				
12.5	0.983	0.991	0.995	0.998	0.999	0.999	1.000			

Table 3

(Continued)

λ \ x	0	1	2	3	4	5	6	7	8	9
13.0	0.000	0.000	0.000	0.001	0.004	0.011	0.026	0.054	0.100	0.166
13.5	0.000	0.000	0.000	0.001	0.003	0.008	0.019	0.041	0.079	0.135
14.0	0.000	0.000	0.000	0.000	0.002	0.006	0.014	0.032	0.062	0.109
14.5	0.000	0.000	0.000	0.000	0.001	0.004	0.010	0.024	0.048	0.088
15.0	0.000	0.000	0.000	0.000	0.001	0.003	0.008	0.018	0.037	0.070

	10	11	12	13	14	15	16	17	18	19
13.0	0.252	0.353	0.463	0.573	0.675	0.764	0.835	0.890	0.930	0.957
13.5	0.211	0.304	0.409	0.518	0.623	0.718	0.798	0.861	0.908	0.942
14.0	0.176	0.260	0.358	0.464	0.570	0.669	0.756	0.827	0.883	0.923
14.5	0.145	0.220	0.311	0.413	0.518	0.619	0.711	0.790	0.853	0.901
15.0	0.118	0.185	0.268	0.363	0.466	0.568	0.664	0.749	0.819	0.875

	20	21	22	23	24	25	26	27	28	29
13.0	0.975	0.986	0.992	0.996	0.998	0.999	1.000			
13.5	0.965	0.980	0.989	0.994	0.997	0.998	0.999	1.000		
14.0	0.952	0.971	0.983	0.991	0.995	0.997	0.999	0.999	1.000	
14.5	0.936	0.960	0.976	0.986	0.992	0.996	0.998	0.999	0.999	1.000
15.0	0.917	0.947	0.967	0.981	0.989	0.994	0.997	0.998	0.999	1.000

Table 4

Critical Values of *t*

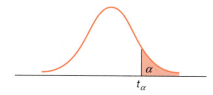

Degrees of Freedom	$t_{0.100}$	$t_{0.050}$	$t_{0.025}$	$t_{0.010}$	$t_{0.005}$
1	3.078	6.314	12.706	31.821	63.657
2	1.886	2.920	4.303	6.695	9.925
3	1.638	2.353	3.182	4.541	5.841
4	1.533	2.132	2.776	3.747	4.604
5	1.476	2.015	2.571	3.365	4.032
6	1.440	1.943	2.447	3.143	3.707
7	1.415	1.895	2.365	2.998	3.499
8	1.397	1.860	2.306	2.896	3.355
9	1.383	1.833	2.262	2.821	3.250
10	1.372	1.812	2.228	2.764	3.169
11	1.363	1.796	2.201	2.718	3.106
12	1.356	1.782	2.179	2.681	3.055
13	1.350	1.771	2.160	2.650	3.012
14	1.345	1.761	2.145	2.624	2.977
15	1.341	1.753	2.131	2.602	2.947
16	1.337	1.746	2.120	2.583	2.921
17	1.333	1.740	2.110	2.567	2.898
18	1.330	1.734	2.101	2.552	2.878
19	1.328	1.729	2.093	2.539	2.861
20	1.325	1.725	2.086	2.528	2.845
21	1.323	1.721	2.080	2.518	2.831
22	1.321	1.717	2.074	2.508	2.819
23	1.319	1.714	2.069	2.500	2.807
24	1.318	1.711	2.064	2.492	2.797
25	1.316	1.708	2.060	2.485	2.787
26	1.315	1.706	2.056	2.479	2.779
27	1.314	1.703	2.052	2.473	2.771
28	1.313	1.701	2.048	2.467	2.763
29	1.311	1.699	2.045	2.462	2.756
∞	1.282	1.645	1.960	2.326	2.576

Source: From M. Merrington, "Table of Percentage Points of the *t*-Distribution," *Biometrika*, 32, 1941, p. 300. Reproduced by permission of the *Biometrika* Trustees.

Table 5

Critical Values of χ^2

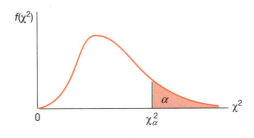

Degrees of Freedom	$\chi^2_{0.995}$	$\chi^2_{0.990}$	$\chi^2_{0.975}$	$\chi^2_{0.950}$	$\chi^2_{0.900}$
1	0.0000393	0.0001571	0.0009821	0.0039321	0.0157908
2	0.0100251	0.0201007	0.0506356	0.102587	0.210720
3	0.0717212	0.114832	0.215795	0.351846	0.584375
4	0.206990	0.297110	0.484419	0.710721	1.063623
5	0.411740	0.554300	0.831211	1.145476	1.61031
6	0.675727	0.872085	1.237347	1.63539	2.20413
7	0.989265	1.239043	1.68987	2.16735	2.83311
8	1.344419	1.646482	2.17973	2.73264	3.48954
9	1.734926	2.087912	2.70039	3.32511	4.16816
10	2.15585	2.55821	3.24697	3.94030	4.86518
11	2.60321	3.05347	3.81575	4.57481	5.57779
12	3.07382	3.57056	4.40379	5.22603	6.30380
13	3.56503	4.10691	5.00874	5.89186	7.04150
14	4.07468	4.66043	5.62872	6.57063	7.78953
15	4.60094	5.22935	6.26214	7.26094	8.54675
16	5.14224	5.81221	6.90766	7.96164	9.31223
17	5.69724	6.40776	7.56418	8.67176	10.0852
18	6.26481	7.01491	8.23075	9.39046	10.8649
19	6.84398	7.63273	8.90655	10.1170	11.6509
20	7.43386	8.26040	9.59083	10.8508	12.4426
21	8.03366	8.89720	10.28293	11.5913	13.2396
22	8.64272	9.54249	10.9823	12.3380	14.0415
23	9.26042	10.19567	11.6885	13.0905	14.8479
24	9.88623	10.8564	12.4011	13.8484	15.6587
25	10.5197	11.5240	13.1197	14.6114	16.4734
26	11.1603	12.1981	13.8439	15.3791	17.2919
27	11.8076	12.8786	14.5733	16.1513	18.1138
28	12.4613	13.5648	15.3079	16.9279	18.9392
29	13.1211	14.2565	16.0471	17.7083	19.7677
30	13.7867	14.9535	16.7908	18.4926	20.5992
40	20.7065	22.1643	24.4331	26.5093	29.0505
50	27.9907	29.7067	32.3574	34.7642	37.6886
60	35.5346	37.4848	40.4817	43.1879	46.4589
70	43.2752	45.4418	48.7576	51.7393	55.3290
80	51.1720	53.5400	57.1532	60.3915	64.2778
90	59.1963	61.7541	63.6466	69.1260	73.2912
100	67.3276	70.0648	74.2219	77.9295	82.3581

Table 5

(Continued)

Degrees of Freedom	$\chi^2_{0.100}$	$\chi^2_{0.050}$	$\chi^2_{0.025}$	$\chi^2_{0.010}$	$\chi^2_{0.005}$
1	2.70554	3.84146	5.02389	6.63490	7.87944
2	4.60517	5.99147	7.37776	9.21034	10.5966
3	6.25139	7.81473	9.34840	11.3449	12.8381
4	7.77944	9.48773	11.1433	13.2767	14.8602
5	9.23635	11.0705	12.8325	15.0863	16.7496
6	10.6446	12.5916	14.4494	16.8119	18.5476
7	12.0170	14.0671	16.0128	18.4753	20.2777
8	13.3616	15.5073	17.5346	20.0902	21.9550
9	14.6837	16.9190	19.0228	21.6660	23.5893
10	15.9871	18.3070	20.4831	23.2093	25.1882
11	17.2750	19.6751	21.9200	24.7250	26.7569
12	18.5494	21.0261	23.3367	26.2170	28.2995
13	19.8119	22.3621	24.7356	27.6883	29.8194
14	21.0642	23.6848	26.1190	29.1413	31.3193
15	22.3072	24.9958	27.4884	30.5779	32.8013
16	23.5418	26.2962	28.8454	31.9999	34.2672
17	24.7690	27.5871	30.1910	33.4087	35.7185
18	25.9894	28.8693	31.5264	34.8053	37.1564
19	27.2036	30.1435	32.8523	36.1908	38.5822
20	28.4120	31.4104	34.1696	37.5662	39.9968
21	29.6151	32.6705	35.4789	38.9321	41.4010
22	30.8133	33.9244	36.7807	40.2894	42.7956
23	32.0069	35.1725	38.0757	41.6384	44.1813
24	33.1963	36.4151	39.3641	42.9798	45.5585
25	34.3816	37.6525	40.6465	44.3141	46.9278
26	35.5631	38.8852	41.9232	45.6417	48.2899
27	36.7412	40.1133	43.1944	46.9630	49.6449
28	37.9159	41.3372	44.4607	48.2782	50.9933
29	39.0875	42.5569	45.7222	49.5879	52.3356
30	40.2560	43.7729	46.9792	50.8922	53.6720
40	51.8050	55.7585	59.3417	63.6907	66.7659
50	63.1671	67.5048	71.4202	76.1539	79.4900
60	74.3970	79.0819	83.2976	88.3794	91.9517
70	85.5271	90.5312	95.0231	100.425	104.215
80	96.5782	101.879	106.629	112.329	116.321
90	107.565	113.145	118.136	124.116	128.299
100	118.498	124.342	129.561	135.807	140.169

Source: From C. M. Thompson, "Tables of the Percentage Points of the χ^2 Distribution," *Biometrika*, 32, 1941, pp. 188–189. Reproduced by permission of the *Biometrika* Trustees.

Table 6

Percentage Points of the *F* Distribution, $\alpha = 0.05$

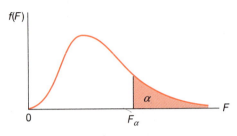

v_1				Numerator Degrees of Freedom					
v_2	1	2	3	4	5	6	7	8	9
1	161.4	199.5	215.7	224.6	230.2	234.0	236.8	238.9	240.5
2	18.51	19.00	19.16	19.25	19.30	19.33	19.35	19.37	19.38
3	10.13	9.55	9.28	9.12	9.01	8.94	8.89	8.85	8.81
4	7.71	6.94	6.59	6.39	6.26	6.16	6.09	6.04	6.00
5	6.61	5.79	5.41	5.19	5.05	4.95	4.88	4.82	4.77
6	5.99	5.14	4.76	4.53	4.39	4.28	4.21	4.15	4.10
7	5.59	4.74	4.35	4.12	3.97	3.87	3.79	3.73	3.68
8	5.32	4.46	4.07	3.84	3.69	3.58	3.50	3.44	3.39
9	5.12	4.26	3.86	3.63	3.48	3.37	3.29	3.23	3.18
10	4.96	4.10	3.71	3.48	3.33	3.22	3.14	3.07	3.02
11	4.84	3.98	3.59	3.36	3.20	3.09	3.01	2.95	2.90
12	4.75	3.89	3.49	3.26	3.11	3.00	2.91	2.85	2.80
13	4.67	3.81	3.41	3.18	3.03	2.92	2.83	2.77	2.71
14	4.60	3.74	3.34	3.11	2.96	2.85	2.76	2.70	2.65
15	4.54	3.68	3.29	3.06	2.90	2.79	2.71	2.64	2.59
16	4.49	3.63	3.24	3.01	2.85	2.74	2.66	2.59	2.54
17	4.45	3.59	3.20	2.96	2.81	2.70	2.61	2.55	2.49
18	4.41	3.55	3.16	2.93	2.77	2.66	2.58	2.51	2.46
19	4.38	3.52	3.13	2.90	2.74	2.63	2.54	2.48	2.42
20	4.35	3.49	3.10	2.87	2.71	2.60	2.51	2.45	2.39
21	4.32	3.47	3.07	2.84	2.68	2.57	2.49	2.42	2.37
22	4.30	3.44	3.05	2.82	2.66	2.55	2.46	2.40	2.34
23	4.28	3.42	3.03	2.80	2.64	2.53	2.44	2.37	2.32
24	4.26	3.40	3.01	2.78	2.62	2.51	2.42	2.36	2.30
25	4.24	3.39	2.99	2.76	2.60	2.49	2.40	2.34	2.28
26	4.23	3.37	2.98	2.74	2.59	2.47	2.39	2.32	2.27
27	4.21	3.35	2.96	2.73	2.57	2.46	2.37	2.31	2.25
28	4.20	3.34	2.95	2.71	2.56	2.45	2.36	2.29	2.24
29	4.18	3.33	2.93	2.70	2.55	2.43	2.35	2.28	2.22
30	4.17	3.32	2.92	2.69	2.53	2.42	2.33	2.27	2.21
40	4.08	3.23	2.84	2.61	2.45	2.34	2.25	2.18	2.12
60	4.00	3.15	2.76	2.53	2.37	2.25	2.17	2.10	2.04
120	3.92	3.07	2.68	2.45	2.29	2.17	2.09	2.02	1.96
∞	3.84	3.00	2.60	2.37	2.21	2.10	2.01	1.94	1.88

Source: From M. Merrington and C. M. Thompson, "Tables of Percentage Points of the Inverted Beta (*F*)-Distribution," *Biometrika*, 33, 1943, pp. 73–88. Reproduced by permission of the *Biometrika* Trustees.

Table 6

(Continued)

v_2	Numerator Degrees of Freedom									
v_1	10	12	15	20	24	30	40	60	120	∞
1	241.9	243.9	245.9	248.0	249.1	250.1	251.1	252.2	253.3	254.3
2	19.40	19.41	19.43	19.45	19.45	19.46	19.47	19.48	19.49	19.50
3	8.79	8.74	8.70	8.66	8.64	8.62	8.59	8.57	8.55	8.53
4	5.96	5.91	5.86	5.80	5.77	5.75	5.72	5.69	5.66	5.63
5	4.74	4.68	4.62	4.56	4.53	4.50	4.46	4.43	4.40	4.36
6	4.06	4.00	3.94	3.87	3.84	3.81	3.77	3.74	3.70	3.67
7	3.64	3.57	3.51	3.44	3.41	3.38	3.34	3.30	3.27	3.23
8	3.35	3.28	3.22	3.15	3.12	3.08	3.04	3.01	2.97	2.93
9	3.14	3.07	3.01	2.94	2.90	2.86	2.83	2.79	2.75	2.71
10	2.98	2.91	2.85	2.77	2.74	2.70	2.66	2.62	2.58	2.54
11	2.85	2.79	2.72	2.68	2.61	2.57	2.53	2.49	2.45	2.40
12	2.75	2.69	2.62	2.54	2.51	2.47	2.43	2.38	2.34	2.30
13	2.67	2.60	2.53	2.46	2.42	2.38	2.34	2.30	2.25	2.21
14	2.60	2.53	2.46	2.39	2.35	2.31	2.27	2.22	2.18	2.13
15	2.54	2.48	2.40	2.33	2.29	2.25	2.20	2.16	2.11	2.07
16	2.49	2.42	2.35	2.28	2.24	2.19	2.15	2.11	2.06	2.01
17	2.45	2.38	2.31	2.23	2.19	2.15	2.10	2.06	2.01	1.96
18	2.41	2.34	2.27	2.19	2.15	2.11	2.06	2.02	1.97	1.92
19	2.38	2.31	2.23	2.16	2.11	2.07	2.03	1.98	1.93	1.88
20	2.35	2.28	2.20	2.12	2.08	2.04	1.99	1.95	1.90	1.84
21	2.32	2.25	2.18	2.10	2.05	2.01	1.96	1.92	1.87	1.81
22	2.30	2.23	2.15	2.07	2.03	1.98	1.94	1.89	1.84	1.78
23	2.27	2.20	2.13	2.05	2.01	1.96	1.91	1.86	1.81	1.76
24	2.25	2.18	2.11	2.03	1.98	1.94	1.89	1.84	1.79	1.73
25	2.24	2.16	2.09	2.01	1.96	1.92	1.87	1.82	1.77	1.71
26	2.22	2.15	2.07	1.99	1.95	1.90	1.85	1.80	1.75	1.69
27	2.20	2.13	2.06	1.97	1.93	1.88	1.84	1.79	1.73	1.67
28	2.19	2.12	2.04	1.96	1.91	1.87	1.82	1.77	1.71	1.65
29	2.18	2.10	2.03	1.94	1.90	1.85	1.81	1.75	1.70	1.64
30	2.16	2.09	2.01	1.93	1.89	1.84	1.79	1.74	1.68	1.62
40	2.08	2.00	1.92	1.84	1.79	1.74	1.69	1.64	1.58	1.51
60	1.99	1.92	1.84	1.75	1.70	1.65	1.59	1.53	1.47	1.39
120	1.91	1.83	1.75	1.66	1.61	1.55	1.50	1.43	1.35	1.25
∞	1.83	1.75	1.67	1.57	1.52	1.46	1.39	1.32	1.22	1.00

Denominator Degrees of Freedom

Table 7

Percentage Points of the F Distribution, $\alpha = 0.01$

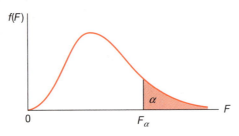

ν_1 ν_2	Numerator Degrees of Freedom								
	1	**2**	**3**	**4**	**5**	**6**	**7**	**8**	**9**
1	4,052	4,999.5	5,403	5,625	5,764	5,859	5,928	5,982	6,022
2	98.50	99.00	99.17	99.25	99.30	99.33	99.36	99.37	99.39
3	34.12	30.82	29.46	28.71	28.24	27.91	27.67	27.49	27.35
4	21.20	18.00	16.69	15.98	15.52	15.21	14.98	14.80	14.66
5	16.26	13.27	12.06	11.39	10.97	10.67	10.46	10.29	10.16
6	13.75	10.92	9.78	9.15	8.75	8.47	8.26	8.10	7.98
7	12.25	9.55	8.45	7.85	7.46	7.19	6.99	6.84	6.72
8	11.26	8.65	7.59	7.01	6.63	6.37	6.18	6.03	5.91
9	10.56	8.02	6.99	6.42	6.06	5.80	5.61	5.47	5.35
10	10.04	7.56	6.55	5.99	5.64	5.39	5.20	5.06	4.94
11	9.65	7.21	6.22	5.67	5.32	5.07	4.89	4.74	4,63
12	9.33	6.93	5.95	5.41	5.06	4.82	4.64	4.50	4.39
13	9.07	6.70	5.74	5.21	4.86	4.62	4.44	4.30	4.19
14	8.86	6.51	5.56	5.04	4.69	4.46	4.28	4.14	4.03
15	8.68	6.36	5.42	4.89	4.56	4.32	4.14	4.00	3.89
16	8.53	6.23	5.29	4.77	4.44	4.20	4.03	3.89	3.78
17	8.40	6.11	5.18	4.67	4.34	4.10	3.93	3.79	3.68
18	8.29	6.01	5.09	4.58	4.25	4.01	3.84	3.71	3.60
19	8.18	5.93	5.01	4.50	4.17	3.94	3.77	3.63	3.52
20	8.10	5.85	4.94	4.43	4.10	3.87	3.70	3.56	3.46
21	8.02	5.78	4.87	4.37	4.04	3.81	3.64	3.51	3.40
22	7.95	5.72	4.82	4.31	3.99	3.76	3.59	3.45	3.35
23	7.88	5.66	4.76	4.26	3.94	3.71	3.54	3.41	3.30
24	7.82	5.61	4.72	4.22	3.90	3.67	3.50	3.36	3.26
25	7.77	5.57	4.68	4.18	3.85	3.63	3.46	3.32	3.22
26	7.72	5.53	4.64	4.14	3.82	3.59	3.42	3.29	3.18
27	7.68	5.49	4.60	4.11	3.78	3.56	3.39	3.26	3.15
28	7.64	5.45	4.57	4.07	3.75	3.53	3.36	3.23	3.12
29	7.60	5.42	4.54	4.04	3.73	3.50	3.33	3.20	3.09
30	7.56	5.39	4.51	4.02	3.70	3.47	3.30	3.17	3.07
40	7.31	5.18	4.31	3.83	3.51	3.29	3.12	2.99	2.89
60	7.08	4.98	4.13	3.65	3.34	3.12	2.95	2.82	2.72
120	6.85	4.79	3.95	3.48	3.17	2.96	2.79	2.66	2.56
∞	6.63	4.61	3.78	3.32	3.02	2.80	2.64	2.51	2.41

Denominator Degrees of Freedom

Table 7

(Continued)

v_2 \ v_1	Numerator Degrees of Freedom									
	10	12	15	20	24	30	40	60	120	∞
1	6,056	6,106	6,157	6,209	6,235	6,261	6,287	6,313	6,339	6,366
2	99.40	99.42	99.43	99.45	99.46	99.47	99.47	99.48	99.49	99.50
3	27.23	27.05	26.87	26.69	26.60	26.50	26.41	26.32	26.22	26.13
4	14.55	14.37	14.20	14.02	13.93	13.84	13.75	13.65	13.56	13.46
5	10.05	9.89	9.72	9.55	9.47	9.38	9.29	9.20	9.11	9.02
6	7.87	7.72	7.56	7.40	7.31	7.23	7.14	7.06	6.97	6.88
7	6.62	6.47	6.31	6.16	6.07	5.99	5.91	5.82	5.74	5.65
8	5.81	5.67	5.52	5.36	5.28	5.20	5.12	5.03	4.95	4.86
9	5.26	5.11	4.96	4.81	4.73	4.65	4.57	4.48	4.40	4.31
10	4.85	4.71	4.56	4.41	4.33	4.25	4.17	4.08	4.00	3.91
11	4.54	4.40	4.25	4.10	4.02	3.94	3.86	3.78	3.69	3.60
12	4.30	4.16	4.01	3.86	3.78	3.70	3.62	3.54	3.45	3.36
13	4.10	3.96	3.82	3.66	3.59	3.51	3.43	3.34	3.25	3.17
14	3.94	3.80	3.66	3.51	3.43	3.35	3.27	3.18	3.09	3.00
15	3.80	3.67	3.52	3.37	3.29	3.21	3.13	3.05	2.90	2.87
16	3.69	3.55	3.41	3.26	3.18	3.10	3.02	2.93	2.84	2.75
17	3.59	3.46	3.31	3.16	3.08	3.00	2.92	2.83	2.75	2.65
18	3.51	3.37	3.23	3.08	3.00	2.92	2.84	2.75	2.66	2.57
19	3.43	3.30	3.15	3.00	2.92	2.84	2.76	2.67	2.58	2.49
20	3.37	3.23	3.09	2.94	2.86	2.78	2.69	2.61	2.52	2.42
21	3.31	3.17	3.03	2.88	2.80	2.72	2.64	2.55	2.46	2.36
22	3.26	3.12	2.98	2.83	2.75	2.67	2.58	2.50	2.40	2.31
23	3.21	3.07	2.93	2.78	2.70	2.62	2.54	2.45	2.35	2.26
24	3.17	3.03	2.89	2.74	2.66	2.58	2.49	2.40	2.31	2.21
25	3.13	2.99	2.85	2.70	2.62	2.54	2.45	2.36	2.27	2.17
26	3.09	2.96	2.81	2.66	2.58	2.50	2.42	2.33	2.23	2.13
27	3.06	2.93	2.78	2.63	2.55	2.47	2.38	2.29	2.20	2.10
28	3.03	2.90	2.75	2.60	2.52	2.44	2.35	2.26	2.17	2.06
29	3.00	2.87	2.73	2.57	2.49	2.41	2.33	2.23	2.14	2.03
30	2.98	2.84	2.70	2.55	2.47	2.39	2.30	2.21	2.11	2.01
40	2.80	2.66	2.52	2.37	2.29	2.20	2.11	2.02	1.92	1.80
60	2.63	2.50	2.35	2.20	2.12	2.03	1.94	1.84	1.73	1.60
120	2.47	2.34	2.19	2.03	1.95	1.86	1.76	1.66	1.53	1.38
∞	2.32	2.18	2.04	1.88	1.79	1.70	1.59	1.47	1.32	1.00

Denominator Degrees of Freedom

Table 8

Factors for Computing Control Chart Lines

Observations in Sample, n	Chart for Averages — Factors for Control Limits			Chart for Standard Deviations — Factors for Central Line		Factors for Control Limits				Chart for Ranges — Factors for Central Line			Factors for Control Limits			
	A	A_2	A_1	c_4	$1/c_4$	B_3	B_4	B_5	B_6	d_2	$1/d_2$	d_3	D_1	D_2	D_3	D_4
2	2.121	1.880	2.659	0.7979	1.2533	0	3.267	0	2.606	1.128	0.8862	0.853	0	3.686	0	3.267
3	1.732	1.023	1.954	0.8862	1.1284	0	2.568	0	2.276	1.693	0.5908	0.888	0	4.358	0	2.575
4	1.500	0.729	1.628	0.9213	1.0854	0	2.266	0	2.088	2.059	0.4857	0.880	0	4.698	0	2.282
5	1.342	0.577	1.427	0.9400	1.0638	0	2.089	0	1.964	2.326	0.4299	0.864	0	4.918	0	2.114
6	1.225	0.483	1.287	0.9515	1.0510	0.030	1.970	0.029	1.874	2.534	0.3946	0.848	0	5.079	0	2.004
7	1.134	0.419	1.182	0.9594	1.0424	0.118	1.882	0.113	1.806	2.704	0.3698	0.833	0.205	5.204	0.076	1.924
8	1.061	0.373	1.099	0.9650	1.0363	0.185	1.815	0.179	1.751	2.847	0.3512	0.820	0.388	5.307	0.136	1.864
9	1.000	0.337	1.032	0.9693	1.0317	0.239	1.761	0.232	1.707	2.970	0.3367	0.808	0.547	5.393	0.184	1.816
10	0.949	0.308	0.975	0.9727	1.0281	0.284	1.716	0.276	1.669	3.078	0.3249	0.797	0.686	5.469	0.223	1.777
11	0.905	0.285	0.927	0.9754	1.0253	0.321	1.679	0.313	1.637	3.173	0.3152	0.787	0.811	5.535	0.256	1.744
12	0.866	0.266	0.886	0.9776	1.0230	0.354	1.646	0.346	1.610	3.258	0.3069	0.778	0.923	5.594	0.283	1.717
13	0.832	0.249	0.850	0.9794	1.0210	0.382	1.618	0.374	1.585	3.336	0.2998	0.770	1.025	5.647	0.307	1.693
14	0.802	0.235	0.817	0.9810	1.0194	0.406	1.594	0.399	1.563	3.407	0.2935	0.763	1.118	5.696	0.328	1.672
15	0.755	0.223	0.789	0.9823	1.0180	0.428	1.572	0.421	1.544	3.472	0.2880	0.756	1.203	5.740	0.347	1.653
16	0.750	0.212	0.763	0.9835	1.0168	0.448	1.552	0.440	1.526	3.532	0.2831	0.750	1.282	5.782	0.363	1.637
17	0.728	0.203	0.739	0.9845	1.0157	0.466	1.534	0.458	1.511	3.588	0.2787	0.744	1.356	5.820	0.378	1.622
18	0.707	0.194	0.718	0.9854	1.0148	0.482	1.518	0.475	1.496	3.640	0.2747	0.739	1.424	5.856	0.391	1.609
19	0.688	0.187	0.698	0.9862	1.0140	0.497	1.503	0.490	1.483	3.689	0.2711	0.733	1.489	5.889	0.404	1.596
20	0.671	0.180	0.680	0.9869	1.0132	0.510	1.490	0.504	1.470	3.735	0.2677	0.729	1.549	5.921	0.415	1.585
21	0.655	0.173	0.663	0.9876	1.0126	0.523	1.477	0.516	1.459	3.778	0.2647	0.724	1.606	5.951	0.425	1.575
22	0.640	0.167	0.647	0.9882	1.0120	0.534	1.466	0.528	1.448	3.819	0.2618	0.720	1.660	5.979	0.435	1.565
23	0.626	0.162	0.633	0.9887	1.0114	0.545	1.455	0.539	1.438	3.858	0.2592	0.716	1.711	6.006	0.443	1.557
24	0.612	0.157	0.619	0.9892	1.0109	0.555	1.445	0.549	1.429	3.895	0.2567	0.712	1.759	6.032	0.452	1.548
25	0.600	0.153	0.606	0.9896	1.0105	0.565	1.435	0.559	1.420	3.931	0.2544	0.708	1.805	6.056	0.459	1.541
Over 25	$3/\sqrt{n}$	...	a	b	c	d	e	f	g	...	...	...	...	...	...	...

Source: "Manual on Presentation of Data and Control Chart Analysis," 6th ed., ASTM, Philadelphia, Pa, p. 91.
Note: Values of all factors in this table were recomputed in 1987 by A. T. A. Holden of the Rochester Institute of Technology. The computed values for d_2 and d_3 as tabulated agree with appropriately rounded values from H. L. Harter, in *Order Statistics and Their Use in Testing and Estimation,* vol. 1, 1969, p. 376.

$a = 3/\sqrt{n} - 0.5,$ $b = (4n-4)/(4n-3),$ $c = (4n-3)/(4n-4),$ $d = 1 - 3/\sqrt{2n-2.5},$ $e = 1 + 3/\sqrt{2n-2.5},$
$f = (4n-4)/(4n-3) - 3/\sqrt{2n-1.5},$ $g = (4n-4)/(4n-3) + 3/\sqrt{2n-1.5}$

Table 9

Sample Size Code Letters: MIL-STD-105D

Lot or Batch Size	Special Inspection Levels				General Inspection Levels		
	S-I	S-2	S-3	S-4	I	II	III
2–8	A	A	A	A	A	A	B
9–15	A	A	A	A	A	B	C
16–25	A	A	B	B	B	C	D
26–50	A	B	B	C	C	D	E
51–90	B	B	C	C	C	E	F
91–150	B	B	C	D	D	F	G
151–280	B	C	D	E	E	G	H
281–500	B	C	D	E	F	H	J
501–1,200	C	C	E	F	G	J	K
1,201–3,200	C	D	E	G	H	K	L
3,201–10,000	C	D	F	G	J	L	M
10,001–35,000	C	D	F	H	K	M	N
35,001–150,000	D	E	G	J	L	N	P
150,001–500,000	D	E	G	J	M	P	Q
500,001 and over	D	E	H	K	N	Q	R

Table 10

Master Table for Normal Inspection (Single Sampling): MIL-STD-105D

Acceptable Quality Levels (normal inspection) — each cell shows Ac Re. Arrows: ↓ = use first sampling plan below arrow; ↑ = use first sampling plan above arrow.

Sample Size Code Letter	Sample Size	0.010	0.015	0.025	0.040	0.065	0.10	0.15	0.25	0.40	0.65	1.0	1.5	2.5	4.0	6.5	10	15	25	40	65	100	150	250	400	650	1.000
A	2	↓	↓	↓	↓	↓	↓	↓	↓	↓	↓	↓	↓	↓	↓	↓	↓	↓	1 2	2 3	3 4	5 6	7 8	10 11	14 15	21 22	30 31
B	3	↓	↓	↓	↓	↓	↓	↓	↓	↓	↓	↓	↓	↓	↓	↓	↓	1 2	2 3	3 4	5 6	7 8	10 11	14 15	21 22	30 31	44 45
C	5	↓	↓	↓	↓	↓	↓	↓	↓	↓	↓	↓	↓	↓	↓	0 1	1 2	2 3	3 4	5 6	7 8	10 11	14 15	21 22	30 31	44 45	↑
D	8	↓	↓	↓	↓	↓	↓	↓	↓	↓	↓	↓	↓	↓	0 1	1 2	2 3	3 4	5 6	7 8	10 11	14 15	21 22	↑	↑	↑	↑
E	13	↓	↓	↓	↓	↓	↓	↓	↓	↓	↓	↓	↓	0 1	1 2	2 3	3 4	5 6	7 8	10 11	14 15	21 22	↑	↑	↑	↑	↑
F	20	↓	↓	↓	↓	↓	↓	↓	↓	↓	↓	↓	0 1	1 2	2 3	3 4	5 6	7 8	10 11	14 15	21 22	↑	↑	↑	↑	↑	↑
G	32	↓	↓	↓	↓	↓	↓	↓	↓	↓	↓	0 1	1 2	2 3	3 4	5 6	7 8	10 11	14 15	21 22	↑	↑	↑	↑	↑	↑	↑
H	50	↓	↓	↓	↓	↓	↓	↓	↓	↓	0 1	1 2	2 3	3 4	5 6	7 8	10 11	14 15	21 22	↑	↑	↑	↑	↑	↑	↑	↑
J	80	↓	↓	↓	↓	↓	↓	↓	↓	0 1	1 2	2 3	3 4	5 6	7 8	10 11	14 15	21 22	↑	↑	↑	↑	↑	↑	↑	↑	↑
K	125	↓	↓	↓	↓	↓	↓	↓	0 1	1 2	2 3	3 4	5 6	7 8	10 11	14 15	21 22	↑	↑	↑	↑	↑	↑	↑	↑	↑	↑
L	200	↓	↓	↓	↓	↓	↓	0 1	1 2	2 3	3 4	5 6	7 8	10 11	14 15	21 22	↑	↑	↑	↑	↑	↑	↑	↑	↑	↑	↑
M	315	↓	↓	↓	↓	↓	0 1	1 2	2 3	3 4	5 6	7 8	10 11	14 15	21 22	↑	↑	↑	↑	↑	↑	↑	↑	↑	↑	↑	↑
N	500	↓	↓	↓	↓	0 1	1 2	2 3	3 4	5 6	7 8	10 11	14 15	21 22	↑	↑	↑	↑	↑	↑	↑	↑	↑	↑	↑	↑	↑
P	800	↓	↓	↓	0 1	1 2	2 3	3 4	5 6	7 8	10 11	14 15	21 22	↑	↑	↑	↑	↑	↑	↑	↑	↑	↑	↑	↑	↑	↑
Q	1,250	↓	↓	0 1	1 2	2 3	3 4	5 6	7 8	10 11	14 15	21 22	↑	↑	↑	↑	↑	↑	↑	↑	↑	↑	↑	↑	↑	↑	↑
R	2,000	↓	0 1	1 2	2 3	3 4	5 6	7 8	10 11	14 15	21 22	↑	↑	↑	↑	↑	↑	↑	↑	↑	↑	↑	↑	↑	↑	↑	↑

↓ = use first sampling plan below arrow. If sample size equals or exceeds lot or batch size, do 100% inspection.

↑ = use first sampling plan above arrow.

Ac = acceptance number.

Re = rejection number.

Table 11

Master Table for Tightened Inspection (Single Sampling): MIL-STD-105D

Acceptable Quality Levels (tightened inspection)

Each cell below is given as **Ac Re** (Ac = acceptance number, Re = rejection number). ↓ = use first sampling plan below arrow. ↑ = use first sampling plan above arrow.

Sample Size Code Letter	Sample Size	0.010	0.015	0.025	0.040	0.065	0.10	0.15	0.25	0.40	0.65	1.0	1.5	2.5	4.0	6.5	10	15	25	40	65	100	150	250	400	650	1.000
A	2	↓	↓	↓	↓	↓	↓	↓	↓	↓	↓	↓	↓	↓	↓	↓	↓	↓	0 1	1 2	2 3	3 4	5 6	8 9	12 13	18 19	27 28
B	3	↓	↓	↓	↓	↓	↓	↓	↓	↓	↓	↓	↓	↓	↓	↓	↓	0 1	1 2	2 3	3 4	5 6	8 9	12 13	18 19	27 28	41 42
C	5	↓	↓	↓	↓	↓	↓	↓	↓	↓	↓	↓	↓	↓	↓	↓	0 1	1 2	2 3	3 4	5 6	8 9	12 13	18 19	27 28	41 42	↑
D	8	↓	↓	↓	↓	↓	↓	↓	↓	↓	↓	↓	↓	↓	↓	0 1	1 2	2 3	3 4	5 6	8 9	12 13	18 19	27 28	41 42	↑	↑
E	13	↓	↓	↓	↓	↓	↓	↓	↓	↓	↓	↓	↓	↓	0 1	1 2	2 3	3 4	5 6	8 9	12 13	18 19	27 28	41 42	↑	↑	↑
F	20	↓	↓	↓	↓	↓	↓	↓	↓	↓	↓	↓	↓	0 1	1 2	2 3	3 4	5 6	8 9	12 13	18 19	27 28	41 42	↑	↑	↑	↑
G	32	↓	↓	↓	↓	↓	↓	↓	↓	↓	↓	↓	0 1	1 2	2 3	3 4	5 6	8 9	12 13	18 19	27 28	41 42	↑	↑	↑	↑	↑
H	50	↓	↓	↓	↓	↓	↓	↓	↓	↓	↓	0 1	1 2	2 3	3 4	5 6	8 9	12 13	18 19	27 28	41 42	↑	↑	↑	↑	↑	↑
J	80	↓	↓	↓	↓	↓	↓	↓	↓	↓	0 1	1 2	2 3	3 4	5 6	8 9	12 13	18 19	27 28	41 42	↑	↑	↑	↑	↑	↑	↑
K	125	↓	↓	↓	↓	↓	↓	↓	↓	0 1	1 2	2 3	3 4	5 6	8 9	12 13	18 19	27 28	41 42	↑	↑	↑	↑	↑	↑	↑	↑
L	200	↓	↓	↓	↓	↓	↓	↓	0 1	1 2	2 3	3 4	5 6	8 9	12 13	18 19	27 28	41 42	↑	↑	↑	↑	↑	↑	↑	↑	↑
M	315	↓	↓	↓	↓	↓	↓	0 1	1 2	2 3	3 4	5 6	8 9	12 13	18 19	27 28	41 42	↑	↑	↑	↑	↑	↑	↑	↑	↑	↑
N	500	↓	↓	↓	↓	↓	0 1	1 2	2 3	3 4	5 6	8 9	12 13	18 19	27 28	41 42	↑	↑	↑	↑	↑	↑	↑	↑	↑	↑	↑
P	800	↓	↓	↓	↓	0 1	1 2	2 3	3 4	5 6	8 9	12 13	18 19	27 28	41 42	↑	↑	↑	↑	↑	↑	↑	↑	↑	↑	↑	↑
Q	1,250	↓	↓	↓	0 1	1 2	2 3	3 4	5 6	8 9	12 13	18 19	27 28	41 42	↑	↑	↑	↑	↑	↑	↑	↑	↑	↑	↑	↑	↑
R	2,000	↓	↓	0 1	1 2	2 3	3 4	5 6	8 9	12 13	18 19	27 28	41 42	↑	↑	↑	↑	↑	↑	↑	↑	↑	↑	↑	↑	↑	↑
S	3,150	↓	0 1	1 2	2 3	3 4	5 6	8 9	12 13	18 19	27 28	41 42	↑	↑	↑	↑	↑	↑	↑	↑	↑	↑	↑	↑	↑	↑	↑

↓ = use first sampling plan below arrow. If sample size equals or exceeds lot or batch size, do 100% inspection.

↑ = use first sampling plan above arrow.

Ac = acceptance number.

Re = rejection number.

Table 12

Master Table for Reduced Inspection (Single Sampling): MIL-STD-105D

Acceptable Quality Levels (reduced inspection)*

Each cell shows the acceptance and rejection numbers as "Ac Re". ↓ = use first sampling plan below arrow. ↑ = use first sampling plan above arrow.

Code	Sample Size	0.010	0.015	0.025	0.040	0.065	0.10	0.15	0.25	0.40	0.65	1.0	1.5	2.5	4.0	6.5	10	15	25	40	65	100	150	250	400	650	1.000
A	2	↓	↓	↓	↓	↓	↓	↓	↓	↓	↓	↓	↓	↓	↓	0 1	0 2	1 3	1 4	2 5	3 6	5 8	7 10	10 13	14 17	21 24	30 31
B	2	↓	↓	↓	↓	↓	↓	↓	↓	↓	↓	↓	↓	↓	0 1	0 2	1 3	1 4	2 5	3 6	5 8	7 10	10 13	14 17	21 24	30 31	↑
C	2	↓	↓	↓	↓	↓	↓	↓	↓	↓	↓	↓	↓	0 1	0 2	1 3	1 4	2 5	3 6	5 8	7 10	10 13	14 17	21 24	30 31	↑	↑
D	3	↓	↓	↓	↓	↓	↓	↓	↓	↓	↓	↓	0 1	0 2	1 3	1 4	2 5	3 6	5 8	7 10	10 13	14 17	21 24	30 31	↑	↑	↑
E	5	↓	↓	↓	↓	↓	↓	↓	↓	↓	↓	0 1	0 2	1 3	1 4	2 5	3 6	5 8	7 10	10 13	14 17	21 24	30 31	↑	↑	↑	↑
F	8	↓	↓	↓	↓	↓	↓	↓	↓	↓	0 1	0 2	1 3	1 4	2 5	3 6	5 8	7 10	10 13	14 17	21 24	30 31	↑	↑	↑	↑	↑
G	13	↓	↓	↓	↓	↓	↓	↓	↓	0 1	0 2	1 3	1 4	2 5	3 6	5 8	7 10	10 13	14 17	21 24	30 31	↑	↑	↑	↑	↑	↑
H	20	↓	↓	↓	↓	↓	↓	↓	0 1	0 2	1 3	1 4	2 5	3 6	5 8	7 10	10 13	14 17	21 24	30 31	↑	↑	↑	↑	↑	↑	↑
J	32	↓	↓	↓	↓	↓	↓	0 1	0 2	1 3	1 4	2 5	3 6	5 8	7 10	10 13	14 17	21 24	30 31	↑	↑	↑	↑	↑	↑	↑	↑
K	50	↓	↓	↓	↓	↓	0 1	0 2	1 3	1 4	2 5	3 6	5 8	7 10	10 13	14 17	21 24	30 31	↑	↑	↑	↑	↑	↑	↑	↑	↑
L	80	↓	↓	↓	↓	0 1	0 2	1 3	1 4	2 5	3 6	5 8	7 10	10 13	14 17	21 24	30 31	↑	↑	↑	↑	↑	↑	↑	↑	↑	↑
M	125	↓	↓	↓	0 1	0 2	1 3	1 4	2 5	3 6	5 8	7 10	10 13	14 17	21 24	30 31	↑	↑	↑	↑	↑	↑	↑	↑	↑	↑	↑
N	200	↓	↓	0 1	0 2	1 3	1 4	2 5	3 6	5 8	7 10	10 13	14 17	21 24	30 31	↑	↑	↑	↑	↑	↑	↑	↑	↑	↑	↑	↑
P	315	↓	0 1	0 2	1 3	1 4	2 5	3 6	5 8	7 10	10 13	14 17	21 24	30 31	↑	↑	↑	↑	↑	↑	↑	↑	↑	↑	↑	↑	↑
Q	500	0 1	0 2	1 3	1 4	2 5	3 6	5 8	7 10	10 13	14 17	21 24	30 31	↑	↑	↑	↑	↑	↑	↑	↑	↑	↑	↑	↑	↑	↑
R	800	0 2	1 3	1 4	2 5	3 6	5 8	7 10	10 13	14 17	21 24	30 31	↑	↑	↑	↑	↑	↑	↑	↑	↑	↑	↑	↑	↑	↑	↑

↓ = use first sampling plan below arrow. If sample size equals or exceeds lot or batch size, do 100% inspection.

↑ = use first sampling plan above arrow.

Ac = acceptance number.

Re = rejection number.

*If the acceptance number has been exceeded but the rejection number has not been reached, accept the lot but reinstate normal inspection.

Table 13

MIL-STD-414

(a) AQL Conversion Table

For Specified AQL Values Falling Within These Ranges	Use This AQL Value
to 0.049	0.04
0.050 – 0.069	0.065
0.070 – 0.109	0.10
0.110 – 0.164	0.15
0.165 – 0.279	0.25
0.280 – 0.439	0.40
0.440 – 0.699	0.65
0.700 – 1.09	1.0
1.10 – 1.64	1.5
1.65 – 2.79	2.5
2.80 – 4.39	4.0
4.40 – 6.99	6.5
7.00 – 10.9	10.0
11.00 – 16.4	15.0

(b) Sample Size Code Letters*

Lot Size	Inspection Levels				
	I	II	III	IV	V
3 – 8	B	B	B	B	c
9 – 15	B	B	B	B	D
16 – 25	B	B	B	C	E
26 – 40	B	B	B	D	F
41 – 65	B	B	C	E	G
66 – 110	B	B	D	F	H
111 – 180	B	C	E	G	I
181 – 300	B	D	F	H	J
301 – 500	C	E	G	I	K
501 – 800	D	F	H	J	L
801 – 1,300	E	G	I	K	L
1,301 – 3,200	F	H	J	L	M
3,201 – 8,000	G	I	L	M	N
8,001 – 22,000	H	J	M	N	O
22,001 – 110,000	I	K	N	O	P
110,001 – 550,000	I	K	O	P	Q
550,001 and over	I	K	P	Q	Q

*Sample size code letters given in body of table are applicable when the indicated inspection levels are to be used.

Table 14

MIL-STD-414 (standard deviation method) Master Table for Normal and Tightened Inspection for Plans Based on Variability Unknown: (single specification limit—form 1)

Sample Size Code Letter	Sample Size	Acceptable Quality Levels (normal inspection)													
		0.04 k	0.065 k	0.10 k	0.15 k	0.25 k	0.40 k	0.65 k	1.00 k	1.50 k	2.50 k	4.00 k	6.50 k	10.00 k	15.00 k
B	3								↓	↓	1.12	0.958	0.765	0.566	0.341
C	4							↓	1.45	1.34	1.17	1.01	0.814	0.617	0.393
D	5							1.65	1.53	1.40	1.24	1.07	0.874	0.675	0.455
E	7					2.00	1.88	1.75	1.62	1.50	1.33	1.15	0.955	0.755	0.536
F	10	↓	↓	↓	2.24	2.11	1.98	1.84	1.72	1.58	1.41	1.23	1.03	0.828	0.611
G	15	2.64	2.53	2.42	2.32	2.20	2.06	1.91	1.79	1.65	1.47	1.30	1.09	0.886	0.664
H	20	2.69	2.58	2.47	2.36	2.24	2.11	1.96	1.82	1.69	1.51	1.33	1.12	0.917	0.695
I	25	2.72	2.61	2.50	2.40	2.26	2.14	1.98	1.85	1.72	1.53	1.35	1.14	0.936	0.712
J	30	2.73	2.61	2.51	2.41	2.28	2.15	2.00	1.86	1.73	1.55	1.36	1.15	0.946	0.723
K	35	2.77	2.65	2.54	2.45	2.31	2.18	2.03	1.89	1.76	1.57	1.39	1.18	0.969	0.745
L	40	2.77	2.66	2.55	2.44	2.31	2.18	2.03	1.89	1.76	1.58	1.39	1.18	0.971	0.746
M	50	2.83	2.71	2.60	2.50	2.35	2.22	2.08	1.93	1.80	1.61	1.42	1.21	1.00	0.774
N	75	2.90	2.77	2.66	2.55	2.41	2.27	2.12	1.98	1.84	1.65	1.46	1.24	1.03	0.804
O	100	2.92	2.80	2.69	2.58	2.43	2.29	2.14	2.00	1.86	1.67	1.48	1.26	1.05	0.819
P	150	2.96	2.84	2.73	2.61	2.47	2.33	2.18	2.03	1.89	1.70	1.51	1.29	1.07	0.841
Q	200	2.97	2.85	2.73	2.62	2.47	2.33	2.18	2.04	1.89	1.70	1.51	1.29	1.07	0.845
		0.065	0.10	0.15	0.25	0.40	0.65	1.00	1.50	2.50	4.00	6.50	10.00	15.00	
		Acceptable Quality Levels (tightened inspection)													

Note: All AQL values are in percent defective.

↓ = use first sampling plan below arrow, that is, both sample size and *k* value. When sample size equals or exceeds lot size, every item in the lot must be inspected.

Table 15

Factors for Two-Sided Tolerance Limits

n \ δ	$1 - \alpha = 0.95$			$1 - \alpha = 0.99$		
	0.90	0.95	0.99	0.90	0.95	0.99
2	32.019	37.674	48.430	160.193	188.491	242.300
3	8.380	9.916	12.861	18.930	22.401	29.055
4	5.369	6.370	8.299	9.398	11.150	14.527
5	4.275	5.079	6.634	6.612	7.855	10.260
6	3.712	4.414	5.775	5.337	6.345	8.301
7	3.369	4.007	5.248	4.613	5.488	7.187
8	3.136	3.732	4.891	4.147	4.936	6.468
9	2.967	3.532	4.631	3.822	4.550	5.966
10	2.839	3.379	4.433	3.582	4.265	5.594
11	2.737	3.259	4.277	3.397	4.045	5.308
12	2.655	3.162	4.150	3.250	3.870	5.079
13	2.587	3.081	4.044	3.130	3.727	4.893
14	2.529	3.012	3.955	3.029	3.608	4.737
15	2.480	2.954	3.878	2.945	3.507	4.605
16	2.437	2.903	3.812	2.872	3.421	4.492
17	2.400	2.858	3.754	2.808	3.345	4.393
18	2.366	2.819	3.702	2.753	3.279	4.307
19	2.337	2.784	3.656	2.703	3.221	4.230
20	2.310	2.752	3.615	2.659	3.168	4.161
25	2.208	2.631	3.457	2.494	2.972	3.904
30	2.140	2.549	3.350	2.385	2.841	3.733
35	2.090	2.490	3.272	2.306	2.748	3.611
40	2.052	2.445	3.213	2.247	2.677	3.518
45	2.021	2.408	3.165	2.200	2.621	3.444
50	1.996	2.379	3.126	2.162	2.576	3.385
55	1.976	2.354	3.094	2.130	2.538	3.335
60	1.958	2.333	3.066	2.103	2.506	3.293
65	1.943	2.315	3.042	2.080	2.478	3.257
70	1.929	2.299	3.021	2.060	2.454	3.225
75	1.917	2.285	3.002	2.042	2.433	3.197
80	1.907	2.272	2.986	2.026	2.414	3.173
85	1.897	2.261	2.971	2.012	2.397	3.150
90	1.889	2.251	2.958	1.999	2.382	3.130
95	1.881	2.241	2.945	1.987	2.368	3.112
100	1.874	2.233	2.934	1.977	2.355	3.096
150	1.825	2.175	2.859	1.905	2.270	2.983
200	1.798	2.143	2.816	1.865	2.222	2.921
250	1.780	2.121	2.788	1.839	2.191	2.880
300	1.767	2.106	2.767	1.820	2.169	2.850
400	1.749	2.084	2.739	1.794	2.138	2.809
500	1.737	2.070	2.721	1.777	2.117	2.783
600	1.729	2.060	2.707	1.764	2.102	2.763
700	1.722	2.052	2.697	1.755	2.091	2.748
800	1.717	2.046	2.688	1.747	2.082	2.736
900	1.712	2.040	2.682	1.741	2.075	2.726
1,000	1.709	2.036	2.676	1.736	2.068	2.718
∞	1.645	1.960	2.576	1.645	1.960	2.576

Source: From C. Eisenart, M. W. Hastay, and W. A. Wallis. *Techniques of Statistical Analysis*, McGraw-Hill Book Company, Inc., 1947. Reproduced with permission of McGraw-Hill.

Table 16

Critical Values for the Kolmogorov-Smirnov One-Sample Test

One-sided test, $\alpha =$	0.10	0.05	0.025	0.01	0.005
Two-sided test, $\alpha =$	0.20	0.10	0.05	0.02	0.01
$n = 1$	0.900	0.950	0.975	0.990	0.995
2	0.684	0.776	0.842	0.900	0.929
3	0.565	0.636	0.708	0.785	0.829
4	0.493	0.565	0.624	0.689	0.734
5	0.447	0.509	0.563	0.627	0.669
6	0.410	0.468	0.519	0.577	0.617
7	0.381	0.436	0.483	0.538	0.576
8	0.358	0.410	0.454	0.507	0.542
9	0.339	0.387	0.430	0.480	0.513
10	0.323	0.369	0.409	0.457	0.489
11	0.308	0.352	0.391	0.437	0.468
12	0.296	0.338	0.375	0.419	0.449
13	0.285	0.325	0.361	0.404	0.432
14	0.275	0.314	0.349	0.390	0.418
15	0.266	0.304	0.338	0.377	0.404
16	0.258	0.295	0.327	0.366	0.392
17	0.250	0.286	0.318	0.355	0.381
18	0.244	0.279	0.309	0.346	0.371
19	0.237	0.271	0.301	0.337	0.361
20	0.232	0.265	0.294	0.329	0.352
21	0.226	0.259	0.287	0.321	0.344
22	0.221	0.253	0.281	0.314	0.337
23	0.216	0.247	0.275	0.307	0.330
24	0.212	0.242	0.269	0.301	0.323
25	0.208	0.238	0.264	0.295	0.317
26	0.204	0.233	0.259	0.290	0.311
27	0.200	0.229	0.254	0.284	0.305
28	0.197	0.225	0.250	0.279	0.300
29	0.193	0.221	0.246	0.275	0.295
30	0.190	0.218	0.242	0.270	0.290
31	0.187	0.214	0.238	0.266	0.285
32	0.184	0.211	0.234	0.262	0.281
33	0.182	0.208	0.231	0.258	0.277
34	0.179	0.205	0.227	0.254	0.273
35	0.177	0.202	0.224	0.251	0.269
36	0.174	0.199	0.221	0.247	0.265
37	0.172	0.196	0.218	0.244	0.262
38	0.170	0.194	0.215	0.241	0.258
39	0.168	0.191	0.213	0.238	0.255
40	0.165	0.189	0.210	0.235	0.252
Approximation for $n > 40$:	$\dfrac{1.0730}{\sqrt{n}}$	$\dfrac{1.2239}{\sqrt{n}}$	$\dfrac{1.3581}{\sqrt{n}}$	$\dfrac{1.5174}{\sqrt{n}}$	$\dfrac{1.6276}{\sqrt{n}}$

Source: This table is extracted from "Table of percentage points of Kolmgorov statistics," *J. Amer. Statist. Assoc.*, **51**: 111–121 (1956), with permission of the author, L. H. Miller, and the editor.

References

American Society for Testing and Materials (1992). *Manual on Presentation of Data and Control Chart Analysis*, 6th ed., ASTM, Philadelphia, Pa.

Bowker, A. H., and G. J. Lieberman (1972). *Engineering Statistics*, 2nd ed., Prentice-Hall, Inc., Englewood Cliffs, N.J.

Box, G. E. P., and D. R. Cox (1964). An Analysis of Transformations, *Journal of the Royal Statistical Society*, B 26, 211–243.

Draper, N. R., and H. Smith (1981). *Applied Regression Analysis*, 2nd ed., John Wiley and Sons, New York.

Ford (1987). *Continuing Process Control and Process Capability Improvement*, Ford Motor Company, Dearborn, Mich.

Grant, E. L., and R. S. Leavenworth (1980). *Statistical Quality Control*, 5th ed., McGraw-Hill Book Co., New York.

Guttman, I.; S. S. Wilks; and J. S. Hunter (1971). *Introductory Engineering Statistics*, 2nd ed., John Wiley & Sons, New York.

Koopmans, L. H. (1987). *An Introduction to Contemporary Statistical Methods*, PWS-KENT Publishing Co., Boston.

Kume, H. (1992). *Statistical Methods for Quality Improvement*, The Association for Overseas Technical Scholarship (AOTS), Tokyo.

Kutner, M. H.; C. J. Nachtsheim; and J. Neter (2004). *Applied Linear Regression Models*, 4th ed., McGraw-Hill/Irwin, New York.

Lapin, L. L. (1990). *Probability and Statistics for Modern Engineering*, 2nd ed., PWS-KENT Publishing Co., Boston.

McClave, J. T., and F. H. Dietrich (1985). *Statistics*, 3rd ed., Dellen Publishing Co., San Francisco.

Mendenhall, W.; D. D. Wackerly; and R. L. Scheaffer (1990). *Mathematical Statistics with Applications*, 4th ed., PWS-KENT Publishing Co., Boston.

Miller, I., and J. E. Freund (1977). *Probability and Statistics for Engineers*, 2nd ed., Prentice-Hall Inc., Englewood Cliffs, N.J.

Ott, L. (1988). *An Introduction to Statistical Methods and Data Analysis*, 3rd ed., PWS-KENT Publishing Co., Boston.

Ross, S. (1984). *A First Course in Probability*, 2nd ed., Macmillan Publishing Co., New York.

Ryan, T. P. (1989). *Statistical Methods for Quality Improvement*, John Wiley & Sons, New York.

Stephens, M. A. (1974). "EDF Statistics for Goodness of Fit and Some Comparisons." *Journal of Am. Stat. Assn.*, vol. 69, no. 347, pp. 730–737.

Thompson, J. R., and J. Koronacki (1993). *Statistical Process Control for Quality Improvement*, Chapman and Hall, New York.

Tukey, J. W. (1977). *Exploratory Data Analysis*, Addison-Wesley Publishing Co.

Wadsworth, H. M.; K. S. Stephens; and A. B. Godfrey (1986). *Modern Methods for Quality Control and Improvement*, John Wiley & Sons, New York.

Walpole, R. E., and R. H. Myers (1985). *Probability and Statistics for Engineers and Scientists*, 3rd ed., Macmillan Publishing Co., New York.

Chapter 1

1.1
a. Qualitative **f.** Quantitative
b. Quantitative **g.** Quantitative
c. Quantitative **h.** Qualitative
d. Quantitative **i.** Quantitative
e. Qualitative **j.** Qualitative

1.3
a.

Income Category	Years of Formal Education				
	12	14	16	18	20+
0–29,999	0.62	0.47	0.32	0.22	0.15
30,000–59,999	0.31	0.41	0.40	0.42	0.30
60,000–89,999	0.05	0.09	0.17	0.23	0.27
90,000 and above	0.01	0.03	0.11	0.14	0.28

b. The relative frequency of higher-income categories increased with the increasing number of years of formal education.

1.5
a. Cleaning of public highways, screening/fencing, and working hours.
b. Cleaning of public highways, working hours, screening/fencing, water courses affected by construction, blue routes & restricted times of use, temporary and permanent diversions, TMP, and property damage.

1.7
a.

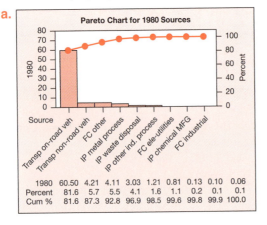

	1980								
1980	60.50	4.21	4.11	3.03	1.21	0.81	0.13	0.10	0.06
Percent	81.6	5.7	5.5	4.1	1.6	1.1	0.2	0.1	0.1
Cum %	81.6	87.3	92.8	96.9	98.5	99.6	99.8	99.9	100.0

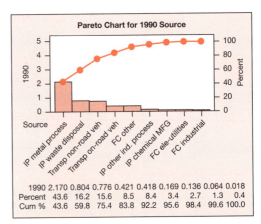

	1990								
1990	2.170	0.804	0.776	0.421	0.418	0.169	0.136	0.064	0.018
Percent	43.6	16.2	15.6	8.5	8.4	3.4	2.7	1.3	0.4
Cum %	43.6	59.8	75.4	83.8	92.2	95.6	98.4	99.6	100.0

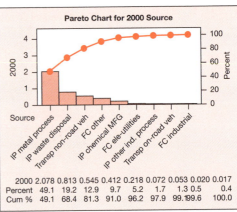

	2000								
2000	2.078	0.813	0.545	0.412	0.218	0.072	0.053	0.020	0.017
Percent	49.1	19.2	12.9	9.7	5.2	1.7	1.3	0.5	0.4
Cum %	49.1	68.4	81.3	91.0	96.2	97.9	99.1	99.6	100.0

b. In 1980, on-road vehicles were the major contributor of lead emission pollution, but in 1990 and 2000 they are a very minor contributor.

c. Yes. The contribution by on-road vehicles has decreased.

1.9
a. The largest number were employed by the information technology services, followed by engineering and RFD services, and missile space vehicle manufacturing.

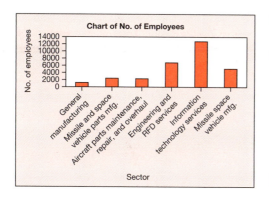

b. Although information technology services employed the largest number of employees, they were not large employers. Engineering RFD services and missile space vehicle manufacturing employed fewer people than the information technology services; they employed far more people per company.

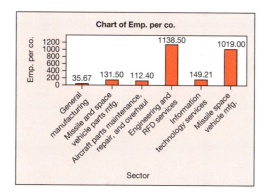

1.11 a.

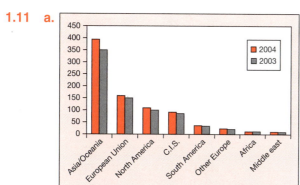

b. In general, the crude oil production has increased from 2003 to 2004.

1.13 The data are grouped together in a histogram, losing identity of individual observations, which are still retained by dotplot. A small number of observations makes it difficult to notice any patterns. Gaps in the data are visible from a dotplot but are not identified from a histogram.

1.15 a. 7.8%

b. A gap just below the LCL

1.17 a. Yes. People are living longer.

b. 1890: about 64%; 2005: about 41%

c. The percentage of older population has increased. In 1890, the percentage of population in different age categories decreased steadily with the increasing age. In 2005, it is fairly evenly distributed across different age groups except for the two oldest age groups.

1.19 a. There is a considerable increase in average index from 1990 to 1998, indicating an increase in industrial product by most of the countries in general. The indexes were more spread out in 1990 compared to 1998. In 1990, there were two outliers on the lower end and one on the higher end. In 1998, there were two outliers on the higher end.

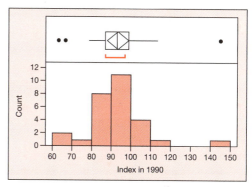

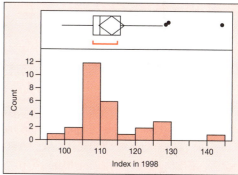

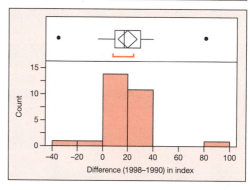

b. Most countries showed improvement (an increase of up to 40 points) in the industrial production; one country in particular showed a tremendous amount of improvement. Only two countries showed a decrease.

1.21 **a.**

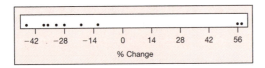

% Change

b. Mean = −0.7, StDev = 49.9
c. Median = −24.5, IQR = 93.6
d. No, because the distribution is skewed with outliers on the higher end that affect the values of mean and standard deviation.

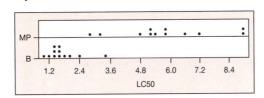

% Change

Mean = −11.0, StDev = 39.3,
Median = −28.6, IQR = 59.8

1.25 **a.** 4.105
     b. 3.428

1.27 **a.** The distribution of MP is more spread out than that of B. The distribution of B is slightly right-skewed. The distribution of MP is fairly symmetric.

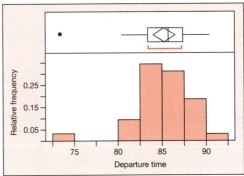

LC50

b. MP: Mean = 5.775, StDev = 1.923
B: Mean = 1.683, StDev = 0.190
Yes. In case of B, mean and standard deviation may be affected slightly by the highest measurement, which may be an outlier.

1.29 **a.** Structurally deficient: Mean = 10.75, StDev = 4.18
Functionally obsolete: Mean = 15.00, StDev = 4.07
b. On the average, there are more functionally obsolete bridges among the states in the study than structurally deficient bridges. The variation in the number of bridges per state seems to be comparable.

1.31 **a.** Both, the arrival and departure time distributions are left-skewed, arrival times more so than the departure times. The median percentage of on-time departures is higher than the

median percentage of on-time arrivals. Both the distributions have about the same range. Both the distributions have outliers on the lower end, indicating a low-performing airport (or airports).

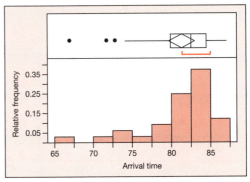

Arrival time

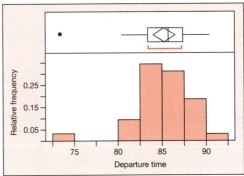

Departure time

b. One
c. None
d. None. The data are skewed with outliers.
e. Arrival: 78%
f. Departure: 94%
g. Arrival: Chicago O'Hare, Newark Int, and New York LaGuardia
Departure: Chicago O'Hare
h. Atlanta: $Z_{Arrival} = -0.1820$,
$Z_{Departure} = -0.5058$. Arrivals.
Chicago O'Hare: $Z_{Arrival} = -3.1425$,
$Z_{Departure} = -3.4591$. Arrivals.

1.33 **a.** Median and IQR

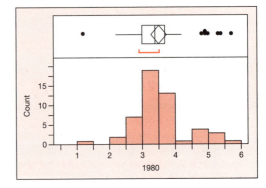

1980

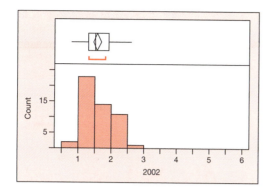

2002

b. Washington, DC; Idaho; Montana; West Virginia; Wyoming; Arizona; New Mexico; Louisiana; and Nevada. Except for Washington, DC, which is heavily populated, all others are less populated large states, mostly rural areas. Drivers tend to speed and medical help is not received fast enough because of distances involved.

c. The median motor vehicle death rate has decreased. The maximum death rate has also reduced to half of what it was in 1980.

1.35 a.

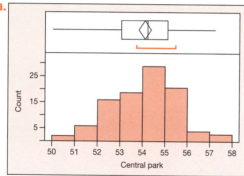

Central park

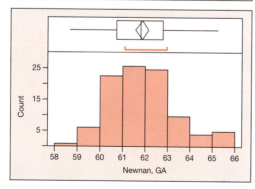

Newnan, GA

b. The distribution of annual temperatures in Central Park is slightly left-skewed. The temperatures ranged from about 50°F to 57°F with a mean about 54°F. There are no outliers. The distribution of temperatures at Newnan is slightly right-skewed. The temperatures ranged from about 58°F to 66°F with a mean about 62°F. There are no outliers.

c. The shapes of the two distributions indicate that Central Park has seen more years with warmer temperatures and Newnan more years with cooler temperatures during the last century. On the average, Newnan is warmer than the Central Park. The range of temperatures is about the same at both locations.

1.37 The percentage of eligible voters who voted in the 2000 presidential election increased steadily with the household income group. From the lowest income group, the lowest percentage of voters voted, whereas from the highest income group the highest percentage of voters voted in this election.

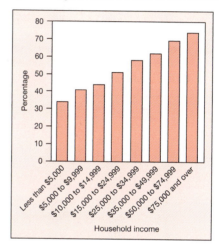

Household income

1.41 a.

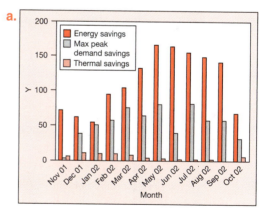

Month

b. Every month the energy savings are the highest and the thermal savings are the lowest. The energy savings show a cycle with highest savings during the summer months and lowest savings during the winter months. On the other hand, thermal savings are highest during the winter months and lowest during the summer months, showing exactly opposite cycles. The maximum peak demand savings are higher in general during summer months and lower in the winter months.

c. Mean = 223.16, StDev = 107.97

d. Although July 2002 showed higher total savings compared to the other months in the study, it was not an outlier.

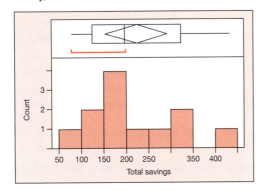

1.43 Fuel combustion is the largest contributor of sulfur dioxide emissions. Although the amount of contribution decreased over the years, it is still a major contributor. Amount of contribution by industrial processes decreased over the years, but the percentage of total emission increased over the years. The percent contribution by transportation increased slightly.

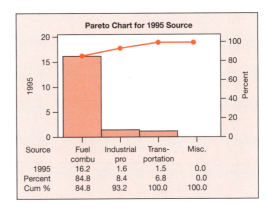

	Fuel combu	Industrial pro	Transportation	Misc.
Source				
1995	16.2	1.6	1.5	0.0
Percent	84.8	8.4	6.8	0.0
Cum %	84.8	93.2	100.0	100.0

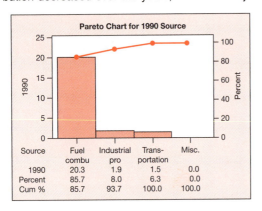

	Fuel combu	Industrial pro	Transportation	Misc.
Source				
1990	20.3	1.9	1.5	0.0
Percent	85.7	8.0	6.3	0.0
Cum %	85.7	93.7	100.0	100.0

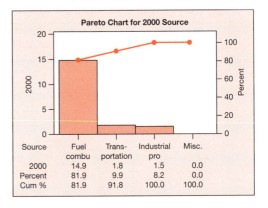

	Fuel combu	Transportation	Industrial pro	Misc.
Source				
2000	14.9	1.8	1.5	0.0
Percent	81.9	9.9	8.2	0.0
Cum %	81.9	91.8	100.0	100.0

Chapter 2

2.1   **a.**

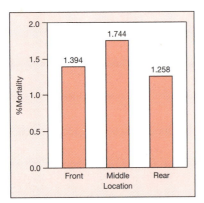

b. The mortality rate is highest among the middle-seat passengers compared to the front- or rear-seat passengers. The mortality rate is lowest among the rear-seat passengers.

2.3

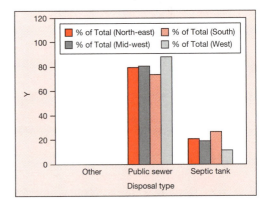

A public sewer is used most commonly in all four geographical areas. Other methods of sewage disposal are the least used in all four geographical areas. The western region has the highest percentage of public sewage users and the southern region the lowest.

2.5 **a.**

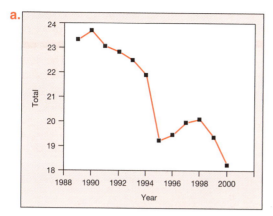

The total sulfur dioxide estimates decreased steadily till 1994; then a sudden drop was observed from 1994 to 1995. After 1995, they increased steadily till 1998 and then decreased.

b.

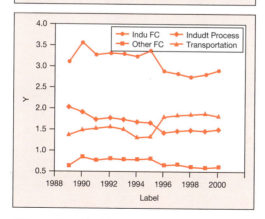

The sulfur dioxide emission estimates from fuel combustion, particularly from electrical utilities, decreased steadily till 1994 and dropped considerably in 1995. They increased for the next three years and then decreased. All other categories are minor contributors of sulfur dioxide emissions, and their amounts

decreased over the years, except for transportation, which showed an increase from 1995 to 1996 and increased slightly thereafter.

2.7 **a.**

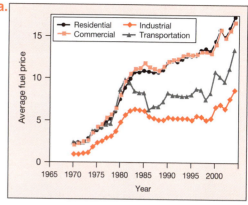

b. The average fuel price showed an increasing trend in residential and commercial sectors, with sudden increases in the early 1980s. Both sectors showed very similar trends with similar prices. The transportation sector has higher average fuel prices. Although the industrial and transportation sectors showed similar trends in average fuel prices over the years, the difference in the average fuel prices increased slightly. In both of these sectors, the prices increased till the mid-80s and then experienced a drop followed by a decade of stable prices. Then the prices showed a steady increasing trend.

2.9 **a.**

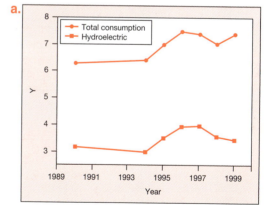

b. Both total consumption and hydroelectric consumption showed a somewhat similar trend of steady increase and then decline, except in 1999, when total consumption increased.

c. Hydroelectric consumption as a percentage of total consumption decreased from 1990 to 1994, then increased steadily for the next three years, and then decreased considerably within two years, bringing it to the level of 1994.

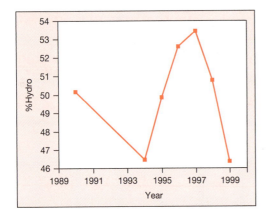

2.11 **a.**

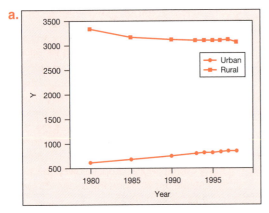

b. From 1980 to 1998, the highways constructed in urban areas (in thousand miles) have increased steadily and those in the rural areas have decreased steadily. However, rural area highways still represent a major portion of federally funded highways (rural area about four times that of urban area).

2.13

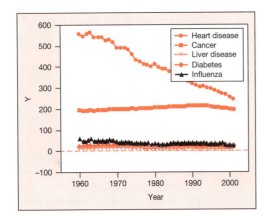

The age-adjusted death rate per 100,000 population due to heart disease has decreased steadily till the year 2001, reducing it to about half of what it was in 1960. The death rate due to cancer, on the other hand, has shown a slight increase till 1995 and then a steady decrease. The death rate due to diabetes has shown a slight decrease with some fluctuations. The death rates due to influenza and liver disease showed a slight decrease over the years.

2.15 **a.**

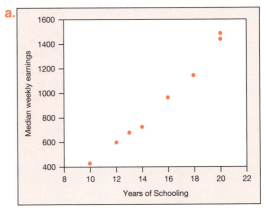

There is extremely strong positive correlation between median weekly earnings and years of schooling. As the years of schooling increase so do the median earnings. Although there is a slight curvature, a linear fit seems like a very good fit.

b.

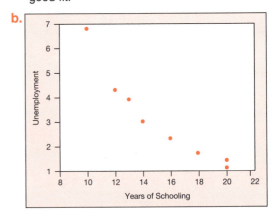

There is very strong negative correlation between the years of schooling and unemployment rate. Higher the number of years of schooling, lower is the unemployment rate. Linear will be a good fit but quadratic will be even better.

2.17 a.

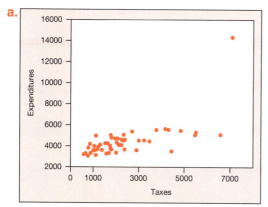

b. There seems to be an increasing relation between expenditure and tax revenue, that is expenditure increased as tax revenue increased. However, there is one outlier data point which is distorting the relation.

c.

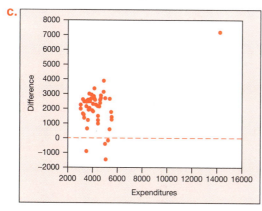

There does not seem to be any relation between expenditure and the differences. There is one outlier state that gives a wrong impression of positive relation.

d.

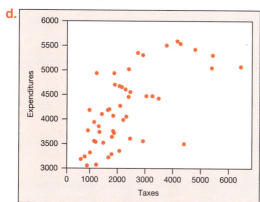

The scatterplot shows a positive relation between expenditure and tax revenue.

2.19 a.

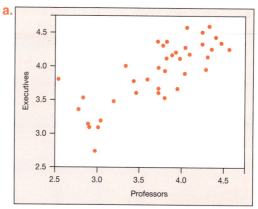

b. There is a strong positive relation between ranking by executives and professors. In general skills ranked by professors are also ranked higher by executives.

2.21 a. $r = 0.708$

b. There is strong positive correlation between the number of motor vehicle deaths in two years. In general the states with the higher death rates in 1980 had the higher death rates in 2002 too.

c. The linear association and presence of possible outliers is easily visible from scatterplot but cannot be assessed from the correlation coefficient.

d. Strength of relation is easily determined from the value of correlation but not necessarily from the scatterplot.

2.23 a.

	Revenue	**Taxes**	**Expenditures**
Revenue	1.0000	−0.0906	−0.0203
Taxes	−0.0906	1.0000	0.6628
Expenditures	−0.0203	0.6628	1.0000

b. There is extremely weak negative correlation between revenue and taxes & revenue and expenditure. There is fairly strong positive correlation between expenditure and taxes (strongest correlation).

c. States Alaska, California, Maryland, and New York are possibly outliers affecting the relation. They are probably weakening the strength of relation.

2.25 a. $r = 0.789$

2.27 a. $r = 0.9837$ **b.** $r = 0.9928$ **c.** $r = 0.9652$

2.29 a. Expenditures $= 3396.85 + 0.346*Taxes$

b. For every dollar per capita of tax revenue collected, the per capita expenditure increased by about 35 cents.

2.31   **a.** Executives = 1.361 + 0.683*Professors
       b. For every unit increase in a rank by professors, the rank by executives increased on the average by only 0.683.
       c. 4.093

2.33   **a.** Extremely strong negative relation
       b. %Fat = 505.253 − 460.678*Density
       c. 21.54

2.35   **a.** flow-through = −0.7110 + 0.6553 Static
       b. Possible presence of three separate groups with different relations.

2.37   **a.** $r = -0.99968$, $r^2 = 0.99936$
       b. There is an extremely strong positive linear association between the % body fat and the body density. Almost all (99.94%) of variation in the % body fat can be explained using a linear relation between the % body fat and body density.

2.39   **a.**

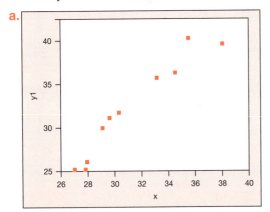

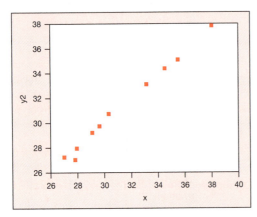

       b. Better after calibration
       c. Before: $r = 0.965$, After: $r = 0.996$

2.41   **a.** $y = -235.116 + 1.273*x^2$
       b. $t = 18.30$, Reject null at $\alpha = 0.05$.
       c. $r^2 = 0.9767$

2.43   Since x_1 and x_2 both have a strong positive correlation, we would expect the slope of the regression line for x_1 and y to have the same sign as the slope of the regression line for x_2 and y.

2.45   **a.** Residual plot

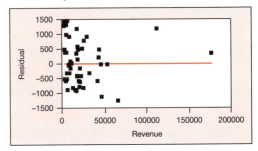

       b. Residuals are fairly randomly distributed around zero without any specific pattern, except for residuals for two states that have extremely high revenue compared to the rest of the states.
       c. A linear model may not be appropriate. After removing two outlier states, the fit must be examined.

2.47   **a.** Residual plot

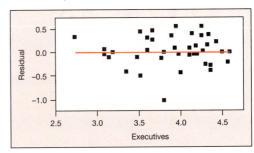

       b. The scatterplot indicates the possibility of two groups. The first part indicates a decreasing trend among the residuals. The latter part shows no pattern.
       c. A linear model may not be appropriate. The possibility of groups must be investigated.

2.49   **a.** Residual plot

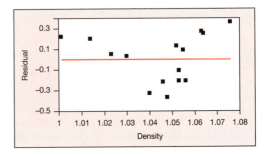

b. The residual plot clearly shows a U-shaped pattern among residuals.

c. The linear model is not appropriate.

2.51 **a.**

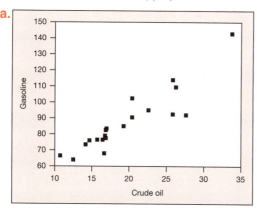

There is a fairly strong positive linear relation between crude oil and gasoline prices.

b. Gasoline = 29.28 + 2.95 crude oil

c.

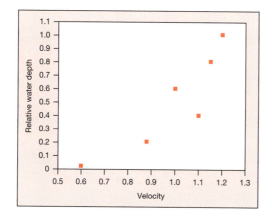

d. Fairly randomly distributed residuals indicate a linear model may be appropriate.

2.53 **a.** Stress = $372.61394 - 247.0381(x/a)$

b. $r^2 = 0.83$

c. Nonlinear

2.55 **a.** Station = 16

Station = 18

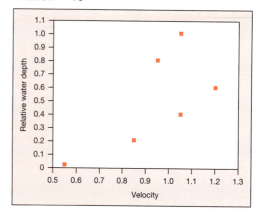

b. A slight curvature is noticed.

c. Station 16: Log(Relative Water Depth) = $-7.417 + 6.295$ Velocity

Station 18: Log(Relative Water Depth) = $-6.599 + 5.738$ Velocity

2.57 **a.** Scatterplots and correlation coefficients indicate increasing relations of varying strengths.

b. Log(Pb) = $-5.6400 + 0.5844$ CO
$NO_2 = 0.0126 + 0.0013$ CO
Ozone = $0.0820 + 0.0057$ CO
$SO_2 = 0.0017 + 0.0010$ CO
Ozone = $0.0334 + 3.9486$ NO_2
$SO_2 = -0.0056 + 0.6382$ NO_2
$SO_2 = -0.0084 + 0.1434$ Ozone $- 4.1392$ (Ozone $- 0.1156)^2$

c.

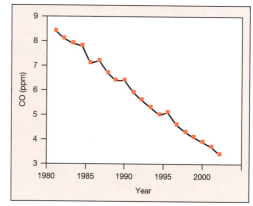

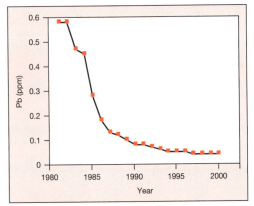

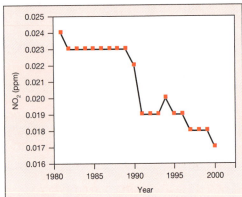

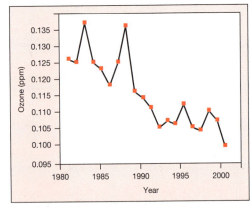

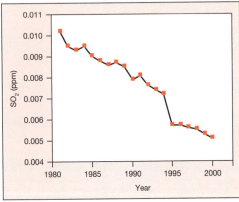

CO and SO_2 show steady decline. Pb declined sharply during the first decade and then declined slowly over the next decade. NO_2 remained fairly constant during the first decade and declined over the next. Overall, ozone level declined, with a few ups and downs.

2.59 Yes. The number of semiactive dampers used increased with the height as well as the number of stories.

No. semiactive dampers $= -93.591 + 5.011$ Stories $+ 0.270$ (Stories $- 28.777)^2$
No. semiactive dampers $= -85.722 + 1.042$ Height (m) $+ 0.014$ (Height (m) $- 130.067)^2$

2.61 **a.** A very strong linear decreasing relation
b. PDA (%) $= 100.760 - 118.660$ Limit load

2.63 **a.**

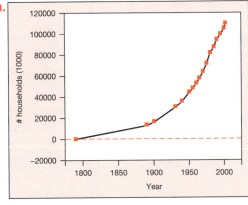

The number of households in the United States has increased over this time period at an increasing rate.

b.

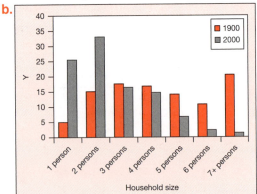

c. The number of households has increased, but the number of persons per household has decreased.

2.65 **a.** Solution = 0.001M NaNO$_3$ Solution = 0.01M NaNO$_3$

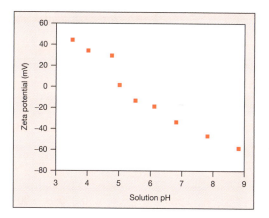
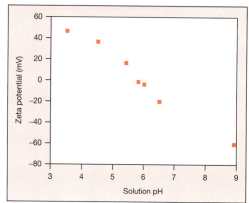

b. Both scatterplots show strong negative linear relations.
c. Zeta potential = 115.65 − 20.69 Solution pH

Chapter 3

3.1 **a.** Observational

3.3 **a.** Experiment

3.5 **a.** Experiment

3.19 **a.** Yield
b. Temperatures 100°F, 150°F, and 200°F   **d.** 4

3.21 **a.** Observational

3.23 **a.** Observational   **b.** No   **d.** No; No

3.25 **a.** Sample survey   **b.** No

Chapter 4

4.1 **a.**

Gender	Engineers	Computer Science
Male	2,578,500	483,000
Female	121,500	216,300
Total	2,700,000	700,000

b.

Race	Engineers	Computer Science
White	2,430,00	618,100
Black	43,200	25,900
Asian	151,200	46,200
Native American	10,800	700
Female	64,800	9,100
Total	2,700,000	700,000

4.3

	Modem	No modem	Total
Printer	3	7	10
No Printer	2	13	15
Total	5	20	25

4.7 Discussion Answer.

4.9 **a.** S = {Turn left, Continue straight ahead}
b. P(Turn left) = 0.5,
   P(Continue straight ahead) = 0.5
c. P(Turns) = 0.5

4.11 **a.** $\frac{1}{3}$ **b.** $\frac{6}{15}$ **c.** $\frac{19}{48}$ **d.** $\frac{2}{3}$

4.13 **a.** 0.08   **b.** 0.16   **c.** 0.14   **d.** 0.84

4.15 **a.** S = {(C1d1, C2d1), (C1d1, C2d2),
   (C1d2, C2d1), (C1d2, C2d2)}

b. P(both customers use door I) = $\frac{1}{4}$

   P(both customers use the same door) = $\frac{1}{2}$

4.17 $\frac{6}{20}$

4.19 $\frac{1}{5}$

4.21 **a.** 168   **b.** $\frac{1}{168}$

4.23 **a.** $\frac{6}{10}$ **b.** $\frac{4}{10}$ **c.** $\frac{3}{5}$

4.25 **a.** $\frac{60}{125}$ **b.** $\frac{5}{125}$ **c.** $\frac{20}{125}$

4.27 **a.** 240   **b.** $\frac{1}{12}$

4.29 0.4 .0833

4.31 **a.** $\dfrac{234}{319}$ **b.** $\dfrac{55}{234}$ **c.** No

4.33 **a.** 0.5   **b.** No

4.35 No

4.37 **a.** $\dfrac{2}{3}$ **b.** $\dfrac{1}{9}$ **c.** $\dfrac{1}{6}$

4.39 **a.** 0.16   **b.** 0.84   **c.** 0.96   **d.** No

4.41 **a.** 0.805   **b.** 0.9947   **c.** 0.8093

4.43 **a.** 0.7225   **b.** 0.99

4.45 **a.** 0.81   **b.** 0.99

4.47 0.8235

4.49 **b.** 2,385,000
   c. 70.1, 13.5, 2.2, 8.2, 0.8, 3.7, 1.6
   d. 795,000

4.51 **a.** No
   b.

≤159	M.I.	No M.I.
Aspirin	2	390
Placebo	9	397

160–209	M.I.	No M.I.
Aspirin	12	575
Placebo	37	474

210–259	M.I.	No M.I.
Aspirin	26	1409
Placebo	43	1401

≥260	M.I.	No M.I.
Aspirin	14	168
Placebo	23	547

The corresponding OR are 0.226, 0.267, 0.601, 1.982.

c. The odds of M.I. for the aspirin group increased as the cholesterol level increased.

4.53 **a.** $\dfrac{24}{40}$ **b.** $\dfrac{24}{60}$ **c.** Yes   **d.** Yes   **e.** 0

4.55 **a.** {dd, nd, dn, nn}   **b.** A = {nn}   **c.** 1/7

4.57 **a.** 36   **b.** $\dfrac{1}{6}$

4.59 **a.** 0.57   **b.** 0.18   **c.** 0.90   **d.** 0.32

4.61 9,000,000

4.63 18

4.65 **a.** 0.0362   **b.** 0.00045, 0.0018

4.67 **a.** $\dfrac{1}{8}$ **b.** $\dfrac{1}{64}$

4.69 0.92

4.71 **a.** $\dfrac{1}{16384}$ **b.** $\dfrac{1}{64}$

4.73 0.8704

4.77 No

4.79 $\dfrac{1}{7}$

4.81 Design A

4.87 **a.** $\dfrac{376}{566}$ **b.** $\dfrac{179}{376}$

Chapter 5

5.1

x	0	1	2	3
p(x)	$\dfrac{1}{30}$	$\dfrac{9}{30}$	$\dfrac{15}{30}$	$\dfrac{9}{30}$

5.3

x	0	1	2	3
p(x)	0.2585	0.4419	0.2518	0.0478

5.5

x	0	1	2	3	4
p(x)	0.3895	0.4142	0.1651	0.0293	0.0019

5.7

x	0	1	2	3
p(x)	$\dfrac{8}{27}$	$\dfrac{12}{27}$	$\dfrac{6}{27}$	$\dfrac{1}{27}$

y	0	1	2	3
p(y)	$\dfrac{2,744}{3,375}$	$\dfrac{588}{3,375}$	$\dfrac{42}{3,375}$	$\dfrac{1}{3,375}$

x + y	0	1	2	3
p(x + y)	0.2409	0.41297	0.26179	0.07445

x + y	4	5	6
p(x + y)	0.00935	0.00053	0.00001

5.9 **a.**

x	0	1	2
p(x)	$\dfrac{1}{6}$	$\dfrac{4}{6}$	$\dfrac{1}{6}$

b.

x	0	1
p(x)	$\dfrac{1}{2}$	$\dfrac{1}{2}$

c.

x	0
p(x)	1

5.11 **a.** $0, \dfrac{2}{3}$ **b.** $0, \dfrac{12}{5}$ **c.** Box II

5.13 Mean = 37.17, Standard deviation = 11.45, and Median = 30–39

5.15 0.4, 0.44, 0.6633

5.17   **a.** (1.4702, 6.5298)   **b.** Yes

5.19   (84.1886, 115.8114)

5.21   **a.** 0.5136   **b.** 0.1808   **c.** 0.9728
      d. 0.8   **e.** 0.64

5.23   **a.** 0.537   **b.** 0.098

5.25   **a.** 16   **b.** 3.2

5.27   8

5.29   0.5931

5.31   **a.** 0.99   **b.** 0.9999

5.33   **a.** 0.1536   **b.** 0.9728

5.35   3.96

5.37   840

5.39   **a.** 0.648   **b.** 1

5.41   **a.** 0.04374   **b.** 0.99144

5.43   **a.** $\dfrac{10}{9}, \dfrac{10}{81}$ **b.** $\dfrac{10}{3}, \dfrac{30}{81}$

5.45   150; 4,500; No

5.47   **a.** 0.128   **b.** 0.03072

5.49   **a.** 0.06561   **b.** $\dfrac{40}{9}, \dfrac{40}{81}$

5.51   **a.** $\dfrac{3}{16}$ **b.** $\dfrac{3}{16}$ **c.** $\dfrac{1}{8}$ **d.** $\dfrac{1}{2}$

5.53   **a.** 0.0183   **b.** 0.908   **c.** 0.997

5.55   **a.** 0.467   **b.** 0.188

5.57   **a.** 0.8187   **b.** 0.5488

5.59   **a.** 0.140   **b.** 0.042   **c.** 0.997

5.61   **a.** 128 exp(−16)   **b.** 128 exp(−16)

5.63   320; 56.5685

5.65 λ^2

5.67   **a.** 0.001   **b.** 0.000017

5.69 $\dfrac{1}{42}$

5.71   **a.** $\dfrac{4}{5}$ **b.** $\dfrac{1}{5}$

5.73   **a.**

y	0	1	2
$p(y)$	$\dfrac{7}{15}$	$\dfrac{7}{15}$	$\dfrac{1}{15}$

     b.

x	0	1	2	3
$p(x)$	$\dfrac{1}{6}$	$\dfrac{1}{2}$	$\dfrac{3}{10}$	$\dfrac{1}{30}$

5.75 $\dfrac{3}{14}$

5.77   **a.** $\dfrac{9}{14}$ **b.** $\dfrac{13}{14}$ **c.** 1

5.79   **a.** 1   **b.** 1   **c.** $\dfrac{18}{19}$ **d.** $\dfrac{49}{57}$ **e.** $\dfrac{728}{969}$

5.81 $\dfrac{10}{21}$

5.83 $E(Y) = np$ and $V(Y) = np(1 - p)$

5.89   0.32805; 0.99999

5.91   **a.** 1   **b.** 0.5905   **c.** 0.1681   **d.** 0.03125   **e.** 0

5.93   **a.** $n = 25, a = 5$ **b.** $n = 25, a = 5$

5.95   **a.** 0.758   **b.** 12; 12   **c.** (5.0718, 18.9282)

5.97   **a.** 5   **b.** 0.007   **c.** 0.384

5.99   0.993

5.103   0.18522

5.105   0.01536; 0.0256

5.107 $E(Y) = 900$; $V(Y) = 90$; $K = 2$; (881.026, 918.974)

5.109   **a.** $p(y) = \dbinom{4}{y}\left(\dfrac{1}{3}\right)^{y}\left(\dfrac{2}{3}\right)^{4-y}$ **b.** $\dfrac{1}{9}$
     c. $\dfrac{4}{3}$ **d.** $\dfrac{8}{9}$

5.111   **a.** 0.1192   **b.** 0.117; yes

5.113   3

5.115   No

Chapter 6

6.1   **a.** Discrete   **b.** Discrete
     c. Discrete   **d.** Continuous

6.3   **a.** $\dfrac{27}{32}$ **b.** 4

6.5   **b.** $F(x) = \begin{cases} 0 & x \le 5 \\ \dfrac{(x-7)^3}{8} + 1 & 5 < x \le 7 \\ 1 & x > 7 \end{cases}$
     c. $\dfrac{7}{8}$ **d.** $\dfrac{37}{56}$

6.7   **a.** $\dfrac{3}{4}$ **b.** $\dfrac{4}{5}$ **c.** 1
     d. $F(x) = \begin{cases} 0 & x \le 0 \\ x^2 & 0 < x \le 1 \\ 1 & x > 1 \end{cases}$

6.9   60; $\dfrac{1}{3}$

6.11   4

6.13   **a.** 5.5; 0.15   **b.** $k = 2$, (4.7254, 6.2746)
     c. About 58% of the time

6.17 a. $\dfrac{1}{20}$ b. $\dfrac{1}{20}$ c. $\dfrac{1}{2}$

6.19 $\dfrac{3}{4}$

6.21 a. $\dfrac{1}{8}$ b. $\dfrac{1}{8}$ c. $\dfrac{1}{4}$

6.23 a. $\dfrac{2}{7}$ b. $\dfrac{3}{200}; \dfrac{49}{120{,}000}$

6.25 $\dfrac{1}{6}$

6.27 a. $\dfrac{1}{2}$ b. $\dfrac{1}{4}$

6.29 a. $60; \dfrac{100}{3}$ b. 4

6.31 a. $\exp\left(\dfrac{-5}{4}\right)$ b. $\exp\left(\dfrac{-5}{6}\right) - \exp\left(\dfrac{-5}{4}\right)$

6.33 a. $\exp(-2)$ b. 460.52

6.35 a. 1,100; 2,920,000
b. No; $p(C > 2{,}000) = \exp(1.94)$

6.37 a. $\exp(-1)$ b. $\exp\left(\dfrac{-1}{2}\right)$

6.39 a. $\exp\left(\dfrac{-1}{2}\right)$ b. $\exp\left(\dfrac{-1}{4}\right)$ c. No

6.41 a. $\exp\left(\dfrac{-5}{4}\right)$ b. $\left[1 - \exp\left(\dfrac{-5}{4}\right)\right]^2$

6.43 a. $1 - \exp\left(\dfrac{-5}{11}\right)$
b. $\exp\left(\dfrac{-15}{11}\right) - \exp\left(\dfrac{-30}{11}\right)$
c. 50.66

6.45 a. 3.2; 6.4 b. (0, 8.26)

6.47 a. 276; 47.664 b. (0, 930.963)

6.49 a. 20; 200 b. 10; 50

6.51 a. 240; 189.74 b. (0, 809.21)

6.53 a. 9.6; 30.72; $f(y) = \begin{cases} 0 & y \le 0 \\ \dfrac{y^2 \exp\,(-y/3.2)}{\Gamma(3)(3.2)^3} & y > 0 \end{cases}$

6.55 a. 0.3849 b. 0.3159 c. 0.3227
d. 0.1586 e. 0.0366

6.57 0.0062

6.59 0.0730

6.61 a. 0.9544 b. 0.8297

6.63 a. 0.6170 b. 0.3174

6.65 a. 0.0062 b. 225.6

6.67 13.67

6.69 a. Yes b. Yes c. No; No d. 4

6.73 a. 13.80; 54.13 b. 0.1607 c. 33.95

6.75 a. 59.55% b. 32.59% c. 171.57

6.77 a. $p(\text{lifetime} > 217) = 0.98$
b. 217E11

6.81 a. 0.8208 b. 4.7; 0.01

6.83 a. 0.75 b. $\dfrac{1}{3}; \dfrac{1}{18}$

6.85 a. 0.875 b. 0.002127

6.87 a. 0.5547 b. 0.6966

6.89 0.03091

6.91 0.09813

6.93 Approximately 6.22 and 43.52

6.95 a. $2\sqrt{\dfrac{2KT}{m\pi}}$ b. $\sqrt{\dfrac{3KT}{2}}$

6.97 $R_n(t) = \exp\left(\dfrac{-nt}{\theta}\right); \dfrac{\theta}{n}$

6.99 3 in parallel

6.103 $\text{Gamma}\left(\dfrac{1}{2}, 2\right)$

6.105 a. $\dfrac{-3}{8}$

b. $F(y) = \begin{cases} 0 & y < 0 \\ \dfrac{1}{2}\left(y^2 - \dfrac{y^3}{4}\right) & 0 \le y \le 2 \\ 1 & y > 2 \end{cases}$

c. Male: Mean = 66.8, SD = 3.1; Female: Mean = 62, SD = 10.9 (Note that values read from the graphs are approximate).

d. $0; 0; \dfrac{3}{8}$ e. 0.1094 f. $\dfrac{7}{6}; 0.2389$

6.107 0.1151

6.109 0.001525

6.111 0.3155

6.113 $\dfrac{\alpha}{\alpha + \beta}; \dfrac{\alpha\beta}{(\alpha + \beta + 1)(\alpha + \beta)^2}$

6.115 0.04996

6.117 a. 105 b. $\dfrac{3}{8}; \dfrac{5}{192}$

6.119 0.8088

6.121 113.33

6.123 0.8753

6.127 0

Chapter 7

Solutions to 7.1 through 7.23 same as those for 5.1 through 5.23 in Edition 4.

7.1   **a.**

x_2 \ x_1	0	1	2
0	$\frac{1}{9}$	$\frac{2}{9}$	$\frac{1}{9}$
1	$\frac{2}{9}$	$\frac{2}{9}$	0
2	$\frac{1}{9}$	0	0

b.

x_1	0	1	2
$p(x_1)$	$\frac{4}{9}$	$\frac{4}{9}$	$\frac{1}{9}$

c. $\frac{1}{2}$

7.3   **a.**

	x_1	
	0	1
x_2 0	.063	.078
1	.101	.056
2	.163	.065
3	.169	.055
4	.193	.057

b.

x_1	
0	1
.45	.55
.64	.36
.71	.29
.75	.25
.77	.23

c.

x_2		
0	.092	.250
1	.146	.179
2	.236	.210
3	.245	.178
4	.280	.185

7.5   **a.** $f(x_1) = 1, 0 \leq x_1 \leq 1$
     $= 0$, otherwise

b. $\frac{1}{2}$ **c.** Yes

7.7   **a.** $\frac{7}{8}$ **b.** $\frac{1}{2}$ **c.** $\frac{2}{3}$

7.9   **a.** $\frac{21}{64}$ **b.** $\frac{1}{3}$ **c.** No

7.11 **a.** Yes   **b.** $\frac{3e^{-1}}{2}$

7.13 $\frac{11}{36}$

7.15 $\frac{23}{144}$

7.17 **a.** 0.2; 0.16; 0.2; 0.16
      b. -0.04 **c.** 0.4; 0.24

7.19 **a.** $\frac{2}{3}; \frac{1}{18}$ **b.** $\left(\frac{1}{3}, 1\right)$

7.21 **a.** 0.0972   **b.** $\frac{1}{2}$ **c.** e^{-1}
      d. $f_1(y_1) = y_1 e^{-y_1}, 0 \leq y_1 < \infty$
        $= 0$, otherwise
      $f_2(y_2) = e^{-y_2}, 0 \leq y_2 < \infty$
        $= 0$, otherwise

7.23 **a.** 1   **b.** 1   **c.** e^{-2}

7.25 0.0895

7.27 **a.** 0.0459   **b.** 0.22622

7.29 0.0935

7.31 0.8031 .4095

7.33 **a.** 0.2759   **b.** 0.8031

7.37 *Hint*: Use the moment-generating function.

7.39 0.0227

7.41 **a.** $f_1(x_1) = \begin{cases} 3x_1^2 & 0 \leq x_1 \leq 1 \\ 0 & \text{Otherwise} \end{cases}$

  $f_2(x_2) = \begin{cases} \frac{3}{2}(1 - x_2^2) & 0 \leq x_2 \leq 1 \\ 0 & \text{Otherwise} \end{cases}$

b. $\frac{23}{64}$ **c.** 0

7.43 Yes

7.45 **a.** $f(x_1, x_2) = \begin{cases} \frac{1}{x_1} & 0 \leq x_2 \leq x_1 \leq 1 \\ 0 & \text{Otherwise} \end{cases}$

b. $\frac{1}{2}$ **c.** $\frac{\ln(2)}{\ln(4)}$

7.47 **a.** $f(x_1) = \begin{cases} x_1 + \frac{1}{2} & 0 \leq x_1 \leq 1 \\ 0 & \text{Otherwise} \end{cases}$

  $f(x_2) = \begin{cases} x_2 + \frac{1}{2} & 0 \leq x_2 \leq 1 \\ 0 & \text{Otherwise} \end{cases}$

b. No

c. $f(x_1 \mid x_2) = \begin{cases} (x_1 + x_2)/(x_2 + 0.5) & 0 \le x_1, x_2 \le 1 \\ 0 & \text{Otherwise} \end{cases}$

7.49 **a.** $f(x_2) = \begin{cases} 2(1 - x_2) & 0 \le x_2 \le 1 \\ 0 & \text{Otherwise} \end{cases}$

b. $f(x_1) = \begin{cases} 1 - |x_1| & |x_1| \le 1 \\ 0 & \text{Otherwise} \end{cases}$

c. $\dfrac{1}{4}$

7.51 $\dfrac{3}{160}$

7.53 **a.** $\dfrac{1}{144}$ **b.** $\dfrac{1}{12}$ **c.** 1.0764

7.55 $\dfrac{1}{4}$

7.57

$f(x_1, x_2, x_3)$

$= \begin{cases} \left(\dfrac{1}{\theta}\right)^3 \exp\left\{-\left(\dfrac{1}{\theta}\right)(x_1 + x_2 + x_3)\right\} & \theta > 0; X_1, X_2, X_3 > 0 \\ 2 & \text{Otherwise} \end{cases}$

7.59 $\dfrac{3}{8}$

7.61 **a.** $M_{X_1 + X_2 + X_3}(t)$

b. $M_{X_1 + X_2}(t)$

c. $E\left(X_1^{k_1} X_2^{k_2} X_3^{k_3}\right)$

Chapter 8

8.1 **b.** 1.6633; 0.2281   **c.** (0.7081, 2.6185)

8.3 **b.** 2.5; 3.0172

8.5 111.4; 323.2; 169.3; 129.9

8.7 385

8.9 153

8.11 0.0132

8.13 **a.** Almost 1   **b.** 0.1230
c. Independence of daily downtimes

8.15 4.4653

8.17 88

8.19 0.0062

8.21 0.6626

8.23 0.3224

8.25 0.2025

8.27 0.0287

8.29 0.1949

8.31 Almost 0

8.33 **a.** 0.0630   **b.** 0.0630

8.35 **a.** 0.0329   **b.** 0.029
c. Contracts are awarded independently.

8.37 0.0023

8.39 0.7498

8.41 1.0017; 3.10435

8.43 **a.** Between 0.05 and 0.10
b. 0.025
c. (12.20, 87.80)
d. Population of resistors is approximately normal.

8.45 About 0.01

8.47 0.1604

8.49 **a.** 0.8815
b. Approximately normal populations

8.51 Almost 0

8.53 Almost 0

8.55 **a.** (14.776, 15.281)
b. No; (14.7613, 15.2595)

8.57 No; (14.5162, 15.5404)

8.59 **a.** (0, 0.1443)   **b.** No; (0, 0.1377)

8.61 **a.** (0, 0.1906)   **b.** No

8.63 (0, 11.8015)

8.65 For Poisson distribution mean = variance.
A separate method is not necessary.

8.67 C_{pk} will increase; proportion will decrease.

8.69 U is distributed chi-square(2).

8.71 0.0062

8.73 **a.** $\mu_1 - \mu_2$ **b.** $\left(\dfrac{\sigma_1^2}{n}\right) + \left(\dfrac{\sigma_2^2}{m}\right)$

8.75 0.1587

8.77 0.0274

Chapter 9

9.1 **a.** $\hat{\theta}_1, \hat{\theta}_2, \hat{\theta}_3, \hat{\theta}_4$ **b.** $\hat{\theta}_4$

9.3 **a.** $\overline{X}$ **c.** $3\overline{X} + (1/n)\sum_{i=1}^{n} X_i^2$

9.5 $\dfrac{1}{12n} + \dfrac{1}{4}$

9.7 $\overline{X} - 1.645\, S\sqrt{\dfrac{n-1}{2}}\, \dfrac{\Gamma((n-1)/2)}{\Gamma(n/2)}$

9.9 Not significantly.

9.11 (45.07, 46.93)

9.13 7,459

9.15 196

9.17 163

9.19 131

9.21 (5.6692, 12.3308)

9.23 **a.** (176.4235, 183.5765)   **b.** (179.02, 180.98)

9.25 (20.7117, 118.9805)

9.27 (54.3207, 89.6793)

9.29 (560.8558, 609.1442)

9.31 (30,498.0235, 173,071.4620)

9.33 We cannot conclude that a majority of voters agree with the statement.

9.35 (−8.1019, −3.8981)

9.37 We cannot conclude that coating B is better than coating A.

9.39 193

9.41 We cannot conclude that intermittent training gives more variable results.

9.43 (0.0733, 0.3067)

9.45 (−0.1900, −0.08996)

9.47 (−0.2147, −0.0653)

9.49 (2.737, 44.6018)

9.51 (3.834, 24.1666)

9.53 (0.0566, 0.1434)

9.55 (8.6923, 10.9077)

9.57 (32.8771, 51.1229)

9.59 (35.0804, 43.7196)

9.61 (0.1496, 16.8504)

9.63 **a.** (1.0706, 1.1294)   **b.** (0.9825, 1.2175)

9.65 **a.** 0.0861   **b.** 0.6242

9.67 $\hat{\lambda} = \bar{x}$

9.69 $\bar{x};\ (n-1)\dfrac{s^2}{n}$

9.71 (9.6781, 26.1255)

9.73 0.06

9.75 0.3125

9.79 (2.0507, 2.1494)

9.81 (−13.4318, −0.5682)

9.83 (−1.5153, −0.5647)

9.85 (3.0091, 3.0709)

9.87 97

9.89 386

9.91 (−4.8176, −1.1824); yes

Chapter 10

10.1 $H_0: \mu = 130$ and $H_a: \mu < 130$

10.3 $H_0: \mu = 64$ and $H_a: \mu < 64$

10.5 $H_0: \mu = 65$ and $H_a: \mu < 65$

10.7 $H_0: \mu = 2500$ and $H_a: \mu < 2500$

10.9 $H_0: \mu = 30$ and $H_a: \mu < 30$

10.11 $H_0: \mu = 490$ and $H_a: \mu > 490$

10.13 Reject null at 5% level of significance.

10.15 70

10.17 0.1131

10.19 p-value = 0.2224

10.21 Fail to reject null at 1% level.

10.23 Fail to reject null at 1% level.

10.25 Reject null at 5% level.

10.27 Fail to reject null at 5% level.

10.29 p-value = 0.0384, reject null at 5% level.

10.31 Reject null at 1% level.

10.33 Fail to reject null at 5% level.

10.35 Fail to reject null at 5% level.

10.37 Reject null at 5% level.

10.39 Reject null at 1% level.

10.41 Fail to reject null at 5% level.

10.43 Reject null at 5% level.

10.45 Reject null at 5% level.

10.47 Fail to reject null at 1% level.

10.49 Reject null at 10% level.

10.51 Reject null at 5% level.

10.53 Fail to reject null at 5% level.

10.55 Fail to reject null at 5% level.

10.57 Fail to reject null at 10% level.

10.59 $p < 0.0001$

10.61 $p = 0.4586$

10.63 $p < 0.0001$

10.65 $p < 0.0001$

10.67 Fail to reject at 10% level.

10.69 Fail to reject at 5% level.

10.71 **a.** Reject null at 5% level.
b. Reject null at 5% level.

10.73 Fail to reject at 5% level.

10.75 Fail to reject at 5% level.

10.77 Fail to reject at 5% level.

10.79 Fail to reject null at 5% level.

10.81 Reject null at 5% level. The Poisson model is inadequate.

10.83 **a.** Reject null at 5% level. The exponential model is inadequate.
b. Reject null at 5% level. The exponential model is inadequate.

10.85 Reject null at 5% level. The Weibull model is inadequate.

10.87 Reject null at 5% level. The exponential model is inadequate.

10.89 **a.** 125; 10   **b.** 125; 8   **c.** 50; 5

10.91 Accept the lot.

10.93 p-value almost 0

10.95 p-value $= 0.025$

10.97 Reject null at 10% level.

10.99 Fail to reject at 10% level.

10.101 Reject null at 5% level.

10.103 Fail to reject at 5% level.

10.105 Fail to reject null at 5% level.

10.107 $H_0: p = \dfrac{1}{3}$ and $H_0: p > \dfrac{1}{3}$

10.109 p-value $= 0.0802$

10.111 **a.** $p = 0.7561$
b. Slab A: p-value $= 0.0065$;
Slab B: p-value $= 0.0058$

10.113 p-value almost 0

Chapter 11

11.1 $\hat{\beta}_0 = 0$, $\hat{\beta}_1 = \dfrac{6}{7}$

11.3 $\hat{y}_i = -0.06 + 2.78x_i$

11.5 $\hat{y}_i = 480.2 - 3.75x_i$

11.7 $\hat{y}_i = 10.20833 + 1.925x_i$

11.9 $SSE = 1.1429$; $s^2 = 0.2857$

11.11 $SSE = 0.6544$; $s^2 = 0.02424$

11.13 $SSE = 87.7855$; $s^2 = 10.9732$

11.15 $t - 6.71$, Reject null at 5% level.

11.17 (2.2618, 3.2982)

11.19 $t = -17.94$; Reject null at 5% level.

11.21 (5.752, 13.498)

11.23 **a.** $\hat{y}_i = 36.68 - 0.3333x$
b. $SSE = 1.3469$; $s^2 = 0.1684$
c. $t = 7.38$; Reject null at 5% level.

11.25 (4.6623, 5.2257)

11.27 (194.1520, 203.5855)

11.29 (30.1164, 36.5002)

11.31 $t = -0.17$; Fail to reject null at 5% level.

11.33 **a.** Rate $= -0.235 + 0.000339$ Temp
c. (0.05938, 0.08109)
d. (0.01651, 0.12396)

11.35 **a.** Yes   **b.** $F = -38.84$; Reject null.

11.37 $F = 1.056$; Do not reject null.

11.39 **a.** $\hat{y}_i = 20.0911 - 0.6705x_i + 0.009535x_i^2$
c. $t = 1.507$; Fail to reject null.
d. $\hat{y}_i = 19.2791 - 0.4449x_i$
e. (-0.5177, -0.3722)

11.41 $t = 3.309$; Reject null.

11.45 **a.** $y = \beta_0 + \beta_1 x_1 + \beta_2 x_1^2 + \varepsilon$
b. $F = 44.50$; reject null
c. p-value $= 0.0001$
d. $t = -4.683$; Reject null
e. p-value $= 0.0001$

11.47 Height $= 0.0790688 - 0.0227425$ Velocity; $R^2 = 0.92$

11.49 **b.** 2 and 20 degrees of freedom

11.51 **a.** $H_0: \beta_1 = \cdots = \beta_5 = 0$;
H_a: at least one of the β_is is nonzero

b. $H_0: \beta_3 = \beta_4 = \beta_5 = 0$;
H_a: at least one of $\beta_3, \beta_4, \beta_5$ is nonzero

c. $F = 18.29$; Reject null.

d. $F = 8.46$; Reject null.

11.53 $F = 1.409$; Do not reject null.

11.55 **c.** $CO = -1.339409 + 937.91235$ SO_2

11.57 **a.** Nonlinear and increasing

b. Dampers $= -85.72201 + 1.0423461$ Height $+ 0.0146712$ (Height $- 130.067)^2$

c. Yes; $F = 102.916$

11.59 **a.** Very strong, linear, negative relation

b. Limit Load $= 0.8402778 - 0.00825$ PDA

c. Yes, $t = -18.04$

11.61 **a.** $\hat{y} = 1157.1905 + 308.5696$ C1 $+ 3.7777$ C4 $+ 1.3249$ C6 $+ 742.4852$ C8

b. $\hat{y} = 264.9791 + 231.8635$ C1 $+ 3.4824$ C4 $+ 684.7012$ C5 $+ 1.4672$ C6 $+ 537.5143$ C8

c. $\hat{y} = 1293.4935 + 326.0507$ C1 $+ 3.9380$ C4 $+ 1.419549$ C6 $+ 799.8290$ C8

d. $\hat{y} = 1584.6891 + 211.7367$ C1 $+ 4.4464$ C4 $+ 2232.2567$ C5

11.63 **b.** Deviation at 1400 $= 8.9636364 - 0.0062727$ temp; $R^2 = 0.943$
Deviation at 1500 $= 8.6681818 - 0.0058636$ temp; $R^2 = 0.937$

c. 1400: about 1425; 1500: about 1425

11.65 Core Velocity $= 0.3885989 - 0.3987871 \times$ Impeller Diameter

11.67 **a.** $\hat{y} = 0.04565 + 0.000785x_1 + 0.23737x_2 - 0.0000381x_1x_2$

b. 2.7152; 0.1697

c. SSE is minimized by the choice of $\hat{\beta}_0, \hat{\beta}_1, \hat{\beta}_2,$ and $\hat{\beta}_3$

11.69 ($7.32, $9.45)

11.71 **a.** $\hat{y} = 0.6013 + 0.5952x_i - 3.7254x_2$ $- 16.2319x_3 - 0.2349x_1x_2 + 0.381x_1x_3$

b. $R^2 = 0.928$; $F = 139.42$; Reject null.

d. $F = 6.866$; Reject null.

11.73 $t = -1.104$; Do not reject null.

11.75 $F = 3.306$; Reject null.

11.77 **a.** $Y = \beta_0 + \beta_1x_1 + \beta_2x_2 + \beta_3x_3 + \varepsilon$;
x_{1-} = area, x_{2-} = number of baths,
x_{3-} = 1 for central air, and 0 otherwise

b. $Y = \beta_0 + \beta_1x_1 + \beta_2x_2 + \beta_3x_3 + \beta_4x_1x_2$ $+ \beta_5x_1x_3 + \beta_6x_2x_3 + \beta_7x_1^2 + \beta_8x_2^2 + \varepsilon$

c. $H_0: \beta_4 = \cdots = \beta_8 = 0$;
H_a: at least one of the β_i, $i = 4, \ldots, 8$ is nonzero

11.79 **a.** $\hat{y} = 1.679 + 0.444x_1 - 0.079x_2$

b. $F = 150.62$; Reject null.

c. $R^2 = 0.9773$; Yes   **d.** (−0.0907, −0.0680)

e. 1.2 hours   **f.** 12.04 hours

g. (1.6430, 1.9785)

11.81 **a.** $R^2 = 0.937$; $F = 100.41$; Reject null.

c. $t = 1.67$; Do not reject null.

11.83 **b.** $\hat{y} = -93.128 + 0.445x$

c. $\hat{y} = 204.46031 - 0.63800x + 0.0000956x^2$

d. $t = 3.64$; Reject null.

11.85 $t = -6.6$; Reject null.

11.87 $F = 6.712$; Yes

11.89 $F = 9.987$; Reject null.

11.91 **a.** $y = 1.1x$ **b.** $y = 36.2 - 2.34x + 0.08x^2$

c. $y = 36.2 - 1.24x + 0.08x^2$

11.93
B_aA: $y = 0.286 + 1.114x$;
B_aP: $y = 0.202 + 1.024x$; Phe: $y = 0.411 + 1.10x$

Chapter 12

12.1 **a.** 4   **b.** 24   **c.** 6

12.3 **a.**

Sources	df	SS	MS	F-ratio
Treatment	4	24.7	6.175	4.914
Error	30	37.7	1.257	
Total	**34**	**62.4**		

b. 5

12.5 **a.**

Sources	df	SS	MS	F-ratio
Treatment	2	11.0752	5.5376	3.151
Error	7	12.3008	1.7573	
Total	**9**			

b. $F = 3.151$; Do not reject null.

12.7 $F = 3.151$; Do not reject null.

12.9 **b.**

Sources	df	SS	MS	F-ratio
Treatment	3	6.816	2.272	8.0283
Error	105	29.715	0.283	
Total	**108**			

  c. $F = 8.0283$; Yes

  d. $(-0.5302, 0.1562)$ **e.** No

12.11 **a.**

Sources	df	SS	MS	F-ratio
Treatment	1	3,237.2	3,237.2	19.622
Error	98	16,167.7	164.9765	
Total	**99**			

  b. $F = 19.622$; Yes

12.13 $F = 6.907$; Yes

12.15 $F = 29.915$; Yes

12.17 $\hat{y} = 1.433 + 1.433x_1 + 2.5417x_2$,

  where for $i = 1, 2$ $x_i = \begin{cases} 1 & \text{Sample } i \\ 0 & \text{Otherwise} \end{cases}$

  $F = 3.15$; Do not reject null.

12.19 $\hat{y} = 0.275 - 0.275x_1 - 0.05x_2$,

  where for $i = 1, 2$ $x_i = \begin{cases} 1 & \text{Thermometer } i \\ 0 & \text{Otherwise} \end{cases}$

  $F = 3.15$; Do not reject null.

12.21 **a.** $t = -0.6676$; No   **b.** $(-1.3857, 0.5857)$
  c. $(3.003, 4.397)$

12.23 **a.** $(-3.3795, -0.8205)$
  b. $i = 1, j = 2$ $(5.7321, 7.7169)$
  $i = 1, j = 3$ $(7.1231, 9.1169)$
  $i = 2, j = 3$ $(7.8231, 9.8169)$

12.25 **a.**

Sources	df	SS	MS	F-ratio
Treatment	2	23.1667	11.5883	12.636
Block	3	14.25	4.75	5.181
Error	6	5.5	0.9167	
Total	**11**	**42.9167**		

  b. $F = 12.636$; Yes
  c. $F = 5.181$; Yes

12.27 $F = 6.3588$; Yes

12.29 **a.** $F = 39.4626$; Yes

12.31 $F = 83.8239$; Yes

12.33 **a.** $F = 2.320$; No   **b.** $F = 4.6761$; No
  c. Do not block.

12.35 **a.** $F = 9.2174$; p-value $= 0.0012$; Yes
  b. Normal distributions with equal variances
  c. Yes

12.37 $F = 39.4626$; Reject null at 5% level.

12.39 **a.** $(-8.551, -1.049)$
  b. $(-8.9331, 0.6669)$, $(-3.2188, 5.0474)$ and
  $(1.5812, 9.8474)$

12.41 1 & 2: $(-9.4421, 9.9421)$;
  1 & 3: $(-36.6921, -17.3079)$;
  2 & 3: $(-36.9421, -17.5579)$

12.43 **a.** $(-2.128, -1.152)$
  b. Normal distributions with equal variances

12.45 **a.** $F = 32.4$; Significant interaction
  b. 1 & 2: $(-6.934, 1.934)$
  1 & 3: $(-8.934, -0.066)$
  1 & 4: $(-2.434, 6.434)$
  2 & 3: $(-6.434, 2.434)$
  2 & 4: $(-0.066, 8.934)$
  3 & 4: $(2.066, 10.934)$

12.47 **a.** Let A_i = weight I total and B_j = sex j total
  $A_1 = 250.4$, $A_2 = 214$, $B_1 = 262.4$,
  $B_2 = 202$, $T_{11} = 146.4$, $T_{12} = 116$,
  $T_{21} = 104$, $T_{22} = 98$

  b. $SST = 174.015$, $SS(A) = 41.405$,
  $SS(B) = 114.106$, and $SS(A \times B) = 18.504$

  c. L & F: 324.63   H & F: 60.0943
  L & M: 177.8112   H & M: 227.43

  d. $SSE = 789.9655$

  e. $TSS = 963.9805$

  f.

Sources	df	SS	MS	F-ratio
Treatment	3	174.015		
A	1	41.405	41.405	1.4676
B	1	114.106	114.106	4.0444
$A \times B$	1	18.605	18.605	0.6594
Error	28	789.9655	28.2131	
Total	**31**	**963.9805**		

  g. Interaction is not significant.
  h. $(-0.139, 10.739)$ **i.** $(-3.189, 7.689)$

12.49 $F = 5.918$; No

12.51 **a.**

Sources	df	SS	MS	F-ratio
Treatment	3	60.9137		
A	1	41.8612	41.8612	29.6625
B	1	12.7512	12.7512	9.0354
$A \times B$	1	6.3013	6.3013	4.4650
Error	4	5.6450	1.4113	
Total	**7**	**66.5588**		

  b. The treatment combination of 0.5% carbon
  and 1.0% manganese

12.53 **a.** Completely randomized design
  b. $F = 7.0245$; Yes
  c. $(7.701, 8.599)$

12.55 a. $F = 0.1925$; No
b.

Sources	df	SS	MS	F-ratio
Treatment	2	0.63166	0.31583	0.9125
Block	3	0.85583	0.28528	0.1739
Error	6	9.84167	1.64028	
Total	**11**	**11.32916**		

c. $F = 0.1739$; no   **d.** $(-2.807, 3.907)$
e. $(-2.865, 2.115)$

12.57 a.

Sources	df	SS	MS	F-ratio
A	3	74.3333	24.7778	16.5185
B	2	4.0833	2.0417	1.3613
$A \times B$	6	168.9167	28.1527	18.7685
Error	12	18.0	1.5	
Total	**23**	**265.3333**		

b. $F = 18.7685$; yes

12.59 a. Completely randomized design
b.

Sources	df	SS	MS	F-ratio
Treatment	2	57.6042	28.8021	0.3375
Error	13	1,109.3333	85.3333	
Total	**15**			

c. $F = 0.3375$; no   **d.** $(73.684, 88.316)$

12.61 a.

Sources	df	SS	MS	F-ratio
A	1	1,444	1,444	29.4694
B	1	361	361	7.3673
$A \times B$	1	1	1	.0204
Error	12	588	49	
Total	**15**	**2,394**		

b. $F = .0204$; no
c. $(-50.854, -25.146)$, $(-31.854, -6.146)$. Use hourly and piece rates and worker-modified schedule.

12.63 a. Randomized block design
b. $F = 10.4673$; yes   **c.** $F = 3.5036$; no

12.65 a. $F = 12.6139$; yes   **b.** Batch 2

12.67 Thickness and diameter affect AE rates significantly.

12.69 The plume samples differ from the rain samples on the elements As, Sb, and Zn.

12.73

Sources	df	SS	MS	F	P
Treatments	12	412237.75			
Gender	1	110224.00	110224.00	242.07	0.0006
Body part	1	247506.25	247506.25	543.57	0.0002
Task	3	42864.75	14288.25	31.38	0.0091
Gender*part	1	3136.00	3136.00	6.89	0.0787
Gender*task	3	5398.00	1799.33	3.95	0.1444
Part*task	3	3108.75	1036.25	2.28	0.2585
Error	3	1366.00			
Total	**15**				

At the 0.05 level, all the main effects are significant.

12.75 Hint: Standardize all means and test for outliers.

12.77 a. $F = 1.4241$; p-value $= 0.3118$; Do not reject null
b. $F = 0.3092$; p-value $= 0.3118$; Do not reject null
c. $F = 10.3933$; p-value $= 0.0112$; Reject null
d. $F = 4.2136$; p-value $= 0.0296$; Reject null
e. $F = 16.3389$; p-value $= 0.0002$; Reject null
f. $F = 9.6188$; p-value $= 0.0019$; Reject null

12.79 a. Stain: $F = 67.7528$; p-value < 0.0001; Reject null
Material: $F = 1.5841$; p-value $= 0.2603$; Do not reject null

Index